AF618862

DIE GRUNDLEHREN DER

MATHEMATISCHEN WISSENSCHAFTEN

IN EINZELDARSTELLUNGEN MIT BESONDERER BERÜCKSICHTIGUNG DER ANWENDUNGSGEBIETE

VOLUME X

RICCI-CALCULUS

AN INTRODUCTION TO TENSOR ANALYSIS AND ITS GEOMETRICAL APPLICATIONS

BY

J. A. SCHOUTEN

SECOND EDITION

SPRINGER-VERLAG BERLIN HEIDELBERG GMBH

1954

RICCI-CALCULUS

AN INTRODUCTION TO TENSOR ANALYSIS AND ITS GEOMETRICAL APPLICATIONS

BY

J. A. SCHOUTEN

EMERITUS PROFESSOR OF MATHEMATICS IN THE UNIVERSITY OF AMSTERDAM

DIRECTOR OF THE MATHEMATICAL CENTRE AT AMSTERDAM

SECOND EDITION

WITH 16 FIGURES

SPRINGER-VERLAG BERLIN HEIDELBERG GMBH

1954

ISBN 978-3-642-05692-5 ISBN 978-3-662-12927-2 (eBook)
DOI 10.1007/978-3-662-12927-2

URSPRÜNGLICH ERSCHIENEN BEI SPRINGER-VERLAG OHG. IN BERLIN, GÖTTINGEN AND HEIDELBERG 1954
SOFTCOVER REPRINT OF THE HARDCOVER 2ND EDITION 1954

This book is dedicated to the memory of

DR. GREGORIO RICCI CURBASTRO

in life Professor of Mathematics in the
University of Padua,
who laid the foundations
of tensor calculus.

Preface to the second edition.

This is an entirely new book. The first edition appeared in 1923 and at that time it was up to date. But in 1935 and 1938 the author and Prof. D. J. STRUIK published a new book, their Einführung I and II, and this book not only gave the first systematic introduction to the kernel-index method but also contained many notions that had come into prominence since 1923. For instance densities, quantities of the second kind, pseudo-quantities, normal coordinates, the symbolism of exterior forms, the LIE derivative, the theory of variation and deformation and the theory of subprojective connexions were included. Now since 1938 there have been many new developments and so a book on RICCI calculus and its applications has to cover quite different ground from the book of 1923. Though the purpose remains to make the reader acquainted with RICCI's famous instrument in its modern form, the book must have quite a different methodical structure and quite different applications have to be chosen.

The first chapter contains algebraical preliminaries but the whole text is modernized and there is a section on hybrid quantities (quantities with indices of the first and of the second kind) and one on the many abridged notations that have been developed by several authors.

In the second chapter the most important analytical notions that come before the introduction of a connexion are dealt with in full. The theory of integrability and PFAFF's problem are treated here and do not get a chapter of their own this time as they did in the first edition because SCHOUTEN and v. D. KULK's book on PFAFF's problem of 1949 can be referred to. A special section pays attention to CARTAN's symbolism.

In the first edition the chapters IV–VI contained the theory of affine, riemannian and WEYL connexions and in these chapters problems of projective and conformal transformation and of imbedding were also dealt with. In the light of modern developments this mixing up of quite different topics could not be tolerated. So there is now a chapter III dealing with linear connexions, curvature and normal coordinates, also with respect to anholonomic systems of reference and in different notations. Then there is a chapter V on imbedding and curvature, in the last section of which the theory of higher curvatures is developed. Chapter VI is dedicated to projective and conformal transformations of connexions and in this chapter subprojective connexions and concircular transformations of a V_n find their natural place.

Between chapter III and these latter chapters it has been necessary to insert a chapter IV on the theory of LIE groups. This theory not only gives a very beautiful example of special linear connexions but it also forms the necessary foundation for the following chapters, especially for chapter VII.

This chapter VII on the theory of variations, deformations and movements leads in §§ 4, 5 to the theory of the holonomy groups that is used in § 6 to give an application of CARTAN's method.

The last chapter in the first edition dealt with the invariant decomposition of tensors and in 1923 this was a new and interesting topic. But now we can refer to D. E. LITTLEWOOD's book of 1951 and it does not seem any longer justifiable to treat such a special purely algebraic problem in a book on RICCI calculus and differential geometry.

Instead a chapter VIII has been added containing miscellaneous examples, mostly taken from recent literature. Here the harmonic V_n (§ 1), the many different connexions for hybrid quantities and their properties with respect to imbedding and transformation (§§ 2–8) and the spaces of recurrent curvature (§ 9) are considered. The choice of these subjects was entirely free. Of course this chapter could have been made many times longer. But there must be a limit to the size of a book of this kind and this is also the reason why all those topics contained in the following long list, however interesting they are, could at most be mentioned very shortly:

1. properties in the large as for instance properties of compact manifolds;
2. theory of geometric objects; fibre bundles etc.;
3. theory of VITALI;
4. geometries of FINSLER, CARTAN, A. KAWAGUCHI;
5. extensors of CRAIG and generalizations of M. KAWAGUCHI;
6. linear elements and surface elements of higher order;
7. connexions of higher order;
8. projective and conformal geometry with supernumerary coordinates either in the tangent spaces or in the whole space;
9. path spaces; K-spreads;
10. natural families of curves;
11. non-linear connexions;
12. special geometries for small values of the dimensions;

13. contact transformations and their connexions;

14. treatment of differential equations by means of linear connexions belonging to them;

15. spaces with a fundamental tensor of rank $< n$;

16. G. KRON's applications of tensor analysis to electric circuits.

In the first edition there was an extensive literature list. With its 196 titles it covered nearly all the literature on many dimensional differential geometry. The literature list of the Einführung of 1935–1938 with its 488 titles represents only a selection of the literature at that time. This is even more true of the literature list presented here which is by no means exhaustive though it contains about 1400 titles and refers to about 350 authors. By selecting the titles in such a way that the reader interested in some topic will always find at least a few titles that can lead him to more references, the author has tried to retain something of the encyclopedic character of the first edition.

The author owes many thanks to the Mathematical Institute of Amsterdam University and the Mathematical Centre at Amsterdam for the valuable help of many of their collaborators. In the first place I mention Dr. NIJENHUIS (M. I. and M. C.) who has given much time to the study of literature and the collection of exercises and who proposed valuable improvements to the text in nearly every chapter. Mr. BARNING (M. I.) and Mrs. v. ROOTSELAAR (M. I.) did important work by studying literature and checking formulae. For the correction of proofs I had the valuable help of Mr. BARNING (M. I.) and Mr. VERHOEFF (M. C.). Mr. WOOD of the Clarendon Press, Oxford, has kindly given advice on the language. This is the third book with which he has helped me and I am very grateful to him for all the English idiom he has taught me. Last but not least I have to thank my wife Mrs. SCHOUTEN-BYLSMA who spent much time on the administration of all literature references.

I wish to express my best thanks to the publishers for their most agreeable collaboration.

Epe (Holland), im Juli 1954. **J. A. SCHOUTEN.**

Contents.

I. Algebraic preliminaries (1).

§ 1. The E_n (1).

§ 2. Quantities in E_n (6).

§ 3. Invariant processes and relations (13).

§ 4. Section and reduction with respect to an E_m in E_n (16).

§ 5. Rank, domain and support of domain with respect to one or more indices (20).

§ 6. Symmetric tensors (21).

II. Analytic preliminaries (61).

§ 1. The arithmetic n-dimensional manifold $\mathfrak{A}_n$ (61).

§ 2. The geometric n-dimensional manifold X_n (62).

§ 3. Geometric objects and quantities in X_n (67).

§ 4. The X_m in X_n (72).

§ 5. The E_m-field or X_n^m in X_n. Systems of linear partial differential equations (78).

§ 6. The invariant operators Rot and Div (83).

§ 3. A linear connexion expressed in terms of $S_{\mu\lambda}^{\cdot\cdot\varkappa}$, an auxiliary symmetric tensor field $g_{\lambda\varkappa}$ of rank n and $\nabla_\mu g_{\lambda\varkappa}$ (131).

§ 4. Curvature (138).

§ 5. The identities for the curvature tensor (144).

§ 6. Integrability conditions in L_n (153).

§ 7. Geodesics and normal coordinates (155).

§ 8. FERMI coordinates (166).

§ 9. Linear connexions expressed in anholonomic coordinates (169).

I. Algebraic preliminaries.

§ 1. The E_n.

In this chapter we consider the "ordinary" n-dimensional *affine space* E_n, i.e. the space in which the coordinates of a point are subject to transformations of the *affine group* G_a.[1] If the $\xi^{\varkappa}$; $\varkappa = 1, \ldots, n$[2] are the "old" coordinates of a point, we can get "new" coordinates $\xi^{\varkappa'}$; $\varkappa' = 1', \ldots, n'$ by means of the coordinate transformation

$$\xi^{\varkappa'} = A^{\varkappa'}_{\varkappa}\, \xi^{\varkappa} + a^{\varkappa'}; \qquad \Delta = \mathrm{Det}\,(A^{\varkappa'}_{\varkappa}) \neq 0\,^{3)} \tag{1.1}$$

where the $A^{\varkappa'}_{\varkappa}$ and $a^{\varkappa'}$ are constants. The coordinate systems are usually denoted by $(\varkappa)$ and $(\varkappa')$ respectively. From (1.1) it follows that there exists a transformation of $\xi^{\varkappa'}$ into $\xi^{\varkappa}$, called the inverse transformation

$$\xi^{\varkappa} = A^{\varkappa}_{\varkappa'}\, \xi^{\varkappa'} + a^{\varkappa}; \qquad \mathrm{Det}\,(A^{\varkappa}_{\varkappa'}) = \Delta^{-1} \tag{1.2}$$

with constant coefficients. For every value of ϱ and σ' the coefficient $A^{\varrho}_{\sigma'}$ equals the minor[4] of the element $A^{\sigma'}_{\varrho}$ in the matrix of the $A^{\varkappa'}_{\varkappa}$, divided by Δ and $a^{\varkappa'}$ and $a^{\varkappa}$ are related by the equation

$$a^{\varkappa'} = -\, A^{\varkappa'}_{\varkappa}\, a^{\varkappa}. \tag{1.3}$$

No mention has been made so far of the meaning of the words "point" and "space". We leave this to Ch. II where the E_n will be defined as a special case of more general manifolds. For the moment it is sufficient to know that each point has certain coordinates, i.e. a set of n real or complex numbers assigned to it and that the linear transformations (1.1, 2) play an important role. The transformations (1.1, 2) will be called *allowable coordinate transformations* and the coordinates obtainable by them *rectilinear coordinates*.

---

[1]) A set of transformations is said to form a *group* if 1° any two transformations of the set performed after each other yield a transformation belonging to the set, 2° the inverse of every transformation of the set belongs to the set, 3° the set contains the identical transformation.

[2]) Unless otherwise specified the indices $\varkappa, \lambda, \mu, \nu, \varrho, \sigma, \tau$ and sometimes ω always take the values $1, \ldots, n$ (in italics); $\varkappa', \lambda', \mu', \nu', \varrho', \sigma', \tau'$ and sometimes ω' take the values $1', \ldots, n'$ (in italics) etc.

[3]) We adopt the summation convention: if an index appears twice in the same term, *once as a subscript and once as a superscript,* the sign $\sum$ will be omitted.

[4]) The minor of $A^{\sigma'}_{\varrho}$ is the determinant remaining after dropping the row and the column containing $A^{\sigma'}_{\varrho}$ in the matrix of $A^{\varkappa'}_{\varkappa}$ and multiplying by $(-1)^{\varrho+\sigma}$.

If in (1.1) we take $a^{\varkappa'}=0$ we get the transformations of the *homogeneous linear group* G_{ho}. The E_n is then called a *centred* E_n. In a centred E_n the origin has the coordinates zero with respect to all allowable coordinate systems.

There are also other transformations called *point transformations*, for instance the linear transformation

$$\eta^{\varkappa} = P^{\varkappa}_{\cdot\lambda}\xi^{\lambda} + p^{\varkappa} \tag{1.4}$$

with constant coefficients $P^{\varkappa}_{\cdot\lambda}$ and $p^{\varkappa}$. Here the point $\xi^{\varkappa}$ is transformed into the point $\eta^{\varkappa}$ but the coordinates are not changed.

To every coordinate transformation $(\varkappa) \rightarrow (\varkappa')$ there belongs a point transformation such that the *old* coordinates of every point are numerically equal to the *new* coordinates of the transformed point. We may express this by the equation

$$\eta^{\varkappa'} \overset{*}{=} \delta^{\varkappa'}_{\varkappa}\xi^{\varkappa} \;{}^{1)} \tag{1.5}$$

valid for the coordinate systems $(\varkappa)$ and $(\varkappa')$ only. The sign $\overset{*}{=}$ (speak: star equals) will always be used if we wish to emphasize the fact that an equation is only valid or that its validity is only asserted for the coordinate system or coordinate systems occuring explicitly in the formula itself. From (1.4, 5) we get

$$A^{\varkappa'}_{\mu} P^{\mu}_{\cdot\lambda}\xi^{\lambda} + A^{\varkappa'}_{\lambda} p^{\lambda} + a^{\varkappa'} \overset{*}{=} \delta^{\varkappa'}_{\varkappa}\xi^{\varkappa} \tag{1.6}$$

valid for every choice of $\xi^{\varkappa}$, and this is only possible if

$$\text{a) } P^{\varkappa}_{\cdot\lambda} \overset{*}{=} A^{\varkappa}_{\lambda'}\delta^{\lambda'}_{\lambda}; \qquad \text{b) } p^{\varkappa} \overset{*}{=} -A^{\varkappa}_{\lambda'}a^{\lambda'}. \tag{1.7}$$

If a point transformation belongs to the coordinate transformation $(\varkappa) \rightarrow (\varkappa')$ we say that $(\varkappa)$ is *dragged along* by this point transformation. Hence the numerical values of the coordinates of a point remain invariant if any point transformation is applied and if at the same time the coordinate system is dragged along by the same point transformation.

In all formulae there are *kernel letters* like A, ξ, P [2]), *running indices* like $\varkappa$, $\varkappa'$ and *fixed indices* like $1, \ldots, n$; $1', \ldots, n'$. With coordinate transformations the kernel letters do not change but we get a new set of running indices and fixed indices. But with point transformations the kernel letters change and the running and fixed indices remain the same.

---

[1]) The symbol $\delta^{\varkappa}_{\lambda}$ used hereafter is the KRONECKER *symbol*. Its value is $+1$ if $\varkappa=\lambda$ (not to summarize) and zero if $\varkappa \neq \lambda$. It occurs also in the forms $\delta^{\varkappa\lambda}$ and $\delta_{\varkappa\lambda}$. The symbol $\delta^{\varkappa'}_{\lambda}$ is the *generalized* KRONECKER *symbol*. Its value is $+1$ if $\varkappa$ and λ' take corresponding values from $1, \ldots, n$ and $1', \ldots, n'$ and zero if they do not.

[2]) A kernel can also be symbolized by a more complicated symbol, cf. for instance (VII 1, 24).

This is the principle of the *kernel-index method* used in this book and in many other publications.[1])

A kernel letter can be changed in the following ways:

1. By changing the letter itself, e.g. $\xi \rightarrow \eta$;

2. by adjoining an accent or another sign, e.g. $*$ preferably *to the left* in order to avoid collision with upper running or fixed indices to the right;

3. by adjoining an index *over* or *under* the kernel[2]). The upper and lower places to the right are reserved for running or fixed indices;

4. by putting upper or lower indices, without moving them, between brackets as will be shown later.

A *flat* or *linear m-dimensional submanifold* or E_m in E_n can be given by a set of $n-m$ linear equations

$$C^x_\lambda \xi^\lambda + C^x = 0; \qquad x = \mathrm{m}+1, \dots, \mathrm{n}\ ^{3)} \tag{1.8}$$

with the conditions that the C^x_λ and the C^x are constants and that the matrix of the C^x_λ has the rank[4]) $n-m$. But it can also be given by a set of m parametric equations

$$\xi^\varkappa = B^\varkappa_b \eta^b + B^\varkappa; \qquad b = 1, \dots, \mathrm{m} \tag{1.9}$$

with m parameters η^b and the condition that the $B^\varkappa_b$ and the $B^\varkappa$ are constants and that the matrix of $B^\varkappa_b$ has rank m. From (1.9) we see that the η^b can be used as rectilinear coordinates in the submanifold and that accordingly this submanifold is really an E_m, because every other set of parameters that could be used in (1.9) must be linearly connected with the η^b. From (1.8, 9) it follows that

$$C^x_\lambda B^\lambda_b = 0; \qquad b = 1, \dots, \mathrm{m}; \qquad x = \mathrm{m}+1, \dots, \mathrm{n} \tag{1.10}$$

and that

$$C^x = -C^x_\lambda B^\lambda. \tag{1.11}$$

An E_0 is a *point*, an E_1 a *straight line*, an E_2 a *plane* and an E_{n-1} a *hyperplane*. To every rectilinear coordinate system there belong n E_1's

[1]) The kernel-index method was sketched for the first time in the authors lectures on continuous groups at Leiden University 1926/27 (opalographic syllabus 1927, 1). The first printed exposition of the method appeared in Schouten and van Kampen 1930, 2 and Golab (thesis) 1930, 1.

[2]) If possible it is advisable to avoid using upper and lower indices of this kind simultaneously.

[3]) In reasonings like this we always suppose that m and m and also n and n stand for the same integer written in italics and vertical.

[4]) A matrix has rank p if it contains at least one non zero subdeterminant with p rows but none with more rows.

the *coordinate axes*, $\binom{n}{2}$ *coordinate planes* and $\binom{n}{p}$ *coordinate* E_p's, $p=1, \ldots, n-1$. These flat manifolds form the *net* of the coordinate system. If a coordinate system is dragged along by a point transformation, this means that the point transformation is applied to its net.

E_p is said to *lie in* E_q and E_q is said to *contain* E_p if every point of E_p is a point of E_q. In order to discuss the common points of an E_p and an E_q we first define parallelism in E_n. A *translation* is a point transformation of the form

$$(1.12) \qquad {}'\xi^\varkappa = \xi^\varkappa + c^\varkappa; \qquad c^\varkappa = \text{constant}$$

with respect to some rectilinear coordinate system and thus with respect to all these systems. Two E_p's are said to be *parallel* if they can be transformed into each other by a translation. The set of all E_p's parallel to a given E_p is called a *p-direction* or an *improper* E_{p-1}.[1]) In a centred E_n there is a one to one correspondence between all improper E_{p-1}'s and all E_p's containing the origin. A *1*-direction is called a *direction* for short.[2]) An E_p is said to *contain* an improper E_q and the E_q is said *to lie in* the E_p if the E_p contains one of the E_{q+1}'s of the E_q. The equation

$$(1.13) \qquad C_\lambda^x \xi^\lambda = \zeta^x; \qquad x = \mathrm{m}+1, \ldots, \mathrm{n}$$

with variable ζ^x represents all E_m's parallel to the E_m fixed by C_λ^x through the origin. In the manifold of these E_m's the ζ^x can be used as rectilinear coordinates. Accordingly this manifold is an E_{n-m} each point of which represents an E_m. We call this E_{n-m} the *reduction* of E_n *with respect to the* E_m through the origin or *with respect to its m-direction.*

The following statements are valid both for proper and for improper flat submanifolds.

1. An E_p and an E_q; $0 < p \leq q$, are called *t/p-parallel*, $t > 0$, and for $t=p$ *parallel* if they contain the same t-direction but no common $(t+1)$-direction.

2. The *section*[3]) of an E_p and an E_q consists of all common E_0's. This section is an E_s; $p \geq s \geq p+q-n$, $q \geq s$.

3. An E_p and an E_q *span* an E_r if they are both contained in E_r and if there is no E_{r-1} containing both E_p and E_q. E_r is called the *join*) of E_p and E_q. r and s are related by the equation $r=p+q-s$.

[1]) Formerly often called an E_{p-1} "at infinity".

[2]) A direction in every day language is according to our definition a direction with a sense (arrow).

[3]) Some authors use the terms *intersection* and *meet*.

[4]) In P. P. 1949, 1 we used the term *junction*.

If in E_n a proper E_p and a q-direction, $q = n - p$, which have no direction in common, are given, then every geometric figure in E_n can be subjected to the following processes:

1. *Section with E_p:* All points of the figure not lying in E_p are dropped. Only the E_p is used.

2. *Reduction with respect to the q-direction:* All points of the figure lying in an E_q with this q-direction are identified. The resulting figure lies in the E_p that arises from reduction of the E_n with respect to the q-direction. Only the q-direction is used.

3. *Projection on E_p in the q-direction*: Through every point of the figure an E_q is laid with the given q-direction. The section of this E_q with E_p, is the projection of this point. Both the E_p and the q-direction are used.

An n-dimensional *screwsense* in E_n is determined by n directions, with a sense (arrow), which are not contained in the same $(n-1)$-direction and are given in a definite order. For $n = 2$ the screwsense is often called *sense of rotation* and for $n = 1$ *sense*. Two screwsenses are *equal* if the defining figures can be transformed into each other by a point transformation (1.4) with Det $(P^{\varkappa}_{.\lambda}) > 0$ and *opposite* if they can not. Hence there exist only two screwsenses in an E_n. The screwsense can also be given by a part of a general curve, not lying in an E_{n-1}, which is provided with a sense (arrow). If $\frac{1}{2} n(n+1)$ is even, e.g. for $n = 3$, the sense of the curve can be omitted. In this case it makes no difference if the senses of all directions are changed and their order is inverted at the same time. A screwsense is invariant for all linear point transformations (1.4) with Det $(P^{\varkappa}_{.\lambda}) > 0$. To every coordinate system $(\varkappa)$ there belongs a definite screwsense fixed by the directions of the axes, their $+$-senses and the order $1, \ldots, n$. This screwsense is invariant for rectilinear coordinate transformations if and only if $\Delta > 0$.

An E_p with a p-dimensional screwsense fixes a screwsense in every E_p parallel to it. Hence a p-direction may have a screwsense. An E_p or a p-direction with screwsense are said to be *oriented* with *inner orientation*. If to a given E_p or p-direction an $(n-p)$-dimensional screwsense is given in some E_{n-p} which has no direction in common with E_p or the p-direction, this screwsense fixes a screwsense in every E_{n-p} with this property. Then the E_p or the p-direction are said to be *oriented* with *outer orientation*. For instance an E_1 in E_3 has inner orientation if there is an arrow *in* it and it has outer orientation if there is an arrow *around* it.[1])

---

[1]) The idea of orientation has been dealt with from a very general and highly interesting point of view by E. CARTAN 1941, 1.

Exercises.

I 1,1. Prove that (cf. Exerc. II 2,1)

$$A^{\varrho}_{\sigma'} = \frac{\partial \log \Delta}{\partial A^{\sigma'}_{\varrho}}. \qquad \text{I 1,1 } \alpha)$$

I 1,2. An E_p in E_n can be fixed by $(p+1)(n-p)$ numbers. An E_p through a given E_q, $q < p$, can be fixed by $(p-q)(n-p)$ numbers.

I 1,3. If an E_p and an E_q, having no direction in common, are given in a definite order, and if in each of them a screwsense is given, a screwsense in the join E_{p+q} is determined. If the order of E_p and E_q is changed the screwsense in E_{p+q} changes if and only if pq is odd.

§ 2. Quantities in E_n.

A *quantity* in E_n is a correspondence between the rectilinear coordinate systems and the ordered sets of N numbers, satisfying the conditions:

1. To every coordinate system $(\varkappa)$ there corresponds one and only one set of N numbers;

2. if Φ_Λ; $\Lambda = 1, \ldots, N$ corresponds to $(\varkappa)$ and $\Phi_{\Lambda'}$; $\Lambda' = 1', \ldots, N'$ to $(\varkappa')$ the $\Phi_{\Lambda'}$ are functions of the Φ_Λ and the $A^{\varkappa'}_{\varkappa}$, homogeneous linear in Φ_Λ and homogeneous algebraic in $A^{\varkappa'}_{\varkappa}$. The Φ_Λ are called the *components of the quantity with respect to* $(\varkappa)$.[1] Quantities with the same number of components and the same manner of transformation are said to be *of the same kind*. According to the definition, by addition of corresponding components of two quantities of the same kind we get components of another quantity of the same kind called the *sum* of both.

Quantities are distinguished by the *manner* of transformation of their components. We give here a list of some of the most important quantities in E_n. Other quantities will be defined later.

1. A *scalar* with one component, invariant for the transformations (1.1).

2. A *contravariant vector*[2]) with n components $v^\varkappa$ and the transformation

$$v^{\varkappa'} = A^{\varkappa'}_{\varkappa} v^{\varkappa}. \qquad (2.1)$$

---

[1]) It is often convenient to denote a quantity in the text by its kernel letter only. This must not of course be done if the same kernel letter is being used for two or more different quantities.

[2]) In the classical theory of invariants two transformations of the kind (2.1) and (2.4) were called *contragredient* with respect to each other. Then the two transformations were distinguished by the words covariant and contravariant. But the reasons for calling (2.1) contravariant and (2.4) covariant (and not the other way round) are purely historical.

A contravariant vector $v^\varkappa$ is represented by two points $\xi^\varkappa$ and $\xi^\varkappa + v^\varkappa$ in definite order fixed to within a common translation. The order can be represented by numbering the points or by a curve or straight line with an arrow joining the points. In a centred E_n a contravariant vector $v^\varkappa$ can be represented by one point with the coordinates $v^\varkappa$, the origin playing the role of first point. Addition of two contravariant vectors can be illustrated by the wellknown figure called "parallelogram of forces" in mechanics. $|v^1|$ is the projection of $v^\varkappa$ on the 1-axis in the $(n-1)$-direction of the other axes measured by the unit in the 1-axis. v^1 is positive if the projected sense of $v^\varkappa$ on the 1-axis is the same as the $+$-sense of the 1-axis and negative otherwise.

To every coordinate system $(\varkappa)$ there belong n contravariant vectors

$$\underset{\lambda}{e}^\varkappa \overset{*}{=} \delta_\lambda^\varkappa = \begin{cases} 1 \text{ for } \varkappa = \lambda \\ 0 \text{ for } \varkappa \neq \lambda \end{cases} \tag{2.2}$$

called the *contravariant basis vectors* of $(\varkappa)$.[1] The components of these vectors with respect to another coordinate system $(\varkappa')$ in general no longer have the values 1 or 0 because

$$\underset{\lambda}{e}^{\varkappa'} = A_\varkappa^{\varkappa'} \underset{\lambda}{e}^\varkappa \overset{*}{=} A_\lambda^{\varkappa'} \tag{2.3}$$

and to $(\varkappa')$ there belongs another contravariant basis $\underset{\lambda'}{e}^\varkappa$. Here we have an example of kernels with indices *under* the letter. Such indices are not subject to the transformation of coordinates and are called *dead indices* in contradistinction to the indices to the right that are called *living*.[2]

3. A *covariant vector* with n components w_λ and the transformation

$$w_{\lambda'} = A_{\lambda'}^\lambda w_\lambda . \tag{2.4}$$

The transformations (2.4) and (2.1) are said to be *contragredient*[3] to each other. A covariant vector w_λ is represented by two parallel hyperplanes

$$w_\lambda \xi^\lambda = c; \qquad w_\lambda \xi^\lambda = c + 1; \qquad c = \text{const.} \tag{2.5}$$

---

[1]) In former publications we used the term measuring vectors instead of basis vectors and called a basis a set of measuring vectors.

[2]) The KRONECKER symbol $\delta_\lambda^\varkappa$ stands for n^2 scalars and accordingly its indices are dead. So it ought to be written $\underset{\lambda}{\overset{\varkappa}{\delta}}$ but we shall not do so for historical reasons and also to avoid such towerlike constructions as much as possible. The same holds for $\delta_{\lambda'}^{\varkappa'}$, $\delta_{\varkappa\lambda}$, $\delta^{\varkappa\lambda}$ etc. All these symbols represent n^2 scalars with the values 1 or 0 and not, as might be expected, different components of some geometric object δ. This is an exception to the general rule for the use of kernels.

[3]) See footnote 2 on p. 6.

given in this order and fixed to within a common translation, i.e. addition of the same arbitrary constant to c. [1]) The order can be represented by a (preferably curved) [2]) arrow in a curve through the hyperplanes. Fig. 1 represents a covariant vector in E_3. In a centred E_n a covariant vector w_λ can be represented by one hyperplane $w_\lambda \xi^\lambda = 1$, the parallel E_{n-1} through the origin playing the role of first hyperplane. Addition of two covariant vectors is illustrated in Fig. 2, representing a section with an arbitrary E_2. [3])

$|w_1|$ is the inverse of the segment on the 1-axis cut out by the E_{n-1}'s of w_λ, measured by the unit on this axis. w_1 is positive if the

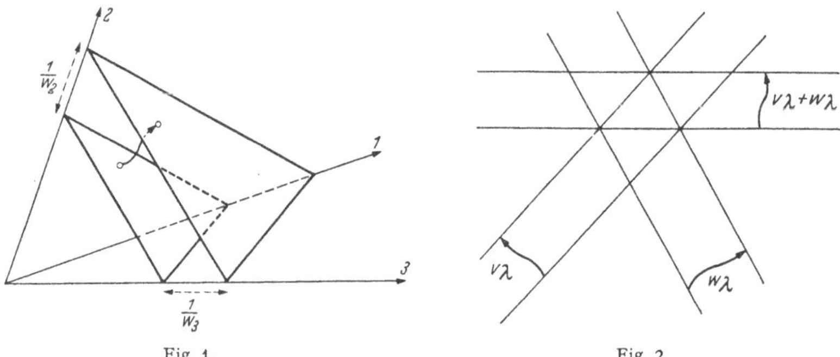

Fig. 1. Fig. 2.

sense of w_λ corresponds to the $+$-sense of the 1-axis and negative otherwise.

To every coordinate system $(\varkappa)$ there belong n covariant vectors

$$\overset{\varkappa}{e}_\lambda \overset{*}{=} \delta^\varkappa_\lambda \tag{2.6}$$

called the *covariant basis vectors* [4]) of $(\varkappa)$. Their components with respect to $(\varkappa')$ no longer have in general the values 1 and 0, and to $(\varkappa')$ there belongs another covariant basis $\overset{\varkappa'}{e}_\lambda$.

The combination of a contravariant and a covariant vector $v^\varkappa w_\varkappa$ is an invariant

$$v^{\varkappa'} w_{\varkappa'} = v^\varkappa A^{\varkappa'}_\varkappa A^\lambda_{\varkappa'} w_\lambda = v^\varkappa w_\varkappa \tag{2.7}$$

and is called the *transvection* of these vectors. $|v^\varkappa w_\varkappa|$ is the length of $v^\varkappa$ measured by the segment cut out from the line of $v^\varkappa$ by the hyperplanes of w_λ. $v^\varkappa w_\varkappa$ is positive if the senses of $v^\varkappa$ and w_λ correspond and

[1]) Cf. SCHOUTEN, 1923, 1, p. 164.

[2]) A straight arrow could suggest that a special direction were given.

[3]) If v_λ and w_λ are parallel the construction can be effected by means of an auxiliary vector u_λ: $v_\lambda + w_\lambda = (v_\lambda + u_\lambda) + (w_\lambda - u_\lambda)$.

[4]) See footnote 1 on p. 7.

negative otherwise. If $v^\varkappa w_\varkappa = \pm 1$ the vector $v^\varkappa$ fits exactly between the hyperplanes of w_λ. If $v^\varkappa w_\varkappa = 0$ the direction of $v^\varkappa$ lies in the $(n-1)$-direction of w_λ.

In Fig. 3 we see the six basis vectors belonging to a coordinate system in E_3.

If n linearly independent contravariant vectors $\underset{i}{v}^\varkappa$ are given, there always exist n covariant vectors $\overset{h}{w}_\lambda$; $h, i = 1, \ldots, n$ such that

$$\underset{i}{v}^\varkappa \overset{h}{w}_\varkappa = \delta_i^h . \tag{2.8}$$

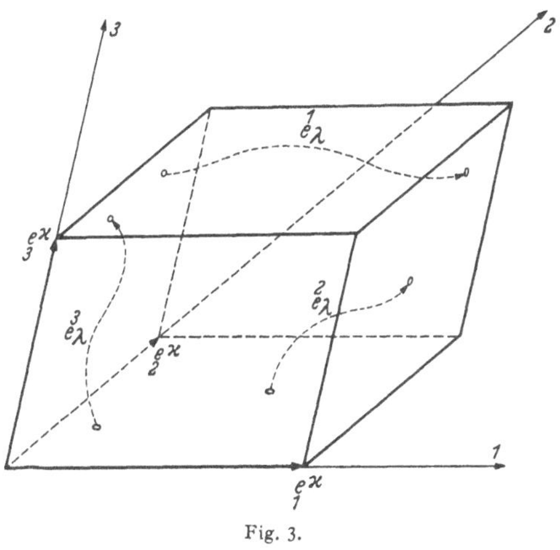

Fig. 3.

The sets $\underset{i}{v}^\varkappa$ and $\overset{h}{w}_\lambda$ are said to be *reciprocal* to each other. For each value of $i, \underset{i}{v}^\varkappa$ fits between the hyperplanes of $\overset{i}{w}_\lambda$ and lies in the $(n-1)$-direction of every other covariant vector $\overset{h}{w}_\lambda$; $h \neq i$. The co- and contravariant basis vectors of a rectilinear coordinate system are reciprocal sets and conversely two reciprocal sets always constitute the basis vectors of some rectilinear coordinate system.

If a vector $v^\varkappa$ is transvected by the n covariant basis vectors $\overset{\varkappa}{e}_\lambda$, we get n scalars

$$\overset{\varkappa}{v} = v^\lambda \overset{\varkappa}{e}_\lambda \tag{2.9}$$

each representing the result of measuring the projection on an axis by the unit on this axis. The equation

$$v^\varkappa = \overset{\lambda}{v} \underset{\lambda}{e}^\varkappa \tag{2.10}$$

gives the construction of $v^\varkappa$ from these scalars and the contravariant basis vectors. The two equations (2.9, 10) are numerically equal if we use the coordinate system $(\varkappa)$, but by introducing $(\varkappa')$ we get the equations

$$\text{a) } \overset{\varkappa}{v} = v^{\lambda'} \overset{\varkappa}{e}_{\lambda'} \qquad \text{b) } v^{\varkappa'} = \overset{\lambda}{v} \underset{\lambda}{e}^{\varkappa'} \tag{2.11}$$

that are no longer numerically equal. The $\varkappa$ in $\overset{\varkappa}{v}$ is a dead index and $\overset{\varkappa}{v}$ stands for a set of n scalars, but the $\varkappa$ in $v^\varkappa$ is a living index transforming with transformations of coordinates. Accordingly we must write

$$\overset{\varkappa}{v} \stackrel{*}{=} v^{\varkappa} \tag{2.12}$$

but $\overset{\varkappa}{v} \neq v^{\varkappa}$. Also $\overset{\varkappa}{e}_{\lambda} \stackrel{*}{=} \underset{\lambda}{e}^{\varkappa}$ but $\overset{\varkappa}{e}_{\lambda} \neq \underset{\lambda}{e}^{\varkappa}$. This illustrates the fact that it is not sufficient to give only the components of a quantity with respect to some coordinate system, and that we must also give the manner of transformation because the components of two quantities of different kind with respect to some special coordinate system may happen to be equal.

The process in (2.9) that leads from a living index to a dead one by means of transvection with the basis vectors is called the *strangling* of an index. It is clear that strangling of the same index or indices at both sides of an equation leaves the equation invariant. But if the strangled indices are not the same on both sides the sign $=$ must be changed into $\stackrel{*}{=}$. Moreover it is clear that in an equation with $\stackrel{*}{=}$ that has the invariant form, both sides have the same living running indices and it is therefore allowable to introduce $=$ instead of $\stackrel{*}{=}$.

4. A *tensor*[1]) of *contravariant valence* p and *covariant valence* q with n^{p+q} components $P^{\varkappa_1 \ldots \varkappa_p}_{\cdot \;\;\cdot\;\; \lambda_1 \ldots \lambda_q}$ and the transformation

$$P^{\varkappa'_1 \ldots \varkappa'_p}_{\cdot \;\;\cdot\;\; \lambda'_1 \ldots \lambda'_q} = A^{\varkappa'_1 \ldots \varkappa'_p \lambda_1 \ldots \lambda_q}_{\varkappa_1 \ldots \varkappa_p \lambda'_1 \ldots \lambda'_q} P^{\varkappa_1 \ldots \varkappa_p}_{\cdot \;\;\cdot\;\; \lambda_1 \ldots \lambda_q}. \tag{2.13}$$ [2])

The place of the indices is essential, for instance we must not write sometimes $P^{\varkappa\lambda}_{\cdot\cdot\mu}$ and at another time $P^{\varkappa \cdot \lambda}_{\cdot \mu \cdot}$ for the same quantity. If only lower (upper) indices occur the tensor is called *co-(contra-)variant*. If both occur it is often called *cocontravariant* (or *mixed*).

The tensor of valence 1,1

$$A^{\varkappa}_{\lambda} \stackrel{*}{=} \delta^{\varkappa}_{\lambda} \tag{2.14}$$

is called the *unity tensor*. Transvection of a quantity over one index with the unity tensor does not change the quantity, for instance

$$A^{\varkappa}_{\lambda} P^{\lambda\mu}_{\cdot\cdot\nu} = P^{\varkappa\mu}_{\cdot\cdot\nu}. \tag{2.15}$$

Tensors with more than one index also have components with respect to two or more different coordinate systems, e.g.

$$P^{\varkappa'\lambda}_{\cdot\cdot\nu} = A^{\varkappa'}_{\varkappa} P^{\varkappa\lambda}_{\cdot\cdot\nu} = A^{\lambda\nu'}_{\lambda'\nu} P^{\varkappa'\lambda'}_{\cdot\cdot\nu'}. \tag{2.16}$$

---

[1]) In former publications we used the word affinor instead of tensor and the word tensor was used in a more restricted sense (cf. footnote 1, p. 21).

[2]) $A^{\varkappa'\lambda'}_{\varkappa\lambda}$ stands for $A^{\varkappa'}_{\varkappa} A^{\lambda'}_{\lambda}$. Generally we do not put upper and lower indices in the same vertical line so as to avoid ambiguity if the process of raising and lowering of indices (to be defined later on) is to be applied. But there are a few exceptions.

These components are called *intermediate*. From this we see that $A_{\varkappa}^{\varkappa'}$ and $A_{\lambda'}^{\lambda}$ in (2.1) and (2.4) are intermediate components of $A_{\lambda}^{\varkappa}$. This justifies the use of the same kernel letter A in all three cases. The equations (2.1) and (2.4) now also express the fact that a vector is invariant for transvection with the unity tensor. The same holds mutatis mutandis for (2.13).

Two quantities of the same kind are said to be *equiform* if there exist two coordinate systems $(\varkappa)$ and $(\varkappa')$ such that the components of the first quantity with respect to $(\varkappa)$ are equal to the corresponding components of the second with respect to $(\varkappa')$, e.g.

$$P^{\varkappa\lambda}{}_{\cdot\cdot\mu} \overset{*}{=} \delta^{\varkappa}_{\varkappa'}{}^{\lambda}_{\lambda'}{}^{\mu'}_{\mu} Q^{\varkappa'\lambda'}{}_{\cdot\cdot\mu'}. \tag{2.17}$$

5. A tensor may be determined to within a scalar factor. Then it is called a *pseudo-tensor* (*pseudo-scalar*, *pseudo-vector* if the valence is 0 or 1). If more pseudo-tensors occur it may be that the unknown scalar factors in them are not entirely independent. For instance a direction and an $(n-1)$-direction fix a contravariant and a covariant pseudo-vector whose scalar factors are entirely free and independent. But if the scalar factors in $v^{\varkappa}$ and w_{λ} are free, the scalar factor in the pseudo-tensor $P^{\varkappa}_{\cdot\lambda} = v^{\varkappa} w_{\lambda}$ depends on the scalar factors in $v^{\varkappa}$ and w_{λ}. In the first case a pseudo-tensor can be distinguished if necessary by the sign $\lfloor\ \rfloor$. For instance $\lfloor P_{\varkappa\lambda} \rfloor$ is the pseudo-tensor belonging to the tensor $P_{\varkappa\lambda}$ and $\lfloor\ \rfloor$ only means that there is an entirely unknown scalar factor. In the second case it is most convenient to introduce next to the $\xi^{\varkappa}$ an auxiliary variable ξ^{0} (or if necessary more of these variables) [1]) with the transformation

$$\xi^{0'} = \sigma\, \xi^{0}. \tag{2.18}$$

A pseudo-tensor of *class c* is then defined by the transformation formula of the form

$$\mathfrak{P}^{\varkappa'\lambda'}{}_{\cdot\cdot\mu'} = \sigma^{c} A^{\varkappa'\lambda'\mu}_{\varkappa\lambda\mu'} \mathfrak{P}^{\varkappa\lambda}{}_{\cdot\cdot\mu}. \tag{2.19}$$

For pseudo-tensors defined in this way we usually prefer gothic kernel letters. Pseudo-scalars have only one component and no indices. It is therefore necessary to introduce an extra notation in order to distinguish between the different values that the pseudo-scalar can take. If we write $\overset{(0)}{\mathfrak{P}}$ or $\mathfrak{P}[0]$ for the component belonging to ξ^{0} we get the transformation formula

$$\overset{(0')}{\mathfrak{P}} = \sigma^{c} \overset{(0)}{\mathfrak{P}} \quad \text{or} \quad \mathfrak{P}[0'] = \sigma^{c} \mathfrak{P}[0]. \tag{2.20}$$

The second notation is used if there is some reason for avoiding an upper index, for instance if there is already another upper or lower index.

1) SCHOUTEN and HLAVATY 1929, 2.

6. If a quantity transforms like a tensor but with an extra factor τ dependent on Δ it is called

a) *tensor Δ-density of weight w and antiweight w'* if $\tau = \Delta^{-w}\bar{\Delta}^{-w'}$; $\bar{\Delta}$ = complex conjugate of Δ;

b) *tensor density of weight w* if $\tau = |\Delta|^{-w}$ (cf. I § 10);

c) *W-tensor* if $\tau = \Delta/|\Delta|$.

Δ-densities frequently occur in differential geometry but densities and W-tensors are more used in physics. For densities we prefer a gothic kernel and for Δ-densities a gothic kernel with the sign $\sim$ (snake) above it, for instance $\tilde{\mathfrak{v}}^{\varkappa\lambda}_{\cdot\cdot\mu}$. For non-scalar W-tensors we prefer roman kernel letters and for W-scalars a small greek letter, both with a snake. But there are exceptions. For instance the mass or charge per parallelepiped of basis vectors is a scalar density of weight $+1$ and is generally denoted by a small greek letter.

If we consider real coordinate systems only, there are two kinds of coordinate systems, transforming into themselves by coordinate transformations with $\Delta > 0$ and into each other by transformations with $\Delta < 0$. They can be distinguished by calling them *right* and *left* systems if some geometric object suitable for comparison is introduced in E_n.[1]) We call this the introduction of a screwsense in E_n. The screwsense can be fixed algebraically by the W-scalar $\tilde{\omega}$, whose component is $+1$ (-1) with respect to right (left) systems.

After the introduction of $\tilde{\omega}$ there is a one to one correspondence between Δ-densities and densities and between W-tensors and tensors. For instance

$$\tilde{\mathfrak{P}}^{\varkappa\lambda}_{\cdot\cdot\mu} = \tilde{\omega}\,\mathfrak{P}^{\varkappa\lambda}_{\cdot\cdot\mu}; \qquad \tilde{Q}^{\varkappa}_{\cdot\lambda} = \tilde{\omega}\,Q^{\varkappa}_{\cdot\lambda}. \tag{2.21}$$

If, instead of G_a, we use only the group of all real affine transformations with $\Delta > 0$, corresponding quantities have the same manner of transformation and can be identified. This is the first example of identification of quantities after introducing a subgroup of G_a.[2])

Analogous to the case of pseudo-scalars we denote the value of a scalar density or Δ-density or a W-scalar with respect to the coordinate system $(\varkappa)$ by writing $(\varkappa)$ right over the kernel or $[\varkappa]$ behind it, for instance

$$\overset{(\varkappa')}{\tilde{\mathfrak{p}}} = \Delta^{-w}\overset{(\varkappa)}{\tilde{\mathfrak{p}}} \quad\text{or}\quad \tilde{\mathfrak{p}}[\varkappa'] = \Delta^{-w}\tilde{\mathfrak{p}}[\varkappa] \tag{2.22}$$

$$\overset{(\varkappa')}{\tilde{p}} = \frac{\Delta}{|\Delta|}\overset{(\varkappa)}{\tilde{p}} \quad\text{or}\quad \tilde{p}[\varkappa'] = \frac{\Delta}{|\Delta|}\tilde{p}[\varkappa]. \tag{2.23}$$

---

[1]) This is just what we do in ordinary space if we call a system right or left after comparison with our own right hand.

[2]) We get another example if instead of G_a we use the so called special affine group G_{sa} consisting of all transformations of G_a with $\Delta = +1$. Then the differences between tensors, tensor densities and tensor Δ-densities vanish.

7. If a quantity has different kinds of indices belonging to different spaces it is called a *connecting quantity*. These quantities occur frequently. E.g. $B_b^\varkappa$ in (1.9) is a connecting quantity, $\varkappa$ belongs to E_n and b to E_m. C_λ^x in (1.8) is another connecting quantity. λ belongs to E_n and x belongs to the E_{n-m} that arises from reducing the E_n with respect to the m-direction of E_m. C_λ^x and $B_b^\varkappa$ are called the *covariant* and the *contravariant connecting quantity* of the E_m in E_n.

Exercises.

I 2,1. n^p numbers $P^{\varkappa_1 \dots \varkappa_p}$ are components of a contravariant tensor of valence p if and only if for some definitely chosen value of $q \leq p$

I 2,1 α) $$P^{\varkappa_1 \dots \varkappa_p} Q_{\varkappa_1 \dots \varkappa_q}$$

is a contravariant tensor for every choice of the tensor $Q_{\varkappa_1 \dots \varkappa_q}$.

I 2,2. $P^{\varkappa}_{\cdot\lambda} = \alpha A^\varkappa_\lambda$ if and only if $P^{\varkappa}_{\cdot\lambda} v^\lambda = \alpha v^\varkappa$ for every choice of $v^\varkappa$.

I 2,3. $P^{\varkappa}_{\cdot\lambda} = \alpha A^\varkappa_\lambda + p_\lambda q^\varkappa$ if and only if $P^{\varkappa}_{\cdot\lambda} v^\lambda = \alpha v^\varkappa$ for every $v^\varkappa$ satisfying the equation $p_\lambda v^\lambda = 0$.

I 2,4. Let $P^{\varkappa}_{\cdot\lambda}$ be the tensor of the linear vector transformation transforming the $\underset{\lambda}{e}^\varkappa$ into the $\underset{\lambda'}{e}^\varkappa$

I 2,4α) $$\underset{\lambda'}{e}^\varkappa = \delta^\lambda_{\lambda'} P^{\varkappa}_{\cdot\mu} \underset{\lambda}{e}^\mu .$$

Prove that

$$P^{\varkappa'}_{\cdot\lambda} \overset{*}{=} \delta^{\varkappa'}_{\lambda} .$$

I 2,5. Prove that two equiform quantities can always be transformed into each other by a homogeneous linear vector transformation.

§ 3. Invariant processes and relations.

The following processes and relations are invariant for all rectilinear coordinate transformations.

1. The *addition* of two quantities of the same kind (cf. I § 2).

2. The construction of an *isomer* by changing the location of the upper indices and/or of the lower indices, e.g.

(3.1) $$Q_{\lambda\varkappa}^{\cdot\cdot\,\nu\mu} = P_{\varkappa\cdot\lambda}^{\cdot\,\mu\,\cdot\,\nu} .$$

3. The (*general*) *multiplication* of two quantities, e.g.

(3.2) $$P^{\varkappa\lambda}_{\cdot\cdot\mu} Q_{\nu\cdot\sigma}^{\cdot\varrho} = R^{\varkappa\varrho\cdot\cdot\lambda}_{\cdot\cdot\sigma\mu\cdot\nu} .$$

4. The *contraction* (Faltung, Verjüngung) with respect to an upper and a lower index or with respect to several such pairs of indices, e.g.

(3.3) $$P^{\varkappa\lambda\cdot\cdot\cdot\sigma}_{\cdot\cdot\varkappa\sigma\varrho} = Q^{\lambda}_{\cdot\varrho} .$$

5. The *transvection* (Überschiebung) of two quantities by application of the process of contraction to the general product, e.g.

$$P^{\varkappa\lambda}_{\cdot\cdot\mu}\,Q^{\mu}_{\cdot\nu\lambda} = R^{\varkappa}_{\cdot\nu}. \tag{3.4}$$

The indices used for summation in contractions and transvections are called *saturated* (also *dummy*), the other *free* indices. The transvection is a kind of multiplication, i.e. it is distributive with respect to addition.

6. The process of *mixing* over p upper or p lower indices by constructing all $p!$ isomers that can be obtained by permutation of these indices, addition of these isomers and division by $p!$. The process is denoted by round brackets ( ).[1]) If indices have to be singled out the sign | | is used, e.g.

$$\begin{cases} P^{\cdot\cdot\cdot\cdot\omega}_{(\varkappa\lambda|\mu|\nu)} = \frac{1}{6}(P^{\cdot\cdot\cdot\cdot\omega}_{\varkappa\lambda\mu\nu} + P^{\cdot\cdot\cdot\cdot\omega}_{\lambda\nu\mu\varkappa} + P^{\cdot\cdot\cdot\cdot\omega}_{\nu\varkappa\mu\lambda} + \\ \qquad + P^{\cdot\cdot\cdot\cdot\omega}_{\varkappa\nu\mu\lambda} + P^{\cdot\cdot\cdot\cdot\omega}_{\nu\lambda\mu\varkappa} + P^{\cdot\cdot\cdot\cdot\omega}_{\lambda\varkappa\mu\nu}) \end{cases} \tag{3.5}$$

where μ stays put and the three other lower indices permute in the remaining places. The effect of ( ) is not intercepted by any form of ordinary brackets, e.g.

$$(P_{(\varkappa|\lambda} + Q_{(\varkappa|\lambda)})\,R_{\mu|\nu)} = P_{(\varkappa|\lambda}\,R_{\mu|\nu)} + Q_{(\varkappa|\lambda}\,R_{\mu|\nu)}. \tag{3.6}$$

A quantity is called *symmetric* in p upper or p lower indices if it is invariant for mixing over these indices. It is then also invariant if two of these indices are interchanged. $P^{(\varkappa_1\ldots\varkappa_p)}$ is called the *symmetric part* of $P^{\varkappa_1\ldots\varkappa_p}$.

7. The process of *alternation* over p upper or p lower indices, effected in the same way as the mixing process with the only difference that all isomers obtained by odd permutations get a negative sign. This process is denoted by square brackets [ ][1]), e.g.

$$\begin{cases} P^{\cdot\cdot\cdot\cdot\omega}_{[\varkappa\lambda|\mu|\nu]} = \frac{1}{6}(P^{\cdot\cdot\cdot\cdot\omega}_{\varkappa\lambda\mu\nu} + P^{\cdot\cdot\cdot\cdot\omega}_{\lambda\nu\mu\varkappa} + P^{\cdot\cdot\cdot\cdot\omega}_{\nu\varkappa\mu\lambda} - \\ \qquad - P^{\cdot\cdot\cdot\cdot\omega}_{\varkappa\nu\mu\lambda} - P^{\cdot\cdot\cdot\cdot\omega}_{\nu\lambda\mu\varkappa} - P^{\cdot\cdot\cdot\cdot\omega}_{\lambda\varkappa\mu\nu}). \end{cases} \tag{3.7}$$

As far as interception by other brackets is concerned [ ] has the same property as ( ). A quantity is called *alternating* in p upper or p lower indices if it is invariant for alternation over these indices. It then changes sign if two of these indices are interchanged. Alternation over more than n indices always leads to zero. $P^{[\varkappa_1\ldots\varkappa_p]}$ is called the *alternating part* of $P^{\varkappa_1\ldots\varkappa_p}$.

Of these seven processes the first four are fundamental, the others being only combinations of them. In order to see what can be done

[1]) The use of ( ) and [ ] for mixing and alternating was proposed by BACH 1921, 1, p. 113.

by these four fundamental processes we need a theorem concerning concomitants. If a set of quantities is given, every quantity whose components can be expressed as functions of the components of the quantities of the set is called a (simultaneous) *concomitant* of the set. For a scalar the term *invariant* also is used. If all these functions are rational integral the concomitant is called *rational integral*. The theorem in question is:

All rational integral concomitants of a given set of quantities can be derived from the quantities of the set by means of addition, construction of isomers, multiplication and contraction.

This theorem is an immediate consequence of the well known first fundamental theorem of the symbolic method used in the theory of invariants.[1], [2]

Exercises.

I 3,1. Prove that $P^{\varkappa_1 \dots \varkappa_p}$ vanishes if it is symmetric in $\varkappa_1 \dots \varkappa_s$ and if $P^{\varkappa_1 \dots \varkappa_p} w_{\varkappa_1} \dots w_{\varkappa_s} = 0$ for every choice of the vector w_λ.

I 3,2. If $u_{\lambda\varkappa} v^\lambda v^\varkappa = 0$ for every vector $v^\varkappa$ for which $v^\lambda w_\lambda = 0$, $u_{(\lambda\varkappa)}$ has the form

I 3,2 α) $$u_{(\lambda\varkappa)} = w_{(\lambda} p_{\varkappa)}$$ [3])

I 3,3. If $w_\varkappa P^{\varkappa}_{\cdot\lambda} = \alpha w_\lambda$ for every vector w_λ for which $v^\lambda w_\lambda = 0$, $P^{\varkappa}_{\cdot\lambda}$ has the form

I 3,3 α) $$P^{\varkappa}_{\cdot\lambda} = \alpha A^{\varkappa}_{\lambda} + v^\varkappa p_\lambda$$ [3])

and α is independent of the choice of w_λ.

I 3,4. If $P^{\varkappa}_{\cdot\lambda\mu} w_\varkappa v^\lambda v^\mu = 0$ for every choice of $v^\varkappa$ and w_λ satisfying $v^\lambda w_\lambda = 0$, $P^{\varkappa}_{\cdot(\lambda\mu)}$ has the form

I 3,4 α) $$P^{\varkappa}_{\cdot(\lambda\mu)} = A^{\varkappa}_{(\lambda} p_{\mu)}$$ [4]), [3]).

---

[1]) Cf. e.g. WEITZENBÖCK, 1923, 1, p. 93.

[2]) There is a large amount of literature on the problem of finding algebraical concomitants of a quantity or a set of quantities. Cf. for instance WEITZENBÖCK 1923, 1; 1939, 1; DROST 1931, 1; OBERTI 1936, 1; WEYL 1939, 1; CISOTTI 1940, 1; 1941, 1; 1943, 1; 1944, 1; WADE 1941, 1; 1943, 1; 1944, 1; 1945, 1; DOYLE 1941, 1; DUBNOV 1941, 1; BRUCK and WADE 1943, 1; BRUCK 1944, 1; LITTLEWOOD 1944, 1; 1950, 1; RUTHERFORD 1948, 1; RACAH 1949, 1.

[3]) Propositions of this kind can always be proved by means of the following well known algebraic theorem. If a polynomial $f(x^1, \dots, x^n)$ does not contain multiple irreducible factors and if the polynomial $F(x^1, \dots, x^n)$ is zero for all values for which $f = 0$, then F is divisible by f.

[4]) From a letter of A. FRIEDMANN.

I 3,5. Verify that $v^{[\varkappa_1 \dots \varkappa_p]}$ can be written out in two ways

I 3,5 α) $$v^{[\varkappa_1 \dots \varkappa_p]} = \frac{1}{p} \sum_{s}^{1,\dots,p} (-1)^{s-1} v^{\varkappa_s [\varkappa_1 \dots \varkappa_{s-1} \varkappa_{s+1} \dots \varkappa_p]}.$$

I 3,5 β) $$v^{[\varkappa_1 \dots \varkappa_p]} = \frac{1}{p} \sum_{s}^{1,\dots,p} (-1)^{s-1} v^{[\varkappa_2 \dots \varkappa_s |\varkappa_1| \varkappa_{s+1} \dots \varkappa_p]}.$$

I 3,6. If $P^{\varkappa}_{\cdot\lambda\mu} = P^{\varkappa}_{\cdot[\lambda\mu]}$ and if $P^{\varkappa}_{\cdot\lambda\mu} w_\varkappa v^\mu \propto w_\lambda$ for every choice of $v^\varkappa$ and w_λ satisfying $v^\lambda w_\lambda = 0$, $P^{\varkappa}_{\cdot\lambda\mu}$ has the form

I 3,6 α) $$P_{\cdot\lambda\mu} = A^{\varkappa}_{[\lambda} p_{\mu]}.$$

§ 4. Section and reduction with respect to an E_m in E_n. Decomposition with respect to a rigged E_m.

Let an E_m in E_n through the origin be given by the equations

(4.1) $$\begin{cases} \text{a) } \xi^\varkappa = B^\varkappa_b \eta^b \quad \text{or} \quad \text{b) } C^x_\lambda \xi^\lambda = 0 \\ b = 1, \dots, m; \qquad x = m+1, \dots, n. \end{cases}$$

The E_m's parallel to it are given by the parametric equation (1.13) with the parameters ζ^x. The η^b are rectilinear coordinates in the E_m and the ζ^x rectilinear coordinates in the $E_{m'}$; $m' = n - m$, arising from the reduction of E_n with respect to the m-direction of E_m.

The points of the E_m which lie in the hyperplanes of a vector w_λ satisfy the equations [cf. (2.5)]

(4.2) $$w_\lambda B^\lambda_b \eta^b = c; \qquad w_\lambda B^\lambda_b \eta^b = c + 1.$$

Hence

(4.3) $$'w_b \stackrel{\text{def}}{=} B^\lambda_b w_\lambda$$

is the *section* of w_λ with the E_m. Conversely w_λ is not uniquely determined by $'w_b$. It is only determined to within an additive vector whose $(n-1)$-direction contains the m-direction of E_m. The section of a *covariant* tensor with valence >1 is formed in the same way, e.g.

(4.4) $$'w_{cba} \stackrel{\text{def}}{=} B^{\mu\lambda\varkappa}_{cba} w_{\mu\lambda\varkappa}.$$

To every *contravariant* vector p^a in E_m there belongs one and only one contravariant vector in E_n

(4.5) $$p^\varkappa \stackrel{\text{def}}{=} B^\varkappa_a p^a$$

$p^\varkappa$ vanishes if it is transvected with $C^x_\varkappa$. Conversely to every contravariant vector in E_n which satisfies this condition there belongs one and only one contravariant vector in E_m. In fact p^a and $p^\varkappa$ have the same geometric interpretation, viz. an arrow lying in the m-direction of E_m. Therefore we consider $p^\varkappa$ and p^a as two different sets of components of the *same* quantity and accordingly we use the same kernel letter. The E_n-components of a *contravariant* tensor of E_m with valence >1 are formed in the same way, e.g.

$$p^{\varkappa\lambda\mu} \overset{\text{def}}{=} B^{\varkappa\lambda\mu}_{abc}\, p^{abc}; \qquad a, b, c = 1, \ldots, \mathrm{m}. \tag{4.6}$$

The E_m's parallel to the given E_m and passing through the points $\xi^\varkappa$, $\xi^\varkappa + v^\varkappa$ have the coordinates $\zeta^x = C^x_\varkappa \xi^\varkappa$, $\zeta^x = C^x_\varkappa \xi^\varkappa + C^x_\varkappa v^\varkappa$. From this we see that to the *contravariant* vector $v^\varkappa$ of E_n there corresponds the vector

$$''v^x \overset{\text{def}}{=} C^x_\lambda v^\lambda, \qquad x = \mathrm{m}+1, \ldots, \mathrm{n} \tag{4.7}$$

in the $E_{m'}$. We call $''v^x$ the *reduction* of $v^\varkappa$ with respect to the given E_m. Conversely $v^\varkappa$ is not uniquely determined by $''v^x$. It is only determined to within an additive vector lying in the m-direction of E_m. The reduction of a *contravariant* tensor with valence >1 is formed in the same way, e.g.

$$''v^{xyz} \overset{\text{def}}{=} C^{xyz}_{\varkappa\lambda\mu}\, v^{\varkappa\lambda\mu}. \tag{4.8}$$

To every *covariant* vector s_y in $E_{m'}$ there belongs one and only one covariant vector in E_n

$$s_\lambda \overset{\text{def}}{=} C^y_\lambda s_y. \tag{4.9}$$

s_λ vanishes if it is transvected with B^λ_b. Conversely to every covariant vector in E_n satisfying this condition there belongs one and only one covariant vector in $E_{m'}$. In fact s_y is represented by the two $E_{m'-1}$'s in $E_{m'}$ resulting from reduction of the two E_{n-1}'s of s_λ. Therefore we consider s_y and s_λ as two different sets of components of the *same* quantity and express this by using the same kernel letter. The E_n-components of a *covariant* tensor of $E_{m'}$ with valence >1 can be formed in the same way, e.g.

$$S_{\mu\lambda\varkappa} \overset{\text{def}}{=} C^{zyx}_{\mu\lambda\varkappa}\, S_{zyx}. \tag{4.10}$$

If in E_n an $E_{m'}$ through the origin is given which has no direction in common with the E_m, the E_m is called *rigged* (eingespannt). Conversely the $E_{m'}$ is rigged by the E_m. This $E_{m'}$ is not identical with the $E_{m'}$ arising from reduction of the E_n with respect to the E_m. But there is a one to one correspondence between the points of these two $E_{m'}$'s because every point of the $E_{m'}$ arising from reduction represents an E_m and this E_m has just one point in common with the rigging $E_{m'}$. Hence we may now, *after having carried out the rigging, identify* both $E_{m'}$'s. From now on we have only one $E_{m'}$ in which the ζ^x in (1.13) can be used as rectilinear coordinates and this $E_{m'}$ can be interpreted geometrically in two ways. Firstly it may be interpreted as an $E_{m'}$ in E_n through the origin with the equations [cf. (4.1 a)]

$$\xi^\varkappa = C^\varkappa_x \zeta^x \tag{4.11}$$

and the contravariant connecting quantity $C^\varkappa_x$. Secondly it can be interpreted as the manifold of all E_m's parallel to the given E_m in which every E_m is considered as a point. Substituting (4.11) in (1.13) we get

$$C^x_y\, C^\lambda_y\, \zeta^y = \zeta^x \tag{4.12}$$

and this proves that $C^x_y \stackrel{\text{def}}{=} C^x_\lambda\, C^\lambda_y$ is the unity tensor of $E_{m'}$.

The $E_{m'}$ is rigged by the E_m. Hence the E_m can be identified with the E_m arising from reduction of the E_n with respect to the $E_{m'}$. The $E_{m'}$'s constituting this latter E_m are the $E_{m'}$'s parallel to the rigging $E_{m'}$ and are given by the equations [cf. (1.13)]

$$B^a_\lambda\, \xi^\lambda = \eta^a. \tag{4.13}$$

B^a_λ is the covariant connecting quantity of the rigging $E_{m'}$. Substituting (4.1 a) in (4.13) we see that $B^a_b \stackrel{\text{def}}{=} B^a_\lambda\, B^\lambda_b$ is the unity tensor of E_m.

After the rigging has been performed we are able to form

1. the *sections* of covariant quantities of E_n with E_m and $E_{m'}$, for instance

$$\text{a)}\ \ 'w_b = B^\lambda_b\, w_\lambda; \qquad \text{b)}\ \ ''w_y = C^\lambda_y\, w_\lambda. \tag{4.14}$$

2. the *reductions* of contravariant quantities of E_n with respect to E_m and to $E_{m'}$, for instance

$$\text{a)}\ \ ''v^x = C^x_\varkappa\, v^\varkappa; \qquad \text{b)}\ \ 'v^a = B^a_\varkappa\, v^\varkappa. \tag{4.15}$$

These reductions can also be considered as projections on $E_{m'}(E_m)$ in the m-(m'-)direction of $E_m(E_{m'})$.

3. the E_n-*components* of covariant and contravariant quantities of E_m and $E_{m'}$, for instance

$$\text{a)}\ \ p^\varkappa = B^\varkappa_a\, p^a; \qquad \text{b)}\ \ q_\lambda = B^b_\lambda\, q_b \tag{4.16}$$

$$\text{a)}\ \ s_\lambda = C^y_\lambda\, s_y; \qquad \text{b)}\ \ r^\varkappa = C^\varkappa_x\, r^x. \tag{4.17}$$

In (4.14–17) the (a) formulae hold for every E_m in E_n but the (b) formulae are only valid if the E_m is rigged. Applying the processes (4.16) and (4.17) to B^a_b and C^x_y respectively we see that B^a_b, $B^\varkappa_b$ and B^a_λ and in the same way C^x_y, C^x_λ and $C^\varkappa_y$ are three different sets of components of the same quantity and that

$$B^{\varkappa}_{\lambda} \overset{\text{def}}{=} B^{\varkappa}_{b} B^{b}_{\lambda} \quad \text{and} \quad C^{\varkappa}_{\lambda} \overset{\text{def}}{=} C^{\varkappa}_{x} C^{x}_{\lambda} \tag{4.18}$$

form a fourth set. This justifies the use of the same kernel letter B and C respectively in all four cases.

In the following table results are gathered together for the rigged E_m.

$$\left\{\begin{array}{ccc}
E_m & E_n & E_{m'} \\
p^a \xrightarrow[\text{components}]{\text{change of}} & \boxed{p^{\varkappa} = B^{\varkappa}_{a} p^a} \quad r^{\varkappa} = C^{\varkappa}_{x} r^x & \xleftarrow[\text{components}]{\text{change of}} r^x \\
v^a \xleftarrow[\text{components}]{\text{change of}} & v^{\varkappa} \text{ with } C^{x}_{\varkappa} v^{\varkappa} = 0; \quad v^{\varkappa} \text{ with } B^{a}_{\varkappa} v^{\varkappa} = 0 & \xrightarrow[\text{components}]{\text{change of}} v^x \\
{}'v^a = B^{a}_{\varkappa} v^{\varkappa} & \xleftarrow[\text{(projection)}]{\text{reduction}} v^{\varkappa} \xrightarrow[\text{(projection)}]{\text{reduction}} & \boxed{{}''v^x = C^{x}_{\varkappa} v^{\varkappa}} \\
{}'v^{\varkappa} = B^{\varkappa}_{a}\, {}'v^a = B^{\varkappa}_{\lambda} v^{\lambda} & & {}''v^{\varkappa} = C^{\varkappa}_{x}\, {}''v^x = C^{\varkappa}_{\lambda} v^{\lambda} \\
\boxed{{}'w_b = B^{\lambda}_{b} w_{\lambda}} & \xleftarrow{\text{section}} w_{\lambda} \xrightarrow{\text{section}} & {}''w_y = C^{\lambda}_{y} w_{\lambda} \\
{}'w_{\lambda} = B^{b}_{\lambda}\, {}'w_b = B^{\varkappa}_{\lambda} w_{\varkappa} & & {}''w_{\lambda} = C^{y}_{\lambda}\, {}''w_y = C^{\varkappa}_{\lambda} w_{\varkappa} \\
w_b \xleftarrow[\text{components}]{\text{change of}} & w_{\lambda} \text{ with } C^{\lambda}_{y} w_{\lambda} = 0; \quad w_{\lambda} \text{ with } B^{\lambda}_{b} w_{\lambda} = 0 & \xrightarrow[\text{components}]{\text{change of}} w_y \\
q_b \xrightarrow[\text{components}]{\text{change of}} & q_{\lambda} = B^{b}_{\lambda} q_b \quad \boxed{s_{\lambda} = C^{y}_{\lambda} s_y} & \xleftarrow[\text{components}]{\text{change of}} s_y
\end{array}\right. \tag{4.19}$$

If the E_m is not rigged the $E_{m'}$ is the reduced E_n. Then only the formulae put in a frame remain valid and the reduction of $v^{\varkappa}$ can no longer be interpreted as a projection. From (4.18), (1.10) and the corresponding formulae for B^{a}_{λ} and C^{λ}_{y} it follows that for a rigged E_m

$$(B^{\varkappa}_{\lambda} + C^{\varkappa}_{\lambda})\, B^{\lambda}_{b} = B^{\varkappa}_{b}; \qquad (B^{\varkappa}_{\lambda} + C^{\varkappa}_{\lambda})\, C^{\lambda}_{y} = C^{\varkappa}_{y} \tag{4.20}$$

and

$$B^{\varkappa}_{\lambda} + C^{\varkappa}_{\lambda} = A^{\varkappa}_{\lambda}. \tag{4.21}$$

Hence by means of $B^{\varkappa}_{\lambda}$ and $C^{\varkappa}_{\lambda}$ a vector can be split up into two parts in E_m and in $E_{m'}$ called the *E_m-part* and the *$E_{m'}$-part* respectively. Besides its components in E_n, each part has also components in the submanifold to which it belongs. Tensors with a valence >1 split up into more parts, for instance

$$P^{\varkappa}_{\cdot\lambda} = B^{\varkappa\sigma}_{\varrho\lambda} P^{\varrho}_{\cdot\sigma} + B^{\varkappa}_{\varrho} C^{\sigma}_{\lambda} P^{\varrho}_{\cdot\sigma} + C^{\varkappa}_{\varrho} B^{\sigma}_{\lambda} P^{\varrho}_{\cdot\sigma} + C^{\varkappa\sigma}_{\varrho\lambda} P^{\varrho}_{\cdot\sigma}. \tag{4.22}$$

The first is the E_m-part and the last the $E_{m'}$-part.

§ 5. Rank, domain and support of domain with respect to one or more indices.

From a co- or contravariant or mixed tensor of valence p we choose one index at some definite place and write μ, say, for this index. If the tensor is then transvected over *all other* indices with all basis vectors belonging to some arbitrary coordinate system $(\varkappa)$ (that is, if $1, \ldots, n$ are substituted for all other indices with respect to this system) we get n^{p-1} co- or contravariant vectors according as μ is an upper or a lower index. The number r of linearly independent ones among them is an arithmetic invariant and called the *μ-rank* of the tensor. This rank is always ≥ 0 and $\leq n$. The set of all vectors which are linearly dependent on these r is called the *μ-domain*. If μ is an upper (lower) index the r vectors span an r-direction $[(n-r)$-direction$]$ called the *support of the μ-domain*. The μ-domain and its support are invariants of the given tensor. The same holds mutatis mutandis for every quantity of valence p. Instead of vectors we then have pseudo-vectors or vector densities or W-vectors.

The rank and the domain with respect to more than one index is defined in the same way.[1]) Obviously the rank with respect to *all* indices is always equal to one.

If the valence is 2 the rank with respect to one index is equal to the rank with respect to the other one. Hence in this case there exists only one non-trivial rank and this rank is equal to the rank of the matrix of the components.

Exercises.

I 5,1. If a symmetric tensor with valence p has μ-rank r, it can be written as a sum of r products of a vector in the μ-domain with a symmetric tensor of valence $p-1$. Hence every symmetric tensor can be written as a sum of products of vectors. (Cf. I § 6.)

I 5,2. Prove that the rank with respect to some indices is always equal to the rank with respect to the other indices.

I 5,3. If a tensor $v^{\varkappa_1 \cdots \varkappa_p}$ is symmetric or alternating in all indices its rank is the same with respect to every index. This rank is r if and only if the equation $v^{\varkappa_1 \cdots \varkappa_p} w_{\varkappa_1} = 0$ has exactly $n-r$ linearly independent solutions w_λ.

I 5,4. Prove that the $v^{\varkappa_1 \cdots \varkappa_p}$ from Exerc. I 5,3 has rank r if and only if [2])

---

[1]) Cf. Alexander 1925, 1; Rice 1928, 1.

[2]) In (I 5,4α) the alternation is only over the last index of each factor. In (I 5,4β) the last indices are alternated, and independently the last but one of each factor are also alternated. The notation [[ ]] is very often used in this sense.

$$\text{I 5,4}\alpha) \qquad v^{\varkappa_1 \ldots [\mu_1} \ldots v^{|\varkappa_s \ldots |\mu_s]} \begin{cases} \neq 0 & \text{for} \quad s \leq r \\ = 0 & \text{for} \quad s > r \end{cases}$$

and also if and only if

$$\text{I 5,4}\beta) \qquad v^{\varkappa_1 \ldots [\lambda_1[\mu_1} \ldots v^{|\varkappa_s \ldots |\lambda_s]\mu_s]} \begin{cases} \neq 0 & \text{for} \quad s \leq r \\ = 0 & \text{for} \quad s > r. \end{cases}$$

§ 6. Symmetric tensors.

A co- or contravariant tensor that is symmetric in *all* indices is called a *symmetric tensor.*[1]) Hence a symmetric tensor is invariant for interchange of any two indices. Symmetric pseudo-tensors, tensor densities and W-tensors are defined in the same way. The number of the linearly independent components of a symmetric tensor of valence p is $\binom{n+p-1}{p}$. In a centred E_n, a covariant symmetric tensor $P_{\lambda_1 \ldots \lambda_p}$ is represented by the hypersurface with the equations

$$(6.1) \qquad P_{\lambda_1 \ldots \lambda_p} \xi^{\lambda_1} \ldots \xi^{\lambda_p} = \pm 1$$

in point coordinates and a contravariant symmetric tensor $Q^{\varkappa_1 \ldots \varkappa_q}$ is represented by the hypersurface with the equations

$$(6.2) \qquad Q^{\varkappa_1 \ldots \varkappa_p} u_{\varkappa_1} \ldots u_{\varkappa_p} = \pm 1$$

in hyperplane coordinates. If we put zero instead of ± 1 in (6.1, 2) we get the equation of the hypercone belonging to the symmetric tensor. The hypercone determines only the symmetric pseudo-tensor $\lfloor P_{\lambda_1 \ldots \lambda_p} \rfloor$ or $\lfloor Q^{\varkappa_1 \ldots \varkappa_p} \rfloor$ respectively. From two covariant (contravariant) symmetric tensors of valences p and q a symmetric tensor of valence $p+q$ can be derived by general multiplication and mixing over all indices. This process is called *symmetric multiplication* of symmetric tensors. It can also be applied to a greater number of symmetric tensors. The polynominal at the left hand side of the equation of the hypersurface of the symmetric product is the product of the polynomials on the left hand side of the equations belonging to the factors. Hence the notions *division* and *divisor* known from the theory of homogeneous polynomials can be extended to symmetric tensors. A symmetric tensor Q is a divisor of a symmetric tensor P if and only if there exists a symmetric tensor R such that P is the symmetric product of Q and R. A symmetric tensor without a divisor is called *irreducible.* It follows from the theory of homogeneous polynomials that every symmetric tensor is the

---

[1]) In former publications we used the term tensor instead of symmetric tensor (cf. footnote 1 on p. 10). The word "tenseur" in CARTAN 1938, 1, I 30 has quite another meaning.

symmetric product of a finite number of irreducible symmetric tensors and that these factors are uniquely determined to within a scalar factor. If all irreducible divisors have valence 1 the hypercone of a covariant symmetric tensor of valence p consists of p hyperplanes and the hypercone of a contravariant symmetric tensor of valence p consists of p straight lines.

Exercises.

I 6,1 [1]). The domain of a symmetric tensor is the same with respect to all its indices. If the rank is r the tensor can be written as a sum of symmetric products of vectors from this domain. The support of the domain of a symmetric tensor $w_{\lambda_1 \dots \lambda_p}$ of rank r is an $(n-r)$-direction. Its hypersurface is a cylinder consisting of $\infty^{r-1} E_{n-r}$'s with this $(n-r)$-direction. The hypercone consists for $r > 1$ of $\infty^{r-2} E_{n-r+1}$'s all of them containing the E_{n-r} with the equation

$$w_{\lambda_1 \dots \lambda_p} \xi^{\lambda_p} = 0. \tag{I 6,1α}$$

I 6,2 [2]). The rank of the symmetric tensor $w_{\lambda_1 \dots \lambda_p}$ is $< n$ if and only if

$$w_{(\varkappa_1 \dots \lambda_1 [\mu_1 [\nu_1} \cdots w_{\varkappa_n \dots \lambda_n) \mu_n] \nu_n]} = 0. \tag{I 6,2α}$$

In this expression the mixing has to be effected over all indices except the last two of every factor. The homogeneous polynomial of degree $n(p-2)$ belonging to the left hand side of (I 6,2α) is the hessian covariant of the polynomial belonging to $w_{\lambda_1 \dots \lambda_p}$.

I 6,3 [3]). If the symmetric tensors $P_{\lambda\varkappa}$ and $Q_{\lambda\varkappa}$ satisfy the equation $P_{[\varkappa (\lambda} Q_{\mu] \nu)} = 0$ they differ only by a scalar factor.

I 6,4. A symmetric tensor of valence 2 is for $n \leq 2$ always the symmetric product of two vectors.

I 6,5. A symmetric tensor $w_{\lambda_1 \dots \lambda_p}$ is the symmetric product of p equal vectors if and only if

$$w_{[\lambda_1 [\lambda_2 | \lambda_3 \dots \lambda_p |} w_{\varkappa_1] \varkappa_2] \varkappa_3 \dots \varkappa_p} = 0. \tag{I 6,5α}$$

§ 7. Multivectors.

A co- or contravariant tensor that is alternating in *all* indices is called a *multivector* or *alternating tensor*. It is also called a *p-vector* if its valence is p. [4]) The valence can not be $> n$. Pseudo-multivectors, multivector densities and W-multivectors are defined in the same way. A multivector changes its sign if two indices are interchanged. The number of linearly independent components of an alternating quantity

[1]) P. P. 1949, 1, p. 12.

[2]) SINIGALLIA 1905, 1, p. 371.

[3]) SCHOUTEN 1924, 1, p. 205; EISENHART 1926, 1, p. 32.

[4]) A multivector is also called antisymmetric or skew symmetric tensor by some authors.

of valence p is $\binom{n}{p}$. It is usual to call a 2-vector *bivector*, a 3-vector *trivector* and a 4-vector *quadrivector*.

From two covariant (contravariant) multivectors of valences p and q; $p+q\leq n$ a multivector of valence $p+q$ can be derived by general multiplication and alternation over all indices. This process is called *alternating multiplication* of multivectors. It can also be applied to a greater number of multivectors, provided that the sum of the valences is $\leq n$. For this alternating product we often use a notation without indices. Instead of $u^{[\varkappa_1 \cdots \varkappa_p} v^{\lambda_1 \cdots \lambda_q]}$ we write shortly $[uv]$.[1]) Note that

$$(7.1) \qquad [u\, v] = (-1)^{pq}\, [v\, u].$$

A multivector Q is called a *divisor* of a multivector P if and only if there exists a multivector R such that P is the alternating product of Q and R. A vector $v^{\varkappa}$ is a divisor of $v^{\varkappa_1 \cdots \varkappa_p}$ if and only if $v^{[\varkappa_1 \cdots \varkappa_p} v^{\varkappa]} = 0$. To prove this we take $\underset{1}{e}^{\varkappa} = v^{\varkappa}$. Then $v^{\varkappa_1 \cdots \varkappa_p}$ cannot have non vanishing components with all indices $\neq 1$. Hence $\underset{1}{e}^{\varkappa}$ is a divisor. If $v^{\varkappa_1 \cdots \varkappa_p}$ has s linearly independent vector divisors this p-vector can be written as the alternating product of these vectors and a $(p-s)$-vector. This can be proved by taking the s vectors as the first s contravariant basis vectors.

A p-vector that can be written as the alternating product of p vectors is called *simple*. The necessary and sufficient condition that $v^{\varkappa_1 \cdots \varkappa_p}$ is simple is

$$(7.2) \qquad v^{[\varkappa_1 \cdots \varkappa_p} v^{\lambda_1] \cdots \lambda_p} = 0.$$[2])

The same holds mutatis mutandis for covariant p-vectors. The necessity of the condition is obvious. In order to prove the sufficiency we take the coordinate system in such a way that $v^{1 \cdots p} \neq 0$. Then (7.2) states that the p vectors $v^{\varkappa 2 \cdots p}$, $v^{\varkappa 1 3 \cdots p}$, $v^{\varkappa 1 2 4 \cdots p}$, $\ldots$, $v^{\varkappa 1 \cdots (p-1)}$ are divisors of $v^{\varkappa_1 \cdots \varkappa_p}$. But these p vectors can not be linearly dependent because the i-th component $\pm v^{1 \cdots p}$ of the i-th vector is $\neq 0$ and the i-th components of all other vectors are zero. Hence $v^{\varkappa_1 \cdots \varkappa_p}$ has p independent vector factors.[3])

Every n-vector is simple. Hence any two n-vectors differ only by a scalar factor. But every $(n-1)$-vector is also simple.[4]) In order

---

[1]) Cf. P. P. 1949, 1, I § 7,10. This notation is intimately connected with CARTAN's notations for differential forms (cf. II § 12).

[2]) Givens proved 1937, 1 that $v^{[\varkappa_1 \cdots \varkappa_p} v^{\lambda_1 \lambda_2] \cdots \lambda_p} = 0$ is another necessary and sufficient condition and moreover that from $v^{[\varkappa_1 \cdots \varkappa_p} v^{\lambda_1 \cdots \lambda_{s+1}] \cdots \lambda_p} = 0$ it always follows that $v^{[\varkappa_1 \cdots \varkappa_s} v^{\lambda_1 \cdots \lambda_s] \cdots \lambda_p} = 0$ if s is odd. This proposition was proved by WEITZENBÖCK 1908, 1 for $p=3$, $n=5$ and formulated 1923, 1, p. 87 for the general case Cf. BOMPIANI 1952, 2.

[3]) If $v^{\varkappa_1 \cdots \varkappa_p}$ is written as the product of these p vectors and a scalar, we get the relations of VAHLEN. Cf. WEITZENBÖCK 1923, 1, p. 118.

[4]) GRASSMANN 1862, 1, p. 61.

to prove this we suppose that $v^{[\varkappa_1\dots\varkappa_{n-1}}\, v^{\lambda_1]\dots\lambda_{n-1}} \neq 0$. From this it would follow that at least one of the vectors obtained by replacing in $v^{\lambda_1\dots\lambda_{n-1}}$ the last $n-2$ indices by $n-2$ of the indices $1, \dots n$ were not zero. By interchanging indices we can always arrange that $v^{\lambda_1 2\dots(n-1)}$ is this vector. But then it would follow that

$$v^{1\dots(n-1)}\, v^{n2\dots(n-1)} - v^{n2\dots(n-1)}\, v^{1\dots(n-1)} \neq 0 \tag{7.3}$$

and this is impossible.

The component $v^{1\dots p}$ of the simple p-vector

$$v^{\varkappa_1\dots\varkappa_p} = p!\, \underset{1}{v}^{[\varkappa_1}\dots\underset{p}{v}^{\varkappa_p]} \tag{7.4}$$

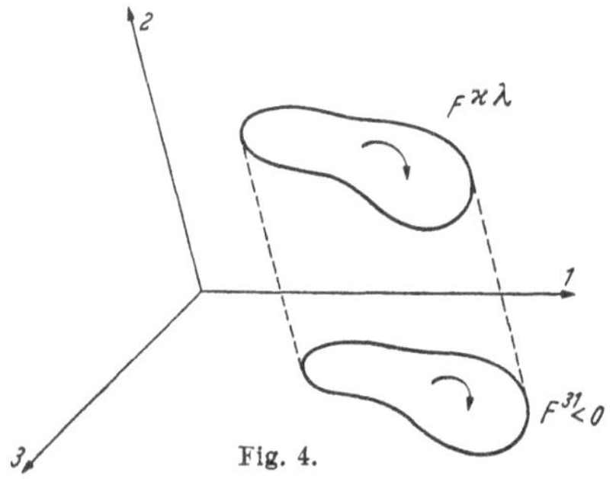

Fig. 4.

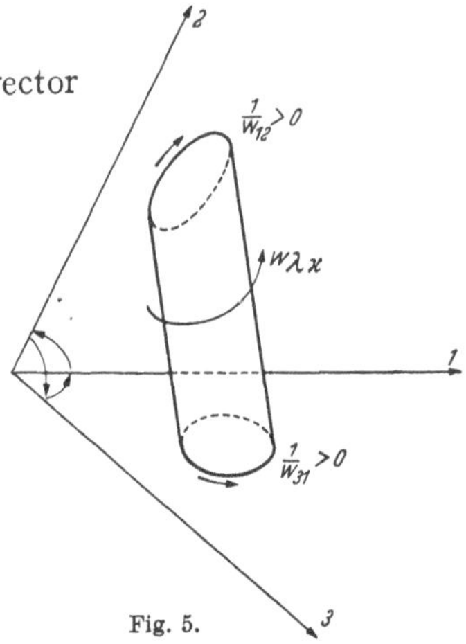

Fig. 5.

is the p-dimensional volume of the projection of the parallelotop of $\underset{1}{v}^{\varkappa}, \dots, \underset{p}{v}^{\varkappa}$ on the E_p of $\underset{1}{e}^{\varkappa}, \dots, \underset{p}{e}^{\varkappa}$ in the $(n-p)$-direction of $\underset{p+1}{e}^{\varkappa}, \dots, \underset{n}{e}^{\varkappa}$, measured by the parallelotop of $\underset{1}{e}^{\varkappa}, \dots, \underset{p}{e}^{\varkappa}$. Its sign is $+$ if the screwsense of the projections of $\underset{1}{v}^{\varkappa}, \dots, \underset{p}{v}^{\varkappa}$ in this order is the same as the screwsense of $\underset{1}{e}^{\varkappa}, \dots, \underset{p}{e}^{\varkappa}$ and $-$ if it is not. From this it follows that a simple contravariant p-vector is represented geometrically by a definite p-dimensional volume with an arbitrary form but a definite screwsense in a definite E_p. Fig. 4 shows a simple contravariant bivector in E_3 with its projection on the 31-plane.

A simple covariant p-vector is represented geometrically by a cylinder of ∞^{p-1} parallel E_{n-p}'s with an *outer* orientation, that is a p-dimensional screwsense *around* the cylinder.[1]) The component $w_{1\dots p}$ of $w_{\lambda_1\dots\lambda_p}$ is the inverse of the p-dimensional volume of the section of this cylinder with the E_p of $\underset{1}{e}^{\varkappa}, \dots, \underset{p}{e}^{\varkappa}$ measured by the parallelotop of these vectors. Its sign is $+$ if the outer orientation of the p-vector corresponds to the screwsense of $\underset{1}{e}^{\varkappa}, \dots, \underset{p}{e}^{\varkappa}$ and $-$ if it does not. Fig. 5 shows a simple covariant bivector in E_3 with its sections with the 31- and 12-plane.[2]) To every rectilinear coordinate system in E_n there belongs

[1]) SCHOUTEN-STRUIK 1922, 1. [2]) $w_{12} > 0$ should be $w_{12} < 0$.

for every value of p from 1 to n, a set of $\binom{n}{p}$ simple contravariant p-vectors and a set of $\binom{n}{p}$ simple covariant p-vectors, uniquely determined for $p=1$ and $p=n$ and determined except for the sign for $1<p<n$. These p-vectors are alternating products of p basis vectors multiplied by $p!$. Their p-directions and $(n-p)$-directions are the $\binom{n}{p}$ coordinate-E_p's and coordinate-E_{n-p}'s respectively. The n-vectors

$$(7.5)\qquad \begin{cases} \underset{(\varkappa)}{E}^{\varkappa_1\ldots\varkappa_n} \overset{\text{def}}{=} n!\, \underset{1}{e}^{[\varkappa_1}\ldots \underset{n}{e}^{\varkappa_n]} \\ \overset{(\varkappa)}{e}_{\lambda_1\ldots\lambda_n} \overset{\text{def}}{=} n!\, \overset{1}{e}_{[\lambda_1}\ldots \overset{n}{e}_{\lambda_n]} \end{cases}$$

are most important among them. They are both represented by the parallelotop of the basis vectors with the screwsense fixed by these vectors in the order $1, \ldots, n$. If the choice of the coordinate system is free, they are determined to within a scalar factor. Their components satisfy the identities (not summarize over $\varkappa$)

$$(7.6)\qquad \underset{(\varkappa)}{E}^{1\ldots n} = \overset{(\varkappa)}{e}_{1\ldots n} = +1$$

$$(7.7)\qquad \underset{(\varkappa)}{E}^{\mu_1\ldots\mu_n} \quad \overset{(\varkappa)}{e}_{\mu_1\ldots\mu_n} = n!$$

$$(7.8)\qquad \underset{(\varkappa)}{E}^{\mu_1\ldots\mu_m\varkappa_{m+1}\ldots\varkappa_n} \quad \overset{(\varkappa)}{e}_{\mu_1\ldots\mu_m\lambda_{m+1}\ldots\lambda_n} = m!\,(n-m)!\; A^{[\varkappa_{m+1}\ldots\varkappa_n]}_{[\lambda_{m+1}\ldots\lambda_n]}\,.$$

These can easily be verified.

From (7.5) it follows that

$$(7.9)\qquad \underset{(\varkappa)}{E}^{1'\ldots n'} = A^{1'\ldots n'}_{\lambda_1\ldots\lambda_n} \underset{(\varkappa)}{E}^{\lambda_1\ldots\lambda_n} = n!\; A^{1'\ldots n'}_{[1\ldots n]} \underset{(\varkappa)}{E}^{1\ldots n} = \Delta \underset{(\varkappa)}{E}^{1\ldots n}.$$

Hence $\underset{(\varkappa)}{E}^{1\ldots n}$ is the component with respect to $(\varkappa)$ of a scalar Δ-density of weight -1. In the same way $\overset{(\varkappa)}{e}_{1\ldots n}$ is the component of a scalar Δ-density of weight $+1$. This is the first example of sets of numbers that can be looked upon in two different ways as components of a quantity.

The n-vectors $\underset{(\varkappa)}{E}$ and $\overset{(\varkappa)}{e}$ determine a one to one correspondence between contravariant pseudo-p-vectors and covariant pseudo-$(n-p)$-vectors:

$$(7.10)\qquad \begin{cases} \lfloor v^{\varkappa_1\ldots\varkappa_p}\rfloor \propto \lfloor w_{\lambda_1\ldots\lambda_q}\rfloor\, \underset{(\varkappa)}{E}^{\lambda_1\ldots\lambda_q\varkappa_1\ldots\varkappa_p}\,; \qquad q=n-p \\ \lfloor w_{\lambda_1\ldots\lambda_q}\rfloor \propto \overset{(\varkappa)}{e}_{\lambda_1\ldots\lambda_q\varkappa_1\ldots\varkappa_p} \lfloor v^{\varkappa_1\ldots\varkappa_p}\rfloor.\,^{1)} \end{cases}$$

---

1) The sign ∞ means proportional to.

If $v^{\varkappa_1 \dots \varkappa_p}$ is simple, $\lfloor v^{\varkappa_1 \dots \varkappa_p} \rfloor$ can be represented in a centred E_n by an E_p through the origin.[1]) Now this E_p can be given by $p(n-p)$ numbers, hence $v^{\varkappa_1 \dots \varkappa_p}$ has only $p(n-p)+1$ independent components and among the equations (7.2) there must be $\binom{n}{p} - p(n-p) - 1$ independent ones. In order to obtain a convenient set of $p(n-p)$ independent components of $\lfloor v^{\varkappa_1 \dots \varkappa_p} \rfloor$ we choose the coordinate system in such a way that $v^{1 \dots p} \neq 0$ and put

$$(7.11) \qquad B^{\varkappa}_{\beta} \stackrel{\text{def}}{=} \frac{v^{1 \dots (\beta-1)\, \varkappa\, (\beta+1) \dots p}}{v^{1 \dots p}}; \qquad \beta = 1, \dots, p.^{2)}$$

Then it is easily proved that

$$(7.12) \qquad \frac{v^{\varkappa_1 \dots \varkappa_p}}{v^{1 \dots p}} = p!\; B^{[\varkappa_1}_{1} \;\; B^{\varkappa_p]}_{p}; \qquad B^{\alpha}_{\beta} \stackrel{\text{def}}{=} \delta^{\alpha}_{\beta}$$

or

$$(7.13) \quad \begin{cases} \dfrac{v^{\xi_1 \dots \xi_s \alpha_{s+1} \dots \alpha_p}}{v^{1 \dots p}} = s!\; B^{[\xi_1}_{\alpha_1} \dots B^{\xi_s]}_{\alpha_s}; \qquad \xi_1, \dots, \xi_s = p+1, \dots, n, \\ \qquad\qquad \alpha_1, \dots, \alpha_p = \text{even permutation of } 1, \dots, p. \end{cases}$$

The B^{ξ}_{α} are the $p(n-p)$ components required. They are very often used in problems concerning systems of partial differential equations.[3])

The alternating quantities $\tilde{\mathfrak{E}}^{\varkappa_1 \dots \varkappa_n}$ and $\tilde{\mathfrak{e}}_{\lambda_1 \dots \lambda_n}$ defined by

$$(7.14) \qquad \tilde{\mathfrak{E}}^{1 \dots n} = +1 \qquad \tilde{\mathfrak{e}}_{1 \dots n} = +1$$

with respect to *every* rectilinear coordinate system in E_n are n-vector $\varDelta$-densities of weight $+1$ and -1 respectively. They exist a priori in E_n just like $A^{\varkappa}_{\lambda}$ and they establish a one to one correspondence between contra- (co-) variant p-vectors and co- (contra-) variant $(n-p)$-vector $\varDelta$-densities of weight -1 $(+1)$:

$$(7.15) \quad \begin{cases} \text{a)} \;\; p!\, \tilde{\mathfrak{v}}_{\lambda_1 \dots \lambda_q} = (-1)^{\alpha_q}\, \tilde{\mathfrak{e}}_{\lambda_1 \dots \lambda_q \varkappa_1 \dots \varkappa_p}\, v^{\varkappa_1 \dots \varkappa_p} \\ \text{b)} \;\; q!\, v^{\varkappa_1 \dots \varkappa_p} = (-1)^{-\alpha_q}\, \tilde{\mathfrak{v}}_{\lambda_1 \dots \lambda_q}\, \tilde{\mathfrak{E}}^{\lambda_1 \dots \lambda_q \varkappa_1 \dots \varkappa_p} \\ \text{c)} \;\; p!\, \tilde{\mathfrak{w}}^{\varkappa_1 \dots \varkappa_p} = (-1)^{\beta_p}\, w_{\lambda_1 \dots \lambda_p}\, \tilde{\mathfrak{E}}^{\lambda_1 \dots \lambda_p \varkappa_1 \dots \varkappa_q} \\ \text{d)} \;\; q!\, w_{\lambda_1 \dots \lambda_p} = (-1)^{-\beta_p}\, \tilde{\mathfrak{e}}_{\lambda_1 \dots \lambda_p \varkappa_1 \dots \varkappa_q}\, \tilde{\mathfrak{w}}^{\varkappa_1 \dots \varkappa_q}; \; q = n - p \end{cases}$$

---

[1]) A contra-(co-)variant pseudo-p-vector can also be represented by an $E_{p-1}(E_{n-p-1})$ in an E_{n-1} if it is simple. The components are then the homogeneous coordinates of this $E_{p-1}(E_{n-p-1})$. If it is not simple it can be represented by an E_{n-p-1}-(E_{p-1}-)complex in E_{n-1}. Hence for $n=4$ a pseudo-bivector can be represented by a line complex in E_3 that degenerates into all straight lines intersecting the same straight line if and only if the pseudo-bivector is simple. Cf. e.g. WEITZENBÖCK 1923, 1, p. 69, 83.

[2]) KÄHLER 1934, 1, p. 17.

[3]) Cf. e.g. KÄHLER 1934, 1; P. P. 1949, 1, Ch. X.

or in another form

$$(7.16)\quad \begin{cases} \text{a)}\ \tilde{\mathfrak{v}}_{\lambda_1\ldots\lambda_q} = (-1)^{\alpha_q} v^{\varkappa_1\ldots\varkappa_p}; \\ \text{b)}\ \tilde{\mathfrak{w}}^{\varkappa_1\ldots\varkappa_p} = (-1)^{\beta_p} w_{\lambda_1\ldots\lambda_q}; \\ \lambda_1\ldots\lambda_q\varkappa_1\ldots\varkappa_p = \text{even permutation of } 1,\ldots n. \end{cases}$$

The choice of the exponents α and β is free. We may take them all equal to zero. But if subgroups of G_a are introduced and if we wish to make use of the possibility of identification of quantities with respect to such a group (cf. e.g. I § 2) it is more convenient to give the α's and β's values suitable for this special purpose.[1])

From (7.16) we see that the components of a contra- (co-) variant p-vector, if labelled in another way, transform as the components of a co- (contra-) variant $(n-p)$-vector Δ-density of weight -1 $(+1)$. Hence there is no geometric difference between these quantities. As an example we prove explicitly that the components v^{23}, v^{31}, v^{12} of a contravariant bivector in E_3 transform like the components $\mathfrak{v}_1$, $\mathfrak{v}_2$, $\mathfrak{v}_3$ of a covariant vector Δ-density of weight (-1). In fact

$$(7.17)\quad \begin{cases} v^{2'3'} = A^{2'3'}_{\varkappa\lambda} v^{\varkappa\lambda} = (A^{2'3'}_{23} - A^{2'3'}_{32})\, v^{23} + \text{cycl. } 1, 2, 3 \\ \qquad = \Delta\,(A^{1}_{1'} v^{23} + A^{2}_{1'} v^{31} + A^{3}_{1'} v^{12}) \end{cases}$$

according to the construction of the $A^{\varkappa}_{\lambda'}$ from the $A^{\varkappa'}_{\lambda}$ given in I § 1.

The densities most frequently occurring in physics are contra- (co-) variant multivectordensities of weight $+1$ (-1) which transform with $|\Delta|$ instead of Δ. Hence for real coordinate transformations they transform in the same way as Δ-densities if $\Delta > 0$ and change the sign for $\Delta < 0$. This implies that their geometric interpretation is the same as that of the corresponding Δ-densities to within the orientation. Instead of an inner (outer) orientation they get an outer (inner) one. If transvected over all indices with $\tilde{\mathfrak{e}}$ or $\tilde{\mathfrak{E}}$ they give rise to W-multivectors, transforming like ordinary multivectors but with a factor[2]) $\Delta/|\Delta|$. The value of a W-scalar[3]) is invariant for $\Delta > 0$ and changes its sign for $\Delta < 0$ (transformation of a right hand system into a left hand system).

The alternating quantities in E_3 with their geometric interpretations and their notations are to be seen in the following tables.[4])

---

[1]) These identifications are elaborately dealt with in T. P. 1951, 1, Ch. III.

[2]) SCHOUTEN 1938, 1. The two different kinds of p-vectors were used in DE RHAM and KODAIRA 1950, 1.

[3]) Often called pseudo-scalar in physical publications.

[4]) Cf. SCHOUTEN 1938, 1; SCHOUTEN and VAN DANTZIG 1940, 1 (note that in these papers the sign $\sim$ is used in another less satisfactory way); T.P. 1951, 1, Ch. II.

Interpretation	Ordinary notation	Second notation	Weight	Orientation
Number	s scalar	$\tilde{\mathfrak{s}}_{\mu\lambda\varkappa}$ cov. triv. Δ-dens.	-1	$\pm$ sign
		$\tilde{\mathfrak{S}}^{\varkappa\lambda\mu}$ contr. triv. Δ-dens.	$+1$	
	$v^{\varkappa}$ contr. vect.	$\tilde{\mathfrak{v}}_{\lambda\varkappa}$ cov. biv. Δ-dens.	-1	inner
	w_{λ} cov. vect.	$\tilde{\mathfrak{w}}^{\varkappa\lambda}$ contr. biv. Δ-dens.	$+1$	outer
	$f^{\varkappa\lambda}$ contr. biv.	$\tilde{\mathfrak{f}}_{\lambda}$ cov. vect. Δ-dens.	-1	inner
	$h_{\lambda\varkappa}$ cov. biv.	$\tilde{\mathfrak{h}}^{\varkappa}$ contr. vect. Δ-dens.	$+1$	outer
	$p^{\varkappa\lambda\mu}$ contr. triv.	$\tilde{\mathfrak{p}}$ scalar Δ-dens.	-1	screw
	$q_{\mu\lambda\varkappa}$ cov. triv.	$\tilde{\mathfrak{q}}$ scalar Δ-dens.	$+1$	

Interpretation	Ordinary notation	Weight	Second notation	Orientation
Number + screwsense	$\mathfrak{s}_{\mu\lambda\varkappa}$ cov. triv. dens.	-1	$\tilde{\mathfrak{s}}$ W-scalar	screw
	$\mathfrak{S}^{\varkappa\lambda\mu}$ contr. triv. dens.	$+1$		
	$\mathfrak{v}_{\lambda\varkappa}$ cov. biv. dens.	-1	$\tilde{v}^{\varkappa}$ contr. W-vect.	outer
	$\mathfrak{w}^{\varkappa\lambda}$ contr. biv. dens.	$+1$	$\tilde{w}_{\lambda}$ cov. W-vect.	inner
	$\mathfrak{f}_{\lambda}$ cov. vect. dens.	-1	$\tilde{f}^{\varkappa\lambda}$ contr. W-biv.	outer
	$\mathfrak{h}^{\varkappa}$ contr. vect. dens.	$+1$	$\tilde{h}_{\lambda\varkappa}$ cov. W-biv.	inner
	$\mathfrak{p}$ scalar dens.	-1	$\tilde{p}^{\varkappa\lambda\mu}$ contr. W-triv.	$\pm$ sign
	$\mathfrak{q}$ scalar dens.	$+1$	$\tilde{q}_{\mu\lambda\varkappa}$ cov. W-triv.	

Exercise.

I 7,1. Prove that for every m-vector $v^{\varkappa_1\ldots\varkappa_m}$:

I 7,1 α) $$v^{\varkappa_1\ldots\varkappa_m}\underset{(\varkappa)}{E}{}^{\lambda_1\ldots\lambda_m\varkappa_{m+1}\ldots\varkappa_n} = \binom{n}{m} v^{[\lambda_1\ldots\lambda_m}\underset{(\varkappa)}{E}{}^{\varkappa_{m+1}\ldots\varkappa_n]\varkappa_1\ldots\varkappa_m}.$$

§ 8. Tensors of valence 2.

A. Cocontravariant tensors.

A tensor $P^{\varkappa}_{\cdot\lambda}$ is represented by the homogeneous linear vector transformation (which is in a centred E_n also a point transformation)

(8.1.) $$'v^{\varkappa} = P^{\varkappa}_{\cdot\lambda}v^{\lambda}.$$

If the rank of $P^{\varkappa}_{\cdot\lambda}$ is n this transformation is invertible

$$(8.2)\qquad v^{\varkappa} = \overset{-1}{P}{}^{\varkappa}_{\cdot\lambda}\, 'v^{\lambda};\quad P^{\varkappa}_{\cdot\mu}\, \overset{-1}{P}{}^{\mu}_{\cdot\lambda} = \overset{-1}{P}{}^{\varkappa}_{\cdot\mu}\, P^{\mu}_{\cdot\lambda} = A^{\varkappa}_{\lambda};\quad \mathrm{Det}\,(P^{\varkappa}_{\cdot\lambda}) \neq 0.$$

For every value of ϱ and σ, $\overset{-1}{P}{}^{\varrho}_{\cdot\sigma}$ is equal to the minor of the element $P^{\sigma}_{\cdot\varrho}$ in the matrix of the $P^{\varkappa}_{\cdot\lambda}$ divided by $\mathrm{Det}\,(P^{\varkappa}_{\cdot\lambda})$ (cf. Exerc. I 1,1 and I 8,1; 2).

The components of

$$(8.3)\qquad s!\, P^{[\varkappa_1}_{\cdot\,[\lambda_1} \ldots P^{\varkappa_s]}_{\cdot\,\lambda_s]} = s!\, P^{[\varkappa_1}_{\cdot\,\lambda_1} \ldots P^{\varkappa_s]}_{\cdot\,\lambda_s} = s!\, P^{\varkappa_1}_{\cdot\,[\lambda_1} \ldots P^{\varkappa_s}_{\cdot\,\lambda_s]}$$

are the s-rowed subdeterminants of the matrix of $P^{\varkappa}_{\cdot\lambda}$. Hence the rank is r if and only if

$$(8.4)\qquad P^{[\varkappa_1}_{\cdot\,\lambda_1} \ldots P^{\varkappa_s]}_{\cdot\,\lambda_s} \begin{cases} \neq 0 & \text{for } s = r \text{ (hence for } s < r) \\ = 0 & \text{for } s > r. \end{cases}$$

The determinant of P is the scalar

$$(8.5)\qquad n!\, P^{[1}_{\cdot\,1} \ldots P^{n]}_{\cdot\,n}.$$

Another concomitant of P is the trace (Spur, spur) $P^{\lambda}_{\cdot\lambda}$.

A vector $v^{\varkappa}$ is said to be an *eigenvector* of P if it satisfies an equation of the form

$$(8.6)\qquad P^{\varkappa}_{\cdot\lambda}\, v^{\lambda} = \lambda\, v^{\varkappa}$$

and λ is called the *eigenvalue* of P belonging to this eigenvector. From (8.6) the rank of $P^{\varkappa}_{\cdot\lambda} - \lambda A^{\varkappa}_{\lambda}$ is $< n$, hence the possible values of λ are the solutions of the algebraic equation of degree n

$$(8.7)\qquad \begin{cases} \varphi(\lambda) \overset{\text{def}}{=} (-1)^n\, \mathrm{Det}\,(P^{\varkappa}_{\cdot\lambda} - \lambda A^{\varkappa}_{\lambda}) \\ \qquad = (-1)^n\, n!\, (P^{[1}_{\cdot\,[1} - \lambda A^{[1}_{[1}) \ldots (P^{n]}_{\cdot\,n]} - \lambda A^{n]}_{n]}) = 0 \end{cases}$$

the so called *characteristic equation* of P. From this we see that if in (8.6) we take the transvection with a covariant vector instead of with a contravariant vector we get the same eigenvalues. Writing (8.7) in the form

$$(8.8)\qquad \varphi(\lambda) = \lambda^n - \underset{1}{p}\, \lambda^{n-1} + \underset{2}{p}\, \lambda^{n-2} - \cdots + (-1)^n\, \underset{n}{p} = 0$$

we have

$$(8.9)\qquad \begin{cases} \underset{1}{p} = n\, n!\, A^{[1}_{[1} \ldots A^{n-1}_{n-1}\, P^{n]}_{\cdot\,n]} = P^{\varkappa}_{\cdot\varkappa} \\ \vdots \\ \underset{i}{p} = \binom{n}{i}\, n!\, A^{[1}_{[1} \ldots A^{n-i}_{n-i}\, P^{n-i+1}_{\cdot\,n-i+1} \ldots P^{n]}_{\cdot\,n]} = P^{[\varkappa_1}_{\cdot\,\varkappa_1} \ldots P^{\varkappa_i]}_{\cdot\,\varkappa_i}\,; \\ \vdots \\ \underset{n}{p} = n!\, P^{[1}_{\cdot\,[1} \ldots P^{n]}_{\cdot\,n]} = P^{[\varkappa_1}_{\cdot\,\varkappa_1} \ldots P^{\varkappa_n]}_{\cdot\,\varkappa_n} = \mathrm{Det}\,(P^{\varkappa}_{\cdot\lambda}). \end{cases}$$

Using the short notation of matrix calculus

(8.10) $$P\,Q \stackrel{\text{def}}{=} P^{\varkappa}_{\cdot\varrho}\,Q^{\varrho}_{\cdot\lambda}\,; \qquad P^2 = P\,P, \text{ etc.}$$

we prove now that P itself satisfies the equation (8.8):

(8.11) $$\varphi(P) \stackrel{\text{def}}{=} P^n - \underset{1}{p}\,P^{n-1} + \underset{2}{p}\,P^{n-2} - \cdots + (-1)^n \underset{n}{p}\,A = 0.$$

Equation (8.11) is easily verified by writing out the identity

(8.12) $$P^{[\varkappa_1}_{\cdot\ [\varkappa_1} \cdots P^{\varkappa_n}_{\cdot\ \varkappa_n} A^{\varkappa_{n+1}]}_{\lambda]} = 0$$

and making use of (8.9).

If the roots of (8.8) are determined, the eigenvectors belonging to some root $\underset{1}{\lambda}$ can be found by solving the equation

(8.13) $$(P^{\varkappa}_{\cdot\lambda} - \underset{1}{\lambda}\,A^{\varkappa}_{\lambda})\,v^{\lambda} = 0.$$

But if we wish to compute eigenvectors for all roots it is more convenient to determine the following auxiliary quantities first.[1])

(8.14) $$\begin{cases} \underset{1}{P} \stackrel{\text{def}}{=} \underset{1}{p}\,A - P; & \underset{1}{p} \stackrel{\text{def}}{=} \operatorname{trace} P;\quad \underset{2}{p} \stackrel{\text{def}}{=} \tfrac{1}{2}\operatorname{trace} P\underset{1}{P} \\ \underset{2}{P} \stackrel{\text{def}}{=} \underset{2}{p}\,A - P\underset{1}{P}; & \underset{3}{p} \stackrel{\text{def}}{=} \tfrac{1}{3}\operatorname{trace} P\underset{2}{P} \\ \vdots & \vdots \\ \underset{n-1}{P} \stackrel{\text{def}}{=} \underset{n-1}{p}\,A - P\underset{n-2}{P}; & \underset{n}{p} \stackrel{\text{def}}{=} \frac{1}{n}\operatorname{trace} P\underset{n-1}{P} \\ \underset{n}{P} \stackrel{\text{def}}{=} \underset{n}{p}\,A - P\underset{n-1}{P}. & \end{cases}$$

It will be found that $\underset{n}{P} = 0$ and this can be used as a check on the correctness of the operations. We now define an auxiliary function of λ

(8.15) $$\begin{cases} (-1)^{n-1}Q(\lambda) \stackrel{\text{def}}{=} -A\,\lambda^{n-1} + \\ \qquad + \underset{1}{P}\,\lambda^{n-2} - \underset{2}{P}\,\lambda^{n-3} + \cdots + (-1)^{n-1}\underset{n-2}{P}\,\lambda + (-1)^n \underset{n-1}{P}.\,^{2)} \end{cases}$$

Then we have

(8.16) $$\lambda\,Q(\lambda) = A \operatorname{Det}(P - \lambda A) + P\,Q(\lambda).$$

Hence

(8.17) $$P\,Q(\underset{1}{\lambda}) = \underset{1}{\lambda}\,Q(\underset{1}{\lambda})$$

and this proves that $Q^{\varkappa}_{\cdot\lambda}(\underset{1}{\lambda})v^{\lambda}$ is an eigenvector of P belonging to the eigenvalue $\underset{1}{\lambda}$ for every choice of $v^{\varkappa}$. This gives an easy method for the computing of eigenvectors if the auxiliary quantities (8.14) and the eigenvalues are known. But if a root $\underset{1}{\lambda}$ has a multiplicity >1 it may

[1]) FETTIS 1950, 1; Souriau 1950, 1.

[2]) FETTIS uses $\underset{1}{\lambda}\,Q$ instead of Q but this can only be done for $\underset{1}{\lambda} \neq 0$.

happen that $Q(\underset{1}{\lambda})$ vanishes and in this case the method can not be used[1]).

The matrix of a given quantity $P^{\varkappa}_{\cdot\lambda}$ can be brought into a canonical form by means of transformation of coordinates. We give here briefly the results of the theory of elementary divisors[2]).

a) For every eigenvalue $\underset{p}{\lambda}$ with frequency r_p there exist r_p linear independent vectors from which σ_{pi} are annihilated by $(P-\underset{p}{\lambda}A)^i$ but not by $(P-\underset{p}{\lambda}A)^{i-1}$. We call them vectors *of order* i with respect to $\underset{p}{\lambda}$. i takes the values $1, \ldots, s_p$ and the numbers σ_{pi} satisfy the conditions

$$\left\{\begin{array}{l}\sigma_{pi} \leq \sigma_{p(i-1)} \\ \sigma_{p1}+\cdots+\sigma_{ps_p}=r_p.\end{array}\right. \tag{8.18}$$

The numbers r_p, σ_{pi} are arithmetic invariants of P. Also the $E_{\sigma_{p1}}$ spanned by the first σ_{p1} vectors, the $E_{\sigma_{p1}+\sigma_{p2}}$ spanned by the first $\sigma_{p1}+\sigma_{p2}$ vectors etc. are invariant but the directions in these flat spaces are not individually uniquely determined by P.

b) The r_p+r_q vectors belonging to two different eigenvalues $\underset{p}{\lambda}$ and $\underset{q}{\lambda}$ are linearly independent.

c) The coordinates in E_n can be chosen in such a way that the matrix of P splits up into a number of square matrices arranged along the main diagonal each of which belongs to one eigenvalue. See fig. 6 for $n=14$, $r_1=5$, $r_2=4$, $r_3=2$, $r_4=2$, $r_5=1$.

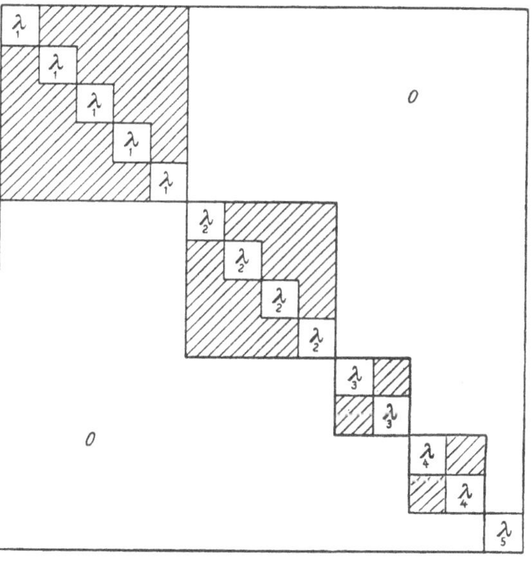

Fig. 6.

If the rank of P is r, the last square has $n-r$ rows and only zero's in the main diagonal. The quantity P is now split up into a number of

1) It can in fact be proved that $Q(\underset{1}{\lambda})$ is always zero if, in the canonical form fig. 6 of the square belonging to $\underset{1}{\lambda}$, there is at least one gap in the secondary diagonal.

2) Muth 1899, 1; Bôcher 1910, 1, p. 282; Pascal Répertorium I, 1, p. 102; Weyl 1922, 1, p. 88; Schreier in Klein 1926, 1, p. 379ff.; Dickson 1926, 1; Van der Waerden 1931, 1, p. 120; E I, 1935, 1, p. 38ff.

quantities in an E_{r_1}, E_{r_2}, etc. and each of these quantities has only one eigenvalue $\underset{1}{\lambda}, \underset{2}{\lambda}$, etc., in its own space. Hence, in order to get a further re-arrangement of the matrix of P we may suppose that P has only one eigenvalue λ. Then the result can be applied later to all subspaces E_{r_1}, E_{r_2} etc.

d) Let P have only one eigenvalue λ. For σ_{pi} and s_p we now write σ_i and s [cf. (8.18)]

$$\begin{cases} \sigma_i \leqq \sigma_{i-1} \\ \sigma_1 + \cdots + \sigma_s = n. \end{cases} \tag{8.20}$$

The basis vectors in E_n can now be chosen as follows.

Let $\underset{1}{e}^{\varkappa}$ be a vector of order s with respect to λ. That means that it is *annihilated by* $(P-\lambda A)^s$ *but not by* $(P-\lambda A)^{s-1}$. Then we choose the first s basis vectors as follows

$$\begin{cases} \underset{1}{e}^{\varkappa} \\ \underset{2}{e}^{\varkappa} = (P-\lambda A)^{\varkappa}_{\lambda} \underset{1}{e}^{\lambda} \\ \vdots \\ \underset{s}{e}^{\varkappa} = \{(P-\lambda A)^{s-1}\}^{\varkappa}_{\lambda} \underset{1}{e}^{\lambda}. \end{cases} \tag{8.21}$$

The last vector $\underset{s}{e}^{\varkappa}$ is an eigenvector. For $\underset{s+1}{e}^{\varkappa}$ we choose another vector of order s which is linearly independent of $\underset{1}{e}^{\varkappa}, \ldots, \underset{s}{e}^{\varkappa}$ and another set of s vectors $\underset{s+1}{e}^{\varkappa}, \ldots, \underset{2s}{e}^{\varkappa}$ is constructed. This process can be repeated σ_s times, the first vectors of every set and also the $\sigma_s\, s$ vectors of all sets being linearly independent. In the $E_{\sigma_{s-1}}$ of all vectors of order $s-1$ there are still $\sigma_{s-1}-\sigma_s$ vectors which are linearly independent of each other and of the σ_s vectors of this order already chosen. From these $\sigma_{s-1}-\sigma_s$ vectors arise $\sigma_{s-1}-\sigma_s$ sets of $s-1$ vectors and among them there are exactly $\sigma_{s-1}-\sigma_s$ vectors of each order from 1 to $s-1$. Proceeding in this way we get at last n linearly independent vectors and among them σ_s of order s, σ_{s-1} of order $s-1$ and so on down to σ_1 vectors of order 1. It may happen that some of the differences $\sigma_{s-1}-\sigma_s$, $\sigma_{s-2}-\sigma_{s-1}$, etc. vanish. Now if $z_1=s, z_2, \ldots, z_{\sigma_1}$ are the numbers of vectors in the sets constructed above, in the order in which they arise we have

$$z_1 = s \geqq z_2 \geqq z_3 \geqq \cdots \geqq z_{\sigma_1} \tag{8.22}$$

and the relations between the numbers $\sigma_1, \ldots, \sigma_{z_1}$ and $z_1, \ldots, z_{\sigma_1}$ can be made clear in a diagram, e.g. for $\sigma_1=5$, $\sigma_2=4$, $\sigma_3=3$, $\sigma_4=1$, $n=13$:

$$
(8.23)\quad \left\{
\begin{array}{llllll}
\circ & \circ & \circ & \circ & \circ & \sigma_1 = 5 \\
\circ & \circ & \circ & \circ & & \sigma_2 = 4 \\
\circ & \circ & \circ & & & \sigma_3 = 3 \\
\circ & & & & & \sigma_4 = 1 \\
s = z_1 = 4 & z_2 = 3 & z_3 = 3 & z_4 = 2 & z_5 = 1 &
\end{array}
\right.
$$

The numbers z are of course also invariants of P. If the basis vectors in E_n are chosen in this way the matrix of P takes the form

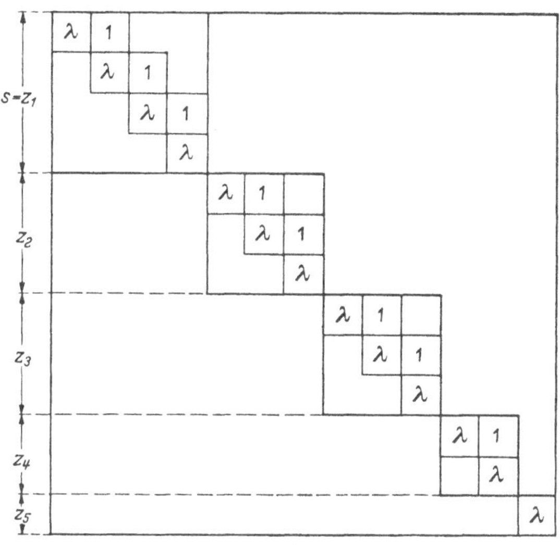

Fig. 7.

If this process is applied to each part of a general quantity P, which comes from splitting up the quantity into parts in E_{r_1}, E_{r_2}, etc. then every square matrix in a diagram like fig. 6 can be brought into a form like fig. 7. We then get the *first canonical form* of the matrix of P.[1])

The determinant Det $(P - \lambda A)$ can be written in the form

$$
(8.24)\qquad \operatorname{Det}(P - \lambda A) = \prod_{p} (\lambda - \underset{p}{\lambda})^{z_{p1}} \ldots (\lambda - \underset{p}{\lambda})^{z_{p\sigma_{p1}}}.
$$

The factors on the right hand side, for instance $(\lambda - \underset{p}{\lambda})^{z_{p1}}$, are called the *elementary divisors* of the matrix of P. Obviously two mixed tensors of valence 2 are equiform (cf. I § 2) if and only if their matrices have the same canonical form, in other words if they have the same invariants z, that is the same elementary divisors.

[1]) Another canonical form was given by WEYL 1922, 1, p. 99.

B. Co- and contravariant tensors.

Every tensor $P^{\varkappa\lambda}$ of rank r can of course be written in the form

$$P^{\varkappa\lambda} = \underset{1}{u}^{\varkappa} \underset{1}{v}^{\lambda} + \cdots + \underset{r}{u}^{\varkappa} \underset{r}{v}^{\lambda}. \tag{8.25}$$

The vectors $u^{\varkappa}$ span the $\varkappa$-domain and the vectors v^{λ} the λ-domain. The same holds mutatis mutandis for the covariant case.

The components of

$$s!\, P^{[\varkappa_1 [\lambda_1} \ldots P^{\varkappa_s] \lambda_s]} = s!\, P^{[\varkappa_1 |\lambda_1|} \ldots P^{\varkappa_s] \lambda_s} = s!\, P^{\varkappa_1 [\lambda_1} \ldots P^{|\varkappa_s| \lambda_s]} \tag{8.26}$$

are the s-rowed subdeterminants of the matrix of $P^{\varkappa\lambda}$. Hence the rank of $P^{\varkappa\lambda}$ is r if and only if the quantity (8.26) vanishes for $s > r$ but not for $s = r$. [1])

If $P^{\varkappa\lambda}$ is symmetric the quantity (8.26) is invariant for the operation of interchanging the set of indices $\varkappa_1, \ldots, \varkappa_s$ with the set of indices $\lambda_1, \ldots, \lambda_s$. Such a quantity is called a symmetric *s-vector-tensor*. It plays the same role with respect to s-vectors as a symmetric tensor of valence *2* does with respect to vectors.

It is possible to define to a co- or contravariant tensor of valence *2* an *adjoint* (cf. Exerc. I 8,2). If the determinant of the matrix is not zero this adjoint is equal to the product of the determinant and the inverse. Relations like (Exerc. I 8,2 β) can be established for the adjoint. [2]) Note that $P^{\varkappa\mu} \overset{-1}{P}_{\mu\lambda} = A_{\lambda}^{\varkappa}$ and that generally $P^{\varkappa\mu} \overset{-1}{P}_{\lambda\mu} \neq A_{\lambda}^{\varkappa}$.

C. Symmetric co- and contravariant tensors.

If we consider real and complex symmetric tensors and real and complex coordinate transformations the following theorem holds:

If the symmetric tensor $P_{\varkappa\lambda}(Q^{\varkappa\lambda})$ has rank r the coordinate system $(\varkappa)$ can always be chosen in such a way that the matrix of $P_{\varkappa\lambda}(Q^{\varkappa\lambda})$ has the diagonal form with r numbers $+1$ and $n-r$ numbers 0 in the main diagonal.

If only *real* symmetric tensors and *real* coordinate transformations are considered we have the theorem:

If the symmetric tensor $P_{\varkappa\lambda}(Q^{\varkappa\lambda})$ is ***real*** *and of rank r, there always exists a* ***real*** *coordinate system such that the matrix of $P_{\varkappa\lambda}(Q^{\varkappa\lambda})$ has the diagonal form with s numbers -1, $r-s$ numbers $+1$ and $n-r$ numbers 0 in the main diagonal.* [3])

The number s, called the *index*, is an invariant of the symmetric tensor. The sequence $- - - \cdots + + + \cdots$ with s $-$-signs and $(r-s)$ $+$-signs is called the *signature*. The signature is said to be *even* (*odd*) if s

[1]) WADE 1943, 1, calls therefore (8.26) the *rank tensors*.

[2]) P. P. 1949, 1, p. 28.

[3]) Cf. for the proof of these two theorems DICKSON 1926, 1, p. 70.

is even (odd). The symmetric tensor is called *positive (negative) definite* if $s = 0\,(n)$ and $r = n$ and *indefinite* in all other cases. It is called *positive (negative) semi-definite* if $n - r > 0$ and $s = 0\,(r)$.

It follows from these theorems that two symmetric tensors are equiform for general transformations if and only if they have the same rank and that two real symmetric tensors are equiform for real transformations if and only if they have the same rank and the same index.

D. Co- and contravariant bivectors.[1]

An antisymmetric matrix always has an even rank. Hence the rank of a bivector is always even. If $v^{\varkappa\lambda}$ has the rank r it can be proved that[2]

$$r!\, v^{[\varkappa_1[\lambda_1} \dots v^{\varkappa_r]\lambda_r]} = \left(\frac{r!}{2^{\frac{1}{2}r}\,(\frac{1}{2}r)!}\right)^2 v^{[\varkappa_1\varkappa_2} \dots v^{\varkappa_{r-1}\varkappa_r]}\, v^{[\lambda_1\lambda_2} \dots v^{\lambda_{r-1}\lambda_r]}. \tag{8.27}$$

Hence, if $r = n$

$$\operatorname{Det}\,(v^{\varkappa\lambda}) = \left(\frac{n!}{2^{\frac{1}{2}n}\,(\frac{1}{2}n)!}\, v^{[12} \dots v^{n-1\,n]}\right)^2. \tag{8.28}$$

The same holds mutatis mutandis for a covariant bivector.

Let $F_{\lambda\varkappa}$ be a bivector of rank r and let $u^\varkappa$ and $v^\varkappa$ be two vectors such that

$$F_{\lambda\varkappa}\, u^\lambda v^\varkappa = 1. \tag{8.29}$$

Then, if

$$w_{\lambda\varkappa} \overset{\text{def}}{=} 2\, F_{[\lambda|\mu|}\, v^\mu F_{\varkappa]\nu}\, u^\nu \tag{8.30}$$

we will prove that the rank $'r$ of

$$'F_{\lambda\varkappa} \overset{\text{def}}{=} F_{\lambda\varkappa} - w_{\lambda\varkappa} \tag{8.31}$$

is $r - 2$. Because the rank of $F_{\lambda\varkappa}$ is r, there exist $n - r$ linearly independent contravariant vectors whose transvections with $F_{\lambda\varkappa}$ vanish. Because of (8.29 to 31) their transvections with $'F_{\lambda\varkappa}$ also vanish. Hence $'r \leq r$. Now, according to (8.30, 31) the transvections of $u^\varkappa$ and $v^\varkappa$ with $'F_{\lambda\varkappa}$ vanish and these vectors are linearly independent of each other and of the $n - r$ vectors just mentioned. Hence $'r \leq r - 2$. Now we have

$$\begin{cases} F_{[\lambda_1[\varkappa_1} \dots F_{\lambda_r]\varkappa_r]} = {'F}_{[\lambda_1[\varkappa_1} \dots {'F}_{\lambda_r]\varkappa_r]} - r\, w_{[\lambda_1[\varkappa_1}\, {'F}_{\lambda_2\varkappa_2} \dots {'F}_{\lambda_r]\varkappa_r]} + \\ \qquad + \binom{r}{2} w_{[\lambda_1[\varkappa_1}\, w_{\lambda_2\varkappa_2}\, {'F}_{\lambda_3\varkappa_3} \dots {'F}_{\lambda_r]\varkappa_r]} \end{cases} \tag{8.32}$$

and from this we see that the rank of $F_{\lambda\varkappa}$ could not be r if the rank of $'F_{\lambda\varkappa}$ were $< r - 2$. Hence $'r = r - 2$. If this process is carried out

---

[1]) There is a great number of papers on special properties of tensors, especially of symmetric tensors and multivectors. We mention only a few: Ruse 1936, 1; Wong 1940, 1; Gurewitsch 1950, 2; 3; 4; Moreau 1950, 1.

[2]) P. P. 1949, 1, p. 28.

$\frac{1}{2}r-1$ times we get a decomposition of $F_{\lambda\varkappa}$ into $\frac{1}{2}r$ simple bivectors. This is called the splitting up of $F_{\lambda\varkappa}$ into $\frac{1}{2}r$ *blades.*[1]) The same can be done with a contravariant bivector. Note that the blades are by no means uniquely determined by the bivector. If $F_{\lambda\varkappa}$ is real, $u^\varkappa$ and $v^\varkappa$ can be chosen to be real and the process can be carried out in such a way that all blades are real.

If

$$F_{\lambda\varkappa}=\overset{1}{f}_{\lambda\varkappa}+\cdots+\overset{\varrho}{f}_{\lambda\varkappa};\quad \varrho=\tfrac{1}{2}r \tag{8.33}$$

is such a decomposition, the r-dimensional domain of $F_{\lambda\varkappa}$ is spanned by the $\frac{1}{2}r$ two-dimensional domains of the blades. Hence these domains can have no vector in common and from this it follows that

$$F_{[\lambda_1\lambda_2}\ldots F_{\lambda_{r-1}\lambda_r]}\neq 0. \tag{8.34}$$

Hence for a bivector, besides the necessary and sufficient condition

$$F_{[\lambda_1[\varkappa_1}\ldots F_{\lambda_s]\varkappa_s]}\begin{cases}\neq 0 & \text{for}\quad s=r\\ =0 & \text{for}\quad s>r\end{cases} \tag{8.35}$$

for the rank to be equal to r, we still have another necessary and sufficient condition[2])

$$F_{[\lambda_1\lambda_2}\ldots F_{\lambda_{s-1}\lambda_s]}\begin{cases}\neq 0 & \text{for}\quad s=r\\ =0 & \text{for}\quad s>r.\end{cases} \tag{8.36}$$

This also holds mutatis mutandis for a contravariant bivector.

If we start from (8.33) we can always choose the coordinate system $(\varkappa)$ in such a way that

$$\overset{1}{f}_{\lambda\varkappa}=2\,\overset{1}{e}_{[\lambda}\overset{2}{e}_{\varkappa]};\quad \overset{2}{f}_{\lambda\varkappa}=2\,\overset{3}{e}_{[\lambda}\overset{4}{e}_{\varkappa]};\ \text{etc.} \tag{8.37}$$

That proves the theorem:

If a ***real*** *bivector has rank r, there is always a* ***real*** *coordinate system such that the matrix of the components with respect to this system consists of exactly $\frac{1}{2}r$ matrices of the form $\begin{Vmatrix}0&1\\-1&0\end{Vmatrix}$, arranged along the main diagonal and zero's at all other places.*[3])

---

[1]) There is a mistake in the proof in E I 1935, 1, p. 46 which is corrected here.

[2]) Cf. (8.27) and GRASSMANN 1844, 1, p. 206; 1862, 1, p. 56; ROTHE 1912, 1, p. 1039.

[3]) The results obtained under A, C and D of this section give a possibility of classification for cocontravariant and symmetric tensors of valence 2 and for bivectors. The problem of the classification of quantities with a valence >2 is only solved in very special cases. For trivectors the classification was given for $n=6$ by REICHEL 1907, 1, for $n=7$ by SCHOUTEN 1931, 1 and for $n=8$ by GUREWITSCH 1935, 1. Cf. W. KÄMMERER 1927, 1; GUREWITSCH 1933, 1; 2; 1934, 1; 2; 3; 1935, 2; 1950, 1; WEITZENBÖCK 1937, 1; 2; 1938, 1; PAPY 1946, 1; PAPY and TOURNAY 1946, 2; HUTCHINSON 1948, 1.

If a covariant bivector $F_{\lambda\varkappa}$ and p linearly independent covariant vectors $\overset{1}{u}_\lambda, \ldots, \overset{p}{u}_\lambda$ are given, we may require the section of $F_{\lambda\varkappa}$ (cf. I §4) with the E_{n-p} spanned by these p vectors. Taking $(\varkappa)$ in such a way that the vectors are $\overset{q+1}{e}_\lambda, \ldots, \overset{n}{e}_\lambda$; $q=n-p$, the ξ^α; $\alpha=1,\ldots,q$ can be taken as coordinates in the E_q (cf. I §4) and we have for the connecting quantity

$$B^\varkappa_\beta \overset{*}{=} \underset{\beta}{e}^\varkappa; \qquad \beta = 1, \ldots, q \tag{8.38}$$

and for the section

$$'F_{\beta\alpha} \overset{*}{=} F_{\lambda\varkappa} \underset{\beta}{e}^\lambda \underset{\alpha}{e}^\varkappa \overset{*}{=} F_{\beta\alpha}. \tag{8.39}$$

This means that with this special choice of the coordinate system we have only to drop all components of $F_{\lambda\varkappa}$ with indices $q+1, \ldots, n$. In order to find the rank of $'F$ we have to consider the set of multivectors (8.36)

$$\left\{\begin{array}{l} \overset{2}{I}_{\lambda\varkappa} \overset{\text{def}}{=} F_{\lambda\varkappa} \\ \overset{4}{I}_{\lambda_1\varkappa_1\lambda_2\varkappa_2} \overset{\text{def}}{=} F_{[\lambda_1\varkappa_1}F_{\lambda_2\varkappa_2]} \\ \vdots \end{array}\right. \quad \text{in short (cf. I § 7):} \left\{\begin{array}{l} \overset{2}{I} \overset{\text{def}}{=} F \\ \overset{4}{I} \overset{\text{def}}{=} [FF] \\ \vdots \end{array}\right. \tag{8.40}$$

the last non vanishing multivector of this set being $\overset{r}{I}$. The multivectors $'\overset{2}{I}, '\overset{4}{I}, \ldots$ belonging to $'F$ are the sections of $\overset{2}{I}, \overset{4}{I}, \ldots$ with the E_{n-p} and these sections can be found in the same way as $'F$, i.e. by dropping all components with indices $q+1, \ldots, n$ with respect to the special coordinate system. Now consider for instance the multivector

$$[\overset{4}{I}\, \overset{1}{u} \ldots \overset{p}{u}] = [\overset{4}{I}\, \overset{q+1}{e} \ldots \overset{n}{e}]. \tag{8.41}$$

Components of $\overset{4}{I}$ with indices $q+1, \ldots, n$ can not occur in this quantity. Hence it is zero if and only if $'\overset{4}{I}$ vanishes. This proves the following theorem:

The section of the covariant bivector $F_{\lambda\varkappa}$ with the E_{n-p} spanned by the covariant vectors $\overset{1}{u}_\lambda, \ldots, \overset{p}{u}_\lambda$ has rank $'r$ if and only if

$$[\overset{2s}{I}\, \overset{1}{u} \ldots \overset{p}{u}] \left\{\begin{array}{ll} \neq 0 & \text{for} \quad 2s = {}'r \\ = 0 & \text{for} \quad 2s > {}'r. \end{array}\right. \tag{8.42}$$

We call $r - {}'r$ the *reduction number* of the vectors $\overset{1}{u}, \ldots, \overset{p}{u}$ with respect to the bivector F.

It can be proved that the reduction number of $\overset{1}{u}, \ldots, \overset{p}{u}$ with respect to F is $\varkappa$ if and only if the matrix

$$(8.43)\qquad \begin{Vmatrix} F_{\mu\lambda} & \overset{1}{u}_\lambda & \ldots & \overset{p}{u}_\lambda \\ -\overset{1}{u}_\mu & 0 & \ldots & 0 \\ \vdots & \vdots & & \vdots \\ -\overset{p}{u}_\mu & 0 & \ldots & 0 \end{Vmatrix}$$

has the rank $R = r + 2p - \varkappa$.[1])

Exercises.

I 8,1. (Cf. Exerc. I 1,1.) Prove that for every tensor $P^{\varkappa}_{\cdot\lambda}$ of rank n (cf. Exerc. II 2,2)

$$\text{I 8,1 }\alpha)\qquad \overset{-1}{P}{}^{\varkappa}_{\cdot\lambda} = \frac{\partial \log \operatorname{Det}(P^{\varrho}_{\cdot\sigma})}{\partial P^{\lambda}_{\cdot\varkappa}}.$$

I 8,2. The quantity

$$\text{I 8,2 }\alpha)\qquad \overset{*}{P}{}^{\varkappa}_{\cdot\lambda} \overset{\text{def}}{=} \frac{\partial \operatorname{Det}(P^{\varrho}_{\cdot\sigma})}{\partial P^{\lambda}_{\cdot\varkappa}}$$

is often called the *adjoint* of P. It exists always, also if $\operatorname{Det}(P) = 0$. The elements of the adjoint are the minors of the elements of P. The adjoint of the adjoint of P does in general not equal P. Prove that for every tensor $P^{\varkappa}_{\cdot\lambda}$ of rank n (cf. Exerc. II 2,2)

$$\text{I 8,2 }\beta)\qquad \overset{*}{P}{}^{\varkappa}_{\cdot\lambda} = n.n!\, A^{\varkappa}_{[1} A^{[1}_{|\lambda|} P^{2}_{\cdot 2} \ldots P^{n]}_{\cdot n]}.$$

I 8,3. Prove that $Q(\lambda)$ defined in (8.15) is the adjoint (cf. Exerc. I 8,2) of $P - \lambda A$.

I 8,4. If p, q and r are the ranks of P, Q and $R = PQ$, prove that $r \leq p$, $r \leq q$ and that $r = p$ if $q = n$.

I 8,5. Prove that the trace of $PQ - QP$ is always zero.

I 8,6. There exist always two coordinate systems $(\varkappa)$ and $(\varkappa')$ such that the matrix of the intermediate components $P^{\varkappa\lambda'}$ has the diagonal form. There exist always two tensors $Q^{\varkappa}_{\cdot\lambda}$, $R^{\varkappa}_{\cdot\lambda}$ such that the matrix of $Q^{\varkappa}_{\cdot\varrho} R^{\lambda}_{\cdot\sigma} P^{\varrho\sigma}$ has the diagonal form.

I 8,7. If $P_{\lambda\varkappa}$ and $Q_{\lambda\varkappa}$ satify the equation

$$\text{I 8,7 }\alpha)\qquad P_{[\varkappa[\lambda} Q_{\mu]\nu]} = 0$$

and if the rank of $P_{\lambda\varkappa}$ is 1 or 2, then the rank of $Q_{\lambda\varkappa}$ is 0, 1 or 2. If the rank of $P_{\lambda\varkappa}$ is > 2, $Q_{\lambda\varkappa}$ is zero.

---

[1]) Cf. P. P. 1949, 1, p. 27.

I 8,8. If $P_{\lambda\varkappa}$ is a tensor of rank n and D the determinant of its matrix, prove that

I 8,8 α)
$$\overbrace{}^{\leftarrow n \rightarrow}\ \overbrace{}^{\leftarrow 1 \rightarrow}\quad \begin{vmatrix} P_{\lambda\varkappa} & u_\varkappa \\ v_\lambda & 0 \end{vmatrix} = -\overset{-1}{P}{}^{\varkappa\lambda} u_\varkappa v_\lambda D.$$

I 8,9. $P^{\varkappa\lambda}$ is determined to within a factor $\sqrt[s]{+1}$ by the quantity (8.26) for $s = 1, \ldots r-1$ but not for $s = r$. The same holds mutatis mutandis for covariant and mixed tensors.

I 8,10. The reduction number of one vector with respect to a bivector is always 0, 1 or 2. It is zero if and only if the vector belongs to the domain of the bivector.

I 8,11. Prove the inequalities [cf. (8.43)]

I 8,11 α)
$$\begin{cases} 0 \leq \varkappa \leq r & 2p \leq R \leq r + 2p \\ 2p \geq \varkappa & R \geq r \\ r - \varkappa \leq n - p & R \leq n + p. \end{cases}$$

I 8,12. The $m \leq n$ vectors $\overset{1}{v}_\lambda, \ldots, \overset{m}{v}_\lambda$ are linearly independent and the m vectors $\overset{1}{w}_\lambda, \ldots, \overset{m}{w}_\lambda$ satisfy the equation

I 8,12 α)
$$\overset{1}{v}_{[\lambda} \overset{1}{w}_{\mu]} + \cdots + \overset{m}{v}_{[\lambda} \overset{m}{w}_{\mu]} = 0.$$

Prove that there exist m^2 coefficients $\underset{uv}{\alpha}$ such that $\overset{u}{w}_\lambda = \underset{uv}{\alpha}\, \overset{v}{v}_\lambda$ and that $\underset{uv}{\alpha} = \underset{vu}{\alpha}$; $u, v = 1, \ldots, m$. [1])

I 8,13. If $P^{\varkappa}_{\cdot\lambda}$ has rank n and if $P^{\alpha}_{\cdot\eta} = 0$; $\alpha = 1, \ldots, m$; $\eta = m+1, \ldots, n$ prove that

I 8,13 α)
$$\overset{-1}{P}{}^{\alpha}_{\cdot\eta} = 0;$$

I 8,13 β)
$$P^{\alpha}_{\cdot\gamma} \overset{1}{P}{}^{\gamma}_{\cdot\beta} = A^{\alpha}_{\beta}; \quad \alpha, \beta, \gamma = 1, \ldots, m$$

I 8,13 γ)
$$P^{\xi}_{\cdot\zeta} \overset{-1}{P}{}^{\zeta}_{\cdot\eta} = A^{\xi}_{\eta}; \quad \xi, \eta, \zeta = m+1, \ldots, n.$$

I 8,14. An E_m is imbedded in E_n. For the tensor $P_{\mu\lambda}$ of E_n the rank of $B^{\mu}_{c} P_{\mu\lambda}$ is r and the rank of $B^{\mu\lambda}_{cb} P_{\mu\lambda}$ is r'. The support of the λ-domain of $B^{\mu}_{c} P_{\mu\lambda}$ intersects E_m in an E_s. Prove that,

I 8,14 α)
$$r - (n - m) \leq r' = m - s \leq r.$$

I 8,15 [2]). If $c_{\lambda\varkappa} = a_{\lambda\varkappa} + i\, b_{\lambda\varkappa}$, where $a_{\lambda\varkappa}$ and $b_{\lambda\varkappa}$ are real and symmetric, $c_{\lambda\varkappa}$ has a rank $< n$ and $a_{\lambda\varkappa}$ is positive definite or positive semi-definite, then there exists a real vector $s^{\varkappa}$ such that $c_{\lambda\varkappa} s^{\varkappa} = 0$.

---

[1]) CARTAN 1919, 1, p. 151.

[2]) PEREMANS, DUPARC and LEKKERKERKER 1952, 1; QUADE 1953, 1.

§ 9. Introduction of a metric in an E_n.

In an E_n the magnitude of two contravariant vectors can only be compared if they are parallel. If we wish to establish a comparison of magnitudes of non-parallel vectors we have to introduce a metric, i.e. a definition of "length". This definition must satisfy the conditions that the length of a real vector is always positive and that multiplication of a vector by a real number k alters its length by a factor $|k|$. Hence, if $F(\underset{2}{\xi}^{\varkappa}-\underset{1}{\xi}^{\varkappa})$ is the *length* of $\underset{2}{\xi}^{\varkappa}-\underset{1}{\xi}^{\varkappa}$ or the *distance* of the points $\underset{1}{\xi}^{\varkappa}$ and $\underset{2}{\xi}^{\varkappa}$, we have

$$F\left(k\left(\underset{2}{\xi}^{\varkappa}-\underset{1}{\xi}^{\varkappa}\right)\right)=|k|\,F\left(\underset{2}{\xi}^{\varkappa}-\underset{1}{\xi}^{\varkappa}\right). \tag{9.1}$$

We assume that the function F is single valued and continuous and that it has continuous derivatives up to a certain order. The equation

$$F\left(\xi^{\varkappa}-\underset{0}{\xi}^{\varkappa}\right)=1 \tag{9.2}$$

represents the *indicatrix* of the metric with respect to the point $\underset{0}{\xi}^{\varkappa}$. In a centred E_n we have $\underset{0}{\xi}^{\varkappa}=0$ and $\xi^{\varkappa}$ is the radius vector. According to (9.1) the indicatrix is symmetric with respect to the point $\underset{0}{\xi}^{\varkappa}$. To every direction there belongs one and only one $(n-1)$-direction viz. the tangent $(n-1)$-direction of the indicatrix at its section with the straight line through $\underset{0}{\xi}^{\varkappa}$ in the given direction. If the equation of the indicatrix is algebraic, its degree is even because $\underset{0}{\xi}^{\varkappa}$ must be the centre.

The projection of a vector $u^{\varkappa}$ on a vector $v^{\varkappa}$ can now be defined as the projection in the $(n-1)$-direction that belongs to the direction of $v^{\varkappa}$. If u is the length of $u^{\varkappa}$ and $'u$ the length of the projection, the angle between $u^{\varkappa}$ and $v^{\varkappa}$ (in this order) can be defined by

$$\cos(u^{\varkappa}, v^{\varkappa}) \overset{\text{def}}{=} \frac{'u}{u}\,. \tag{9.3}$$ [1])

But in general $\cos(u^{\varkappa}, v^{\varkappa}) \neq \cos(v^{\varkappa}, u^{\varkappa})$.

If we wish to define a "rotation" in a centred E_n in a natural way we will have to define it as a linear homogeneous transformation which leaves invariant both an n-dimensional screwsense and the indicatrix.[2]) That excludes all forms of indicatrix which are not invariant for some subgroup of G_{ho}. This subgroup must be chosen big enough to guarantee some freedom of "motion". For instance it could be required that

[1]) FINSLER 1918, 1, p. 39.

[2]) Cf. WEYL 1921, 1, p. 125ff.

an E_1 through the centre, an E_2 through E_1, an E_3 through E_2, etc. being given, there always exists a rotation turning E_1 in a given direction, E_2 in a 2-direction containing this direction, etc. Then it can be proved[1]) that the only possible indicatrix is a non-degenerate real quadratic hypersurface. Following another line of thought it would be possible to require that the angle between two directions be independent of their order. This condition also leads to a real quadratic indicatrix.[2])

Accordingly, let us take the equation of the indicatrix in the form

$$g_{\lambda\varkappa}(\xi^\lambda - \underset{0}{\xi^\lambda})(\xi^\varkappa - \underset{0}{\xi^\varkappa}) = \pm 1 \tag{9.4}$$

where $g_{\lambda\varkappa}$ is a real symmetric tensor of rank n and index $s \leq n$. $g_{\lambda\varkappa}$ and its inverse $g^{\varkappa\lambda}$ are then called the *covariant* and *contravariant fundamental tensor*. An E_n with a symmetric fundamental tensor is called an R_n and is said to be *ordinary* if $g_{\lambda\varkappa}$ is positive definite. The geometry in R_n is called *euclidean* if $g_{\lambda\varkappa}$ is *definite* and *minkowskian* if $g_{\lambda\varkappa}$ is *indefinite*. The *length* of a vector $v^\varkappa$

$$\left|\sqrt{g_{\lambda\varkappa} v^\lambda v^\varkappa}\right| \tag{9.5}$$

is always positive or zero but $g_{\lambda\varkappa} v^\lambda v^\varkappa$ can be >0 or <0 or zero. In the first case $v^\varkappa$ is called *time-like* and in the second *space-like*.[3]) The time-like vectors through the origin fill the $+$-*region* and the space-like vectors the $-$-*region* of a centred R_n. The vectors with zero length fill the *nullcone*. A real nullcone only exists in the indefinite case $0 < s < n$. A vector with length $+1$ is called a *unitvector*. The *angle* between two *real* vectors, both of which lie in the $+$-region or both in the $-$-region is defined by

$$\cos(u^\varkappa, v^\varkappa) = \cos(v^\varkappa, u^\varkappa) \overset{\text{def}}{=} \frac{\pm g_{\lambda\varkappa} u^\lambda v^\varkappa}{\left|\sqrt{g_{\varrho\sigma} u^\varrho u^\sigma g_{\tau\omega} v^\tau v^\omega}\right|} \quad \text{for the } \pm\text{-region.} \tag{9.6}$$

The possible values of the cos depend on the form of the section of the plane of the vectors with the nullcone:

(9.7)

Section	cos
2 real E_1's	$\geq +1$ or ≤ -1
1 real E_1	± 1
2 imaginary E_1's	$\leq +1$ and ≥ -1.

1) Helmholz 1868, 1; Lie 1890, 1; Weyl 1921, 1, p. 26.

2) Finsler 1918, 1, p. 40.

3) The space of special relativity is an R_4 with signature $---+$ and a minkowskian metric. For time-like vectors the term *duration* (Dauer) is sometimes used instead of the term length.

If we take a unitvector as first vector we see from this that the length of its projection on the other vector is ≥ 1, $=1$ and ≤ 1 in these three cases. In the first case the geometry in the plane is minkowskian and in the third case it is euclidean. In the second case the plane is not an R_2 because it does not contain a symmetric tensor of rank 2. Such an E_2 in R_n is called *isotropic.*

An E_p in R_n is called *isotropic* if the symmetric tensor of its metric has a rank $<p$ and *full isotropic* or a *null* E_p if this rank is zero.

Any two vectors $u^\varkappa$ and $v^\varkappa$ for which $g_{\lambda\varkappa} u^\lambda v^\varkappa = 0$ are said to be *mutually perpendicular.* With this exception the angle between a vector in the $+$-region and a vector in the $-$-region is not defined.

An E_p and an E_q in R_n are mutually perpendicular if every direction in E_p is perpendicular to every direction of E_q. All directions perpendicular to E_p span an E_{n-p}. The section of E_p and this E_{n-p} is an E_t with the property that all its directions are mutually perpendicular. Such an E_t is a null-E_t. Its t-direction lies in the nullcone. A null-E_t can never be an R_t. For $t>0$ the E_p must be isotropic because it contains a direction perpendicular to all its directions.

By using the second theorem of I § 8.C, a real coordinate system (h); $h = 1, \ldots, n$ may be introduced such that

$$(9.8)\quad \left.\begin{cases} \text{a)}\ g_{11}=-1;\ldots;g_{\mathrm{s\,s}}=-1;\ g_{\mathrm{s+1\,s+1}}=+1;\ldots;g_{\mathrm{n\,n}}=+1;\ g_{ih}=0 \\ \text{b)}\ g^{11}=-1;\ldots;g^{\mathrm{s\,s}}=-1;\ g^{\mathrm{s+1\,s+1}}=+1;\ldots;g^{\mathrm{n\,n}}=+1;\ g^{hi}=0 \end{cases}\right\}\text{for } h\neq i.$$

Such a coordinate system is called *cartesian.* We write $\underset{i}{i}^h$, $\overset{h}{i}_i$ for its basis vectors. These vectors are mutually perpendicular unitvectors. We always use *roman* running indices and *vertical* fixed indices for cartesian coordinate systems in R_n.

Fig. 8 shows the minkowskian geometry in two dimensions. The signature is $-+$. The indicatrix consists of the hyperbolas

$$(9.9)\qquad -(x^1)^2+(x^2)^2=\pm 1$$

or in parametric form

$$(9.10)\qquad x^1=\cosh\varphi;\ x^2=\sinh\varphi;\qquad x^1=\sinh\varphi;\ x^2=\cosh\varphi.$$

The fact that, if the figure is looked at from a euclidean point of view, the 1-axis seems to be perpendicular to the 2-axis and that $\underset{1'}{i}$ and $\underset{2'}{i}$ seem to have the same length, has no significance at all. The whole figure could be subjected to an affine transformation and would then represent just the same thing. In a minkowskian metric, length is fixed by the indicatrix, hence $\underset{1}{i}$, $\underset{2}{i}$, $\underset{1'}{i}$, and $\underset{2'}{i}$ all have the same length.

The area of any part of the plane happens to be the same for both metrics. The angle φ is real and $\frac{1}{2}\varphi$ is the shaded area. According to (9.6) we have

$$\cos(\underset{1}{i}, \underset{1'}{i}) = \cosh\varphi \geq 1 \tag{9.11}$$

for the angle between 1-axis and 1'-axis. The vectors $\underset{1}{i}$ and $\underset{2}{i}$ are perpendicular to each other and so are $\underset{1'}{i}$ and $\underset{2'}{i}$. But the angle between arbitrary directions that do not belong to the same region is not defined.

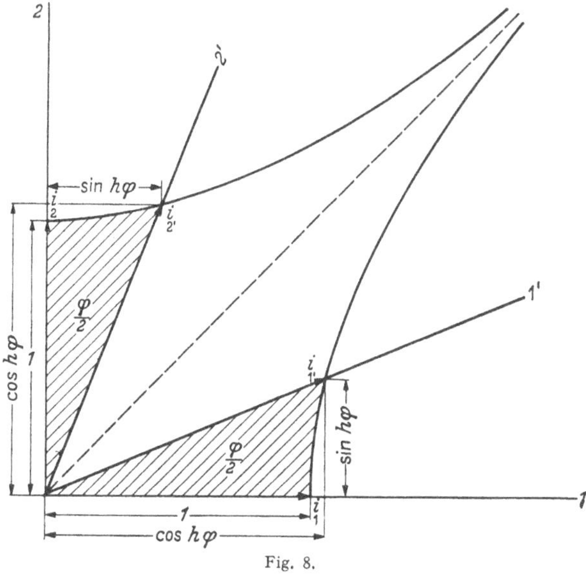

Fig. 8.

The +-region and the −-region each consist of two parts. Every orthogonal transformation leaves each region invariant as a whole, but it may happen that it interchanges its parts. So we have four different kinds of orthogonal transformations in a minkowskian R_2

(9.12)

	+-region	−-region	Δ	example
A)	not interchanged	not interchanged	$+1$	$\underset{1}{i} \to \underset{1'}{i}$; $\underset{2}{i} \to \underset{2'}{i}$
B)	not interchanged	interchanged	-1	$\underset{1}{i} \to -\underset{1'}{i}$; $\underset{2}{i} \to \underset{2'}{i}$
C)	interchanged	not interchanged	-1	$\underset{1}{i} \to \underset{1'}{i}$; $\underset{2}{i} \to -\underset{2'}{i}$
D)	interchanged	interchanged	$+1$	$\underset{1}{i} \to -\underset{1'}{i}$; $\underset{2}{i} \to -\underset{2'}{i}$

The transformations (A) form a group and are called *proper minkowskian rotations* and in the theory of relativity, LORENTZ *transformations*. (B) and (C) are *minkowskian reflexions* and (D) are *improper minkowskian rotations*. They do not form a group neither separately nor together. But (A), (B), (C) and (D) together form the group of all minkowskian orthogonal transformations.

For all values of n, the fundamental tensor establishes a one to one correspondence between co- and contravariant vectors

$$v_\lambda = g_{\lambda\varkappa} v^\varkappa; \qquad v^\varkappa = g^{\varkappa\lambda} v_\lambda. \tag{9.13}$$

We use the same kernel for corresponding vectors and consider $v^\varkappa$ and v_λ as two different sets of components of the *same* quantity, called *vector*, that can be represented geometrically either by an arrow or by an oriented set of two parallel hyperplanes.[1] The distance between the two hyperplanes is the inverse of the length of $v^\varkappa$. The orientations of $v^\varkappa$ and v_λ arc equal if $v^\varkappa$ lies in the $+$-region and opposite if $v^\varkappa$ lies in the $-$-region. The process (9.13) is called *raising* and *lowering of indices.*[2] It can also be applied to one or more indices of a quantity of valence >1, e.g.

$$P^{\varrho\cdot\nu}_{\cdot\lambda} = g^{\varrho\varkappa} P^{\cdot\cdot\nu}_{\varkappa\lambda}; \qquad P_{\varkappa\lambda\sigma} = P^{\cdot\cdot\nu}_{\varkappa\lambda} g_{\nu\sigma}. \tag{9.14}$$

The relation

$$g_{\lambda\varkappa} = g_{\lambda\varrho}\, g_{\varkappa\sigma}\, g^{\varrho\sigma} \tag{9.15}$$

justifies the use of the same kernel in $g_{\lambda\varkappa}$ and $g^{\varkappa\lambda}$.

Because $\underset{j}{i}^\varkappa$ and $\overset{j}{i}_\lambda$ have the same orientation for every value of j, we have

$$\underset{1}{i}^\varkappa = -\overset{1}{i}{}^\varkappa; \ldots; \underset{\mathrm{s}}{i}^\varkappa = -\overset{\mathrm{s}}{i}{}^\varkappa;\ \underset{\mathrm{s}+1}{i}^\varkappa = +\overset{\mathrm{s}+1}{i}{}^\varkappa; \ldots; \underset{\mathrm{n}}{i}^\varkappa = +\overset{\mathrm{n}}{i}{}^\varkappa \tag{9.16}$$

and accordingly

$$v^h = \begin{cases} -v_h; & h = 1, \ldots, \mathrm{s} \\ +v_h; & h = \mathrm{s}+1, \ldots, \mathrm{n}. \end{cases} \tag{9.17}$$

The length of $v^\varkappa$ in cartesian coordinates is

$$\left|\sqrt{(-v^1 v^1 - \cdots - v^{\mathrm{s}} v^{\mathrm{s}} + v^{\mathrm{s}+1} v^{\mathrm{s}+1} + \cdots + v^{\mathrm{n}} v^{\mathrm{n}})}\right| \tag{9.18}$$

and for the equation of the nullcone in these coordinates we get

$$-x^1 x^1 - \cdots - x^{\mathrm{s}} x^{\mathrm{s}} + x^{\mathrm{s}+1} x^{\mathrm{s}+1} + \cdots + x^{\mathrm{n}} x^{\mathrm{n}} = 0. \tag{9.19}$$ [3]

---

[1]) Identifications of this kind and their influence on the choice of the coefficients α, β in I § 7 are dealt with elaborately in T. P. 1951, 1, Ch. III.

[2]) This process will never be applied to geometric objects which are not quantities. Cf. footnote 2 on p. 10.

[3]) If cartesian coordinates are used we often write x instead of ξ for the kernel of the radius vector in a centred R_n.

For the n-vectors belonging to a cartesian coordinate system (h) we use the kernels $\underset{(h)}{I}$ and $\overset{(h)}{i}$ (cf. 7.5)

$$(9.20)\qquad \begin{cases} \underset{(h)}{I}^{\varkappa_1 \ldots \varkappa_n} \overset{\text{def}}{=} n!\, \underset{1}{i}^{[\varkappa_1} \ldots \underset{n}{i}^{\varkappa_n]}; & \underset{(h)}{I}^{1 \ldots n} = +1 \\ \overset{(h)}{i}_{\lambda_1 \ldots \lambda_n} \overset{\text{def}}{=} n!\, \overset{1}{i}_{[\lambda_1} \ldots \overset{n}{i}_{\lambda_n]}; & \overset{(h)}{i}_{1 \ldots n} = +1. \end{cases}$$

They change sign if we pass from a right hand system to a left hand system and are invariant if (h) and (h') have the same screwsense. Geometrically they are represented by a *unit* volume with the screwsense of (h). The component $\overset{(h)}{i}_{1 \ldots n}$ may be considered as the component of a scalar $\varDelta$-density $\tilde{\mathfrak{i}}$ of weight $+1$

$$(9.21)\qquad \mathfrak{i}\,[\varkappa] = \overset{(h)}{i}_{1 \ldots n} = \frac{1}{\underset{(h)}{I}^{1 \ldots n}}\,.$$

If $\tilde{\omega}$ is a W-scalar with the component $+1$ with respect to all rectilinear coordinate systems with the same screwsense as (h), the product

$$(9.22)\qquad \mathfrak{i} \overset{\text{def}}{=} \tilde{\omega}\, \tilde{\mathfrak{i}}$$

is a scalar density of weight $+1$ with a component $+1$ with respect to *all* cartesian coordinate systems. This latter density is intimately connected with $g_{\lambda\varkappa}$. In fact

$$(9.23)\qquad \mathfrak{g}\,[\varkappa] \overset{\text{def}}{=} |\mathrm{Det}\,(g_{\lambda\varkappa})| = n!\, |g_{[1[1} \cdots g_{n]n]}|$$

is a scalar density of weight $+2$ and $\mathfrak{g}\,[h] = +1$, hence

$$(9.24)\qquad \mathfrak{i} = |\mathfrak{g}^{\frac{1}{2}}|.$$

This density has no screwsense and this is in accordance with the fact that $g_{\lambda\varkappa}$ fixes length and angles but not a screwsense.

After introduction of a fundamental tensor we may ask whether any vectors exist which satisfy the equation

$$(9.25)\qquad T_{\lambda\varkappa}\, v^{\varkappa} = \sigma\, v_{\lambda}$$

where $T_{\lambda\varkappa}$ is some given *real* symmetric tensor and σ a suitably chosen scalar. If we write

$$(9.26)\qquad (T_{\lambda\varkappa} - \sigma\, g_{\lambda\varkappa})\, v^{\varkappa} = 0$$

we see that the rank of $T_{\lambda\varkappa} - \sigma\, g_{\lambda\varkappa}$ must be $< n$. Hence σ must be a solution of

$$(9.27)\qquad \mathrm{Det}\,(T_{\lambda\varkappa} - \sigma\, g_{\lambda\varkappa}) = 0\,.$$

This algebraic equation of degree n in σ is the so-called S-equation, well known from the theory of quadrics in ordinary R_3. If $g_{\lambda\varkappa}$ is *definite* it can be proved that the solutions of (9.27) are all real.[1]) If $\underset{1}{\sigma}$ and $\underset{2}{\sigma}$ are two *different* solutions and $\underset{1}{v}^{\varkappa}$ and $\underset{2}{v}^{\varkappa}$ vectors satisfying (9.26) for $\sigma = \sigma_1$ and $\sigma = \sigma_2$, we have

$$(9.28) \qquad \underset{1}{v}^{\lambda} T_{\lambda\varkappa} \underset{2}{v}^{\varkappa} = \underset{1}{\sigma} \underset{1}{v}^{\lambda} g_{\lambda\varkappa} \underset{2}{v}^{\varkappa} = \underset{2}{\sigma} \underset{1}{v}^{\lambda} g_{\lambda\varkappa} \underset{2}{v}^{\varkappa};$$

hence $\underset{1}{v}^{\varkappa}$ and $\underset{2}{v}^{\varkappa}$ are perpendicular to each other. If $\underset{1}{\sigma}$ is a solution of (9.27) with multiplicity m it can be proved that the vectors satisfying (9.26) for $\sigma = \sigma_1$ are the vectors that lie in a definite real m-direction called the *principal multidirection* belonging to $\underset{1}{\sigma}$, and that principal multidirections belonging to different solutions are mutually perpendicular. If the contravariant basis vectors of a cartesian coordinate system (h) are chosen in these multidirections, we have

$$(9.29) \qquad T_{ih} \overset{*}{=} \underset{i}{i}^{\lambda} \underset{h}{i}^{\varkappa} T_{\lambda\varkappa} = 0 \quad \text{for} \quad i \neq h$$

and this proves the

Theorem of principal axes of a symmetric tensor in R_n [2]):

If $T_{\lambda\varkappa}$ is a **real** *symmetric tensor and if the fundamental tensor is* **definite**, *the realways exists a* **real** *cartesian coordinate system (h) such that*

$$(9.30) \qquad T_{ih} \overset{*}{=} 0; \qquad h \neq i.$$

The axes of this coordinate system are called *principal axes* of the symmetric tensor. They are uniquely determined if and only if the solutions of the S-equation are all different.

There is a corresponding theorem for *real* bivectors.[1])

Theorem of principal blades of a bivector in R_n [3]):

If $F^{\varkappa\lambda}$ is a **real** *bivector of rank r and if the fundamental tensor is* **definite** *there always exists a* **real** *cartesian system (h) such that all components F^{hi} vanish except*

$$(9.31) \qquad F^{12} = -F^{21};\; F^{34} = -F^{43};\; \ldots;\; F^{r-1,r} = -F^{r,r-1}.$$

The $\frac{1}{2}r$ simple bivectors with the 2-directions 12; 34; ...; $r-1, r$ are called the *principal blades* of the bivector. They are uniquely determined if and only if the components F^{12}, F^{34}, ..., $F^{r-1,r}$ all have different absolute values.

[1]) Cf. e.g. Dickson 1926, 1.

[2]) See p. 34.

[3]) See p. 36.

The theorems of principal axes and of principal blades do not hold if the fundamental tensor is indefinite because in this case a real symmetric tensor or bivector may possibly have a special position with respect to the real nullcone. But if the symmetric tensor or bivector does not have any such a special position the theorems remain valid.

Orthogonal *coordinate* transformations are transformations from one cartesian coordinate system (h) to another (h'). Because

$$A_{h'i'}^{hi}\, g_{hi} = g_{h'i'} \tag{9.32}$$

we have

$$A_{i'}^{i}\, g_{hi} = A_{h}^{h'}\, g_{h'i'} \tag{9.33}$$

and, because $A_{i'}^{i} \overset{*}{=} \underset{i'}{i}^{i}$; $A_{h}^{h'} \overset{*}{=} \underset{h}{i}^{h'}$ this leads to the reciprocal relation between the covariant (contravariant) components of the contravariant (covariant) basis vectors of both systems

$$\underset{i'}{i}_h \overset{*}{=} \underset{h}{i}_{i'}; \qquad \overset{i'}{i}{}^h \overset{*}{=} \overset{h}{i}{}^{i'}. \tag{9.34}$$

Orthogonal *point* transformations in a centred R_n are those that leave the fundamental tensor invariant. Hence if $T^{\varkappa}_{\cdot\lambda}$ is such a transformation we have

$$T^{\lambda}_{\cdot\sigma}\, g^{\varkappa\sigma} = \overset{-1}{T}{}^{\varkappa}_{\cdot\varrho}\, g^{\varrho\lambda} \tag{9.35}$$

hence

$$T^{\lambda\varkappa} = \overset{-1}{T}{}^{\varkappa\lambda}; \qquad T_{\lambda\varkappa} = \overset{-1}{T}_{\varkappa\lambda} \tag{9.36}$$

from which we see that $\mathrm{Det}\,(T^{\varkappa}_{\cdot\lambda}) = \pm 1$. Orthogonal point transformations are called *rotations* (*reflexotations*) if $\mathrm{Det}\,(T^{\varkappa}_{\cdot\lambda})$ is equal to $+1$ (-1).

An infinitesimal point transformation is a transformation of the form $A^{\varkappa}_{\lambda} + F^{\varkappa}_{\cdot\lambda}\,dt$. Its inverse is $A^{\varkappa}_{\lambda} - F^{\varkappa}_{\cdot\lambda}\,dt$. Hence an infinitesimal point transformation is orthogonal if and only if $F^{\varkappa\lambda}$ is a *bivector*. Of course such a transformation is always a rotation. If the fundamental tensor is definite, $F^{\varkappa\lambda}$ can be split up into $\frac{1}{2}r$ mutually perpendicular blades, r being the rank of $F^{\varkappa\lambda}$. From this it follows that in this case the transformation consists of $\frac{1}{2}r$ rotations in the $\frac{1}{2}r$ R_2's of the blades, leaving these R_2's and the R_{n-r} perpendicular to them invariant. The R_{n-r} plays the part of an "axis of rotation". Hence, if n is odd, there is always such an "axis".

In the indefinite case with index s we consider two real flat manifolds R_s and $'R_s$ through the origin O both of which lie wholly in the $-$-region and who intersect in an R_z. Let a hypersphere with radius 1

and centre in O in R_s be projected orthogonally on $'R_s$. Then we get a quadratic hypersurface in $'R_s$. It always has a set of s principal axes and these are the projections of s mutually perpendicular unitvectors in R_s exactly z of which lie in R_z. The s angles between these unitvectors and their projections are called the *principal angles* between R_s and $'R_s$. Exactly z of them are zero. We now choose (h) in such a way that $\underset{1}{i}^\varkappa, \ldots, \underset{s}{i}^\varkappa$ lie in the direction of the s unitvectors and $\underset{1}{i}^\varkappa, \ldots, \underset{z}{i}^\varkappa$ lie in R_z. Then if $\underset{a}{p}^\varkappa$, $a = 1, \ldots, s$ is the projection of $\underset{a}{i}^\varkappa$ we have the equations

$$\underset{a}{i}^\varkappa = \underset{a}{p}^\varkappa + \underset{a}{q}^\varkappa; \qquad a = 1, \ldots, s \tag{9.37}$$

where $\underset{1}{q}^\varkappa, \ldots, \underset{z}{q}^\varkappa$ are zero and $\underset{z+1}{q}^\varkappa, \ldots, \underset{s}{q}^\varkappa$ are perpendicular to $'R_s$ and consequently lie in the $+$-region. Now all non-vanishing vectors $\underset{a}{p}^\varkappa$ are mutually perpendicular and also perpendicular to all vectors $\underset{a}{q}^\varkappa$. Hence, according to (9.37)

$$0 = (\underset{a}{p}^\varkappa + \underset{a}{q}^\varkappa)(\underset{b}{p}_\varkappa + \underset{b}{q}_\varkappa) = \underset{a}{q}^\varkappa \underset{b}{q}_\varkappa; \qquad a, b = 1, \ldots, s;\ a \neq b \tag{9.38}$$

and from this it follows that the $s - z$ planes of projection of $\underset{z+1}{i}^\varkappa, \ldots, \underset{s}{i}^\varkappa$ are mutually perpendicular. All these planes have a minkowskian geometry and the length of $\underset{a}{p}^\varkappa$ is $|\cos \underset{a}{\varphi}| \geq 1$, where $\underset{a}{\varphi}$ is the angle between $\underset{a}{p}^\varkappa$ and $\underset{a}{i}^\varkappa$. Hence the principal angles between R_s and $'R_s$ are all imaginary or zero and R_s can be transformed into $'R_s$ by a continuous rotation consisting of $s - z$ real minkowskian rotations in the $s - z$ planes of projection of $\underset{z+1}{i}^\varkappa, \ldots, \underset{s}{i}^\varkappa$. Note that during this rotation R_s *remains in the* $-$*-region* and that $\underset{a}{i}^\varkappa$ is transformed into $\underset{a}{p}^\varkappa \cos^{-1} \underset{a}{\varphi}$.

If originally an s-dimensional screwsense is given in R_s the process of orthogonal projection fixes a screwsense in $'R_s$. We call two screwsenses in R_s and $'R_s$ *the same* if each is the orthogonal projection of the other and *opposite* if not. But we still have to prove that if R_s, $'R_s$ and $''R_s$ all lie in the $-$-region and each is provided with a screwsense and if the screwsenses of R_s and $'R_s$ and also those of $'R_s$ and $''R_s$ are the same, then the screwsenses of R_s and $''R_s$ are also the same. The screwsenses in R_s and $''R_s$ are both orthogonal projections of the screwsense in $'R_s$. But as we have seen already, R_s can be transformed into $''R_s$ by a real rotation. Since this process is continuous the projection of the screwsense in $'R_s$ on the moving R_s during its way to $''R_s$ can never change except if there is a moment where there is no screwsense at all, i.e. where R_s contains a direction perpendicular to $'R_s$. But at that moment R_s would contain a direction in the $+$-region. This is impossible. Hence it is possible to fix a screwsense in all real R_s's in the $-$-region simultaneously by fixing a screwsense in one of them. The same can of course be proved for the real R_{n-s}'s in the $+$-region.

Now let $'\underset{1}{i}^{\varkappa}, \ldots, '\underset{s}{i}^{\varkappa}$ be unitvectors in the direction and with the sense of $\underset{1}{p}^{\varkappa}, \ldots, \underset{s}{p}^{\varkappa}$:

$$(9.39)\qquad '\underset{a}{i}^{\varkappa}\,|\cos\underset{a}{\varphi}| = \underset{a}{p}^{\varkappa}; \qquad a = 1, \ldots, s.$$

We know that

$$(9.40)\qquad s!\, '\underset{1}{i}^{[\varkappa_1} \ldots '\underset{s}{i}^{\varkappa_s]}\, '\underset{1}{i}_{[\varkappa_1} \ldots '\underset{s}{i}_{\varkappa_s]} = (-1)^s$$

and from this and (9.37) it follows that

$$(9.41)\qquad \begin{cases} s!\, \underset{1}{i}^{[\varkappa_1} \ldots \underset{s}{i}^{\varkappa_s]}\, '\underset{1}{i}_{[\varkappa_1} \ldots '\underset{s}{i}_{\varkappa_s]} \\ \quad = s!\, \underset{1}{p}^{[\varkappa_1} \ldots \underset{s}{p}^{\varkappa_s]}\, '\underset{1}{i}_{[\varkappa_1} \ldots '\underset{s}{i}_{\varkappa_s]} = (-1)^s\, |\cos\underset{z+1}{\varphi} \ldots \cos\underset{s}{\varphi}| \end{cases}$$

and in this formula instead of $\underset{a}{i}^{\varkappa}$ and $'\underset{a}{i}^{\varkappa}$ we may take two arbitrarily chosen sets of s mutually perpendicular unitvectors in R_s and $'R_s$, provided that the screwsenses fixed by their orders are the same in the sense defined above.

Now let a coordinate system (h) be chosen in such a way that $\underset{1}{i}^{\varkappa}, \ldots, \underset{s}{i}^{\varkappa}$ lie in an arbitrary way in R_s and let there be another coordinate system (h') such that $\underset{1'}{i}^{\varkappa}, \ldots, \underset{s'}{i}^{\varkappa}$ lie in an arbitrary way in $'R_s$. Then the transvection

$$(9.42)\qquad (-1)^s\, s!\, \underset{1}{i}^{[\varkappa_1} \ldots \underset{s}{i}^{\varkappa_s]}\, \underset{1'}{i}_{[\varkappa_1} \ldots \underset{s'}{i}_{\varkappa_s]}$$

is ≥ 1 (≤ -1) if the two screwsenses fixed in R_s and $'R_s$ by the order of these two sets of unitvectors are the same (opposite).

If $P^{\varkappa}{}_{\lambda}$ represents the real point transformation belonging to the coordinate transformation $(h) \to (h')$, i.e. the transformation transforming $\underset{1}{i}^{\varkappa}, \ldots, \underset{n}{i}^{\varkappa}$ into $\underset{1'}{i}^{\varkappa}, \ldots, \underset{n'}{i}^{\varkappa}$, we have

$$(9.43)\qquad P^{\varkappa}_{\cdot\lambda}\, \underset{i}{i}^{\lambda} = \underset{i'}{i}^{\varkappa}; \qquad i = 1, \ldots, n;\ i' = 1', \ldots, n'$$

or

$$(9.44)\qquad P^{h}_{\cdot i} \overset{*}{=} \underset{i'}{i}^{h} \overset{*}{=} A^{h}_{i'}.$$

Accordingly (9.41) can now be written as

$$(9.45)\quad \begin{cases} (-1)^s s!\, \underset{1}{i}^{[j_1} \dots \underset{s}{i}^{j_s]} P^{h_1}_{\cdot j_1} \dots P^{h_s}_{\cdot j_s} \underset{1}{i}_{h_1} \dots \underset{s}{i}_{h_s} \\ \qquad = s!\, \underset{1}{i}^{[j_1} \dots \underset{s}{i}^{j_s]} P^{h_1}_{\cdot j_1} \dots P^{h_s}_{\cdot j_s} \underset{1}{i}_{h_1} \dots \underset{s}{i}_{h_s} \\ \qquad \overset{*}{=} s!\, P^{[1}_{\cdot [1} \dots P^{s]}_{\cdot s]} = \mathrm{Det}\,(P^{\mathfrak{a}}_{\cdot \mathfrak{b}}); \qquad \mathfrak{a}, \mathfrak{b} = 1, \dots, s. \end{cases}$$

Hence the screwsense in R_s and its transform in $'R_s$ are the same (opposite) if and only if $\mathrm{Det}\,(P^{\mathfrak{a}}_{\cdot\mathfrak{b}}) \geq +1$ (≤ -1). But we still have to prove that if, instead of R_s, we take another R_s in the $-$-region not in any way connected with the coordinate system (h), $\mathrm{Det}\,(P^{\mathfrak{a}}_{\cdot\mathfrak{b}}) \geq 1$ is also the necessary and sufficient condition for the screwsense in this new R_s to be the same as its transform. Let all R_s's in the $-$-region be provided with a screwsense and let all these screwsenses be the same in the sense defined before. Now let $P^{\varkappa}_{\cdot\lambda}$ be applied to all these R_s's and let there be, among all transforms, one with the same and one with the opposite screwsense. Because every R_s in the $-$-region can be transformed into every other R_s, with the same screwsense, in that region by a continuous rotation, there must be a transform where the screwsense changes and this transform could not have any screwsense at all which is impossible. Hence we may state the theorem[1]):

If $P^{\varkappa}_{\cdot\lambda}$ represents a real orthogonal transformation in an R_n with index s and if a real coordinate system (h) is chosen in such a way that $\underset{1}{i}^{\varkappa}, \dots, \underset{s}{i}^{\varkappa}$ lie in the $-$-region and the other unitvectors in the $+$-region, $P^{\varkappa}_{\lambda}$ transforms every real R_s, with a given screwsense in the $-$-region into an R_s with the same (opposite) screwsense in that region, if the subdeterminant $\mathrm{Det}\,(P^{\mathfrak{a}}_{\cdot\mathfrak{b}})$; $\mathfrak{a}, \mathfrak{b} = 1, \dots, s$ *is $\geq +1$ (≤ -1) and every real R_{n-s} with a given screwsense in the $+$-region, into an R_{n-s} with the same (opposite) screwsense in that region if the subdeterminant* $\mathrm{Det}\,(P^{\mathfrak{x}}_{\cdot\mathfrak{y}})$; $\mathfrak{x}, \mathfrak{y} = s+1, \dots, n$ *is $\geq +1$ (≤ -1).*

If we call a real orthogonal transformation *$-$-reflexional* (*$+$-reflexional*) if it changes the screwsense of a real R_s (R_{n-s}) in the $-$-region ($+$-region), we have the following possibilities for real orthogonal transformations in R_n

A) non $+$-reflexional; non $-$-reflexional: *proper rotation,* $\Delta = +1$
B) non $+$-reflexional; $-$-reflexional: reflexotation, $\Delta = -1$
C) $+$-reflexional; non $-$-reflexional: reflexotation, $\Delta = -1$
D) $+$-reflexional; $-$-reflexional: *improper rotation,* $\Delta = +1$.

The group of all real orthogonal transformations splits up into four sets of transformations. Obviously a transformation of one set cannot be transformed by a continuous proper rotation into a trans-

[1]) Cf. BRAUER and WEYL 1935, 1; GIVENS 1940, 1 gave an elementary proof.

formation of another set. From these sets only the set (A) and the four sets together form a group.

Exercises.

I 9,1. Prove that for a definite fundamental tensor

I 9,1α) $$I_{\varkappa_1\ldots\varkappa_p\mu_1\ldots\mu_{n-p}} I^{\cdot\ \ \cdot\ \mu_1\ldots\mu_{n-p}}_{\lambda_1\ldots\lambda_p} = p!\,(n-p)!\, g_{[\varkappa_1[\lambda_1} \cdots g_{\varkappa_p]\lambda_p]}\,.$$ [1]

I 9,2. An E_p and an E_q, $p \leq q$, in E_n were defined in I § 1 to be t/p-parallel if the E_p contains an E_t and no E_{t+1} parallel to E_q.[2]

An E_p and an E_q, $p \leq q$, in R_n are said to be *t/p-perpendicular*, if the E_p contains an E_t and no E_{t+1} perpendicular to E_q.[3]

If $v^{\varkappa_1\ldots\varkappa_p}$ and $w^{\varkappa_1\ldots\varkappa_q}$ are simple multivectors in E_p and E_q prove that E_p and E_q are t/p-parallel if and only if

I 9,2α) $$v^{\varkappa_1\ldots\varkappa_u[\lambda_1\ldots\lambda_{p-u}} w^{\nu_1\ldots\nu_q]} \begin{cases} \neq 0 & \text{for} \quad u = t \\ = 0 & \text{for} \quad u = t-1. \end{cases}$$

Prove that E_p and E_q are t/p-perpendicular if and only if

I 9,2β) $$v^{\varkappa_1\ldots\varkappa_u}_{\cdot\ \ \cdot\ \ \lambda_1\ldots\lambda_{p-u}} w^{\lambda_1\ldots\lambda_q} \begin{cases} \neq 0 & \text{for} \quad u = t \\ = 0 & \text{for} \quad u = t-1. \end{cases}$$

I 9,3. The real symmetric tensors $P_{\lambda\varkappa}$ and $Q_{\lambda\varkappa}$ in an ordinary R_n have the same principal multidirections if and only if

I 9,3α) $$P_{[\varkappa}^{\cdot\,\lambda} Q_{\mu]\lambda} = 0.$$

I 9,4.[4] If $g_{\lambda\varkappa}$ is definite and $v^{\varkappa_1\ldots\varkappa_p}$ real, the equations $v^{\varkappa_1\ldots\varkappa_p} v_{\varkappa_1\ldots\varkappa_p} = 0$ and $v^{\varkappa_1\ldots\varkappa_p} = 0$ are equivalent.

§ 10. Hybrid quantities.

Every set of N independent functions of the components of a quantity with N components could be considered as a kind of components of this quantity. But generally such components are not used because their transformation formulae are too complicated. There is one exception. In dealing with a quantity with not only real components, for instance a vector $v^\varkappa$, the complex conjugates of the $v^\varkappa$ may be considered as another set of components of the same vector. But we may also look upon them as components of another quantity, a quantity *of the second kind* in contradistinction to the ordinary quantities that will be called *of the first kind*. If we write $v^{\bar{\varkappa}}$ for these complex conjugates, they transform as follows

(10.1) $$v^{\bar{\varkappa}'} = A^{\bar{\varkappa}'}_{\bar{\varkappa}} v^{\bar{\varkappa}}$$

[1] LIPKA 1922, 1, p. 243.
[2] SCHOUTEN 1902, 1, p. 34.
[3] SCHOUTEN 1902, 1, p. 49.
[4] NARLIKAR 1938, 1 for $p = 2$.

where the $A_{\bar{\varkappa}}^{\bar{\varkappa}'}$ stand for the complex conjugates of the $A_{\varkappa}^{\varkappa'}$. In the same way to every tensor for instance $P^{\varkappa\lambda}_{\cdot\cdot\mu}$ a tensor of the second kind with components $P^{\bar{\varkappa}\bar{\lambda}}_{\cdot\cdot\bar{\mu}} = \overline{P^{\varkappa\lambda}_{\cdot\cdot\mu}}$ can be constructed. But it is not possible to construct in this way components $P^{\bar{\varkappa}\lambda}_{\cdot\cdot\mu}$, because $P^{\varkappa\lambda}_{\cdot\cdot\mu}$ can be written in several ways as a sum of products of vectors and these several ways would lead to different values for $P^{\bar{\varkappa}\lambda}_{\cdot\cdot\mu}$. Notwithstanding this it has a definite meaning to consider for instance a quantity with components $Q^{\bar{\varkappa}}_{\cdot\bar{\lambda}\mu}$ transforming as follows

$$Q^{\bar{\varkappa}'}_{\cdot\bar{\lambda}'\mu'} = A^{\bar{\varkappa}'\bar{\lambda}\mu}_{\bar{\varkappa}\bar{\lambda}'\mu'} Q^{\bar{\varkappa}}_{\cdot\bar{\lambda}\mu}. \tag{10.2}$$

Such quantities we call *hybrid*.[1]) The indices with a bar are called indices *of the second kind* in contradistinction to those *of the first kind* without a bar. Hybrid Δ-densities can be defined in the same way as ordinary Δ-densities. If they get the factor $\Delta^{-w}\bar{\Delta}^{-w'}$, w is the *weight* and w' the *antiweight*. If $w = w'$ we have an ordinary density of weight $2w$. Hence ordinary densities as defined in § 2 are really a kind of hybrid quantities. Ordinary tensors and their complex conjugates may be considered as special cases of hybrid quantities but as a rule we use this latter term only if there are really two different kinds of indices.

To a co- or contravariant hybrid quantity of valence 2 for instance $T^{\bar{\varkappa}\lambda}$ there belongs the complex conjugate $T^{\varkappa\bar{\lambda}}$. Now if

$$T^{\bar{\varkappa}\lambda} = +T^{\lambda\bar{\varkappa}} \tag{10.3}$$

$T^{\varkappa\bar{\lambda}}$ (and its conjugate) is called a ***hermitian symmetric tensor*** or ***hermitian tensor*** for short. But if

$$T^{\bar{\varkappa}\lambda} = -T^{\lambda\bar{\varkappa}} \tag{10.4}$$

$T^{\bar{\varkappa}\lambda}$ (and its conjugate) is called a ***hermitian alternating tensor*** or a ***hybrid bivector***. The difference between hermitian tensors and hybrid bivectors is not so great as in the ordinary case, because if $T^{\bar{\varkappa}\lambda}$ is hermitian, $i\,T^{\bar{\varkappa}\lambda}$ is a hybrid bivector. The property is obviously invariant for coordinate transformations and also for multiplication with a *real* scalar.

A mixed hybrid quantity of valence 2, $V^{\bar{\varkappa}}_{\cdot\lambda}$ has the complex conjugate $V^{\varkappa}_{\cdot\bar{\lambda}}$. If its determinant is not zero the inverses $\overset{-1}{V}{}^{\varkappa}_{\cdot\bar{\lambda}}$, $\overset{-1}{V}{}^{\bar{\varkappa}}_{\cdot\lambda}$ exist. Now if

$$V^{\bar{\varkappa}}_{\cdot\lambda} = \pm\overset{-1}{V}{}^{\bar{\varkappa}}_{\cdot\lambda} \tag{10.5}$$

[1]) SCHOUTEN 1929, 1; SCHOUTEN and VAN DANTZIG 1930, 1; E I 1935, 1, p. 8. In these publications we wrote $\bar{v}^{\bar{\varkappa}}$ and $\bar{A}_{\bar{\varkappa}}^{\bar{\varkappa}'}$ instead of $v^{\bar{\varkappa}}$ and $A_{\bar{\varkappa}}^{\bar{\varkappa}'}$. This is useful if at any time we wish to drop the indices as for instance in matrix calculus. Because otherwise, where the indices had been dropped, we could no longer make a difference between a quantity and its complex conjugate. But as long as we have no intention of dropping indices and no ambiguity arises we can simplify all notations considerably by dropping the bar over the kernels. It can always be reintro-

$V^{\bar{\varkappa}}_{\cdot\lambda}$ (and its conjugate) is called *positive* or *negative invertible.* [1]) This property is invariant for coordinate transformations and also for multiplication with a factor of the form $e^{i\varphi}$ where φ is real. [2])

For hermitian tensors the following theorem holds [3]) (cf. I § 8):

If r is the rank of $T_{\lambda\bar{\varkappa}} = T_{\bar{\varkappa}\lambda}$ $(U^{\bar{\varkappa}\lambda} = U^{\lambda\bar{\varkappa}})$, the coordinate system can always be chosen such that the matrix of $T_{\lambda\bar{\varkappa}}(U^{\bar{\varkappa}\lambda})$ has the diagonal form with s numbers -1, $r-s$ numbers $+1$ and $n-r$ numbers 0 in the main diagonal.

The number s, called the *index*, is an invariant of the hermitian tensor. The sequence $- - - \cdots + + + \cdots$ with s $-$-signs and $(r-s)$ $+$-signs is called the *signature*. The signature is said to be even (odd) if s is even (odd). The hermitian tensor is called *positive* (*negative*) *definite* if $s=0(n)$ and $r=n$ and *indefinite* in all other cases. It is called *positive* (*negative*) *semi-definite* if $n-r>0$ and $s=0(r)$ (cf. I § 8).

As a corollary we get that two hermitian tensors are equiform (cf. I § 2) if and only if they have the same rank and the same index.

In a centred E_n a hermitian tensor $T_{\lambda\bar{\varkappa}}$ is represented by a subspace satisfying the equations

$$T_{\lambda\bar{\varkappa}}\, x^{\lambda}\, x^{\bar{\varkappa}} = \pm 1. \tag{10.6}$$

The nullcone of $T_{\lambda\bar{\varkappa}}$ has the equation

$$T_{\lambda\bar{\varkappa}}\, x^{\lambda}\, x^{\bar{\varkappa}} = 0. \tag{10.7}$$

If $T_{\lambda\bar{\varkappa}}$ is definite its nullcone degenerates into the origin because in this case there are no other points satisfying (10.7).

In dealing with hybrid quantities it is often convenient to consider an auxiliary E_{2n} with coordinates $x^{\varkappa}$, $x^{\bar{\varkappa}}$ that are now supposed to be linearly independent and each to take all complex values. The allowable coordinate transformations in this E_{2n} are the linear homogeneous transformations of the $x^{\varkappa}$ and $x^{\bar{\varkappa}}$ allowable in the original E_n, hence the E_{2n} has a restricted transformation group. It contains the original E_n as a subspace given by the n equations

$$x^{\bar{\varkappa}} = \overline{x^{\varkappa}}. \tag{10.8}$$

This subspace, called the *principal* E_n, is invariant for the allowable coordinate transformations and the same holds for the subspaces re-

duced where necessary (cf. T. P. 1951, 1, Ch. X, also for the shorter notations of matrix calculus arising by dropping all indices of quantities with valences 1 and 2).

[1]) Canonical forms for $V^{\bar{\varkappa}}_{\cdot\lambda}$ were given for the general case by Haantjes 1935, 1 and for the invertible case by Jacobsthal 1934, 1. Cf. Veblen, v. Neumann and Givens 1936, 1, p. 4. 23; Schouten 1949, 3, p. 218.

[2]) Invertible quantities were introduced by Schouten and Haantjes 1935, 2, p. 179 and by Veblen, v. Neumann and Givens 1936, 1, p. 4.23. For the projective case these latter authors called them anti-involutions. They are important in spinspace. Cf. Schouten 1949, 2; 3; 4; 1950,1.

[3]) Cf. for the proof Dickson 1926, 1, p. 70.

presented by each of the two sets of n equations

(10.9) a) $x^{\bar{\varkappa}} = \underset{1}{c}^{\bar{\varkappa}} = \text{const.}$ b) $x^{\varkappa} = \underset{2}{c}^{\varkappa} = \text{const.}$

Each set of equations represents a set of ∞^n E_n's called the *invariant* E_n's of the *first* and of the *second set* respectively. The $x^{\varkappa}$ can be used as coordinates in every E_n of the first set and the $x^{\bar{\varkappa}}$ play the same role for the E_n's of the second set. Every E_n of one set is mapped on each E_n of the same set by means of the E_n's of the other set. According to KLEIN's principle instead of $x^{\varkappa}$ and $x^{\bar{\varkappa}}$ we may now also introduce general rectilinear coordinate systems subject to the full group G_{ho} in E_{2n}, provided that we adjoin the two sets of invariant E_n's and the principal E_n (10.8).

Instead of an ordinary symmetric tensor $g_{\lambda\varkappa}$ a hermitian tensor $a_{\lambda\bar{\varkappa}}$ of rank n can be introduced as fundamental tensor. Lowering and raising of indices can then be effected by means of $a_{\lambda\bar{\varkappa}}$ and its inverse denoted by $a^{\bar{\varkappa}\lambda}$. The determinant of $a_{\lambda\bar{\varkappa}}$ is a scalar density of weight $+2$. An E_n with a fundamental tensor $a_{\lambda\bar{\varkappa}}$ is called an $\widetilde{R}_n$ [1]) and it is said to be *ordinary* if $a_{\lambda\bar{\varkappa}}$ is positive definite. In an $\widetilde{R}_n$, $v^{\varkappa}$ and $v^{\bar{\varkappa}}$ each get two kinds of components, viz. $v^{\varkappa}$, $v_{\bar{\lambda}}$ and $v^{\bar{\varkappa}}$, v_{λ}. So in $\widetilde{R}_n$ the difference between co- and contravariant vectors does not vanish, as it does in R_n. There remain in $\widetilde{R}_n$ two kinds of vectors, called *kets* and *bras* by DIRAC. The ket (bra) is a contravariant vector of the first (second) kind, equivalent to a covariant vector of the second (first) kind[2]). Two vectors $v^{\varkappa}$ and $w^{\varkappa}$ are said to be *(unitary) perpendicular* to each other if $v^{\lambda} w^{\bar{\varkappa}} a_{\lambda\bar{\varkappa}} = 0$. This is equivalent to $w^{\lambda} v^{\bar{\varkappa}} a_{\lambda\bar{\varkappa}} = 0$. The *norm* of a vector $v^{\varkappa}$ is the expression $v^{\lambda} v^{\bar{\varkappa}} a_{\lambda\bar{\varkappa}}$ *that is always real*. The vectors with a positive (negative) norm fill the $+$ $(-)$-*region* of the $\widetilde{R}_n$. A vector is called *unitvector* if its norm equals ± 1. A unitvector remains unitvector if it is multiplied by $e^{i\varphi}$ with real φ.

According to the theorem stated above there exists at least one coordinate system (h) with basis vectors $\underset{i}{u}^{\varkappa}$; $\overset{h}{u}_{\lambda}$ such that

$$a_{\lambda\bar{\varkappa}} = -\overset{1}{u}_{\lambda}\overset{1}{u}_{\bar{\varkappa}} - \cdots - \overset{s}{u}_{\lambda}\overset{s}{u}_{\bar{\varkappa}} + \overset{s+1}{u}_{\lambda}\overset{s+1}{u}_{\bar{\varkappa}} + \cdots + \overset{n}{u}_{\lambda}\overset{n}{u}_{\bar{\varkappa}}. \tag{10.10}$$

The $\overset{h}{u}_{\lambda}$ are mutually perpendicular unitvectors and they satisfy the equations

$$\begin{cases} \underset{i}{u}^{\varkappa} = -\overset{i}{u}^{\varkappa} = -\overset{i}{u}_{\bar{\lambda}} a^{\bar{\lambda}\varkappa} & i = 1, \ldots, s; \\ \underset{j}{u}^{\varkappa} = +\overset{j}{u}^{\varkappa} = +\overset{j}{u}_{\bar{\lambda}} a^{\bar{\lambda}\varkappa} & j = s+1, \ldots, n. \end{cases} \tag{10.11}$$

They form a *unitary cartesian* coordinate system. All these systems are transformed into each other by the transformations of the *unitary*

[1]) In E I 1935, 1, p. 60 we wrote U_n instead of $\widetilde{R}_n$ (cf. Ch. VIII).

[2]) DIRAC 1947, 1; cf. T. P. 1951, 1, p. 243.

orthogonal group. This is the group of homogeneous linear transformations characterized by the invariance of $a_{\lambda\bar{\varkappa}}$ (cf. 9.36):

$$(10.12) \quad \text{a)}\ U^{\varrho}_{\cdot\lambda} U^{\bar{\sigma}}_{\cdot\bar{\varkappa}} a_{\varrho\bar{\sigma}} = a_{\lambda\bar{\varkappa}} \quad \text{or} \quad \text{b)}\ U_{\varrho\bar{\varkappa}} = \overset{-1}{U}_{\bar{\varkappa}\varrho} \quad \text{or} \quad \text{c)}\ U^{\varkappa}_{\cdot\lambda} = \overset{-1}{U}{}_{\lambda}^{\cdot\varkappa}.$$

From this we see that Det $(U^{\varkappa}_{\cdot\lambda})$ has the form $e^{i\varphi}$ with real φ and that if $U^{\varkappa}_{\cdot\lambda}$ represents a unitary orthogonal transformation, the same holds for $e^{i\psi} U^{\varkappa}_{\cdot\lambda}$ with real ψ. [1]

The subspace represented by

$$(10.13) \qquad a_{\lambda\bar{\varkappa}}\, x^{\lambda} x^{\bar{\varkappa}} = \pm 1 \ (0)$$

is called the *fundamental figure* (the *nullcone*) in the $\widetilde{R}_n$. But (10.13) can also be interpreted as the equation of a hyperquadric in the auxiliary E_{2n} that makes this E_{2n} to an R_{2n}. In this R_{2n} the $a_{\lambda\bar{\varkappa}}$ are the components of a fundamental tensor whose components $a_{\lambda\varkappa}$ and $a_{\bar{\lambda}\bar{\varkappa}}$ are zero. Because of this the hyperquadric has a very peculiar position with respect to the two sets of invariant E_n's. In fact, the section of the hyperquadric (nullcone) with an E_n of the first set is an E_{n-1} with the $n+1$ linear equations

$$(10.14) \qquad \text{a)}\ x^{\bar{\varkappa}} = \underset{1}{c}^{\bar{\varkappa}} \qquad \text{b)}\ 2\, a_{\lambda\bar{\varkappa}}\, x^{\lambda} \underset{1}{c}^{\bar{\varkappa}} = 1 \ (0)$$

in which n independent parameters occur. Hence the hyperquadric is built up of two sets of ∞^n E_{n-1}'s and each E_{n-1} of one set has at most one point in common with an E_{n-1} of the other set. [2]

If $T_{\lambda\bar{\varkappa}}$ is an arbitrary hermitian tensor, we may ask whether any vectors exist satisfying the relation

$$(10.15) \qquad v^{\lambda} T_{\lambda\bar{\varkappa}} = \sigma\, v_{\bar{\varkappa}}.$$

σ is a solution of the equation

$$(10.16) \qquad \operatorname{Det}(T_{\lambda\bar{\varkappa}} - \sigma\, a_{\lambda\bar{\varkappa}}) = 0.$$

If $a_{\lambda\bar{\varkappa}}$ is *definite*, it can be proved that the solutions are all *real* and that to every solution with multiplicity m there belongs an m-direction perpendicular to all multidirections belonging to other solutions.

---

[1]) Instead of $a_{\lambda\bar{\varkappa}}$ a positive or negative invertible hybrid quantity (cf. 10.5) could be introduced as fundamental quantity in E_n. This would lead to a geometry with a group leaving this quantity invariant. This geometry seems to have escaped attention and even its group seems not to have got a name.

[2]) It is well known that a quadric in E_{2n} contains $\infty^{\binom{n+1}{2}}$ E_{n-1}'s. But a quadric in E_{2n} does not fix two preferred sets of E_n's. This means that the E_{2n} is not an ordinary R_{2n} but an R_{2n} that contains besides a quadric also two sets of invariant E_n's. If these E_n's are fixed it contains moreover the principal E_n with the equation (10.8). This is the real reason why there is something new in the geometry of an $\widetilde{R}_n$ and why this geometry is not the ordinary geometry of an R_{2n} described in a queer way.

These multidirections are called the *principal multidirections* of the hermitian tensor.[1] The unitvectors of (h) can be chosen in these multidirections, hence (cf. I § 9)

Theorem of principle axes of a hermitian tensor in $\widetilde{R}_n$:

If $T_{\lambda\bar{\varkappa}}$ is a hermitian tensor and if the fundamental tensor is ***definite*** *there always exists a unitary cartesian system (h) such that*

$$T_{i\bar{h}} \stackrel{*}{=} 0; \qquad h \neq i. \tag{10.17}$$

The axes of this coordinate system are called *principal axes* of the hermitian tensor. They are uniquely determined if and only if the solutions of the equation (10.16) are all different. The theorem of principal axes does not hold if the fundamental tensor is indefinite because in this case the tensor $T_{\lambda\bar{\varkappa}}$ may possibly have a special position with respect to the nullcone. If this is not the case the theorem remains valid.

Exercises.

I 10,1. The infinitesimal transformation

$$'x^{\varkappa} = x^{\varkappa} + P^{\varkappa}_{\cdot\lambda}\, x^{\lambda}\, dt; \qquad t \text{ real} \tag{I 10,1 α}$$

belongs to the unitary orthogonal group if and only if $P_{\bar{\varkappa}\lambda}$ is hermitian alternating. Then $iP_{\bar{\varkappa}\lambda}$ is hermitian symmetric and every principal direction of this tensor is invariant for the transformation.

I 10,2[2]. If $T_{\lambda\bar{\varkappa}}$ is hermitian symmetric (alternating) and $S_{\lambda}^{\cdot\varkappa}$ arbitrary, prove that

$$S_{\lambda}^{\cdot\sigma}\, T_{\sigma\bar{\varrho}}\, S_{\bar{\varkappa}}^{\cdot\bar{\varrho}} \tag{I 10,2 α}$$

is also hermitian symmetric (alternating).

If $T^{\bar{\varrho}}_{\cdot\lambda}$ is positive (negative) invertible and $S_{\lambda}^{\cdot\varkappa}$ arbitrary, prove that

$$S_{\lambda}^{\cdot\sigma}\, \overset{-1}{T}{}^{\bar{\varkappa}}_{\cdot\sigma}\, S_{\bar{\varkappa}}^{\cdot\bar{\varrho}} \tag{I 10,2 β}$$

is also positive (negative) invertible. If $'x^{\varkappa} = S_{\lambda}^{\cdot\varkappa}\, x^{\lambda}$ is a transformation of the unitary orthogonal group, in both formulae $S_{\bar{\varkappa}}^{\cdot\bar{\varrho}}$ can be replaced by $\overset{-1}{S}{}^{\bar{\varrho}}_{\cdot\bar{\varkappa}}$.

I 10,3[2]. In an ordinary $\widetilde{R}_2$ the most general unitary orthogonal transformation can be written in the form

$$\left\{\begin{aligned} 'x &= x\, e^{i\alpha} \cos\varphi + y\, e^{i\beta} \sin\varphi \\ 'y &= -\, x\, e^{i\gamma} \sin\varphi + y\, e^{i(\gamma-\alpha+\beta)} \cos\varphi \end{aligned}\right. \tag{I 10,3 α}$$

with four real parameters α, β, γ, φ with respect to a unitary cartesian coordinate system.

---

[1]) Cf. e.g. Dickson 1926, 1.

[2]) Cf. E I 1935, 1, § 5.

§ 11. Abridged notations.

In tensor calculus there sometimes appears a great number of indices. Many authors therefore have tried to get rid of them by introducing abridged notations. Most of these attempts failed because an abridged notation always means a new kind of stenography that has to be learned by the reader, and because most readers are overbusy and seldom willing. It should be held in mind that as a rule abbreviations make the reading more difficult and the calculus less foolproof. Authors should therefore be very careful and they should introduce them only when necessary. We mention here some of the most interesting abbreviations.

1. Collecting indices. Some of the indices of a quantity can be replaced by one upper or lower index. For instance $P^{A}_{\cdot\,\nu}$ or $P_{A\nu}$ may be written for $P^{\varkappa\lambda}_{\cdot\,\cdot\,\mu\nu}$. Even transvections over collected indices can be made possible, for instance $P^{A}_{\cdot\,\nu}\, Q^{\cdot\,\cdot\,\sigma}_{A\,\varrho}$ for $P^{\varkappa\lambda}_{\cdot\,\cdot\,\mu\nu}\, Q^{\cdot\,\cdot\,\cdot\,\mu\sigma}_{\varkappa\lambda\varrho}$. This abbreviation can be very useful if during an investigation or even during some part of it only the indices not collected are important. It is often used and there seems to be no objection provided its use is not exaggerated.

2. Representative indices. Let $\lambda_p\,\lambda_q$ where $q<p$, always stand for $\lambda_p \ldots \lambda_{q+1}\,\lambda_q \ldots \lambda_1$.[1] Then we are able to write $P_{[\lambda_p]\,\lambda_q}$ for $P_{[\lambda_p \ldots \lambda_{q+1}]\,\lambda_q \ldots \lambda_1}$. This notation is useful in investigations where repeated differentiations occur often, for instance $\partial_{\lambda_p}\, w_{\lambda_q}$ stands for $\partial_{\lambda_p} \ldots \partial_{\lambda_{q+1}}\, w_{\lambda_q \ldots \lambda_1}$. But it needs to be handled carefully and this may be the reason that it has not been accepted. The same holds for the notation $v^{\varkappa_1|p}_{\cdot\;\;\lambda_1|q}$ for $v^{\varkappa_1 \ldots \varkappa_p}_{\cdot\;\;\;\;\cdot\;\lambda_1 \ldots \lambda_q}$ [2]) that has the same disadvantages and is less practical. A less developed version of this latter notation that has come out recently[3]) is to write $P^{\varkappa \ldots}_{\cdot\;\;\lambda \ldots}$ instead of $P^{\varkappa_1 \ldots \varkappa_p}_{\cdot\;\;\;\;\cdot\;\lambda_1 \ldots \lambda_q}$. If this version should find approval this would demonstrate the fact that only very uncomplicated abbreviations have any chance of being accepted.

3. The notation of GIVENS.[4]) GIVENS proposed to write (λ) instead of p alternated neighbouring indices $[\lambda_1 \ldots \lambda_p]$ wherever they might occur and $|(\lambda)|$ for their number p. This notation makes it easy to write down transvections of multivectors. Though GIVENS' results were interesting his notation was not accepted.

4. VITALI'S *method.* VITALI[5]) introduced a kind of indices that can take the values *1*, ..., *n*, *11*, *12*, ..., *1n*, *21*, *22*, ..., *2n*, ..., *nn*, *111*, ..., *nnn*, ... and so on till all combinations (with repetition) from the numbers *1*, ..., *n*. The notation is to be used in connexion

1) SCHOUTEN 1925, 1.

2) E I 1935, 1, p. 33.

3) CRAIG 1943, 1.

4) VEBLEN, V. NEUMANN and GIVENS 1936, 1; GIVENS 1937, 1.

5) VITALI 1923, 1.

with higher differentiations. Though strongly advocated by BORTOLOTTI[1]) it never came in general use. The method is outside of the realm of this book.

5. CRAIG'S *extensors*. Another notation, also connected with higher differentiations is the method of extensors introduced by CRAIG.[2]) Though the double-indices used are in the beginning somewhat bewildering the notation is often used especially by japanese authors and it seems to have a fair chance of being accepted for those investigations where higher differentiations play an important role. A. KAWAGUCHI[3]) and M. KAWAGUCHI[4]) gave generalizations. The method is outside of the realm of this book.

6. *Ideal vectors* (cf. III § 4). A tensor, for instance $P_{\kappa\lambda}^{\cdot\,\cdot\,\nu}$, is generally not a product of vectors. But we may write $P_{\kappa\lambda}^{\cdot\,\cdot\,\nu} = u_\kappa v_\lambda w^\nu$ and look upon u_κ, v_λ and w^ν as symbolic factors. If this process is applied to a symmetric tensor the factors can be chosen equal, for instance $w_{\lambda_1 \ldots \lambda_q} = w_{(\lambda_1 \ldots \lambda_q)} = w_{\lambda_1} \ldots w_{\lambda_q}$ and the transvection $w_{\lambda_1 \ldots \lambda_q} x^{\lambda_1} \ldots x^{\lambda_q}$ can then be written $(w_\lambda x^\lambda)^q$. So we get the CLEBSCH-ARONHOLD symbolism well known from the theory of algebraic invariants. If $w_{\lambda_1 \ldots \lambda_q}$ is a q-vector we may write $w_{\lambda_1 \ldots \lambda_q} = w_{\lambda_1} \ldots w_{\lambda_q}$ if we assume that $w_{\lambda_1} w_{\lambda_2} = -w_{\lambda_2} w_{\lambda_1}$ for every choice of λ_1 and λ_2. Then we arrive at WEITZENBÖCK'S "Komplexsymbolik".[5]) Of course in both cases in all terms that contain a non-symbolical factor more than once, separate and distinguishable ideal vectors must be introduced for every factor in order to avoid ambiguity. This is all well known from algebra. Now both symbolisms are very useful in purely algebraic investigations as so many publications on algebraic invariants prove and they can of course be equally useful in tensor algebra. But as soon as differentiations are introduced, as in differential geometry, many difficulties arise with respect to the differentiation of ideal vectors. In some earlier publications[6]) an attempt was made to introduce ideal vectors in differential geometry. But the simplifications to be got in this way did not appear important enough to compensate for the introduction of symbolic differentiations of ideal vectors and so ideal vectors never got a proper place outside of the purely algebraic domain.

7. *Systems of "direct" calculus*. An invariant equation between tensors does not depend on the choice of the coordinates or the letters

---

[1]) BORTOLOTTI 1930, 1; 1937, 1; also for more literature.

[2]) CRAIG 1943, 1.

[3]) A. KAWAGUCHI 1939, 1.

[4]) M. KAWAGUCHI 1952, 1.

[5]) WEITZENBÖCK 1908, 1.

[6]) SCHOUTEN 1918, 1; 1921, 1; SCHOUTEN and STRUIK 1919, 1; 1921, 2; 3; 1922, 1; 2; STRUIK 1922, 1. Cf. III § 4.

used for the indices. It is a message from the author to the reader concerning certain properties of tensors independent of any coordinate system and this message is readable as soon as we know the *skeleton* of the formula that is the combination of the *kernel*, the *places* of the indices for every manifold concerned and the *places* where the transvections act. Hence it must be possible to invent notations that represent in some way this skeleton without the use of indices. Such a notation is often called a "direct" system of calculus. Well known examples are GRASSMANN's "Ausdehnungslehre", HAMILTON's quaternion calculus, ordinary "vectoranalysis", "tensoranalysis", matrix calculus etc. A direct system must have a notation for the kernels, mostly a letter of some special alphabet, in order to distinguish between tensors and scalars, a special method of marking the places of upper and lower indices that makes it possible to drop these indices and a number of special signs for the most frequently used multiplications and transvections. For instance in the most developed direct system[1]) the vectors $v^{\varkappa}$ and w_{λ} were written $\mathbf{v}'$ and $\mathbf{w}$ and the tensor $P^{\varkappa\cdot\mu\nu}_{\cdot\lambda}$ was represented by $\mathbf{P}'{}^{\cdot}{}''$. The transvections $v^{\varkappa} w_{\varkappa}$, $P^{\varkappa\cdot\mu\nu}_{\cdot\lambda} w_{\nu}$, $w_{\varkappa} P^{\varkappa\cdot\mu\nu}_{\cdot\lambda}$ were written $\mathbf{v}'\,\underset{\cdot}{1}\,\mathbf{w}$, $\mathbf{P}'{}^{\cdot}{}''\,\underset{\cdot}{1}\,\mathbf{w}$, $\mathbf{w}\,\underset{\cdot}{1}\,\mathbf{P}'{}^{\cdot}{}''$ and other multiplication signs were given for a number of other transvections. This system worked rather efficiently and it was used not only for the translation of well known results (as many of these systems were) but especially for the establishment of new theorems. From the beginning there was a great difficulty, notwithstanding the great number of multiplication symbols the use of ideal vectors for the representation of all necessary transvections could not be avoided. For instance $w_{\mu} P^{\varkappa\cdot\mu\nu}_{\cdot\lambda}$ could only be represented by $(\mathbf{w}\,\underset{\cdot}{1}\,\mathbf{r}')\,\mathbf{p}'\mathbf{q}\mathbf{s}'$ after introduction of ideal vectors: $\mathbf{P}'{}^{\cdot}{}'' = \mathbf{p}'\mathbf{q}\mathbf{r}'\mathbf{s}'$. Now in view of this difficulty one could get the idea to improve the direct calculus by introducing instead of the two signs and ' to mark the places of the indices, as many signs as there are places. Then for instance a tensor $P^{\varkappa\cdot\mu\nu}_{\cdot\lambda}$ could be written $\mathbf{P}'{}^{\cdot}_{\cdot\times}{}^{\circ *}$ and it would be easy to write all possible transvections, for instance $\mathbf{w}_{,}\mathbf{P}'{}^{\cdot}_{\cdot\times}{}^{\circ *}$; $\mathbf{w}_{\circ}\mathbf{P}'{}^{\cdot}_{\cdot\times}{}^{\circ *}$; $\mathbf{P}'{}^{\cdot}_{\cdot\times}{}^{\circ *}\mathbf{v}^{\times}$ without the necessity of introducing ideal vectors. This would certainly be an improvement but then the next step would be the remark: "why not $\varkappa\lambda\mu\nu$ instead of $, \times \circ *$?" and this would lead us back to RICCI's calculus. The real cause of this remarkable situation is that tensor calculus has two different aspects. The *running* indices are on the one hand only marks labelling certain places and constituting together with the kernels the skeleton of the formula. From this point of view tensor calculus is a highly efficient "direct" calculus because it enables us

[1]) SCHOUTEN 1918, 1; STRUIK 1922, 1. In SCHOUTEN and STRUIK 1922, 1 (the first edition of E I 1935, 1) all formulae were given in duplo, once indexfree and once with indices.

to represent all kinds of multiplications and transvections without the help of any auxiliary device. But on the other hand as soon as a definite coordinate system is chosen, the running indices may be replaced by *fixed* ones, for instance λ by $1, \ldots, n$ and if this is done the equations can be read as equations between components. The "directness" of the calculus lies in the fact that the formulae *can* be read (and *should* be read) as relations between quantities and not between components.[1]) The process leading from the direct formulae to the formulae in components (in most kinds of "direct" calculus rather a difficult process!) happens automatically here because the same formulae *can* also be read (and *should* be read were desirable) as relations between components. Of course this is only exactly true if the kernel-index method is used, that is, if every object gets its own kernel and if every coordinate system gets its own kind of running indices, for instance $\varkappa$, $\varkappa'$, h with their corresponding fixed indices, $1, \ldots, n$; $1', \ldots, n'$; $1, \ldots, n$.

The greatest difficulties for all systems without indices lie in the construction of the necessary equivalent for the places of these indices and in the representation of all possible multiplications and transvections. These difficulties increase strongly if the valences of the quantities occurring become higher. Therefore indexfree systems only succeeded for valences *0*, *1* and *2*. In fact vector analysis in R_3 with valences *0* and *1* only, and matrix calculus[2]) in R_n with valences *0*, *1* and *2* only, are rather satisfactory. But the vector analysis that can be constructed for E_3 or R_4 is already much more complicated and systems of this kind for E_n and R_n are very amusing for the inventor himself and perhaps for some of his friends but they never will find general application.[3])

8. The method of the radiusvector in E_n and R_n. For differential geometry in E_n and R_n often a middlecourse between tensor calculus and indexfree calculus is followed by writing $\mathfrak{r}$ (instead of $r^\varkappa$) for the radiusvector and conserving all other indices. In this way quantities

---

[1]) All students of tensor calculus should be encouraged from the beginning to try to read all invariant formulae as formulae between geometric objects and not as formulae between components only.

[2]) Matrix calculus in connexion with tensor calculus was dealt with elaborately in E I 1935, 1, p. 35ff. and in T. P. 1951, 1, p. 39ff. and Ch. X (DIRAC's matrix calculus).

[3]) The present author has some experience in this matter. He wrote his paper 1922, 3 on different kinds of connexions in differential geometry, in his own indexfree system. But following the advice of F. KLEIN who warned him that the paper would never be read, he translated it in terms of RICCI calculus. The advice was a good one, the paper was much referred to and it had its influence on the development of modern differential geometry.

of valence p derived from $\mathfrak{r}$ may get $p-1$-indices. The method is certainly efficient and it is very often used, of course only in euclidean spaces.

9. CARTAN's *method.* Though from the point of view of RICCI calculus CARTAN's method could be considered as an abbreviation we prefer to consider it as a full sized method in its own right. It is not necessary to discuss it here because it will be dealt with elaborately at many places in this book.

10. Abbreviations ad hoc. Sometimes it is convenient to introduce for a short time during an investigation some abbreviation that is specially adapted to this part of the calculation but will be dropped later on. For instance in II § 8 we write Ω with suppressed indices for any quantity and $-\circ$ for any operation of the kind considered there. This is certainly justified because only in this way could such a general formula be written down. So abbreviations ad hoc may have their advantages. But if a book contains too many of them it becomes unreadable. Young authors especially, who can not yet foresee all the consequences, *should abstain as far as possible from introducing ad hoc abbreviations.* It is inexusable to spend valuable time on the writing of a paper that will not actually be read.

II. Analytic preliminaries.

§ 1. The arithmetic n-dimensional manifold $\mathfrak{A}_n$. [1]

A set of n real or complex values of n ordered variables $\overset{\varkappa}{\xi}$ is called an *arithmetic point* and the totality of all these points an *arithmetic manifold* $\mathfrak{A}_n$. The $\overset{\varkappa}{\xi}$ are called the *components* (Bestimmungszahlen) of the arithmetic point. "Point $\overset{\varkappa}{\xi}$" means "point with components $\overset{\varkappa}{\xi}$".

The set of all arithmetic points satisfying the inequalities

$$|\overset{\varkappa}{\xi} - \underset{0}{\overset{\varkappa}{\xi}}| < \overset{\varkappa}{\beta}, \tag{1.1}$$

where $\underset{0}{\overset{\varkappa}{\xi}}$ are n arbitrarily given numbers and $\overset{\varkappa}{\beta}$ are n arbitrarily given positive numbers, is called a *polycylinder* (or *box*) of $\mathfrak{A}_n$.

If a set of arithmetic points is given, satisfying the conditions:

1. every point of the set belongs to at least one polycylinder consisting of points of the set;

[1]) For § 1 and § 2 of this Chapter cf. VEBLEN and WHITEHEAD 1932, 1; BEHNKE and THULLEN 1934, 1; BOCHNER and MARTIN 1948, 1.

2. for two points of the set there always exists a finite chain of polycylinders, all consisting of points of the set, such that the first point lies in the first and the second point in the last polycylinder and that consecutive polycylinders have at least one point in common, this set is called a *region* of $\mathfrak{A}_n$. Every polycylinder is a region and the whole $\mathfrak{A}_n$ is a region. But not every region is a polycylinder. Every region is called a *neighbourhood* of each of its points. For "neighbourhood of $\overset{\varkappa}{\xi}$" we write $\mathfrak{N}(\overset{\varkappa}{\xi})$.

§ 2. The geometric n-dimensional manifold X_n.

Let the elements of a set M be in one to one correspondence with the points of a region $\underset{0}{\mathfrak{R}}$ of $\mathfrak{A}_n$. We make no presuppositions concerning the "nature" of these elements. They may be, for instance, homogeneous linear forms in n variables or polynomials of degree $n-1$ in one variable but they may be also just elements not connected in any way with other notions. We call the one to one correspondence between the elements of M and the points of $\underset{0}{\mathfrak{R}}$ a *coordinate system defined over M*. If the point $\overset{\varkappa}{\xi}$ corresponds to a certain element of M, we call the $\overset{\varkappa}{\xi}$ also *coordinates* of this element *with respect to the coordinate system* $(\varkappa)$. If the $\overset{\varkappa}{\xi}$ are considered in this way we write $\xi^\varkappa$ instead of $\overset{\varkappa}{\xi}$ in order to emphasize this fact.

We consider now a point transformation in $\mathfrak{A}_n$

$$\overset{\varkappa'}{\xi} = \overset{\varkappa'}{f}(\overset{\varkappa}{\xi}) \tag{2.1}$$

satisfying the conditions that the $\overset{\varkappa'}{f}$ are analytic [1)] at $\underset{0}{\overset{\varkappa}{\xi}}$ and that the functional determinant

$$\Delta \overset{\text{def}}{=} \operatorname{Det}(\partial_\lambda \overset{\varkappa'}{\xi}); \quad \partial_\lambda \overset{\text{def}}{=} \frac{\partial}{\partial \overset{\lambda}{\xi}} \tag{2.2}$$

is $\neq 0$ for $\overset{\varkappa}{\xi} = \underset{0}{\overset{\varkappa}{\xi}}$. Then it is a well established fact that there is an inverse transformation

$$\overset{\varkappa}{\xi} = \overset{\varkappa}{f}(\overset{\varkappa'}{\xi}) \tag{2.3}$$

1) A function defined in an $\mathfrak{N}(\underset{0}{\overset{\varkappa}{\xi}})$ is said to be *analytic* at $\underset{0}{\overset{\varkappa}{\xi}}$ if there exists an $\mathfrak{N}(\underset{0}{\overset{\varkappa}{\xi}})$ where it can be expanded into a power series in $\overset{\varkappa}{\xi} - \underset{0}{\overset{\varkappa}{\xi}}$, convergent in this latter $\mathfrak{N}(\underset{0}{\overset{\varkappa}{\xi}})$. Instead we may only require that the functions $\overset{\varkappa'}{f}$ are *of class u* in some $\mathfrak{N}(\underset{0}{\overset{\varkappa}{\xi}})$. A function is said to be of *class u* in some region if it is continuous and has continuous derivatives up to the order u at each point of that region. This case has been dealt with by VEBLEN and WHITEHEAD 1932, 1, p. 36.

with functions analytic at $\underset{0}{\overset{\varkappa'}{\xi}} \overset{\text{def}}{=} \overset{\varkappa'}{f}(\underset{0}{\overset{\varkappa}{\xi}})$ and that there exists a neighbourhood $\mathfrak{N}$ of $\underset{0}{\overset{\varkappa}{\xi}}$ and a neighbourhood $\mathfrak{N}'$ of $\underset{0}{\overset{\varkappa'}{\xi}}$ for the points of which (2.1) and (2.3) establish a one to one correspondence. Suppose now that $\mathfrak{N}$ lies in $\underset{0}{\mathfrak{N}}$. Then there is a subset R of M whose points are in one to one correspondence with the points of $\mathfrak{N}$ and also with the points of $\mathfrak{N}'$. But this latter correspondence is another coordinate system $(\varkappa')$ over R. For the $\overset{\varkappa'}{\xi}$, considered as coordinates of the elements of R, we write $\xi^{\varkappa'}$.

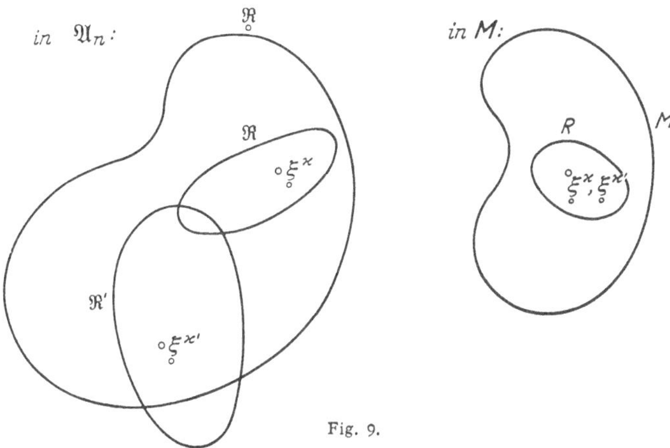

Fig. 9.

Fig. 9[1]) illustrates what is going on in $\mathfrak{A}_n$ and in M. In $\mathfrak{A}_n$ we have only one kind of variables and a transformation means a point transformation $\overset{\varkappa}{\xi} \to \overset{\varkappa'}{\xi}$; $\underset{0}{\overset{\varkappa}{\xi}} \to \underset{0}{\overset{\varkappa'}{\xi}}$; $\mathfrak{N} \to \mathfrak{N}'$. In M we have two kinds of variables, the coordinates $\xi^{\varkappa}$ and $\xi^{\varkappa'}$ and the transformation

$$(2.4) \qquad \xi^{\varkappa'} = \overset{\varkappa'}{f}(\xi^{\varkappa}); \qquad \xi^{\varkappa} = \overset{\varkappa}{f}(\xi^{\varkappa'}); \qquad \operatorname{Det}(A_{\varkappa}^{\varkappa'}) \neq 0; \qquad A_{\varkappa}^{\varkappa'} \overset{\text{def}}{=} \partial \xi^{\varkappa'} / \partial \xi^{\varkappa}$$

corresponding to (2.1, 3) transforms the coordinates only and leaves the elements of M unchanged. As we see, our notation agrees with the principles of the kernel-index method (cf. I § 1). In $\mathfrak{A}_n$ the transformation (2.1, 3) is considered as a point transformation and accordingly the kernel changes. But in M the same transformation is a coordinate transformation and accordingly in (2.4) the index changes and the kernel is invariant.

Collecting results we see that a coordinate system over a set of elements R is a one to one correspondence between the elements of R and the points

[1]) It is not necessary that $\underset{0}{\overset{\varkappa'}{\xi}}$ lies in $\underset{0}{\mathfrak{N}}$.

of a region of $\mathfrak{A}_n$ *and that transformation of coordinates in* R *means passing to another one to one correspondence between these elements and the points of another region of* $\mathfrak{A}_n$.[1])

Until now we had only the set R with a coordinate system over it and the possibility of introducing other coordinate systems by means of point transformations in $\mathfrak{A}_n$. Now everything depends on the choice of these point transformations. If we allow all analytic transformations (or transformations of class u) with $\mathfrak{R} = \mathfrak{R}' = \underset{0}{\mathfrak{R}}$ then these transformations form a group (cf. I § 1). But if we allow the set $\mathfrak{P}_n$ of all transformations of points of $\mathfrak{A}_n$ that are analytic (or of class u) each in a certain region, but without the condition that these regions coincide, the set does not constitute a group. Because, if $\mathfrak{R} \to \mathfrak{R}'$ and $\mathfrak{R}'' \to \mathfrak{R}'''$ are two transformations of $\mathfrak{P}_n$, the second can only be performed after the first if $\mathfrak{R}'$ and $\mathfrak{R}''$ have some region in common. Such a set of transformations, satisfying the first condition for groups (cf. I § 1) only if the second transformation can really be effected after the first is called a *pseudo-group*. [2]), [3])

The set M, provided with the pseudo-group $\mathfrak{P}_n$ and with all *allowable coordinate systems*, viz. all systems that can be derived from one originally specified system by transformations of $\mathfrak{P}_n$ is called a *general geometric n-dimensional manifold* or X_n. The elements of M are called *geometric points* or *points of* X_n. In the following "point" means always "geometric point" and "point $\xi^\varkappa$" means "point with coordinates $\xi^\varkappa$ with respect to $(\varkappa)$".

Instead of $\mathfrak{P}_n$ some subgroup or sub-pseudo-group of $\mathfrak{P}_n$ can be taken. If for instance $\underset{0}{\mathfrak{R}} = \mathfrak{A}_n$ and if $\mathfrak{P}_n$ is replaced by the group G_a or G_{ho} we get instead of X_n, the E_n or the centred E_n (cf. I § 1). If we take $\underset{0}{\mathfrak{R}} = \mathfrak{A}_n$ and the group G_{or} of all orthogonal transformations or the group G_{ro} of all proper rotations we get the R_n or the R_n with a fixed screwsense (cf. I § 9). In all these cases the geometry is fixed wholly by the choice of the group or pseudo-group of allowable coordinate transformations. As was pointed out by F. KLEIN in 1872[4]) the geometry of a figure in some space is the theory of the invariants of this figure

---

[1]) VEBLEN and WHITEHEAD 1932, 1, p. 22. We consider here only ordinary coordinates. For supernumerary coordinates cf. R. S. 1947, 1; P. P. 1949, 1, II § 9.

[2]) VEBLEN and WHITEHEAD 1932, 1, p. 38.

[3]) For a discussion of the foundation of pseudo-groups cf. WAGNER 1950, 1; PENSOV 1950, 1. NIJENHUIS 1952, 1, succeeded in replacing the pseudo-group by a groupoid in the sense of BRANDT, by taking as the elements the transformations of the neighbourhoods of points instead of the transformations of regions.

[4]) KLEIN 1872, 1.

with respect to the underlying set of allowable coordinate transformations.[1])

If we wish to make an E_n from a given X_n, we have to choose one coordinate system and consider all coordinate systems that can be derived from this chosen one by means of transformations of G_a. Then the X_n provided with this preferred set of coordinate systems is an E_n and the geometry of a figure in this E_n is, in X_n, the geometry of this figure together with the set of all preferred coordinate systems. This is an application of the principle of F. KLEIN[2]):

If in any space with a group G_1 the subgroup G_2 is introduced, consisting of all transformations of G_1 which leave a figure F_1 invariant, then the geometry of a figure F_2 with respect to G_2 is identical with the geometry of the set of figures F_1, F_2 with respect to G_1.

Starting from X_n we took for F_1 the set of all preferred coordinate systems in order to get E_n. But this is not always necessary, we could also have chosen some figure from which these preferred coordinate systems might be derived and that is uniquely determined by these systems. For instance if we start from E_n in order to get R_n, we can use for F_1 the set of all orthogonal coordinate systems, but we can as well use (cf. I § 9) the quadratic hypersurface representing the fundamental tensor.

The allowable coordinate systems in E_n are the rectilinear systems used in Ch. I. This does not mean in any way that we are not allowed to introduce more general coordinate systems in E_n. But in doing this we consider the E_n as an X_n in which the allowable coordinate systems of E_n play the part of preferred coordinate systems. In the same way in an R_n besides the allowable orthogonal coordinate systems, general rectilinear or curvilinear systems can be used, but then we look upon the R_n as an E_n or X_n with preferred coordinate systems.

In an X_n we can not use the notion of polycylinder because contrary to $\mathfrak{A}_n$ there is no preferred coordinate system in X_n. Instead we use cells, a *cell* being defined as a set of points given by the inequalities

$$|\xi^\varkappa| < 1 \tag{2.5}$$

with respect to some allowable coordinate system. A pointset R in X_n is called a *region*, if there exists an allowable coordinate system, determining a one to one correspondence between the points of R and the arithmetic points of a region $\mathfrak{R}$ in $\mathfrak{A}_n$.[3]) Every cell is a region, but not

[1]) Cf. for more general views on geometry SCHOUTEN and v. DANTZIG 1934, 1 and NIJENHUIS 1952, 1.

[2]) See footnote 4, p. 64.

[3]) A region in X_n can also be defined in the same way as a region in $\mathfrak{A}_n$ by using cells instead of polycylinders.

every region need be a cell. Also the X_n is itself a region and every region in X_n is itself an X_n. Every X_n which is a region of another X_n is said to be *imbedded* in the latter. Every region in X_n is called a *neighbourhood* of every one of its points. For "neighbourhood of $\underset{0}{\xi^{\varkappa}}$" we write $\mathfrak{N}(\underset{0}{\xi^{\varkappa}})$.

If $\mathfrak{P}_n$ consists of analytical transformations only, a field p is said to be *analytic in the region* R if for some choice of $(\varkappa)$ we have $p = f(\xi^{\varkappa})$ and if $f(\overset{\varkappa}{\xi})$ is analytic at all arithmetic points of the region $\mathfrak{R}$ of $\mathfrak{A}_n$ corresponding to R by means of $(\varkappa)$. Because in this case $\overset{\varkappa}{f}(\overset{\varkappa'}{f}(\xi))$ is analytic at all points of the region $\mathfrak{R}'$ of $\mathfrak{A}_n$ corresponding to R by means of $(\varkappa')$, the analytic nature of p is an invariant notion.[1])

If $\mathfrak{P}_n$ consists of transformations of class u an analytic function in a region of X_n can not be defined. But with the same reasoning as above we may call p a function of *class* u in some region R of X_n, if with respect to some allowable coordinate system $(\varkappa)$ we have $p = f(\xi^{\varkappa})$ and if $f(\overset{\varkappa}{\xi})$ is of class u in the region $\mathfrak{R}$ corresponding to R by means of $(\varkappa)$. In both cases the choice of the allowable coordinate system is not important.

Exercises.

II 2,1. Prove that (cf. Exerc. I 1,1)

II 2,1 α) $$A^{\lambda}_{\varkappa'}\, \partial_\mu A^{\varkappa'}_{\lambda} = \partial_\mu \log \mathrm{Det}\,(A^{\varkappa'}_{\lambda}).$$

II 2,2 (cf. Exerc. I 8,1). If $P_{\lambda\varkappa}$ is a tensor of rank n, $\overset{-1}{P}{}^{\varkappa\lambda}$ its inverse and $\mathfrak{P}$ its determinant (scalar density of weight $+2$), prove that

II 2,2 α) $$\overset{-1}{P}{}^{\varkappa\lambda}\, \partial_\mu P_{\lambda\varkappa} = \partial_\mu \log \mathfrak{P}.$$

II 2,3 (cf. V § 5). If $h_{\lambda}^{\cdot\varkappa}$ is a tensor, prove that

II 2,3 α) $$H_{\mu\lambda}^{\cdot\cdot\varkappa} = 2\, h_{[\mu}^{\cdot\varrho}\, \partial_{|\varrho|}\, h_{\lambda]}^{\cdot\varkappa} - 2\, h_{\varrho}^{\cdot\varkappa}\, \partial_{[\mu}\, h_{\lambda]}^{\cdot\varrho}$$

is also a tensor.[2])

II 2,4. If the eigenvalues of $h_{\lambda}^{\cdot\varkappa}$ of Exerc. II 2,3 are all different and if (h) is chosen in such a way that $\underset{1}{e^{\varkappa}}, \ldots, \underset{n}{e^{\varkappa}}$ are eigenvectors of $h_{\lambda}^{\cdot\varkappa}$, prove that any p of them are X_p-forming (cf. II § 5) if and only if $H_{ji}^{\cdot\cdot h} = 0;\ h, i, j \neq$.[2])

[1]) In general in this book we only consider fields in a neighbourhood of a point and not "in the large" though these fields play a very important role in "global" differential geometry, cf. e.g. Bochner 1950, 1; 1951, 1.

[2]) Nijenhuis 1951, 1; 2.

II 2,5. If $P^{\varkappa\lambda}$ and $Q^{\varkappa\lambda}$ are tensors, prove that

$$\text{II 2,5 } \alpha) \quad P^{\mu[\varkappa_1}\partial_\mu Q^{\varkappa_2\varkappa_3]} + P^{[\varkappa_1|\mu|}\partial_\mu Q^{\varkappa_3\varkappa_2]} - Q^{\mu[\varkappa_1}\partial_\mu P^{\varkappa_3\varkappa_2]} - Q^{[\varkappa_1|\mu|}\partial_\mu P^{\varkappa_2\varkappa_3]}$$

is a tensor and that the same holds for ( ) instead of [ ].[1])

§ 3. Geometric objects and quantities in X_n.

A *geometric object*[2]), [3]) *at a point* $\xi^\varkappa$ *of* X_n is a correspondence between the allowable coordinate systems in regions containing $\xi^\varkappa$ and the ordered sets of N numbers, satisfying the conditions:

1. To every coordinate system $(\varkappa)$ there belongs one and only one ordered set of N numbers.

2. If Φ_Λ; $\Lambda = 1, \ldots, N$ corresponds to $(\varkappa)$ and $\Phi_{\Lambda'}$; $\Lambda' = 1', \ldots, N'$ to $(\varkappa')$, the $\Phi_{\Lambda'}$ are functions of the Φ_Λ and functionals of the functions $f^{\varkappa'}$ in a neighbourhood of the point $\xi^\varkappa$. Instead we may also require that the $\Phi_{\Lambda'}$ are functions of the Φ_Λ and of the values of the $\xi^{\varkappa'}$ and of the derivatives $A^{\varkappa'}_\varkappa = \partial_\varkappa \xi^{\varkappa'}$, $\partial_\mu A^{\varkappa'}_\varkappa$, $\partial_\nu \partial_\mu A^{\varkappa'}_\varkappa, \ldots$ up to a certain order u at the point $\xi^\varkappa$. If the $f^{\varkappa'}$ are analytic and $u = \infty$ both definitions are equivalent, but in all other cases the first is more general. But we prefer here the second since it is sufficiently general for our purposes and avoids the use of functionals.[4]) Then the transformation formula of a general geometric object takes the form

$$(3.1\,\mathrm{a}) \qquad \Phi_{\Lambda'}\{\xi\} = \delta^\Lambda_{\Lambda'} F_\Lambda(\Phi_\Lambda\{\xi\}, \xi^\varkappa, \xi^{\varkappa'}, A^{\varkappa'}_\varkappa\{\xi\}, \partial_\mu A^{\varkappa'}_\varkappa\{\xi\}, \ldots)$$

where $\{\xi\}$ means that the values of the whole foregoing expression must be taken at the point $\xi^\varkappa$. Note that in expressions as $\Phi_\Lambda\{\xi\}$ it would not be better to write $\xi^\varkappa$ instead of ξ, because $\Phi_\Lambda\{\xi^\varkappa\}$ would have the same meaning as $\Phi_\Lambda\{\xi^{\varkappa'}\}$. The form of the functions F_Λ does not depend on the values at $\xi^\varkappa$ mentioned between the parentheses. Hence the inversion of (3.1 a)

$$(3.1\,\mathrm{b}) \qquad \Phi_\Lambda\{\xi\} = F_\Lambda(\Phi_{\Lambda'}\{\xi\}, \xi^{\varkappa'}, \xi^\varkappa, A^\varkappa_{\varkappa'}\{\xi\}, \partial_{\mu'} A^\varkappa_{\lambda'}\{\xi\}, \ldots)$$

contains the *same* functions F_Λ.

1) Schouten 1940, 2 also for general valences of P and Q.

2) Cf. for more general definitions Schouten and Haantjes 1937, 1, and P. P. 1949, 1.

3) Further literature: Wundheiler 1937, 1; Golab 1938, 1; 1946, 1; 1947, 1; 1948, 1; 1950, 1; 2; Nakae 1944, 1; Wagner 1945, 1; 1949, 1; 1950, 1; Pensov 1946, 1; 1948, 1; Georghiu 1949, 1; Liber 1951, 1; Vasiliev 1951, 1; Ehresmann 1952, 1; Haantjes and Laman 1953, 1; Nijenhuis 1952, 1, gave a development from the point of view of the representation theory of a groupoid.

4) Of course the pseudo-group $\mathfrak{P}_n$ used in the definition of the X_n must only contain transformations of class $\geq u$.

The Φ_Λ are called the *components of the geometric object* with respect to $(\varkappa)$.

If we wish to express that a scalar p is a function of the $\xi^\varkappa$ we must write $p = f(\xi^\varkappa)$ and not $p = p(\xi^\varkappa)$. In fact, if a new coordinate system is introduced we can write $p = f'(\xi^{\varkappa'})$ but we could not write $p = p(\xi^{\varkappa'})$ because the same letter p can not be used as a function symbol for two different functions. The same holds for more general geometric objects as Φ_Λ and this is the reason why we introduced the notation $\Phi_\Lambda\{\xi\}$ for the values of Φ_Λ at the point $\xi^\varkappa$.

For all geometric objects we may make use of a notation first introduced by NIJENHUIS.[1]) If the components Φ_Λ of a geometric object with respect to the coordinate system $(\varkappa)$ are expressed as functions of the coordinates $\xi^{\varkappa'}$ he uses ${}_{(\varkappa')}\Phi_\Lambda$ as function symbol. Hence

$$(3.2)\quad \Phi_\Lambda = {}_{(\varkappa')}\Phi_\Lambda(\xi^{\varkappa'}) = {}_{(\varkappa)}\Phi_\Lambda(\xi^\varkappa); \qquad \Phi_{\Lambda'} = {}_{(\varkappa')}\Phi_{\Lambda'}(\xi^{\varkappa'}) = {}_{(\varkappa)}\Phi_{\Lambda'}(\xi^\varkappa).$$

It should be remarked that ${}_{(\varkappa')}\Phi_\Lambda$ is not a kernel in our sense because the change of a kernel always means a change of the object. If the object is not a scalar it is convenient to write for shortage Φ_Λ instead of ${}_{(\varkappa)}\Phi_\Lambda$ and $\Phi_{\Lambda'}$ instead of ${}_{(\varkappa')}\Phi_{\Lambda'}$, though by doing this we use a symbol for components as a function symbol.

It may happen that two or more geometric objects have the same kernel letter. If, in some investigation, this is *not* the case it is often convenient to denote a geometric object in the text by its kernel letter only. Objects with the same number of components and the same manner of transformation are said to be *of the same kind*.

If, for every choice of $(\varkappa)$ and $(\varkappa')$, the $\Phi_{\Lambda'}$ are functions of the Φ_Λ and the $A_\varkappa^{\varkappa'}$ only, linear homogeneous in the Φ_Λ and homogeneous algebraic in the $A_\varkappa^{\varkappa'}$ the geometric object is called a *geometric quantity* or briefly a *quantity*.

Geometric objects may be defined with respect to two or more spaces[2]), for instance an X_n with the coordinates $\xi^\varkappa$ and an X_m with the coordinates η^a; $a = 1, \ldots, m$. The components may be functions of $\xi^\varkappa$ and η^a. Objects of this kind are often called *connecting objects*. There may be relations between the X_m and the X_n, for instance the $\xi^\varkappa$ may be functions of the η^a. If a connecting object is a quantity it often happens that its components have two different kinds of indices each belonging to one of the underlying spaces. For instance $B_b^\varkappa = \partial\xi^\varkappa/\partial\eta^b$.

In dealing with objects with components with respect to more than one coordinate system, in most cases it is more convenient to introduce

---

[1]) NIJENHUIS 1952, 1, p. 44ff.

[2]) Such quantities were introduced incidentally by BOMPIANI 1921, 1 and more systematically by VAN DER WAERDEN 1927, 1. Cf. SCHOUTEN 1927, 2; DUSCHEK and MAYER 1930, 1, Ch. VII; TUCKER 1931, 1; RUSE 1931, 1; E I 1935, 1, p. 9.

special function symbols, for instance $A_\lambda^{\varkappa'} = F_\lambda^{\varkappa'}(\xi^\nu) = G_\lambda^{\varkappa'}(\xi^{\nu'})$. But if symbols for components are used as function symbols in this more general case, it must always be made very clear to which coordinate system or systems the function symbols introduced in this way belong.

Pseudo-quantities are a special kind of connecting quantities. They are defined with respect to the X_n of the $\xi^\varkappa$ and the X_1 of a variable ξ^0 with the transformation group (cf. I § 2 under 5):

$$\left\{\begin{array}{l} \xi^{\varkappa'} = f^{\varkappa'}(\xi^\varkappa) \\ \xi^{0'} = \sigma(\xi^\varkappa)\,\xi^0. \end{array}\right. \tag{3.3}$$

Their components may be functions of the $\xi^\varkappa$ and of ξ^0. A *pseudo-scalar of class c* has one component, depending only on the $\xi^\varkappa$ and has the transformation

$$\overset{(0')}{\mathfrak{p}} = \sigma^c \overset{(0)}{\mathfrak{p}}. \tag{3.4}$$

Hence ξ^0 itself is a pseudo-scalar of class 1. Other pseudo-quantities arise if a pseudo-scalar is multiplied by an ordinary quantity. All these pseudo-quantities can be multiplied and transvected. Addition is only possible if the pseudo-quantities are defined with respect to the same $(n+1)$st variable and if the valences, the weights and the class are the same.

If we have some quantity or some set of quantities each defined to within an entirely free scalar factor, we might consider them as pseudo-quantities, each defined with respect to its own $(n+1)$st auxiliary variable. But as there are no connections between these auxiliary variables there is in general no use in introducing them. In this case we use the sign $\lfloor\ \rfloor$ (cf. I § 2). For instance $\lfloor v^\varkappa \rfloor$ is a vector defined to within a scalar factor. Note that it is no use asking whether $\lfloor v^\varkappa \rfloor$ is a vector or a vector density because the influence of a factor Δ is nullified by the entirely free scalar factor. $\lfloor v^\varkappa \rfloor$ is neither a vector nor a vector density but a pseudo-vector though an auxiliary variable ξ^0 is not mentioned.

The $\xi^\varkappa$ represent a geometric object which is no quantity. But according to (2.4) the $d\xi^\varkappa$ undergo a homogeneous linear transformation $d\xi^{\varkappa'} = A_\varkappa^{\varkappa'}\,d\xi^\varkappa$. Hence the $d\xi^\varkappa$ at the point $\xi^\varkappa$ represent a quantity at that point. But from the same equation we see that every allowable coordinate transformation in an $\mathfrak{N}(\xi^\varkappa)$ induces one and only one transformation of the group G_{ho} at $\xi^\varkappa$ and that to every transformation of this group at least one allowable coordinate transformation in X_n can be found inducing this very transformation of G_{ho} at $\xi^\varkappa$. This means that to every point $\xi^\varkappa$ of X_n there belongs a centred E_n, the *tangent space* or *tangent* E_n of the point $\xi^\varkappa$.[1])

[1]) In former publications we used the term *local* E_n.

From the definition it follows that a quantity of X_n at $\xi^\varkappa$ is the same as a quantity as defined in I § 2 in the tangent E_n of $\xi^\varkappa$. Hence all quantities of E_n considered in Ch. I may occur as quantities of X_n at every point of X_n. All equations of transformation like (I 2.1), (I 2.4), (I 2.13), (I 2.19) and (I 2.22, 23) are valid in X_n, but the $A_\lambda^{\varkappa'}$, $A_{\lambda'}^{\varkappa}$ and Δ are no longer constants.

If the components of a geometric object are defined in some region of X_n as functions of the $\xi^\varkappa$, the object is said to form an *object field*. A field is said to be *analytic* (*of class u*) in a region if all its components are analytic (of class u) in that region.

As a first example we take a scalar field $p = f(\xi^\varkappa)$. If p is analytic or at least of class 1 in the region considered we may form the derivatives $\partial_\lambda p$. Because of

$$\partial_{\lambda'} p = A_{\lambda'}^{\lambda} \partial_\lambda p \tag{3.5}$$

these derivatives are the components of a covariant vector, the ***natural derivative*** or ***gradient*** of p. If we use the abridged notation of I § 7, we write Grad p or Dp for the gradient of p and also $\overline{p}$ if the vector occurs in the abridged form (I 7.1) of an alternating product.

Now we take a vector field $v^\varkappa = f^\varkappa(\xi^\lambda)$ and form its derivatives $\partial_\lambda v^\varkappa$. Because of [1])

$$\partial_{\lambda'} v^{\varkappa'} = A_{\lambda'}^{\lambda} \partial_\lambda A_\varkappa^{\varkappa'} v^\varkappa = A_{\lambda'\varkappa}^{\lambda\varkappa'} \partial_\lambda v^\varkappa + A_{\lambda'}^{\lambda} v^\varkappa \partial_\lambda A_\varkappa^{\varkappa'} \tag{3.6}$$

these derivatives are neither the components of a tensor nor of any geometric object, since the $\partial_{\lambda'} v^{\varkappa'}$ depend not only on $\partial_\lambda v^\varkappa$, $A_\varkappa^{\varkappa'}$ and $\partial_\lambda A_\varkappa^{\varkappa'}$ but also on $v^\varkappa$. But the $n+n^2$ expressions $v^\varkappa$, $\partial_\lambda v^\varkappa$ are the components of a geometric object whose transformation involves the $\partial_\lambda A_\varkappa^{\varkappa'}$. The same is true for the derivatives of all geometric quantities except scalars and W-scalars.

To every coordinate system ($\varkappa$) in X_n there belong n scalar fields $\overset{\varkappa}{\xi}$. The transformation $\xi^\varkappa \to \xi^{\varkappa'}$ is a coordinate transformation in X_n, accompanied by the point transformations $\overset{\varkappa}{\xi} \to \overset{\varkappa'}{\xi}$ in $\mathfrak{A}_n$. But this latter transformation can be considered also as an object transformation in X_n, viz. the transformation of the scalar fields $\overset{\varkappa}{\xi}$ into the scalar fields $\overset{\varkappa'}{\xi}$. The gradients of the scalar fields $\overset{\varkappa}{\xi}$ are the vectors of the *covariant basis*

[1]) In dealing with operators like ∂_μ (and also ∇_μ in the next chapter, cf. footnote 3 on p. 132) we agree that the operator works by differentiation on all quantities to the right of it in the same term till the first closing bracket, whose corresponding opening bracket stands at the left of the operator. If these operators are applied several times in succession we usually prefer to write the kernel letter ∂ (or ∇) only once, for instance $\partial_{\mu_p \ldots \mu_1}$ instead of $\partial_{\mu_p} \ldots \partial_{\mu_1}$.

belonging to $(\varkappa)$.[1]) Their components with respect to $(\varkappa)$ have the values 1 and 0 only

(3.7) $$\overset{\varkappa}{e}_\lambda \overset{\text{def}}{=} \partial_\lambda \overset{\varkappa}{\xi} \overset{*}{=} \delta_\lambda^\varkappa$$

but the components with respect to $(\varkappa')$ may have quite other values

(3.8) $$\overset{\varkappa}{e}_{\lambda'} = \partial_{\lambda'} \overset{\varkappa}{\xi} \overset{*}{=} A_{\lambda'}^{\varkappa}.$$

The vectors of the *contravariant basis* belonging to $(\varkappa)$ arise from differentiation of the coordinates $\xi^\varkappa$ in X_n with respect to the $\overset{\varkappa}{\xi}$

(3.9) $$\underset{\lambda}{e}^\varkappa \overset{\text{def}}{=} \frac{\partial \xi^\varkappa}{\partial \underset{\lambda}{\xi}} \overset{*}{=} \delta_\lambda^\varkappa; \qquad \underset{\lambda}{e}^{\varkappa'} = \frac{\partial \xi^{\varkappa'}}{\partial \underset{\lambda}{\xi}} \overset{*}{=} A_\lambda^{\varkappa'}.$$

The equations (3.7–9) and

(3.10) $$A_\lambda^\varkappa = \partial_\lambda \xi^\varkappa; \qquad A_\lambda^{\varkappa'} = \partial_\lambda \xi^{\varkappa'}$$

(3.11) $$\delta_\lambda^\varkappa = \frac{\partial \overset{\varkappa}{\xi}}{\partial \underset{\lambda}{\xi}}$$

show clearly the difference between the tensor $A_\lambda^\varkappa$, the two bases each consisting of n vectors $\underset{\lambda}{e}^\varkappa$ and $\overset{\varkappa}{e}_\lambda$ and the set of n^2 scalars $\delta_\lambda^\varkappa$. We also see the difference between the living indices, transforming only with coordinate transformations in X_n, and the dead indices, transforming only with the object transformations in X_n corresponding to point transformations in $\mathfrak{A}_n$.[2])

At every point every quantity field has a rank with respect to each set of indices, as defined in I § 5. If such a rank is r at $\underset{0}{\xi}^\varkappa$, this means that a certain matrix formed from the components of the quantity has rank r at that point. Hence all subdeterminants with more than r rows vanish at $\underset{0}{\xi}^\varkappa$ and there is at least one non vanishing subdeterminant with r rows. Now let the quantity be analytic in a region R containing $\underset{0}{\xi}^\varkappa$ and let $\underset{0}{\xi}^\varkappa$ be a point where r has a maximum value. Because all subdeterminants are analytic, there exists an $\mathfrak{N}(\underset{0}{\xi}^\varkappa)$ such that the subdeterminants not vanishing at $\underset{0}{\xi}^\varkappa$ can not vanish at any point of $\mathfrak{N}(\underset{0}{\xi}^\varkappa)$. Hence the rank has to be r at all points of $\mathfrak{N}(\underset{0}{\xi}^\varkappa)$. We call $\mathfrak{N}(\underset{0}{\xi}^\varkappa)$ a region of *constant rank* with respect to the index or the indices concerned.

[1]) The $\underset{\lambda}{e}^\varkappa$ and $\overset{\varkappa}{e}_\lambda$ were formerly called the contra- and covariant *measuring vectors* belonging to $(\varkappa)$.

[2]) All these differences should not be neglected. Neglecting them leads to many difficulties especially after the introduction of special kinds of processes of differentiation.

Now let there be two subregions of constant rank in R in which the rank, always taken with respect to the same indices, takes the values r_1 and $r_2 \geq r_1$. Then in the first region all subdeterminants with r_1+1 rows vanish. But this implies that they vanish at all points of R. Hence $r_1 = r_2$. This proves that a quantity analytic in R has the same rank with respect to some given indices in all subregions of constant rank in R and that there is no point where it can have a higher rank. This does however not imply that there are no points in R where the rank is lower than in the subregions of constant rank. But these points can not form a region. For instance a vector field $v^\varkappa$ analytic in R and not vanishing at all points of R has rank zero at all points where $v^\varkappa = 0$ and rank 1 at all other points. Points where $v^\varkappa = 0$ may occur but they can not form a region because in that case $v^\varkappa$ would vanish all over R. If $v^\varkappa = 0$ at $\underset{0}{\xi^\varkappa}$, in every $\mathfrak{N}(\underset{0}{\xi^\varkappa})$ there is at least one point where $v^\varkappa \neq 0$.

Exercise.

II 3,1[1]). Let Φ_Λ be a tensor field with the collecting index Λ. If Ψ_Ω is a tensor field with the collecting index Ω whose components are functions of the $\xi^\varkappa$ and the Φ_Λ, prove that the $\partial\Psi_\Omega/\partial\Phi_\Lambda$, where the $\xi^\varkappa$ are supposed to be left constant, are components of a tensor field. The same holds mutatis mutandis for tensor densities.

§ 4. The X_m in X_n.[2])

Let N functions $F^\mathfrak{a}(\xi^\varkappa)$; $\mathfrak{a} = 1, \ldots, \mathrm{N}$ be analytic in $\mathfrak{N}(\underset{0}{\xi^\varkappa})$. The matrix

$$(4.1)\qquad \| \partial_\lambda F^\mathfrak{a} \|$$

is the *functional matrix* of the set of functions and its rank r at $\underset{0}{\xi^\varkappa}$ is the *rank* of the set at that point. r is also the maximum number of linearly independent differentials among the $dF^\mathfrak{a}$, hence $r \leq n$; $r \leq N$. It is well known that the functions $F^\mathfrak{a}$ are independent in $\mathfrak{N}(\underset{0}{\xi^\varkappa})$ if and only if $r = N$ at at least one point of $\mathfrak{N}(\underset{0}{\xi^\varkappa})$.

Let $r = N \leq n$. Then we can take $m = n - N$ auxiliary functions $F^{\mathrm{N}+1}(\xi^\varkappa), \ldots, F^\mathrm{n}(\xi^\varkappa)$ analytic in $\mathfrak{N}(\underset{0}{\xi^\varkappa})$, such that the rank of the set F^h; $h = 1, \ldots, \mathrm{n}$ equals n at all points of $\mathfrak{N}(\underset{0}{\xi^\varkappa})$. Then the transformation

$$(4.2)\qquad \xi^h = F^h(\xi^\varkappa); \quad h = 1, \ldots, \mathrm{n}$$

is an allowable coordinate transformation in $\mathfrak{N}(\underset{0}{\xi^\varkappa})$. That proves the

1) Cf. Horak 1929, 1.

2) Cf. for a more elaborate treatment P. P. 1949, 1, Ch. II; R. S. 1947, 1, § 2.

Theorem of adaptation.

If the functions $F^{\mathfrak{a}}(\xi^{\varkappa})$; $\mathfrak{a} = 1, \ldots, \mathrm{N}$, *are analytic in* $\mathfrak{N}(\underset{0}{\xi^{\varkappa}})$ *and if* $r = N \leq n$, *there exists in* $\mathfrak{N}(\underset{0}{\xi^{\varkappa}})$ *an allowable coordinate system* ξ^{h}; $h = 1, \ldots, \mathrm{n}$, *such that* $\xi^{\mathfrak{a}} = F^{\mathfrak{a}}(\xi^{\varkappa})$; $\mathfrak{a} = 1, \ldots, \mathrm{N}$.

Let the system of equations

$$F^{\mathfrak{a}}(\xi^{\varkappa}) = 0 \tag{4.3}$$

be consistent. We call r the *rank of the system*. Every point $\xi^{\varkappa}$ of $\mathfrak{N}(\underset{0}{\xi^{\varkappa}})$ satisfying (4.3) is called a *null point* of (4.3) and the set of all null points is called the *null manifold* of (4.3). (4.3) is said to be a *null form* of this manifold. Two systems of equations having the same null points in $\mathfrak{N}(\underset{0}{\xi^{\varkappa}})$ are said to be *equivalent in that region*. Two equivalent systems need not have the same rank at all null points. E.g. $\xi^1 = 0$; $\xi^2 = 0$ and $\xi^1 \xi^1 = 0$; $\xi^2 = 0$ at the point $\xi^{\varkappa} = 0$.

Now we consider a system of this kind

$$C^{x}(\xi^{\varkappa}) = 0; \quad x = \mathrm{m}+1, \ldots, \mathrm{n}; \quad (C^{x} \text{ analytic}) \tag{4.4}$$

with rank $n-m$ at the null point $\underset{0}{\xi^{\varkappa}}$. Then there exists an $\mathfrak{N}(\underset{0}{\xi^{\varkappa}})$ such that the rank is $n-m$ at all its points. In this case (4.4) is called *minimal regular of dimension m at* $\underset{0}{\xi^{\varkappa}}$. The number m is called the *dimension* of the null manifold and we have $0 \leq m \leq n$. From the definition we see that every subsystem of a system which is minimal regular at $\underset{0}{\xi^{\varkappa}}$ is itself minimal regular at $\underset{0}{\xi^{\varkappa}}$ and we see also that there exists an $\mathfrak{N}(\underset{0}{\xi^{\varkappa}})$ such that the system is minimal regular and the null manifold m-dimensional at all null points in $\mathfrak{N}(\underset{0}{\xi^{\varkappa}})$. The notions "minimal regular" and "dimension" are invariant for all allowable coordinate transformations. Of the following four examples of systems of equations only the last one is minimal regular at $\xi^{\varkappa} = 0$.

$$\begin{cases} \text{a)} & \xi^1 \xi^2 = 0; & \xi^1 \xi^3 = 0 \\ \text{b)} & \xi^1 \xi^1 = 0; & \xi^2 = 0 \\ \text{c)} & \xi^1 \xi^1 = 0; & \xi^1 = 0; & \xi^2 = 0 \\ \text{d)} & \xi^1 \quad = 0; & \xi^2 = 0 \end{cases} \tag{4.5}$$

and only (4.5b) and (4.5c) are equivalent to a minimal regular system, viz. (4.5d). The null manifold of (4.5a) has no dimension at all at $\xi^{\varkappa} = 0$. If the $\xi^{\varkappa}$ are rectilinear coordinates in E_n, (4.5a) represents a figure consisting of an E_{n-1} and an E_{n-2} through the origin.

If (4.3) is minimal regular of dimension m at $\mathfrak{N}(\overset{0}{\xi^\varkappa})$, according to the theorem of adaptation there exists a coordinate system (h); $h = 1, \ldots, n$ such that (4.3) is equivalent to the system

$$\xi^x = 0; \quad x = m+1, \ldots, n. \tag{4.6}$$

Accordingly the ξ^a; $a = 1, \ldots, m$ can be used as coordinates in the null manifold. Now the pseudo-group $\mathfrak{P}_n$ contains as a sub-pseudo-group the pseudo-group $\mathfrak{P}_m$ of all analytic and invertible transformations of the ξ^a only, the ξ^x being left invariant. Hence the null manifold is an X_m. Such an X_m is said to be *imbedded in* an X_n or an X_m *in* X_n. An X_m in X_n is often called an *m-dimensional surface* and for $m = 1, 2$ and $n-1$ it is also called *curve, surface* and *hypersurface* respectively.[1])

The quantity

$$C_\lambda^x \overset{\text{def}}{=} \partial_\lambda C^x; \quad x = m+1, \ldots, n \tag{4.7}$$

with rank $n-m$ is the *covariant connecting quantity* of the X_m in X_n (cf. I § 2 point 7 and II § 3).

If a function $s(\xi^\varkappa)$ is analytic in $\mathfrak{N}(\overset{0}{\xi^\varkappa})$ and zero at all null points of (4.4) in $\mathfrak{N}(\overset{0}{\xi^\varkappa})$, there always exists an $\mathfrak{N}(\overset{0}{\xi^\varkappa})$ such that s satisfies in this $\mathfrak{N}(\overset{0}{\xi^\varkappa})$ an equation of the form

$$s = \Phi_x(\xi^\varkappa)\, C^x; \quad x = m+1, \ldots, n \tag{4.8}$$

with analytic functions Φ_x. This is the so-called *first base theorem.*[2]) A set of $n-m$ functions $C^{x'}$; $x' = (m+1)', \ldots, n'$ which take the value zero at every null point of (4.4) and whose differentials $dC^{x'}$ are linearly independent at $\overset{0}{\xi^\varkappa}$, constitutes a *base* of the X_m at $\overset{0}{\xi^\varkappa}$ and consequently at all null points in some $\mathfrak{N}(\overset{0}{\xi^\varkappa})$.[3]) Hence the C^x in (4.4) constitute a base. As a consequence of the first base theorem we have equations of the form

$$C^{x'} = C_x^{x'} C^x; \quad x = m+1, \ldots, n; \quad x' = (m+1)', \ldots, n' \tag{4.9}$$

valid in an $\mathfrak{N}(\overset{0}{\xi^\varkappa})$ with coefficients $C_x^{x'}$ analytic in this region. Hence

$$dC^{x'} = C_x^{x'}\, dC^x; \quad x = m+1, \ldots, n; \quad x' = (m+1)', \ldots, n' \tag{4.10}$$

---

[1]) Note that in consequence of the definition an X_m in X_n never has singularities. An X_2 or surface in our sense in ordinary space is never a surface in the ordinary sense with singular points and curves, but it can be a part of such a surface, free from singularities.

[2]) Cf. for a proof of this well known theorem for instance P. P. 1949, 1, p. 37; R. S. 1947, 1, § 3.

[3]) We define a base in a different way as KÄHLER, who also allows sets of more than $n-m$ functions. Cf. 1934, 1, p. 13.

at all null points in this region. We see from this that $C^{\varkappa'} = 0$ is a null form of the X_m in some $\mathfrak{N}(\underset{0}{\xi}^{\varkappa})$ which is minimal regular in this region. This is the *second base theorem.*[1]) It is easy to see that $\operatorname{Det}(C^{\varkappa'}_{\varkappa}) \neq 0$ in $\mathfrak{N}(\underset{0}{\xi}^{\varkappa})$.

Every transformation of the form (4.9) with $\operatorname{Det}(C^{\varkappa'}_{\varkappa}) \neq 0$ is called a *base transformation.* If a coordinate transformation and a base transformation are effected simultaneously we have

$$C^{\varkappa'}_{\lambda'} = A^{\lambda}_{\lambda'} C^{\varkappa'}_{\varkappa} C^{\varkappa}_{\lambda} \tag{4.11}$$

valid *at the null points* of the region considered.

There is another way of defining an X_m in X_n. Let the η^a; $a = 1, \ldots, m$ be allowable coordinates of some X_m which has no point in common with X_n and let there be given a system of equations

$$\xi^{\varkappa} = B^{\varkappa}(\eta^a); \quad a = 1, \ldots, m \tag{4.12}$$

where the $B^{\varkappa}$ are analytic in an $\mathfrak{N}(\underset{0}{\eta}^a)$ and the matrix of

$$B^{\varkappa}_b \stackrel{\text{def}}{=} \partial_b \xi^{\varkappa}; \quad \partial_b \stackrel{\text{def}}{=} \frac{\partial}{\partial \eta^b}; \quad b = 1, \ldots, m \tag{4.13}$$

has rank m in this region. Then (4.12) fixes a one to one correspondence between the points of X_m in an $\mathfrak{N}(\underset{0}{\eta}^a)$ and certain points of X_n in an $\mathfrak{N}(\underset{0}{\xi}^{\varkappa})$; $\underset{0}{\xi}^{\varkappa} \stackrel{\text{def}}{=} B^{\varkappa}(\underset{0}{\eta}^a)$. Hence every point of this $\mathfrak{N}(\underset{0}{\eta}^a)$ can be identified with its corresponding point in $\mathfrak{N}(\underset{0}{\xi}^{\varkappa})$. This process is called the *imbedding of the X_m in X_n.* (4.12) is called the *parametric form* (or *parametric representation*) of the X_m in X_n, *minimal regular of dimension m at $\underset{0}{\eta}^a$*, and $B^{\varkappa}_b$ is the *contravariant connecting quantity of the X_m in X_n.*

In order to prove that this second definition of an X_m in X_n is in agreement with the first one, we start from the null form (4.4) and use the theorem of adaptation. According to this theorem the coordinate system $(\varkappa)$ can always be chosen in such a way that the null form takes the form $\xi^{\zeta} = 0$; $\zeta = m+1, \ldots, n$. Now, writing these equations in the form

$$\xi^{\alpha} = \xi^{\alpha}; \quad \xi^{\zeta} = 0; \quad \alpha = 1, \ldots, m; \quad \zeta = m+1, \ldots, n \tag{4.14}$$

and considering the ξ^{α} as parameters we see that we have got a minimal regular parametric form. Now let us start from the parametric form (4.12). Since $B^{\varkappa}_b$ has rank m, the indices $1, \ldots, n$ can always be rearranged in such a way that B^{α}_b; $\alpha = 1, \ldots, m$; $b = 1, \ldots, m$ has rank m.

---

[1]) Cf. for a proof of this well known theorem for instance P. P. 1949, 1, p. 37; R. S. 1947, 1, § 3.

Then the transformation $\xi^{\varkappa} \to \xi^{h}$; $h = 1, \ldots, n$, defined by

$$(4.15) \quad \xi^{\varkappa} = B^{\varkappa}(\xi^{a}) + \delta^{\varkappa}_{x}\,\xi^{x}; \qquad a = 1, \ldots, m; \qquad x = m+1, \ldots, n$$

is an allowable coordinate transformation in the region concerned. If this transformation is applied to (4.12) we get the system

$$(4.16) \quad \begin{cases} \text{a)} \quad B^{\alpha}(\xi^{a}) = B^{\alpha}(\eta^{a}) \qquad \text{b)} \quad B^{\zeta}(\xi^{a}) + \delta^{\zeta}_{x}\,\xi^{x} = B^{\zeta}(\eta^{a}); \\ \qquad a = 1, \ldots, m; \qquad x = m+1, \ldots, n \\ \qquad \alpha = 1, \ldots, m; \qquad \zeta = m+1, \ldots, n. \end{cases}$$

From (4.16a) it follows that $\xi^{a} = \eta^{a}$ and consequently that (4.16b) is equivalent to the system $\xi^{x} = 0$, constituting a minimal regular null form.[1])

To every allowable coordinate system $(\varkappa)$ there belong for every value of $m < n$ just $\binom{n}{m}$ sets of ∞^{n-m} X_m's with equations of the form

$$(4.17) \qquad \xi^{\zeta} = c^{\zeta}; \qquad \zeta = m+1, \ldots, n$$

where the c^{ζ} are $n-m$ parameters. These X_m's are called the *coordinate-X_m's* (*-curves, -surfaces, -hypersurfaces* for $m = 1, 2, n-1$) of $(\varkappa)$. Hence the theorem of adaptation proves that every X_m in X_n is coordinate-X_m of some suitable allowable coordinate system. ∞^{n-m} X_m's with equations of the form (4.17) are said to constitute a *normal system of X_m's*. The most important property of a normal system is *that every point of the region considered lies in one and only one X_m of the system*. From this we see once more that an X_m in our sense must always be considered in a certain region only. For instance the straight lines through the origin of a centred E_n form a normal system of X_1's for all regions not containing the origin and for these regions only.

The coordinate-X_m's are said to constitute a *net*, the *net of the coordinate system*. A net contains for instance n normal systems of X_1's whose directions at every point are linearly independent. Not every net can belong to a coordinate system (cf. II § 9).

The ∞^{n-m} X_m's of a normal system themselves constitute an $(n-m)$-dimensional manifold. If the equations have the form (4.17) the ξ^{ζ} can be used as coordinates in this manifold. Now the pseudo-group $\mathfrak{P}_n$ in $\mathfrak{N}(\underset{0}{\xi^{\varkappa}})$ contains a sub-pseudo-group $\mathfrak{P}_{n-m}$ of all analytic and invertible transformations of the ξ^{ζ} only, the ξ^{α}; $\alpha = 1, \ldots, m$, being left invariant. Hence the manifold is an X_{n-m}. It is called the *reduction* (Zusammenlegung) *of the X_n with respect to the normal system of*

1) In P. P. 1949, 1, II § 4 it has been proved that it is also possible to derive a minimal regular parametric form from a minimal regular null form and vice versa without using coordinate transformations.

X_m's. This process of reduction is a generalization of the process defined in Ch. I § 1.

If an X_m in X_n is given by its null form and its parametric form, the connecting quantities $B_b^\varkappa$ and C_λ^x are known at every point of X_m. In the tangent E_n of each of these points they fix an E_m called the *tangent* E_m *of* X_m. It is a sub-space of the tangent E_n at that point. It can be identified with the tangent E_m of X_m, as defined in II § 3. $B_b^\varkappa$ and C_λ^x can now be used in the way described in I § 4 to form

1° the section of any covariant tensor of X_n, called here the *section with* X_m, e.g. $'w_b = B_b^\lambda w_\lambda$ (cf. I 4.3, 14a);

2° the reduction of any contravariant tensor of X_n, called here the *reduction with respect to* X_m, e.g. $''v^x = C_\varkappa^x v^\varkappa$ (cf. I 4.7, 15a);

3° the X_n*-components* of a contravariant tensor of X_m, e.g. $p^\varkappa = B_a^\varkappa p^a$ (cf. I 4.5, 16a);

4° the X_n*-components* of a covariant tensor in the E_{n-m} arising from the tangent E_n by reduction with respect to the tangent E_m, e.g. $s_\lambda = C_\lambda^y s_y$ (cf. I 4.9, 17a).

An X_m in X_n is called *rigged* if all its tangent E_m's are rigged (cf. I § 4), that is if, at each of its points, an E_{n-m} in the tangent E_n is given having no direction in common with the tangent E_m. In this case which occurs frequently in differential geometry the unity tensors B_b^a and C_y^x of the X_m and of the local E_{n-m} also have components $B_b^\varkappa$, B_λ^a, $B_\lambda^\varkappa$ and C_λ^x, $C_y^\varkappa$, $C_\lambda^\varkappa$. The quantities can be used in the way described in I § 4 to form:

1° the *section with* E_{n-m} of any covariant tensor of X_n, e.g. $''w_y = C_y^\lambda w_\lambda$ (cf. I 4.14b);

2° the *reduction with respect to* E_{n-m} of any contravariant tensor of X_n, e.g. $'v^a = B_\varkappa^a v^\varkappa$ (cf. I 4.15b);

3° the X_n*-components* of a covariant tensor of X_m, e.g. $q_\lambda = B_\lambda^b q_b$ (cf. I 4.16b);

4° the X_n*-components* of a contravariant tensor in the E_{n-m}, e.g. $r^\varkappa = C_x^\varkappa r^x$ (cf. I 4.17b);

5° the X_m*-part* and the E_{n-m}*-part* of any tensor of X_n, e.g. (cf. I 4.21)

$$\text{(4.18)} \quad \begin{cases} 'v^\varkappa = B_\lambda^\varkappa v^\lambda; & 'w_\lambda = B_\lambda^\varkappa w_\varkappa \\ ''v^\varkappa = C_\lambda^\varkappa v^\lambda; & ''w_\lambda = C_\lambda^\varkappa w_\varkappa \\ v^\varkappa = B_\lambda^\varkappa v^\lambda + C_\lambda^\varkappa v^\lambda; & w_\lambda = B_\lambda^\varkappa w_\varkappa + C_\lambda^\varkappa w_\varkappa. \end{cases}$$

If a field, for instance a scalar field p, is defined in an $\mathfrak{N}(\underset{0}{\eta^a})$ over the X_m with the parametric equations $\xi^\varkappa = B^\varkappa(\eta^a)$, it is often useful to consider a field, analytic (or of a certain class u) at all points of an $\mathfrak{N}(\underset{0}{\xi^\varkappa})$;

$\underset{0}{\xi}^{\varkappa} = \underset{0}{B}^{\varkappa}(\eta^a)$, having at all points of $\underset{0}{\mathfrak{N}}(\eta^a)$ the same values as the given field. This process is called a *prolongation* of the field over X_n and it is convenient to denote the prolonged field with the same kernel. Then $\partial_\lambda p$ also has a sense and the equation $\partial_b p = B_b^\mu \partial_\mu p$ holds. This enables us to use the operator ∂_λ freely. Of course all those results obtained in this way, that can be expressed in terms of quantities and operators of X_m only, are independent of the manner of prolongation.

§ 5. The E_m-field or X_n^m in X_n. Systems of linear partial differential equations.

If we take the tangent E_m's of the X_m's belonging to a normal system we get an E_m-field. But this is a very special field because its E_m's can be arranged in ∞^{n-m} sets of ∞^m E_m's each set consisting of the tangent E_m's of an X_m.

Instead of this special field we may now define a general E_m-field by giving, in each tangent E_n of a point of a region, an E_m through the contactpoint by means of its connecting quantities $B_b^\varkappa$ and (or) C_λ^x. Such a general field is called an X_n^m in X_n.[1]) The E_m at a point is called tangent E_m because it lies in the tangent E_n. If a b-transformation and an x-transformation is effected in all E_n's simultaneously we have

$$(5.1)\qquad \left\{\begin{array}{ll} \text{a)}\ \ B_{b'}^\varkappa = B_{b'}^b B_b^\varkappa; & \text{b)}\ \ C_\lambda^{x'} = C_x^{x'} C_\lambda^x \\ b\ = 1, \ldots, \mathrm{m}\,; & x\ = \mathrm{m}+1, \ldots, \mathrm{n} \\ b' = 1', \ldots, \mathrm{m}'; & x' = (\mathrm{m}+1)', \ldots, \mathrm{n}' \end{array}\right.$$

but the $B_{b'}^b$ and the $C_x^{x'}$ are now functions of the $\xi^\varkappa$. The E_m at each point is spanned by the contravariant vectors $B_1^\varkappa, \ldots, B_{\mathrm{m}}^\varkappa$ and also by the covariant vectors $C_\lambda^{\mathrm{m}+1}, \ldots, C_\lambda^{\mathrm{n}}$. A b-transformation can be interpreted as a linear homogeneous transformation of the m vectors $B_b^\varkappa$; $b = 1, \ldots, \mathrm{m}$ and an x-transformation as a transformation of the same kind of the $n-m$ vectors C_λ^x; $x = \mathrm{m}+1, \ldots, \mathrm{n}$. But the b-transformations here are not connected in any way with a coordinate transformation in some variables η^a (cf. II § 4).

By means of $B_b^\varkappa$ and C_λ^x we are now able to define the *section with* X_n^m of any covariant tensor of X_n, the *reduction with respect to* X_n^m of

[1]) Cf. for the elder literature on the X_n^m and anholonomic coordinate systems VRANCEANU 1926, 1; 2; 1927, 1; 2; 1928, 1; 2; 3; 4; 1929, 1; 2; 1930, 1; 1931, 1; 1932, 1; 1933, 1; 1934, 1; 1936, 1; HLAVATY 1924, 1; 1928, 1; 1930, 1; 2; T. Y. THOMAS 1926, 1; HORAK 1927, 1; 2; SCHOUTEN 1928, 1; 1929, 3; SYNGE 1928, 1; SCHLESINGER 1929, 1; SCHOUTEN and v. KAMPEN 1930, 2; BORTOLOTTI 1931, 1; 2; 1932, 1; 1933, 1; 1937, 2; 3; 1940, 1; DIENES 1932, 1. Cf. for the newer literature Ch. V § 7.

any contravariant tensor of X_n and the X_n-*components* of any contravariant tensor of X_n^m and any covariant tensor in the local E_{n-m} arising from reduction of the tangent E_n with respect to the X_n^m. The formulae are the same as (I 4.14a, 15a, 16a, 17a). If, besides the X_n^m we give an X_n^{n-m} such that the tangent E_m and E_{n-m} have no direction in common, the X_n^m (and also the X_n^{n-m}) are said to be *rigged*. In this case we have B_λ^a, B_b^a, $B_b^\varkappa$, $B_\lambda^\varkappa$ and C_λ^x, C_y^x, $C_y^\varkappa$, $C_\lambda^\varkappa$ at every point. By means of these we are now able to construct also the section with X_n^{n-m} of any covariant tensor of X_n, the reduction with respect to X_n^{n-m} of any contravariant tensor of X_n, the X_n-components of every tensor of X_n^m and of X_n^{n-m} and the X_n^m-*part* and X_n^{n-m}-*part* of every tensor of X_n.

Now the question arises whether a given E_m-field belongs to some normal system of X_m's. But this is only a special case of a far more general question. An X_p is said to *envelop* an E_m-field if its tangent E_p contains, at every point, the tangent E_m of the field and it is said to be *enveloped by* an E_m-field if its tangent E_p is contained at every point in the tangent E_m. Hence there are two problems, the *outer problem* concerning all enveloping X_p's and the *inner problem* concerning all enveloped X_p's. [1]) We require especially all enveloped normal systems of X_p's for the maximum value of p and also all enveloping normal systems of X_p's for the minimum value of p. According to the definition an enveloping X_m is at the same time enveloped, hence the two problems are identical for $p=m$. But for all other values of p they are essentially different and require quite different mathematical methods.

The inner problem is much more difficult than the outer problem. Only the case $m=n-1$, known as PFAFF's *problem*, can be dealt with by fairly elementary means as will be shown in II § 7.

The outer problem is closely connected with the theory of systems of linear partial differential equations of the first order. We give first a definition of totally integrable systems of general partial differential equations.

If a system of N partial differential equations of the first order with n independent variables $\xi^\varkappa$ and M unknown variables $\underset{\mathfrak{b}}{z}$; $\mathfrak{b}=1, \ldots, M$

$$\overset{i}{F}(\xi^\varkappa, \underset{\mathfrak{b}}{z}, \partial_\lambda \underset{\mathfrak{b}}{z}) = 0; \quad i=1, \ldots, N; \quad \mathfrak{b}=1, \ldots, M \tag{5.2}$$

is given, every set of functions $\underset{\mathfrak{b}}{z}$ satisfying (5.2) is called a *solution* of (5.2) and every set of values $\underset{*}{\xi^\varkappa}$, $\underset{\mathfrak{b}}{\overset{*}{\eta}}$, $\underset{\mathfrak{b}}{\overset{*}{\eta}_\lambda}$ satisfying the equations

$$\overset{i}{F}(\underset{*}{\xi^\varkappa}, \underset{\mathfrak{b}}{\overset{*}{\eta}}, \underset{\mathfrak{b}}{\overset{*}{\eta}_\lambda}) = 0 \tag{5.3}$$

[1]) Cf. P. P. 1949, 1 Ch. III, VIII, also for literature.

is called a *null point* of (5.2). Now the system (5.2) is said to be *totally integrable*[1]) in an $\mathfrak{N}(\underset{0}{\xi^\varkappa}, \overset{0}{\underset{\mathfrak{b}}{\eta}}, \overset{0}{\underset{\mathfrak{b}}{\eta_\lambda}})$ if for every null point $\underset{*}{\xi^\varkappa}, \overset{*}{\underset{\mathfrak{b}}{\eta}}, \overset{*}{\underset{\mathfrak{b}}{\eta_\lambda}}$ in this region there exists at least one solution of (5.2) satisfying the condition that for $\xi^\varkappa = \underset{*}{\xi^\varkappa}$

$$(5.4) \qquad \underset{\mathfrak{b}}{z} = \overset{*}{\underset{\mathfrak{b}}{\eta}}; \qquad \partial_\lambda \underset{\mathfrak{b}}{z} = \overset{*}{\underset{\mathfrak{b}}{\eta_\lambda}}.$$

The necessary and sufficient conditions for a system to be totally integrable are called its first *conditions of integrability.*

Now let p be the smallest number for which a normal system of X_p's exists enveloping the E_m-field $B_b^\varkappa$, C_λ^x and let

$$(5.5) \qquad F^{\mathfrak{x}}(\xi^\varkappa) = c^{\mathfrak{x}}; \qquad \xi^\varkappa = \Phi^\varkappa(\eta^{\mathfrak{a}}, c^{\mathfrak{x}}); \qquad \begin{array}{l}\mathfrak{a} = \overline{1}, \ldots, \overline{\mathfrak{p}};\\ \mathfrak{x} = \overline{\mathfrak{p}+1}, \ldots, \overline{\mathfrak{n}}\end{array}$$

be the null form and the parametric form of the enveloping normal system. Then the $F^{\mathfrak{x}}$ are $n-p$ independent solutions (i.e. solutions with linearly independent gradients) of the homogeneous linear equations

$$(5.6) \qquad B_b^\mu \partial_\mu f = 0; \qquad b = 1, \ldots, \mathrm{m}$$

and every linear element $d\xi^\varkappa$ satisfying the equations

$$(5.7) \qquad C_\mu^x d\xi^\mu = 0; \qquad x = \mathrm{m}+1, \ldots, \mathrm{n}$$

also satisfies the equations

$$(5.8) \qquad d\xi^\mu \partial_\mu F^{\mathfrak{x}} = 0; \qquad \mathfrak{x} = \overline{\mathfrak{p}+1}, \ldots, \overline{\mathfrak{n}}.$$[2])

The equations (5.6) and (5.7) are said to be *adjoint to each other.* The first gives the condition for the tangent E_p's of an enveloping X_p and the second the condition for its linear elements. If F is a solution of (5.6), F is called an *integral function* of (5.7) and the equation $F =$ const. an *integral* of (5.7).

If the operator $B_b^\mu \partial_\mu$ is applied twice in succession we get, after alternation,

$$(5.9) \qquad (B_{[c}^\nu \partial_{|\nu|} B_{b]}^\mu) \partial_\mu f = 0$$

because $\partial_{[\nu} \partial_{\mu]} f = 0$. Hence every solution of (5.6) must satisfy the $\binom{m}{2}$ linear homogeneous equations (5.9). If, among these equations, there are exactly μ_1 linearly independent of (5.6) and of each other, we get

[1]) For the definition of total integrability for systems of higher order see for instance P. P. 1949, 1, p. 89.

[2]) Cf. for the conditions for one single enveloping X_p, P. P. 1949, 1, p. 85f.

a system of $m_1 \overset{\text{def}}{=} m + \mu_1$ equations

$$B^{\mu}_{b_1} \partial_\mu f = 0; \qquad b_1 = 1, \ldots, \mathrm{m}_1 \tag{5.10}$$

called the *first derived system of* (5.6). This system has the same solutions as (5.6). $\mu_1 = 0$ if and only if

$$C_{cb}^{\cdot\cdot\,\varkappa} \overset{\text{def}}{=} - B^{\nu}_{[c} (\partial_{|\nu|} B^{\mu}_{b]}) C^{\varkappa}_{\mu} = + B^{\nu\mu}_{c\,b} \partial_{[\nu} C^{\varkappa}_{\mu]} = 0. \tag{5.11}$$

It is remarkable that $C_{cb}^{\cdot\cdot\,\varkappa}$ is a tensor though $B^{\nu}_{[c} \partial_{|\nu|} B^{\varkappa}_{b]}$ does not have this property. If $\mu_1 = 0$ the system (5.6) is said to be *complete.* It can be proved that a complete system of m equations is totally integrable, that it has just $n-m$ independent solutions and that its integrability conditions[1]) are identically satisfied. Hence if (5.11) is satisfied and only in that case there exists an enveloped and enveloping normal system of X_m's. The E_m-field is then called *X_m-forming.*

The solution of the complete system is equivalent to the determination of this set of ∞^{n-m} X_m's. If the X_m's are known, every function of the $\xi^\varkappa$ that is constant on each of these X_m's is a solution.

If the first derived system is complete there exists a normal system of $X_{m+\mu_1}$'s. If it is not complete a second derived system with $m+\mu_1+\mu_2$ equations can be found, and so on. At last we get a complete system having the same solutions as (5.6). If this last system consists of n equations, (5.6) has no solutions except the trivial one $f = \text{const}$. The numbers $\mu_1, \mu_2, \ldots$ are arithmetic invariants of the E_m-field.

By

$$v^{\varkappa_1 \ldots \varkappa_m} \propto B^{[\varkappa_1}_{\mathrm{1}} \ldots B^{\varkappa_m]}_{\mathrm{m}} \tag{5.12}$$

a contravariant pseudo-m-vector is given with the m-direction of the tangent E_m. In the same way by

$$w_{\lambda_1 \ldots \lambda_{n-m}} \propto C^{\mathrm{m}+1}_{[\lambda_1} \ldots C^{\mathrm{n}}_{\lambda_{n-m}]} \tag{5.13}$$

a covariant pseudo-$(n-m)$-vector is introduced with the same m-direction. The $n.a.s.$ conditions (5.11) for the E_m-field to be X_m-forming can now be written as

$$(\partial_{\varkappa_1} v^{\varkappa_1 \ldots \varkappa_m}) w_{\varkappa_2 \lambda_2 \ldots \lambda_{n-m}} = 0 \text{ [2])}, \tag{5.14}$$

from which we may easily derive the following equivalent forms

$$(\partial_{\varkappa_1} v^{\varkappa_1 \varkappa_2 \ldots [\varkappa_m}) v^{\nu_1 \ldots \nu_m]} = 0 \text{ [3])} \tag{5.15}$$

$$(\partial_{\varkappa_1} w_{\varkappa_2 \lambda_2 \ldots \lambda_{n-m}}) v^{\varkappa_1 \ldots \varkappa_m} = 0 \text{ [4]), [2])} \tag{5.16}$$

$$(\partial_{[\varkappa_1} w_{\varkappa_{m+1} \ldots \varkappa_n}) w_{\varkappa_2] \lambda_2 \ldots \lambda_{n-m}} = 0. \tag{5.16a}$$

---

[1]) Cf. for the exact formulation and the proof of the existence theorem for instance P. P. 1949, 1, p. 108.

[2]) Schouten 1918, 2; Schouten and Struik 1919, 1; Struik 1922, 1, p. 53ff.

[3]) v. Weber 1900, 1, p. 73ff. has an equivalent form.

[4]) v. Weber 1900, 1, p. 103 has an equivalent form.

To every simple contravariant pseudo-vector field $v^{\varkappa_1 \dots \varkappa_m}$ that is X_m-forming there belongs a complete system with the equations

$$(5.17) \qquad v^{\varkappa_1 \dots \varkappa_m} \partial_{\varkappa_1} f = 0 \quad \text{or} \quad w_{[\lambda_1 \dots \lambda_{n-m}} \partial_{\lambda]} f = 0,$$

expressing the fact that the $(n-1)$-direction of $\partial_\lambda f$ contains the tangent m-direction, and the adjoint system with the equations

$$(5.18) \qquad v^{[\varkappa_1 \dots \varkappa_m} d\xi^{\varkappa]} = 0 \quad \text{or} \quad w_{\lambda_1 \dots \lambda_{n-m}} d\xi^{\lambda_1} = 0$$

expressing the fact that the direction of $d\xi^\varkappa$ lies in the tangent m-direction.

If the E_m-field is X_m-forming and if

$$(5.19) \qquad F^x(\xi^\varkappa) = c^x; \quad x = \mathrm{m}+1, \dots, \mathrm{n}$$

are the equations of the normal system of X_m's, the tangent E_m at every point lies in the tangent E_{n-1} of each of the $n-m$ X_{n-1}'s (5.19). Hence

$$(5.20) \qquad w_{\lambda_1 \dots \lambda_{n-m}} \propto (\partial_{[\lambda_1} F^{\mathrm{m}+1}) \dots \partial_{\lambda_{n-m}]} F^{\mathrm{n}}$$

and this equation expresses the fact that $w_{\lambda_1 \dots \lambda_{n-m}}$ is a *gradient product* (alternating product of gradients) to within a scalar factor. This proves that (5.16a) is the necessary and sufficient condition for $w_{\lambda_1 \dots \lambda_{n-m}}$ to be a gradient product to within a scalar factor.

The theory of homogeneous linear equations of the first order can be generalized for non homogeneous linear equations of the form

$$(5.21) \qquad B_b^\mu \partial_\mu f = \varphi_b(\xi^\varkappa); \quad b = 1, \dots, \mathrm{m},$$

where B_b^μ has rank m. It can be proved[1]) that such a system is totally integrable and that it has just $n-m+1$ independent solutions if and only if (5.6), called the *reduced system of* (5.21), is complete, that is, if there exist functions $\sigma_{cb}^a(\xi^\varkappa)$ such that

$$(5.22) \qquad B_{[c}^\mu \partial_{|\mu|} B_{b]}^\varkappa = \sigma_{cb}^a B_a^\varkappa$$

and if moreover the φ_b satisfy the equations

$$(5.23) \qquad B_{[c}^\mu \partial_{|\mu|} \varphi_{b]} = \sigma_{cb}^a \varphi_a.$$

If the integrability conditions (5.22, 23) are satisfied, (5.21) is called *complete*. If $\overset{1}{f}, \dots, \overset{n-m}{f}$ are $n-m$ independent solutions of the reduced system, and if $\overset{0}{f}$ is one solution of the complete system (5.21), the general solution is $\overset{0}{f} + \psi(\overset{1}{f}, \dots, \overset{n-m}{f})$ where ψ denotes an arbitrary analytic function.[2])

[1]) Cf. for instance P. P. 1949, 1, III § 14.

[2]) Cf. for the exact formulation and the proof of the existence theorem for instance P. P. 1949, 1, p. 111.

We apply this theory first to the gradient equation

$$\partial_\mu p = w_\mu(\xi^\varkappa). \tag{5.24}$$

If this system is written in the form

$$A^\mu_\lambda \partial_\mu p = w_\lambda \tag{5.25}$$

we see that the reduced system is complete and that the coefficients σ are all zero. Hence here the only integrability conditions are

$$\partial_{[\mu} w_{\lambda]} = 0. \tag{5.26}$$

In fact, if a solution $\overset{0}{p}$ of (5.24) is known, the general solution is $p = \overset{0}{p} + c$ and this means that the value of p at any point $\underset{*}{\xi^\varkappa}$ can be given arbitrarily.

Exercise.

II 5,1. If a vector field $v^\varkappa$ is given in $\mathfrak{N}(\overset{0}{\xi^\varkappa})$, there always exists a coordinate system $(\varkappa)$ in an $\mathfrak{N}(\underset{0}{\xi^\varkappa})$ such that $v^\varkappa = \underset{1}{e^\varkappa}$. Every covariant $(n-1)$-vector in X_n is product of a scalar and a gradient product.[1])

§ 6. The invariant operators Rot and Div.

As we have seen in II § 3, the derivatives of the components of a tensor with valence >0 with respect to the $\xi^\varkappa$ in an X_n are not components of a tensor. But if we form the *alternated* derivative of a *covariant* q-vector, the additional term in

$$\begin{cases} \partial_{[\mu'} w_{\lambda'_1 \ldots \lambda'_q]} = A^{\mu\lambda_1 \ldots \lambda_q}_{\mu'\lambda'_1 \ldots \lambda'_q} \partial_{[\mu} w_{\lambda_1 \ldots \lambda_q]} + w_{\lambda_1 \ldots \lambda_q} \partial_{[\mu'} A^{\lambda_1}_{\lambda'_1} \ldots A^{\lambda_q}_{\lambda'_q]} \\ \qquad = A^{\mu\lambda_1 \ldots \lambda_q}_{\mu'\lambda'_1 \ldots \lambda'_q} \partial_{[\mu} w_{\lambda_1 \ldots \lambda_q]} \end{cases} \tag{6.1}$$

vanishes because $\partial_{[\mu'} A^\varkappa_{\lambda']} = 0$. We call this alternated derivative *multiplied by* $q+1$ the *natural derivative* (ct. II § 3) and for $q>0$ also the *rotation* of $w_{\lambda_1 \ldots \lambda_q}$. If we use the abridged notation of I § 7, we write Rot w and also $(q+1)\, D w$. Rot w vanishes identically if $q=n$. As a consequence of the definition the process of forming the natural derivative applied twice in succession, always leads to zero:

$$\begin{cases} \text{Rot Grad}\, p = 2\, D\, D\, p = 0 \\ \text{Rot Rot}\, w = (q+2)(q+1)\, D\, D\, w = 0. \end{cases} \tag{6.2}$$

All this is also true for W-scalars and covariant W-q-vectors. If the corresponding $(n-q)$-vector Δ-density $\tilde{\mathfrak{w}}^{\varkappa_1 \ldots \varkappa_p}$; $p = n-q$ is formed from

[1]) GOURSAT 1922, 1, p. 117.

w by means of $\tilde{\mathfrak{E}}^{\varkappa_1 \ldots \varkappa_n}$ (cf. I § 7) it follows that for $p \geqq 1$

$$(6.3)\quad \begin{cases} \partial_\mu \tilde{\mathfrak{w}}^{\mu \varkappa_2 \ldots \varkappa_p} = (-1)^{-\beta_q} \dfrac{1}{q!} \partial_\mu w_{\lambda_1 \ldots \lambda_q} \tilde{\mathfrak{E}}^{\lambda_1 \ldots \lambda_q \mu \varkappa_2 \ldots \varkappa_p} \\ \qquad = (-1)^{q-\beta_q} \dfrac{1}{(q+1)!} (q+1) \partial_{[\mu} w_{\lambda_1 \ldots \lambda_q]} \tilde{\mathfrak{E}}^{\mu \lambda_1 \ldots \lambda_q \varkappa_2 \ldots \varkappa_p}. \end{cases}$$

Hence, $\partial_\mu \tilde{\mathfrak{w}}^{\mu \varkappa_2 \ldots \varkappa_p}$ is a $(p-1)$-vector Δ-density of weight $+1$. It is called the *natural derivative* or the *divergence* of $\tilde{\mathfrak{w}}^{\varkappa_1 \ldots \varkappa_p}$ and it is written $\mathrm{Div}\, \tilde{\mathfrak{w}}$ in abridged notation. $\mathrm{Div}\, \tilde{\mathfrak{w}}$ is by definition zero if $p = 0$. From (6.3) we see that if

$$(6.4)\qquad W = \mathrm{Rot}\, w; \qquad \tilde{\mathfrak{W}} = \mathrm{Div}\, \tilde{\mathfrak{w}},$$

$\tilde{\mathfrak{W}}$ corresponds to $(-1)^{q-\beta_q+\beta_{q+1}} W$. From the definition it follows that the rule concerning the result of two natural derivations in succession holds also in this case

$$(6.5)\qquad \mathrm{Div}\,\mathrm{Div}\, \tilde{\mathfrak{w}} = 0.$$

All this is also true for contravariant p-vector densities of weight $+1$.

Let $W_{\lambda_1 \ldots \lambda_{q+1}}$ be a *simple* $(q+1)$-vector whose rotation vanishes in an $\mathfrak{N}(\overset{}{\underset{0}{\xi}}{}^{\varkappa})$. Then according to (5.16a) the E_{p-1}-field with the $(p-1)$-direction of W is X_{p-1}-forming and this means that there exists an $\mathfrak{N}(\underset{0}{\xi}{}^{\varkappa})$ where W is a gradient product to within a scalar factor

$$(6.6)\quad W_{\lambda_1 \ldots \lambda_{q+1}} = s\, (\partial_{[\lambda_1} \underset{1}{s}) \ldots \partial_{\lambda_{q+1}]} \underset{q+1}{s} \quad \text{or} \quad W = s\,[\underset{1}{\bar{s}} \ldots \underset{q+1}{\bar{s}}].$$

Taking the rotation we get

$$(6.7)\quad (\partial_{[\lambda} s)(\partial_{\lambda_1} \underset{1}{s}) \ldots (\partial_{\lambda_{q+1}]} \underset{q+1}{s}) = 0 \quad \text{or} \quad [\bar{s}\, \underset{1}{\bar{s}} \ldots \underset{q+1}{\bar{s}}] = 0$$

and this equation expresses the fact that s can be written as a function of $\underset{1}{s}, \ldots, \underset{q+1}{s}$. If we take $\underset{1}{s}, \ldots, \underset{q+1}{s}$ as the first $q+1$ coordinates $\xi^1, \ldots, \xi^{q+1}$ of $(\varkappa)$, we have in the region considered

$$(6.8)\qquad W_{\lambda_1 \ldots \lambda_{q+1}} = s\, \overset{1}{e}_{[\lambda_1} \ldots \overset{q+1}{e}_{\lambda_{q+1}]}$$

where s is a function of $\xi^1, \ldots, \xi^{q+1}$ only. Now in the X_{q+1} of these variables we take a solution of the non homogeneous equation

$$(6.9)\qquad \underset{1}{e}{}^\beta \partial_\beta z = s; \qquad \beta = 1, \ldots, q+1.$$

Then

$$(6.10)\qquad W_{\lambda_1 \ldots \lambda_{q+1}} = z_{[\lambda_1} \overset{2}{e}_{\lambda_2} \ldots \overset{q+1}{e}_{\lambda_{q+1}]}; \qquad z_\lambda \overset{\text{def}}{=} \partial_\lambda z$$

and this proves that a simple covariant $(q+1)$-vector is a gradient product and consequently the rotation of some q-vector in an $\mathfrak{N}(\underset{0}{\xi}{}^{\varkappa})$

if and only if its rotation vanishes in an $\mathfrak{N}(\overset{}{\underset{0}{\xi^\varkappa}})$. As a corollary it follows that every n-vector given in an $\mathfrak{N}(\underset{0}{\xi^\varkappa})$ is a gradient product in an $\mathfrak{N}(\underset{0}{\xi^\varkappa})$.

If the $(q+1)$-vector is not simple it is still true that it is the rotation of a q-vector if and only if its rotation vanishes. This can be proved by induction. Let the theorem be proved for every value of q, $(1 \leq q+1 \leq n-1)$ in an X_{n-1}. Then it can be proved that it is also true in an X_n.[1]) Now if $n \leq 3$ the $(q+1)$-vector is always simple and the theorem has been proved for simple $(q+1)$-vectors. Hence it is true for all values of n. The q-vector is determined to within an additive term whose rotation vanishes and it can be found by quadratures only.[2]) We give here the exact formulation of the theorem for $(q+1)$-vectors and for $(p-1)$-vector Δ-densities:[3])

If $W_{\lambda_1 \ldots \lambda_{q+1}}$ is a covariant $(q+1)$-vector, $1 \leq q+1 \leq n$, analytic in an $\mathfrak{N}(\underset{0}{\xi^\varkappa})$, it can be written in some $\mathfrak{N}(\underset{0}{\xi^\varkappa})$ as the gradient of a scalar for $q=0$ and as the rotation of a q-vector for $q>0$, if and only if there exists an $\mathfrak{N}(\underset{0}{\xi^\varkappa})$ where Rot $W=0$. *This condition is identically satisfied for $q+1=n$.*

If $\mathfrak{W}^{\varkappa_1 \ldots \varkappa_{p-1}}$ is a contravariant $(p-1)$-vector Δ-density of weight $+1$; $0 \leq p-1 \leq n-1$, analytic in an $\mathfrak{N}(\underset{0}{\xi^\varkappa})$, it can be written in some $\mathfrak{N}(\underset{0}{\xi^\varkappa})$ as the divergence of a p-vector Δ-density of weight $+1$, if and only if there exists an $\mathfrak{N}(\underset{0}{\xi^\varkappa})$ where Div $\mathfrak{W}=0$. *This condition is identically satisfied for $p-1=0$.*

There are of course theorems of exactly the same form for covariant W-$(q+1)$-vectors and $(p-1)$-vector densities of weight $+1$.

Both theorems express conditions of integrability. For the equations

$$\partial_{[\mu}\, w_{\lambda_1 \ldots \lambda_q]} = W_{\mu \lambda_1 \ldots \lambda_q} \tag{6.11}$$

[1]) P. P. 1949, 1, II § 11. The whole theorem is proved there without using the theory of linear partial differential equations.

[2]) GOURSAT 1922, 1, p. 105 ff. Using the homotopy operator known from the theory of graded rings (H. CARTAN 1949, 1, p. 107) a very elegant expression can be found. If

$$w_{\lambda_1 \ldots \lambda_q} = f_{\lambda_1 \ldots \lambda_q}(\xi^\varkappa)$$

and

$$u_{\lambda_1 \ldots \lambda_{q-1}} \overset{\text{def}}{=} \frac{1}{q}\, \xi^{\lambda_q} \int_{t=0}^{t=1} f_{\lambda_1 \ldots \lambda_q}(\xi^\varkappa t)\, s\, t^{s-1}\, dt$$

it follows immediately that $w=$ Rot u. If we write $u=$ Op w it is proved easily that

$$\text{Rot Op } w - \text{Op Rot } w = w$$

for every field w. (From a personal communication of N. H. KUIPER.)

[3]) VOLTERRA 1889, 1 for R_n; BROUWER 1906, 1, p. 22 for R_n.

the conditions are

(6.12) $$\partial_{[\nu} W_{\mu\lambda_1\ldots\lambda_q]} = 0$$

and for the equations

(6.13) $$\partial_\mu \tilde{\mathfrak{w}}^{\mu\varkappa_2\ldots\varkappa_p} = \mathfrak{W}^{\varkappa_2\ldots\varkappa_p}$$

we have the conditions

(6.14) $$\partial_\nu \mathfrak{W}^{\nu\varkappa_3\ldots\varkappa_p} = 0.$$

We remark that these integrability conditions are independent of the manner of transformation of the objects concerned. For instance if $W_{\mu\lambda_1\ldots\lambda_q}$, alternating in $\mu\,\lambda_1\ldots\lambda_q$, is any geometric object whose components with respect to $(\varkappa)$ satisfy (6.12), in general no such relation holds for the components with respect to any other coordinate system $(\varkappa')$. But nevertheless there exist functions $w_{\lambda_1\ldots\lambda_q}$ of the $\xi^\varkappa$ such that (6.11) holds.

Here the question arises whether there exist solutions of equations of the form

(6.15) $$\partial_{[\mu} \pi_{\lambda_1\ldots\lambda_q]A} = \Pi_{\mu\lambda_1\ldots\lambda_q A}$$

where A is a collective index standing for any set of co- and contravariant indices and where $\pi_{\lambda_1\ldots\lambda_q A}$ and $\Pi_{\mu\lambda_1\ldots\lambda_q A}$ are alternating in the first q and $q+1$ indices respectively. The equations are *not* invariant and π and Π need not be tensors. But they are closely connected with invariant equations to be dealt with in the next chapter. In order to show what is happening we take the simple example

(6.16) $$\partial_{[\mu} \pi_{\lambda]}{}^{\varkappa}{}_{\varrho} = \Pi_{\mu\lambda}{}^{\varkappa}{}_{\varrho}.$$

We may consider (6.16) as a set of n^2 equations by giving $\varkappa$ and ϱ all values from 1 to n. For instance if we take $\varkappa = 1$, $\varrho = 2$, the integrability conditions of this special equation are

(6.17) $$\partial_{[\nu} \Pi_{\mu\lambda]}{}^{1}{}_{2} = 0$$

and this proves that the integrability conditions of (6.16) are

(6.18) $$\partial_{[\nu} \Pi_{\mu\lambda]}{}^{\varkappa}{}_{\varrho} = 0.$$

It can be proved in the same way that the equation

(6.19) $$\partial_\nu \pi^{\nu\varkappa_2\ldots\varkappa_p A} = \Pi^{\varkappa_2\ldots\varkappa_p A}$$

where A is a collective index and where π and Π are alternating in the first p and $p-1$ indices respectively, has the integrability conditions

(6.20) $$\partial_\nu \Pi^{\nu\varkappa_3\ldots\varkappa_p A} = 0.$$

If $q = 0$ we can go a step further and consider the case

(6.21) $$\partial_\mu \pi_B = \Pi_{\mu B}(\xi^\varkappa, \pi_C)$$

where the right hand side depends also on the π_B (the π_B need not be components of a geometric object!).[1], [2]) (6.21) can be written in the form of a system of total differential equations

$$d\pi_B - \Pi_{\mu B}(\xi^\varkappa, \pi_C)\, d\xi^\mu = 0. \tag{6.22}$$

Now let $\Omega(\xi^\varkappa, \pi_B)$ be an integral function (hence Ω = const. an integral) of this system, then Ω is a solution (II § 5) of the homogeneous linear system

$$\frac{\partial \Omega}{\partial \xi^\mu} + \frac{\partial \Omega}{\partial \pi_B} \Pi_{\mu B} = 0 \tag{6.23}$$

with the independent variables $\xi^\varkappa$ and π_B. This system is complete if and only if

$$(\bar{\partial}_{[\nu} \Pi_{\mu] B}) \frac{\partial}{\partial \pi_B} + \Pi_{[\nu|C|} \frac{\partial \Pi_{\mu] B}}{\partial \pi_C} \frac{\partial}{\partial \pi_B} \tag{6.24}$$

(where $\bar{\partial}_\nu$ denotes differentiation with respect to ξ^ν, the π_B being left constant) depends linearly on the operators $\bar{\partial}_\mu + \Pi_{\mu B} \dfrac{\partial}{\partial \pi_B}$ in the left hand side of (6.23). But this is only possible if

$$\partial_{[\nu} \Pi_{\mu] B} = \bar{\partial}_{[\nu} \Pi_{\mu] B} + \Pi_{[\nu|C|} \frac{\partial \Pi_{\mu] B}}{\partial \pi_C} = 0. \tag{6.25}$$

Now let these equations, which do not contain any derivatives of π_B, be symbolized by

$$\overset{1}{F}(\pi_B, \xi^\varkappa) = 0. \tag{6.26}$$

If we apply ∂_ω to (6.26) and eliminate $\partial_\omega \pi_A$ by means of (6.21) we get a second set of the same kind, symbolized by

$$\overset{2}{F}(\pi_B, \xi^\varkappa) = 0 \tag{6.27}$$

and this process can be continued ad infinitum. We need not consider the case when a relation between the $\xi^\varkappa$ can be derived from all these equations, because in that case we could diminish the number of independent variables. So there are only two possibilities. Either after $i+1$ steps the $(i+1)$th set contains an equation which is not consistent with the former ones, in which case (6.21) has no solutions. Or the $(j+1)$th set depends on the foregoing sets. For this latter case VEBLEN and J. M. THOMAS[3]) have proved that then all further sets depend on the first i ones. Collecting results we get the rule:

In order to find the integrability conditions of equations of the form (6.21), *where the* $\Pi_{\mu A}$ *depend on the* $\xi^\varkappa$ *and the* π_B, *we have to form the*

---

[1]) GUREWITSCH considered 1933, 1 equations of the form $\partial_{[\mu} w_{\lambda_1 \ldots \lambda_q]} = p_{[\mu} w_{\lambda_1 \ldots \lambda_q]}$.

[2]) Cf. for literature SCHOUTEN 1925, 1, p. 442ff.

[3]) VEBLEN and J. M. THOMAS 1926, 1; cf. T. Y. THOMAS and LEVINE 1934, 1; T. Y. THOMAS 1934, 2, Ch. X.

integrability conditions (6.26), (6.27), *and so on. If these conditions are inconsistent,* (6.21) *has no solutions. If the first i sets do not contain inconsistent equations and if the* $(i+1)$*th set depends on the foregoing sets, the equation* (6.21) *together with the i first sets form a totally integrable system.*[1]) *This means here that if* $\underset{0}{\pi}_B$ *and* $\underset{0}{\Pi}_{\mu B}$ *satisfy* $\underset{0}{\Pi}_{\mu B} = \Pi_{\mu B}(\underset{0}{\pi}_A, \underset{0}{\xi}^\varkappa)$ *and the first i conditions of integrability for* $\xi^\varkappa = \underset{0}{\xi}^\varkappa$, *there exists exactly one solution of* (6.21) *such that* $\pi_B = \underset{0}{\pi}_B$ *and* $\partial_\mu \pi_B = \underset{0}{\Pi}_{\mu B}$ *at* $\underset{0}{\xi}^\varkappa$.

If besides (6.21) a set of equations is given in π_B and $\xi^\varkappa$ which does not contain derivatives of π_B, this set must be adjoined to the first integrability conditions.[2])

As we see the construction of the integrability conditions depends always on the possibility of eliminating at each step the derivatives of the π_B. Hence the same method can in general not be used for an equation of the form (6.15) if the right hand side depends on the $\xi^\varkappa$ and the π, because the derivatives of π arising at every step do not in general appear in the form $\partial_{[\mu} \pi_{\lambda_1 \ldots \lambda_q] B}$. Only if they happen to appear in this form the elimination can be effected and in that case the integrability conditions can be constructed in the way described. But in the more general case the problem whether a set of equations has solutions can only be solved either by the theory of CARTAN or by the theory of RIQUIER.[3])

It often happens that the $\Pi_{\mu B}$ in (6.21) are linear homogeneous in the π_B

$$\Pi_{\mu B} = \Pi_{\mu B}^{A}(\xi^\varkappa)\, \pi_A . \tag{6.28}$$

Then for total integrability the equation

$$\{\partial_{[\mu} \Pi_{\lambda] B}^{A} + \Pi_{[\mu |C|}^{A} \Pi_{\lambda] B}^{C}\} \pi_A = 0 \tag{6.29}$$

must be satisfied identically. Accordingly the conditions of integrability are in this case

$$\partial_{[\mu} \Pi_{\lambda] B}^{A} + \Pi_{[\mu |C|}^{A} \Pi_{\lambda] B}^{C} = 0. \tag{6.30}$$

If an X_m in X_n is given, the section of a field $w_{\lambda_1 \ldots \lambda_q}$; $q \leq m$ with X_m is

$$'w_{b_1 \ldots b_q} = B_{b_1 \ldots b_q}^{\lambda_1 \ldots \lambda_q} w_{\lambda_1 \ldots \lambda_q}; \qquad B_b^\lambda \stackrel{\text{def}}{=} \partial_b \xi^\lambda \tag{6.31}$$

and the section of its rotation is

$$'W_{c b_1 \ldots b_q} = (q+1)\, B_{c\, b_1 \ldots b_q}^{\mu \lambda_1 \ldots \lambda_q} \partial_{[\mu} w_{\lambda_1 \ldots \lambda_q]} = (q+1)\, B_{[c\, b_1 \ldots b_q]}^{\mu \lambda_1 \ldots \lambda_q} \partial_\mu w_{\lambda_1 \ldots \lambda_q} . \tag{6.32}$$

[1]) Cf. II § 5.

[2]) J. M. THOMAS 1927, 1.

[3]) Cf. also for literature P. P. 1949, 1, Ch. X.

We prove that *the rotation of the section equals the section of the rotation.* In fact

$$(6.33)\qquad \begin{cases} (q+1)\,\partial_{[c}\,{}'w_{b_1\ldots b_q]} = (q+1)\,\partial_{[c}\,B^{\lambda_1\ldots\lambda_q}_{b_1\ldots b_q]}\,w_{\lambda_1\ldots\lambda_q} \\ \qquad = (q+1)\,B^{\mu\,\lambda_1\ldots\lambda_q}_{c\,b_1\ldots b_q}\,\partial_{[\mu}\,w_{\lambda_1\ldots\lambda_q]} \end{cases}$$

because $\partial_{[c}\,B^{\lambda}_{b]} = 0$.

Exercises.

II 6,1. If $w_{\lambda_1\ldots\lambda_{p-1}n} = 0$ and $Dw = 0$ the field w does not depend on ξ^n.[1])

II 6,2.[2]) If for every choice of the q-vector $P_{\lambda_1\ldots\lambda_q}$ the rotation of $P_{\lambda_1\ldots\lambda_q}$ vanishes always if the rotation of $Q_{[\lambda}^{\cdot\cdot\lambda_q}P_{\lambda_1\ldots\lambda_q]}$ is zero, it follows that

II 6,2 α) $$Q_{\lambda}^{\cdot\cdot\varkappa} = \alpha\,A^{\varkappa}_{\lambda};\quad \alpha = \text{constant}.$$

§ 7. PFAFF's problem.[3])

A field w_λ in X_n, analytic in some $\mathfrak{N}(\overset{0}{\xi^\varkappa})$, fixes an E_{n-1}-field or X_n^{n-1}. For such a field the outer problem is trivial because it is immediately clear that the E_{n-1}-field is X_{n-1}-forming if and only if w_λ is a gradient to within a scalar factor. Only the inner problem is interesting, in fact we have here the most simple form of this problem.

We form the rotation $W_{\mu\lambda} = 2\,\partial_{[\mu}w_{\lambda]}$ and the multivectors

$$(7.1\,\text{a})\quad \overset{1}{I}_\lambda = w_\lambda;\ \ \overset{2}{I}_{\mu\lambda} = W_{\mu\lambda};\ \ \overset{3}{I}_{\nu\mu\lambda} = w_{[\nu}\,W_{\mu\lambda]};\ \ \overset{4}{I}_{\omega\nu\mu\lambda} = W_{[\omega\nu}\,W_{\mu\lambda]};\ \text{etc.}$$

or, using the abridged notation of I § 7

$$(7.1\,\text{b})\qquad \overset{1}{I} = w;\ \ \overset{2}{I} = W;\ \ \overset{3}{I} = [w\,W];\ \ \overset{4}{I} = [W\,W];\ \text{etc.}$$

Then if

$$(7.2)\qquad \overset{K}{I} \neq 0;\qquad \overset{K+1}{I} = 0,$$

and consequently $\overset{s}{I} \neq 0$ for $s \leq K$ and $\overset{s}{I} = 0$ for $s > K$, we call K the *class* of w. K is the λ-rank of the set w_λ, $W_{\mu\lambda}$. The rank 2ϱ of W is called the *rotation class* and the λ-rank k of w_λ, $w_{[\nu}\,W_{\mu]\lambda}$ the *similarity class* of w. There are two cases:

In case I w does not belong to the domain of W (cf. I § 5). The λ-domain of the set w_λ, $W_{\mu\lambda}$ contains both the domain of W and the vector w outside it, hence $K = 2\varrho+1$ and $k = K$. $\overset{2\varrho}{I}$ and $\overset{2\varrho+1}{I}$ are both

---

1) CARTAN 1922, 1, p. 72.

2) NAGABHUSHANAM 1949, 1.

3) We give here only a brief summary of results. Cf. for a detailed treatment for instance P. P. 1949, 1, Ch. IV, also for literature.

simple multivectors and gradient products because their rotations vanish (cf. II § 6). Hence $\overset{2\varrho}{I}$ is $X_{n-2\varrho}$-forming and $\overset{K}{I}$ is X_{n-K}-forming and the corresponding complete systems are

$$(S_1'): [\overset{2\varrho}{I}\bar{f}] = 0 \tag{7.3}$$

$$(S_4'): [\overset{k}{I}\bar{f}] = 0; \qquad (k = K)\ ^{1)}. \tag{7.4}$$

The $X_{n-2\varrho}$'s are called the *supports of rotation* of w and the X_{n-K}'s the *supports* and also the *characteristics* of w. Because $\overset{2\varrho}{I}$ and $\overset{K}{I}$ are both gradient products and $\overset{2\varrho}{I}$ lies in the domain of $\overset{K}{I}$, there exists a scalar p such that

$$\overset{K}{I} = [w\overset{2\varrho}{I}] = [\bar{p}\overset{2\varrho}{I}]. \tag{7.5}$$

In case II w belongs to the domain of W, hence $K = 2\varrho$ and $k = 2\varrho - 1$. To prove this take the coordinate system at a definite point such that $w = \overset{1}{e}$, $W = [\overset{1}{e}\overset{2}{e}] + \cdots + [\overset{k}{e}\overset{k+1}{e}]$. The rotation of $\overset{k}{I}$ does not vanish because $\overset{2\varrho}{I} = 2D\overset{k}{I}$. But now it follows from (5.16a) that $\overset{k}{I}$ is X_{n-k}-forming. $\overset{2\varrho}{I}$ is simple and a gradient product, hence $X_{n-2\varrho}$-forming, because its rotation vanishes. The corresponding complete systems are

$$(S_4'): [\overset{k}{I}\bar{f}] = 0 \tag{7.6}$$

$$(S_1'): [\overset{2\varrho}{I}\bar{f}] = 0. \tag{7.7}$$

The $X_{n-2\varrho}$'s are called the *supports of rotation* and (because $K = 2\varrho$) also the *supports* of w and the X_{n-k}'s are called the *characteristics* of w. Because $\overset{k}{I}$ lies in the domain of $\overset{2\varrho}{I}$ there must exist a vector u_λ such that $\overset{2\varrho}{I} = [u\overset{k}{I}]$. Moreover we can prove that it is always possible to make u_λ a gradient. Let

$$\overset{\mathfrak{a}}{F}(\xi^\varkappa) = \text{const}; \qquad \mathfrak{a} = \bar{1}, \ldots, \bar{\mathfrak{k}} \tag{7.8}$$

be the equations of the X_{n-k}'s formed by $\overset{k}{I}$. Then every tangent E_{n-k} of them has the $(n-k)$-direction of $\overset{k}{I}$. Hence there exists an equation of the form

$$\overset{k}{I} = \tau[\overset{\bar{1}}{F}\ldots\overset{\bar{\mathfrak{k}}}{F}] \tag{7.9}$$

from which it follows that

$$\overset{2\varrho}{I} = 2[\bar{\tau}\overset{\bar{1}}{F}\ldots\overset{\bar{\mathfrak{k}}}{F}] = 2[\overline{\log\tau}\,\overset{k}{I}]. \tag{7.10}$$

---

1) As in P. P. 1949, 1, IV § 2 we denote the complete systems (7.3, 4) by S_1' and S_4' and their adjoint systems by S_1 and S_4.

Note that in both cases 2ϱ is the greatest even number $\leq K$ and k the greatest odd number $\leq K$.

Transformations of the form

$$'w_\lambda = \sigma w_\lambda; \qquad \sigma \neq 0 \tag{7.11}$$

are called *similarity transformations* of the field w. Such a transformation transforms W into

$$'W = \sigma W + 2[\bar{\sigma} w] \tag{7.12}$$

and accordingly $\overset{2\varrho}{I}$ and $\overset{k}{I}$ into

$$'\overset{2\varrho}{I} = \sigma^\varrho \overset{2\varrho}{I} + 2\varrho\,\sigma^{\varrho-1}[\bar{\sigma}\overset{2\varrho-1}{I}] \tag{7.13}$$

$$'\overset{k}{I} = \sigma^{\frac{k+1}{2}} \overset{k}{I}. \tag{7.14}$$

Hence k *and the characteristics are invariant for similarity transformations.*

In case I $'\overset{K}{I} \neq 0$ and $'\overset{K+2}{I} = 0$. For $'\overset{K+1}{I}$ we get

$$'\overset{K+1}{I} = (2\varrho + 2)\,\sigma^\varrho [\bar{\sigma}\overset{K}{I}]. \tag{7.15}$$

If σ is a solution of (S_4') we get $'\overset{K+1}{I} = 0$, but if it is not a solution, $'\overset{K+1}{I}$ does not vanish and we pass from case I to case II. Of course this latter possibility can never occur for $K = n$.

In case II $'\overset{k}{I} \neq 0$ and $'\overset{k+2}{I} = 0$. From (7.13) we see that $'\overset{2\varrho}{I}$ vanishes if and only if σ is a solution of the non homogeneous linear equation

$$2\varrho\,[\overline{\log\sigma}\,\overset{2\varrho-1}{I}] = -\overset{2\varrho}{I}. \tag{7.16}$$

This equation implies that σ is a solution of (S_1') and not a solution of (S_4'), but these conditions are only necessary and not sufficient. Comparing (7.16) with (7.10), it follows that $\sigma = \tau^{-\frac{1}{\varrho}}$ is a solution of (7.16). Collecting results we see that by means of a similarity transformation it is always possible to pass from case II to case I with $'K = K - 1$ and for $K < n$ also from case I to case II with $'K = K + 1$.

A transformation of the form

$$'w_\lambda = w_\lambda + \partial_\lambda u; \qquad (u \text{ not constant}) \tag{7.17}$$

is called a *gradient transformation* of the field w. Such a transformation leaves W invariant and accordingly transforms $\overset{2\varrho}{I}$ and $\overset{k}{I}$ into

$$'\overset{2\varrho}{I} = \overset{2\varrho}{I} \tag{7.18}$$

$$'\overset{k}{I} = \overset{k}{I} + [\bar{u}\overset{k-1}{I}]. \tag{7.19}$$

Hence 2ϱ *and the supports of rotation are invariant for gradient transformations.*

In case I $'\overset{2\varrho}{I} \neq 0$ and $'\overset{2\varrho+2}{I} = 0$. Because $K = k$ we get $'\overset{K}{I} = 0$ if and only if u is a solution of the non homogeneous linear equation

$$(7.20) \qquad [\bar{u}\overset{2\varrho}{I}] = -\overset{K}{I}.$$

This equation implies that u is a solution of (S_4') and not a solution of (S_1'), but these conditions are only necessary and not sufficient. In this case it follows $'K = 2\varrho = K - 1$ so that we get from case I to case II. Comparing (7.20) with (7.5) it follows that $u = -\varrho$ is a solution of (7.20).

In case II $'\overset{K}{I} \neq 0$ and $'\overset{K+2}{I} = 0$. For $'\overset{K+1}{I}$ we get

$$(7.21) \qquad '\overset{K+1}{I} = [\bar{u}\overset{K}{I}].$$

If u is a solution of (S_1') we get $'\overset{k+1}{I} = 0$ but if u is not a solution, $'\overset{k+1}{I}$ does not vanish and we pass from case II to case I. Of course this latter possibility can never occur for $K = n$. Collecting results we see that by means of a gradient transformation it is always possible to pass from case I to case II with $'K = K - 1$ and, for $K < n$, also to pass from case II to case I with $'K = K + 1$.

Here is a table of results[1]:

Similarity transformations

$$(7.22) \quad K = 2\varrho + 1 \begin{cases} 'K = K + 1; & \sigma \text{ is not a solution of } (S_4') \text{ (never for } K = n) \\ 'K = K; & \sigma \text{ is a solution of } (S_4') \text{ (always for } K = n) \end{cases}$$

$$(7.23) \quad K = 2\varrho \qquad \begin{cases} 'K = K; & \sigma \text{ is not a solution of (7.16)} \\ 'K = K - 1; & \sigma \text{ is a solution of (7.16)} \end{cases}$$

Gradient transformations

$$(7.24) \quad K = 2\varrho \qquad \begin{cases} 'K = K + 1; & u \text{ is not a solution of } (S_1') \text{ (never for } K = n) \\ 'K = K; & u \text{ is a solution of } (S_1') \text{ (always for } K = n) \end{cases}$$

$$(7.25) \quad K = 2\varrho + 1 \begin{cases} 'K = K; & u \text{ is not a solution of (7.20)} \\ 'K = K - 1; & u \text{ is a solution of (7.20).} \end{cases}$$

If a field w is given, K can be found by means of differentiations and algebraic operations only. Then by alternately applying a gradient transformation and a similarity transformation the class can be reduced

[1]) FROBENIUS 1879, 1.

to zero after K steps. But this proves that w can always be written in the form

$$(7.26)\quad \begin{cases} w_\lambda = \varepsilon \overset{0}{s}_\lambda + \overset{1}{z}\overset{1}{s}_\lambda + \cdots + \overset{\varrho}{z}\overset{\varrho}{s}_\lambda; \quad \varepsilon = \begin{cases} 1 \text{ for } K \text{ odd} \\ 0 \text{ for } K \text{ even} \end{cases} \\ \overset{\mathfrak{a}}{s}_\lambda \overset{\text{def}}{=} \partial_\lambda \overset{\mathfrak{a}}{s}; \qquad \mathfrak{a} = 1, \ldots, \varrho. \end{cases}$$

This form is called a *canonical form* for w. There can not be a canonical form with less then $\varrho + \varepsilon$ terms, because in that case we see, by forming the rotation, that the class had to be smaller than K. The K variables occurring in (7.26) must be independent functions of the $\xi^\varkappa$. Because otherwise w could be expressed in terms of less than K variables and that would imply that $\overset{K}{I} = 0$. But these K variables are not uniquely determined. For instance for K odd

$$(7.27)\qquad w_\lambda = (\overset{0}{s}_\lambda + \partial_\lambda(\overset{1}{z}\overset{1}{s})) - \overset{1}{s}\overset{1}{z}_\lambda + \overset{2}{z}\overset{2}{s}_\lambda + \cdots + \overset{\varrho}{z}\overset{\varrho}{s}_\lambda$$

is another canonical form for w.

Taking the K variables in (7.26) as the first K coordinates of $(\varkappa)$ we get

$$(7.28)\qquad \begin{cases} w_\lambda = \overset{1}{e}_\lambda + \xi^2 \overset{3}{e}_\lambda + \cdots + \xi^{K-1}\overset{K}{e}_\lambda \quad \text{for } K \text{ odd} \\ w_\lambda = \xi^1 \overset{2}{e}_\lambda + \cdots + \xi^{K-1}\overset{K}{e}_\lambda \quad \text{for } K \text{ even.} \end{cases}$$

We now describe the process leading to a canonical form in more detail. If K is odd we have first to find a gradient transformation $'w_\lambda = w_\lambda - \partial_\lambda \overset{0}{s}$, diminishing K by 1. According to (7.25) $-\overset{0}{s}$ is an arbitrary solution of the non-homogeneous equation (7.20). If K is even we have first to find a similarity transformation $'w_\lambda = \overset{1}{z}{}^{-1} w_\lambda$ diminishing K by 1. According to (7.23) $\overset{1}{z}{}^{-1}$ is a solution of the non-homogeneous equation (7.16). Now it seems that (7.20) and (7.16) are each a set of $\binom{n}{K}$ equations instead of one equation. But this is not true. Because this is a property independent of the choice of the coordinate system we may suppose that $(\varkappa)$ is chosen such that (7.28) holds. Then for K odd (7.20) takes the form

$$(7.29)\qquad \partial_1 \overset{0}{s} = 1; \qquad \partial_\mathfrak{b} \overset{0}{s} = 0; \qquad \mathfrak{b} = K+1, \ldots, n$$

with the general solution

$$(7.30)\qquad \overset{0}{s} = \xi^1 + \Phi(\xi^2, \ldots, \xi^K)$$

with the arbitrary function Φ. If K is even (7.16) takes the form

$$(7.31)\qquad \begin{cases} 1 - \overset{1}{z}{}^{-1}(\xi^1 \partial_1 \overset{1}{z} + \xi^3 \partial_3 \overset{1}{z} + \cdots + \xi^{K-1}\partial_{K-1}\overset{1}{z}) = 0; \\ \partial_\mathfrak{b} \overset{1}{z} = 0; \qquad \mathfrak{b} = K+1, \ldots, n \end{cases}$$

whose general solution is any function of $\xi^1, \ldots, \xi^K$ which is homogeneous of degree one in $\xi^1, \xi^3, \ldots, \xi^{K-1}$. Of course the equations (7.29) and (7.31) do not require any integrations. But we must remember that in order to get the special coordinate system we need a canonical form and that this form can not be formed without integrations.

If w is X_m-enveloping, there exists in the region considered an enveloped normal system of X_m's with equations of the form

$$F^x(\xi^\varkappa) = c^x; \qquad x = m+1, \ldots, n \tag{7.32}$$

and because w_λ is, at every point, tangent to the local X_m, it can be expressed linearly in the $\partial_\lambda F^x$

$$w_\lambda - \alpha_x \partial_\lambda F^x; \qquad x = m+1, \ldots, n. \tag{7.33}$$

From this equation it follows that $K = 2\varrho + \varepsilon \leq 2(n-m)$ or $m \leq n - \varrho - \varepsilon$. But from (7.26) we see that

$$\varepsilon \overset{0}{s} = \varepsilon \overset{0}{c}; \quad \overset{1}{s} = \overset{1}{c}; \ldots; \quad \overset{\varrho}{s} = \overset{\varrho}{c} \tag{7.34}$$

represents an enveloped normal system of $X_{n-\varrho-\varepsilon}$'s. Hence $\nu \overset{\text{def}}{=} n - \varrho - \varepsilon$ is the maximum value of m. This proves

Every field w_λ of class $K = 2\varrho + \varepsilon$ is X_ν-enveloping but not $X_{\nu+1}$-enveloping; $\nu = n - \varrho - \varepsilon$.

If (7.28) holds we have

$$\left\{\begin{array}{ll} \varepsilon = 1 & \varepsilon = 0 \\ \overset{K}{I} = 2^p p! \, [\xi^1 \ldots \xi^K] & \overset{2\varrho}{I} = 2^p p! \, [\xi^1 \ldots \xi^K] \\ \overset{2\varrho}{I} = 2^p p! \, [\xi^2 \ldots \xi^K] & \overset{k}{I} = 2^{p-1}(p-1)! \{\xi^1 [\xi^2 \ldots \xi^K] + \\ & \quad + \xi^3 [\xi^1 \xi^2 \xi^4 \ldots \xi^K] + \cdots \\ & \quad + \xi^{K-1} [\xi^1 \ldots \xi^{K-2} \xi^K]\} \end{array}\right. \tag{7.35}$$

and

(7.36)

for $\varepsilon = 1$

supports = characteristics: $\xi^1 = \text{const.}; \ldots; \xi^K = \text{const.}$

supports of rotation: $\xi^2 = \text{const.}; \ldots; \xi^K = \text{const.}$

for $\varepsilon = 0$

supports = supports of rotation: $\xi^1 = \text{const.}; \ldots; \xi^K = \text{const.}$

characteristics: $\begin{cases} \xi^1 : \xi^3 : \ldots : \xi^{K-1} = \text{const.}; \\ \xi^2 = \text{const.}; \xi^4 = \text{const.}; \ldots; \xi^K = \text{const.} \end{cases}$

Exercise.

II 7,1[1]). Prove that the class of

a) $x^1 x^3 dx^2 + x^1 x^2 dx^3 + (x^1 + x^3 x^5) dx^4 + x^3 x^4 dx^5$

b) $x^2 dx^1 + x^1 dx^2 - x^3 x^5 dx^4 - x^3 x^4 dx^5 + x^2 dx^6$

is 4 and 5 respectively.

§ 8. The theorem of STOKES.

In X_n we consider an X_{q+1} with the equations

$$\xi^{q+2} = 0; \ldots; \quad \xi^n = 0 \tag{8.1}$$

with respect to some coordinate system $(\varkappa)$ and in this X_{q+1} we consider a $(q+1)$-dimensional volume τ_{q+1} and its boundary τ_q, such that from the curves on X_{q+1}

$$\xi^1 = \text{const.}; \ldots; \xi^{\mathfrak{a}-1} = \text{const.}; \xi^{\mathfrak{a}+1} = \text{const.}; \ldots; \xi^{q+1} = \text{const.} \tag{8.2}$$

for every value of $\mathfrak{a}$ from 1 up to $q+1$ each intersects τ_q in at most two points. Let an *inner* orientation be fixed on τ_{q+1} by $\underset{1}{e}^{\varkappa}, \ldots, \underset{q+1}{e}^{\varkappa}$ in this order. Then the volume element $df^{\varkappa_1 \ldots \varkappa_{q+1}}$ with this orientation may be written in the form

$$df^{\varkappa_1 \ldots \varkappa_{q+1}} = (q+1)! \underset{1}{d} \xi^{[\varkappa_1} \ldots \underset{q+1}{d} \xi^{\varkappa_{q+1}]}. \tag{8.3}$$

If the $q+1$ vectors $\underset{1}{d} \xi^{\varkappa}, \ldots, \underset{q+1}{d} \xi^{\varkappa}$ are chosen in such a way that

$$\left\{ \begin{array}{llll} \underset{1}{d} \xi^1 = d\xi^1; & \underset{1}{d} \xi^2 = 0; \ldots; & \underset{1}{d} \xi^n = 0 & \\ \underset{2}{d} \xi^1 = 0; & \underset{2}{d} \xi^2 = d\xi^2; & \underset{2}{d} \xi^3 = 0; \ldots; & \underset{2}{d} \xi^n = 0 \\ \vdots & & & \\ \underset{q+1}{d} \xi^1 = 0; \ldots; & \underset{q+1}{d} \xi^q = 0; & \underset{q+1}{d} \xi^{q+1} = d\xi^{q+1}; & \underset{q+1}{d} \xi^{q+2} = 0; \ldots; \underset{q+1}{d} \xi^n = 0 \end{array} \right. \tag{8.4}$$

we have

$$df^{1 \ldots q+1} \overset{*}{=} d\xi^1 \ldots d\xi^{q+1} \tag{8.5}$$

and all components of $df^{\varkappa_1 \ldots \varkappa_q}$ which have an index $q+2, \ldots, n$ vanish. Let one of the curves (8.2) for $\mathfrak{a} = 1$ intersect the boundary τ_q at the two points

$$\left\{ \begin{array}{l} P_1\colon\ \xi^1 = \underset{1}{\xi}^1 \\ P_2\colon\ \xi^1 = \underset{2}{\xi}^1 \end{array} \right\} \xi^2 = \underset{0}{\xi}^2; \ldots; \xi^{q+1} = \underset{0}{\xi}^{q+1}; \ \underset{2}{\xi}^1 > \underset{1}{\xi}^1. \tag{8.6}$$

---

[1]) CARTAN 1899, 1, p. 259 and 265.

The points of τ_{q+1} satisfying the equations

$$\underset{1}{\xi^1} \leq \xi^1 \leq \underset{2}{\xi^1}; \quad \underset{0}{\xi^a} \leq \xi^a \leq \underset{0}{\xi^a} + d\xi^a; \quad a = 2, \ldots, q+1 \tag{8.7}$$

constitute a $(q+1)$-dimensional tube in τ_{q+1} cutting of a q-dimensional element from τ_q at each of the points P_1 and P_2.

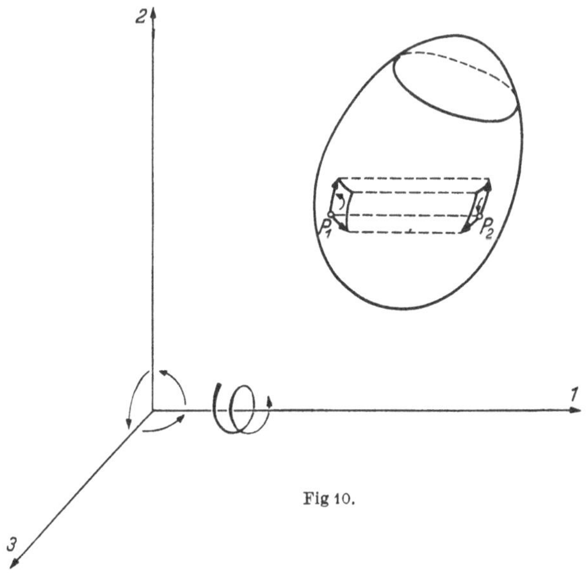

Fig 10.

Fig. 10 illustrates the case $n=3$, $q=2$. For convenience the coordinate-X_1's are drawn in this figure as straight lines but this has no real significance. If we fix now an inner orientation all over τ_q by $\underset{2}{e^\varkappa}, \ldots, \underset{q+1}{e^\varkappa}$ in this order at P_2, then this orientation is opposite to the orientation of $\underset{2}{e^\varkappa}, \ldots, \underset{q+1}{e^\varkappa}$ at P_1. The orientations in τ_{q+1} and τ_q are now chosen in such a way that the direction from a point of τ_{q+1} towards the boundary followed by the orientation of τ_q gives the orientation of τ_{q+1}. Accordingly we have for the q-dimensional element of τ_q at P_2

$$\overset{2}{d} f^{2 \ldots q+1} \stackrel{*}{=} d\xi^2 \ldots d\xi^{q+1} \tag{8.8}$$

and at P_1

$$\overset{1}{d} f^{2 \ldots q+1} \stackrel{*}{=} - d\xi^2 \ldots d\xi^{q+1}. \tag{8.9}$$

After these preliminaries we now consider a covariant q-vector field $v_{\lambda_1 \ldots \lambda_q}$ satisfying the following conditions:

a) the $v_{\lambda_1 \ldots \lambda_q}$ are continuous in τ_{q+1} and on τ_q;
b) the derivatives which occur in $\partial_{[\mu} v_{\lambda_1 \ldots \lambda_q]}$ exist at all points of τ_{q+1};

c) these derivatives are continuous at all points of τ_{q+1} except at the points of a finite number of X_q's.

We consider the integral of $\partial_1 v_{2\ldots q+1}$ over τ_{q+1} (cf. 8.5)

$$\int_{\tau_{q+1}} \partial_1 v_{2\ldots q+1}\, d f^{1\ldots q+1} \overset{*}{=} \int_{\tau_{q+1}} \partial_1 v_{2\ldots q+1}\, d\xi^1 \ldots d\xi^{q+1}. \tag{8.10}$$

As a consequence of the conditions for the field $v_{\lambda_1\ldots\lambda_q}$, the part of the integral over the tube is

$$\left\{ \begin{array}{l} (\overset{2}{v}_{2\ldots q+1} - \overset{1}{v}_{2\ldots q+1})\, d\xi^2 \ldots d\xi^{q+1} \\ \qquad \overset{*}{=} \overset{2}{v}_{2\ldots q+1}\, \overset{2}{d} f^{2\ldots q+1} + \overset{1}{v}_{2\ldots q+1}\, \overset{1}{d} f^{2\ldots q+1} \end{array} \right. \tag{8.11}$$

where $\overset{2}{v}_{2\ldots q+1}$ and $\overset{1}{v}_{2\ldots q+1}$ are the values of $v_{2\ldots q+1}$ at P_2 and at P_1 respectively. Hence

$$\int_{\tau_{q+1}} \partial_1 v_{2\ldots q+1}\, d f^{1\ldots q+1} \overset{*}{=} \int_{\tau_q} v_{2\ldots q+1}\, d f^{2\ldots q+1}. \tag{8.12}$$

By using the other q sets of coordinate-X_1's, equations similar to (8.12) but with $\partial_2, \ldots, \partial_{q+1}$ instead of ∂_1, can be derived. By addition we get

$$\boxed{\int_{\tau_{q+1}} \partial_{[\mu} v_{\lambda_1\ldots\lambda_q]}\, d f^{\mu\lambda_1\ldots\lambda_q} = \int_{\tau_q} v_{\lambda_1\ldots\lambda_q}\, d f^{\lambda_1\ldots\lambda_q}} \tag{8.13}$$ [1]).

In order to get rid of the condition of intersection imposed on the form of τ_q we consider the case where τ_{q+1} can be split up into a finite number of τ_{q+1}'s such that the condition of intersection holds for the boundary of each of them. Then (8.13) holds for each of these τ_{q+1}'s and by summation of the integrals all integrals over boundaries between two adjacent τ_{q+1}'s cancel each other out, because the orientations in the common boundary are opposite. Hence (8.13) is proved now for this more general case.

(8.13) is a form of the generalized theorem of STOKES (in Germany also named after GAUSS and in France after OSTROGRADSKI). Many other forms can be deduced from it. First $\tilde{\mathfrak{v}}^{\varkappa_1\ldots\varkappa_p}$, $\tilde{\mathfrak{f}}_{\lambda_1\ldots\lambda_{p-1}}$ and $\tilde{\mathfrak{f}}_{\lambda_1\ldots\lambda_p}$ $(p = n - q)$ can be introduced instead of $v_{\lambda_1\ldots\lambda_q}$, $f^{\mu\varkappa_1\ldots\varkappa_q}$ and $f^{\varkappa_1\ldots\varkappa_q}$. Then we get four other forms

$$p \int_{\tau_{q+1}} \partial_{[\mu} v_{\lambda_1\ldots\lambda_q}\, d\tilde{\mathfrak{f}}_{\varkappa_1\ldots\varkappa_{p-1}]} = \int_{\tau_q} v_{[\lambda_1\ldots\lambda_q}\, d\tilde{\mathfrak{f}}_{\varkappa_1\ldots\varkappa_{p-1}\mu]} \tag{8.14}$$

$$p \int_{\tau_{q+1}} \partial_\mu \tilde{\mathfrak{v}}^{\lambda_1\ldots\lambda_{p-1}\mu}\, d\tilde{\mathfrak{f}}_{\lambda_1\ldots\lambda_{p-1}} = \int_{\tau_q} \tilde{\mathfrak{v}}^{\lambda_1\ldots\lambda_p}\, d\tilde{\mathfrak{f}}_{\lambda_1\ldots\lambda_p} \tag{8.15}$$

[1]) POINCARÉ 1887, 1; 1895, 1; BROUWER 1906, 1; 2; 1919, 1; WEITZENBÖCK 1923, 1, p. 398ff., also for literature; WEYSSENHOFF 1937, 1; SCHOUTEN and v. DANTZIG 1940, 1; LEVI CIVITA 1940, 1.

$$(8.16)\quad \begin{cases} \dfrac{n-q}{q+1} \displaystyle\int_{\tau_{q+1}} \partial_\mu \tilde{\mathfrak{v}}^{[\lambda_1 \ldots \lambda_{p-1}|\mu|} \, d\, f^{\lambda_p \varkappa_1 \ldots \varkappa_q]} \\ \qquad = \displaystyle\int_{\tau_{q+1}} \partial_\mu \tilde{\mathfrak{v}}^{[\lambda_1 \ldots \lambda_p} \, d\, f^{|\mu| \varkappa_1 \ldots \varkappa_q]} = \int_{\tau_q} \tilde{\mathfrak{v}}^{[\lambda_1 \ldots \lambda_p} \, d\, f^{\varkappa_1 \ldots \varkappa_q]} \end{cases}$$

which all have the same geometrical signification as (8.13). If the elements of τ_q and τ_{q+1} have *outer* orientation (this is the case that occurs most frequently in physics) and if these orientations are chosen in such a way that the orientation of τ_{q+1} followed by the direction from a point of τ_{q+1} towards the boundary gives the orientation of τ_q we get the following five equations for a field $\tilde{v}_{\lambda_1 \ldots \lambda_q}$

$$(8.17)\quad \int_{\tau_{q+1}} \partial_{[\mu} \tilde{v}_{\lambda_1 \ldots \lambda_q]} \, d\, \tilde{f}^{\mu \lambda_1 \ldots \lambda_q} = \int_{\tau_q} \tilde{v}_{\lambda_1 \ldots \lambda_q} \, d\, \tilde{f}^{\lambda_1 \ldots \lambda_q}$$

$$(8.18)\quad p \int_{\tau_{q+1}} \partial_{[\mu} \tilde{v}_{\lambda_1 \ldots \lambda_q} \, d\, \mathfrak{f}_{\varkappa_1 \ldots \varkappa_{p-1}]} = \int_{\tau_q} \tilde{v}_{[\lambda_1 \ldots \lambda_q} \, d\, \mathfrak{f}_{\varkappa_1 \ldots \varkappa_{p-1} \mu]}$$

$$(8.19)\quad p \int_{\tau_{q+1}} \partial_\mu \mathfrak{v}^{\lambda_1 \ldots \lambda_{p-1} \mu} \, d\, \mathfrak{f}_{\lambda_1 \ldots \lambda_{p-1}} = \int_{\tau_q} \mathfrak{v}^{\lambda_1 \ldots \lambda_p} \, d\, \mathfrak{f}_{\lambda_1 \ldots \lambda_p}$$

$$(8.20)\quad \begin{cases} \dfrac{n-q}{q+1} \displaystyle\int_{\tau_{q+1}} \partial_\mu \mathfrak{v}^{[\lambda_1 \ldots \lambda_{p-1}|\mu|} \, d\, \tilde{f}^{\lambda_p \varkappa_1 \ldots \varkappa_q]} \\ \qquad = \displaystyle\int_{\tau_{q+1}} \partial_\mu \mathfrak{v}^{[\lambda_1 \ldots \lambda_p} \, d\, \tilde{f}^{|\mu| \varkappa_1 \ldots \varkappa_q]} = \int_{\tau_q} \mathfrak{v}^{[\lambda_1 \ldots \lambda_p} \, d\, \tilde{f}^{\varkappa_1 \ldots \varkappa_q]}. \end{cases}$$ [1]

All integrals which have occurred up till now in this section have been scalars or n-vector Δ-densities of weight ± 1. In fact only these quantities can be added even if they belong to different points. But other forms of STOKES' theorem may arise if we restrict ourselves to less general transformations of coordinates. For instance, if only those coordinate transformations with $\Delta > 0$ are allowed, W-scalars also can be added and thus besides (8.13) we have in that case the invariant formula

$$(8.21)\quad \int_{\tau_{q+1}} \partial_{[\mu} \tilde{v}_{\lambda_1 \ldots \lambda_q]} \, d\, f^{\mu \lambda_1 \ldots \lambda_q} = \int_{\tau_q} \tilde{v}_{\lambda_1 \ldots \lambda_q} \, d\, f^{\lambda_1 \ldots \lambda_q}.$$

If transformations of G_a only are allowed, that is, if we are in an E_n, also quantities that belong to *different* points can be added, provided that they have the same manner of transformation. Accordingly in E_n the following equations hold

$$(8.22)\quad \int_{\tau_{q+1}} d\, f^{\mu \lambda_1 \ldots \lambda_q} \, \partial_\mu \!-\!\circ\, \Omega = \int_{\tau_q} d\, f^{\lambda_1 \ldots \lambda_q} \!-\!\circ\, \Omega$$

$$(8.23)\quad p \int_{\tau_{q+1}} d\, \tilde{\mathfrak{f}}_{[\lambda_1 \ldots \lambda_{p-1}} \, \partial_{\lambda_p]} \!-\!\circ\, \Omega = \int_{\tau_q} d\, \tilde{\mathfrak{f}}_{\lambda_1 \ldots \lambda_p} \!-\!\circ\, \Omega$$

[1] Cf. T. P. 1951, 1, p. 71 f. for the formulae occurring most frequently in physics.

(8.24) $$\int_{\tau_{q+1}} d\tilde{f}^{\mu\lambda_1\ldots\lambda_q}\,\partial_\mu -\!\circ\, \Omega = \int_{\tau_q} d\tilde{f}^{\lambda_1\ldots\lambda_q} -\!\circ\, \Omega$$

(8.25) $$p\int_{\tau_{q+1}} d\mathfrak{f}_{[\lambda_1\ldots\lambda_{p-1}}\,\partial_{\lambda_p]} -\!\circ\, \Omega = \int_{\tau_q} d\mathfrak{f}_{\lambda_1\ldots\lambda_p} -\!\circ\, \Omega$$

where Ω is a symbol with suppressed indices standing for any quantity and where $-\!\circ$ symbolizes an operation performed on the indices $\lambda_1, \ldots, \lambda_p$ or $\lambda_1, \ldots, \lambda_q$ and the indices of Ω, which is composed of alternation, mixing and transvection[1]).

For instance if

(8.26) $$\mathfrak{v}_\lambda = \partial_\mu \mathfrak{P}^{\mu}_{\cdot\,\lambda}$$

we have in E_3

(8.27) $$\int_{\tau_3} \mathfrak{v}_\lambda\, d\mathfrak{f} = \int_{\tau_2} \mathfrak{P}^{\mu}_{\cdot\,\lambda}\, d\mathfrak{f}_\mu$$

but this equation has no invariant meaning in X_3.

The STOKES integrals may be looked upon from a quite different point of view, which is very important for many investigations. Take for instance (8.13). The $v_{\lambda_1\ldots\lambda_q}$ may be considered as $\binom{n}{q}$ functions of the $\xi^\varkappa$ that need not be components of a q-vector or of any geometric object. Then we are not able to give a corresponding formula for another coordinate system $(\varkappa')$ but the formula (8.13) remains true. This enables us to use STOKES' formulae in all cases where the manner of transformation of a set of variables is not known or not yet known. Of course in this case the integrals are not scalars because transformation of coordinates is not allowed.

§ 9. Anholonomic coordinates (cf. II § 5).

If at every point of X_n we introduce a set of n linearly independent contravariant vectors $\underset{i}{e}^\varkappa$ and the reciprocal set $\overset{h}{e}_\lambda$; $h, i = 1, \ldots, n$, in such a way that these fields are analytic (or of class u) in the region considered, these vectors are contra- and covariant basis vectors of a coordinate system ξ^h if and only if the rotations $\partial_{[\mu}\overset{h}{e}_{\lambda]}$ vanish at all points. In fact $\partial_{[\mu}\overset{h}{e}_{\lambda]} = 0$ are the integrability conditions of the system $\partial_\lambda \overset{h}{\xi} = \overset{h}{e}_\lambda$. If these conditions are *not* satisfied, the system of reference formed by the fields $\underset{i}{e}^\varkappa$, $\overset{h}{e}_\lambda$ is called an *anholonomic* coordinate system (h) [2]) in contradistinction to the coordinate systems used before, that we call

[1]) E I 1935, 1, p. 130.

[2]) Cf. for literature the footnote 1 on p. 78 and Ch. V § 7. Though we often prefer latin running indices for anholonomic coordinate systems, this is not obligatory; also $(\varkappa)$ can be taken anholonomic.

holonomic. The $\underset{i}{e}^{\varkappa}$ determine a *net* of X_1's (cf. II § 4), called the *net of the anholonomic coordinate system*. But in general an anholonomic coordinate system has no coordinate-X_m's, $1 < m < n$. The components of any quantity with respect to (h) can be derived from the components with respect to $(\varkappa)$ by using the intermediate components A_λ^h, $A_i^\varkappa$ of the unity tensor

$$(9.1) \qquad A_\lambda^h \overset{\text{def}}{=} \underset{i}{e}^h \overset{i}{e}_\lambda \overset{*}{=} \overset{h}{e}_\lambda; \qquad A_i^\varkappa \overset{\text{def}}{=} \overset{h}{e}_i \underset{h}{e}^\varkappa \overset{*}{=} \underset{i}{e}^\varkappa$$

just in the same way as if (h) were holonomic. But the components of a more general geometric object (as defined in II § 3) with respect to (h) can not be derived by using A_λ^h, $A_i^\varkappa$ and $\partial_j \overset{\text{def}}{=} A_j^\mu \partial_\mu$ just as if (h) were holonomic. In some way the components with respect to (h) must first be fixed by the introduction of a condition.[1]

From now on we sometimes use the sign $\overset{h}{=}$ in order to emphasize the fact that an equation is only valid with respect to holonomic coordinate systems. We write $(d\xi)^h$ instead of $d\xi^h$ for the components of the linear element $d\xi^\varkappa$ with respect to an anholonomic coordinate system (h) because variables ξ^h do not exist.

The expression

$$(9.2) \qquad \Omega_{ji}^{\cdot\cdot h} \overset{\text{def}}{=} A_{j\,i}^{\mu\lambda} \partial_{[\mu} A_{\lambda]}^h \overset{*}{=} \underset{j}{e}^\mu \underset{i}{e}^\lambda \partial_{[\mu} \overset{h}{e}_{\lambda]}$$

is called the *object of anholonomity*. It is a geometric object all of whose components vanish with respect to all holonomic systems and only with respect to such systems. For two holonomic or anholonomic systems (h) and (h') it follows from this definition that

$$(9.3) \qquad \Omega_{j'i'}^{\cdot\cdot h'} = A_{j'i'h}^{j\,i\,h'} \Omega_{ji}^{\cdot\cdot h} + A_{[j'}^{j} A_{i']}^{i} \partial_j A_i^{h'}$$

(9.2) is a special case of this formula.

If any expression with respect to holonomic coordinates is transformed with respect to an anholonomic system, correction terms appear, all containing the object of anholonomity. For instance[2])

$$(9.4) \quad \partial_{[j} \partial_{i]} p = A_{[j}^{\mu} \partial_{|\mu|} A_{i]}^{\lambda} \partial_\lambda p = (A_{[j}^{\mu} \partial_{|\mu|} A_{i]}^{\lambda}) A_\lambda^h \partial_h p = -\Omega_{ji}^{\cdot\cdot h} \partial_h p.$$

If an X_m in X_n is given by its parametric equations

$$(9.5) \qquad \xi^\varkappa = f^\varkappa(\eta^\alpha); \qquad \alpha, \beta, \gamma, \delta = \dot{1}, \ldots, \dot{m}$$

we may introduce besides $(\varkappa)$ and (α) the anholonomic coordinate systems (h); $h, i, j, k = 1, \ldots, n$ and (a); $a, b, c, d = \dot{1}, \ldots, \dot{m}$ in X_n and X_m respectively. If $'\Omega_{cb}^{\cdot\cdot a}$ is the object of anholonomity of the X_m and if the X_m is rigged, we have

[1]) For an example cf. III § 9. Here the condition for the geometric object Γ is that the equations (III 9.1) hold for an anholonomic coordinate system.

[2]) Cf. Exerc. II 9,1.

(9.6) $B^{jia}_{cbh}\Omega^h_{ji} = B^{\mu\lambda a}_{cbh}\partial_{[\mu}A^h_{\lambda]} = B^a_h\partial_{[c}B^h_{b]} - B^{\mu}_{[c}B^{a}_{|\lambda}\partial_{\mu|}B^{\lambda}_{b]} = B^a_h\partial_{[c}B^h_{b]} + '\Omega^a_{cb}$

where

(9.6a) $B^{\mu}_{c} \overset{\text{def}}{=} B^{\beta}_{c}B^{\mu}_{\beta};\quad B^{j}_{c} \overset{\text{def}}{=} B^{\beta}_{c}B^{\mu}_{\beta}A^{j}_{\mu};\quad B^{a}_{\lambda} \overset{\text{def}}{=} B^{a}_{\alpha}B^{\alpha}_{\lambda};\quad B^{a}_{i} \overset{\text{def}}{=} B^{a}_{\alpha}B^{\alpha}_{\lambda}A^{\lambda}_{i}.$

If in this case we choose at all points of X_m the first m vectors $\underset{i}{e}^{\varkappa}$ coinciding with the vectors $\underset{b}{e}^{\alpha}$ and the last $n-m$ vectors $\underset{i}{e}^{\varkappa}$ in the E_{n-m} of the rigging, we have, taking $a, b, c, d = 1, \ldots, \mathrm{m}$ and $x, y, z, u = \mathrm{m}+1, \ldots, \mathrm{n}$

$$(9.7)\quad \begin{cases} A^h_\lambda \overset{*}{=} \overset{h}{e}_\lambda;\quad A^{\varkappa}_{i} \overset{*}{=} \underset{i}{e}^{\varkappa};\quad B^{a}_{\beta} \overset{*}{=} \overset{a}{e}_{\beta};\quad B^{\alpha}_{b} \overset{*}{=} \underset{b}{e}^{\alpha};\quad B^{j}_{c} \overset{*}{=} \underset{c}{e}^{\mu}A^{j}_{\mu} = \underset{c}{e}^{j} \overset{*}{=} \delta^{j}_{c};\\ B^{\mu}_{c} \overset{*}{=} \underset{c}{e}^{\beta}B^{\mu}_{\beta} = \underset{c}{e}^{\mu};\quad B^{a}_{\lambda} \overset{*}{=} \overset{a}{e}_{\alpha}B^{\alpha}_{\lambda} = \overset{a}{e}_{\lambda};\quad B^{a}_{i} \overset{*}{=} \overset{a}{e}_{\lambda}A^{\lambda}_{i} = \overset{a}{e}_{i} \overset{*}{=} \delta^{a}_{i}, \end{cases}$$

hence

$$(9.8)\qquad '\Omega^a_{cb} \overset{*}{=} \Omega^a_{cb}.$$

Anholonomic coordinate systems are frequently used when dealing with an X^m_n in X_n[1]). If the X^m_n is rigged it is often convenient to take at each point the $\underset{b}{e}^{\varkappa}$; $a, b, c, d = 1, \ldots, \mathrm{m}$ in the tangent E_m and the $\underset{y}{e}^{\varkappa}$; $x, y, z, u = \mathrm{m}+1, \ldots, \mathrm{n}$ in the tangent E_{n-m}. For this choice we prove that the X^m_n is tangent to $\infty^{n-m}\, X_m$'s if and only if $\Omega^x_{cb} \overset{*}{=} 0$. In fact, the fields $\underset{b}{e}^{\varkappa}$ are X_m-forming if and only if the $\underset{[c}{e}^{\mu}\,\partial_{|\mu|}\underset{b]}{e}^{\varkappa}$ depend linearly on the $\underset{b}{e}^{\varkappa}$, that is if

$$(9.9)\qquad 0 = \underset{[c}{e}^{\mu}\left(\partial_{|\mu|}\underset{b]}{e}^{\lambda}\right)\overset{x}{e}_{\lambda} = -\underset{[c}{e}^{\mu}\underset{b]}{e}^{\lambda}\partial_{\mu}\overset{x}{e}_{\lambda} \overset{*}{=} -\Omega^x_{cb}.$$

All other applications of anholonomic coordinates will be postponed until after the introduction of a linear connexion in Ch. III.

Besides the Ω^h_{ji}, components of another kind are often used, these are the intermediate components ($(\varkappa)$ holonomic)

$$(9.10)\qquad \Omega^h_{\mu\lambda} \overset{\text{def}}{=} A^{ji}_{\mu\lambda}\Omega^h_{ji} = \partial_{[\mu}A^h_{\lambda]} \overset{*}{=} \partial_{[\mu}\overset{h}{e}_{\lambda]}.$$

The n objects $\Omega^{(h)}_{\mu\lambda}$ resulting from strangling the index h are bivectors.

Exercises.

II 9,1. Prove the formulae for $(\varkappa)$ holonomic

II 9,1 α) $\qquad \partial_{[\mu}w_{\lambda]} = A^{ji}_{\mu\lambda}(\partial_{[j}w_{i]} + w_h\Omega^h_{ji});$

II 9,1 β) $\qquad \partial_\mu \mathfrak{v}^\mu = \Delta(\partial_j\mathfrak{v}^j - 2\Omega^j_{ji}\mathfrak{v}^i);\qquad \Delta = \mathrm{Det}\,(A^h_\lambda)$

where w_λ is a covariant vector and $\mathfrak{v}^{\varkappa}$ a vector Δ-density of weight $+1$.

II 9,2[2]). Prove the identity

II 9,2 α) $\qquad \partial_{[k}\Omega^h_{ji]} - 2\Omega^l_{[kj}\Omega^h_{i]l} = 0.$

1) Cf. for applications to anholonomic dynamical systems and references to literature T. P. 1951, 1, Ch. VIII.

2) Cf. (II 12.10b).

§ 10. The LIE derivative.[1], [2]

Let the points of a region R be subject to a point transformation

$$\eta^{\varkappa} = f^{\varkappa}(\xi^{\lambda}); \qquad \mathrm{Det}\,(\partial_{\lambda} f^{\varkappa}) \neq 0. \tag{10.1}$$

The functions $f^{\varkappa}$ are supposed to be analytic in the region considered and to establish a one to one correspondence between the points of R and those of some other region R'. Another coordinate system $(\varkappa')$ is now introduced such that the transform in R' has the same coordinates with respect to $(\varkappa')$ as the original point in R had with respect to $(\varkappa)$. Hence if

$$\xi^{\varkappa} = \varphi^{\varkappa}(\eta^{\lambda}) \tag{10.2}$$

is the inverse of (10.1) and if $\eta^{\varkappa}$ and $\eta^{\varkappa'}$ are the coordinates with respect to $(\varkappa)$ and to $(\varkappa')$ of the transform of $\xi^{\varkappa}$, we have

$$\eta^{\varkappa'} = \delta^{\varkappa'}_{\varkappa}\, \xi^{\varkappa} = \delta^{\varkappa'}_{\varkappa}\, \varphi^{\varkappa}(\eta^{\lambda}). \tag{10.3}$$

Accordingly the transformation of $\xi^{\varkappa}$ into $\xi^{\varkappa'}$ and vice versa is given by the equations

$$\xi^{\varkappa'} = \delta^{\varkappa'}_{\varkappa}\, \varphi^{\varkappa}(\xi^{\lambda}); \qquad \xi^{\varkappa} = f^{\varkappa}(\xi^{\lambda'}). \tag{10.4}$$

This process is called the *dragging along*[3] *of the coordinate system* $(\varkappa)$ by the point transformation (10.1) and $(\varkappa')$ is called the *coordinate system dragged along*. The dragging along of a coordinate system $(\varkappa)$ can be interpreted geometrically in the following way. Let the region considered be filled up with gelatine and let a coordinate net of $(\varkappa)$ with sufficiently small meshes be made visible by red curves in the gelatine. Then, if the material points of the gelatine suffer the point transformation (10.1), the red curves in their new position form a coordinate net of the system dragged along $(\varkappa')$.

Now let the components of a geometric object Φ_{Λ} as defined in II § 3 be analytic in R. Then we form a field $\overset{m}{\Phi}_{\Lambda}$ in R' whose components $\overset{m}{\Phi}_{\Lambda'}$ with respect to $(\varkappa')$ at each point of R' are equal to the Φ_{Λ} at the corresponding point of R. This process is called the *dragging along of a field* by the point transformation (10.1) and the field $\overset{m}{\Phi}_{\Lambda}$ is called the

---

[1]) This section was written in collaboration with A. NIJENHUIS. For literature cf. SLEBODZINSKI 1931, 1; SCHOUTEN and v. KAMPEN 1933, 1; E I 1935, 1, p. 142; E II 1938, 2, p. 161; P. P. 1949, 1, Ch. II; T. P. 1951, 1, Ch. IV; NIJENHUIS 1952, 1.

[2]) The LIE derivative is defined here for an X_n. It can be defined for more general spaces for instance path-space, that is an X_n in which a system of curves is given such that through any two points there passes one and only one curve of the system. See YANO 1945, 1; KOSAMBI 1952, 1.

[3]) "Mitschleppen" in SCHOUTEN and v. KAMPEN 1933, 1 and E I 1935, 1.

field dragged along[1]). The dragging along of a geometric object can be interpreted geometrically. Every geometric object as defined in II § 3 can be represented by some figure or figures in X_n, for instance a contravariant vector in a point by an arrow on an infinitesimal scale. Let the points of these figures be marked blue in the gelatine filling up the region. Then, if the gelatine is transformed, the transformed figures represent the object dragged along.

From the definition of dragging along it follows that (cf. II § 3)

$$\overset{m}{\Phi}_{\Lambda'}\{\eta\} = \delta^{\Lambda}_{\Lambda'}\Phi_{\Lambda}\{\xi\}, \tag{10.5}$$

hence, making use of (3.1b)

$$\begin{cases} \overset{m}{\Phi}_{\Lambda}\{\eta\} = F_{\Lambda}(\overset{m}{\Phi}_{M'}\{\eta\}, \eta^{\varkappa'}, \eta^{\varkappa}, A^{\varkappa}_{\lambda'}\{\eta\}, \partial_{\mu'} A^{\varkappa}_{\lambda'}\{\eta\}, \ldots) \\ \qquad = F_{\Lambda}(\Phi_M\{\xi\}, \xi^{\varkappa}, f^{\varkappa}\{\xi\}, \partial_{\lambda} f^{\varkappa}\{\xi\}, \partial_{\mu}\partial_{\lambda} f^{\varkappa}\{\xi\}, \ldots). \end{cases} \tag{10.6}$$

If the transformation is infinitesimal: $\xi^{\varkappa} \to \xi^{\varkappa} + v^{\varkappa}\, dt$, we use the term *dragging along over* $v^{\varkappa}\, dt$. The process of dragging along of a contravariant vector field $u^{\varkappa}\, ds$ over $v^{\varkappa}\, dt$ is illustrated in Fig. 11. The vector $\overset{m}{u}{}^{\varkappa}\, ds$ in $\xi^{\varkappa} + v^{\varkappa}\, dt$ results from the displacement of beginning-point and endpoint of the vector $u^{\varkappa}\, ds$ in $\xi^{\varkappa}$ over $v^{\varkappa}\, dt$. In general $\overset{m}{u}{}^{\varkappa}\, ds$ and $u^{\varkappa}\, ds$ in $\xi^{\varkappa} + v^{\varkappa}\, dt$ do not coincide and so there arises a pentagon.

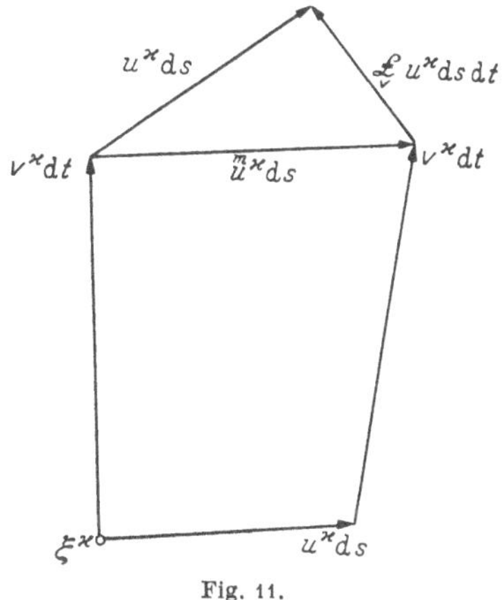

Fig. 11.

We now consider a vector field $v^{\varkappa} = \psi^{\varkappa}(\xi^{\nu})$ and the system of n ordinary differential equations

$$\frac{d\eta^{\varkappa}}{dt} = \psi^{\varkappa}(\eta^{\nu}) \tag{10.7}$$

with the independent variable t and the unknown variables $\eta^{\varkappa}$, with functions $\psi^{\varkappa}$ analytic in the region considered and with the initial conditions

$$\eta^{\varkappa} = \xi^{\varkappa} \quad \text{for } t = 0. \tag{10.8}$$

The solution is

$$\begin{cases} \eta^{\varkappa} = \xi^{\varkappa} + t\, v^{\varkappa} + \dfrac{t^2}{2!} v^{\mu}\partial_{\mu} v^{\varkappa} + \dfrac{t^3}{3!}(v^{\mu}\partial_{\mu})^2 v^{\varkappa} + \cdots \\ \quad = \left(1 + t\, v^{\mu}\partial_{\mu} + \dfrac{t^2}{2!}(v^{\mu}\partial_{\mu})^2 + \cdots\right)\xi^{\varkappa} \end{cases} \tag{10.9}$$

[1]) "Mitschleppen" in SCHOUTEN and v. KAMPEN 1933, 1 and E I 1935, 1.

or symbolically

(10.10) $$\eta^\varkappa = e^{tX}\xi^\varkappa; \qquad X \stackrel{\text{def}}{=} v^\mu \partial_\mu.$$

This solution represents a point transformation, conveniently indicated by $\xi^\varkappa \xrightarrow{tv} \eta^\varkappa$ or $\eta^\varkappa = \overset{tv}{T}\xi^\varkappa$. Its geometrical significance follows from (10.7). According to this equation t is a parameter fixed by the field $v^\varkappa$ on every streamline of this field if the point $t=0$ on each streamline is chosen arbitrarily. Now $\xrightarrow{\tau v}$ indicates the point transformation that on every streamline transforms a point with parameter t into the point with parameter $t+\tau$. We call this transformation the *displacement over* $\tau v^\varkappa$. From this it follows immediately that

(10.11) $$\overset{t_1 v}{T}\overset{t_2 v}{T} = \overset{t_3 v}{T}; \qquad t_3 \stackrel{\text{def}}{=} t_1 + t_2$$

and that accordingly the transformations $\overset{tv}{T}$ form a one-parameter abelian group (cf. Ch. IV).

Neglecting second and higher powers of t and writing dt instead of t we get the infinitesimal transformation

(10.12) $$\eta^\varkappa \stackrel{*}{=} \xi^\varkappa + v^\varkappa dt \quad \text{or} \quad \eta^\varkappa = \overset{v\,dt}{T}\xi^\varkappa; \quad \xi^\varkappa \stackrel{*}{=} \eta^\varkappa - v^\varkappa dt \quad \text{or} \quad \xi^\varkappa = \overset{-v\,dt}{T}\eta^\varkappa$$

which is said to *generate* the group of finite transformations (10.10). According to (10.1, 2, 4) the corresponding transformations of coordinates are

(10.13) $$\begin{cases} \xi^{\varkappa'} \stackrel{*}{=} \delta^{\varkappa'}_\varkappa (\xi^\varkappa - v^\varkappa dt) & \text{or} \quad \xi^{\varkappa'} = \overset{-v\,dt}{T}\xi^\varkappa \\ \xi^\varkappa \stackrel{*}{=} \delta^\varkappa_{\varkappa'}\xi^{\varkappa'} + v^\varkappa dt & \text{or} \quad \xi^\varkappa = \overset{v\,dt}{T}\xi^{\varkappa'} \end{cases}$$

and $(\varkappa')$ is the coordinate system *dragged along over* $v^\varkappa dt$. By differentiation of (10.13) we get

(10.14) $$\begin{cases} \text{a)} & A^{\varkappa'}_\lambda \stackrel{*}{=} \delta^{\varkappa'}_\varkappa (A^\varkappa_\lambda - \partial_\lambda v^\varkappa dt) \\ \text{b)} & \partial_\mu A^{\varkappa'}_\lambda \stackrel{*}{=} -\delta^{\varkappa'}_\varkappa \partial_\mu \partial_\lambda v^\varkappa dt \end{cases}$$

(10.15) $$\begin{cases} \text{a)} & A^\varkappa_{\lambda'} \stackrel{*}{=} \delta^\lambda_{\lambda'} (A^\varkappa_\lambda + \partial_\lambda v^\varkappa dt) \\ \text{b)} & \partial_{\mu'} A^\varkappa_{\lambda'} \stackrel{*}{=} \delta^{\mu\lambda}_{\mu'\lambda'} \partial_\mu \partial_\lambda v^\varkappa dt. \end{cases}$$

The expression

(10.16) $$\underset{v}{\pounds}\,\Phi \stackrel{\text{def}}{=} \lim_{t\to 0} \frac{\Phi - \overset{m}{\Phi}}{t}$$

will be called the LIE *derivative of the field Φ with respect to the field $v^\varkappa$ and $\underset{v}{\pounds}\,\Phi\,dt$ the* LIE *differential of Φ with respect to the infinitesimal displacement $v^\varkappa dt$*[1]. This can be expressed in another way:

[1]) SCHOUTEN and VAN KAMPEN 1933, 1; YANO 1946, 5; 1949, 1; YANO and TAKANO 1946, 6; SCHOUTEN 1949, 5 and for more general geometric objects LAP-

If a field Φ is dragged along over $v^\varkappa\, dt$ the variation of the field at $\xi^\varkappa$ is the negative LIE *differential of Φ at that point.*

If Φ_A is a geometric object, also $(\Phi_A, \partial_\mu \Phi_A)$ are components of a geometric object. From the definition of dragging along it follows that $\overbrace{\Phi_A, \partial_\mu \Phi_A}^{m} = \overset{m}{\Phi}_A, \partial_\mu \overset{m}{\Phi}_A$. But this implies that $\underset{v}{\pounds}(\Phi_A, \partial_\mu \Phi_A) = (\underset{v}{\pounds}\Phi_A, \partial_\mu \underset{v}{\pounds}\Phi_A)$. Now on the other hand this same LIE derivative can also be written $(\underset{v}{\pounds}\Phi_A, \underset{v}{\pounds}\,\partial_\mu \Phi_A)$, and this proves that in the expression of the LIE derivative of the object consisting of a geometric object and its first derivative, $\underset{v}{\pounds}\,\partial_\mu \Phi_A$ may be replaced by $\partial_\mu \underset{v}{\pounds}\Phi_A$:

$$\underset{v}{\pounds}\,\partial_\mu \Phi_A = \partial_\mu \underset{v}{\pounds}\Phi_A \quad (10.17)$$ [1].

Of course the expression $\underset{v}{\pounds}\,\partial_\mu \Phi_A$ can not be considered as the LIE derivative of $\partial_\mu \Phi_A$ but only as a set of components of the LIE derivative of $(\Phi_A, \partial_\mu \Phi_A)$, because $\partial_\mu \Phi_A$ is not a geometric object. In the same way $\underset{v}{\pounds}\, u^1$ is not the LIE derivative of the component u^1 but only the *1*-component of the LIE derivative $\underset{v}{\pounds}\, u^\varkappa$. Obviously (10.17) is also true if $(\varkappa)$ is anholonomic [2].

According to the definition we have for a contravariant vector field $u^\varkappa$ and the field $\overset{m}{u}{}^\varkappa$ dragged along over $v^\varkappa\, dt$

$$\overset{m}{u}{}^{\varkappa'}\{\xi + v\, dt\} \overset{*}{=} \delta^{\varkappa'}_{\varkappa}\, u^\varkappa \{\xi\} \quad (10.18\text{a})$$

or

$$\overset{m}{u}{}^{\varkappa'}\{\xi\} \overset{*}{=} \delta^{\varkappa'}_{\varkappa}\, u^\varkappa\{\xi - v\, dt\} = \delta^{\varkappa'}_{\varkappa}\left(u^\varkappa\{\xi\} - v^\mu\, \partial_\mu u^\varkappa\{\xi\}\, dt\right), \quad (10.18\text{b})$$

hence, according to (10.15)

$$\left\{\begin{aligned} \overset{m}{u}{}^\varkappa &= (A^\varkappa_\lambda + \partial_\lambda v^\varkappa\, dt)\,(u^\lambda - v^\mu\, \partial_\mu u^\lambda dt) \\ &= u^\varkappa - v^\mu\, \partial_\mu u^\varkappa\, dt + u^\mu\, \partial_\mu v^\varkappa\, dt \end{aligned}\right. \quad (10.19)$$

and, according to (10.16)

$$\underset{v}{\pounds}\, u^\varkappa = v^\mu\, \partial_\mu u^\varkappa - u^\mu\, \partial_\mu v^\varkappa. \quad (10.20\text{a})$$

Returning now to Fig. 9 we see that $\underset{v}{\pounds}\, u^\varkappa\, ds\, dt$ is the fifth side of the pentagon and that $\underset{v}{\pounds}\, u^\varkappa$ vanishes if and only if the fields $u^\varkappa$ and $v^\varkappa$ are X_2-forming (cf. II § 5).

TEW 1938, 1; DE DONDER and GÉHÉNIAU 1942, 1; RACHEVSKI 1949, 1; TASHIRO 1950, 1; WAGNER 1950, 1; NIJENHUIS 1952, 1, Ch. II. The name LIE derivative occurred first in v. DANTZIG 1932, 1; 2.

[1]) NIJENHUIS 1952, 1, p. 95.

[2]) Cf. for LIE derivatives in anholonomic coordinates NIJENHUIS 1952, 1, II § 6 and Exerc. II 10,6; 7.

From the definition of dragging along it follows that for any product or transvection of geometric objects we have $\overset{m}{\widehat{\Phi\Psi}} = \overset{m}{\Phi}\overset{m}{\Psi}$ and from this we see that the rule of LEIBNIZ

$$\underset{v}{\pounds}\Phi\Psi = \Phi\underset{v}{\pounds}\Psi + (\underset{v}{\pounds}\Phi)\Psi \tag{10.21}$$

holds for $\underset{v}{\pounds}$ applied to products and transvections provided that Φ, Ψ and $\Phi\Psi$ are geometric objects. It can be proved in a similar way that

1) the LIE derivative of a sum of geometric objects equals the sum of the LIE derivatives of the summands, provided that the sum is a geometric object;

2) the LIE derivative of a contraction equals the contraction of the LIE derivative.

If Φ is a quantity, $\overset{m}{\Phi}$ is a quantity of the same kind and therefore $\underset{v}{\pounds}\Phi$ is also a quantity of the same kind. The same is not true for every geometric object.

According to the definition we have for a scalar

$$\underset{v}{\pounds}\, s = v^\mu \partial_\mu s \tag{10.22}$$

and from this and the rule of LEIBNIZ for transvections it can be proved that for a covariant vector

$$\underset{v}{\pounds}\, w_\lambda = v^\mu \partial_\mu w_\lambda + w_\mu \partial_\lambda v^\mu \tag{10.20b}$$

because $v^\lambda w_\lambda$ is a scalar.

Because every tensor can be written as a sum of products of vectors we get for the LIE derivative of a tensor, for instance $P^{\varkappa}_{\cdot\lambda}$:

$$\underset{v}{\pounds}\, P^{\varkappa}_{\cdot\lambda} = v^\mu \partial_\mu P^{\varkappa}_{\cdot\lambda} - P^{\mu}_{\cdot\lambda}\partial_\mu v^\varkappa + P^{\varkappa}_{\cdot\mu}\partial_\lambda v^\mu . \tag{10.23}$$

For a tensor with higher valences the LIE derivative is constructed in the same way. Each upper (lower) index gives rise to a term with a negative (positive) sign. As a special case we mention the contra- and covariant p-vector field

$$\left\{\begin{array}{ll} \text{a)} & \underset{v}{\pounds}\, u^{\varkappa_1\ldots\varkappa_p} = v^\mu \partial_\mu u^{\varkappa_1\ldots\varkappa_p} - p\, u^{\mu[\varkappa_2\ldots\varkappa_p}\partial_\mu v^{\varkappa_1]} \\ \text{b)} & \underset{v}{\pounds}\, w_{\lambda_1\ldots\lambda_p} = v^\mu \partial_\mu w_{\lambda_1\ldots\lambda_p} + p\, w_{\mu[\lambda_2\ldots\lambda_p}\partial_{\lambda_1]} v^\mu . \end{array}\right. \tag{10.24}$$

From (10.24b) we get for $p = n$

$$\left\{\begin{array}{l} \underset{v}{\pounds}\, w_{\lambda_1\ldots\lambda_n} = v^\mu \partial_\mu w_{\lambda_1\ldots\lambda_n} + n\, w_{\mu[\lambda_2\ldots\lambda_n}\partial_{\lambda_1]} v^\mu \\ \qquad = v^\mu \partial_\mu w_{\lambda_1\ldots\lambda_n} + w_{\lambda_1\ldots\lambda_n}\partial_\mu v^\mu = \partial_\mu v^\mu w_{\lambda_1\ldots\lambda_n} \end{array}\right. \tag{10.25}$$

and because $w_{1\ldots n}$ may be considered as the component of a scalar Δ-density of weight $+1$, it follows that for such a quantity $\tilde{\mathfrak{q}}$

$$\underset{v}{\pounds}\,\tilde{\mathfrak{q}} = v^\mu \partial_\mu \tilde{\mathfrak{q}} + \tilde{\mathfrak{q}}\,\partial_\mu v^\mu = \partial_\mu \tilde{\mathfrak{q}}\, v^\mu = \operatorname{Div}(\tilde{\mathfrak{q}}\, v) . \tag{10.26}$$

Since every scalar density $\mathfrak{q}$ of weight $+1$ is a product of the W-scalar $\widetilde{\omega}$ (cf. I § 2) with a scalar Δ-density of weight $+1$ we have

$$(10.27)\qquad \underset{v}{\pounds}\, \mathfrak{q} = v^\mu \partial_\mu \mathfrak{q} + \mathfrak{q}\, \partial_\mu v^\mu = \partial_\mu \mathfrak{q}\, v^\mu = \operatorname{Div}(\mathfrak{q}\, v).$$

From this we get for a scalar density $\mathfrak{s}$ of weight w

$$(10.28)\qquad \underset{v}{\pounds}\, \mathfrak{s} = v^\mu \partial_\mu \mathfrak{s} + w\, \mathfrak{s}\, \partial_\mu v^\mu$$

and for a tensor density of weight w for instance $\mathfrak{P}^{\varkappa}_{\cdot\lambda}$,

$$(10.29)\qquad \underset{v}{\pounds}\, \mathfrak{P}^{\varkappa}_{\cdot\lambda} = v^\mu \partial_\mu \mathfrak{P}^{\varkappa}_{\cdot\lambda} - \mathfrak{P}^{\mu}_{\cdot\lambda} \partial_\mu v^\varkappa + \mathfrak{P}^{\varkappa}_{\cdot\mu} \partial_\lambda v^\mu + w\, \mathfrak{P}^{\varkappa}_{\cdot\lambda} \partial_\mu v^\mu.$$

Formulae of the same form hold for scalar Δ-densities and tensor Δ-densities.

It can be proved for more general geometric objects as defined in II § 3 that the LIE derivative of a geometric object together with the object itself form a geometric object and that the LIE derivative itself is always a geometric object if and only if the transformation of the given geometric object is *linear* (not necessary homogeneous) in the components with coefficients depending on $\xi^\varkappa$, $\xi^{\varkappa'}$, $A^{\varkappa'}_{\varkappa}$, $\partial_\mu A^{\varkappa'}_{\varkappa}$, ...[1])

A field is said to be *absolutely invariant with respect to the field* $v^\varkappa$ if its LIE derivative with respect to $v^\varkappa$ vanishes. The necessary and sufficient condition is that the field does not change if it is dragged along by the point transformations of the one-parameter groups generated by $v^\varkappa\, dt$. Obviously sums, products and transvections of absolutely invariant fields are absolutely invariant.

A field is said to be *absolutely invariant with respect to the streamlines of* $v^\varkappa$ if it is absolutely invariant with respect to $\sigma v^\varkappa$ for every choice of the scalar σ.

The additional condition for this stronger invariance for an absolutely invariant covariant p-vector field $w_{\lambda_1 \ldots \lambda_p}$ is

$$(10.30)\qquad v^\mu w_{\mu \lambda_2 \ldots \lambda_p} = 0.$$

This follows immediately from (10.24b).[2]) The LIE derivative of every geometric object Φ_Λ as defined in II § 3 can be found from the definition (10.16) and the form $\overset{m}{\Phi}_\Lambda$ in (10.6). But the computing may be rather cumbersome if the transformation of the object is complicated. As an example we deal here with a rather simple geometric object without using (10.6). Let the transformation of the object be[3])

$$(10.31)\qquad \Pi^{\varkappa'}_{\mu'\lambda'} = A^{\mu\,\lambda\,\varkappa'}_{\mu'\lambda'\varkappa} \Pi^{\varkappa}_{\mu\lambda} - A^{\mu\,\lambda}_{\mu'\lambda'} \partial_\mu A^{\varkappa'}_{\lambda}.$$

---

[1]) NIJENHUIS 1952, 1, Ch. II. Cf. SCHOUTEN and VAN KAMPEN 1933, 1; E I 1935, 1; TASHIRO 1950, 1.

[2]) Cf. for more particulars concerning absolutely and relatively invariant fields P. P. 1949, 1, p. 43ff. and T. P. 1951, 1, p. 78.

[3]) Objects of this kind occur frequently in differential geometry, cf. Ch. III.

For $\overset{m}{\Pi}{}^{\varkappa}_{\mu\lambda}$ we get

(10.32) $$\overset{m}{\Pi}{}^{\varkappa'}_{\mu'\lambda'}\{\xi\} \overset{*}{=} \delta^{\varkappa'\mu\lambda}_{\varkappa\mu'\lambda'} \Pi^{\varkappa}_{\mu\lambda}\{\xi - v\,dt\} = \delta^{\varkappa'\mu\lambda}_{\varkappa\mu'\lambda'} (\Pi^{\varkappa}_{\mu\lambda}\{\xi\} - v^{\nu}\partial_{\nu}\Pi^{\varkappa}_{\mu\lambda}\{\xi\}\,dt)$$

hence, according to (10.14, 15)

(10.33) $$\left\{\begin{array}{l}\overset{m}{\Pi}{}^{\varkappa}_{\mu\lambda} = (A^{\tau}_{\mu} - \partial_{\mu} v^{\tau}\,dt)\,(A^{\sigma}_{\lambda} - \partial_{\lambda} v^{\sigma}\,dt)\,(A^{\varkappa}_{\varrho} + \partial_{\varrho} v^{\varkappa}\,dt)\times \\ \quad \times (\Pi^{\varrho}_{\tau\sigma} - v^{\omega}\partial_{\omega}\Pi^{\varrho}_{\tau\sigma}\,dt) - (A^{\tau}_{\mu} - \partial_{\mu} v^{\tau}\,dt)\,(A^{\sigma}_{\lambda} - \partial_{\lambda} v^{\sigma}\,dt)\,\partial_{\tau}\partial_{\sigma} v^{\varkappa}\,dt\,.\end{array}\right.$$

(10.34) $$\underset{v}{£}\,\Pi^{\varkappa}_{\mu\lambda} = v^{\nu}\partial_{\nu}\Pi^{\varkappa}_{\mu\lambda} - \Pi^{\varrho}_{\mu\lambda}\partial_{\varrho} v^{\varkappa} + \Pi^{\varkappa}_{\mu\sigma}\partial_{\lambda} v^{\sigma} + \Pi^{\varkappa}_{\varrho\lambda}\partial_{\mu} v^{\varrho} + \partial_{\mu}\partial_{\lambda} v^{\varkappa}.$$

We see from (10.31) that the difference of two objects of this kind is always a tensor, because the second term to the right in (10.31) does not contain $\Pi^{\varkappa}_{\mu\lambda}$. Now $\overset{m}{\Pi}{}^{\varkappa}_{\mu\lambda}$ is an object of the same kind as $\Pi^{\varkappa}_{\mu\lambda}$ and this implies that $\underset{v}{£}\,\Pi^{\varkappa}_{\mu\lambda}$ is a tensor.

If a field $v^{\varkappa}$ is given, the LIE operator $\underset{v}{£}$ can be used to find the field $\overset{m}{\Phi}_{\Lambda}$ that results from a field of a *quantity* Φ_{Λ} if this field is dragged along by a finite point transformation $\overset{tv}{T}$ belonging to the one-parameter group generated by the infinitesimal point transformation $\xi^{\varkappa} \to \xi^{\varkappa} + v^{\varkappa}\,dt$. Let $(\varkappa)$ be chosen in such a way that $v^{\varkappa} = \underset{1}{e}{}^{\varkappa}$ (cf. Exerc. II 5,1). Then $\underset{v}{£}\,\underset{\lambda}{e}{}^{\varkappa} = 0$ for every value of λ and this has as a consequence that the transformation $\overset{tv}{T}$ takes the form

(10.35) $$'\xi^{\varkappa} \overset{*}{=} \xi^{\varkappa} + t\,\underset{1}{e}{}^{\varkappa}$$

and that

(10.36) $$\overset{m}{\Phi}_{\Lambda}\{\xi\} \overset{*}{=} \Phi_{\Lambda}\{\xi - t\,\underset{1}{e}\}.$$

Now for this special choice of $(\varkappa)$ we have

(10.37) $$\underset{v}{£}\,\Phi_{\Lambda} \overset{*}{=} \partial_1 \Phi_{\Lambda},$$

hence

(10.38) $$\overset{m}{\Phi}_{\Lambda} = e^{-t\underset{v}{£}}\,\Phi_{\Lambda} = \Phi_{\Lambda} - t\,\underset{v}{£}\,\Phi_{\Lambda} + \frac{t^2}{2!}\,\underset{v}{£}{}^2\,\Phi_{\Lambda} - \cdots\text{[1]}.$$

Because this equation has the invariant form it is independent of the choice of $(\varkappa)$. Every term represents the values at $\xi^{\varkappa}$ of the components of a quantity of the same kind as Φ_{Λ}.

It can be proved that (10.38) also holds for all *geometric objects* whose transformation is *linear*[2]).

---

[1]) This series occurs in KOSAMBI 1946, 1 but no proof is given. SHABBAR's justification of (10.38) in 1942, 1 really shows only the correctness of the first two terms. In 1951, 1 KOSAMBI gave for the first time a complete proof. The proof given here is from NIJENHUIS 1952, 1.

[2]) NIJENHUIS 1952, 1. GOLAB 1950, 1 called these objects "purely differential". The $\Pi^{\varkappa}_{\mu\lambda}$ in (10.31) are components of such an object.

In Ch. IV we need the following formula[1])

$$(10.39)\qquad \boxed{e^{-s\underset{u}{\pounds}}\, e^{-t\underset{v}{\pounds}}\, e^{+s\underset{u}{\pounds}}\, \Phi = e^{-t\underset{w}{\pounds}}\, \Phi; \qquad w^\varkappa \overset{\text{def}}{=} e^{-s\underset{u}{\pounds}}\, v^\varkappa}$$

valid for all geometric objects whose transformation is linear. First we prove for objects of this kind

$$(10.40)\qquad {}^*\psi \overset{\text{def}}{=} e^{-s\underset{u}{\pounds}}\, \underset{v}{\pounds}\, e^{+s\underset{u}{\pounds}}\, \psi = \underset{w}{\pounds}\, \psi .$$

For the derivatives of ${}^*\psi$ we get (cf. Exerc. II 10,9)

$$(10.41)\qquad \begin{cases} \dfrac{d\,{}^*\psi}{d\,s} = -\, e^{-s\underset{u}{\pounds}} \left(\underset{u}{\pounds}\,\underset{v}{\pounds} - \underset{v}{\pounds}\,\underset{u}{\pounds}\right) e^{+s\underset{u}{\pounds}}\, \psi = -\, e^{-s\underset{u}{\pounds}}\, \underset{\underset{1}{w}}{\pounds}\, e^{+s\underset{u}{\pounds}}\, \psi \\[2ex] \dfrac{d^2\,{}^*\psi}{d\,s^2} = e^{-s\underset{u}{\pounds}}\, \underset{\underset{2}{w}}{\pounds}\, e^{+s\underset{u}{\pounds}}\, \psi; \qquad \underset{1}{w}{}^\varkappa \overset{\text{def}}{=} \underset{u}{\pounds}\, v^\varkappa; \qquad \underset{2}{w}{}^\varkappa \overset{\text{def}}{=} \underset{u}{\pounds}\, \underset{1}{w}{}^\varkappa = \pounds^2\, v^\varkappa . \end{cases}$$

Hence we have

$$(10.42)\qquad {}^*\psi = \underset{v}{\pounds}\, \psi - s\, \underset{\underset{1}{w}}{\pounds}\, \psi + \frac{s^2}{2!}\, \underset{\underset{2}{w}}{\pounds}\, \psi - \cdots$$

and because of

$$(10.43)\qquad \underset{u+v}{\pounds}\, \psi = \underset{u}{\pounds}\, \psi + \underset{v}{\pounds}\, \psi$$

and the definition of $w^\varkappa$, it follows that ${}^*\psi = \underset{w}{\pounds}\, \psi$. Now writing χ for the left hand side of (10.39) we have according to (10.40)

$$(10.44)\qquad \frac{d\chi}{d\,t} = -\, e^{-s\underset{u}{\pounds}}\, \underset{v}{\pounds}\, e^{-t\underset{v}{\pounds}}\, e^{+s\underset{u}{\pounds}}\, \Phi = -\, e^{-s\underset{u}{\pounds}}\, \underset{v}{\pounds}\, e^{+s\underset{u}{\pounds}}\, \chi = -\, \underset{w}{\pounds}\, \chi$$

and if this differential equation is solved by means of a series, we get (10.39).

Exercises.

II 10,1. Prove that $\underset{v}{\pounds}\, v^\varkappa = 0$ and that the LIE derivative of $A_\lambda^\varkappa$, $\tilde{\mathfrak{E}}^{\varkappa_1 \ldots \varkappa_n}$ and $\tilde{\mathfrak{e}}_{\lambda_1 \ldots \lambda_n}$ vanish for every choice of $v^\varkappa$.

II 10,2. Prove that the additional conditions for absolute invariance of a p-vector density (or Δ-density) $\mathfrak{w}^{\varkappa_1 \ldots \varkappa_p}$ of weight $+1$ is

$$\text{II 10,2 }\alpha)\qquad \mathfrak{w}^{[\varkappa_1 \ldots \varkappa_p}\, v^{\varkappa]} = 0 .$$

II 10,3[2]). Prove the identities for the geometrical object $\xi^\varkappa$ and for the scalar fields $\overset{\varkappa}{\xi}$

$$\text{II 10,3 }\alpha)\qquad \underset{v}{\pounds}\, \xi^\varkappa = 0$$

$$\text{II 10,3 }\beta)\qquad \underset{v}{\pounds}\, \overset{\varkappa}{\xi} = v^\mu\, \partial_\mu \overset{\varkappa}{\xi} = v^\mu\, \overset{\varkappa}{e}_\mu = \overset{\varkappa}{v} .$$

---

1) NIJENHUIS 1952, 1, II § 8.

2) NIJENHUIS 1952, 1, II § 4.

II 10,4[1]). Prove that the operators $\underset{v}{\pounds}$ and D are commutative, for instance

II 10,4 α) $$\underset{v}{\pounds}\, \partial_{[\mu}\, w_{\lambda_1 \ldots \lambda_q]} = \partial_{[\mu}\, \underset{v}{\pounds}\, w_{\lambda_1 \ldots \lambda_q]}$$

II 10,4 β) $$\underset{v}{\pounds}\, \partial_\mu\, \mathfrak{w}^{\mu \varkappa_2 \ldots \varkappa_p} = \partial_\mu\, \underset{v}{\pounds}\, \mathfrak{w}^{\mu \varkappa_2 \ldots \varkappa_p}.$$

II 10,5[2]). Prove that

II 10,5 α) $$\underset{v}{\pounds}\, \mathbf{w} = \mathbf{v}.\, \mathrm{Rot}\, \mathbf{w} + \mathrm{Rot}\, \mathbf{v}.\, \mathbf{w}$$

II 10,5 β) $$\underset{v}{\pounds}^k\, \mathbf{w} = (\mathbf{v}.\, \mathrm{Rot})^k\, \mathbf{w} + (\mathrm{Rot}\, \mathbf{v}.)^k\, \mathbf{w}$$

II 10,5 γ) $$\underset{v}{\pounds}\, \boldsymbol{\mathfrak{w}} = \mathrm{Div}\, \mathbf{v}.\, \boldsymbol{\mathfrak{w}} - \mathbf{v}.\, \mathrm{Div}\, \boldsymbol{\mathfrak{w}}$$

II 10,5 δ) $$\underset{v}{\pounds}^k\, \boldsymbol{\mathfrak{w}} = (\mathrm{Div}\, \mathbf{v}.)^k\, \boldsymbol{\mathfrak{w}} + (-\, \mathbf{v}.\, \mathrm{Div})^k\, \boldsymbol{\mathfrak{w}},$$

where $\mathbf{w}$ is a covariant q-vector and $\mathfrak{w}$ a contravariant p-vector density of weight $+1$. Rot has to be changed into Grad if working on a scalar and $\mathbf{v}.$ is defined by:

II 10,5 ε) $$\mathbf{v}.\, \mathbf{w}: \qquad v^\mu\, w_{\mu \lambda_2 \ldots \lambda_q}$$

II 10,5 ζ) $$\mathbf{v}.\, \boldsymbol{\mathfrak{w}}: \qquad (p+1)\, v^{[\varkappa}\, \mathfrak{w}^{\varkappa_1 \ldots \varkappa_p]}.$$

II 10,6. Prove that[3])

II 10,6 α) $$\underset{v}{\pounds}\, u^h = v^j\, \partial_j\, u^h - (\partial_j\, v^h - 2 v^i\, \Omega_{ji}^{h})\, u^j$$

II 10,6 β) $$\underset{v}{\pounds}\, w_i = v^j\, \partial_j\, w_i + (\partial_i\, v^h - 2 v^j\, \Omega_{ij}^{h})\, w_h$$

II 10,6 γ) $$\underset{v}{\pounds}\, \Omega_{ji}^{h} = 0.$$

II 10,6 δ) $$\underset{v}{\pounds}\, \overset{(h)}{\mathfrak{p}} = v^j\, \partial_j\, \overset{(h)}{\mathfrak{p}}\, (\partial_j\, v^j + 2 v^j\, \Omega_{ji}^{i}).$$

II 10,7. Prove that[3])

II 10,7 α) $$\underset{j}{\pounds}\, \underset{i}{e}^h \overset{*}{=} -\, 2\, \Omega_{ji}^{h}$$

II 10,7 β) $$\underset{j}{\pounds}\, \underset{i}{e}^\varkappa \overset{*}{=} -\, 2\, \Omega_{ji}^{h}\, A_h^\varkappa.$$

where $\underset{j}{\pounds}$ denotes the Lie derivative with respect to $\underset{j}{e}^\varkappa$.

II 10,8. Prove that for any three vector fields $t^\varkappa$, $u^\varkappa$, $v^\varkappa$ [4])

II 10,8 α) $$\underset{t}{\pounds}\, \underset{u}{\pounds}\, v^\varkappa + \underset{u}{\pounds}\, \underset{v}{\pounds}\, t^\varkappa + \underset{v}{\pounds}\, \underset{t}{\pounds}\, u^\varkappa = 0.$$

II 10,9. Prove that for any quantity Φ (indices suppressed) and any vector fields $u^\varkappa$, $v^\varkappa$ [5])

---

[1]) P. P. 1949, 1, p. 76. This is only a special case of the commutativity of $\underset{v}{\pounds}$ and ∂_μ [cf. (10.17)].

[2]) E. J. Post, personal communication.

[3]) Nijenhuis 1952, 1, II § 6.

[4]) Davies 1938, 1.

[5]) Nijenhuis 1952, 1, II § 8; Taschiro 1952, 1.

II 10,9 α) $(\underset{u}{£}\underset{v}{£} - \underset{v}{£}\underset{u}{£})\Phi = \underset{w}{£}\Phi; \quad w^\varkappa \overset{\text{def}}{=} \underset{u}{£} v^\varkappa = - \underset{v}{£} u^\varkappa.$

II 10,10. Prove that $\underset{u}{£} v^\varkappa = 0$ if and only if [1])

II 10,10 α) $$\overset{su}{T}\overset{tv}{T}\xi^\varkappa = \overset{tv}{T}\overset{su}{T}\xi^\varkappa.$$

II 10,11 [2]). In an X_n a covariant q-vector field $w_{\lambda_1 \ldots \lambda_q}$ and a set of ∞^1 X_q's depending on a parameter t are given. Let a set of curves intersect each X_q in one point, such that t is also a parameter on each of these curves. If then τ_q is a q-dimensional part of one X_q there is a one to one correspondence between the points of this τ_q and ∞^1 τ_q's on the ∞^1 X_q's and

II 10,11 α) $$\Phi(t) \overset{\text{def}}{=} \int_{\tau_q} w_{\lambda_1 \ldots \lambda_q} d f^{\lambda_1 \ldots \lambda_q}$$

is a function of t. Prove that

II 10,11 β) $$\frac{d\Phi}{dt} = \int_{\tau_q} \underset{v}{£} w_{\lambda_1 \ldots \lambda_q} d f^{\lambda_1 \ldots \lambda_q}; \qquad v^\varkappa \overset{\text{def}}{=} \frac{d\xi^\varkappa}{dt}.$$

II 10,12. Prove that $£\, \partial_j u^h = \partial_j £\, u^h$ for (h) anholonomic.

§ 11. The LAGRANGE derivative.

Let Φ_Λ; $\Lambda = 1, \ldots, N$ be a set of functions of the $\xi^\varkappa$. They may be scalars or components of geometric objects or just quite arbitrary functions for which no manner of transformation is defined. Let the index Λ be suppressed and the derivatives of the Φ_Λ with respect to the $\xi^\varkappa$ be denoted by $\Phi_\mu, \Phi_{\nu\mu}$, etc.

If $\mathfrak{L}$ is a function of the $\Phi, \Phi_\mu, \Phi_{\nu\mu}$, etc. up to a certain definite order, then, in a certain region of X_n, $\mathfrak{L}$ is a function of the $\xi^\varkappa$. We consider the volume integral (cf. II 8.5)

(11.1) $$\int_{\tau_n} \mathfrak{L}\, d\xi^1 \ldots d\xi^n$$

over an arbitrary region τ_n where the Φ are analytic.

Let the field Φ be subjected to a variation $\overset{v}{d}\Phi$ such that the variations of Φ and of all its derivatives occurring in $\mathfrak{L}$ vanish at the boundary τ_{n-1} of τ_n.

Then we have for the variation of $\mathfrak{L}$

(11.2) $$\overset{v}{d}\mathfrak{L} = \frac{\partial \mathfrak{L}}{\partial \Phi} \overset{v}{d}\Phi + \frac{\partial \mathfrak{L}}{\partial \Phi_\mu} \overset{v}{d}\Phi_\mu + \cdots$$

---

[1]) A. NIJENHUIS, personal communication.

[2]) E. J. POST, personal communication.

and consequently

$$\overset{v}{d}\int_{\tau_n} \mathfrak{L}\, d\xi^1 \ldots \xi^n = \int_{\tau_n} \left(\frac{\partial \mathfrak{L}}{\partial \Phi} \overset{v}{d}\Phi + \frac{\partial \mathfrak{L}}{\partial \Phi_\mu} \overset{v}{d}\Phi_\mu + \cdots\right) d\xi^1 \ldots d\xi^n. \tag{11.3}$$

Let now τ_{n-1} have a form such that the theorem of STOKES is valid for τ_n and τ_{n-1}. Then we have, integrating by parts and using this theorem in the form (8.20) for $p=1$ [1])

$$\left\{\begin{aligned} \overset{v}{d}\int_{\tau_n} \mathfrak{L}\, d\xi^1 \ldots d\xi^n &= \int_{\tau_n} \left(\frac{\partial \mathfrak{L}}{\partial \Phi} \overset{v}{d}\Phi + \frac{\partial \mathfrak{L}}{\partial \Phi_\mu} \partial_\mu \overset{v}{d}\Phi + \cdots\right) d\xi^1 \ldots d\xi^n \\ &= \int_{\tau_n} \left(\frac{\partial \mathfrak{L}}{\partial \Phi} - \partial_\mu \frac{\partial \mathfrak{L}}{\partial \Phi_\mu} + \partial_\nu \partial_\mu \frac{\partial \mathfrak{L}}{\partial \Phi_{\nu\mu}} - \cdots\right) \overset{v}{d}\Phi\, d\xi^1 \ldots d\xi^n \end{aligned}\right. \tag{11.4}$$

hence

$$\overset{v}{d}\int_{\tau_n} \mathfrak{L}\, d\xi^1 \ldots d\xi^n = \int_{\tau_n} [\mathfrak{L}]\, \overset{v}{d}\Phi\, d\xi^1 \ldots d\xi^n \tag{11.5}$$

where

$$[\mathfrak{L}] \overset{\text{def}}{=} \frac{\partial \mathfrak{L}}{\partial \Phi} - \partial_\mu \frac{\partial \mathfrak{L}}{\partial \Phi_\mu} + \partial_\nu \partial_\mu \frac{\partial \mathfrak{L}}{\partial \Phi_{\nu\mu}} - \cdots. \tag{11.6}$$

$[\mathfrak{L}]$ is called the LAGRANGE *derivative* of $\mathfrak{L}$. $[\mathfrak{L}]$ has of course a suppressed upper index Λ just as Φ has a suppressed lower index Λ. If $[\mathfrak{L}]=0$ the variation (11.3) vanishes for every choice of τ_n, provided that the variation of Φ satisfies the boundary conditions.

The equation $[\mathfrak{L}]=0$ is called the LAGRANGE *equation*. In classical dynamics LAGRANGE equations with $n=1$ occur. The $\xi^\varkappa$ reduce then to one variable t and the Φ_Λ are the coordinates of the dynamical system. [2])

If the Φ_Λ are components of a geometric object with a given manner of transformation, and if $\mathfrak{L}$ is computed from the Φ_Λ and their derivatives *in the same way for all coordinate systems* it may happen that $\mathfrak{L}$ is a scalar Δ-density of weight $+1$. In that case the integral (11.3) is a scalar and the process of forming the LAGRANGE derivative is invariant for all coordinate transformations. If moreover Φ is a tensor with the valences p, q, $[\mathfrak{L}]$ is a tensor Δ-density of weight $+1$ with the valences q, p. If only coordinate transformations with $\Delta>0$ are considered we may replace Δ-density by density in the above statements.

If Φ is a tensor and $\mathfrak{L}$ a scalar Δ-density (or density for $\Delta>0$) of weight $+1$, a very important relation exists between Φ, $[\mathfrak{L}]$ and their first derivatives. In order to obtain this relation we suppose

---

1) Cf. the remark at the end of II § 8.

2) Cf. T. P. 1951, 1, p. 198.

that the variation $\overset{v}{d}\Phi$ is due to the dragging along of the field Φ over $v^\varkappa\, dt$, hence

$$\overset{v}{d}\Phi = -\underset{v}{\pounds}\Phi\, dt \tag{11.7}$$

and that $v^\varkappa$, $\partial_\mu v^\varkappa$, $\partial_\mu \partial_\lambda v^\varkappa$, ... are zero on τ_{n-1}. Let Φ be the tensor $P^{\varkappa_1 \ldots \varkappa_p}{}_{\lambda_1 \ldots \lambda_q}$. Then we have

$$\left\{\begin{aligned} \overset{m}{P}{}^{\varkappa_1 \ldots \varkappa_p}{}_{\lambda_1 \ldots \lambda_q} = P^{\varkappa_1 \ldots \varkappa_p}{}_{\lambda_1 \ldots \lambda_q} - v^\mu \partial_\mu P^{\varkappa_1 \ldots \varkappa_p}{}_{\lambda_1 \ldots \lambda_q}\, dt\, + \\ + P^{\mu \varkappa_2 \ldots \varkappa_p}{}_{\lambda_1 \ldots \lambda_q} \partial_\mu v^{\varkappa_1}\, dt + \cdots \\ - P^{\varkappa_1 \ldots \varkappa_p}{}_{\mu \lambda_2 \ldots \lambda_q} \partial_{\lambda_1} v^\mu\, dt - \cdots \end{aligned}\right. \tag{11.8}$$

and thus

$$\partial_\nu \overset{m}{P}{}^{\varkappa_1 \ldots \varkappa_p}{}_{\lambda_1 \ldots \lambda_q} = \partial_\nu P^{\varkappa_1 \ldots \varkappa_p}{}_{\lambda_1 \ldots \lambda_q} + * \tag{11.9}$$

where $*$ denotes terms each of which contains $v^\varkappa$ or a first or second derivative of $v^\varkappa$ as a factor. Hence $\overset{v}{d}P$ and $\overset{v}{d}\,\partial_\mu P$ vanish on τ_{n-1} and the same can be proved for the variations of all higher derivatives of P.

By substituting (11.7) in (11.4) we get using STOKES' theorem

$$\left\{\begin{aligned} \overset{v}{d}\int_{\tau_n} \mathfrak{L}\, d\xi^1 \ldots d\xi^n &= -\int_{\tau_n} [\mathfrak{L}]_{\varkappa_1 \ldots \varkappa_p}{}^{\lambda_1 \ldots \lambda_q} \underset{v}{\pounds} P^{\varkappa_1 \ldots \varkappa_p}{}_{\lambda_1 \ldots \lambda_q}\, dt\, d\xi^1 \ldots d\xi^n \\ &= -\int_{\tau_n} [\mathfrak{L}]_{\varkappa_1 \ldots \varkappa_p}{}^{\lambda_1 \ldots \lambda_q} \{v^\mu \partial_\mu P^{\varkappa_1 \ldots \varkappa_p}{}_{\lambda_1 \ldots \lambda_q} - P^{\mu \varkappa_2 \ldots \varkappa_p}{}_{\lambda_1 \ldots \lambda_q} \partial_\mu v^{\varkappa_1} - \cdots \\ &\qquad + P^{\varkappa_1 \ldots \varkappa_p}{}_{\mu \lambda_2 \ldots \lambda_q} \partial_{\lambda_1} v^\mu + \cdots\}\, dt\, d\xi^1 \ldots d\xi^n \\ &= -\int_{\tau_n} \Big[[\mathfrak{L}]_{\varkappa_1 \ldots \varkappa_p}{}^{\lambda_1 \ldots \lambda_q} \{\partial_\mu P^{\varkappa_1 \ldots \varkappa_p}{}_{\lambda_1 \ldots \lambda_q} + A_\mu^{\varkappa_1} \partial_\varrho P^{\varrho \varkappa_2 \ldots \varkappa_p}{}_{\lambda_1 \ldots \lambda_q} + \cdots \\ &\qquad - \partial_{\lambda_1} P^{\varkappa_1 \ldots \varkappa_p}{}_{\mu \lambda_2 \ldots \lambda_q} - \cdots\} + \\ &\qquad + P^{\varrho \varkappa_2 \ldots \varkappa_p}{}_{\lambda_1 \ldots \lambda_q} \partial_\varrho [\mathfrak{L}]_{\mu \varkappa_2 \ldots \varkappa_p}{}^{\lambda_1 \ldots \lambda_q} + \cdots \\ &\qquad - P^{\varkappa_1 \ldots \varkappa_p}{}_{\mu \lambda_2 \ldots \lambda_q} \partial_{\lambda_1} [\mathfrak{L}]_{\varkappa_1 \ldots \varkappa_p}{}^{\lambda_1 \ldots \lambda_q} - \cdots \Big] v^\mu\, dt\, d\xi^1 \cdots d\xi^n \end{aligned}\right. \tag{11.10}$$

but this variation must be zero because the field dragged along has exactly the same components with respect to $(\varkappa')$ as the original field with respect to $(\varkappa)$. Hence

$$\left\{\begin{aligned} &[\mathfrak{L}]_{\varkappa_1 \ldots \varkappa_p}{}^{\lambda_1 \ldots \lambda_q} \{\partial_\mu P^{\varkappa_1 \ldots \varkappa_p}{}_{\lambda_1 \ldots \lambda_q} + A_\mu^{\varkappa_1} \partial_\varrho P^{\varrho \varkappa_2 \ldots \varkappa_p}{}_{\lambda_1 \ldots \lambda_q} + \cdots \\ &\quad - \partial_{\lambda_1} P^{\varkappa_1 \ldots \varkappa_p}{}_{\mu \lambda_2 \ldots \lambda_q} - \cdots\} + P^{\varrho \varkappa_2 \ldots \varkappa_p}{}_{\lambda_1 \ldots \lambda_q} \partial_\varrho [\mathfrak{L}]_{\mu \varkappa_2 \ldots \varkappa_p}{}^{\lambda_1 \ldots \lambda_q} + \cdots \\ &\quad - P^{\varkappa_1 \ldots \varkappa_p}{}_{\mu \lambda_2 \ldots \lambda_q} \partial_{\lambda_1} [\mathfrak{L}]_{\varkappa_1 \ldots \varkappa_p}{}^{\lambda_1 \ldots \lambda_q} - \cdots = 0. \end{aligned}\right. \tag{11.11}$$

For $p = 0$, $q = 2$ the identity takes the simple form

$$[\mathfrak{L}]^{\lambda\varkappa} (\partial_\mu P_{\lambda\varkappa} - \partial_\lambda P_{\mu\varkappa} - \partial_\varkappa P_{\lambda\mu}) - P_{\mu\varkappa} \partial_\lambda [\mathfrak{L}]^{\lambda\varkappa} - P_{\lambda\mu} \partial_\varkappa [\mathfrak{L}]^{\lambda\varkappa} = 0 \tag{11.12}$$

which is of special importance in the theory of relativity. (11.11) is not the only relation between a quantity, its LAGRANGE derivative and the derivatives of these two, as has already been pointed out by BUCHDAHL[1]) and SCHRÖDINGER.[2]) BUCHDAHL derived identities of this kind for a density as a function of the fundamental tensor and its derivatives up to a certain order. SCHRÖDINGER considered a function of the parameters of a linear connexion and their first derivatives.

We give here only an example, the generalization being then obvious.[3]) Let $P^{\varkappa}_{\cdot\lambda}$ be a tensor field and $\mathfrak{L}$ a density of weight $+1$ depending on the $P^{\varkappa}_{\cdot\lambda}$ and their first and second derivatives

$$(11.13) \qquad \overset{(\varkappa)}{\mathfrak{L}} = F(P^{\varkappa}_{\cdot\lambda}, \partial_\mu P^{\varkappa}_{\cdot\lambda}, \partial_\nu \partial_\mu P^{\varkappa}_{\cdot\lambda}).$$

Then we have

$$(11.14) \quad F(P^{\varkappa'}_{\cdot\lambda'}, \partial_{\mu'} P^{\varkappa'}_{\cdot\lambda'}, \partial_{\nu'} \partial_{\mu'} P^{\varkappa'}_{\cdot\lambda'}) = \Delta^{-1} F(P^{\varkappa}_{\cdot\lambda}, \partial_\mu P^{\varkappa}_{\cdot\lambda}, \partial_\nu \partial_\mu P^{\varkappa}_{\cdot\lambda})$$

and in this equation the expressions

$$(11.15) \quad \begin{cases} \text{a)} \quad P^{\varkappa'}_{\cdot\lambda'} = A^{\varkappa'\lambda}_{\varkappa\lambda'} P^{\varkappa}_{\cdot\lambda} \\ \text{b)} \quad \partial_{\mu'} P^{\varkappa'}_{\cdot\lambda'} = (\partial_{\mu'} A^{\varkappa'\lambda}_{\varkappa\lambda'}) P^{\varkappa}_{\cdot\lambda} + A^{\varkappa'\lambda\mu}_{\varkappa\lambda'\mu'} \partial_\mu P^{\varkappa}_{\cdot\lambda} \\ \text{c)} \quad \partial_{\nu'} \partial_{\mu'} P^{\varkappa'}_{\cdot\lambda'} = (\partial_{\nu'} \partial_{\mu'} A^{\varkappa'\lambda}_{\varkappa\lambda'}) P^{\varkappa}_{\cdot\lambda} + (\partial_{\mu'} A^{\varkappa'\lambda}_{\varkappa\lambda'}) A^{\nu}_{\nu'} \partial_\nu P^{\varkappa}_{\cdot\lambda} + \\ \qquad + (\partial_{\nu'} A^{\varkappa'\lambda\mu}_{\varkappa\lambda'\mu'}) \partial_\mu P^{\varkappa}_{\cdot\lambda} + A^{\varkappa'\lambda\mu\nu}_{\varkappa\lambda'\mu'\nu'} \partial_\nu \partial_\mu P^{\varkappa}_{\cdot\lambda} \end{cases}$$

can be substituted. After this substitution (11.14) is an identity in the variables $P^{\varkappa}_{\cdot\lambda}$, $\partial_\mu P^{\varkappa}_{\cdot\lambda}$, $\partial_\nu \partial_\mu P^{\varkappa}_{\cdot\lambda}$, $A^{\varkappa'}_{\lambda}$, $\partial_\mu A^{\varkappa'}_{\lambda}$, $\partial_\nu \partial_\mu A^{\varkappa'}_{\lambda}$ and these may be considered as *independent*. If this identity is differentiated with respect to $\partial_\nu \partial_\mu P^{\varkappa}_{\cdot\lambda}$ we get

$$(11.16) \qquad A^{\varkappa'\lambda\mu\nu}_{\varkappa\lambda'\mu'\nu'} \frac{\partial \overset{(\varkappa')}{\mathfrak{L}}}{\partial(\partial_{\nu'} \partial_{\mu'} P^{\varkappa'}_{\cdot\lambda'})} = \Delta^{-1} \frac{\partial \overset{(\varkappa)}{\mathfrak{L}}}{\partial(\partial_\nu \partial_\mu P^{\varkappa}_{\cdot\lambda})}$$

and this proves that

$$(11.17) \qquad \mathfrak{L}^{\nu\mu\lambda}_{\cdot\cdot\cdot\varkappa} \overset{\text{def}}{=} \frac{\partial \mathfrak{L}}{\partial(\partial_\nu \partial_\mu P^{\varkappa}_{\cdot\lambda})}$$

is a *tensor density of weight* $+1$. Similarly by differentiating the identity with respect to $\partial_\mu P^{\varkappa}_{\cdot\lambda}$ and with respect to $P^{\varkappa}_{\cdot\lambda}$ it can be proved once more that $[\mathfrak{L}]^{\cdot\lambda}_{\varkappa}$ is a tensor density of weight $+1$. But new identities arise if we *first* differentiate with respect to $A^{\varkappa'}_{\lambda}$, $\partial_\mu A^{\varkappa'}_{\lambda}$ and $\partial_\nu \partial_\mu A^{\varkappa'}_{\lambda}$ and *afterwards* substitute $A^{\varkappa'}_{\lambda} = \delta^{\varkappa'}_{\lambda}$, $\partial_\mu A^{\varkappa'}_{\lambda} = 0$, $\partial_\nu \partial_\mu A^{\varkappa'}_{\lambda} = 0$. This treatment, which is in reality equivalent to effecting an infinitesimal coordinate transformation, is rather complicated and it is much simpler to replace

[1]) BUCHDAHL 1948, 1; 1951, 1; 2; 3.

[2]) SCHRÖDINGER 1948, 2; 1950, 1; 1951, 1.

[3]) NIJENHUIS 1952, 1.

the coordinate transformation by a point transformation. Then we can use the LIE derivative. The LIE derivative of $\mathfrak{L}$ can be computed in two ways (cf. 10.27 and 10.17)

$$(11.18)\begin{cases} \text{a)}\ \underset{v}{\pounds}\,\mathfrak{L} = \partial_\mu\, \mathfrak{L}\, v^\mu \\ \text{b)}\ \underset{v}{\pounds}\,\mathfrak{L} = \dfrac{\partial \mathfrak{L}}{\partial P^{\varkappa}_{\cdot\lambda}}\, \underset{v}{\pounds}\, P^{\varkappa}_{\cdot\lambda} + \dfrac{\partial \mathfrak{L}}{\partial(\partial_\mu P^{\varkappa}_{\cdot\lambda})}\, \partial_\mu \underset{v}{\pounds}\, P^{\varkappa}_{\cdot\lambda} + \dfrac{\partial \mathfrak{L}}{\partial(\partial_\nu \partial_\mu P^{\varkappa}_{\cdot\lambda})}\, \partial_\nu \partial_\mu \underset{v}{\pounds}\, P^{\varkappa}_{\cdot\lambda}. \end{cases}$$

Equating these expressions we get

$$(11.19)\begin{cases} \left(\dfrac{\partial \mathfrak{L}}{\partial P^{\varkappa}_{\cdot\lambda}} - \partial_\mu \dfrac{\partial \mathfrak{L}}{\partial(\partial_\mu P^{\varkappa}_{\cdot\lambda})} + \partial_\nu \partial_\mu \dfrac{\partial \mathfrak{L}}{\partial(\partial_\nu \partial_\mu P^{\varkappa}_{\cdot\lambda})}\right) \underset{v}{\pounds}\, P^{\varkappa}_{\cdot\lambda} + \\ \qquad + \partial_\mu \left\{\dfrac{\partial \mathfrak{L}}{\partial(\partial_\mu P^{\varkappa}_{\cdot\lambda})}\, \underset{v}{\pounds}\, P^{\varkappa}_{\cdot\lambda} - 2\left(\partial_\nu \dfrac{\partial \mathfrak{L}}{\partial(\partial_\nu \partial_\mu P^{\varkappa}_{\cdot\lambda})}\right) \underset{v}{\pounds}\, P^{\varkappa}_{\cdot\lambda} - \mathfrak{L}\, v^\mu\right\} + \\ \qquad + \partial_\nu \partial_\mu \dfrac{\partial \mathfrak{L}}{\partial(\partial_\nu \partial_\mu P^{\varkappa}_{\cdot\lambda})}\, \underset{v}{\pounds}\, P^{\varkappa}_{\cdot\lambda} = 0 \end{cases}$$

or [cf. (11.17)]

$$(11.20)\begin{cases} [\mathfrak{L}]_{\varkappa}^{\cdot\lambda}\, \underset{v}{\pounds}\, P^{\varkappa}_{\cdot\lambda} + \partial_\mu \left\{\left(\dfrac{\partial \mathfrak{L}}{\partial(\partial_\mu P^{\varkappa}_{\cdot\lambda})} - 2\,\partial_\nu\, \mathfrak{L}^{\nu\mu\lambda}_{\cdot\cdot\cdot\varkappa}\right) \underset{v}{\pounds}\, P^{\varkappa}_{\cdot\lambda} - \mathfrak{L}\, v^\mu\right\} + \\ \qquad + \partial_\nu \partial_\mu\, \mathfrak{L}^{\nu\mu\lambda}_{\cdot\cdot\cdot\varkappa}\, \underset{v}{\pounds}\, P^{\varkappa}_{\cdot\lambda} = 0 \end{cases}$$

and after some calculation

$$(11.21)\begin{cases} v^\mu \left([\mathfrak{L}]_{\varkappa}^{\cdot\lambda}\, \partial_\mu P^{\varkappa}_{\cdot\lambda} - \partial_\lambda [\mathfrak{L}]_{\varkappa}^{\cdot\lambda}\, P^{\varkappa}_{\cdot\mu} + \partial_\lambda P^{\lambda}_{\cdot\varkappa} [\mathfrak{L}]_{\mu}^{\cdot\varkappa}\right) + \\ \qquad + \partial_\lambda \left([\mathfrak{L}]_{\varkappa}^{\cdot\lambda}\, P^{\varkappa}_{\cdot\mu}\, v^\mu - P^{\lambda}_{\cdot\varkappa}\, [\mathfrak{L}]_{\mu}^{\cdot\varkappa}\, v^\mu\right) + \\ \qquad + \partial_\mu \Big[-\mathfrak{L}\, v^\mu + v^\nu \Big\{\left(\dfrac{\partial \mathfrak{L}}{\partial(\partial_\mu P^{\varkappa}_{\cdot\lambda})} - 2\,\partial_\varrho\, \mathfrak{L}^{\varrho\mu\lambda}_{\cdot\cdot\cdot\varkappa}\right) \partial_\nu P^{\varkappa}_{\cdot\lambda} - \\ \qquad - \partial_\lambda P^{\varkappa}_{\cdot\nu} \left(\dfrac{\partial \mathfrak{L}}{\partial(\partial_\mu P^{\varkappa}_{\cdot\lambda})} - 2\,\partial_\varrho\, \mathfrak{L}^{\varrho\mu\lambda}_{\cdot\cdot\cdot\varkappa}\right) \,| \\ \qquad + \partial_\lambda P^{\lambda}_{\cdot\varkappa} \left(\dfrac{\partial \mathfrak{L}}{\partial(\partial_\mu P^{\nu}_{\cdot\varkappa})} - 2\,\partial_\varrho\, \mathfrak{L}^{\varrho\mu\varkappa}_{\cdot\cdot\cdot\nu}\right)\Big\} + \\ \qquad + \partial_\lambda \Big\{\left(\dfrac{\partial \mathfrak{L}}{\partial(\partial_\mu P^{\varkappa}_{\cdot\lambda})} - 2\,\partial_\varrho\, \mathfrak{L}^{\varrho\mu\lambda}_{\cdot\cdot\cdot\varkappa}\right) P^{\varkappa}_{\cdot\nu}\, v^\nu - \\ \qquad - P^{\lambda}_{\cdot\varkappa} \left(\dfrac{\partial \mathfrak{L}}{\partial(\partial_\mu P^{\nu}_{\cdot\varkappa})} - 2\,\partial_\varrho\, \mathfrak{L}^{\varrho\mu\varkappa}_{\cdot\cdot\cdot\nu}\right) v^\nu\Big\}\Big] + \\ \qquad + \partial_\nu \partial_\mu \Big[v^\varrho \left(\mathfrak{L}^{\nu\mu\lambda}_{\cdot\cdot\cdot\varkappa}\, \partial_\varrho P^{\varkappa}_{\cdot\lambda} - \partial_\lambda\, \mathfrak{L}^{\nu\mu\lambda}_{\cdot\cdot\cdot\varkappa}\, P^{\varkappa}_{\cdot\varrho} + \partial_\lambda P^{\lambda}_{\cdot\varkappa}\, \mathfrak{L}^{\nu\mu\varkappa}_{\cdot\cdot\cdot\varrho}\right) + \\ \qquad + \partial_\lambda \left(\mathfrak{L}^{\nu\mu\lambda}_{\cdot\cdot\cdot\varkappa}\, P^{\varkappa}_{\cdot\varrho}\, v^\varrho - P^{\lambda}_{\cdot\varkappa}\, \mathfrak{L}^{\nu\mu\varkappa}_{\cdot\cdot\cdot\varrho}\, v^\varrho\right)\Big] = 0. \end{cases}$$

This is an identity of the form

$$(11.22)\qquad \mathfrak{p}_\varkappa\, v^\varkappa + \partial_\lambda\, \mathfrak{q}^{\lambda}_{\cdot\varkappa}\, v^\varkappa + \partial_\mu \partial_\lambda\, \mathfrak{r}^{\mu\lambda}_{\cdot\cdot\varkappa}\, v^\varkappa + \partial_\nu \partial_\mu \partial_\lambda\, \mathfrak{s}^{\nu\mu\lambda}_{\cdot\cdot\cdot\varkappa}\, v^\varkappa = 0$$

valid for all values of the field $v^\varkappa$ and its derivatives. Now such an identity has as a consequence that

$$(11.23)\quad \begin{cases} (\mathfrak{p}_\varkappa + \partial_\lambda \mathfrak{q}^{\lambda}_{\cdot\varkappa} + \partial_\mu \partial_\lambda \mathfrak{r}^{(\mu\lambda)}_{\cdot\cdot\varkappa} + \partial_\nu \partial_\mu \partial_\lambda \mathfrak{s}^{(\nu\mu\lambda)}_{\cdot\cdot\cdot\varkappa})\, v^\varkappa + \\ \quad + (\mathfrak{q}^{\lambda}_{\cdot\varkappa} + 2\,\partial_\mu \mathfrak{r}^{(\mu\lambda)}_{\cdot\cdot\varkappa} + 3\,\partial_\nu \partial_\mu \mathfrak{s}^{(\nu\mu\lambda)}_{\cdot\cdot\cdot\varkappa})\, \partial_\lambda v^\varkappa + \\ \quad + (\mathfrak{r}^{\mu\lambda}_{\cdot\cdot\varkappa} + 3\,\partial_\nu \mathfrak{s}^{(\nu\mu\lambda)}_{\cdot\cdot\cdot\varkappa})\, \partial_\mu \partial_\lambda v^\varkappa + \mathfrak{s}^{(\nu\mu\lambda)}_{\cdot\cdot\cdot\varkappa}\, \partial_\nu \partial_\mu \partial_\lambda v^\varkappa = 0, \end{cases}$$

hence

$$(11.24)\qquad \mathfrak{s}^{(\nu\mu\lambda)}_{\cdot\cdot\cdot\varkappa} = 0; \qquad \mathfrak{r}^{(\mu\lambda)}_{\cdot\cdot\varkappa} = 0; \qquad \mathfrak{q}^{\lambda}_{\cdot\varkappa} = 0; \qquad \mathfrak{p}_\varkappa = 0.$$

Applying this to (11.22) we get four identities

$$(11.25)\quad \begin{cases} \text{a)}\ [\mathfrak{L}]^{\cdot\lambda}_{\varkappa}\, \partial_\mu P^{\varkappa}_{\cdot\lambda} - \partial_\lambda [\mathfrak{L}]^{\cdot\lambda}_{\varkappa} P^{\varkappa}_{\cdot\mu} + \partial_\lambda P^{\lambda}_{\cdot\varkappa} [\mathfrak{L}]^{\cdot\varkappa}_{\mu} = 0 \\ \text{b)}\ -\mathfrak{L} A^{\lambda}_{\nu} + [\mathfrak{L}]^{\cdot\lambda}_{\varkappa} P^{\varkappa}_{\cdot\nu} - P^{\lambda}_{\cdot\varkappa} [\mathfrak{L}]^{\cdot\varkappa}_{\nu} + \left(\dfrac{\partial\mathfrak{L}}{\partial(\partial_\lambda P^{\varkappa}_{\cdot\mu})} - 2\,\partial_\varrho \mathfrak{L}^{\varrho\lambda\mu}_{\cdot\cdot\cdot\varkappa}\right) \partial_\nu P^{\varkappa}_{\cdot\mu} - \\ \quad - \partial_\mu P^{\varkappa}_{\cdot\nu} \left(\dfrac{\partial\mathfrak{L}}{\partial(\partial_\lambda P^{\varkappa}_{\cdot\mu})} - 2\,\partial_\varrho \mathfrak{L}^{\varrho\lambda\mu}_{\cdot\cdot\cdot\varkappa}\right) + \partial_\mu P^{\mu}_{\cdot\varkappa} \left(\dfrac{\partial\mathfrak{L}}{\partial(\partial_\lambda P^{\nu}_{\cdot\varkappa})} - 2\,\partial_\varrho \mathfrak{L}^{\varrho\lambda\varkappa}_{\cdot\cdot\cdot\nu}\right) = 0 \\ \text{c)}\ \left(\dfrac{\partial\mathfrak{L}}{\partial(\partial_\mu P^{\varkappa}_{\cdot\lambda})} - 2\,\partial_\varrho \mathfrak{L}^{\varrho(\mu\lambda)}_{\cdot\cdot\cdot\varkappa}\right) P^{\varkappa}_{\cdot\sigma} - A^{(\mu}_{\nu} P^{\lambda)}_{\cdot\varkappa} \left(\dfrac{\partial\mathfrak{L}}{\partial(\partial_\nu P^{\sigma}_{\cdot\varkappa})} - 2\,\partial_\varrho \mathfrak{L}^{\varrho\nu\varkappa}_{\cdot\cdot\cdot\sigma}\right) + \\ \quad + \mathfrak{L}^{(\lambda\mu)\nu}_{\cdot\cdot\cdot\varkappa}\, \partial_\sigma P^{\varkappa}_{\cdot\nu} - \partial_\nu \mathfrak{L}^{(\lambda\mu)\nu}_{\cdot\cdot\cdot\varkappa} P^{\varkappa}_{\cdot\sigma} + \partial_\nu P^{\nu}_{\cdot\varkappa} \mathfrak{L}^{(\lambda\mu)\varkappa}_{\cdot\cdot\cdot\sigma} = 0 \\ \text{d)}\ \mathfrak{L}^{(\nu\mu\lambda)}_{\cdot\cdot\cdot\varrho} P^{\varrho}_{\cdot\varkappa} - \mathfrak{L}^{(\nu\mu|\varrho|}_{\cdot\cdot\cdot\varkappa} P^{\lambda)}_{\cdot\varrho} = 0. \end{cases}$$

Of these (11.25a) is a special case of (11.11). (11.25d) is an algebraic identity and the others are differential identities.

Similarly, if a density $\mathfrak{L}$ of weight $+1$ is formed from a tensor P (with arbitrary valences) and its derivatives up to the order p, an equation of the form (11.23) arises with $p+2$ terms and thus we get $p+2$ identities whose total number of equations is

$$(11.26)\qquad n + n^2 + n\binom{n+1}{2} + \cdots + n\binom{n+p}{p+1} = n\binom{n+p+1}{p+1}.$$

The first n equations can be found in the same way in which (11.11) is found and the last $n\binom{n+p}{p+1}$ are algebraic identities linear homogeneous in P and in the derivatives of $\mathfrak{L}$ with respect to $\partial_{\mu_p} \ldots \partial_{\mu_1} P$.

If instead of $P^{\varkappa}_{\cdot\lambda}$ a tensor $g_{\lambda\varkappa}$ of rank n is taken, the second and third term of (11.25b) are

$$(11.27)\qquad -[\mathfrak{L}]^{\varkappa\nu} g_{\varkappa\lambda} - [\mathfrak{L}]^{\nu\varkappa} g_{\lambda\varkappa} = -2\,[\mathfrak{L}]^{\varkappa\nu} g_{\varkappa\lambda}.$$

This is also true when higher derivatives of $g_{\lambda\varkappa}$ occur. Hence by transvection with the inverse of $g_{\lambda\varkappa}$ the LAGRANGE derivative can be expressed in terms of $\mathfrak{L}$, $g_{\lambda\varkappa}$ and their derivatives. [1])

[1]) BUCHDAHL 1951, 3.

If P is a geometric object but not a quantity [for instance the $\Pi^{\varkappa}_{\mu\lambda}$ in (10.31)] the LIE derivative of Π contains not only first but also *second* derivatives of $v^{\varkappa}$. If for instance Π and its first derivatives occur in $\mathfrak{L}$, one finds $n\binom{n+3}{3}$ equations of which the last $n\binom{n+2}{3}$ are algebraic.[1])

Exercise.

II 11,1. Prove that the left hand side of (11.11) is a covariant vector Δ-density of weight $+1$ if $[\mathfrak{L}]$ is replaced by an arbitrary tensor Δ-density $\mathfrak{Q}$ of valences q and p and weight $+1$. For $q=1$, $p=0$ and $q=0$, $p=1$ this gives the differential concomitants

II 11,1 α) $$2\mathfrak{Q}^{\lambda}\partial_{[\mu}P_{\lambda]} - P_{\mu}\partial_{\lambda}\mathfrak{Q}^{\lambda}$$

II 11,1 β) $$\mathfrak{Q}_{\lambda}\partial_{\mu}P^{\lambda} + \partial_{\lambda}(\mathfrak{Q}_{\mu}P^{\lambda})\,.$$

In an R_3 with signature $+++$ these are the concomitants written in ordinary vector notation:

II 11,1 γ) $$\mathbf{Q}\times\operatorname{rot}\mathbf{P} - \mathbf{P}\operatorname{div}\mathbf{Q}$$

II 11,1 δ) $$\operatorname{grad}(\mathbf{P}\,.\,\mathbf{Q}) - \mathbf{P}\times\operatorname{rot}\mathbf{Q} + \mathbf{Q}\operatorname{div}\mathbf{P}\,.$$

§ 12. CARTAN's symbolical method.

CARTAN's symbolical method is intimately connected with anholonomic coordinate systems. It is in fact an abridged notation specially suited to dealing with these systems. If the A^h_λ of an anholonomic system (h) are fixed, these components may be considered as n sets of components of the n basis vectors $\overset{1}{e}_{\lambda}, \ldots, \overset{n}{e}_{\lambda}$ of (h) or (if we like) of the corresponding pfaffians $\overset{h}{e}_{\lambda}\,d\xi^{\lambda}$. Then any geometry introduced into X_n can be expressed in terms of these vectors or these pfaffians. This is the point of view of CARTAN. In his thesis of 1894[2]) he developed a very ingenious method to deal with pfaffians and their generalizations and this method was used in connection with some kind of tensor calculus in nearly all his papers on differential geometry. The introduction of a tensor calculus will always be necessary if non alternating quantities are considered, because the symbolical method is made for alternating quantities only. In 1934 KÄHLER[3]) gave a systematic account of the possibilities of the method as far as alternating quantities only were concerned and under his influence some notations were slightly changed.

---

[1]) SCHRÖDINGER 1951, 1.

[2]) CARTAN 1894, 1; 1901, 1.

[3]) KÄHLER 1934, 1.

In this section we try to give an exposition of the abridged notation in such a way that it becomes fully adapted to the notations of tensor calculus.[1] Using the kernel-index method this can be attained very easily by *using the same kernels in both sets of formulae*. Then a simple rule for translation can be established.

If $\underset{1}{d}\xi^\varkappa, \ldots, \underset{p}{d}\xi^\varkappa$ are linearly independent linear elements and $w_{\lambda_1 \ldots \lambda_p}$ a covariant p-vector, all at the same point, the expression

$$[d\xi^{\varkappa_1} \ldots d\xi^{\varkappa_p}] \overset{\text{def}}{=} \alpha_p \underset{1}{d}\xi^{[\varkappa_1} \ldots \underset{p}{d}\xi^{\varkappa_p]}, \tag{12.1}$$

where α_p is a constant that will be fixed later on, is a contravariant p-vector and

$$\begin{cases} w \overset{\text{def}}{=} w_{\lambda_1 \ldots \lambda_p} [d\xi^{\lambda_1} \ldots d\xi^{\lambda_p}] = \alpha_p w_{\lambda_1 \ldots \lambda_p} \underset{1}{d}\xi^{\lambda_1} \ldots \underset{p}{d}\xi^{\lambda_p} \\ \quad = \alpha_p w_{\lambda'_1 \ldots \lambda'_p} \underset{1}{d}\xi^{\lambda'_1} \ldots \underset{p}{d}\xi^{\lambda'_p} = \alpha_p w_{i_1 \ldots i_p} (\underset{1}{d}\xi)^{i_1} \ldots (\underset{p}{d}\xi)^{i_p} \end{cases} \tag{12.2}$$

is a generalization of a pfaffian, called an *alternating differential form*. For $p = 1$ we take $\alpha_1 = 1$ and

$$w = w_\lambda d\xi^\lambda = w_{\lambda'} d\xi^{\lambda'} = w_i (d\xi)^i. \tag{12.3}$$

The product of two forms $\overset{1}{w}$ and $\overset{2}{w}$ of valence p and q may be defined as follows

$$[\overset{1}{w}\overset{2}{w}] \overset{\text{def}}{=} \alpha_{p+q} \overset{1}{w}_{[\lambda_1 \ldots \lambda_p} \overset{2}{w}_{\varkappa_1 \ldots \varkappa_q]} \underset{1}{d}\xi^{\lambda_1} \ldots \underset{p}{d}\xi^{\lambda_p} \underset{p+1}{d}\xi^{\varkappa_1} \ldots \underset{p+q}{d}\xi^{\varkappa_q}; \; p+q \leq n. \tag{12.4}$$

Then we have (cf. I 7.1)

$$[\overset{1}{w}\overset{2}{w}] = (-1)^{pq} [\overset{2}{w}\overset{1}{w}]. \tag{12.5}$$

In all these expressions it is quite unimportant whether we consider w as an alternating form or as an abridged notation of a multivector. The results are the same and the interpretation is only a question of personal predilection.[2]

---

[1] This is very important. All authors who are using, where convenient, CARTAN's abridged calculus, always use also notations of tensor calculus and they seem to have no difficulties with the latter. But as a matter of fact most authors using tensor calculus have the greatest difficulties with CARTAN's abridged calculus, even in translating results. This is a great pity. Both methods have their advantages (CARTAN himself performed miracles with his method!) and both should be used by every author, each on the place where it suits best. But this ideal can only be attained if there is a convenient bridge between them. We try to build this bridge here.

[2] Just the same thing occurs in linear algebra. We may look upon w as an abridged notation for a covariant vector w_λ or as a linear form $w_\lambda x^\lambda$. The old theory of invariants preferred the latter point of view but nowadays many people think more easily in terms of vectors.

If s is a scalar, an invariant differentiation of the p-vector $s\,[d\xi^{\varkappa_1}\dots d\xi^{\varkappa_p}]$ can be defined as follows

$$(12.6)\quad \left\{\begin{aligned} [d\{s\,[d\xi^{\varkappa_1}\dots d\xi^{\varkappa_p}]\}] &\overset{\text{def}}{=} \alpha_{p+1}\,(\partial_\lambda s)\, d\xi^{[\lambda} \underset{1}{d}\xi^{\varkappa_1}\dots \underset{p}{d}\xi^{\varkappa_p]} \\ &= (\partial_\lambda s)\,[d\xi^\lambda d\xi^{\varkappa_1}\dots d\xi^{\varkappa_p}]. \end{aligned}\right.$$

From this definition and (12.2) it follows that

$$(12.7)\quad [dw] = (\partial_\lambda w_{\lambda_1\dots\lambda_p})\,[d\xi^\lambda d\xi^{\lambda_1}\dots d\xi^{\lambda_p}] = \alpha_{p+1}\,\partial_{[\lambda} w_{\lambda_1\dots\lambda_p]}\, d\xi^\lambda \underset{1}{d}\xi^{\lambda_1}\dots \underset{p}{d}\xi^{\lambda_p}.$$

The differentiation of a product we easily find then to be

$$(12.8)\quad [d\overset{1}{w}\overset{2}{w}] = [[d\overset{1}{w}]\overset{2}{w}] + (-1)^p\,[\overset{1}{w}\,[d\overset{2}{w}]]; \qquad p = \text{valence of } \overset{1}{w}.$$

CARTAN always took $\alpha_p = p!$ and he used the product notation $[\overset{1}{w}\overset{2}{w}]$. But instead of $[dw]$, at first he wrote w' and more recently under the influence of KÄHLER dw. Now this latter notation may lead to ambiguity because $[dw]$ is certainly not an ordinary differential. KÄHLER wrote $d(\xi^{\varkappa_1}, \dots, \xi^{\varkappa_p})$ for what we have written $\alpha_p\, \underset{1}{d}\xi^{[\varkappa_1}\dots \underset{p}{d}\xi^{\varkappa_p]}$ and $[d\xi^{\varkappa_1}\dots d\xi^{\varkappa_p}]$, and he did not fix the values of α_p. In fact, if only symbolic expressions are used, and if we never wish to translate into notations of tensor calculus, the constants α_p never occur. For $[\overset{1}{w}\overset{2}{w}]$ KÄHLER wrote $\overset{1}{w}\overset{2}{w}$ without brackets. But in dealing with less simple problems the notation with brackets should be preferred. As to the choice of the α_p we remark that, according to (12.4), $[\overset{1}{w}\overset{2}{w}]$ does not correspond to the ordinary alternating product of the multivectors corresponding to $\overset{1}{w}$ and $\overset{2}{w}$ but to $\alpha_{p+q}\,\alpha_p^{-1}\,\alpha_q^{-1}$ times this product. So, if we wish to establish a reasonable adaptation we have to take $\alpha_p = 1$ for all values of p. Doing this and *always using the same kernels in both notations* we get the following simple scheme of translation. Let $\Pi_{\lambda_1\dots\lambda_p A}$ be components of a geometric object, alternating in the first p indices and carrying a collective index A standing for any number of co- and contravariant indices, each belonging to any holonomic or anholonomic coordinate system in X_n or any other space. Let the objects $\Pi_{\lambda_1\dots\lambda_p(A)}$ with p living indices $\lambda_1\dots\lambda_p$, resulting from strangling all indices symbolized by A, be covariant p-vectors. As an example we take the intermediate components A^h_λ and $\Omega^h_{\mu\lambda}$ (cf. II § 9). In fact $A^{(h)}_\lambda = \overset{h}{e}_\lambda$ are n vectors and $\Omega^{(h)}_{\mu\lambda}$ are n bivectors. Now Π_A denotes the p-vectors, resulting in this way:

$$(12.9)\quad \boxed{\Pi_A:\ \left\{\begin{array}{l} \Pi_{\lambda_1\dots\lambda_p(A)} \\ \text{and also} \\ \Pi_{\lambda_1\dots\lambda_p(A)}\, \underset{1}{d}\xi^{\lambda_1}\dots \underset{p}{d}\xi^{\lambda_p} \end{array}\right.}\,.$$

The product $[\overset{1}{\Pi}_A \overset{2}{\Pi}_B]$ stands for

$$(12.10)\qquad [\overset{1}{\Pi}_A \overset{2}{\Pi}_B]: \begin{cases} \overset{1}{\Pi}_{[\lambda_1 \dots \lambda_p|(A)|} \overset{2}{\Pi}_{\varkappa_1 \dots \varkappa_q]\,(B)} \\ \text{and also for} \\ \overset{1}{\Pi}_{[\lambda_1 \dots \lambda_p|(A)|} \overset{2}{\Pi}_{\varkappa_1 \dots \varkappa_q]\,(B)} \underset{1}{d}\,\xi^{\lambda_1} \dots \underset{p}{d}\,\xi^{\lambda_p} \underset{p+1}{d}\,\xi^{\varkappa_1} \dots \underset{p+q}{d}\,\xi^{\varkappa_q} \end{cases}$$

and the expression $[d\Pi_A]$ symbolizes

$$(12.11)\qquad [d\Pi_A]: \begin{cases} \partial_{[\mu} \Pi_{\lambda_1 \dots \lambda_p]\,(A)} \\ \text{and also} \\ \partial_{[\mu} \Pi_{\lambda_1 \dots \lambda_p]\,(A)}\, d\xi^\mu \underset{1}{d}\,\xi^{\lambda_1} \dots \underset{p}{d}\,\xi^{\lambda_p} \end{cases}.$$

With this scheme every abridged formula can immediately be translated into a formula of tensor calculus. There is one point that is at first bewildering for one accustomed to tensor calculus only. For instance w does not stand for w_λ only, but also for $w_{\lambda'}$ or w_i and even for $w_\lambda\, d\xi^\lambda$, $w_{\lambda'}\, d\xi^{\lambda'}$ and $w_i (d\xi)^i$, and $[dw]$ denotes not only $\partial_{[\mu} w_{\lambda]}$ but also $\partial_{[\mu'} w_{\lambda']}$ and $\partial_{[j} w_{i]} + w_h \Omega_{ji}^{h}$ (cf. Exerc. II 9,1) and even the alternating differential forms obtained by transvecting these expressions by $\underset{1}{d}\xi^\mu \underset{2}{d}\xi^\lambda$, $\underset{1}{d}\xi^{\mu'} \underset{2}{d}\xi^{\lambda'}$ and $(\underset{1}{d}\xi)^j (\underset{2}{d}\xi)^i$ respectively. This means that in *passing to the abridged notation the coordinate system belonging to the alternating part of the object is set free and that at the same time the indices not belonging to this part are all strangled.*

From now on we mark all formulae in abridged notation where they appear together with the non abridged formula by the sign $\mathbb{C}$ (a sign taken from music) and give both formulae the same number. We give here two examples, concerning the object of anholonomity [cf. (9.10) and Exerc. II 9,2]:

$$(12.12)\qquad \partial_{[\mu} A_{\lambda]}^{h} = A_{\mu\lambda}^{ji} \Omega_{ji}^{h} = \Omega_{\mu\lambda}^{h}; \qquad ((\varkappa)\ \text{holonomic})$$

$$(12.12)\quad \mathbb{C} \qquad [dA^h] = \Omega^h = \Omega_{ji}^{h} [A^j A^i]$$

$$(12.13)\qquad 0 = A_{\nu\mu\lambda}^{kji} \{\partial_{[k} \Omega_{ji]}^{h} - 2\Omega_{[kj}^{l} \Omega_{i]l}^{h}\}$$

$$(12.13)\quad \mathbb{C} \qquad 0 = [(d\Omega_{ji}^{h}) A^j A^i] - 2\Omega_{kj}^{l} \Omega_{il}^{h} [A^k A^j A^i].$$

So far we have seen that certain (not all) formulae of tensor calculus can be written in the abridged form and that by doing this a lot of indices can be saved. On the other hand the abridged calculus requires more attention during the calculation and so it is merely a question of personal inclination whether one or the other calculus will be preferred. But in order to give abridged calculus fair play we draw the attention to some cases where it has certainly advantages.

Let us take a vector field w_λ and its rotation $W_{\mu\lambda}$:

$$(12.14) \qquad 2\,\partial_{[\mu}\, w_{\lambda]} = W_{\mu\lambda}$$

$$(12.14) \quad \mathfrak{C} \qquad 2\,[d\,w] = W$$

and let us suppose that we wish to consider w_λ not only as a function of the $\xi^\varkappa$ but also of certain parameters η^α; $\alpha = \dot{1}, \ldots, \dot{r}$. Then we introduce in the X_{n+r} of $\xi^\varkappa, \eta^\alpha$ a field w_B; $B = 1, \ldots, n, \dot{1}, \ldots, \dot{r}$ with the $n+r$ components $w_\lambda, w_\beta = 0$. If this X_{n+r} is reduced with respect to the normal system of X_r's $\xi^\varkappa =$ const. the field w_B is reduced to the original field w_λ in the X_n, depending on the parameters η^α. In tensor-calculus we have now

$$(12.15) \qquad \begin{cases} W_{\mu\lambda} = 2\,\partial_{[\mu}\, w_{\lambda]}; & \mu, \lambda = 1, \ldots, n \\ W_{\gamma\lambda} \overset{\text{def}}{=} \partial_\gamma\, w_\lambda; & \beta, \gamma = \dot{1}, \ldots, \dot{r} \\ W_{\gamma\beta} \overset{\text{def}}{=} 2\,\partial_{[\gamma}\, w_{\beta]} = 0 \end{cases}$$

or

$$(12.16) \qquad W_{CB} = 2\,\partial_{[C}\, w_{B]}; \qquad B, C = 1, \ldots, n, \dot{1}, \ldots, \dot{r}$$

but in the abridged calculus this equation is symbolized by the *same* equation (12.14 $\mathfrak{C}$). Hence *the abridged calculus gives the possibility of introducing at any moment auxiliary variables without changing the form of the formulae.*

Some other examples of the use of the abridged calculus will be found in Ch. III.

III. Linear connexions.

§ 1. Parallel displacement in an E_n.[1]

Let $v^\varkappa$ be a vector field in an X_n. As we have already seen in II § 3, $dv^\varkappa$ is not a vector and $\partial_\mu v^\varkappa$ not a tensor. But if the X_n is an E_n and if (h) is a rectilinear coordinate system, dv^h and $\partial_i v^h$ are a vector and a tensor because if (h') is another rectilinear system, the $A_h^{h'}$ are constants.

[1]) Tensor calculus was founded, developed and applicated by Ricci in a great number of papers. We mention here 1884, 1 and 1886, 1; 2 (all preliminary); 1887, 1; 2; 1888, 1; 2; 1889, 1; 2; 1892, 1; 1893, 1; 2; 1894, 1; 2; 1895, 1; 2; 1897,

dv^h is the difference between the vector $v^h + dv^h$ at $\xi^h + d\xi^h$ and a vector with the components v^h at the same point. This latter vector can be derived from the vector v^h at ξ^h by the well known process of parallel displacement.

Now let $(\varkappa)$ be a general coordinate system in E_n. Then the $A_h^\varkappa$ are not necessarily constants and we have

$$(1.1) \quad dv^h = d(v^\varkappa A_\varkappa^h) = A_\varkappa^h dv^\varkappa + v^\lambda \partial_\mu A_\lambda^h d\xi^\mu = A_\varkappa^h (dv^\varkappa + \Gamma_{\mu\lambda}^\varkappa v^\lambda d\xi^\mu)$$

where

$$(1.2) \quad \Gamma_{\mu\lambda}^\varkappa \stackrel{\text{def}}{=} A_h^\varkappa \partial_\mu A_\lambda^h = -A_\lambda^h \partial_\mu A_h^\varkappa.$$

Hence the components of the vector dv^h with respect to $(\varkappa)$ are

$$(1.3) \quad \delta v^\varkappa \stackrel{\text{def}}{=} A_h^\varkappa dv^h = dv^\varkappa + \Gamma_{\mu\lambda}^\varkappa v^\lambda d\xi^\mu.$$

We call $\delta v^\varkappa$ the *covariant differential* of $v^\varkappa$ and the $\Gamma_{\mu\lambda}^\varkappa$ the *parameters* of the parallel displacement in E_n with respect to $(\varkappa)$. If we write $\delta v^\varkappa = \nabla_\mu v^\varkappa d\xi^\mu$, we get

$$(1.4) \quad \nabla_\mu v^\varkappa \stackrel{\text{def}}{=} \partial_\mu v^\varkappa + \Gamma_{\mu\lambda}^\varkappa v^\lambda.$$

$\nabla_\mu v^\varkappa$ is called the *covariant derivative* of $v^\varkappa$. If $(\varkappa)$ is rectilinear, the $\Gamma_{\mu\lambda}^\varkappa$ are all zero and we have $\delta v^\varkappa \stackrel{*}{=} dv^\varkappa$ and $\nabla_\mu v^\varkappa \stackrel{*}{=} \partial_\mu v^\varkappa$.

Instead of v^h we can take a covariant vector field w_i. Then for the components of w_i and $\partial_j w_i$ with respect to $(\varkappa)$ we get

$$(1.5) \quad \delta w_\lambda \stackrel{\text{def}}{=} dw_\lambda - \Gamma_{\mu\lambda}^\varkappa w_\varkappa d\xi^\mu,$$

$$(1.6) \quad \nabla_\mu w_\lambda \stackrel{\text{def}}{=} \partial_\mu w_\lambda - \Gamma_{\mu\lambda}^\varkappa w_\varkappa.$$

But these formulae can also be derived by using the fact that the transvection $v^\varrho w_\varrho$ is invariant if both vectors undergo the same parallel displacement. Hence we have $d(v^\varrho w_\varrho) = \delta(v^\varrho w_\varrho)$ and the equation

$$(1.7) \quad w_\varrho dv^\varrho + v^\varrho dw_\varrho = (dv^\varrho + \Gamma_{\mu\lambda}^\varrho v^\lambda d\xi^\mu) w_\varrho + v^\varrho \delta w_\varrho$$

can only be true for every choice of the field $v^\varkappa$ if (1.5) holds.

1; 2; 3; 1898, 1; 2; 1902, 1; 2; 1903, 1; 1904, 1; 1905, 1; 1910, 1; 2; 1912, 1; 2; 1918, 1; 2. It found its first systematic representation in RICCI and LEVI CIVITA 1901, 1. His "derivazione covariante" got a geometric interpretation in the "pseudo-parallel" or "geodesic" displacement (now called parallel displacement) introduced by LEVI CIVITA 1917, 1 and independently by SCHOUTEN 1918, 1. Cf. also BROUWER 1906, 2, p. 121 who gave a first example of a displacement of a vector in an S_n (cf. III § 5).

§ 2. Parallel displacements in X_n [1], [2].

We have seen in III § 1 that the construction of the covariant differential in E_n only depends on the existence of parallel displacements in E_n. We do not need all displacements but only those from one point to a "neighbouring" point. Hence if in an X_n a definition of a displacement of quantities were given for every point of the region considered and for every linear element in such a point, it must be possible to construct some kind of a covariant differential.

Let Φ (indices suppressed) be a field of *quantities* defined in some region of X_n. Then at $\xi^\varkappa$ and $\xi^\varkappa + d\xi^\varkappa$ we have the field values Φ and $\Phi + d\Phi$. Let a parallel displacement for these quantities be defined for every choice of $\xi^\varkappa$ and $d\xi^\varkappa$ in the region considered and let $\Phi + \overset{*}{d}\Phi$ be the field value at $\xi^\varkappa + d\xi^\varkappa$ after a parallel displacement from $\xi^\varkappa$ to $\xi^\varkappa + d\xi^\varkappa$. Then $d\Phi - \overset{*}{d}\Phi$ is a covariant differential $\delta\Phi$ of Φ belonging to this displacement. Of course we make the condition that if the components of $\Phi + \overset{*}{d}\Phi$ are given for one coordinate system $(\varkappa)$ its com-

[1]) Parallel displacements (formerly often called pseudo-parallel displacements) are considered here only with respect to the tangent E_n's of an X_n. König generalized 1920, 1 this idea for a set of E_N's each of which is associated with a point of X_n. Cf. König 1932, 1; Sibata 1934, 1; Hlavaty 1934, 1; 1935, 1; König and Peschl 1934, 1; König and Weise 1935, 1. As we proved 1924, 2; 3 the conformal and projective connexions introduced by Cartan 1923, 2 and 1924, 1 can be looked upon as special cases of König connexions. There is much literature on these general connexions. We mention here only a few papers where also other literature can be found: Schouten 1924, 1, p. 62; 1926, 1; Schlesinger 1928, 1; Morinaga 1934, 2; 3; Tucker 1935, 1; Yano and Muto 1935, 1; 1936, 1; Wagner 1943, 1; Yano 1945, 2; 3.

[2]) There is a large literature on connexions and parallelism. We mention here only Hessenberg 1916, 1 (preliminary); Levi Civita 1917, 1; Schouten 1918, 1; 1922, 3; 1923, 1; 1926, 1 (also non linear) 1928, 1; Frieseke 1925, 1 (non linear); Eisenhart 1927, 1; Struik 1928, 1 (literature); Schlesinger 1928, 1; 1929, 1; Wundheiler 1929, 1 (non linear); Duschek 1930, 2 (axiomatic); Bortolotti 1930, 3 (literature); 1930, 2 and 1931, 3 (degenerate fundamental tensor); 1931, 4 (also non linear and parallel directions) 1935, 1; 1936, 1 (non linear); 1937, 3; 1939, 1 (also for literature) 1942, 1; Conforto 1931, 1; 2 (in function space); A. Kawaguchi 1932, 1 (also non linear) 1933, 1; 1933, 2 (also non linear); 1934, 1; 1935, 1 (in function space and literature); 1952, 1 (non linear, general theory and literature); Vanderslice 1934, 1; Hokari 1934, 1 (for bivectors); Bortolotti and Hlavaty 1936, 2 (also literature); Urabi 1940, 1; Mikami 1941, 1; Wagner 1943, 1 and 1945, 2 (general geom. obj.); Norden 1945, 1 and 1950, 1 (space with two connexions); Dalla Volta 1946, 1 and 2 (asymmetric); Bompiani 1945, 1; 1946, 1 (for tensors); 1952, 1 (conn. non posizionali); Rozenfel'd 1947, 1 and 1948, 1 (generalizations); Johnson 1948, 1; Liber 1950, 1 (classification of connexions in L_2); Ehresmann 1950, 1 (in fibre space); 1952, 2 (higher order); Knebelmann 1951, 1 (parallelism relative to a continuous curve); Winogradski 1951, 1 (axiomatic); Takasu 1953, 1.

ponents with respect to $(\varkappa')$ can be derived by the rules valid for the transformation of $\Phi + d\Phi$. That means that we introduce as a first condition

C I) *The covariant differential of a quantity Φ transforms in the same way as Φ.*

Now there are many possibilities for the definition of a displacement. Here we consider only those displacements that satisfy (C I) and the following four conditions (all quantities considered are supposed to be given as fields in a certain region):

C II) *The covariant differential of a sum of quantities is the sum of the covariant differentials of the terms.*

C III) *For the covariant differential of a product or a transvection of quantities the rule of* LEIBNIZ *holds, for instance* (indices suppressed)

$$\delta(\Phi\Psi) = (\delta\Phi)\Psi + \Phi\,\delta\Psi .$$

C IV) *The covariant differential of a quantity is linear homogeneous in the $d\xi^\varkappa$:*

$$\delta\Psi = \nabla_\mu \Psi\, d\xi^\mu .$$

C V) *The covariant differential of a scalar or W-scalar is identical with the ordinary differential.*

For a contravariant vector $v^\varkappa$ we have $\delta v^\varkappa = d v^\varkappa - \overset{*}{d} v^\varkappa$, hence according to (C IV) $\overset{*}{d} v^\varkappa$ is linear homogeneous in $d\xi^\varkappa$

$$\overset{*}{d} v^\varkappa = \varphi^\varkappa_\mu (v^\lambda)\, d\xi^\mu . \tag{2.1}$$

But then it follows from (C II) that $\varphi^\varkappa_\mu(v^\lambda)$ is linear homogeneous in $v^\varkappa$. Hence $\delta v^\varkappa$ and $\nabla_\mu v^\varkappa$ have the form

$$\delta v^\varkappa = d v^\varkappa + \Gamma^\varkappa_{\mu\lambda} v^\lambda d\xi^\mu , \tag{2.2}$$

$$\nabla_\mu v^\varkappa = \partial_\mu v^\varkappa + \Gamma^\varkappa_{\mu\lambda} v^\lambda . \tag{2.3}$$

According to (C I) we get for the transformation $(\varkappa) \to (\varkappa')$

$$A^{\varkappa'}_\varkappa (d v^\varkappa + \Gamma^\varkappa_{\mu\lambda} v^\lambda d\xi^\mu) = d v^{\varkappa'} + \Gamma^{\varkappa'}_{\mu'\lambda'} v^{\lambda'} d\xi^{\mu'} \tag{2.4}$$

from which we easily deduce the transformation formulae

$$\Gamma^{\varkappa'}_{\mu'\lambda'} = A^{\mu\,\lambda\,\varkappa'}_{\mu'\lambda'\varkappa} \Gamma^\varkappa_{\mu\lambda} + A^{\varkappa'}_\lambda \partial_{\mu'} A^\lambda_{\lambda'} = A^{\mu\,\lambda\,\varkappa'}_{\mu'\lambda'\varkappa} \Gamma^\varkappa_{\mu\lambda} - A^{\mu\,\lambda}_{\mu'\lambda'} \partial_\mu A^{\varkappa'}_\lambda \tag{2.5}$$

showing that the $\Gamma^\varkappa_{\mu\lambda}$ do not constitute a quantity but a more general geometric object of the kind already considered in II § 10.

The n^3 parameters $\Gamma^\varkappa_{\mu\lambda}$ can be given arbitrarily for any system $(\varkappa)$. Then by (2.5) the parameters with respect to every allowable coordinate system are known. They are said to fix a *linear displacement* or a *linear connexion* in X_n.[1]) An X_n provided with a linear connexion is called an L_n. If a vector in an L_n is displaced in such a way that $\delta v^\varkappa = 0$ the displacement is said to be *parallel*.

Using (C V) it can be proved in the same way as in III § 1 that for a covariant field w_λ

$$\delta w_\lambda = d w_\lambda - \Gamma^\varkappa_{\mu\lambda} w_\varkappa d\xi^\mu, \tag{2.6}$$

$$\nabla_\mu w_\lambda = \partial_\mu w_\lambda - \Gamma^\varkappa_{\mu\lambda} w_\varkappa. \tag{2.7}$$

Every tensor can be written as a sum of general products of vectors (cf. I § 3). Hence as a consequence of (C III) we have for instance for $P^{\varkappa\lambda}_{\cdot\cdot\mu}$

$$\delta P^{\varkappa\lambda}_{\cdot\cdot\mu} = d P^{\varkappa\lambda}_{\cdot\cdot\mu} + \Gamma^\varkappa_{\nu\varrho} P^{\varrho\lambda}_{\cdot\cdot\mu} d\xi^\nu + \Gamma^\lambda_{\nu\varrho} P^{\varkappa\varrho}_{\cdot\cdot\mu} d\xi^\nu - \Gamma^\varrho_{\nu\mu} P^{\varkappa\lambda}_{\cdot\cdot\varrho} d\xi^\nu, \tag{2.8}$$

$$\nabla_\nu P^{\varkappa\lambda}_{\cdot\cdot\mu} = \partial_\nu P^{\varkappa\lambda}_{\cdot\cdot\mu} + \Gamma^\varkappa_{\nu\varrho} P^{\varrho\lambda}_{\cdot\cdot\mu} + \Gamma^\lambda_{\nu\varrho} P^{\varkappa\varrho}_{\cdot\cdot\mu} - \Gamma^\varrho_{\nu\mu} P^{\varkappa\lambda}_{\cdot\cdot\varrho}. \tag{2.9}$$

For every upper (lower) index there is a term with a +-(—-)sign. Every W-tensor is the product of a tensor with a W-scalar. Hence, according to (C V) the formulae for the covariant differentiation of tensors hold also for W-tensors.

The covariant differential is the difference between the field value at $\xi^\varkappa + d\xi^\varkappa$ and the field value of the field that has been displaced parallel from $\xi^\varkappa$ to $\xi^\varkappa + d\xi^\varkappa$. This can be said in another way; *the covariant differential is the ordinary differential from the point of view of a local coordinate system that is displaced parallel* from $\xi^\varkappa$ to $\xi^\varkappa + d\xi^\varkappa$.

A quantity is called *allround covariant stationary at a point* of an L_n if its covariant differential vanishes at that point for every direction. It is called *covariant stationary at a point in an m-direction* if the covariant differential vanishes at that point for all directions lying in this m-direction. It is said to be *covariant constant over* L_n or *covariant constant* for short if it is allround covariant stationary at all points of L_n.[2])

If a quantity is defined over an X_m in L_n it is called *covariant constant over* X_m if at all points of X_m it is covariant stationary in the tangent m-direction of X_m. If the quantity is also defined in some

[1]) GALVANI has shown that connexions often can be realized by taking for points certain subspaces of a flat space, 1942, 1; 2; 1943, 1; 2; 1945, 1; 1946, 1. Cf. also for other choices of the elements of space ROZENFEL'D 1947, 1; 1948, 1; YANO and HIRAMATU (sets of hypersurfaces) 1951, 1.

[2]) The existence of covariant constant fields *in the large* is quite another problem that does not belong to the realm of this book. We mention here only a few papers where also other literature can be found: BORTOLOTTI 1928, 1; 1929, 1; T. Y. THOMAS 1936, 1; MAYER and T. Y. THOMAS, 1937, 1.

neighbourhood of X_m it may happen that it is allround covariant stationary in L_n at all points of X_m. In this case it is called *allround covariant stationary* in L_n over X_m. Of course it is then also covariant constant over X_m. The following scheme gives the relations between these five cases

	$\supset$	allround cov. stat. at point of X_m	$\supset$	cov. stat. in m-direction at point of X_m
cov. constant over L_n		$\cap$		$\cap$
	$\supset$	allround cov. stat. in L_n over X_m	$\supset$	cov. constant over X_m [1])

A field that is at every point and for every direction proportional to its covariant differential is called *recurrent*. A recurrent vector field is also called *parallel*.

If we wish to write out a covariant derivative in terms of its intermediate components (cf. I § 2) for instance between $(\varkappa)$ and $(\varkappa')$ it is convenient to introduce the notations

$$\Gamma^{\varkappa}_{\mu'\lambda} \overset{\text{def}}{=} A^{\mu}_{\mu'} \Gamma^{\varkappa}_{\mu\lambda}; \qquad \Gamma^{\varkappa'}_{\mu\lambda'} \overset{\text{def}}{=} A^{\mu'}_{\mu} \Gamma^{\varkappa'}_{\mu'\lambda'}. \tag{2.10}$$

Then for $P^{\varkappa\lambda}_{\cdot\cdot\mu}$ for instance we have the formulae

$$\nabla_\nu P^{\varkappa'\lambda}_{\cdot\cdot\mu} = \partial_\nu P^{\varkappa'\lambda}_{\cdot\cdot\mu} + \Gamma^{\varkappa'}_{\nu\varrho'} P^{\varrho'\lambda}_{\cdot\cdot\mu} + \Gamma^{\lambda}_{\nu\varrho} P^{\varkappa'\varrho}_{\cdot\cdot\mu} - \Gamma^{\varrho}_{\nu\mu} P^{\varkappa'\lambda}_{\cdot\cdot\varrho}, \tag{2.11}$$

$$\nabla_{\nu'} P^{\varkappa\lambda'}_{\cdot\cdot\mu} = \partial_{\nu'} P^{\varkappa\lambda'}_{\cdot\cdot\mu} + \Gamma^{\varkappa}_{\nu'\varrho} P^{\varrho\lambda'}_{\cdot\cdot\mu} + \Gamma^{\lambda'}_{\nu'\varrho'} P^{\varkappa\varrho'}_{\cdot\cdot\mu} - \Gamma^{\varrho}_{\nu'\mu} P^{\varkappa\lambda'}_{\cdot\cdot\varrho} \tag{2.12}$$

equivalent to (2.9). They each contain only intermediate components of the same kind.[2]) The $\Gamma^{\varkappa}_{\mu'\lambda}$ and $\Gamma^{\varkappa'}_{\mu\lambda'}$ might be considered as a kind of intermediate components of the geometric object $\Gamma^{\varkappa}_{\mu\lambda}$. But in using them we have always to remember that they can only be formed by a transvection over the index μ and that $\Gamma^{\varkappa}_{\mu'\lambda} \neq \Gamma^{\varkappa'}_{\mu'\lambda'} A^{\varkappa\lambda'}_{\varkappa'\lambda}$. That is the reason why, in general, we use intermediate components of quantities only.

From (2.5) it follows that

$$S_{\mu\lambda}^{\cdot\cdot\varkappa} \overset{\text{def}}{=} \Gamma^{\varkappa}_{[\mu\lambda]} \tag{2.13}$$

is a tensor. It is called the *tensor of asymmetry* or *torsion*[3]) of the L_n. The connexion is said to be *symmetric* and the L_n is called an A_n if $S_{\mu\lambda}^{\cdot\cdot\varkappa} = 0$ at all points of the region considered. It is said to be *semi-symmetric* if $S_{\mu\lambda}^{\cdot\cdot\varkappa}$ has the form

$$S_{\mu\lambda}^{\cdot\cdot\varkappa} = S_{[\mu} A^{\varkappa}_{\lambda]}\ ^{4}). \tag{2.14}$$

[1]) Cf. E I 1935, 1, p. 77f. The term "allround" introduced here is new.

[2]) Formulae of this kind were introduced first by R. LAGRANGE 1926, 1, p. 10.

[3]) McCONNELL used 1928, 2; 3 the word torsion in V_n in another sense connected with a geodesic.

[4]) Cf. FRIEDMANN and SCHOUTEN 1924, 1.

From (2.14) we easily deduce

$$S_\mu = \frac{2}{n-1} S_{\mu\lambda}^{\cdot\cdot\lambda} \tag{2.15}$$

and this equation will also be taken as the definition of the concomitant S_μ if (2.14) does not hold. $S_{\mu\lambda}^{\cdot\cdot\varkappa}$ is intimately connected with the operation Rot (cf. II § 6). The covariant derivative of a q-vector $w_{\lambda_1 \ldots \lambda_q}$ is

$$\begin{cases} \nabla_\mu w_{\lambda_1 \ldots \lambda_q} = \partial_\mu w_{\lambda_1 \ldots \lambda_q} - \Gamma_{\mu\lambda_1}^{\varkappa} w_{\varkappa\lambda_2 \ldots \lambda_q} - \cdots - \Gamma_{\mu\lambda_q}^{\varkappa} w_{\lambda_1 \ldots \lambda_{q-1}\varkappa} \\ \qquad = \partial_\mu w_{\lambda_1 \ldots \lambda_q} - q\, \Gamma_{\mu[\lambda_q}^{\varkappa} w_{\lambda_1 \ldots \lambda_{q-1}]\varkappa}, \end{cases} \tag{2.16}$$

hence

$$\nabla_{[\mu} w_{\lambda_1 \ldots \lambda_q]} = \partial_{[\mu} w_{\lambda_1 \ldots \lambda_q]} - q\, S_{[\mu\lambda_q}^{\cdot\cdot\varkappa} w_{\lambda_1 \ldots \lambda_{q-1}]\varkappa} \tag{2.17}$$

and for the semi-symmetric case

$$\nabla_{[\mu} w_{\lambda_1 \ldots \lambda_q]} = \partial_{[\mu} w_{\lambda_1 \ldots \lambda_q]} - q\, S_{[\mu} w_{\lambda_1 \ldots \lambda_q]} \tag{2.18}$$

from which it follows that only in an A_n the symbols ∂ and ∇ can be interchanged in the expression $\nabla_{[\mu} w_{\lambda_1 \ldots \lambda_q]}$.

If two linear elements $\underset{1}{d}\xi^\varkappa$ and $\underset{2}{d}\xi^\varkappa$ at $\xi^\varkappa$ are displaced parallel along each other, at $\xi^\varkappa + \underset{1}{d}\xi^\varkappa$ we get the vector $\underset{2}{d}\xi^\varkappa - \Gamma_{\mu\lambda}^{\varkappa} \underset{2}{d}\xi^\lambda \underset{1}{d}\xi^\mu$ and at $\xi^\varkappa + \underset{2}{d}\xi^\varkappa$ the vector $\underset{1}{d}\xi^\varkappa - \Gamma_{\mu\lambda}^{\varkappa} \underset{1}{d}\xi^\lambda \underset{2}{d}\xi^\mu$, neglecting all terms of third and higher orders. The figure obtained is in general not a parallelogram but a pentagon with the closing vector

$$2 S_{\mu\lambda}^{\cdot\cdot\varkappa} \underset{1}{d}\xi^\mu \underset{2}{d}\xi^\lambda. \tag{2.19}$$

Hence in an L_n with an asymmetric connection, in general no infinitesimal parallelograms exist. In the semi-symmetrical case the closing vector is

$$2 S_\mu A_\lambda^\varkappa \underset{1}{d}\xi^{[\mu} \underset{2}{d}\xi^{\lambda]} = 2 S_\mu \underset{1}{d}\xi^{[\mu} \underset{2}{d}\xi^{\varkappa]} \tag{2.20}$$

and accordingly infinitesimal parallelograms only exist in the 2-directions contained in the $(n-1)$-direction of S_λ. In the symmetric case there exist infinitesimal parallelograms in every 2-direction. [1])

There is another geometric interpretation of $S_{\mu\lambda}^{\cdot\cdot\varkappa}$ due to CARTAN. Thus far we have only displaced the *vectors* and not the *points* of the tangent E_n. This comes to the same thing as if we had assumed that

[1]) Cf. CARTAN 1922, 2; FRIEDMANN and SCHOUTEN 1924, 1; J. M. THOMAS 1926, 1. Cf. for other geometric interpretations of $S_{\mu\lambda}^{\cdot\cdot\varkappa}$, PINL 1951, 1; BOMPIANI 1951, 1.

the contact-point always remained a contact-point. But another displacement of points can be fixed by the assumption that the contact-point of the tangent E_n of $\xi^\varkappa$ has after its arrival in the tangent E_n of $\xi^\varkappa + d\xi^\varkappa$ the radiusvector $-d\xi^\varkappa$. This we call a CARTAN *displacement*. If the X_n is an E_n all tangent E_n's coincide and the CARTAN displacement is the displacement zero. The covariant differential $\delta v^\varkappa$ of a vector $v^\varkappa$ is the differential with respect to a parallel displaced local coordinate system. But the covariant differential of a point is the differential with respect to a local coordinate system that is undergoing a CARTAN displacement. Hence, if $v^\varkappa$ is the radiusvector of a field of points, each of which lies in a tangent E_n, its covariant differential is $d\xi^\varkappa + \delta v^\varkappa$. Vice versa, if a vector is displaced parallel over $d\xi^\varkappa$, its ordinary differential is $-d\xi^\mu \Gamma^\varkappa_{\mu\lambda} v^\lambda$, but if the point with radiusvector $v^\varkappa$ undergoes a CARTAN displacement, its ordinary differential is $-d\xi^\varkappa - d\xi^\mu \Gamma^\varkappa_{\mu\lambda} v^\lambda$.[1]) Now let us suppose that there is a closed curve in X_n through $\underset{0}{\xi^\varkappa}$ and that the local E_n at $\underset{0}{\xi^\varkappa}$ undergoes a CARTAN displacement along this curve. The points of the curve may be considered as a point field having in each local coordinate system along the curve the radiusvector zero. During the motion the moving E_n coincides at every moment with one of the tangent E_n's of the points of the curve. Hence at every moment the place of this point can be marked in the moving E_n. Then in this E_n we get a curve and it can be proved that if the original curve is contracted into a point and if terms of third and higher orders in the differentials are neglected, the image in the moving E_n is always closed in the symmetric case and is closed in the semi-symmetric case if and only if the 2-direction of the enclosed surface element is lying in the $(n-1)$-direction of S_λ. In an E_n the curve and its image coincide. A rigorous proof of theorems of this kind always requires a rather detailed discussion. Therefore we only give an illustration here, that, without being a rigorous proof, may have some heuristic value. Let the curve be the quadrilateral of the points $\xi^\varkappa$, $\xi^\varkappa + \underset{1}{d}\xi^\varkappa$, $\xi^\varkappa + \underset{1}{d}\xi^\varkappa + \underset{2}{d}\xi^\varkappa$ and $\xi^\varkappa + \underset{2}{d}\xi^\varkappa$. Starting from the first contact-point $\xi^\varkappa$ the radiusvector of this point with respect to the second contact-point $\xi^\varkappa + \underset{1}{d}\xi^\varkappa$ is after the arrival of the E_n at this latter point equal to $-\underset{1}{d}\xi^\varkappa$. Going from $\xi^\varkappa + \underset{1}{d}\xi^\varkappa$ to the third contact-point $\xi^\varkappa + \underset{1}{d}\xi^\varkappa + \underset{2}{d}\xi^\varkappa$ the point $\xi^\varkappa + \underset{1}{d}\xi^\varkappa$

[1]) From the large amount of literature on the CARTAN displacement in various spaces we mention here only a few titles: DIENES 1934, 1; WAGNER 1942, 1; TACHIBANA 1949, 1; SASAKI and YANO 1949, 1; ÔTSUKI 1950, 1; KANITANI 1950, 1. Of course the CARTAN displacement can also be defined for the more general linear connexions mentioned in footnote 1, 2 on p. 123 and especially for conformal and projective connexions (cf. CARTAN 1937, 1, p. 163ff., 278ff.).

gets the radiusvector $-\underset{2}{d}\xi^\varkappa$ with respect to this third contact-point but at the same time the vector $-\underset{1}{d}\xi^\varkappa$ is displaced parallel over $\underset{2}{d}\xi^\varkappa$. Hence $\delta \underset{1}{d}\xi^\varkappa = d\,\underset{1}{d}\xi^\varkappa + \underset{2}{d}\xi^\mu\, \Gamma^\varkappa_{\mu\lambda}\, \underset{1}{d}\xi^\lambda = 0$ and this has as a consequence that the point that had a radiusvector zero with respect to the first contact-point and after the first CARTAN displacement a radiusvector $-\underset{1}{d}\xi^\varkappa$ with respect to the second contact-point gets after the second displacement a radiusvector $-\underset{1}{d}\xi^\varkappa - \underset{2}{d}\xi^\varkappa + \underset{2}{d}\xi^\mu\, \Gamma^\varkappa_{\mu\lambda}\, \underset{1}{d}\xi^\lambda$ with respect to the third contact-point. Going via $\xi^\varkappa + \underset{2}{d}\xi^\varkappa$ we get in the same way a radiusvector $-\underset{2}{d}\xi^\varkappa - \underset{1}{d}\xi^\varkappa + \underset{1}{d}\xi^\mu\, \Gamma^\varkappa_{\mu\lambda}\, \underset{2}{d}\xi^\lambda$. Hence, if we go from $\xi^\varkappa$ round the quadrilateral back to $\xi^\varkappa$, the original contact-point has after its return in $\xi^\varkappa$ the radiusvector $-\,2\,\underset{1}{d}\xi^\mu\, \underset{2}{d}\xi^\lambda\, S_{\mu\lambda}{}^{\cdot\cdot\varkappa}$.

Applying (2.16) for $q = n$ we get

$$(2.21)\qquad \begin{cases} \nabla_\mu w_{1\ldots n} = \partial_\mu w_{1\ldots n} - n\,\Gamma^\varkappa_{\mu[n}\, w_{1\ldots n-1]\varkappa} \\ \phantom{\nabla_\mu w_{1\ldots n}} = \partial_\mu w_{1\ldots n} - \Gamma^\varkappa_{\mu\varkappa}\, w_{1\ldots n}. \end{cases}$$

Hence, if $w_{\lambda_1\ldots\lambda_n}$ is displaced parallel we have

$$(2.22)\qquad d\,w_{1\ldots n} = \overset{*}{d}\,w_{1\ldots n} = \Gamma_\mu\, w_{1\ldots n}\, d\xi^\mu; \qquad \Gamma_\mu \overset{\text{def}}{=} \Gamma^\varkappa_{\mu\varkappa}.$$

Γ_μ is a geometric object with the transformation [cf. Exerc. II 2,1 and (2.5)]

$$(2.23)\qquad \Gamma_{\mu'} = A^\mu_{\mu'}\, \Gamma_\mu - A^\lambda_{\varkappa'}\, \partial_{\mu'} A^{\varkappa'}_\lambda = A^\mu_{\mu'}\, \Gamma_\mu - \partial_{\mu'} \log \Delta.$$

Now $w_{1\ldots n}$ may be considered as the component of a scalar Δ-density of weight $+1$. Accordingly we have for such a density $\tilde{\mathfrak{p}}$

$$(2.24)\qquad \delta\tilde{\mathfrak{p}} = d\tilde{\mathfrak{p}} - \Gamma_\mu\, \tilde{\mathfrak{p}}\, d\xi^\mu,$$

$$(2.25)\qquad \nabla_\mu \tilde{\mathfrak{p}} = \partial_\mu \tilde{\mathfrak{p}} - \Gamma_\mu\, \tilde{\mathfrak{p}}$$

and for a Δ-density $\tilde{\mathfrak{q}}$ of weight w

$$(2.26)\qquad \delta\tilde{\mathfrak{q}} = d\,\mathfrak{q} - w\,\Gamma_\mu\, \tilde{\mathfrak{q}}\, d\xi^\mu,$$

$$(2.27)\qquad \nabla_\mu \tilde{\mathfrak{q}} = \partial_\mu \tilde{\mathfrak{q}} - w\,\Gamma_\mu\, \tilde{\mathfrak{q}}.$$

If a Δ-density is multiplied by a W-scalar we get an ordinary density. Hence, according to (C V) the formulae for the covariant differentiation of ordinary scalar densities and scalar Δ-densities are the same.[1])

---

[1]) Cf. for covariant differentiation of densities VEBLEN and T. Y. THOMAS 1924, 1; HLAVATY 1927, 2; SCHOUTEN and HLAVATY 1929, 2.

Because every tensor density can be written as the product of a tensor and a scalar density, in the formulae for the covariant differential and the covariant derivative we get an extra term with Γ_μ, for instance [cf. (2.8, 9)]

$$\text{(2.28)}\quad \left\{ \begin{aligned} \delta \mathfrak{P}^{\varkappa\lambda}_{\cdot\cdot\mu} = d\,\mathfrak{P}^{\varkappa\lambda}_{\cdot\cdot\mu} + \Gamma^{\varkappa}_{\nu\varrho}\,\mathfrak{P}^{\varrho\lambda}_{\cdot\cdot\mu}\,d\xi^\nu + \Gamma^{\lambda}_{\nu\varrho}\,\mathfrak{P}^{\varkappa\varrho}_{\cdot\cdot\mu}\,d\xi^\nu - \\ - \Gamma^{\varrho}_{\nu\mu}\,\mathfrak{P}^{\varkappa\lambda}_{\cdot\cdot\varrho}\,d\xi^\nu - w\,\mathfrak{P}^{\varkappa\lambda}_{\cdot\cdot\mu}\,\Gamma_\nu\,d\xi^\nu, \end{aligned} \right.$$

$$\text{(2.29)}\quad \nabla_\nu \mathfrak{P}^{\varkappa\lambda}_{\cdot\cdot\mu} = \partial_\nu \mathfrak{P}^{\varkappa\lambda}_{\cdot\cdot\mu} + \Gamma^{\varkappa}_{\nu\varrho}\,\mathfrak{P}^{\varrho\lambda}_{\cdot\cdot\mu} + \Gamma^{\lambda}_{\nu\varrho}\,\mathfrak{P}^{\varkappa\varrho}_{\cdot\cdot\mu} - \Gamma^{\varrho}_{\nu\mu}\,\mathfrak{P}^{\varkappa\lambda}_{\cdot\cdot\varrho} - w\,\mathfrak{P}^{\varkappa\lambda}_{\cdot\cdot\mu}\,\Gamma_\nu.$$

For a p-vector density of weight $+1$ we have

$$\text{(2.30)}\quad \left\{ \begin{aligned} \nabla_\mu \mathfrak{w}^{\mu\varkappa_2\ldots\varkappa_p} &= \partial_\mu \mathfrak{w}^{\mu\varkappa_2\ldots\varkappa_p} + \Gamma^{\mu}_{\mu\lambda}\,\mathfrak{w}^{\lambda\varkappa_2\ldots\varkappa_p} + \Gamma^{\varkappa_2}_{\mu\lambda}\,\mathfrak{w}^{\mu\lambda\varkappa_3\ldots\varkappa_p} + \cdots \\ &\qquad \cdots + \Gamma^{\varkappa_p}_{\mu\lambda}\,\mathfrak{w}^{\mu\varkappa_2\ldots\varkappa_{p-1}\lambda} - \Gamma^{\lambda}_{\mu\lambda}\,\mathfrak{w}^{\mu\varkappa_2\ldots\varkappa_p} \\ &= \partial_\mu \mathfrak{w}^{\mu\varkappa_2\ldots\varkappa_p} + 2\,S_{\nu\mu}^{\cdot\cdot\nu}\,\mathfrak{w}^{\mu\varkappa_2\ldots\varkappa_p} + \\ &\qquad + (p-1)\,S_{\mu\lambda}^{\cdot\cdot[\varkappa_2}\,\mathfrak{w}^{|\mu\lambda|\varkappa_3\ldots\varkappa_q]} \end{aligned} \right.$$

from which we see that in the expression $\nabla_\mu \mathfrak{w}^{\mu\varkappa_2\ldots\varkappa_p}$ in an A_n the symbols ∂ and ∇ can be interchanged but that in a general L_n a correction term containing S_λ occurs and for $p > 1$ also one containing $S_{\mu\lambda}^{\cdot\cdot\varkappa}$.

In order to establish the covariant differentiation of pseudo-quantities as defined in II § 3 with respect to an auxiliary variable ξ^0, we consider a pseudo-scalar $\mathfrak{p}$ of class 1. Its ordinary differential is not a pseudo-scalar because

$$\text{(2.31)}\quad d\overset{(0')}{\mathfrak{p}} = d\sigma\overset{(0)}{\mathfrak{p}} = \sigma\, d\overset{(0)}{\mathfrak{p}} + \overset{(0)}{\mathfrak{p}}\,\partial_\mu\sigma\, d\xi^\mu.$$ [1]

From the conditions (C I–IV) we see that the covariant differential must have the form

$$\text{(2.32)}\quad \delta\mathfrak{p} = d\mathfrak{p} - \Lambda_\mu\,\mathfrak{p}\,d\xi^\mu$$

with parameters Λ_μ subject to the transformation

$$\text{(2.33)}\quad \Lambda_{\mu'} = A^{\mu}_{\mu'}\,\Lambda_\mu + \partial_{\mu'}\log\sigma.$$

This implies that in order to get the covariant differential of a pseudo-quantity Φ (indices suppressed) of class c we have first to write down all terms that would occur if Φ were a quantity and then to add the term $-c\,\Phi\,\Lambda_\mu\,d\xi^\mu$. For instance the covariant derivative of a pseudo-tensor density $\mathfrak{p}^{\varkappa}_{\cdot\lambda}$ of weight w and class c is

$$\text{(2.34)}\quad \nabla_\mu \mathfrak{p}^{\varkappa}_{\cdot\lambda} = \partial_\mu \mathfrak{p}^{\varkappa}_{\cdot\lambda} + \Gamma^{\varkappa}_{\mu\varrho}\,\mathfrak{p}^{\varrho}_{\cdot\lambda} - \Gamma^{\sigma}_{\mu\lambda}\,\mathfrak{p}^{\varkappa}_{\cdot\sigma} - w\,\mathfrak{p}^{\varkappa}_{\cdot\lambda}\,\Gamma_\mu - c\,\mathfrak{p}^{\varkappa}_{\cdot\lambda}\,\Lambda_\mu.$$

If a vector field $v^\varkappa$ is given we may require the field $'\Phi_\Lambda$ that arises from a given field Φ_Λ if this field is displaced parallel along the stream-lines of $v^\varkappa$ over $v^\varkappa t$ (cf. II § 10). In order to find $'\Phi_\Lambda$ we take the co-

[1] If σ also depends on ξ^0 there is another term $\overset{(0)}{\mathfrak{p}}\,\partial_0\sigma\,d\xi^0$.

ordinate system $(\varkappa)$ such that $v^{\varkappa} = \underset{1}{e}^{\varkappa}$ (cf. Exerc. II 5,1) and introduce an anholonomic system (h) such that the $\underset{i}{e}^{\varkappa}$; $h, i = 1, \ldots, n$ are covariant constant along the streamlines of $v^{\varkappa}$. Let the components of Φ_{Λ} with respect to (h) be symbolized by Φ_L. Then we have

$$v^l \nabla_l \Phi_L \overset{*}{=} v^l \partial_l \Phi_L = v^{\mu} \partial_{\mu} \Phi_L \overset{*}{=} \partial_1 \Phi_L. \tag{2.35}$$

Since the $\underset{i}{e}^{\varkappa}$ are covariant constant along the streamlines, $'\Phi_L(\xi^{\varkappa})$ equals $\Phi_L(\xi^{\varkappa} - t\,\underset{1}{e}^{\varkappa})$. Hence

$$'\Phi_L(\xi^{\varkappa}) = \Phi_L(\xi^{\varkappa} - t\,\underset{1}{e}^{\varkappa}) = e^{-t\partial_1}\Phi_L(\xi^{\varkappa}) = \exp(-t\,v^l \nabla_l)\,\Phi_L(\xi^{\varkappa}) \tag{2.36}$$

and, because this result has the invariant form

$$\boxed{'\Phi_{\Lambda}(\xi^{\varkappa}) = e^{-t v^{\mu}\nabla_{\mu}}\,\Phi_{\Lambda}(\xi^{\varkappa})}\;.^{1)} \tag{2.37}$$

Exercises.

III 2,1. Prove that $A_{\lambda}^{\varkappa}$ is covariant constant.

III 2,2. Prove (2.5) starting from $\nabla_{\mu} A_{\lambda}^{\varkappa'} = 0$.

III 2,3. The vector $S_{\mu\lambda}^{\cdot\cdot\varkappa}\,\underset{1}{d}\xi^{\mu}\,\underset{2}{d}\xi^{\lambda}$ lies in the 2-direction of $\underset{1}{d}\xi^{\varkappa}$ and $\underset{2}{d}\xi^{\varkappa}$ for every choice of these latter vectors if and only if the connexion is semi-symmetric (cf. Exerc. I 3,6).

III 2,4. Prove that for every given point there exists always a coordinate system $(\varkappa)$ such that $\underset{(\varkappa)}{E}^{\varkappa_1 \ldots \varkappa_n}$ is covariant stationary at this point.

III 2,5. Prove that $\tilde{\mathfrak{E}}^{\varkappa_1 \ldots \varkappa_n}$ and $\tilde{\mathfrak{e}}_{\lambda_1 \ldots \lambda_n}$ are covariant constant (cf. Exerc. II 10,1).

III 2,6. In the identity (II 11.11) for $[\mathfrak{L}]$ in A_n it is allowed to replace ∂_{μ} by ∇_{μ}. Prove this for the special case (II 11.12).

§ 3. A linear connexion expressed in terms of $S_{\mu\lambda}^{\cdot\cdot\varkappa}$, an auxiliary symmetric tensor field $g_{\lambda\varkappa}$ of rank n and $\nabla_{\mu} g_{\lambda\varkappa}$.

Let $g_{\lambda\varkappa}$ be a symmetric tensor, $g_{\lambda\varkappa} = g_{\varkappa\lambda}$ and let its inverse be written $g^{\varkappa\lambda}$. At first we do not consider $g_{\lambda\varkappa}$ as a fundamental tensor in tangent space, it is just an auxiliary symmetric tensor and nothing more. But we will use it to raise and lower indices in the way described in I § 9. If we write

$$\nabla_{\mu} g^{\varkappa\lambda} \overset{\text{def}}{=} Q_{\mu}^{\cdot\varkappa\lambda} \tag{3.1}$$

---

1) This formula was first found by MORINAGA 1934, 2, the proof given here is by NIJENHUIS (personal communication). Cf. also the footnote 1 on p. 340 on product integrals VII § 1 and DIENES 1934, 1; JOHNSON 1948, 1; NIJENHUIS 1952, 1, p. 174. Φ_{Λ} and $'\Phi_{\Lambda}$ are here used as function symbols, cf. II § 3.

it follows from $g^{\varkappa\varrho} g_{\varrho\lambda} = A_\lambda^\varkappa$ that

$$\nabla_\mu g_{\lambda\varkappa} = - Q_{\mu\lambda\varkappa}. \tag{3.2}$$

By writing out these equations

$$\partial_\mu g_{\lambda\varkappa} = \Gamma_{\mu\lambda}^\varrho g_{\varrho\varkappa} + \Gamma_{\mu\varkappa}^\varrho g_{\lambda\varrho} - Q_{\mu\lambda\varkappa} \tag{3.3}$$

permuting the indices $\mu\,\lambda\,\varkappa$ and taking the sum of the permutations $\mu\,\lambda\,\varkappa$, $\varkappa\,\mu\,\lambda$ diminished by the permutation $\lambda\,\varkappa\,\mu$ we get by addition

$$\left\{\begin{aligned} 2\Gamma_{(\varkappa\mu)}^\varrho g_{\varrho\lambda} = \partial_\mu g_{\lambda\varkappa} + \partial_\varkappa g_{\mu\lambda} - \partial_\lambda g_{\varkappa\mu} + Q_{\mu\varkappa\lambda} + Q_{\varkappa\mu\lambda} - Q_{\lambda\varkappa\mu} - \\ - 2 S_{\mu\lambda}{}^{\cdot\cdot\varrho} g_{\varrho\varkappa} - 2 S_{\varkappa\lambda}{}^{\cdot\cdot\varrho} g_{\mu\varrho} \end{aligned}\right. \tag{3.4}$$

or

$$\boxed{\Gamma_{\mu\lambda}^\varkappa = \left\{{\varkappa \atop \mu\lambda}\right\} + S_{\mu\lambda}{}^{\cdot\cdot\varkappa} - S_{\lambda\cdot\mu}^{\cdot\varkappa} + S^\varkappa{}_{\cdot\mu\lambda} + \tfrac{1}{2}\left(Q_{\mu\lambda}{}^{\cdot\cdot\varkappa} + Q_{\lambda\cdot\mu}^{\cdot\varkappa} - Q^\varkappa{}_{\cdot\mu\lambda}\right)} \tag{3.5}$$

where

$$\left\{{\varkappa \atop \mu\lambda}\right\} \stackrel{\text{def}}{=} \tfrac{1}{2} g^{\varkappa\sigma} \left(\partial_\mu g_{\lambda\sigma} + \partial_\lambda g_{\mu\sigma} - \partial_\sigma g_{\mu\lambda}\right) \tag{3.6}$$

is the so called CHRISTOFFEL *symbol* belonging to the tensor $g_{\lambda\varkappa}$.[1] In fact (3.5) satisfies the equation (3.3).

The $\frac{1}{2} g^{\varkappa\sigma} \partial_\mu g_{\lambda\sigma}$, $S_{\mu\lambda}{}^{\cdot\cdot\varkappa}$ and $\frac{1}{2} Q_{\mu\lambda}{}^{\cdot\cdot\varkappa}$ occur in (3.5) in an analogous way. This can be made clearer by using the abbreviation (the place of the indices is important!)

$$\psi_{\{\mu\lambda\varkappa\}} \stackrel{\text{def}}{=} \psi_{\mu\lambda\varkappa} - \psi_{\lambda\varkappa\mu} + \psi_{\varkappa\mu\lambda} \tag{3.7}$$

for any expression $\psi_{\mu\lambda\varkappa}$. Then, if we write

$$\chi_{\mu\lambda\varkappa} \stackrel{\text{def}}{=} \tfrac{1}{2} \partial_\mu g_{\lambda\varkappa} \tag{3.8}$$

(3.5) takes the form

$$\boxed{\Gamma_{\mu\lambda}^\varkappa = g^{\varkappa\varrho} \left(\chi_{\{\mu\varrho\lambda\}} - S_{\{\mu\varrho\lambda\}} + \tfrac{1}{2} Q_{\{\mu\varrho\lambda\}}\right)}\,. \tag{3.9}$$

If $Q_{\mu\lambda}{}^{\cdot\cdot\varkappa} = 0$ the connexion is called *metric*[2] with respect to $g_{\lambda\varkappa}$. In this case the covariant differential of $g_{\lambda\varkappa}$ vanishes and this implies that the processes of covariant differentiation and of raising and lowering of indices are commutative, for instance

$$\nabla_\mu g_{\lambda\varkappa} v^\varkappa = g_{\lambda\varkappa} \nabla_\mu v^\varkappa = \nabla_\mu v_\lambda.\text{[3]} \tag{3.10}$$

---

[1] CHRISTOFFEL 1869, 1.

[2] The equation (3.9) can also be used to establish relations between two different non-metric connexions. Cf. SEN 1944, 1; 1945, 1; 1946, 1; 1948, 1; 1949, 1; 2; 1950, 1; 2; NORDEN 1945, 1.

[3] Most authors accept the convention that the range of differentiation of ∇ stretches as far as the first closing bracket whose corresponding opening bracket stands at the left hand side of ∇ (cf. footnote 1 on p. 70).

If the connexion is *metric* and if $g_{\lambda\varkappa}$ is *real* and introduced as a fundamental tensor (cf. I § 9) in each of the local E_n's, the X_n is called a U_n and $g_{\lambda\varkappa}$ is said to be the *fundamental tensor* of the U_n. A U_n with a *symmetric* connexion is called a V_n. [1]) This is the case of *riemannian geometry*. A U_n or V_n is called *ordinary* if $g_{\lambda\varkappa}$ is positive definite. In a V_n (3.5) takes the form

$$\Gamma_{\mu\lambda}^{\varkappa} = \left\{ {\varkappa \atop \mu\lambda} \right\} \quad (3.11)$$ [2])

and from this it follows that

$$\Gamma_\mu = \tfrac{1}{2} g^{\varkappa\lambda} \partial_\mu g_{\varkappa\lambda} = \tfrac{1}{2} \partial_\mu \log \mathfrak{g}; \qquad \mathfrak{g} \stackrel{\text{def}}{=} |\operatorname{Det}(g_{\lambda\varkappa})|. \quad (3.12)$$

If

$$Q_\mu^{\cdot\lambda\varkappa} = Q_\mu g^{\lambda\varkappa} \quad (3.13)$$

the connexion is called *semi-metric*. From (3.2) it follows that in this case

$$\nabla_\mu g_{\lambda\varkappa} = - Q_\mu g_{\lambda\varkappa}. \quad (3.14)$$

If the connexion is *semi-metric* and *symmetric*, and if $g_{\lambda\varkappa}$ is *real*, the X_n is called a W_n. This is the case of WEYL's geometry [3]). Let a vector $v^\varkappa$ in a W_n be displaced parallel. Then $\delta v^\varkappa = 0$ and

$$\delta g_{\lambda\varkappa} v^\lambda v^\varkappa = - g_{\lambda\varkappa} v^\lambda v^\varkappa Q_\mu d\xi^\mu. \quad (3.15)$$

Hence in a W_n the length of a vector is not invariant for parallel displacement but the ratio of the length of two vectors at the same point is invariant. If $g_{\lambda\varkappa}$ in a W_n undergoes a conformal transformation

$$'g_{\lambda\varkappa} = \sigma(\xi^\nu) g_{\lambda\varkappa} \quad (3.16)$$

we have

$$\nabla_\mu {}'g_{\lambda\varkappa} = (-\sigma Q_\mu + \partial_\mu \sigma) g_{\lambda\varkappa} = (-Q_\mu + \partial_\mu \log \sigma)\, 'g_{\lambda\varkappa}. \quad (3.17)$$

Hence if we take $'g_{\lambda\varkappa}$ as fundamental tensor instead of $g_{\lambda\varkappa}$, and if at the same time Q_λ is transformed into $'Q_\lambda = Q_\lambda - \partial_\lambda \log \sigma$ we get the

[1]) An A_n may be considered as a V_n if and only if there exists a covariant constant symmetric tensor $g_{\lambda\varkappa}$ of rank n. The conditions were investigated by T. Y. THOMAS and LEVINE 1934, 1; T. Y. THOMAS 1935, 1; 1936, 2; LEVINE 1948, 1. Several authors investigated the geometries with a fundamental tensor of rank $< n$. We mention here only BORTOLOTTI 1930, 2; MOISIL 1940, 1; VRANCEANU 1942, 1; NORDEN 1945, 2.

[2]) Note that $\left\{ {\varkappa \atop \mu\lambda} \right\}$ does not change if $g_{\lambda\varkappa}$ gets a *constant* factor. Introducing such a factor in fact means a change of gauge in the whole space without changing the geometry.

[3]) WEYL 1918, 1, p. 400; R. K. 1924, 1, Ch. 6; HLAVATY 1929, 2. The semi-metric semi-symmetric case appeared in the classification R. K. 1924, 1, p. 75 and was applied to mechanics by LICHNEROWICZ 1941, 1.

same connexion. That implies that in a W_n the fundamental tensor is only fixed to within an arbitrary scalar factor. But a W_n is not the same as a conformal space. In the latter only $\lfloor g_{\lambda\varkappa} \rfloor$ is given (cf. I § 2) but no linear connexion can be fixed by $\lfloor g_{\lambda\varkappa} \rfloor$ alone. Therefore we need the vector field Q_λ, given for one of the possible fields $g_{\lambda\varkappa}$. Of course we must exclude the case when Q_λ is a gradientvector because in that case it would be possible to fix $'g_{\lambda\varkappa}$ by chosing σ in such a way that $'Q_\lambda$ vanishes, and that would mean that we did not really have a W_n but a V_n with an undetermined gauge in which $g_{\lambda\varkappa}$ could have been given in a more practical way (determined to within a *constant* scalar factor). In one dimensional space every vector is a gradientvector. Hence every W_1 is a V_1 with an undetermined gauge. In a W_n (3.5) takes the form

$$\Gamma^{\varkappa}_{\mu\lambda} = \left\{ {\varkappa \atop \mu\lambda} \right\} + \frac{1}{2} (Q_\mu A^{\varkappa}_{\lambda} + Q_\lambda A^{\varkappa}_{\mu} - Q^{\varkappa} g_{\mu\lambda}) \tag{3.18}$$

and for Γ_μ we have the formula

$$\Gamma_\mu = \frac{1}{2} \partial_\mu \log \mathfrak{g} + \frac{n}{2} Q_\mu . \tag{3.19}$$

In a W_n the process of covariant differentiation does not commute with the process of raising and lowering of indices. But, since the fundamental tensor is fixed only to within an arbitrary scalar factor, this latter process is not invariant and can be effected only if a preference is given to one of the possible fundamental tensor fields. If this can be done in an invariant way we go back to the V_n and in the other case it is better not to use the non invariant process at all.

The geometry in W_n is based on two different transformation groups, the group of the coordinate transformations and the group of the conformal transformations of $g_{\lambda\varkappa}$ (the "Umeichung" of WEYL). But in the above these two groups are not dealt with in the same way; the coordinate transformations are fundamental because they form the foundation for the definition of all objects and the conformal transformations come in after the introduction of a fundamental tensor to make this tensor variable. Though quite correct mathematically, this method seems incongruous from a methodical and aesthetic point of view.[1]) It is just the same as if, in dealing with some geometry, we first worked out everything with respect to one definite coordinate system and afterwards introduced coordinate transformations and raised the question of invariance.[2]) Now it is possible to handle both groups in the same

---

[1]) My attention was drawn for the first time to this point by G. LYRA in his letter of 27. 6. 1949. Cf. LYRA 1951, 1. The same point of view occurs in HLAVATY 1949, 1; 1952, 1.

[2]) In many 19th century publications this was actually done!

way by introducing at the beginning a fundamental pseudo-tensor $\mathfrak{g}_{\lambda\varkappa}$ of class $+1$, defined with respect to an auxiliary variable ξ^0 with the transformation $\xi^{0'}=\sigma\,\xi^0$ (cf. II § 3).[1] Then, using the formula [cf. (2.34)] for the covariant differentiation of pseudo-quantities we get

$$\nabla_\mu \mathfrak{g}_{\lambda\varkappa} = \partial_\mu \mathfrak{g}_{\lambda\varkappa} - \Gamma_{\mu\lambda}^{\varrho} \mathfrak{g}_{\varrho\varkappa} - \Gamma_{\mu\varkappa}^{\varrho} \mathfrak{g}_{\varrho\lambda} - \mathfrak{g}_{\lambda\varkappa} A_\mu . \tag{3.20}$$

Equating this expression to zero, taking $S_{\mu\lambda}^{\cdot\cdot\varkappa}=0$ and using $\mathfrak{g}_{\lambda\varkappa}$ and its inverse $\mathfrak{g}^{\varkappa\lambda}$ for raising and lowering of indices we get

$$\Gamma_{\mu\lambda}^{\varkappa} = \left\{ {\varkappa \atop \mu\lambda} \right\}^* - \frac{1}{2}\left(A_\mu A_\lambda^\varkappa + A_\lambda A_\mu^\varkappa - A^\varkappa \mathfrak{g}_{\mu\lambda}\right) \tag{3.21}$$

$$\Gamma_\mu = \frac{1}{2}\,\partial_\mu \log \mathfrak{G} - \frac{n}{2} A_\mu; \qquad \mathfrak{G} \stackrel{\text{def}}{=} |\operatorname{Det}(\mathfrak{g}_{\lambda\varkappa})| \tag{3.22}$$

where

$$\left\{ {\varkappa \atop \mu\lambda} \right\}^* \stackrel{\text{def}}{=} \frac{1}{2}\,\mathfrak{g}^{\varkappa\sigma}\left(\partial_\mu \mathfrak{g}_{\lambda\sigma} + \partial_\lambda \mathfrak{g}_{\mu\sigma} - \partial_\sigma \mathfrak{g}_{\mu\lambda}\right). \tag{3.23}$$

$\mathfrak{G}$ is a pseudo-scalar density of weight $+2$ and class n. From (3.18, 19) and (3.21–23) we see that A_λ plays now the same role as $-Q_\lambda$ before. The only difference is that we have now two definitely fixed geometric objects $\mathfrak{g}_{\lambda\varkappa}$ and A_λ instead of the variable quantities $g_{\lambda\varkappa}$ and Q_λ. $\mathfrak{g}_{\lambda\varkappa}$ being fixed, it can be used for raising or lowering of indices and according to $\nabla_\mu \mathfrak{g}_{\lambda\varkappa} = 0$ this process is commutative with the process of covariant differentiation. Note however that by raising (lowering) of an index the class increases with -1 $(+1)$.[2]

It is very remarkable that in fixing a linear displacement the tensor $g_{\lambda\varkappa}$ plays such a fundamental part. Why not any quantity with another valence and another condition of symmetry? This has been made clear by Weyl. First let an arbitrary linear displacement be given by its parameters $\Gamma_{\mu\lambda}^{\varkappa}$. At a point $\underset{0}{\xi}^\varkappa$ we choose an arbitrary symmetric tensor $\overset{0}{g}_{\lambda\varkappa}$ of rank n. Now we try to form a field $g_{\lambda\varkappa}$ in $\mathfrak{R}\,(\underset{0}{\xi}^\varkappa)$ such that $g_{\lambda\varkappa}=\overset{0}{g}_{\lambda\varkappa}$ at $\underset{0}{\xi}^\varkappa$ and that the given displacement from $\underset{0}{\xi}^\varkappa$ to any point $\underset{0}{\xi}^\varkappa + d\xi^\varkappa$ is the result of

1° the displacement from $\underset{0}{\xi}^\varkappa$ to $\underset{0}{\xi}^\varkappa + d\xi^\varkappa$ belonging to the Riemannian geometry with $g_{\lambda\varkappa}$ as fundamental tensor;

2° a rotation of all vectors at $\underset{0}{\xi}^\varkappa + d\xi^\varkappa$ with respect to $\overset{0}{g}_{\lambda\varkappa} + d g_{\lambda\varkappa}$ as fundamental tensor.

---

[1]) Newman 1927, 1; Schouten and Hlavaty 1929, 2, p. 426; E I 1935, 1, p. 87.

[2]) Note that the geometry of a W_n is not the conformal geometry. In a conformal geometry we have only the pseudo-tensor $\mathfrak{g}_{\lambda\varkappa}$ and no A_λ. That implies that there is no fixed linear connexion.

Such a displacement WEYL calls a "congruent transplantation" (kongruente Verpflanzung). If $v^\varkappa$ is a vector at $\underset{0}{\xi}^\varkappa$, it changes first into

$$v^\varkappa - \left\{{}^{\,\varkappa}_{\mu\lambda}\right\} v^\lambda \, d\xi^\mu \tag{3.24}$$

where $\left\{{}^{\,\varkappa}_{\mu\lambda}\right\}$ is the CHRISTOFFEL symbol belonging to $g_{\lambda\varkappa}$ and afterwards into

$$v^\varkappa - \left\{{}^{\,\varkappa}_{\mu\lambda}\right\} v^\lambda \, d\xi^\mu - F_\mu^{\cdot\varkappa\nu} \overset{0}{g}_{\nu\lambda} v^\lambda \, d\xi^\mu \tag{3.25}$$

where $F_\mu^{\cdot\varkappa\nu} d\xi^\mu$ is the bivector of the rotation (cf. I § 9). Hence at $\underset{0}{\xi}^\varkappa$

$$\Gamma_{\mu\lambda}^\varkappa = \left\{{}^{\,\varkappa}_{\mu\lambda}\right\} - F_{\mu\lambda\nu} g^{\nu\varkappa}. \tag{3.26}$$

Now $F_{\mu\lambda\nu}$ is alternating in $\lambda\nu$ and from this it follows that

$$g_{\varkappa\nu} \Gamma_{\mu\lambda}^\nu + g_{\lambda\nu} \Gamma_{\mu\varkappa}^\nu = \partial_\mu g_{\lambda\varkappa} \tag{3.27}$$

or

$$\nabla_\mu g_{\lambda\varkappa} = 0. \tag{3.28}$$

Now let a field $g_{\lambda\varkappa}$ of rank n be given in $\mathfrak{N}(\underset{0}{\xi}^\varkappa)$. Then all linear connexions representing congruent transplantations with respect to $g_{\lambda\varkappa}$ satisfy equations of the form (3.26) where $F_{\mu\lambda\nu}$ is an arbitrary tensor, alternating in $\lambda\nu$. Among these connexions there exists one and only one symmetric connexion because $F_{\mu\lambda\nu}$ vanishes if $F_{[\mu\lambda]\nu} = 0$. Collecting results we have two theorems:

I. *If a linear connexion and a symmetric tensor* $\overset{0}{g}_{\lambda\varkappa}$ *of rank* n *at a point* $\underset{0}{\xi}^\varkappa$ *be given, it is always possible to find a field* $g_{\lambda\varkappa}$ *in an* $\mathfrak{N}(\underset{0}{\xi}^\varkappa)$ *such that* $g_{\lambda\varkappa} = \overset{0}{g}_{\lambda\varkappa}$ *at* $\underset{0}{\xi}^\varkappa$ *and that the linear displacement from* $\underset{0}{\xi}^\varkappa$ *to every point* $\underset{0}{\xi}^\varkappa + d\xi^\varkappa$ *is a congruent transplantation with respect to* $g_{\lambda\varkappa}$.

II. *If a symmetric tensor field* $g_{\lambda\varkappa}$ *of rank* n *be given there exists one and only one symmetric connexion representing a congruent transplantation with respect to* $g_{\lambda\varkappa}$.

The first theorem was proved by WEYL[1]). The second theorem deals with a special case of the general formula (3.5).

Now one could try to generalize the notion of congruent transplantation by using instead of the group of rotations some other subgroup of the group G_{ho} which left the volume invariant. Then, if the following conditions are introduced:

1. If the subgroup and a linear connexion be given, it is always possible to define the "generalized rotation" in an $\mathfrak{N}(\underset{0}{\xi}^\varkappa)$ in such a

[1]) WEYL 1921, 1, p. 131; cf. WEYL 1922, 2, p. 117.

way that the connexion represents a congruent transplantation from $\overset{}{\underset{0}{\xi}}{}^{\varkappa}$ to $\underset{0}{\xi}{}^{\varkappa}+d\xi^{\varkappa}$ for every $d\xi^{\varkappa}$;

2. if the generalized rotations are given at each point of $\mathfrak{N}(\underset{0}{\xi}{}^{\varkappa})$, there exists one and only one symmetric connexion representing a congruent transplantation with respect to this definition of rotation,

it has been proved[1]) that there exists one and only one subgroup of G_{h_0} that leaves the volume invariant and also satisfies these conditions, and that this subgroup is the group that leaves invariant a symmetric tensor of valence *2* and rank n. This theorem makes clear why in the problem of fixing a symmetric connexion, a symmetric tensor of valence *2* and rank n plays such an important part.

Exercises.

III 3,1[2]). The symmetric part $'\Gamma_{\mu\lambda}^{\varkappa} \overset{\text{def}}{=} \Gamma_{(\mu\lambda)}^{\varkappa}$ depends not only on $Q_{\mu\lambda}^{\cdot\cdot\varkappa}$ but also on $S_{\mu\lambda}^{\cdot\cdot\varkappa}$. The $'\Gamma_{\mu\lambda}^{\varkappa}$ are parameters of another connexion for which

III 3,1 α) $$'\nabla_{\mu} g_{\lambda\varkappa} = 2S_{\mu(\lambda\varkappa)} - Q_{\mu\lambda\varkappa}.$$

III 3,2[2]). If the connexion $\Gamma_{\mu\lambda}^{\varkappa}$ is metric, the symmetric connexion $'\Gamma_{\mu\lambda}^{\varkappa} = \Gamma_{(\mu\lambda)}^{\varkappa}$ is also metric, if and only if $S_{\mu\lambda\varkappa}$ is a trivector.

III 3,3[2]). If $g_{\lambda\varkappa}$ and $\nabla_{\mu} g_{\lambda\varkappa}$ are known, Γ_{μ} is fixed without it being necessary to give $S_{\mu\lambda}^{\cdot\cdot\varkappa}$.

III 3,4. In a metric connexion length and angles are invariant for all parallel displacements.

III 3,5. In a W_n angles are invariant for parallel displacement.

III 3,6. Quantities in L_n form a ring (in the sense of abstract algebra) with respect to addition, multiplication and the operators ∇_{μ} and δ. Prove that pseudo-quantities in W_n defined with respect to ξ^0 with a class equal to one half of the difference between covariant and contravariant valence form a ring with respect to addition, multiplication and the operators $(\xi^0)^{-\frac{1}{2}}\nabla_{\mu}$ and $(\xi^0)^{-\frac{1}{2}}\delta$.

III 3,7[3]). A semi-symmetric linear connexion is for $n>2$ fixed by the covariant derivative $Q_{\mu}^{\cdot\varkappa\lambda}$ of some symmetric tensor $g^{\varkappa\lambda}$ of rank n and the derivative $I^{\varkappa} = \nabla_{\mu} f^{\mu\varkappa}$ of some bivector $f^{\mu\varkappa}$ of rank n. This derivative can be replaced by the derivative $F_{\mu\lambda\varkappa} = 6\nabla_{[\mu} f_{\lambda\varkappa]}$ of some bivector $f_{\lambda\varkappa}$ of rank n.

---

[1]) WEYL 1922, 2; CARTAN gave another proof, 1923, 1; cf. also KOSAMBI 1952 1, p. 6.

[2]) E I 1935, 1, p. 84.

[3]) E I 1935, 1, p. 88.

§ 4. Curvature.

We consider a field $v^\varkappa$ or w_λ in an $\mathfrak{N}(\underset{0}{\xi}^\varkappa)$ and an infinitesimal simple contravariant bivector $df^{\varkappa\lambda}$ at $\underset{0}{\xi}^\varkappa$ represented by a part of an E_2 (with inner orientation) through $\underset{0}{\xi}^\varkappa$ with a definitely fixed boundary curve passing through $\underset{0}{\xi}^\varkappa$. Now let the vector $v^\varkappa$ or w_λ at $\underset{0}{\xi}^\varkappa$ be displaced parallel along this curve till it returns to $\underset{0}{\xi}^\varkappa$. Then it can be proved that the difference between the final value and the initial value can be written in the form

$$\text{(4.1)} \qquad \text{a)}\ -\tfrac{1}{2} R_{\nu\mu\lambda}^{\cdot\cdot\cdot\,\varkappa}\, v^\lambda\, df^{\nu\mu}; \qquad \text{b)}\ +\tfrac{1}{2} R_{\nu\mu\lambda}^{\cdot\cdot\cdot\,\varkappa}\, w_\varkappa\, df^{\nu\mu}$$

where $R_{\nu\mu\lambda}^{\cdot\cdot\cdot\,\varkappa}$ is a tensor depending on the $\Gamma_{\mu\lambda}^\varkappa$ and their first derivatives

$$\text{(4.2)} \qquad R_{\nu\mu\lambda}^{\cdot\cdot\cdot\,\varkappa} \overset{\text{def}}{=} 2\,\partial_{[\nu}\Gamma_{\mu]\lambda}^\varkappa + 2\,\Gamma_{[\nu|\varrho|}^\varkappa\,\Gamma_{\mu]\lambda}^\varrho .$$

The rigorous proof of this theorem will be given after (4.9b).[1] Here we give a short illustration that is not meant to be a proof. For the curve we take the quadrilateral of the points $\underset{0}{\xi}^\varkappa$, $\underset{0}{\xi}^\varkappa + \underset{1}{d}\xi^\varkappa$, $\underset{0}{\xi}^\varkappa + \underset{1}{d}\xi^\varkappa + \underset{2}{d}\xi^\varkappa$ and $\underset{0}{\xi}^\varkappa + \underset{2}{d}\xi^\varkappa$. Then, displacing $v^\varkappa$ from $\underset{0}{\xi}^\varkappa$ to $\underset{0}{\xi}^\varkappa + \underset{1}{d}\xi^\varkappa + \underset{2}{d}\xi^\varkappa$ via the point $\underset{0}{\xi}^\varkappa + \underset{1}{d}\xi^\varkappa$ we get at $\underset{0}{\xi}^\varkappa + \underset{1}{d}\xi^\varkappa$ the field value

$$\text{(4.3)} \qquad v^\varkappa - \Gamma_{\mu\lambda}^\varkappa\, v^\lambda\, \underset{1}{d}\xi^\mu$$

neglecting terms of second and higher orders. At $\underset{0}{\xi}^\varkappa + \underset{1}{d}\xi^\varkappa + \underset{2}{d}\xi^\varkappa$ we get

$$\text{(4.4)} \quad v^\varkappa - \Gamma_{\mu\lambda}^\varkappa v^\lambda \underset{1}{d}\xi^\mu - \Gamma_{\mu\lambda}^\varkappa \underset{2}{d}\xi^\mu (v^\lambda - \Gamma_{\nu\varrho}^\lambda v^\varrho \underset{1}{d}\xi^\nu) - (\partial_\nu \Gamma_{\mu\lambda}^\varkappa)\, v^\lambda \underset{1}{d}\xi^\nu \underset{2}{d}\xi^\mu .$$

Going via $\underset{0}{\xi}^\varkappa + \underset{2}{d}\xi^\varkappa$ we get in the same way

$$\text{(4.5)} \quad v^\varkappa - \Gamma_{\nu\lambda}^\varkappa v^\lambda \underset{2}{d}\xi^\nu - \Gamma_{\nu\lambda}^\varkappa \underset{1}{d}\xi^\nu (v^\lambda - \Gamma_{\mu\varrho}^\lambda v^\varrho \underset{2}{d}\xi^\mu) - (\partial_\mu \Gamma_{\nu\lambda}^\varkappa)\, v^\lambda \underset{2}{d}\xi^\mu \underset{1}{d}\xi^\lambda$$

and the difference of these two values is

$$\text{(4.6)} \qquad -2(\partial_{[\nu}\Gamma_{\mu]\lambda}^\varkappa + \Gamma_{[\nu|\varrho|}^\varkappa \Gamma_{\mu]\lambda}^\varrho)\, v^\lambda \underset{1}{d}\xi^\nu \underset{2}{d}\xi^\mu = -\tfrac{1}{2} R_{\nu\mu\lambda}^{\cdot\cdot\cdot\,\varkappa}\, v^\lambda\, df^{\nu\mu}$$

because $df^{\nu\mu} = 2\underset{1}{d}\xi^{[\nu} \underset{2}{d}\xi^{\mu]}$. If (4.1) is proved to be true for every field $v^\varkappa$ (or w_λ) and every bivector $f^{\nu\mu}$, $R_{\nu\mu\lambda}^{\cdot\cdot\cdot\,\varkappa}$ must be a tensor because every

[1] Cf. TIETZE 1923, 1; SYNGE 1924, 1; T. Y. THOMAS 1925, 1; SCHLESINGER 1928, 1; 1931, 1; 1932, 1; McCONNELL 1928, 1; AGOSTINELLI 1933, 1; MORINAGA 1934, 1; GUGINO 1935, 1; JOHNSON 1948, 1; GRAEUB 1950, 1. If instead of vectors other quantities are displaced, other quantities arise instead of $R_{\nu\mu\lambda}^{\cdot\cdot\cdot\,\varkappa}$, cf. for instance HOKARI 1934, 1; HOMBU 1936, 1; TAKENO 1942, 1; BOMPIANI 1946, 1; PETROV 1948, 1; BOCHNER 1951, 2.

expression of this kind with four indices alternating in $\nu\mu$, whose transvection with every bivector $f^{\nu\mu}$ and every vector $v^{\varkappa}$ (or w_{λ}) is a vector, is necessarily a tensor.

$R_{\nu\mu\lambda}^{\cdot\cdot\cdot\varkappa}$ is called the RIEMANN-CHRISTOFFEL *tensor* or *curvature tensor* of the connexion $\Gamma_{\mu\lambda}^{\varkappa}$. If it is zero at all points, the connexion is said to be *integrable*.

Connecting these results with those obtained in III § 2 we see that if a point with radiusvector $v^{\varkappa}$ at $\underset{0}{\xi}^{\varkappa}$ undergoes a CARTAN displacement along the boundary curve of the infinitesimal bivector $d f^{\nu\mu}$ in the sense of this bivector, the point will after its return have the radiusvector

$$v^{\varkappa} - \tfrac{1}{2} R_{\nu\mu\lambda}^{\cdot\cdot\cdot\varkappa} v^{\lambda} d f^{\nu\mu} - S_{\nu\mu}^{\cdot\cdot\varkappa} d f^{\nu\mu}. \tag{4.7}$$

If the process of covariant differentiation is applied twice in succession we get, after alternation over the indices of differentiation, an expression containing $S_{\mu\lambda}^{\cdot\cdot\varkappa}$ and $R_{\nu\mu\lambda}^{\cdot\cdot\cdot\varkappa}$. For instance

$$\tag{4.8} \left\{ \begin{aligned} \nabla_{[\nu} \nabla_{\mu]} w_{\lambda} &= \nabla_{[\nu} (\partial_{\mu]} w_{\lambda} - \Gamma_{\mu]\lambda}^{\varrho} w_{\varrho}) \\ &= \partial_{[\nu} \partial_{\mu]} w_{\lambda} - (\partial_{[\nu} \Gamma_{\mu]\lambda}^{\varrho}) w_{\varrho} - \Gamma_{[\mu|\lambda|}^{\varrho} \partial_{\nu]} w_{\varrho} - \Gamma_{[\nu\mu]}^{\varrho} \partial_{\varrho} w_{\lambda} - \\ &\quad - \Gamma_{[\nu|\lambda|}^{\varrho} \partial_{\mu]} w_{\varrho} + \Gamma_{[\nu\mu]}^{\varrho} \Gamma_{\varrho\lambda}^{\sigma} w_{\sigma} + \Gamma_{[\nu|\lambda|}^{\varrho} \Gamma_{\mu]\varrho}^{\sigma} w_{\sigma} \end{aligned} \right.$$

or

$$\boxed{\nabla_{[\nu} \nabla_{\mu]} w_{\lambda} = -\tfrac{1}{2} R_{\nu\mu\lambda}^{\cdot\cdot\cdot\varkappa} w_{\varkappa} - S_{\nu\mu}^{\cdot\cdot\varrho} \nabla_{\varrho} w_{\lambda}} \tag{4.9a}$$

and in the same way

$$\boxed{\nabla_{[\nu} \nabla_{\mu]} v^{\varkappa} = \tfrac{1}{2} R_{\nu\mu\lambda}^{\cdot\cdot\cdot\varkappa} v^{\lambda} - S_{\nu\mu}^{\cdot\cdot\varrho} \nabla_{\varrho} v^{\varkappa}}\,. \tag{4.9b}$$

Using (4.9b) we can now give a rigorous proof of (1.1a). Let $t^{\varkappa}$ and $u^{\varkappa}$ be two vector fields with the parameters t and u and let these fields satisfy the equation $\underset{t}{£}\, u^{\varkappa} = 0$.

Then for any field $v^{\varkappa}$ it follows from (4.9b) that

$$(t^{\nu} \nabla_{\nu} u^{\mu} \nabla_{\mu} - u^{\nu} \nabla_{\nu} t^{\mu} \nabla_{\mu}) v^{\varkappa} = t^{\nu} u^{\mu} R_{\nu\mu\lambda}^{\cdot\cdot\cdot\varkappa} v^{\lambda} + (\underset{t}{£}\, u^{\varrho}) \nabla_{\varrho} v^{\varkappa} = t^{\nu} u^{\mu} R_{\nu\mu\lambda}^{\cdot\cdot\cdot\varkappa} v^{\lambda}. \tag{4.10}$$

Now instead of displacing a vector at $\underset{0}{\xi}^{\varkappa}$ parallel along an infinitesimal quadrilateral as was done in the illustration given above we give the whole vector field $v^{\varkappa}$ a finite parallel displacement over $t^{\varkappa} t$ (cf. III § 2). We follow this with another displacement over $u^{\varkappa} u$, then over $-t^{\varkappa} t$ and finally over $-u^{\varkappa} u$. Because $\underset{t}{£}\, u^{\varkappa} = 0$ [1]), after these four displacements every point is back at its original place, and according to (2.37), the new field

[1]) Cf. Exerc. II 10,10.

value $'v^\varkappa$ is

$$(4.11)\quad \begin{cases} 'v^\varkappa = \exp(u\,u^\mu \nabla_\mu)\exp(t\,t^\mu \nabla_\mu)\exp(-u\,u^\mu \nabla_\mu)\exp(-t\,t^\mu \nabla_\mu)\,v^\varkappa \\ \quad = v^\varkappa - t\,u\,(t^\nu \nabla_\nu u^\mu \nabla_\mu - u^\nu \nabla_\nu t^\mu \nabla_\mu)\,v^\varkappa + \cdots \\ \quad = v^\varkappa - t\,u\,t^\nu u^\mu R_{\nu\mu\lambda}^{\cdot\cdot\cdot\varkappa} v^\lambda + \cdots \end{cases}$$

which proves (4.1a). [1])

The operator $\nabla_{[\nu}\nabla_{\mu]}$ has the very important property that it satisfies the rule of LEIBNIZ. In fact, if Φ and Ψ are two quantities with suppressed indices, then we have (cf. footnote 3 on p. 132).

$$(4.12)\quad \begin{cases} \nabla_{[\nu}\nabla_{\mu]}\Phi\Psi = \nabla_{[\nu}(\nabla_{\mu]}\Phi)\Psi + \nabla_{[\nu}\Phi\nabla_{\mu]}\Psi \\ \quad = (\nabla_{[\nu}\nabla_{\mu]}\Phi)\Psi + (\nabla_{[\mu}\Phi)\nabla_{\nu]}\Psi + (\nabla_{[\nu}\Phi)\nabla_{\mu]}\Psi + \Phi\nabla_{[\nu}\nabla_{\mu]}\Psi \\ \quad = (\nabla_{[\nu}\nabla_{\mu]}\Phi)\Psi + \Phi\nabla_{[\nu}\nabla_{\mu]}\Psi. \end{cases}$$

Now every tensor can be written as a sum of products of vectors and from this we get immediately a formula for the operator $\nabla_{[\nu}\nabla_{\mu]}$ applied to a tensor. In this formula there appears a term with $R_{\nu\mu\lambda}^{\cdot\cdot\cdot\varkappa}$ with the +-(−-)sign for each upper (lower) index. For instance, for the tensor $P^{\varkappa\lambda}_{\cdot\cdot\omega}$

$$(4.13)\quad \begin{cases} \nabla_{[\nu}\nabla_{\mu]} P^{\varkappa\lambda}_{\cdot\cdot\omega} = \tfrac{1}{2} R_{\nu\mu\varrho}^{\cdot\cdot\cdot\varkappa} P^{\varrho\lambda}_{\cdot\cdot\omega} + \tfrac{1}{2} R_{\nu\mu\varrho}^{\cdot\cdot\cdot\lambda} P^{\varkappa\varrho}_{\cdot\cdot\omega} - \\ \quad - \tfrac{1}{2} R_{\nu\mu\omega}^{\cdot\cdot\cdot\sigma} P^{\varkappa\lambda}_{\cdot\cdot\sigma} - S_{\nu\mu}^{\cdot\cdot\varrho}\nabla_\varrho P^{\varkappa\lambda}_{\cdot\cdot\omega}. \end{cases}$$

If $\nabla_{[\nu}\nabla_{\mu]}$ is applied to a scalar density or Δ-density $\mathfrak{p}$ of weight $+1$ we may deduce in the same way as (4.9a)

$$(4.14)\quad \nabla_{[\nu}\nabla_{\mu]}\mathfrak{p} = -\mathfrak{p}\,\partial_{[\nu}\Gamma_{\mu]} - S_{\nu\mu}^{\cdot\cdot\lambda}\nabla_\lambda\mathfrak{p}.$$

But, although Γ_μ is not a vector, the expression $\partial_{[\nu}\Gamma_{\mu]}$ is always a bivector because

$$(4.15)\quad V_{\nu\mu} \overset{\text{def}}{=} R_{\nu\mu\lambda}^{\cdot\cdot\cdot\lambda} = 2\partial_{[\nu}\Gamma_{\mu]} + 2\Gamma^{\lambda}_{[\nu|\varrho|}\Gamma^{\varrho}_{\mu]\lambda} = 2\partial_{[\nu}\Gamma_{\mu]}\ ^{2)}.$$

Hence, we get for a scalar density or Δ-density $\mathfrak{p}$ of weight w

$$(4.16)\quad \nabla_{[\nu}\nabla_{\mu]}\mathfrak{p} = -\tfrac{1}{2} w\,\mathfrak{p}\,V_{\nu\mu} - S_{\nu\mu}^{\cdot\cdot\lambda}\nabla_\lambda\mathfrak{p}$$

and for a tensor density or Δ-density of weight w, for instance $\mathfrak{P}^{\varkappa}_{\cdot\lambda}$, according to the rule of LEIBNIZ

$$(4.17)\quad \nabla_{[\nu}\nabla_{\mu]}\mathfrak{P}^{\varkappa}_{\cdot\lambda} = \tfrac{1}{2} R_{\nu\mu\varrho}^{\cdot\cdot\cdot\varkappa}\mathfrak{P}^{\varrho}_{\cdot\lambda} - \tfrac{1}{2} R_{\nu\mu\lambda}^{\cdot\cdot\cdot\sigma}\mathfrak{P}^{\varkappa}_{\cdot\sigma} - S_{\nu\mu}^{\cdot\cdot\varrho}\nabla_\varrho\mathfrak{P}^{\varkappa}_{\cdot\lambda} - \tfrac{1}{2} w\,V_{\nu\mu}\mathfrak{P}^{\varkappa}_{\cdot\lambda}.$$

In dealing with pseudo-quantities for which a covariant differentiation (that is a set of parameters Λ_λ besides the $\Gamma^{\varkappa}_{\mu\lambda}$) has been defined

[1]) Apart from a slight change this is the proof given by MORINAGA 1934, 1.

[2]) Cf. III § 6 for the geometric meaning of the bivector $V_{\nu\mu}$.

(cf. III § 2) we get in the same way an additional term containing the *bivector*

(4.18) $$l_{\nu\mu} \overset{\text{def}}{=} 2\partial_{[\nu} A_{\mu]}$$

instead of $V_{\nu\mu}$ and c (class) instead of w (weight). For instance for a pseudo-vector density $\mathfrak{v}^\varkappa$

(4.19) $$\nabla_{[\nu}\nabla_{\mu]}\mathfrak{v}^\varkappa = \tfrac{1}{2}R_{\nu\mu\lambda}^{\cdot\cdot\cdot\varkappa}\mathfrak{v}^\lambda - S_{\nu\mu}^{\cdot\cdot\lambda}\nabla_\lambda\mathfrak{v}^\varkappa - \tfrac{1}{2}(w\,V_{\nu\mu} + c\,l_{\nu\mu})\,\mathfrak{v}^\varkappa.$$

If a symmetric tensorfield $g_{\lambda\varkappa}$ of rank n, its covariant derivative $-Q_{\mu\lambda\varkappa}$ and $S_{\mu\lambda}^{\cdot\cdot\varkappa}$ are given, the curvature tensor can be expressed in terms of these quantities, their first ordinary derivatives and the second ordinary derivatives of $g_{\lambda\varkappa}$. If we write [cf. (3.9)]

(4.20) $$\Gamma_{\mu\lambda}^{\varkappa} = \left\{{}_{\mu\lambda}^{\varkappa}\right\} + T_{\mu\lambda}^{\cdot\cdot\varkappa}; \qquad T_{\mu\lambda}^{\cdot\cdot\varkappa} \overset{\text{def}}{=} g^{\varkappa\varrho}\left(-S_{\{\mu\varrho\lambda\}} + \tfrac{1}{2}Q_{\{\mu\varrho\lambda\}}\right)$$

and

(4.21) $$K_{\nu\mu\lambda}^{\cdot\cdot\cdot\varkappa} \overset{\text{def}}{=} 2\partial_{[\nu}\left\{{}_{\mu]\lambda}^{\varkappa}\right\} + 2\left\{{}_{[\nu|\varrho|}^{\varkappa}\right\}\left\{{}_{\mu]\lambda}^{\varrho}\right\}.$$

$K_{\nu\mu\lambda}^{\cdot\cdot\cdot\varkappa}$ is the curvature tensor of the riemannian connexion belonging to $g_{\lambda\varkappa}$, and $T_{\mu\lambda}^{\cdot\cdot\varkappa}$ is a tensor representing the difference between the connexion $\Gamma_{\mu\lambda}^{\varkappa}$ and this riemannian connexion. If we indicate the covariant differentiation of the connexion $\left\{{}_{\mu\lambda}^{\varkappa}\right\}$ by $\overset{*}{\nabla}$ we get from (4.2) and (4.20)

(4.22a) $$R_{\nu\mu\lambda}^{\cdot\cdot\cdot\varkappa} = K_{\nu\mu\lambda}^{\cdot\cdot\cdot\varkappa} + 2\overset{*}{\nabla}_{[\nu} T_{\mu]\lambda}^{\cdot\cdot\varkappa} + 2\,T_{[\nu|\varrho|}^{\cdot\cdot\varkappa}\,T_{\mu]\lambda}^{\cdot\cdot\varrho}$$

and also

(4.22b) $$R_{\nu\mu\lambda}^{\cdot\cdot\cdot\varkappa} = K_{\nu\mu\lambda}^{\cdot\cdot\cdot\varkappa} + 2\nabla_{[\nu} T_{\mu]\lambda}^{\cdot\cdot\varkappa} - 2\,T_{[\nu|\varrho|}^{\cdot\cdot\varkappa}\,T_{\mu]\lambda}^{\cdot\cdot\varrho} + 2\,S_{\nu\mu}^{\cdot\cdot\varrho}\,T_{\varrho\lambda}^{\cdot\cdot\varkappa}.$$

From (4.22a) we may derive the relations

(4.23) $$\begin{cases} R_{\mu\lambda} = K_{\mu\lambda} + \overset{*}{\nabla}_\varkappa T_{\mu\lambda}^{\cdot\cdot\varkappa} - \overset{*}{\nabla}_\mu T_{\varkappa\lambda}^{\cdot\cdot\varkappa} + T_{\varkappa\varrho}^{\cdot\cdot\varkappa}\,T_{\mu\lambda}^{\cdot\cdot\varrho} - T_{\mu\varrho}^{\cdot\cdot\varkappa}\,T_{\varkappa\lambda}^{\cdot\cdot\varrho}; \\ R_{\mu\lambda} \overset{\text{def}}{=} R_{\varkappa\mu\lambda}^{\cdot\cdot\cdot\varkappa}; \qquad K_{\mu\lambda} = K_{\lambda\mu} \overset{\text{def}}{=} K_{\varkappa\mu\lambda}^{\cdot\cdot\cdot\varkappa} \end{cases}$$

(4.24) $$V_{\nu\mu} = 2\overset{*}{\nabla}_{[\nu} T_{\mu]\varkappa}^{\cdot\cdot\varkappa} = 2\partial_{[\nu} T_{\mu]\varkappa}^{\cdot\cdot\varkappa}; \quad \text{(cf. 4.15)}.$$

$R_{\mu\lambda}$ ($K_{\mu\lambda}$) is called the RICCI *tensor* of the L_n (V_n).

(4.22) is a special case of a more general formula. If $A_{\mu\lambda}^{\cdot\cdot\varkappa}$ is an arbitrary tensor, $'\Gamma_{\mu\lambda}^{\varkappa} = \Gamma_{\mu\lambda}^{\varkappa} + A_{\mu\lambda}^{\cdot\cdot\varkappa}$ represents another connexion and the relation between the curvature tensors of these two connexions is easily proved to be

(4.25) $$'R_{\nu\mu\lambda}^{\cdot\cdot\cdot\varkappa} = R_{\nu\mu\lambda}^{\cdot\cdot\cdot\varkappa} + 2\nabla_{[\nu} A_{\mu]\lambda}^{\cdot\cdot\varkappa} + 2S_{\nu\mu}^{\cdot\cdot\varrho} A_{\varrho\lambda}^{\cdot\cdot\varkappa} - 2A_{[\nu|\lambda|}^{\cdot\cdot\varrho} A_{\mu]\varrho}^{\cdot\cdot\varkappa}.$$

In a WEYL connexion we have according to (3.18)

(4.26) $$T_{\mu\lambda}^{\cdot\cdot\varkappa} = \tfrac{1}{2}(Q_\mu A_\lambda^\varkappa + Q_\lambda A_\mu^\varkappa - Q^\varkappa g_{\lambda\mu})$$

and from this it follows after some calculation

$$(4.27)\quad \begin{cases} R_{\nu\mu\lambda}^{\cdot\cdot\cdot\varkappa} = K_{\nu\mu\lambda}^{\cdot\cdot\cdot\varkappa} + \nabla_{[\nu} Q_{\mu]} A_\lambda^\varkappa - g_{[\nu[\varrho} (2\nabla_{\mu]} Q_{\lambda]} + Q_{\mu]} Q_{\lambda]} - \\ \qquad - \frac{1}{2} Q_{|\sigma|} Q^\sigma g_{\mu]\lambda]}) g^{\varrho\varkappa}. \end{cases}$$

In a metric and semi-symmetric connexion we have according to (2.14) and (3.5)

$$(4.28)\quad T_{\mu\lambda}^{\cdot\cdot\varkappa} = - S_\lambda A_\mu^\varkappa + S^\varkappa g_{\lambda\mu}$$

and this leads without difficulties to

$$(4.29)\quad R_{\nu\mu\lambda}^{\cdot\cdot\cdot\varkappa} = K_{\nu\mu\lambda}^{\cdot\cdot\cdot\varkappa} + g^{\varkappa\varrho} g_{[\nu[\varrho} (4\nabla_{\mu]} S_{\lambda]} - 4 S_{\mu]} S_{\lambda]} + 2 S_{|\sigma|} S^\sigma g_{\mu]\lambda]}).$$

An L_n is said to possess *teleparallelism*[1], [2]) or *absolute parallelism* (Fernparallelismus) in $\mathfrak{N}(\overset{}{\underset{0}{\xi}}{}^\varkappa)$ if for any two points P and Q of this region the parallel displacement of any quantity from P to Q along a curve of $\mathfrak{N}(\underset{0}{\xi}{}^\varkappa)$ connecting P and Q gives a result at Q which is independent of the choice of the curve. According to (4.1) the curvature tensor vanishes in this case. Naturally an E_n possesses teleparallelism. In order to prove that every A_n with teleparallelism is an E_n, we take n linearly independent covariant vectors $\overset{h}{e}_\lambda$; $h = 1, \ldots, n$ at $\underset{0}{\xi}{}^\varkappa$ and by parallel displacement we construct n covariant constant fields $\overset{h}{e}_\lambda$ in an $\mathfrak{N}(\underset{0}{\xi}{}^\varkappa)$. Because $\partial_{[\mu} \overset{h}{e}_{\lambda]} = \nabla_{[\mu} \overset{h}{e}_{\lambda]} = 0$, the $\overset{h}{e}_\lambda$ are gradientfields. If $\overset{h}{e}_\lambda \overset{*}{=} \partial_\lambda \xi^h$, the ξ^h are independent because their gradients are linearly independent. If we take them as new coordinates we have

$$(4.30)\quad 0 = \nabla_j \overset{h}{e}_i = \partial_j \overset{h}{e}_i - \Gamma_{ji}^k \overset{h}{e}_k \overset{*}{=} - \Gamma_{ji}^h$$

and this proves that (h) is a rectilinear coordinate system in an E_n. The same does not hold for an L_n. In fact an L_n can be endowed with teleparallelism without being an E_n.[3])

Here is another proof showing the relations of $R_{\nu\mu\lambda}^{\cdot\cdot\cdot\varkappa}$ to integrability conditions. We look for a *holonomic* coordinate system (h) in L_n such that the Γ_{ji}^h vanish. The differential equations are [cf. (2.5)]

$$(4.31)\quad \begin{cases} \text{a)} \quad \partial_\mu \xi^h = A_\mu^h \\ \text{b)} \quad \partial_\mu A_\lambda^h = A_\varkappa^h \Gamma_{\mu\lambda}^\varkappa \end{cases}$$

[1]) Cf. WEITZENBÖCK 1923, 1; VITALI 1924, 1; GRISS 1925, 1; CARTAN and SCHOUTEN 1926, 3; 4; CARTAN 1927, 2; BORTOLOTTI 1927, 1; 2; 1928, 1; 1929, 1; EINSTEIN 1928, 1; WEITZENBÖCK 1928, 1; SCHOUTEN 1929, 4; E I 1935, 1, p. 88; HAIMOVICI 1943, 1.

[2]) For literature on absolute parallelism in the large, cf. footnote 2 on p. 125.

[3]) It is proved in Chapter IV that every non-trivial group space is such an L_n.

with the integrability conditions

$$(4.32)\quad \begin{cases} \text{a)} \quad 0 = \partial_{[\nu}\, A^h_{\mu]} = A^h_\varkappa\, S_{\nu\mu}{}^{\cdot\cdot\varkappa} \\ \text{b)} \quad 0 = \partial_{[\nu}\,\partial_\mu\, A^h_{\lambda]} = (\partial_{[\nu}\, A^h_{|\varkappa|})\, \Gamma^\varkappa_{\mu]\lambda} + A^h_\varkappa\, \partial_{[\nu}\, \Gamma^\varkappa_{\mu]\lambda} = \tfrac{1}{2} A^h_\varkappa\, R_{\nu\mu\lambda}{}^{\cdot\cdot\cdot\varkappa}. \end{cases}$$

These conditions are satisfied identically if and only if $S_{\mu\lambda}{}^{\cdot\cdot\varkappa} = 0$ and $R_{\nu\mu\lambda}{}^{\cdot\cdot\cdot\varkappa} = 0$.[1])

If the fundamental tensor in a V_n is written as a product of two equal ideal factors (cf. I § 11), $g_{\lambda\varkappa} = a_\lambda\, a_\varkappa = b_\lambda\, b_\varkappa = \cdots$, we have

$$(4.33)\qquad 0 = \nabla_\mu\, g_{\lambda\varkappa} = \nabla_\mu\, a_\lambda\, a_\varkappa = 2(\nabla_\mu\, a_{(\lambda)}\, a_{\varkappa)}.$$

The expression $(\nabla_\mu\, a_\lambda)\, a_\varkappa$ has no meaning by itself, only the sum of $(\nabla_\mu\, a_\lambda)\, a_\varkappa$ and $(\nabla_\mu\, a_\varkappa)\, a_\lambda$ is known to be zero. This enables us to simplify the calculations by taking $(\nabla_\mu\, a_\lambda)\, a_\varkappa = 0$. Then we get for instance

$$(4.34)\qquad \nabla_\mu\, v^\varkappa = \nabla_\mu\, v^\lambda\, a_\lambda\, a^\varkappa = a^\varkappa\, \partial_\mu\, v^\lambda\, a_\lambda = \partial_\mu\, v^\varkappa + v^\lambda\, a^\varkappa\, \partial_\mu\, a_\lambda,$$

hence

$$(4.35)\qquad \Gamma^\varkappa_{\mu\lambda} = a^\varkappa\, \partial_\mu\, a_\lambda = -\, a_\lambda\, \partial_\mu\, a^\varkappa$$

and

$$(4.36)\qquad 2\,\nabla_{[\nu}\,\nabla_{\mu]}\, v^\varkappa = 2\,\nabla_{[\nu}\,\nabla_{\mu]}\, v^\lambda\, a_\lambda\, a^\varkappa = 2\, v^\lambda\, a_\lambda\, \nabla_{[\nu}\,\nabla_{\mu]}\, a^\varkappa,$$

hence

$$(4.37)\qquad K_{\nu\mu\lambda}{}^{\cdot\cdot\cdot\varkappa} = 2 a_\lambda\, \nabla_{[\nu}\,\nabla_{\mu]}\, a^\varkappa.$$

Using this symbolism many calculations can be shortened considerably (cf. I § 11 under 6, footnote 6, p. 58).

Exercises.

III 4,1[2]). A man is moving on the surface of the earth always facing one definite point, say Jerusalem or Mekka or the North pole. Prove that this displacement is semi-symmetric and metric and compute S_λ.[3]) During the mathematical congress in Moscow in 1934 one evening some of us invented the "Moscow displacement". The streets of Moscow are approximately straight lines through the Kremlin and concentric circles around it. Now let a person walk in the street always facing the Kremlin. Prove that also this displacement is semi-symmetric and metric and compute S_λ.

III 4,2. Prove that for a symmetric connexion the λ-rank of $R_{\nu\mu\lambda}{}^{\cdot\cdot\cdot\varkappa}$ can not be greater than the ν-rank. For a metric and symmetric connexion the rank of $K_{\nu\mu\lambda\varkappa}$ is the same for all indices.

---

[1]) RIEMANN 1861, 1, p. 402, LIPSCHITZ 1869, 1, p. 94; 1874, 1, p. 109; RICCI 1884, 1, p. 142, all for V_n.

[2]) Cf. E I 1935, 1, p. 127, Exerc. 11,9.

[3]) HESSENBERG 1924, 1 gave other examples of teleparallelism on a sphere.

III 4,3. Prove that (cf. 4.20)

III 4,3 α) $\frac{1}{2} R_{\nu\mu\lambda}^{\cdot\cdot\cdot\varkappa} - \nabla_{[\nu} T_{\mu]\lambda}^{\cdot\cdot\varkappa} - S_{\nu\mu}^{\cdot\cdot\varrho} T_{\varrho\lambda}^{\cdot\cdot\varkappa} = \frac{1}{2} K_{\nu\mu\lambda}^{\cdot\cdot\cdot\varkappa} + \overset{*}{\nabla}_{[\nu} T_{\mu]\lambda}^{\cdot\cdot\varkappa}$.

III 4,4. If the vector $v^\varkappa$ is displaced parallel along the boundary of the infinitesimal simple bivector $df^{\nu\mu}$ the difference between the final value and the initial value lies in the plane of $df^{\nu\mu}$ if and only if

III 4,4 α) $(n-1) R_{\nu\mu\lambda}^{\cdot\cdot\cdot\varkappa} = 2 A_{[\nu}^{\varkappa} R_{\mu]\lambda}$ [1]).

III 4,5. Prove that in an L_n teleparallelism for scalar densities exists if and only if $V_{\nu\mu} = 0$ (cf. III § 6). The connexion is called in that case *volume preserving* [2]).

§ 5. The identities for the curvature tensor. [3])

From the definition (4.2) we get the *first identity*

(5.1) $$\boxed{R_{(\nu\mu)\lambda}^{\cdot\cdot\cdot\varkappa} = 0}$$

and by alternation of (4.2) over $\nu\mu\lambda$ we get the *second identity*

(5.2) $$\boxed{R_{[\nu\mu\lambda]}^{\cdot\cdot\cdot\varkappa} = 2\nabla_{[\nu} S_{\mu\lambda]}^{\cdot\cdot\varkappa} - 4 S_{[\nu\mu}^{\cdot\cdot\varrho} S_{\lambda]\varrho}^{\cdot\cdot\varkappa}}$$

taking for *semi-symmetric* connexions the form

(5.3) $$R_{[\nu\mu\lambda]}^{\cdot\cdot\cdot\varkappa} = 2 A_{[\nu}^{\varkappa} \nabla_\mu S_{\lambda]}$$ [4])

and for the *symmetric* case the form

(5.4) $$\boxed{R_{[\nu\mu\lambda]}^{\cdot\cdot\cdot\varkappa} = 0}.$$

If we write (cf. 4.15, 23)

(5.5) $$R_{\mu\lambda} \overset{\text{def}}{=} R_{\nu\mu\lambda}^{\cdot\cdot\cdot\nu}; \qquad V_{\nu\mu} \overset{\text{def}}{=} R_{\nu\mu\lambda}^{\cdot\cdot\cdot\lambda}$$

we get from (5.2) by contraction

(5.6) $$\boxed{2R_{[\mu\lambda]} = -V_{\mu\lambda} + 2\nabla_\sigma S_{\mu\lambda}^{\cdot\cdot\sigma} + 2(n-1)\nabla_{[\mu} S_{\lambda]} + 2(n-1) S_{\mu\lambda}^{\cdot\cdot\sigma} S_\sigma}$$

from which for the *semi-symmetric* case

(5.7) $$2R_{[\mu\lambda]} = -V_{\mu\lambda} + 2(n-2)\nabla_{[\mu} S_{\lambda]}$$

---

[1]) COSSU 1949, 1.

[2]) VEBLEN 1923, 2; SCHOUTEN 1923, 1; EISENHART 1923, 3. Cf. GOLAB 1931, 1.

[3]) Cf. for complete sets of identities for different cases T. Y. THOMAS 1934, 2; LEVINE 1936, 1; SUN 1937, 1; BLUM 1947, 1.

[4]) The case where (5.3) holds for a not semi-symmetric connexion was investigated by V. D. KULK 1939, 1.

and for the *symmetric* case

(5.8) $$\boxed{2R_{[\mu\lambda]} = -V_{\mu\lambda}}\,.$$

By differentiation and alternation we get from (3.2) and (4.13)

(5.9) $$\nabla_{[\nu}\nabla_{\mu]}g_{\lambda\varkappa} = -R_{\nu\mu(\lambda\varkappa)} - S_{\nu\mu}^{\cdot\cdot\varrho}\nabla_\varrho g_{\lambda\varkappa}$$

from which follows the *third identity*

(5.10) $$\boxed{R_{\nu\mu(\lambda\varkappa)} = \nabla_{[\nu}Q_{\mu]\lambda\varkappa} + S_{\nu\mu}^{\cdot\cdot\varrho}Q_{\varrho\lambda\varkappa}}\ ^{1)}\,.$$

This identity takes the following forms for special cases

semi-symmetric connexion

(5.11) $$R_{\nu\mu(\lambda\varkappa)} = \nabla_{[\nu}Q_{\mu]\lambda\varkappa} + S_{[\nu}Q_{\mu]\lambda\varkappa}$$

semi-metric connexion

(5.12) $$R_{\nu\mu(\lambda\varkappa)} = (\nabla_{[\nu}Q_{\mu]} + S_{\nu\mu}^{\cdot\cdot\varrho}Q_\varrho)\,g_{\lambda\varkappa}$$

metric connexion

(5.13) $$\boxed{R_{\nu\mu(\lambda\varkappa)} = 0}\,.$$

Every quantity $Q_{\nu\mu\lambda\varkappa}$, alternating in $\nu\mu$ satisfies the identity

(5.14) $$\begin{cases} Q_{\lambda\varkappa\nu\mu} = Q_{\nu\mu\lambda\varkappa} - \tfrac{3}{2}(Q_{[\varkappa\lambda\nu]\mu} + Q_{[\varkappa\mu\nu]\lambda} + Q_{[\mu\lambda\nu]\varkappa} + Q_{[\lambda\varkappa\mu]\nu}) + \\ \qquad + Q_{\nu\varkappa(\mu\lambda)} + Q_{\lambda\nu(\mu\varkappa)} + Q_{\varkappa\mu(\lambda\nu)} + Q_{\mu\lambda(\varkappa\nu)} + Q_{\mu\nu(\varkappa\lambda)} + Q_{\lambda\varkappa(\mu\nu)}\,. \end{cases}$$

If (5.2) and (5.10) are substituted in (5.14) for $Q_{\nu\mu\lambda\varkappa} = R_{\nu\mu\lambda\varkappa}$ we get $R_{\nu\mu\lambda\varkappa} - R_{\lambda\varkappa\nu\mu}$ expressed in terms of $S_{\mu\lambda}^{\cdot\cdot\varkappa}$, $\nabla_{[\nu}S_{\mu\lambda]}^{\cdot\cdot\varkappa}$, $Q_{\mu\lambda\nu}$ and $\nabla_{[\nu}Q_{\mu]\lambda\varkappa}$. This rather complicated formula is the *fourth identity*. For the *metric* and *symmetric* case it takes a very simple form. For this special case (riemannian connexion) we collect the four identities

(5.15) $$\begin{cases} \text{I)} & K_{(\nu\mu)\lambda\varkappa} = 0 \\ \text{II)} & K_{[\nu\mu\lambda]\varkappa} = 0 \\ \text{III)} & K_{\nu\mu(\lambda\varkappa)} = 0 \\ \text{IV)} & K_{\nu\mu\lambda\varkappa} = K_{\lambda\varkappa\nu\mu}\,. \end{cases}$$

From (II, III, IV) it follows that

(5.16) $$\text{II}')\quad K_{[\nu\mu\lambda\varkappa]} = 0\,.$$

* 1) CASTOLDI 1947, 1 finds only one term, $\overset{*}{\nabla}_{[\nu}Q_{\mu]\lambda\varkappa}$ at the right hand side, where $\overset{*}{\nabla}$ is the operator of the riemannian connexion belonging to $g_{\lambda\varkappa}$. This seems to be a mistake.

It can be proved that (I, II, III, IV) is equivalent to (I, II′, III, IV). The identities (I, II′, III, IV) can be used to evaluate the number of independent components of $K_{\nu\mu\lambda\varkappa}$. According to (I, III, IV) $K_{\nu\mu\lambda\varkappa}$ is a symmetric bivector tensor (I § 8 under B) i.e. a symmetric tensor in the $\binom{n}{2}$-dimensional space of all bivectors.[1]) Now a symmetric bivector tensor has $\frac{1}{2}\binom{n}{2}\left\{\binom{n}{2}+1\right\}=\frac{1}{8}(n^2-n)(n^2-n+2)$ components. But (II′) stands for $\binom{n}{4}$ independent relations. Hence the number of independent components is $\frac{1}{12}n^2(n^2-1)$, that is 1 for $n=2$; 6 for $n=3$ and 20 for $n=4$ (the case of relativity).

From (5.15) it follows that $V_{\nu\mu}$ is always zero in a V_n and that $K_{\mu\lambda}$ is always symmetric there. $K_{\mu\lambda}$ is the RIÇCI tensor of the V_n. The principal directions of $K_{\mu\lambda}$ are called the *principal directions of the* V_n.[2])

The equations (4.2)

$$\partial_{[\nu}\Gamma^{\varkappa}_{\mu]\lambda}=-\Gamma^{\varkappa}_{[\nu|\varrho|}\Gamma^{\varrho}_{\mu]\lambda}+\tfrac{1}{2}R_{\nu\mu\lambda}{}^{\cdot\cdot\cdot\varkappa} \tag{5.17}$$

have the form (II 6.15) with a right hand member that depends on the $\Gamma^{\varkappa}_{\mu\lambda}$. As we have seen in II § 6 the construction of the integrability conditions depends on the possibility of being able to eliminate the derivatives of the $\Gamma^{\varkappa}_{\mu\lambda}$ at each step. Now for (5.17) this possibility exists and we get for the first set of integrability conditions

$$\left\{\begin{aligned}0&=\Gamma^{\varkappa}_{[\omega|\sigma|}\Gamma^{\sigma}_{\nu|\varrho|}\Gamma^{\varrho}_{\mu]\lambda}-\tfrac{1}{2}R_{[\omega\nu|\varrho|}{}^{\cdot\cdot\cdot\varkappa}\Gamma^{\varrho}_{\mu]\lambda}+\Gamma^{\varkappa}_{[\nu|\varrho|}\Gamma^{\varrho}_{\omega|\sigma|}\Gamma^{\sigma}_{\mu]\lambda}-\\&\qquad-\tfrac{1}{2}\Gamma^{\varkappa}_{[\nu|\varrho|}R_{\omega\mu]\lambda}{}^{\cdot\cdot\cdot\varrho}+\tfrac{1}{2}\partial_{[\omega}R_{\nu\mu]\lambda}{}^{\cdot\cdot\cdot\varkappa}\\&=\tfrac{1}{2}\nabla_{[\omega}R_{\nu\mu]\lambda}{}^{\cdot\cdot\cdot\varkappa}+S_{[\omega\nu}{}^{\cdot\cdot\sigma}R_{|\sigma|\mu]\lambda}{}^{\cdot\cdot\cdot\varkappa}.\end{aligned}\right. \tag{5.18}$$

If ∂_{ϱ} is applied to the right hand side and if we alternate over $\varrho\omega\nu\mu$ the expression is identically zero.[3]) Hence the only integrability conditions are

$$\boxed{\nabla_{[\omega}R_{\nu\mu]\lambda}{}^{\cdot\cdot\cdot\varkappa}=2S_{[\omega\nu}{}^{\cdot\cdot\sigma}R_{\mu]\sigma\lambda}{}^{\cdot\cdot\cdot\varkappa}} \tag{5.19}$$

known as BIANCHI'S *identity*.[4]) The identity can be proved in many other ways, for instance by computing $\nabla_{[\omega}\nabla_{\nu}\nabla_{\mu]}w_{\lambda}$ from $\nabla_{[\omega}\nabla_{\nu]}\nabla_{\mu}w_{\lambda}$

---

[1]) The geometric properties of this quantity were for $n=4$ investigated by CHURCHILL 1932, 1; RUSE 1936, 1; 1944, 1; 2; 1946, 1; 2; 1948, 1; LANCZOS 1938, 1; cf. WRONA 1948, 1 for general values of n.

[2]) RICCI 1903, 1; 1904, 1. Cf. EISENHART 1922, 1. Other principal directions were defined by SYNGE 1922, 1.

[3]) This result will be obtained later in a much shorter way.

[4]) An identity of this kind was first considered by VOSS 1880, 1. Then RICCI found it independently and sent it to PADOVA, who published it in 1889, 1. In 1902, 1 it was proved independently by BIANCHI. Therefore italien mathematiciens

and from $\nabla_\omega \nabla_{[\nu} \nabla_{\mu]} w_\lambda$. For a *semi-symmetric* connexion it takes the form

(5.20) $$\nabla_{[\omega} R_{\nu\mu]\lambda}^{\cdot\cdot\cdot\varkappa} = 2 S_{[\omega} R_{\mu\nu]\lambda}^{\cdot\cdot\cdot\varkappa}$$

and for a *symmetric* connexion we have

(5.21) $$\boxed{\nabla_{[\omega} R_{\nu\mu]\lambda}^{\cdot\cdot\cdot\varkappa} = 0}\ ^{1)},\ ^{2)}.$$

By contraction of (5.19) over $\varkappa\lambda$ we get for the general case:

(5.22) a) $\boxed{\nabla_{[\omega} V_{\nu\mu]} = 2 S_{[\omega\nu}^{\cdot\cdot\cdot\varrho} V_{\mu]\varrho}}$; (equivalent to b) $\partial_{[\omega} V_{\nu\mu]} = 0$),

for the *semi-symmetric* case:

(5.23) $$\nabla_{[\omega} V_{\nu\mu]} = 2 S_{[\omega} V_{\mu\nu]}$$

and for the *symmetric* case:

(5.24) $$\boxed{\nabla_{[\omega} V_{\nu\mu]} = 0}\,.$$

By contraction of (5.19) over $\omega\varkappa$ we get for the general case:

(5.25) $$\boxed{\nabla_\varkappa R_{\nu\mu\lambda}^{\cdot\cdot\cdot\varkappa} - 2\nabla_{[\nu} R_{\mu]\lambda} = 4 S_{\varkappa[\nu}^{\cdot\cdot\cdot\varrho} R_{\mu]\varrho\lambda}^{\cdot\cdot\cdot\varkappa} + 2 S_{\nu\mu}^{\cdot\cdot\cdot\varrho} R_{\varrho\lambda}};$$

for the *semi-symmetric* case:

(5.26) $$\nabla_\varkappa R_{\nu\mu\lambda}^{\cdot\cdot\cdot\varkappa} - 2\nabla_{[\nu} R_{\mu]\lambda} = 4 S_\varkappa R_{\mu\nu\lambda}^{\cdot\cdot\cdot\varkappa} + 2 S_{[\nu} R_{\mu]\lambda}$$

and for the *symmetric* case:

(5.27) $$\boxed{\nabla_\varkappa R_{\nu\mu\lambda}^{\cdot\cdot\cdot\varkappa} - 2\nabla_{[\nu} R_{\mu]\lambda} = 0}\,.$$

---

often speak of the identity of RICCI-BIANCHI. All these publications consider the metric symmetric case only. BACH gave 1921, 1 the proof for a WEYL connexion and the author at the Jena congress 1921 did the same for a general symmetric connexion, cf. 1923, 2. In 1922, 1 VEBLEN dealt with the symmetric case and in 1923, 1 WEITZENBÖCK published the formula (5.19). Cf. SCHOUTEN-STRUIK 1923, 3. As a more recent publication we mention LEVINE 1936, 1. WAGNER gave 1945, 2 a generalization for general geometric objects.

1) For an A_n the second identity and BIANCHI's identity can be proved in a very short way by using normal coordinates (cf. III § 7, Exerc. III 7,4). Another method is used in III § 9.

2) BLUM proved 1946, 1 that BIANCHI's identity in a V_n consists of $\frac{1}{24} n^2 (n^2 - 1) \times (n-2)$ independent equations. Cf. BLUM 1947, 1 for complete sets of identities

For a riemannian connexion we get from (5.27) by transvection with $g^{\mu\lambda}$

$$\nabla_\varkappa K_{\nu}^{\cdot\,\varkappa} - \nabla_\nu K + \nabla_\varkappa K_{\nu}^{\cdot\,\varkappa} = 0; \qquad K \stackrel{\text{def}}{=} K_{\mu\lambda}\, g^{\mu\lambda} \tag{5.28}$$

or

$$\boxed{\nabla_\mu G^{\mu}_{\cdot\lambda} = 0; \qquad G_{\varkappa\lambda} \stackrel{\text{def}}{=} K_{\varkappa\lambda} - \tfrac{1}{2} K\, g_{\varkappa\lambda}} \;{}^{1)}. \tag{5.29}$$

The scalar $\varkappa \stackrel{\text{def}}{=} \dfrac{1}{n\,(n-1)} K$ is called the *scalar curvature*[2]) of the V_n. For $n = 2$ we have $G_{\lambda\varkappa} = 0$, and $\varkappa$ is equal to the *gaussian curvature*. For $n = 4$, $G_{\lambda\varkappa}$ is the tensor of momentum and energy in general relativity and (5.29) is the well known equation of conservation of momentum and energy.

An L_n in which $\nabla_\nu R_{\mu\lambda} = 0$ is called a RICCI *space*. If $R_{\mu\lambda} = 0$ it is said to be a *special* RICCI *space*. A V_n, $n > 1$ in which $K_{\mu\lambda}$ differs from $g_{\mu\lambda}$ only by a scalar factor is called an EINSTEIN *space*.[3]) If $K_{\mu\lambda} = 0$ it is said to be a *special* EINSTEIN *space*. According to the definition of K and $\varkappa$ we have in this case

$$K_{\mu\lambda} = \frac{1}{n} K\, g_{\mu\lambda} = (n-1)\,\varkappa\, g_{\mu\lambda}. \tag{5.30}$$

For $n > 2$ it follows from (5.29) that the scalar curvature of an EINSTEIN space is constant. But a V_n with a constant scalar curvature need not be an EINSTEIN space.[4]) This is a theorem of HERGLOTZ.[5]) An EINSTEIN space is for $n > 2$ a special case of a RICCI space.[6])

A V_n, $n > 2$ in which $K_{\nu\mu\lambda\varkappa}$ differs from $g_{[\nu[\lambda}\, g_{\mu]\varkappa]}$ only by a scalar factor is called an S_n or a *space of constant curvature*. In this case we have

$$K_{\nu\mu\lambda\varkappa} = \frac{-2}{n\,(n-1)} K\, g_{[\nu[\lambda}\, g_{\mu]\varkappa]} = -2\varkappa\, g_{[\nu[\lambda}\, g_{\mu]\varkappa]}. \tag{5.31}$$

For $n = 2$ an S_2 is defined as a V_2 in which $\varkappa$ is constant.

---

[1]) BIANCHI 1902, 1; RICCI 1903, 1; for $n = 3$.

[2]) WEYL 1921, 2 used the term "scalar curvature" in another sense. He called an S_n a "space of scalar curvature". Cf. footnote 4 and VI p. 289, footnote 1.

[3]) Cf. E I 1935, 1, p. 125; FIALKOW 1938, 1; 1939, 1; 1942, 1; T. Y. THOMAS 1938, 1 (invariants); HAANTJES and WRONA 1939, 1 (and C_n); YANO 1943, 1 (conf. and concirc); WRONA 1947, 1 (C_n); KUIPER 1950, 1; 2 (conf. and proj.); 1951, 1 (conf.); TACHIBANA 1951, 1 (parallel fields).

[4]) The opposite assertion in LOVELL 1934, 1 is erroneous and perhaps caused by a wrong interpretation of WEYL's term "scalar curvature". Cf. footnote 2 and footnote 1 Ch. VI, p. 289.

[5]) HERGLOTZ 1916, 1, p. 203. Cf. for generalizations WRONA 1941, 1; BOMPIANI 1950, 1; VARGA 1943, 1.

[6]) There are many other examples of spaces whose curvature tensor satisfies special identities. We mention here only a paper of V. D. KULK 1939, 1 where the case $(n+1)\, R_{\nu\,(\mu\lambda)}^{\cdot\;\cdot\;\cdot\;\varkappa} = (V_{\nu(\mu} - R_{\nu(\mu)})\, A^{\varkappa}_{\lambda)}$ and especially $R_{\nu\,(\mu\lambda)}^{\cdot\;\cdot\;\cdot\;\varkappa} = 0$ or $R_{\nu\mu\lambda}^{\cdot\;\cdot\;\cdot\;\varkappa} = R_{[\nu\mu\lambda]}^{\cdot\;\cdot\;\cdot\;\varkappa}$ was investigated.

An S_n is a special case of an EINSTEIN space.

The second identity and the identity of BIANCHI can be put into another form by introducing another invariant differential operator.

Let $P_{\lambda_1 \ldots \lambda_p A}$ be a tensor *alternating in the first p indices* and let A be a collecting index standing for any number of co- and contravariant indices. Then we write

$$(5.32)\qquad \overset{p}{\nabla}_{[\mu} P_{\lambda_1 \ldots \lambda_p] A} \overset{\text{def}}{=} \nabla_{[\mu} P_{\lambda_1 \ldots \lambda_p] A} + p\, S_{[\mu \lambda_1}^{\cdot\cdot\cdot\,\sigma} P_{|\sigma| \lambda_2 \ldots \lambda_p] A}\,.$$

As we see the significance of $\overset{p}{\nabla}_{[\mu\ldots]}$ is that the alternating part $S_{\mu\lambda}^{\cdot\cdot\cdot\,\varkappa}$ of $\Gamma_{\mu\lambda}^{\varkappa}$ does *not* apply to the *alternated* indices $\lambda_1 \ldots \lambda_p$. In an A_n, the only difference between $\overset{p}{\nabla}_\mu$ and ∇_μ is that $\overset{p}{\nabla}_\mu$ can be used only if we alternate over the first p indices. In $\overset{0}{\nabla}_\mu$ there is no alternation, hence $\overset{0}{\nabla}_\mu$ is identical with ∇_μ. For this new operator the rule of LEIBNIZ holds in the form

$$(5.33)\qquad \begin{cases} \overset{p+q}{\nabla}_{[\mu} P_{\lambda_1 \ldots \lambda_p |A|} Q_{\varkappa_1 \ldots \varkappa_q] B} \\ \qquad = (\overset{p}{\nabla}_{[\mu} P_{\lambda_1 \ldots \lambda_p |A|}) Q_{\varkappa_1 \ldots \varkappa_q] B} + P_{[\lambda_1 \ldots \lambda_p |A|} \overset{q}{\nabla}_\mu Q_{\varkappa_1 \ldots \varkappa_q] B} \end{cases}$$

and for the operator $\overset{p+1}{\nabla}_{[\nu} \overset{p}{\nabla}_{\mu \ldots]}$ the rule of LEIBNIZ takes the form

$$(5.34)\qquad \begin{cases} \overset{p+1}{\nabla}_{[\nu} \overset{p}{\nabla}_\mu P_{\lambda_1 \ldots \lambda_s |A|} Q_{\lambda_{s+1} \ldots \lambda_p] B} \\ = (\overset{s+1}{\nabla}_{[\nu} \overset{s}{\nabla}_\mu P_{\lambda_1 \ldots \lambda_s |A|}) Q_{\lambda_{s+1} \ldots \lambda_p] B} + P_{[\lambda_1 \ldots \lambda_s |A|} \overset{p-s+1}{\nabla}_\nu \overset{p-s}{\nabla}_\mu Q_{\lambda_{s+1} \ldots \lambda_p] B}\,. \end{cases}$$

These identities and the identities

$$(5.35)\qquad \boxed{\begin{aligned} \overset{1}{\nabla}_{[\nu} \overset{0}{\nabla}_{\mu]} v^{\varkappa} &= \tfrac{1}{2} R_{\nu\mu\lambda}^{\cdot\cdot\cdot\,\varkappa} v^{\lambda} \\ \overset{1}{\nabla}_{[\nu} \overset{0}{\nabla}_{\mu]} w_{\lambda} &= -\tfrac{1}{2} R_{\nu\mu\lambda}^{\cdot\cdot\cdot\,\varkappa} w_{\varkappa} \end{aligned}}$$

that are valid in L_n (not only in A_n) and simplify the equations (4.9), are easily verified and from them and (5.32) we get the identity

$$(5.36)\qquad \begin{cases} \overset{p+1}{\nabla}_{[\nu} \overset{p}{\nabla}_\mu P_{\lambda_1 \ldots \lambda_p]}^{\cdot\cdot\cdot\cdot\,\varkappa\ldots}{}_{\tau\ldots} \\ = \tfrac{1}{2} R_{[\nu\mu|\varrho|}^{\cdot\cdot\cdot\,\varkappa} P_{\lambda_1 \ldots \lambda_p]}^{\cdot\cdot\cdot\cdot\,\varrho\ldots}{}_{\tau\ldots} + \cdots - \tfrac{1}{2} R_{[\nu\mu|\tau|}^{\cdot\cdot\cdot\,\sigma} P_{\lambda_1 \ldots \lambda_p]}^{\cdot\cdot\cdot\cdot\,\varkappa\ldots}{}_{\sigma\ldots} - \cdots. \end{cases}$$

This is a simplification and generalization of (4.13).[1])

[1]) The new operator was introduced by CARTAN. The general idea was in the first edition 1928, 1, p. 211 and the operator itself, symbolized by D, appeared in the second edition 1946, 1, p. 209.

In order to find the geometrical meaning of the new operator we consider an X_{p+1} through $\underset{0}{\xi}^{\varkappa}$ and an *infinitesimal* (properly chosen) closed X_p in this X_{p+1} surrounding $\underset{0}{\xi}^{\varkappa}$. Taking for simplicity the field $P_{\lambda_1 \dots \lambda_p \cdot \tau}^{\cdot \; \cdot \; \cdot \; \varkappa}$, the field value at a point $\xi^{\varkappa}$ of X_p is

$$\underset{0}{P}_{\lambda_1 \dots \lambda_p \cdot \tau}^{\cdot \; \cdot \; \cdot \; \varkappa} + (\xi^{\mu} - \underset{0}{\xi}^{\mu}) \, \partial_{\mu} \, P_{\lambda_1 \dots \lambda_p \cdot \tau}^{\cdot \; \cdot \; \cdot \; \varkappa}. \tag{5.37}$$

If $d f^{\varkappa_1 \dots \varkappa_p}$ is the infinitesimal volume element of the X_p with a screw-sense that is the same for all points, then

$$d f^{\lambda_1 \dots \lambda_p} \left(\underset{0}{P}_{\lambda_1 \dots \lambda_p \cdot \tau}^{\cdot \; \cdot \; \cdot \; \varkappa} + (\xi^{\mu} - \underset{0}{\xi}^{\mu}) \, \partial_{\mu} \, P_{\lambda_1 \dots \lambda_p \cdot \tau}^{\cdot \; \cdot \; \cdot \; \varkappa}\right) \tag{5.38}$$

is a well defined tensor at $\xi^{\varkappa}$. But the formation of the integral over X_p of this expression has no geometrical sense because addition of tensors at different points is not an invariant process. Now if we displace each of them parallel to $\underset{0}{\xi}^{\varkappa}$ we get

$$\left\{ \begin{aligned} & d f^{\lambda_1 \dots \lambda_p} \{\underset{0}{P}_{\lambda_1 \dots \lambda_p \cdot \tau}^{\cdot \; \cdot \; \cdot \; \varkappa} + (\xi^{\mu} - \underset{0}{\xi}^{\mu}) \, \partial_{\mu} \, P_{\lambda_1 \dots \lambda_p \cdot \tau}^{\cdot \; \cdot \; \cdot \; \varkappa} + \\ & \qquad + (\xi^{\mu} - \underset{0}{\xi}^{\mu}) \, \Gamma_{\mu \varrho}^{\varkappa} \, P_{\lambda_1 \dots \lambda_p \cdot \tau}^{\cdot \; \cdot \; \cdot \; \varrho} - (\xi^{\mu} - \underset{0}{\xi}^{\mu}) \, \Gamma_{\mu \tau}^{\sigma} \, P_{\lambda_1 \dots \lambda_p \cdot \sigma}^{\cdot \; \cdot \; \cdot \; \varkappa}\} \end{aligned} \right. \tag{5.39}$$

and since these tensors can now be added we can form the integral over X_p. Then, using the theorem of STOKES we can form the integral over the part τ_{p+1} of X_{p+1} bordered by X_p. Neglecting terms containing $\xi^{\varkappa} - \underset{0}{\xi}^{\varkappa}$ in the result, we get

$$\left\{ \begin{aligned} & \int_{\tau_{p+1}} \{\partial_{[\nu} \, P_{\lambda_1 \dots \lambda_p] \cdot \tau}^{\cdot \; \cdot \; \cdot \; \varkappa} + \Gamma_{[\nu |\varrho|}^{\varkappa} \, P_{\lambda_1 \dots \lambda_p] \cdot \tau}^{\cdot \; \cdot \; \cdot \; \varrho} - \Gamma_{[\nu |\tau|}^{\sigma} \, P_{\lambda_1 \dots \lambda_p] \cdot \sigma}^{\cdot \; \cdot \; \cdot \; \varkappa}\} \, d f^{\nu \lambda_1 \dots \lambda_p} \\ & \qquad = \int_{\tau_{p+1}} \overset{p}{\nabla}_{[\nu} \, P_{\lambda_1 \dots \lambda_p] \cdot \tau}^{\cdot \; \cdot \; \cdot \; \varkappa} \, d f^{\nu \lambda_1 \dots \lambda_p} \end{aligned} \right. \tag{5.40}$$

if $d f^{\nu \lambda_1 \dots \lambda_p}$ is the infinitesimal volume element of τ_{p+1}. This result also holds *mutatis mutandus* if there are more contravariant and covariant indices $\varkappa$ and τ. The integral is zero for every choice of the $(p+1)$-direction of X_{p+1} at $\underset{0}{\xi}^{\varkappa}$ if and only if $\overset{p}{\nabla}_{[\nu} P_{\lambda_1 \dots \lambda_p] \cdot \tau}^{\cdot \; \cdot \; \cdot \; \varkappa} = 0$ at $\underset{0}{\xi}^{\varkappa}$. In that case we say that the field $P_{\lambda_1 \dots \lambda_p \cdot \tau}^{\cdot \; \cdot \; \cdot \; \varkappa}$ is *in equilibrium at* $\underset{0}{\xi}^{\varkappa}$.

If we compare (5.32) with (5.2) and (5.19) we see that these equations can now be written in the simple form

$$\boxed{R_{[\nu \mu \lambda]}^{\cdot \; \cdot \; \cdot \; \varkappa} = 2 \overset{2}{\nabla}_{[\nu} S_{\mu \lambda]}^{\cdot \; \cdot \; \varkappa}} \qquad \text{(second identity)} \tag{5.41}$$

$$\boxed{\overset{2}{\nabla}_{[\omega} R_{\nu \mu] \lambda}^{\cdot \; \cdot \; \cdot \; \varkappa} = 0} \qquad \text{(identity of BIANCHI).} \tag{5.42}$$

Hence BIANCHI's identity expresses the fact that the curvature tensor is in equilibrium at each point of L_n.

In order to obtain the geometrical meaning of (5.41, 42) we have to make use of CARTAN displacements. We choose an X_3 through $\underset{0}{\xi}^{\varkappa}$ and in this X_3 we choose a closed X_2 surrounding $\underset{0}{\xi}^{\varkappa}$. Let $\underset{0}{v}^{\varkappa}$ be a vector at $\underset{0}{\xi}^{\varkappa}$ and let $\underset{0}{\Gamma}{}_{\mu\lambda}^{\varkappa}$, $\underset{0}{R}{}_{\nu\mu\lambda}^{\cdot\cdot\cdot\,\varkappa}$ and $\underset{0}{S}{}_{\mu\lambda}^{\cdot\cdot\,\varkappa}$ be field values at $\underset{0}{\xi}^{\varkappa}$. Let $df^{\nu\mu}$ be the bivector of an infinitesimal surface-element of X_2 whose bordering X_1 goes through a point $\xi^{\varkappa}$ of X_2. The point with radiusvector $\underset{0}{v}^{\varkappa}$ at $\underset{0}{\xi}^{\varkappa}$ now undergoes a CARTAN displacement from $\underset{0}{\xi}^{\varkappa}$ to $\xi^{\varkappa}$, along the border of $df^{\nu\mu}$ in the sense of $df^{\nu\mu}$ back to $\xi^{\varkappa}$ and finally from $\xi^{\varkappa}$ to $\underset{0}{\xi}^{\varkappa}$. From $\underset{0}{\xi}^{\varkappa}$ to $\xi^{\varkappa}$ the radiusvector changes into

$$\underset{0}{v}^{\varkappa} - (\xi^{\varkappa} - \underset{0}{\xi}^{\varkappa}) - (\xi^{\omega} - \underset{0}{\xi}^{\omega})\, \underset{0}{\Gamma}{}_{\omega\lambda}^{\varkappa}\, \underset{0}{v}^{\lambda}. \tag{5.43}$$

At $\xi^{\varkappa}$ we have the field values $\underset{0}{S}{}_{\mu\lambda}^{\cdot\cdot\,\varkappa} + (\xi^{\omega} - \underset{0}{\xi}^{\omega})\, \partial_{\omega} S_{\mu\lambda}^{\cdot\cdot\,\varkappa}$ and $\underset{0}{R}{}_{\nu\mu\lambda}^{\cdot\cdot\cdot\,\varkappa} + (\xi^{\omega} - \underset{0}{\xi}^{\omega})\, \partial_{\omega} R_{\nu\mu\lambda}^{\cdot\cdot\cdot\,\varkappa}$. Hence the change of the moving point due to the displacement around $df^{\nu\mu}$ is given by the vector (cf. 4.7)

$$\left\{\begin{aligned} &\{-\underset{0}{S}{}_{\nu\mu}^{\cdot\cdot\,\varkappa} - (\xi^{\omega} - \underset{0}{\xi}^{\omega})\, \partial_{\omega} S_{\nu\mu}^{\cdot\cdot\,\varkappa} - \tfrac{1}{2} \underset{0}{R}{}_{\nu\mu\lambda}^{\cdot\cdot\cdot\,\varkappa}\, \underset{0}{v}^{\lambda} + \tfrac{1}{2} \underset{0}{R}{}_{\nu\mu\omega}^{\cdot\cdot\cdot\,\varkappa} (\xi^{\omega} - \underset{0}{\xi}^{\omega}) + \\ &\quad + \tfrac{1}{2} \underset{0}{R}{}_{\nu\mu\lambda}^{\cdot\cdot\cdot\,\varkappa} (\xi^{\omega} - \underset{0}{\xi}^{\omega})\, \underset{0}{\Gamma}{}_{\omega\varrho}^{\lambda}\, \underset{0}{v}^{\varrho} - \tfrac{1}{2} (\xi^{\omega} - \underset{0}{\xi}^{\omega})\, \partial_{\omega} R_{\nu\mu\lambda}^{\cdot\cdot\cdot\,\varkappa}\, \underset{0}{v}^{\lambda}\}\, df^{\nu\mu}. \end{aligned}\right. \tag{5.44}$$

For the integration of this vector over X_2 we must first displace it parallel to $\underset{0}{\xi}^{\varkappa}$ in order to make addition possible. Then at $\underset{0}{\xi}^{\varkappa}$ we get the vector

$$\left\{\begin{aligned} &\{-\underset{0}{S}{}_{\nu\mu}^{\cdot\cdot\,\varkappa} - \tfrac{1}{2} \underset{0}{R}{}_{\nu\mu\lambda}^{\cdot\cdot\cdot\,\varkappa}\, \underset{0}{v}^{\lambda} + (\xi^{\omega} - \underset{0}{\xi}^{\omega}) \big(-\underset{0}{S}{}_{\nu\mu}^{\cdot\cdot\,\varrho}\, \underset{0}{\Gamma}{}_{\omega\varrho}^{\varkappa} - \tfrac{1}{2} \underset{0}{R}{}_{\nu\mu\lambda}^{\cdot\cdot\cdot\,\varrho}\, \underset{0}{v}^{\lambda}\, \underset{0}{\Gamma}{}_{\omega\varrho}^{\varkappa} - \\ &\quad - \partial_{\omega} S_{\nu\mu}^{\cdot\cdot\,\varkappa} + \tfrac{1}{2} \underset{0}{R}{}_{\nu\mu\omega}^{\cdot\cdot\cdot\,\varkappa} + \tfrac{1}{2} \underset{0}{R}{}_{\nu\mu\sigma}^{\cdot\cdot\cdot\,\varkappa}\, \underset{0}{\Gamma}{}_{\omega\lambda}^{\sigma}\, \underset{0}{v}^{\lambda} - \tfrac{1}{2} \underset{0}{v}^{\lambda}\, \partial_{\omega} R_{\nu\mu\lambda}^{\cdot\cdot\cdot\,\varkappa}\big)\}\, df^{\nu\mu}. \end{aligned}\right. \tag{5.45}$$

According to the theorem of STOKES the integral of this vector over X_2 is equal to

$$\left\{\begin{aligned} &\int_{\tau_3} \{-\underset{0}{S}{}_{[\nu\mu}^{\cdot\cdot\,\varrho}\, \underset{0}{\Gamma}{}_{\omega]\varrho}^{\varkappa} - \tfrac{1}{2} \underset{0}{R}{}_{[\nu\mu|\lambda|}^{\cdot\cdot\cdot\,\varrho}\, \underset{0}{v}^{\lambda}\, \underset{0}{\Gamma}{}_{\omega]\varrho}^{\varkappa} - \partial_{[\omega} S_{\nu\mu]}^{\cdot\cdot\,\varkappa} + \\ &\qquad + \tfrac{1}{2} \underset{0}{R}{}_{[\omega\nu\mu]}^{\cdot\cdot\cdot\,\varkappa} + \tfrac{1}{2} \underset{0}{R}{}_{[\nu\mu|\sigma|}^{\cdot\cdot\cdot\,\varkappa}\, \underset{0}{\Gamma}{}_{\omega]\varrho}^{\sigma}\, \underset{0}{v}^{\varrho} - \tfrac{1}{2} \underset{0}{v}^{\lambda}\, \partial_{[\omega} R_{\nu\mu]\lambda}^{\cdot\cdot\cdot\,\varkappa}\}\, df^{\omega\nu\mu} \\ &\quad = \int_{\tau_3} \{-\overset{2}{\nabla}_{[\omega} S_{\nu\mu]}^{\cdot\cdot\,\varkappa} + \tfrac{1}{2} R_{[\omega\nu\mu]}^{\cdot\cdot\cdot\,\varkappa} - \tfrac{1}{2} \underset{0}{v}^{\lambda}\, \overset{2}{\nabla}_{[\omega} R_{\nu\mu]\lambda}^{\cdot\cdot\cdot\,\varkappa}\}_{\xi^{\varkappa} = \underset{0}{\xi}^{\varkappa}}\, df^{\omega\nu\mu}, \end{aligned}\right. \tag{5.46}$$

and this integral is zero in consequence of the second identity (5.41) and BIANCHI's identity (5.42). Hence these two identities together express the following geometric fact:

If an infinitesimal closed X_2 in an X_3 through $\overset{0}{\xi}{}^\varkappa$ in X_n surrounds $\overset{0}{\xi}{}^\varkappa$ and if a point with radiusvector $\overset{0}{v}{}^\varkappa$ in the local E_n of $\overset{0}{\xi}{}^\varkappa$ undergoes a CARTAN *displacement from $\overset{0}{\xi}{}^\varkappa$ to a point $\xi^\varkappa$ of X_2, after that along the border of a surfaceelement $d f^{\nu\mu}$ of X_2 in the sense of $d f^{\nu\mu}$, and finally back from $\xi^\varkappa$ to $\overset{0}{\xi}{}^\varkappa$, the difference between the end value and the beginning value at $\overset{0}{\xi}{}^\varkappa$ be called the* CARTAN-*deviation of the point $\overset{0}{v}{}^\varkappa$ with respect to the surface-element $d f^{\nu\mu}$. The integral of the* CARTAN-*deviation over X_2 is zero for every (proper) choice of X_3 and X_2, to within infinitesimal quantities of an order higher than 1 in $\xi^\varkappa - \overset{0}{\xi}{}^\varkappa$.*

There is an identity connecting the curvature tensor with the LIE derivative of $\Gamma^\varkappa_{\mu\lambda}$

$$\boxed{\underset{v}{£}\,\Gamma^\varkappa_{\mu\lambda} = \nabla_\mu v_{;\lambda}^{\cdot\;\varkappa} + v^\sigma R_{\sigma\mu\lambda}^{\cdot\cdot\cdot\;\varkappa}} \quad ; \quad v_{;\lambda}^{\cdot\;\varkappa} \overset{\text{def}}{=} \nabla_\lambda v^\varkappa - 2 S_{\lambda\varrho}^{\cdot\cdot\;\varkappa} v^\varrho \tag{5.47}$$

that can be derived immediately from (II 10.34). The quantity $v_{;\lambda}^{\cdot\;\varkappa}$ is a covariant derivative of $v^\varkappa$ formed after changing the order of the lower indices of $\Gamma^\varkappa_{\mu\lambda}$. It frequently occurs in problems of deformation. In the general formula for the operation of $\underset{v}{£}$ on a quantity in A_n, for instance (II 10.29), it is allowed to replace ∂_μ by ∇_μ in all terms at the same time. This is clear because in an A_n the coordinates can always be chosen in such a way that $\Gamma^\varkappa_{\mu\lambda} = 0$ at some given point (cf. III § 7). But in an L_n the same rule holds provided that $\nabla_\mu v^\varkappa$ is changed into $v_{;\mu}^{\cdot\;\varkappa}$. For instance

$$(5.48)\quad \begin{cases} \text{a)}\ \underset{v}{£}\, u^\varkappa = v^\mu \partial_\mu u^\varkappa - u^\mu \partial_\mu v^\varkappa = v^\mu \nabla_\mu u^\varkappa - v^\mu \Gamma^\varkappa_{\mu\lambda} u^\lambda - u^\mu \partial_\mu v^\varkappa = v^\mu \nabla_\mu u^\varkappa - u^\mu v_{;\mu}^{\cdot\;\varkappa} \\ \text{b)}\ \underset{v}{£}\, w_\lambda = v^\mu \partial_\mu w_\lambda + w_\mu \partial_\lambda v^\mu = v^\mu \nabla_\mu w_\lambda + w_\mu v_{;\lambda}^{\cdot\;\mu}. \end{cases}$$

Now, making use of the commutability of $\underset{v}{£}$ and ∂_μ (cf. II § 10) we get for $\underset{v}{£}\,\nabla_\mu u^\varkappa$

$$\underset{v}{£}\,\nabla_\mu u^\varkappa = \underset{v}{£}\,\partial_\mu u^\varkappa + \underset{v}{£}\,\Gamma^\varkappa_{\mu\lambda} u^\lambda = \partial_\mu \underset{v}{£}\, u^\varkappa + \Gamma^\varkappa_{\mu\lambda} \underset{v}{£}\, u^\lambda + u^\lambda \underset{v}{£}\,\Gamma^\varkappa_{\mu\lambda}, \tag{5.49}$$

hence

$$\boxed{\underset{v}{£}\,\nabla_\mu u^\varkappa - \nabla_\mu \underset{v}{£}\, u^\varkappa = u^\lambda \underset{v}{£}\,\Gamma^\varkappa_{\mu\lambda}} \tag{5.50}$$

is the commutation formula for $\underset{v}{£}$ and ∇_μ. If this formula is written out we get

$$(5.51)\quad \begin{cases} v^\varrho \nabla_\varrho \nabla_\mu u^\varkappa + (\nabla_\varrho u^\varkappa)\, v_{;\mu}^{\cdot\;\varrho} - (\nabla_\mu u^\varrho)\, v_{;\varrho}^{\cdot\;\varkappa} \\ \quad = (\nabla_\mu v^\varrho)\, \nabla_\varrho u^\varkappa + v^\varrho \nabla_\mu \nabla_\varrho u^\varkappa - (\nabla_\mu u^\varrho)\, v_{;\varrho}^{\cdot\;\varkappa} - u^\varrho \nabla_\mu v_{;\varrho}^{\cdot\;\varkappa} + u^\lambda \underset{v}{£}\,\Gamma^\varkappa_{\mu\lambda} \end{cases}$$

from which (5.47) follows, according to (4.9b).

Exercises.

III 5,1. Prove (5.2) by writing out $\nabla_{[\nu} \nabla_\mu \nabla_{\lambda]} p$ in two ways.

III 5,2. If a vector $v^\varkappa$ in V_2 is displaced parallel along a closed curve, the angle between the beginning position and the end position is equal to the surface integral of $\varkappa$ over the area bounded by the curve.

III 5,3. In a V_n the expression $-\frac{1}{4} K_{\nu\mu\lambda\varkappa} f^{\nu\mu} f^{\lambda\varkappa}$, where $f^{\nu\mu}$ is a simple bivector with the area 1, is called the *riemannian curvature* of the V_n *with respect to the 2-direction* of $f^{\nu\mu}$ at the point considered. For $n=2$ it is equal to $\varkappa$. The principal directions of the symmetric tensor $K_{\mu\lambda}$ are the principal directions of the V_n. If they are determined uniquely and if (h) is an orthogonal system whose basis vectors are the unitvectors $\underset{1}{i}^\varkappa, \ldots, \underset{n}{i}^\varkappa$ in these directions, prove that K_{11} is the sum of the riemannian curvatures with respect to $n-1$ arbitrary mutually perpendicular 2-directions through the direction of $\underset{1}{i}^\varkappa$ and that $-G_{11}$ is the sum of $\binom{n-1}{2}$ arbitrary mutually and to the direction of $\underset{1}{i}^\varkappa$ perpendicular 2-directions.[1])

III 5,4. (Theorem of SCHUR.) A V_n is an S_n if and only if at every point the riemannian curvature with respect to a 2-direction is independent of the choice of this 2-direction.[2])

III 5,5. A V_n is an S_n if and only if the components K_{kjih} with three different indices with respect to every local cartesian coordinate system at every point vanish.[3])

III 5,6. Prove the identity for V_n [4])

III 5,6 α) $$\nabla_{[\varrho\sigma]} K_{\nu\mu\lambda\varkappa} + \nabla_{[\nu\mu]} K_{\lambda\varkappa\varrho\sigma} + \nabla_{[\lambda\varkappa]} K_{\varrho\sigma\nu\mu} = 0 \qquad \nabla_{[\varrho\sigma]} \overset{\text{def}}{=} \nabla_{[\varrho} \nabla_{\sigma]}.$$

III 5,7. Prove (5.41) by applying $\overset{2}{\nabla}$ to the identity

III 5,7 α) $$\overset{1}{\nabla}_{[\mu} A^\varkappa_{\lambda]} = S_{\mu\lambda}^{\cdot\cdot\varkappa}.$$

Check (III 5,7 α). Prove (5.42) by computing $\overset{2}{\nabla}_{[\omega} \overset{1}{\nabla}_\nu \nabla_{\mu]} w_\lambda$ in two ways.

§ 6. Integrability conditions in L_n.

In II § 6 we have developed integrability conditions for sets of partial differential equations in X_n. In an L_n differential equations

1) HERGLOTZ 1916, 1. Cf. E I 1935, 1, p. 117.

2) SCHUR 1886, 1; cf. E I 1935, 1, p. 126; WRONA 1941, 1 and 1947, 1, also for generalizations: EISENHART 1949, 1, p. 83. Cf. footnote 5 on p. 148; BOMPIANI 1950, 1.

3) R. K. 1924, 1, p. 171; E I 1935, 1, p. 126.

4) Cf. WALKER 1950, 1 and (VII 4.45).

occur containing the covariant instead of the ordinary differentiation. We consider equations of the first order formed by covariant differentiation and alternation. A very general form is

$$(6.1)\qquad \overset{p}{\nabla}_{[\nu} P^{\cdot}_{\mu_1\ldots\mu_p]}{}^{\cdot\,\varkappa\ldots}_{\cdot\,\lambda\ldots} = Q^{\cdot\,\cdot}_{\nu\mu_1\ldots\mu_p}{}^{\cdot\,\varkappa\ldots}_{\cdot\,\lambda\ldots}$$

where Q depends on the $\xi^\varkappa$ only and not on the components of P. If this equation is written out, on the left hand side we get $\partial_{[\nu} P^{\cdot}_{\mu_1\ldots\mu_p]}{}^{\cdot\,\varkappa\ldots}_{\cdot\,\lambda\ldots}$ and on the right hand side we get terms which are linear in the components of P and contain a factor $\Gamma^{\varkappa}_{\mu\lambda}$. But these are equations of the form (II 6.15); the only difference being that the right hand side is not independent of the components of P. As we have seen in Chapter II the integrability conditions can only be found easily if the derivatives of P arising at every step can be eliminated by means of (6.1). Now this is just the case here, hence, proceeding in this way we must get at the first step equations containing first derivatives of Q and no derivatives of P. The equations are obtained by applying ∂_ω to (6.1) and alternating over $\omega\nu\mu_1\ldots\mu_p$. But instead of ∂_ω we can just as well use $\overset{p+1}{\nabla}_\omega$ because the difference vanishes by virtue of (6.1). But if we use this operator we have nothing to compute because the result is already known from the preceeding section (5.36):

$$(6.2)\qquad \overset{p+1}{\nabla}_{[\omega} Q^{\cdot\,\cdot}_{\nu\mu_1\ldots\mu_p]}{}^{\cdot\,\varkappa\ldots}_{\cdot\,\lambda\ldots} = \tfrac{1}{2} R_{[\omega\nu|\varrho|}{}^{\cdot\cdot\cdot\varkappa} P^{\cdot}_{\mu_1\ldots\mu_p]}{}^{\cdot\,\varrho}_{\cdot\,\lambda} + \cdots - \tfrac{1}{2} R_{[\omega\nu|\lambda|}{}^{\cdot\cdot\cdot\sigma} P^{\cdot}_{\mu_1\ldots\mu_p]}{}^{\cdot\,\varkappa}_{\cdot\,\sigma} - \cdots.$$

For $p = 0$ we get

$$(6.3)\qquad \overset{0}{\nabla}_\nu P^{\varkappa\ldots}_{\cdot\,\lambda\ldots} = \nabla_\nu P^{\varkappa\ldots}_{\cdot\,\lambda\ldots} = Q^{\cdot\,\varkappa\ldots}_{\nu\,\cdot\,\lambda\ldots}$$

with the first integrability conditions

$$(6.4)\qquad \overset{1}{\nabla}_{[\omega} Q^{\cdot\,\varkappa\ldots}_{\nu]\,\cdot\,\lambda\ldots} = \tfrac{1}{2} R_{\omega\nu\varrho}{}^{\cdot\cdot\cdot\varkappa} P^{\varrho\ldots}_{\cdot\,\lambda\ldots} + \cdots - \tfrac{1}{2} R_{\omega\nu\lambda}{}^{\cdot\cdot\cdot\sigma} P^{\varkappa\ldots}_{\cdot\,\sigma\ldots} - \cdots$$

and if P is a scalar

$$(6.5)\qquad \nabla_\lambda P = \partial_\lambda P = Q_\lambda,$$

(6.4) takes the form

$$(6.6)\qquad \overset{1}{\nabla}_{[\mu} Q_{\lambda]} = \partial_{[\mu} Q_{\lambda]} = 0.$$

Applying the general rule to the equation

$$(6.7)\qquad \nabla_\nu v^{\varkappa_1\ldots\varkappa_n} = 0; \qquad v^{\varkappa_1\ldots\varkappa_n} = v^{[\varkappa_1\ldots\varkappa_n]}$$

we get the first integrability conditions

$$(6.8)\qquad R_{\nu\mu\varrho}{}^{\cdot\cdot\cdot[\varkappa_1} v^{|\varrho|\varkappa_2\ldots\varkappa_n]} = \frac{1}{n} R_{\nu\mu\varrho}{}^{\cdot\cdot\cdot\varrho} v^{\varkappa_1\ldots\varkappa_n} = \frac{1}{n} V_{\nu\mu} v^{\varkappa_1\ldots\varkappa_n} = 0.$$

Hence in an L_n a covariant constant n-vector field (or a scalar density-field) is possible if and only if $V_{\nu\mu}=0$. This can also be proved by using (5.22b). It is in accordance with the property of volume preserving of an L_n with $V_{\nu\mu}=0$, mentioned already in Exerc. III 4,5.[1])

§ 7. Geodesics and normal coordinates[2]).

A curve $\xi^\varkappa=f^\varkappa(t)$ in L_n is called a *geodesic* if the tangent vector $d\xi^\varkappa/dt$ at a point remains tangent if it is displaced parallel along the curve.

The n. a. s. conditions are

$$\frac{\delta}{dt}\frac{d\xi^\varkappa}{dt}=\alpha(t)\frac{d\xi^\varkappa}{dt} \tag{7.1}$$

or

$$\frac{d^2\xi^\varkappa}{dt^2}+\Gamma^\varkappa_{\mu\lambda}\frac{d\xi^\mu}{dt}\frac{d\xi^\lambda}{dt}=\alpha(t)\frac{d\xi^\varkappa}{dt} \tag{7.2}$$

from which we see that the geodesics only depend on $\Gamma^\varkappa_{(\mu\lambda)}$. If a new parameter z is introduced

$$z=z(t); \qquad t=t(z) \tag{7.3}$$

the equation (7.2) takes the form

$$\frac{d^2\xi^\varkappa}{dz^2}+\Gamma^\varkappa_{\mu\lambda}\frac{d\xi^\mu}{dz}\frac{d\xi^\lambda}{dz}=-\frac{d\xi^\varkappa}{dz}\left(\frac{d^2z}{dt^2}-\alpha\frac{dz}{dt}\right)\left(\frac{dz}{dt}\right)^{-2} \tag{7.4}$$

and the right hand side of these equations reduces to zero if z is a solution of the ordinary differential equation of second order

$$\frac{d^2z}{dt^2}-\alpha(t)\frac{dz}{dt}=0. \tag{7.5}$$

If this is the case, z is called *an affine parameter* on the geodesic. The general solution of (7.5) has the form

$$z=C_1\int e^{\int\alpha dt}dt+C_2; \qquad C_1\neq 0; \qquad C_1, C_2 \text{ constants} \tag{7.6}$$

---

1) Cf. for a generalization URABE 1940, 1.

2) General references: RIEMANN 1854, 1 (for V_n); VEBLEN 1922, 1; 1927, 1; BIRKHOFF 1923, 1; VEBLEN and T. Y. THOMAS 1923, 1; 1924, 1; VEBLEN and J. M. THOMAS 1925, 1; T. Y. THOMAS 1925, 2; 1926, 2 and 1927, 3 (projective); 1928, 1; 1936, 3; J. M. THOMAS 1926, 2 (projective); HLAVATY 1926, 1; 2; 3; 1927, 1; 1930, 3; T. Y. THOMAS and MICHAL 1927, 2; EISENHART 1927, 1, § 41 and 1930, 1 (projective); DUBOURDIEU 1927, 1; 1928, 1; NALLI 1928, 1; BORTOLOTTI 1929, 2; 1930, 3 (asymmetric); 1940, 2; MICHAL 1930, 1; 2; 1931, 1; 2; MATTIOLI 1930, 3; LEVY 1930, 1; WHITEHEAD and WILLIAMS 1930, 1; RUSE 1930, 1; 1931, 2; 3; DOUGLAS 1931, 1; WHITEHEAD 1931, 1; E I 1935, 1, p. 100 ff.; SASAKI 1941, 1 (conformal); BOMPIANI 1945, 1; BLUM 1947, 1; NARLIKAR and PRASAD 1948, 1 (relativity); VARGA 1950, 1; PINL 1951, 1 (asymmetric); HARTMANN 1951, 1.

and accordingly the affine parameter is fixed to within the place of the nullpoint and a constant factor. This implies that two segments on the same geodesic have an invariant ratio of "length" that can be measured by means of one parameter chosen arbitrarily from all possible affine parameters.

In an ordinary V_n the length s on a *real* geodesic is always an affine parameter because $d\xi^\varkappa/ds$ is a unitvector and the covariant differential of a unitvector is always perpendicular to the vector. Hence, if we write $i^\varkappa$ for $d\xi^\varkappa/ds$, the equation of a geodesic in V_n takes the form

$$(7.7) \qquad i^\mu \nabla_\mu i^\varkappa = \frac{d^2 \xi^\varkappa}{d s^2} + \Gamma^\varkappa_{\mu\lambda} \frac{d\xi^\mu}{ds} \frac{d\xi^\lambda}{ds} = 0\,^{1)}.$$

We may ask whether it is possible to transform a connexion $\Gamma^\varkappa_{\mu\lambda}$ in such a way that all geodesics remain geodesics. If

$$(7.8) \qquad \Gamma^\varkappa_{\mu\lambda} \to \Gamma^\varkappa_{\mu\lambda} + P_{\mu\lambda}^{\cdot\cdot\varkappa}$$

$P_{\mu\lambda}^{\cdot\cdot\varkappa}$ is necessarily a tensor (cf. II § 10) and we may take $P_{[\mu\lambda]}^{\cdot\cdot\varkappa} = 0$ because the alternating part does not affect the geodesics. According to (7.2) we must have

$$(7.9) \qquad P_{\mu\lambda}^{\cdot\cdot\varkappa} \frac{d\xi^\mu}{dt} \frac{d\xi^\lambda}{dt} = \beta(t) \frac{d\xi^\varkappa}{dt}$$

for every choice of $d\xi^\varkappa/dt$ with a function $\beta(t)$ that depends on this choice. But this is only possible (cf. Exerc. I 3,4) if $P_{\mu\lambda}^{\cdot\cdot\varkappa}$ has the form

$$(7.10) \qquad P_{\mu\lambda}^{\cdot\cdot\varkappa} = p_\mu A^\varkappa_\lambda + p_\lambda A^\varkappa_\mu\,^{2)}.$$

If (7.10) is satisfied the new affine parameter $'z$ has the form

$$(7.11) \qquad 'z = C_1 \int e^{2\int p_\mu d\xi^\mu} dz + C_2; \qquad C_1 \neq 0; \qquad C_1, C_2 \text{ constants}$$

and from this it follows that z is an affine parameter for the transformed connexion if and only if $p_\lambda d\xi^\lambda = 0$ at all points of the geodesic considered, i.e. if at all points p_λ is tangent to the geodesic. z remains affine parameter for all geodesics if and only if $p_\lambda = 0$, hence *a symmetric connexion is wholly determined by its geodesics and the affine parameters on them.*

As we have already seen in III § 4 a coordinate system in an A_n with the $\Gamma^\varkappa_{\mu\lambda}$ vanishing at *all* points exists if and only if the A_n is an E_n. But it is easy to find a coordinate system in A_n for which the $\Gamma^\varkappa_{\mu\lambda}$ vanish at *one* arbitrarily given point $\underset{0}{\xi}^\varkappa$. Such a coordinate system is called

---

[1]) The expression $\nabla_\mu i^\varkappa$ has a sense only if the field $i^\varkappa$ is prolonged (cf. II § 4) but $i^\mu \nabla_\mu i^\varkappa$ always has a sense.

[2]) Cf. Eisenhart 1926, 1, p. 132; T. Y. Thomas 1934, 2, p. 9.

geodesic at $\underset{0}{\xi}^\varkappa$. VEBLEN has given the following general method for the construction of a coordinate system of this kind, that has moreover many other useful properties. At $\underset{0}{\xi}^\varkappa$ we take an arbitrary vector $\underset{0}{t}^\varkappa \neq 0$ depending on $n-1$ parameters, in such a way that to every direction at $\underset{0}{\xi}^\varkappa$ there belongs one of the vectors $\underset{0}{t}^\varkappa$ with that direction. Then we consider all geodesics through $\underset{0}{\xi}^\varkappa$ and on each of these geodesics we fix an affine parameter z such that at $\underset{0}{\xi}^\varkappa$

$$\text{(7.12)} \qquad \frac{d\xi^\varkappa}{dz} = \underset{0}{t}^\varkappa; \qquad z = 0.$$

Now the equation of a geodesic enables us to compute for each geodesic the value of $d^2\xi^\varkappa/dz^2$ at $\underset{0}{\xi}^\varkappa$

$$\text{(7.13)} \qquad \frac{d^2\xi^\varkappa}{dz^2}(\underset{0}{\xi}) = -\Gamma^\varkappa_{\mu\lambda}(\underset{0}{\xi})\,\underset{0}{t}^\mu\,\underset{0}{t}^\lambda$$

and by repeated differentiation of the equation of a geodesic we can find all derivatives of the $\xi^\varkappa$ with respect to z at $\underset{0}{\xi}^\varkappa$

$$\text{(7.14)} \qquad \frac{d^{p+1}\xi^\varkappa}{dz^{p+1}}(\underset{0}{\xi}) = -\Gamma^\varkappa_{\mu_p\ldots\mu_1\lambda}(\underset{0}{\xi})\,\underset{0}{t}^{\mu_p}\ldots\underset{0}{t}^{\mu_1}\,\underset{0}{t}^\lambda$$

where [1])

$$\text{(7.15)} \qquad \Gamma^\varkappa_{\mu_p\ldots\mu_1\lambda} \overset{\text{def}}{=} \partial_{(\mu_p}\Gamma^\varkappa_{\mu_{p-1}\ldots\mu_1\lambda)} - p\,\Gamma^\tau_{(\mu_p\mu_{p-1}}\Gamma^\varkappa_{\mu_{p-2}\ldots\mu_1\lambda)\tau}.$$

Then $\xi^\varkappa$ can be expanded in a series

$$\text{(7.16)} \quad \begin{cases} \xi^\varkappa = \underset{0}{\xi}^\varkappa + \left(\dfrac{d\xi^\varkappa}{dz}\right)_{z=0} z + \dfrac{1}{2!}\left(\dfrac{d^2\xi^\varkappa}{dz^2}\right)_{z=0} z^2 + \cdots \\ \quad = \underset{0}{\xi}^\varkappa + z\,\underset{0}{t}^\varkappa - \dfrac{1}{2!}\Gamma^\varkappa_{\mu\lambda}(\underset{0}{\xi})\, z^2\,\underset{0}{t}^\mu\,\underset{0}{t}^\lambda - \dfrac{1}{3!}\Gamma^\varkappa_{\mu_2\mu_1\lambda}(\underset{0}{\xi})\, z^3\,\underset{0}{t}^{\mu_2}\,\underset{0}{t}^{\mu_1}\,\underset{0}{t}^\lambda - \cdots. \end{cases}$$

If this series is convergent the expressions $z\,\underset{0}{t}^\varkappa$ can be used in an $\mathfrak{N}(\underset{0}{\xi}^\varkappa)$ as new coordinates. Taking $\underset{0}{\xi}^\varkappa = 0$ for simplicity and writing ξ^h; $h = 1, \ldots, n$ for these new coordinates

$$\text{(7.17)} \qquad \xi^h \overset{\text{def}}{=} \delta^h_\varkappa\,\underset{0}{t}^\varkappa z$$

we get

$$\text{(7.18)} \quad \xi^\varkappa = \delta^\varkappa_h\,\xi^h - \frac{1}{2!}\Gamma^\varkappa_{\mu\lambda}(0)\,\delta^{\mu\lambda}_{ji}\,\xi^j\xi^i - \frac{1}{3!}\Gamma^\varkappa_{\mu_2\mu_1\lambda}(0)\,\delta^{\mu_2\mu_1\lambda}_{j_2 j_1 i}\,\xi^{j_2}\xi^{j_1}\xi^i - \cdots,$$

---

[1]) BORTOLOTTI and HLAVATY remarked 1936, 2 that the construction of $\Gamma^\varkappa_{\mu_p\ldots\mu_1}$ is analogous to the construction of a covariant derivative for the covariant indices only.

and the inverse

$$\xi^h = \delta^h_\varkappa \left(\xi^\varkappa + \frac{1}{2!} \Lambda^\varkappa_{\mu\lambda}(0)\, \xi^\mu \xi^\lambda + \frac{1}{3!} \Lambda^\varkappa_{\mu_2\mu_1\lambda}(0)\, \xi^{\mu_2} \xi^{\mu_1} \xi^\lambda + \cdots \right) \tag{7.19}$$

where

$$\left\{ \begin{aligned} \Lambda^\varkappa_{\mu\lambda} &\overset{\text{def}}{=} \Gamma^\varkappa_{\mu\lambda} \\ \Lambda^\varkappa_{\mu_2\mu_1\lambda} &\overset{\text{def}}{=} \Gamma^\varkappa_{\mu_2\mu_1\lambda} + 3\,\Gamma^\varrho_{(\mu_2\mu_1}\Gamma^\varkappa_{\lambda)\varrho} \\ \Lambda^\varkappa_{\mu_3\mu_2\mu_1\lambda} &\overset{\text{def}}{=} \Gamma^\varkappa_{\mu_3\mu_2\mu_1\lambda} + 4\,\Gamma^\varrho_{(\mu_3\mu_2\mu_1}\Gamma^\varkappa_{\lambda)\varrho} + 6\,\Gamma^\varrho_{(\mu_3\mu_2}\Lambda^\varkappa_{\mu_1\lambda)\varrho} - 3\,\Gamma^\varkappa_{\sigma\varrho}\Gamma^\sigma_{(\mu_3\mu_2}\Gamma^\varrho_{\mu_1\lambda)}. \end{aligned} \right. \tag{7.20}$$

By differentiation of (7.18) and (7.19) we get *at* $\underset{0}{\xi}{}^\varkappa$

$$\left\{ \begin{array}{lll} \text{a)} & A^\varkappa_i(\underset{0}{\xi}) = \delta^\varkappa_i; & A^h_\lambda(\underset{0}{\xi}) = \delta^h_\lambda \\ \text{b)} & \partial_j A^\varkappa_i(\underset{0}{\xi}) = -\,\delta^{\mu\lambda}_{j\,i}\,\Gamma^\varkappa_{\mu\lambda}(\underset{0}{\xi}); & \partial_\mu A^h_\lambda(\underset{0}{\xi}) = \delta^h_\varkappa\,\Gamma^\varkappa_{\mu\lambda}(\underset{0}{\xi}) \\ \text{c)} & \partial_k\partial_j A^\varkappa_i(\underset{0}{\xi}) = -\,\delta^{\nu\mu\lambda}_{kji}\,\Gamma^\varkappa_{\nu\mu\lambda}(\underset{0}{\xi}); & \partial_\nu\partial_\mu A^h_\lambda(\underset{0}{\xi}) = \delta^h_\varkappa\,\Lambda^\varkappa_{\nu\mu\lambda}(\underset{0}{\xi}) \end{array} \right. \tag{7.21}$$

and accordingly *at* $\underset{0}{\xi}{}^\varkappa$

$$\Gamma^h_{ji}(\underset{0}{\xi}) = \delta^{\mu\lambda h}_{j\,i\,\varkappa}\,\Gamma^\varkappa_{\mu\lambda}(\underset{0}{\xi}) - A^{\mu\lambda}_{j\,i}\,\partial_\mu A^h_\lambda(\underset{0}{\xi}) = 0. \tag{7.22}$$

Now the choice of the system $(\varkappa)$ is entirely free. Hence, if we take $(\varkappa)$ equal to (h) it follows from (7.18) that *at* $\underset{0}{\xi}{}^\varkappa$

$$\Gamma^h_{j_p\ldots j_1 i}(\underset{0}{\xi}) = 0 \tag{7.23}$$

and in consequence of (7.15) *at* $\underset{0}{\xi}{}^\varkappa$

$$\partial_{(j_{p+1}}\Gamma^h_{j_p\ldots j_1 i)}(\underset{0}{\xi}) = 0; \tag{7.24}$$

$$\partial_{(j_p}\partial_{j_{p-1}}\cdots\partial_{j_1}\Gamma^h_{ji)}(\underset{0}{\xi}) = 0. \tag{7.25}$$

The coordinates (h) defined in this way are called by VEBLEN the *normal coordinates in* A_n *with respect to* $\underset{0}{\xi}{}^\varkappa$ *and to the coordinate system* $(\varkappa)$. They are a generalization of the normal coordinates introduced in a V_n by RIEMANN[1]). If we use normal coordinates, not only the Γ^h_{ji} vanish at $\underset{0}{\xi}{}^\varkappa$ but so do the symmetrized derivatives of Γ^h_{ji} of every order also.

If the coordinates $\xi^\varkappa$ are transformed into $\xi^{\varkappa'}$ we get the new normal coordinates $\xi^{h'}$ belonging to the same point $\underset{0}{\xi}{}^\varkappa$

$$\xi^{h'} \overset{\text{def}}{=} \delta^{h'}_{\varkappa'}\,\underset{0}{t}{}^{\varkappa'} z = \delta^{h'}_{\varkappa'}\,A^{\varkappa'}_\varkappa(\underset{0}{\xi})\,\underset{0}{t}{}^\varkappa z = A^{h'}_h\,\xi^h \tag{7.26}$$

[1]) RIEMANN 1854, 1, p. 279.

where the $A_h^{h'}$ are constants. Hence, if the $\xi^\varkappa$ are transformed, the normal coordinates with respect to $\overset{0}{\xi}{}^\varkappa$ and $(\varkappa)$ undergo a linear homogeneous transformation with *constant* coefficients and these coefficients are the values of $A_\varkappa^{\varkappa'}$ at $\overset{0}{\xi}{}^\varkappa$.

By means of the normal coordinates ξ^h the points of an $\mathfrak{N}(\overset{0}{\xi}{}^\varkappa)$ are mapped on the points of the tangent E_n of $\overset{0}{\xi}{}^\varkappa$ in such a way that the point of $\mathfrak{N}(\overset{0}{\xi}{}^\varkappa)$ that has normal coordinates ξ^h is mapped on the point of E_n with rectilinear coordinates ξ^h with respect to the local coordinate system of E_n belonging to the coordinates $\xi^\varkappa$.

From (7.21a) we see that the basis vectors of $(\varkappa)$ and of (h) at $\overset{0}{\xi}{}^\varkappa$ coincide. But that implies that the components of a *quantity* (not of every geometric object, consider for instance $\Gamma_{\mu\lambda}^\varkappa$) at $\overset{0}{\xi}{}^\varkappa$ with respect to $(\varkappa)$ and to (h) are numerically equal. This proves the following proposition:

If a set of numbers be given at $\overset{0}{\xi}{}^\varkappa$ with respect to (h) and if we know that they transform like the components of a quantity if (h) is transformed into (h'), these numbers are the components of a quantity at $\overset{0}{\xi}{}^\varkappa$ with respect to $(\varkappa)$ and $(\varkappa')$ and numerically equal to the components of this quantity with respect to (h) and (h') respectively.

This proposition has been used by VEBLEN and T. Y. THOMAS[1]) for a very elegant construction of a set of quantities called the normal tensors. The (not symmetrized) derivatives of order p of the Γ_{ij}^h with respect to the ξ^h at $\overset{0}{\xi}{}^\varkappa$

$$N_{j_p \ldots j_1 j i}^{\cdot \ldots \cdot \cdot \cdot h}(\overset{0}{\xi}) \overset{\text{def}}{=} \partial_{j_p} \ldots \partial_{j_1} \Gamma_{ji}^h(\overset{0}{\xi}) \tag{7.27}$$

transform obviously like the components of a tensor with valences $1, p+2$ if (h) is transformed into (h'). Hence the $N_{j_p \ldots j_1 j i}$ are components of a tensor. These tensors are called the *normal tensors* belonging to the connexion $\Gamma_{\mu\lambda}^\varkappa$ and the point $\overset{0}{\xi}{}^\varkappa$. They occur in the series

$$\Gamma_{ji}^h(\xi) = \xi^k N_{kji}^{\cdot\cdot\cdot h}(\overset{0}{\xi}) + \frac{1}{2!}\xi^{k_2}\xi^{k_1} N_{k_2 k_1 j i}^{\cdot \cdot \cdot \cdot h} + \cdots; \quad (\overset{0}{\xi}{}^h = 0)\ ^{2)}. \tag{7.28}$$

They are all symmetric in the *first* p and in the *last* 2 lower indices and vanish identically by mixing over *all* lower indices according to (7.25). The normal tensors belonging to the connexion $\Gamma_{\mu\lambda}^\varkappa$ can be constructed at every point and thus form tensor fields. But if we do this in the way described, at every point we get the components of these fields with respect to a normal coordinate system belonging to *this*

[1]) VEBLEN and T. Y. THOMAS 1923, 1.

[2]) The convergence of this series was discussed by T. Y. THOMAS 1934, 2, p. 126.

point and we should like to have the components with respect to a general coordinate system $(\varkappa)$. These can of course be computed by repeated differentiation of [cf. (2.5)]

$$\Gamma_{ji}^{h} = A_{j\,i\,\varkappa}^{\mu\lambda h} \Gamma_{\mu\lambda}^{\varkappa} - A_{j\,i}^{\mu\lambda} \partial_{\mu} A_{\lambda}^{h} \tag{7.29}$$

using (7.21). In this way we find for the first normal tensor at $(\underset{0}{\xi^{\varkappa}})$

$$N_{kji}^{\cdot\cdot\cdot h} = \delta_{k\,j\,i\,\varkappa}^{\nu\mu\lambda h} (\partial_{\nu} \Gamma_{\mu\lambda}^{\varkappa} - \Gamma_{\nu\mu\lambda}^{\varkappa} - 2\,\Gamma_{\nu(\mu}^{\varrho} \Gamma_{\lambda)\varrho}^{\varkappa}) \tag{7.30}$$

and accordingly for the general components

$$N_{\nu\mu\lambda}^{\cdot\cdot\cdot\varkappa} = \partial_{\nu} \Gamma_{\mu\lambda}^{\varkappa} - \Gamma_{\nu\mu\lambda}^{\varkappa} - 2\,\Gamma_{\nu(\mu}^{\varrho} \Gamma_{\lambda)\varrho}^{\varkappa} \tag{7.31}$$

valid at all points of $\mathfrak{N}(\underset{0}{\xi^{\varkappa}})$. The same could be done for all normal tensors but as we shall see later there is a much shorter way using the curvature tensor and its covariant derivatives.

If we start from the $\Gamma_{j_p \ldots j_1 i}^{h}$ (zero at $\underset{0}{\xi^{\varkappa}}$) other sets of generalized normal tensors can be formed since the $\partial_{k_q} \ldots \partial_{k_1} \Gamma_{j_p \ldots j_1 i}^{h}$ transform for $(h) \to (h')$ like the components of a tensor. All these tensor fields are differential concomitants of the connexion $\Gamma_{\mu\lambda}^{\varkappa}$ and there exists a great number of relations between them and their covariant derivatives. Much information can be found in the publications of VEBLEN and T. Y. THOMAS[1]).

The process applied here to the Γ_{ji}^{h} etc. can also be applied to any quantity. In fact if we take the (h)-components Φ_L (L is a collecting index), the derivatives $\partial_{j_1} \Phi_L$, $\partial_{j_2} \partial_{j_1} \Phi_L$ at $\underset{0}{\xi^{\varkappa}}$ transform for $(h) \to (h')$ like the components of a quantity and this gives rise to the construction of a series of quantities, called by VEBLEN the first, second, etc. *extension* of Φ_L. We denote them by the operators $\overset{e}{\nabla}_{\mu}$, $\overset{e}{\nabla}_{\mu_2\mu_1}$, etc. The first extension is of course identical with the first covariant derivative but this is not true for the higher extensions. For instance for a vector field $v^{\varkappa}$ we have at $\underset{0}{\xi^{\varkappa}}$:

$$(7.32)\begin{cases} \text{a)}\ \nabla_{\mu} v^{\varkappa}(\underset{0}{\xi}) = A_{\mu h}^{j\varkappa} \nabla_{j} v^{h}(\underset{0}{\xi}) = A_{\mu h}^{j\varkappa} \partial_{j} v^{h}(\underset{0}{\xi}) = A_{\mu h}^{j\varkappa} \overset{e}{\nabla}_{j} v^{h}(\underset{0}{\xi}) = \overset{e}{\nabla}_{\mu} v^{\varkappa}(\underset{0}{\xi}), \\ \text{b)}\ \nabla_{\mu_2\mu_1} v^{\varkappa}(\underset{0}{\xi}) = A_{\mu_2\mu_1 h}^{j_2 j_1 \varkappa} \nabla_{j_2} \nabla_{j_1} v^{h}(\underset{0}{\xi}) = A_{\mu_2\mu_1 h}^{j_2 j_1 \varkappa} (\partial_{j_2} \partial_{j_1} v^{h} + v^{i} \partial_{j_2} \Gamma_{j_1 i}^{h})(\underset{0}{\xi}) \\ \qquad = A_{\mu_2\mu_1 h}^{j_2 j_1 \varkappa} (\overset{e}{\nabla}_{j_2 j_1} v^{h} + v^{i} N_{j_2 j_1 i}^{\cdot\cdot\cdot h})(\underset{0}{\xi}) \\ \qquad = \overset{e}{\nabla}_{\mu_2\mu_1} v^{\varkappa}(\underset{0}{\xi}) + v^{\lambda} N_{\mu_2\mu_1\lambda}^{\cdot\cdot\cdot\varkappa}(\underset{0}{\xi}) \end{cases}$$

[1]) VEBLEN and T. Y. THOMAS 1924, 1; T. Y. THOMAS 1934, 2. When dealing with all these sets of quantities the way is open for "orgies of formalism" but it must be said that these authors have restricted themselves in an exemplary way to those relations that are really interesting.

and because these results have the invariant form and $\underset{0}{\xi}^{\varkappa}$ is arbitrary

$$(7.33)\quad \begin{cases} \text{a)} \quad \nabla_{\mu} v^{\varkappa} = \overset{e}{\nabla}_{\mu} v^{\varkappa} \\ \text{b)} \quad \nabla_{\mu_2 \mu_1} v^{\varkappa} = \overset{e}{\nabla}_{\mu_2 \mu_1} v^{\varkappa} + v^{\lambda} N_{\mu_2 \mu_1 \lambda}^{\cdot\cdot\cdot\,\varkappa} \end{cases}$$

valid at all points of $\mathfrak{N}(\underset{0}{\xi})$. Conversely all extensions can be expressed in terms of normal tensors and the covariant derivatives of the same and lower orders.

The most interesting relations are those between the simplest normal tensors defined by (7.27) and the curvature tensor $R_{\nu\mu\lambda}^{\cdot\cdot\cdot\,\varkappa}$ and its derivatives. With respect to (h) we have at $\underset{0}{\xi}^{\varkappa}$ according to (7.31)

$$(7.34)\quad R_{kji}^{\cdot\cdot\cdot\,h}(\underset{0}{\xi}) = 2\partial_{[k} \Gamma_{j]i}^{h}(\underset{0}{\xi}) = 2N_{[kj]i}^{\cdot\cdot\cdot\,h}(\underset{0}{\xi}),$$

hence everywhere

$$(7.35)\quad \boxed{R_{\nu\mu\lambda}^{\cdot\cdot\cdot\,\varkappa} = 2N_{[\nu\mu]\lambda}^{\cdot\cdot\cdot\,\varkappa}}\,.$$

But, because of $N_{\nu[\mu\lambda]}^{\cdot\cdot\cdot\,\varkappa} = 0$ and $N_{(\nu\mu\lambda)}^{\cdot\cdot\cdot\,\varkappa} = 0$, $N_{\nu\mu\lambda}^{\cdot\cdot\cdot\,\varkappa}$ can be solved from (7.35):

$$(7.36)\quad \boxed{N_{\nu\mu\lambda}^{\cdot\cdot\cdot\,\varkappa} = \tfrac{2}{3} R_{\nu(\mu\lambda)}^{\cdot\cdot\cdot\,\varkappa}}\,.$$

If the higher covariant derivatives of $R_{\nu\mu\lambda}^{\cdot\cdot\cdot\,\varkappa}$ are written out we get the general formula

$$(7.37)\quad \nabla_{\nu_p \ldots \nu_1} R_{\nu\mu\lambda}^{\cdot\cdot\cdot\,\varkappa} = 2\partial_{\nu_p} \ldots \partial_{\nu_1} \partial_{[\nu} \Gamma_{\mu]\lambda}^{\varkappa} + *$$

where $*$ stands for terms containing only the $\Gamma_{\mu\lambda}^{\varkappa}$ and their derivatives up to the order p. Using the coordinate system (h) we get at $\underset{0}{\xi}^{\varkappa}$

$$(7.38)\quad \nabla_{k_p \ldots k_1} R_{kji}^{\cdot\cdot\cdot\,h}(\underset{0}{\xi}) = 2N_{k_p \ldots k_1 [kj]i}^{\cdot\;\;\cdot\;\;\cdot\cdot\cdot\,h}(\underset{0}{\xi}) + *$$

where $*$ symbolizes terms containing only normal tensors with a valence lower than $p+4$. Note that for $p=0$ and $p=1$ these additional terms vanish because $\Gamma_{ji}^{h}(\underset{0}{\xi}) = 0$. Going back to general coordinates we get from (7.38)

$$(7.39)\quad \boxed{\nabla_{\nu_p \ldots \nu_1} R_{\nu\mu\lambda}^{\cdot\cdot\cdot\,\varkappa} = 2N_{\nu_p \ldots \nu_1 [\nu\mu]\lambda}^{\cdot\;\;\cdot\;\;\cdot\cdot\cdot\,\varkappa} + *}$$

where $*$ symbolizes the same quantities as in (7.38), valid at all points of $\mathfrak{N}(\underset{0}{\xi}^{\varkappa})$. Using the symmetry of the normal tensor with valence $p+4$ in the first $p+1$ and in the last 2 lower indices and the fact that mixing over all $p+3$ lower indices leads to zero, from (7.38) we get an equation of the form

$$(7.40)\quad \boxed{N_{\nu_p \ldots \nu_1 \nu\mu\lambda}^{\cdot\;\;\cdot\;\;\cdot\cdot\cdot\,\varkappa} = L(\nabla_{\nu_p \ldots \nu_1} R_{\nu\mu\lambda}^{\cdot\cdot\cdot\,\varkappa}) + *}$$

where $L(\,)$ denotes a polynomial linear homogeneous in the isomers of $\nabla_{\nu_p \dots \nu_1} R_{\nu\mu\lambda}^{\cdot\cdot\cdot\,\varkappa}$ with constant scalar coefficients and where $*$ stands for an expression in terms of $R_{\nu\mu\lambda}^{\cdot\cdot\cdot\,\varkappa}$ and its covariant derivatives up to the order $p-1$.

In fact the equation

$$N_{(\nu_p \dots \nu_1 \nu \mu \lambda)}^{\cdot \;\dots\; \cdot\, \cdot\, \cdot\, \cdot\,\varkappa} = 0 \tag{7.41}$$

contains three kinds of terms: 1. those with $\mu\lambda$ as the last two lower indices; 2. those with one ν-index among the last two lower indices and 3. those with two ν-indices as the last two lower indices. Because of the symmetry of $\overset{p+4}{N}$ in the first $p+1$ and in the last 2 lower indices the place of the indices in each of these two sets of indices is not important. Now according to (7.39) a term of the second kind can be written as a sum of $\overset{p+4}{R}$, a term of the first kind and terms containing only N's with lower valences. In the same way a term of the third kind can be written as a sum of an isomer of $\overset{p+4}{R}$, a term of the second kind and terms containing only N's with lower valences. If the term of the second kind arising in this way is expressed in the way described above we have finally in (7.41) only a sum of isomers of $\overset{p+4}{R}$ with constant coefficients, terms of the first kind, and terms containing only N's with lower valences. If these N's are dealt with in the same way, we get (7.40). It is possible to give some more information concerning the form of the equations (7.39, 40). Let $\overset{v}{R}$ and $\overset{v}{N}$ symbolize the $(v-4)$th covariant derivative of $R_{\nu\mu\lambda}^{\cdot\cdot\cdot\,\varkappa}$ and the normal tensor of valence v respectively. Then it is clear that (7.39) expresses $\overset{4}{R}$ in terms of $\overset{4}{N}$, and $\overset{5}{R}$ in terms of $\overset{5}{N}$, but that in the expression for $\overset{6}{R}$ we have not only $\overset{6}{N}$ but also a transvection of two factors $\overset{4}{N}$. The general scheme is as follows

$$\left\{\begin{array}{l} \overset{4}{R}\colon\; \overset{4}{N} \\ \overset{5}{R}\colon\; \overset{5}{N} \\ \overset{6}{R}\colon\; \overset{6}{N},\, \overset{4}{N}\overset{4}{N} \\ \overset{7}{R}\colon\; \overset{7}{N},\, \overset{5}{N}\overset{4}{N} \\ \overset{8}{R}\colon\; \overset{8}{N},\, \overset{6}{N}\overset{4}{N},\, \overset{5}{N}\overset{5}{N},\, \overset{4}{N}\overset{4}{N}\overset{4}{N} \\ \overset{9}{R}\colon\; \overset{9}{N},\, \overset{7}{N}\overset{4}{N},\, \overset{6}{N}\overset{5}{N},\, \overset{5}{N}\overset{4}{N}\overset{4}{N} \\ \qquad\quad \text{etc.} \end{array}\right. \tag{7.42}$$

There is a similar scheme belonging to (7.40) and derived from (7.42) by interchanging the letters R and N.

From these schemes we see 1° that a term with $\overset{v_1}{N}\ldots\overset{v_u}{N}(\overset{v_1}{R}\ldots\overset{v_u}{R})$ occurs in $\overset{v}{R}(\overset{v}{N})$ if and only if the following conditions hold

$$v = v_1 + \cdots + v_u - 2(u-1); \tag{7.43}$$

2° that there are exactly $u-1$ transvections in this term, every $N(R)$ being transvected with every other $N(R)$ over at most one index and 3° that every $N(R)$ in a product is transvected at least once.

The components of the normal tensors at $\underset{0}{\xi}^\varkappa$ with respect to (h) are the derivatives of the Γ_{ij}^h with respect to the ξ^h. Hence (7.40) gives very valuable information about the Γ_{ji}^h as functions of the ξ^h in $\mathfrak{N}(\underset{0}{\xi}^\varkappa)$. We consider the case when all covariant derivatives of $R_{\nu\mu\lambda}^{\cdot\cdot\cdot\varkappa}$ of *odd* orders are zero at $\underset{0}{\xi}^\varkappa$. Then according to (7.40, 43) all N's with an *odd* valence vanish at $\underset{0}{\xi}^\varkappa$ and it follows from (7.28) that all terms with an even number of factors ξ^h vanish at that point and that consequently

$$\Gamma_{ji}^h(\xi^k) = -\Gamma_{ji}^h(-\xi^k). \tag{7.44}$$

But this means that the A_n is invariant for the transformation $\xi^h \to -\xi^h$, that is a reflection at the point $\underset{0}{\xi}^\varkappa$. In fact a vector field with a field value v^h at ξ^h has a field value $-v^h$ at $-\xi^h$ after the reflection. Now a linear element $d\xi^h$ at ξ^h has for reflection the element $-d\xi^h$ at $-\xi^h$, hence (7.44) has as a consequence that the reflection of δv^h at ξ^h is a vector with the same components but with opposite sign at $-\xi^h$. An A_n with this property, is called *symmetric with respect* to $\underset{0}{\xi}^\varkappa$. If an A_n has this property all N's with an odd valence must vanish in (7.28) and then it follows from (7.39, 43) that all covariant derivatives of $R_{\nu\mu\lambda}^{\cdot\cdot\cdot\varkappa}$ of odd orders vanish at $\underset{0}{\xi}^\varkappa$. Hence *the vanishing at $\underset{0}{\xi}^\varkappa$ of all derivatives of $R_{\nu\mu\lambda}^{\cdot\cdot\cdot\varkappa}$ of odd orders is necessary and sufficient for the symmetry of the A_n with respect to $\underset{0}{\xi}^\varkappa$.*

An A_n is called *symmetric* if it is symmetric with respect to all points of an $\mathfrak{N}(\underset{0}{\xi}^\varkappa)$. It follows from the above that this occurs if and only if the first and consequently all covariant derivatives of $R_{\nu\mu\lambda}^{\cdot\cdot\cdot\varkappa}$ vanish in $\mathfrak{N}(\underset{0}{\xi}^\varkappa)$:

An A_n is symmetric if and only if $R_{\nu\mu\lambda}^{\cdot\cdot\cdot\varkappa}$ is covariant constant.[1]

[1] (Cf. also VII § 5.) CARTAN 1926, 1; 2; 1927, 1, p. 87; 1932, 1; LEVY 1926, 1; 2; WHITEHEAD 1932, 1; WALKER 1944, 1; 2; RASCHEVSKI 1950, 1; ROZENFEL'D and ABRAMOV 1950, 1; SCHIROKOV 1950, 1; BOREL and LICHNEROWICZ 1952, 1 (also for $\tilde{U}_n$).

Any *quantity* whose components can be expressed in terms of the $\Gamma^{\varkappa}_{\mu\lambda}$ and their derivatives up to a certain order by means of equations that are invariant for coordinate transformations will be called a *differential concomitant* of the connexion $\Gamma^{\varkappa}_{\mu\lambda}$. For instance $R_{\nu\mu\lambda}^{\cdot\cdot\cdot\varkappa}$, $R_{\mu\lambda}$, $V_{\mu\lambda}$, $\nabla_{\nu} R_{\mu\lambda}$. In general we only consider concomitants whose components can be expressed *algebraically* in terms of the $\Gamma^{\varkappa}_{\mu\lambda}$ and their derivatives. Special cases are scalars and scalar densities (or Δ-densities), often called *absolute differential invariants* and *relative differential invariants* of the connexion respectively. If we use first a normal coordinate system (h) with respect to a point $\underset{0}{\xi}^{\varkappa}$, we get at this point equations containing only the derivatives of the Γ^{h}_{ji} with respect to the ξ^{h} and these derivatives are the components of the normal tensors. Since the expressions are invariant for coordinate transformations we can go back now to a general coordinate system $(\varkappa)$, and by using (7.40) we may change these expressions into expressions in $R_{\nu\mu\lambda}^{\cdot\cdot\cdot\varkappa}$ and its covariant derivatives. Hence we have proved the *first reduction theorem* for differential concomitants (cf. I § 3):

All differential concomitants of a symmetric connexion are ordinary concomitants of $R_{\nu\mu\lambda}^{\cdot\cdot\cdot\varkappa}$ *and its covariant derivatives.*[1])

Any quantity whose components can be expressed in terms of the $\Gamma^{\varkappa}_{\mu\lambda}$, their derivatives up to a certain order, the components of a set of quantities $\underset{1}{\Phi}, \ldots, \underset{M}{\Phi}$ (indices suppressed) and their (ordinary) derivatives up to a certain order by means of equations that are invariant for coordinate transformations, are called *differential concomitants* of the set $\underset{1}{\Phi}, \ldots, \underset{M}{\Phi}$ and the connexion $\Gamma^{\varkappa}_{\mu\lambda}$. As before in general we only consider concomitants whose components can be expressed *algebraically*. *Absolute* and *relative invariants* are defined as before. If we use a normal coordinate system (h) for the point $\underset{0}{\xi}^{\varkappa}$ first, we get at this point equations containing only the derivatives of the Γ^{h}_{ji} with respect to the ξ^{h}, and the (h)-components of $\underset{1}{\Phi}, \ldots, \underset{M}{\Phi}$ and their derivatives with respect to the ξ^{h}. Now these latter derivatives are the components of the extensions of $\underset{1}{\Phi}, \ldots, \underset{M}{\Phi}$ and as we have seen, these can be expressed in terms of the covariant derivatives of $\underset{1}{\Phi}, \ldots, \underset{M}{\Phi}$ and the normal tensors. Since the expressions are invariant for coordinate transformations we can now go back to a general coordinate system $(\varkappa)$ and by means of (7.40) we may change all normal tensors into expressions in $R_{\nu\mu\lambda}^{\cdot\cdot\cdot\varkappa}$ and its covariant derivatives. This proves the *second reduction theorem* for differential concomitants:

1) CHRISTOFFEL 1869, 1.

All differential concomitants of a set of quantities $\underset{1}{\Phi}, \ldots, \underset{M}{\Phi}$ *(indices suppressed) and the symmetric connexion* $\Gamma_{\mu\lambda}^{\varkappa}$ *are ordinary concomitants of* $\underset{1}{\Phi}, \ldots, \underset{M}{\Phi}, R_{\nu\mu\lambda}^{\cdot\cdot\cdot\varkappa}$ *and their covariant derivatives.*[1])

If the connexion $\Gamma_{\mu\lambda}^{\varkappa}$ is not symmetric we have only to consider $S_{\mu\lambda}^{\cdot\cdot\varkappa}$ as one of the set of given quantities. Then all differential concomitants of the connexion $\underset{0}{\Gamma_{\mu\lambda}^{\varkappa}}$ and the set $\underset{1}{\Phi}, \ldots, \underset{M}{\Phi}$ are ordinary concomitants of $\underset{1}{\Phi}, \ldots, \underset{M}{\Phi}$, $S_{\mu\lambda}^{\cdot\cdot\varkappa}$ and $\overset{0}{R}{}_{\nu\mu\lambda}^{\cdot\cdot\cdot\varkappa}$ (the curvature tensor for the connexion $\Gamma_{(\mu\lambda)}^{\varkappa}$) and their covariant derivatives with respect to the connexion $\Gamma_{(\mu\lambda)}^{\varkappa}$. But they are of course also ordinary concomitants of $\underset{1}{\Phi}, \ldots, \underset{M}{\Phi}$, $S_{\mu\lambda}^{\cdot\cdot\varkappa}$ and $R_{\nu\mu\lambda}^{\cdot\cdot\cdot\varkappa}$ (the curvature tensor for the connexion $\Gamma_{\mu\lambda}^{\varkappa}$) and their covariant derivatives with respect to the connexion $\Gamma_{\mu\lambda}^{\varkappa}$.[1])

Exercises.

III 7,1[2]). If an X_m in A_n is formed by ∞^{m-1} geodesics through $\underset{0}{\xi^{\varkappa}}$ tangent to an E_m in the tangent space of that point, its equations in the normal coordinates belonging to $\underset{0}{\xi^{\varkappa}}$ are linear. (Note that an X_m in our sense can not be singular at $\underset{0}{\xi^{\varkappa}}$, cf. Ch. II, footnote 1 on p. 74.)

III 7,2)[2]. If the ξ^h are normal coordinates with respect to $\underset{0}{\xi^{\varkappa}}$, prove that

III 7,2 α) $$\Gamma_{ji}^{h} \xi^j \xi^i \overset{*}{=} 0$$

is true at *all* points of $\mathfrak{N}(\underset{0}{\xi^{\varkappa}})$ (use 7.25).

III 7,3[3]). If the ξ^h are normal coordinates in a V_n with respect to $\underset{0}{\xi^{\varkappa}}$, prove that

III 7,3 α) $$\begin{cases} ds^2 \overset{*}{=} (g_{ih}\{\underset{0}{\xi}\} + N_{kjih}\{\underset{0}{\xi}\}\,\xi^k\xi^j + \cdots)\,d\xi^i d\xi^h \\ \quad = (g_{ih}\{\underset{0}{\xi}\} + \frac{1}{3}K_{kijh}\{\underset{0}{\xi}\}\,\xi^k\xi^j + \cdots)\,d\xi^i d\xi^h. \end{cases}$$

III 7,4. Prove the second identity and the identity of Bianchi using normal coordinates.

---

1) Ricci and Levi Civita 1901, 1; Weitzenböck 1923, 1; T. Y. Thomas 1926, 2. Cf. T. Y. Thomas and Michal 1927, 2; T. Y. Thomas 1929, 1; 1934, 2, Ch. V. Cf. for the projective and the conformal case T. Y. Thomas 1927, 3; Levine 1942, 1 and for a more general case Noether 1918, 1.

2) E I 1935, 1, p. 102.

3) E I 1935, 1, p. 104, 138.

III 7,5. Prove that

III 7,5 α) $$N_{\omega\nu\mu\lambda}^{\cdot\cdot\cdot\cdot\,\varkappa} = \tfrac{5}{6}\nabla_{(\omega} R_{\nu)(\mu\lambda)}^{\cdot\cdot\cdot\,\varkappa} - \tfrac{1}{6}\nabla_{(\mu} R_{\lambda)(\omega\nu)}^{\cdot\cdot\cdot\,\varkappa}$$

using normal coordinates.

III 7,6[1]). Prove that in normal coordinates with respect to $\underset{0}{\xi}^{\varkappa}$

$$(\nabla_k \nabla_j P^h_{\cdot i})_0 \overset{*}{=} (\partial_k \partial_j P^h_{\cdot i} + N_{kjm}^{\cdot\cdot\cdot\, h} P^m_{\cdot i} - N_{kji}^{\cdot\cdot\cdot\, m} P^h_{\cdot m})_0 .$$

§ 8. Fermi coordinates [2]).

If a curve is given in A_n, there exist coordinate systems (h) such that the Γ^h_{ji} vanish at *every* point of the curve. Fermi proved this for a V_n [3]) and Eisenhart gave the generalization for A_n. [4]) Let

(8.1) $$\xi^{\varkappa} = f^{\varkappa}(t); \qquad \underset{0}{\xi}^{\varkappa} = f^{\varkappa}(0)$$

be the equations of a curve and let

(8.2) $$\underset{0}{t}^{\varkappa} = \varphi^{\varkappa}(\alpha_1, \ldots, \alpha_{n-2})$$

be a set of ∞^{n-2} vectors at $\underset{0}{\xi}^{\varkappa}$ in an E_{n-1} that does not contain the tangent of the curve, such that there is one vector in every direction of E_{n-1}. If now the vectors $\underset{0}{t}^{\varkappa}$ are displaced parallel along the curve we get a set of ∞^{n-2} vector fields

(8.3) $$t^{\varkappa} = \psi^{\varkappa}(\alpha_1, \ldots, \alpha_{n-2}, t); \qquad \underset{0}{t}^{\varkappa} = \psi^{\varkappa}(\alpha_1, \ldots, \alpha_{n-2}, 0)$$

whose covariant differentials vanish for displacements along the curve:

(8.4) $$d\xi^{\mu} \nabla_{\mu} t^{\varkappa} = 0 .$$

We only consider points of the curve where the $t^{\varkappa}$ span an E_{n-1} that does not contain the tangent direction. We now take all geodesics that are tangent to one of the vectors $t^{\varkappa}$ at an arbitrary point $\underset{1}{\xi}^{\varkappa}$ of the curve in $\mathfrak{N}(\underset{0}{\xi}^{\varkappa})$, and on each of these geodesics we choose an affine parameter z such that $d\xi^{\varkappa}/dz = t^{\varkappa}$ at $\underset{1}{\xi}^{\varkappa}$. Then according to (7.16)

(8.5) $$\xi^{\varkappa} = \underset{1}{\xi}^{\varkappa} + z\, t^{\varkappa} - \frac{1}{2!}\Gamma^{\varkappa}_{\mu\lambda}\{\underset{1}{\xi}\}\, z^2 t^{\mu} t^{\lambda} - \frac{1}{3!}\Gamma^{\varkappa}_{\mu_2\mu_1\lambda}\{\underset{1}{\xi}\}\, z^3 t^{\mu_2} t^{\mu_1} t^{\lambda} - \cdots$$

expressing $\xi^{\varkappa}$ in terms of $\underset{1}{\xi}^{\varkappa}$ and $z t^{\varkappa}$. Hence, taking the general point $f^{\varkappa}(t)$ of the curve for $\underset{1}{\xi}^{\varkappa}$, we get the expansion in $\mathfrak{N}(\underset{1}{\xi}^{\varkappa})$

(8.6) $$\xi^{\varkappa} = f^{\varkappa}(t) + z\, t^{\varkappa} - \frac{1}{2!}\Gamma^{\varkappa}_{\mu\lambda}\{f^{\varkappa}(t)\}\, z^2 t^{\mu} t^{\lambda} - \cdots$$

---

[1]) E I 1935, 1, p. 105 also for other examples.

[2]) Fermi 1922, 1; Eisenhart 1927, 1, p. 64; Bortolotti 1929, 2; 1930, 4; 1934, 1; Dienes 1933, 1; 2; E I 1935, 1, p. 106.

[3]) Fermi 1922, 1.

[4]) Eisenhart 1927, 1, p. 64.

expressing $\underset{0}{\xi}^\varkappa$ in terms of the $n+1$ variables t, $z t^\varkappa$ and valid in a neighbourhood of the curve. If now a set of n linearly independent vectors $\overset{0}{\underset{i}{e}}{}^\varkappa$; $i=1,\ldots,n$ is introduced at $\underset{0}{\xi}^\varkappa$, such that $\overset{0}{\underset{2}{e}}{}^\varkappa,\ldots,\overset{0}{\underset{n}{e}}{}^\varkappa$ span the E_{n-1} of the $\underset{0}{t}^\varkappa$, we have at $\underset{0}{\xi}^\varkappa$

$$\underset{0}{t}^\varkappa \overset{*}{=} t^x \overset{0}{\underset{x}{e}}{}^\varkappa; \quad x=2,\ldots,n \tag{8.7}$$

where the t^x are functions of the parameters $\alpha_1,\ldots,\alpha_{n-2}$. If the vectors $\overset{0}{\underset{i}{e}}{}^\varkappa$ are displaced parallel along the curve we get at each point of the curve a set $\underset{i}{e}^\varkappa$ such that

$$t^\varkappa \overset{*}{=} t^x \underset{x}{e}^\varkappa; \quad x=2,\ldots,n \tag{8.8}$$

where the t^x are *constant* along the curve. The $\underset{i}{e}^\varkappa$ are known as functions of t. Introducing the values (8.8) in (8.6) we get

$$\left\{\begin{aligned} \xi^\varkappa \overset{*}{=} f^\varkappa(t) + \underset{x}{e}^\varkappa z t^x - \frac{1}{2!}\underset{y}{e}^\mu \underset{x}{e}^\lambda \Gamma^\varkappa_{\mu\lambda}\{t\} z^2 t^x t^y - \\ - \frac{1}{3!}\underset{z}{e}^{\mu_2}\underset{y}{e}^{\mu_1}\underset{x}{e}^\lambda \Gamma^\varkappa_{\mu_2\mu_1\lambda}\{t\} z^3 t^x t^y t^z - \cdots; \quad x,y,z=2,\ldots,n. \end{aligned}\right. \tag{8.9}$$

Now we introduce the new coordinate system (h) of A_n by means of the equations

$$\left\{\begin{array}{lll} \xi^1 \overset{*}{=} \omega^1(t); & \omega^1(0)\overset{*}{=}0; & t=\theta(\xi^1) \\ \xi^x \overset{*}{=} \omega^x(t) + z t^x; & \omega^x(0)\overset{*}{=}0; & x=2,\ldots,n \end{array}\right. \tag{8.10}$$

and try to determine the as yet unknown functions $\omega^h(t)$; $h=1,\ldots,n$ in such a way that the contravariant basis vectors of (h) at all points of the curve are identical with the $\underset{i}{e}^\varkappa$ already defined. The equations of the curve are

$$\xi^x - \omega^x(\theta(\xi^1)) \overset{*}{=} 0 \tag{8.11}$$

in these new coordinates and the $\underset{i}{e}^\varkappa$ and $\Gamma^\varkappa_{\mu\lambda}$ along the curve may now be considered as functions of ξ^1. It follows from (8.9) and (8.10) that

$$\left\{\begin{aligned} \xi^\varkappa \overset{*}{=} f^\varkappa(\theta(\xi^1)) + \underset{x}{e}^\varkappa\{\xi^1\}(\xi^x - \omega^x(\theta(\xi^1))) - \\ - \frac{1}{2!}\underset{y}{e}^\mu\{\xi^1\}\underset{x}{e}^\lambda\{\xi^1\}\Gamma^\varkappa_{\mu\lambda}\{\xi^1\}(\xi^x-\omega^x(\theta(\xi^1)))(\xi^y-\omega^y(\theta(\xi^1))) - \cdots \end{aligned}\right. \tag{8.12}$$

and, by differentiation with respect to ξ^h for all points of the curve, that

$$\partial_1\xi^\varkappa \overset{*}{=} \frac{d f^\varkappa}{dt}\left(\frac{d\omega^1}{dt}\right)^{-1} - \underset{x}{e}^\varkappa \frac{d\omega^x}{dt}\left(\frac{d\omega^1}{dt}\right)^{-1}; \quad \partial_y\xi^\varkappa \overset{*}{=} \underset{y}{e}^\varkappa; \quad x,y=2,\ldots,n. \tag{8.13}$$

Hence the $\underset{y}{e}^{\varkappa}$ are, in fact, basis vectors of (h), and in order to make $\underset{1}{e}^{\varkappa}$ also such a basis vector, $\omega^1(t)$ and $\omega^x(t)$ must be chosen such that they satisfy the equations

$$\frac{d f^{\varkappa}}{d t} - \underset{x}{e}^{\varkappa} \frac{d \omega^x}{d t} \overset{*}{=} \underset{1}{e}^{\varkappa} \frac{d \omega^1}{d t} . \tag{8.14}$$

Now according to our assumptions $d f^{\varkappa}/d t$ and $\underset{1}{e}^{\varkappa}$ are linearly independent vectors and they are known as functions of t along the curve. That means that there exists an equation of the form

$$\frac{d f^{\varkappa}}{d t} = \overset{1}{\alpha}\, \underset{1}{e}^{\varkappa} + \overset{x}{\alpha}\, \underset{x}{e}^{\varkappa}; \qquad \overset{1}{\alpha} \neq 0 \tag{8.15}$$

valid at all points of the curve and that $\overset{1}{\alpha}, \overset{x}{\alpha}$ are known as functions of t. But then it follows from (8.14, 15) that

$$\frac{d \omega^1}{d t} \overset{*}{=} \overset{1}{\alpha}; \qquad \frac{d \omega^x}{d t} \overset{*}{=} \overset{x}{\alpha}, \tag{8.16}$$

from which the functions $\omega^1(t)$ and $\omega^x(t)$ can be found by means of quadratures. If the curve happens to be a geodesic and if for t we choose an affine parameter, $\underset{1}{e}^{\varkappa}$ can be chosen such that $\overset{1}{\alpha} = 1, \overset{x}{\alpha} = 0$. Then we get as a solution $\xi^1 \overset{*}{=} t, \xi^x \overset{*}{=} z t^x$.

In order to prove that the Γ_{ji}^h vanish at all points of the curve we remark that (8.12) holds for every choice of $(\varkappa)$, hence it holds if we take (h) for $(\varkappa)$. From (8.14) we then get $f^h \overset{*}{=} \omega^h$ at all points of the curve, which is in accordance with (8.10). (8.12) changes into

$$\left\{\begin{aligned} \xi^h \overset{*}{=} \omega^h (\theta(\xi^1)) &+ \underset{x}{e}^h (\xi^x - \omega^x(\theta(\xi^1))) - \\ &- \frac{1}{2!} \underset{y}{e}^j \underset{x}{e}^i \Gamma_{ji}^h \{\xi^1\} (\xi^x - \omega^x(\theta(\xi^1))) (\xi^y - \omega^y(\theta(\xi^1))) - \cdots \end{aligned}\right. \tag{8.17}$$

and from this equation we get by differentiation

$$\left\{\begin{aligned} &\delta_y^h \overset{*}{=} \partial_y \xi^h \overset{*}{=} \underset{y}{e}^h - \Gamma_{yx}^h (\xi^x - \omega^x(\theta(\xi^1))) - \cdots \\ &\qquad x, y = 2, \ldots, \mathrm{n}; \; h = 1, \ldots, \mathrm{n}; \end{aligned}\right. \tag{8.18}$$

hence $\Gamma_{yx}^h \overset{*}{=} 0$ at all points of the curve. Moreover we know that the fields $\underset{i}{e}^{\varkappa}$ are covariant constant along the curve. Hence we have on the curve [cf. (8.15)]

$$0 = \frac{d f^j}{d t} \nabla_j \underset{i}{e}^h \overset{*}{=} \overset{1}{\alpha}\, \Gamma_{1i}^h + \overset{y}{\alpha}\, \Gamma_{yi}^h \tag{8.19}$$

or

$$\overset{1}{\alpha}\, \Gamma_{11}^h + \overset{y}{\alpha}\, \Gamma_{1y}^h \overset{*}{=} 0; \qquad \overset{1}{\alpha}\, \Gamma_{1x}^h \overset{*}{=} 0; \qquad x, y = 2, \ldots, \mathrm{n}; \tag{8.20}$$

and because of $\overset{1}{\alpha} \neq 0$ this also proves that $\Gamma^h_{y1} = \Gamma^h_{1y} \overset{*}{=} 0$ and $\Gamma^h_{11} \overset{*}{=} 0$ at all points of the curve. The coordinates (h) defined in this way are called FERMI *coordinates* for the curve considered. By repeated differentiation it can be proved that [cf. (7.15)]

$$(8.21) \qquad \begin{cases} \Gamma^h_{z_p \ldots z_1 y} \overset{*}{=} 0; \qquad y, z_1, \ldots, z_{p+1} = 2, \ldots, n; \quad h = 1, \ldots, n \\ \partial_{(z_{p+1}} \Gamma^h_{z_p \ldots z_1 y)} \overset{*}{=} 0 \end{cases}$$

at all points of the curve.

If we know of an X_m that there exist n linearly independent vector fields $\underset{i}{e}^\varkappa$, *allround covariant stationary* in A_n [1]) over X_m, it can easily be proved that there exists a coordinate system (h) such that the Γ^h_{ji} vanish at all points of X_m. [2]) In the case of an X_1 we have proved more, the fields $\underset{i}{e}^\varkappa$ being only *covariant constant* over X_1. It is not yet known whether a system (h) with Γ^h_{ji} vanishing over X_m can be constructed if only covariant constant fields $\underset{i}{e}^\varkappa$ are supposed to exist over X_m.

§ 9. Linear connexions expressed in anholonomic coordinates (cf. II § 9).

If we write for the covariant derivative of a vector with respect to an anholonomic coordinate system (h)

$$(9.1) \qquad \begin{cases} \text{a)} \quad \nabla_j v^h = \partial_j v^h + \Gamma^h_{ji} v^i; \qquad \partial_j \overset{\text{def}}{=} A^\mu_j \partial_\mu \\ \text{b)} \quad \nabla_j w_i = \partial_j w_i - \Gamma^h_{ji} w_h \end{cases}$$

and if $(\varkappa)$ is a holonomic system, we get for the Γ^h_{ji} the transformation [cf. (2.5)]

$$(9.2) \qquad \Gamma^h_{ji} = A^{\mu\lambda h}_{ji\varkappa} \Gamma^\varkappa_{\mu\lambda} - A^{\mu\lambda}_{ji} \partial_\mu A^h_\lambda,$$

hence (cf. II § 9)

$$(9.3) \qquad \Gamma^h_{[ji]} = S_{ji}{}^{\cdot\cdot h} - \Omega^h_{ji},$$

which implies that $\Gamma^h_{[ji]}$ need not be zero if the connexion is symmetric. From (9.2) follows

$$(9.4) \qquad \Gamma_j = A^\mu_j \Gamma_\mu - A^{\mu\lambda}_{ji} \partial_\mu A^i_\lambda.$$

The equations (9.2) and (9.4) also hold if (h) and $(\varkappa)$ are both anholonomic.

1) Cf. III § 2.

2) E I 1935, 1, p. 106. Read on that page A_n instead of L_n.

From (9.3) we get the general formulae for Rot and Div with ∇ in anholonomic components (cf. 2.17, 2.30)

$$(9.5)\qquad \nabla_{[j} w_{i_1\ldots i_q]} = \partial_{[j} w_{i_1\ldots i_q]} - q\, S_{[j i_q}^{\cdot\cdot\; h} w_{i_1\ldots i_{q-1}]h} + q\, \Omega^h_{[j i_q} w_{i_1\ldots i_{q-1}]h};$$

$$(9.6)\qquad \begin{cases} \nabla_j \mathfrak{w}^{j h_2\ldots h_p} = \partial_j \mathfrak{w}^{j h_2\ldots h_p} - 2 S_{ji}^{\cdot\cdot\; i} \mathfrak{w}^{j h_2\ldots h_p} + (p-1)\, S_{ji}^{\cdot\cdot\;[h_2} \mathfrak{w}^{|ji|h_3\ldots h_p]} + \\ \qquad + 2\Omega^i_{ji} \mathfrak{w}^{j h_2\ldots h_p} - (p-1)\, \Omega^{[h_2}_{ji} \mathfrak{w}^{|ji|h_3\ldots h_p]}; \\ \qquad \mathfrak{w}^{h_1\ldots h_p} = p\text{-vector density of weight } +1. \end{cases}$$

From these equations another very simple proof of the equations in Exerc. II 9,1 can be derived.

If we try to express a linear connexion in terms of $S_{ji}^{\cdot\cdot\; h}$, g_{ih} and $\nabla_j g_{ih}$ the only difference from III § 3 is that (9.3) occurs instead of (2.13). Hence, if we write

$$(9.7)\qquad \Omega_{jih} \overset{\text{def}}{=} \Omega^k_{ji} g_{kh}\ ^{1)}$$

we get

$$(9.8)\qquad \boxed{\Gamma^h_{ji} = g^{hk}\left(\chi_{\{jki\}} - S_{\{jki\}} + \Omega_{\{jki\}} + \tfrac{1}{2} Q_{\{jki\}}\right); \qquad \chi_{jki} \overset{\text{def}}{=} \tfrac{1}{2}\partial_j g_{ki}}$$

instead of (3.9).

This equation takes a very simple form in an ordinary V_n if we introduce an anholonomic system consisting of mutually perpendicular unit-vectors $\underset{i}{i}^{\varkappa}$, $\overset{h}{i}_{\lambda}$ at every point. Then the components of g_{ih} are all 1 or 0 and $S_{ji}^{\cdot\cdot\; h}$ and $Q_{ji}^{\cdot\cdot\; h}$ vanish, hence[2])

$$(9.9)\qquad \Gamma^h_{ji} \overset{*}{=} + g^{hk}\Omega_{\{jki\}} = -\, 2 g^{hk}\Omega_{k(ji)} - \Omega^h_{ji}$$

and

$$(9.10)\qquad \Gamma_j \overset{*}{=} -\,\Omega^i_{ji} + g_{ih}\Omega^h_{jl} g^{il} = 0.$$

In this case the Γ^h_{ji} have a simple geometric meaning. If we write

$$(9.11)\qquad \Gamma_{jih} \overset{\text{def}}{=} \Gamma^k_{ji} g_{kh}$$

it follows from $\nabla_j g_{ih} = 0$ that

$$(9.12)\qquad \Gamma_{j(ih)} \overset{*}{=} 0.$$

Now the covariant differential of $\underset{i}{i}^h$ is for $g_{\lambda\varkappa}$ positive definite

$$(9.13)\qquad \delta \underset{i}{i}^h \overset{*}{=} \Gamma^h_{jk} \underset{i}{i}^k (d\xi)^j \overset{*}{=} \Gamma_{jih} (d\xi)^j \overset{*}{=} \Gamma_{jkh} \underset{i}{i}^k (d\xi)^j$$

---

1) Note that the Ω_{jih} are not components of a geometric object.

2) VRANCEANU 1932, 1.

and this means that for every fixed value of i the vector $\underset{i}{i}^h$ undergoes an infinitesimal rotation with respect to a parallel moving coordinate system, with $\Gamma_{jkh}(d\xi)^j$ as "bivector" of rotation (cf. I § 9). As such the Γ_{jih} were discovered by RICCI[1]) long before anholonomic coordinates came into use. He wrote

$$(9.14) \qquad \gamma_{ihj} \overset{\text{def}}{=} \Gamma_{jih} = \Gamma_{ji}^k g_{kh}$$

and called the γ_{ihj} the *coefficients of rotation* of the orthogonal net consisting of the n congruences of streamlines of the n fields $\underset{i}{i}^\varkappa$. The γ_{ihj} alternate in the *first* two indices.[2])

In an L_n facts are quite analogous. Having introduced the anholonomic system (h) the covariant differential of $\underset{i}{e}^h$ is

$$(9.15) \qquad \delta \underset{i}{e}^h \overset{*}{=} \Gamma_{ji}^h (d\xi)^j \overset{*}{=} \Gamma_{jk}^h \underset{i}{e}^k (d\xi)^j.$$

Hence, for every fixed value of i the vector $\underset{i}{e}^h$ undergoes a homogeneous affine transformation with respect to a parallel moving coordinate system, with $\Gamma_{jk}^h(d\xi)^j$ as "tensor" of the transformation.[3])

The rotation coefficients Γ_{jih} contain a lot of information about the orthogonal net of (h). A congruence $i^\varkappa$ is geodesic (cf. III § 7, V § 5) if and only if $i^\mu \nabla_\mu i^h = 0$. Hence the n.a.s. conditions for the congruence $\underset{\mathrm{n}}{i}^\varkappa$ to be geodesic are

$$(9.16) \qquad \Gamma_{\mathrm{nn}h} \overset{*}{=} -\Gamma_{\mathrm{n}h\mathrm{n}} \overset{*}{=} 0.$$

A congruence is V_{n-1}-normal if and only if (cf. II 5.16, V § 5) $i_{[\nu} \nabla_\mu i_{\lambda]} = 0$, hence the n.a.s. conditions for the congruence $\underset{\mathrm{n}}{i}^\varkappa$ to be V_{n-1}-normal are

$$(9.17) \qquad \Gamma_{[cb]\mathrm{n}} \overset{*}{=} 0; \qquad b, c = 1, \ldots, \mathrm{n}-1.$$

Taking (9.16) and (9.17) together we get for a geodesic and V_{n-1}-normal congruence $\underset{\mathrm{n}}{i}^\varkappa$ the n.a.s. conditions

$$(9.18) \qquad \Gamma_{[ji]\mathrm{n}} \overset{*}{=} 0.$$

All congruences $\underset{i}{i}^\varkappa$ are geodesic and V_{n-1}-normal if and only if $\Gamma_{[ji]h} \overset{*}{=} 0$. But since the Γ_{jih} are alternating in ih this is only possible if $\Gamma_{jih} \overset{*}{=} 0$,

1) RICCI 1895, 1, p. 303. Cf. SCHOUTEN-STRUIK 1919, 1; LEVY 1925, 1; HLAVATY 1928, 1; VRANCEANU 1929, 3; MATTIOLI 1930, 1; 2; DIENES 1933, 1; 2; E I 1935, 1, p. 85; E II 1938, 2, p. 34. The coefficients of rotation are intimately connected with the theory of the "trièdre mobile" of DARBOUX, cf. RICCI and LEVI CIVITA 1901, 1, p. 157.

2) Of course, if we have introduced the anholonomic system (h) and thus the Γ_{ji}^h and Γ_{jih}, it is incorrect from a methodical point of view to introduce *besides* these the rotation γ_{ihj} also, as many authors still do.

3) Cf. SLEBODZINSKI 1930, 1. In this case also it would be incorrect to introduce some kind of "generalized coefficients of rotation" *besides* the Γ_{ji}^h.

that is if the Γ_{ji}^{h} vanish at all points. But this implies according to (9.3) that (h) is holonomic and that consequently $K_{kji}^{\cdot\cdot\cdot h}$ vanishes. Hence *an orthogonal net of geodesic and V_{n-1}-normal congruences is only possible in an R_n.*

In order to find the expression for the anholonomic components $R_{kji}^{\cdot\cdot\cdot h}$ of the curvature tensor we start from (9.1b) and use (9.3) and (II 9.4). Then we find

$$(9.19)\qquad \nabla_{[k}\nabla_{j]}w_i = -\{\partial_{[k}\Gamma_{j]i}^{h} + \Gamma_{[k|l|}^{h}\Gamma_{j]i}^{l} + \Omega_{kj}^{l}\Gamma_{li}^{h}\}w_h - S_{kj}^{\cdot\cdot h}\nabla_h w_i.$$

Comparing this result with (4.9a) we get

$$(9.20)\qquad R_{kji}^{\cdot\cdot\cdot h} = 2\partial_{[k}\Gamma_{j]i}^{h} + 2\Gamma_{[k|l|}^{h}\Gamma_{j]i}^{l} + 2\Omega_{kj}^{l}\Gamma_{li}^{h}$$

instead of (4.2) and, instead of (4.15) we get

$$(9.21)\qquad V_{kj} = 2\partial_{[k}\Gamma_{j]} + 2\Omega_{kj}^{l}\Gamma_{l}.$$

The equations (9.20, 21) could also be derived from (4.2), (4.15) and (9.2).

§ 10. CARTAN's symbolical method used for connexions.

Using the symbolic notation explained in II § 12 we write[1])

$$(10.1)\qquad \begin{cases} A^h \text{ for } A_\lambda^h & \text{and also for } A_\lambda^h\, d\xi^\lambda \\ \Omega^h \text{ for } \Omega_{\mu\lambda}^h = \Omega_{ji}^h A_\mu^j A_\lambda^i & \text{and also for } \Omega_{\mu\lambda}^h \underset{1}{d}\xi^\mu \underset{2}{d}\xi^\lambda \\ \Gamma_i^h \text{ for } \Gamma_{\mu i}^h = \Gamma_{ji}^h A_\mu^j & \text{and also for } \Gamma_{\mu i}^h\, d\xi^\mu \\ S^h \text{ for } S_{\mu\lambda}^{\cdot\cdot h} & \text{and also for } S_{\mu\lambda}^{\cdot\cdot h} \underset{1}{d}\xi^\mu \underset{2}{d}\xi^\lambda \\ R_i^{\cdot h} \text{ for } R_{\mu\lambda i}^{\cdot\cdot\cdot h} & \text{and also for } R_{\mu\lambda i}^{\cdot\cdot\cdot h} \underset{1}{d}\xi^\mu \underset{2}{d}\xi^\lambda. \end{cases}$$

In II § 12 we had already for $\varkappa$ holonomic (cf. II 12.9):

$$(10.2)\qquad \begin{cases} \text{a)} & \partial_{[\mu}A_{\lambda]}^h = \Omega_{ji}^h A_{\mu\lambda}^{ji} = \Omega_{\mu\lambda}^h \\ \text{b)}\ \mathfrak{C} & [dA^h] = \Omega^h. \end{cases}$$

---

1) CARTAN writes A_{ji}^h and A_{ikj}^h or R_{ikj}^h for $S_{ji}^{\cdot\cdot h}$ and $R_{kji}^{\cdot\cdot\cdot h}$ but ω^h, ω_i^h, Ω^h and Ω_i^h for what we have written A^h, Γ_i^h, S^h and $R_i^{\cdot h}$. This gives rise to a duplication of formulae that could have been avoided. The same holds for all publications where the abridged calculus is used, except those dealing with alternating forms only. T. Y. THOMAS gave 1927, 1 a severe criticism and used such expressions as "superabundance of geometrical terminology" and "a lack of content". This is certainly too severe. The first objection can be countered by the change of notation we have proposed here, that makes duplication of formulae unnecessary and the second one is only true where the abridged calculus is used at places where it is not appropriate to do so.

Now we know from (9.3) that

$$(10.3)\quad \begin{cases} \text{a)} & \Omega^h_{\mu\lambda} = - A^{ji}_{\mu\lambda}(\Gamma^h_{[ji]} - S_{ji}^{\cdot\cdot h}) = - \Gamma^h_{[\mu|k|}A^k_{\lambda]} + S_{\mu\lambda}^{\cdot\cdot h}; \\ \text{b)}\ \mathfrak{C} & \Omega^h = [A^i \Gamma^h_i] + S^h. \end{cases}$$

Hence, from (10.2) and (10.3)

$$(10.4)\quad \begin{cases} \text{a)} & \partial_{[\mu} A^h_{\lambda]} = A^i_{[\mu}\Gamma^h_{\lambda]i} + S_{\mu\lambda}^{\cdot\cdot h} \\ \text{b)}\ \mathfrak{C} & [dA^h] = [A^i\Gamma^h_i] + S^h. \end{cases}$$

Moreover we know from (9.20) that

$$(10.5)\quad \begin{cases} \text{a)} & \partial_{[\nu}\Gamma^h_{\mu]i} + \Gamma^h_{[\nu|l|}\Gamma^l_{\mu]i} = \tfrac12 R_{\nu\mu i}^{\cdot\cdot\cdot h} \\ \text{b)}\ \mathfrak{C} & [d\Gamma^h_i] + [\Gamma^h_l \Gamma^l_i] = \tfrac12 R_i^{\cdot h}. \end{cases}$$

CARTAN calls (10.4b) and (10.5b) the *structural equations* (équations de structure) of the L_n. According to the definition, repeated application of $\partial_{[\nu\ldots]}$ or $[d\ldots]$ leads to zero, hence from (10.4), giving the calculation in full for the symbolical method

$$(10.6)\quad \begin{cases} \text{a)} & 0 = -\tfrac12 A^i_{[\omega} R_{\nu\mu]i}^{\cdot\cdot\cdot h} + \partial_{[\omega} S_{\nu\mu]}^{\cdot\cdot h} + S_{[\omega\nu}^{\cdot\cdot i}\Gamma^h_{\mu]i} \\ \text{b)}\ \mathfrak{C} & 0 = [[dA^i]\Gamma^h_i] - [A^i[d\Gamma^h_i]] + [dS^h] \\ & \quad = [A^l\Gamma^i_l\Gamma^h_i] + [S^i\Gamma^h_i] + [A^i\Gamma^h_l\Gamma^l_i] - \tfrac12[A^iR_i^{\cdot h}] + [dS^h] \\ & \quad = -\tfrac12[A^iR_i^{\cdot h}] + [dS^h] + [S^i\Gamma^h_i], \end{cases}$$

or

$$(10.7)\quad \boxed{\begin{array}{ll} \text{a)} & A^i_{[\omega}R_{\nu\mu]i}^{\cdot\cdot\cdot h} - 2\overset{2}{\nabla}_{[\omega} S_{\nu\mu]}^{\cdot\cdot h} = 0 \\ \text{b)}\ \mathfrak{C} & [A^iR_i^{\cdot h}] - 2[\overset{2}{\nabla} S^h] = 0 \end{array}}$$ [1] (second identity, cf. 5.2, 41 and Exerc. III 5,7).

In the same way, we get from (10.5)

$$(10.8)\quad \begin{cases} \text{a)} & 0 = \tfrac12\partial_{[\omega}R_{\nu\mu]i}^{\cdot\cdot\cdot h} - \tfrac12\Gamma^l_{[\omega|i|}R_{\nu\mu]l}^{\cdot\cdot\cdot h} + \tfrac12\Gamma^h_{[\omega|l|}R_{\nu\mu]i}^{\cdot\cdot\cdot l} \\ \text{b)}\ \mathfrak{C} & 0 = \tfrac12[dR_i^{\cdot h}] - [[d\Gamma^h_l]\Gamma^l_i] + [\Gamma^h_l[d\Gamma^l_i]] \\ & \quad = \tfrac12[dR_i^{\cdot h}] + [\Gamma^h_k\Gamma^k_l\Gamma^l_i] - \tfrac12[\Gamma^l_i R_l^{\cdot h}] - \\ & \qquad - [\Gamma^h_l\Gamma^l_k\Gamma^k_i] + \tfrac12[\Gamma^h_l R_i^{\cdot l}] \end{cases}$$

or

$$(10.9)\quad \boxed{\begin{array}{ll} \text{a)} & \overset{2}{\nabla}_{[\omega}R_{\nu\mu]i}^{\cdot\cdot\cdot h} = 0 \\ \text{b)}\ \mathfrak{C} & [\overset{2}{\nabla} R_i^{\cdot h}] = 0 \end{array}}$$ [1] (identity of BIANCHI, cf. 5.19, 42 and Exerc. III 5,7).

[1]) CARTAN calls (10.7b) and (10.9b) "théorème de conservation de la courbure et de la torsion".

In these formulae we have used the operator $[\overset{p}{\nabla}\ldots]$ corresponding to the operator $\overset{p}{\nabla}_{[\omega\ldots]}$ defined in III § 5. The working of the first operator on any quantity $P_{\lambda_1\ldots\lambda_p\cdot}^{\cdot\ldots\cdot h_1\ldots h_q}{}_{\cdot i_1\ldots i_r}$ or its corresponding alternating differential form (cf. II § 12) is defined for $p>0$ by the equation (cf. 5.32)

$$(10.10)\quad \begin{cases} [\overset{p}{\nabla} P^{h_1\ldots h_q}_{\cdot\quad\cdot\, i_1\ldots i_r}] = [d\, P^{h_1\ldots h_q}_{\cdot\quad\cdot\, i_1\ldots i_r}] + [\Gamma_l^{h_1} P^{l h_2\ldots h_q}_{\cdot\cdot\quad\cdot\, i_1\ldots i_r}] + \cdots \\ \qquad\qquad - [\Gamma_{i_1}^{l} P^{h_1\ldots h_q}_{\cdot\quad\cdot\, l i_2\ldots i_r}] - \cdots. \end{cases}$$

In order to make comparison between tensor calculus and CARTAN's notation possible in a more complicated case we now translate the proof of the geometrical meaning of the two identities (10.7, 9) given in III § 5. Writing $d\xi^\varkappa$ for $\xi^\varkappa - \underset{0}{\xi^\varkappa}$, $\underset{1}{d}\,\xi^{[\nu}\underset{2}{d}\,\xi^{\mu]}$ for $df^{\nu\mu}$, $\underset{1}{d}\xi^{[\omega}\,\underset{2}{d}\xi^{\nu}\,\underset{3}{d}\xi^{\mu]}$ for $df^{\omega\nu\mu}$ and passing to general anholonomic coordinates we get for the change of the radiusvector $\underset{0}{v^h}$ from $\underset{0}{\xi^\varkappa}$ to $\xi^\varkappa$

$$\begin{matrix}(10.11)\\(5.43)\end{matrix}\qquad \mathfrak{C}\quad \underset{0}{v^h} - A^h - \underset{0}{\Gamma_i^h}\,\underset{0}{v^i}$$

and for the change after moving around $df^{\nu\mu}$

$$\begin{matrix}(10.12)\\(5.44)\end{matrix}\qquad \mathfrak{C}\quad \underset{0}{S^h} + d\,S^h + \tfrac{1}{2}\underset{0}{R_i^{\cdot h}}\,\underset{0}{v^i} - \tfrac{1}{2}\underset{0}{R_i^{\cdot h}}A^i - \tfrac{1}{2}\underset{0}{R_i^{\cdot h}}\,\underset{0}{\Gamma_j^i}\,\underset{0}{v^j} + \tfrac{1}{2}\underset{0}{v^i}\,d\,R_i^{\cdot h}.$$

After parallel displacement we get

$$\begin{matrix}(10.13)\\(5.45)\end{matrix}\quad \begin{cases} \mathfrak{C}\quad \underset{0}{S^h} + \tfrac{1}{2}\underset{0}{R_i^{\cdot h}}\,\underset{0}{v^i} + \underset{0}{S^i}\,\underset{0}{\Gamma_i^h} + \tfrac{1}{2}\underset{0}{R_i^{\cdot j}}\,\underset{0}{v^i}\,\underset{0}{\Gamma_j^h} + d\,S^h - \\ \qquad\qquad - \tfrac{1}{2}\underset{0}{R_i^{\cdot h}}A^i - \tfrac{1}{2}\underset{0}{R_j^{\cdot h}}\,\underset{0}{\Gamma_i^j}\,\underset{0}{v^i} + \tfrac{1}{2}\underset{0}{v^i}\,d\,R_i^{\cdot h}, \end{cases}$$

and after applying the theorem of STOKES

$$\begin{matrix}(10.14)\\(5.46)\end{matrix}\quad \begin{cases} \mathfrak{C}\quad \int_{\tau_3} [\underset{0}{S^i}\,\underset{0}{\Gamma_i^h}] + \tfrac{1}{2}\underset{0}{v^i}\,[\underset{0}{R_i^{\cdot j}}\,\underset{0}{\Gamma_j^h}] + [d\,S^h] - \tfrac{1}{2}[A^i \underset{0}{R_i^{\cdot h}}] - \\ \qquad\qquad - \tfrac{1}{2}[\underset{0}{R_j^{\cdot h}}\,\underset{0}{\Gamma_i^j}]\,\underset{0}{v^i} + \tfrac{1}{2}\underset{0}{v^i}\,[d\,R_i^{\cdot j}] \\ \quad = \int_{\tau_3} \{[\overset{2}{\nabla} S^h] - \tfrac{1}{2}[A^i R_i^{\cdot h}] + \tfrac{1}{2}\underset{0}{v^i}\,[\overset{2}{\nabla} R_i^{\cdot h}]\}_{\xi^\varkappa = \underset{0}{\xi^\varkappa}}. \end{cases}$$

The abridged calculus has often been used by CARTAN to solve problems concerning point transformations transforming fields into each other. We consider first the question whether n given linearly independent vector fields $\overset{h}{w}_\lambda$ in an X_n with coordinates $\xi^\varkappa$ can be transformed by a transformation $\xi^\varkappa \to \xi^{\varkappa'}$ into n given linearly independent vector fields $'\overset{h}{w}_{\lambda'}$ in another X_n with coordinates $\xi^{\varkappa'}$

(10.15)
$$\begin{cases} \text{a)} & \overset{h}{w}_{\lambda}\, d\xi^{\lambda} = '\overset{h}{w}_{\lambda'}\, d\xi^{\lambda'} \text{ [1]} \\ \text{b)}\ \mathfrak{C} & \overset{h}{w} = '\overset{h}{w}. \end{cases}$$

By differentiation we get

(10.16 a)
$$\begin{cases} \alpha) & \partial_{[\mu}\, \overset{h}{w}_{\lambda]} = u_{j\,i}^{\cdot\cdot h}\, \overset{j}{w}_{[\mu}\, \overset{i}{w}_{\lambda]}\,; \\ \beta) & \partial_{[\mu'}\, '\overset{h}{w}_{\lambda']} = 'u_{j\,i}^{\cdot\cdot h}\, '\overset{j}{w}_{[\mu'}\, '\overset{i}{w}_{\lambda']}\,; \end{cases}$$

(10.16 b) $\mathfrak{C}$
$$\begin{cases} \alpha) & [d\overset{h}{w}] = u_{j\,i}^{\cdot\cdot h}\, [\overset{j}{w}\, \overset{i}{w}]\,; \\ \beta) & [d\, '\overset{h}{w}] = 'u_{j\,i}^{\cdot\cdot h}\, ['\overset{j}{w}\, '\overset{i}{w}]\,; \end{cases}$$

where $u_{j\,i}^{\cdot\cdot h}$ and $'u_{j\,i}^{\cdot\cdot h}$ are known functions of $\xi^{\varkappa}$ and $\xi^{\varkappa'}$ respectively. From (10.15) and (10.16) there follows as a first condition the set of non-differential equations between $\xi^{\varkappa}$ and $\xi^{\varkappa'}$

(10.17)
$$u_{j\,i}^{\cdot\cdot h}(\xi^{\varkappa}) = 'u_{j\,i}^{\cdot\cdot h}(\xi^{\varkappa'})\,.$$

If these equations are inconsistent, no transformation $\xi^{\varkappa} \rightarrow \xi^{\varkappa'}$ transforming the fields $\overset{h}{w}$ into the fields $'\overset{h}{w}$ is possible. If they are consistent we get by differentiation (defining the operator ∂_j in both X_n's by $\partial_{\mu} = \overset{j}{w}_{\mu}\, \partial_j$; $\partial_{\mu'} = '\overset{j}{w}_{\mu'}\, \partial_j$):

(10.18 a)
$$\begin{cases} \alpha) & \partial_{\mu}\, u_{j\,i}^{\cdot\cdot h} = \overset{k}{w}_{\mu}\, \partial_k\, u_{j\,i}^{\cdot\cdot h}\,; \\ \beta) & \partial_{\mu'}\, 'u_{j\,i}^{\cdot\cdot h} = '\overset{k}{w}_{\mu'}\, \partial_k\, 'u_{j\,i}^{\cdot\cdot h}\,; \end{cases}$$

(10.18 b) $\mathfrak{C}$
$$\begin{cases} \alpha) & d u_{j\,i}^{\cdot\cdot h} = \overset{k}{w}\, \partial_k u_{j\,i}^{\cdot\cdot h}\,; \\ \beta) & d\, 'u_{j\,i}^{\cdot\cdot h} = '\overset{k}{w}\, \partial_k\, 'u_{j\,i}^{\cdot\cdot h}\,; \end{cases}$$

and from this we get as a second set of non-differential equations between $\xi^{\varkappa}$ and $\xi^{\varkappa'}$

(10.19)
$$\partial_k u_{j\,i}^{\cdot\cdot h} = \partial_k\, 'u_{j\,i}^{\cdot\cdot h}.$$

In the same way we can derive

(10.20)
$$\partial_l\, \partial_k u_{j\,i}^{\cdot\cdot h} = \partial_l\, \partial_k\, 'u_{j\,i}^{\cdot\cdot h}$$

and so on till we get inconsistent equations or a set depending on the former sets. In this latter case all following sets depend on these former sets and we have got a finite number of equations from which the $\xi^{\varkappa'}$ can be solved as functions of the $\xi^{\varkappa}$ and a finite number of parameters. These functions give the transformation $\xi^{\varkappa} \rightarrow \xi^{\varkappa'}$ required.

[1]) CARTAN 1946, 1, p. 315ff.

The result just obtained can be applied to the question of whether there exists a transformation $\xi^\varkappa \to \xi^{\varkappa'}$ mapping an L_n with coordinates $\xi^\varkappa$ and a connexion $\Gamma^\varkappa_{\mu\lambda}$ on another L_n, called hereafter $'L_n$ with coordinates $\xi^{\varkappa'}$ and a connexion $'\Gamma^{\varkappa'}_{\mu'\lambda'}$.[1]) In L_n we take an arbitrary anholonomic coordinate system (h) and in $'L_n$ an arbitrary anholonomic coordinate system (h). Then at each point of L_n the A^h_λ depend on r $(=n^2)$ parameters η^α, $\alpha = \dot{1}, \ldots, \dot{r}$ and at each point of $'L_n$ the $A^h_{\lambda'}$ depend on r parameters $\eta^{\alpha'}$, $\alpha' = \dot{1}', \ldots, \dot{r}'$.[2]) Then we have to investigate whether it is possible to choose the systems (h) in L_n and $'L_n$ such that $\Gamma^h_{ji} = {'\Gamma^h_{ji}}$ at corresponding points. We remark that, though $\Gamma^\varkappa_{\mu\lambda}$, $S_{\mu\lambda}^{\cdot\cdot\varkappa}$ and $R_{\nu\mu\lambda}^{\cdot\cdot\cdot\varkappa}$ are independent of the η^α, the components Γ^h_{ji}, $S_{ji}^{\cdot\cdot h}$ and $R_{kji}^{\cdot\cdot\cdot h}$ also depend on these parameters, and that the same holds mutatis mutandis in $'L_n$. In L_n the equations (10.4) and (10.5) are valid but A^h_λ, $\Gamma^h_{\lambda i}$, $S_{\mu\lambda}^{\cdot\cdot h}$ and $R_{\nu\mu i}^{\cdot\cdot\cdot h}$ now depend not only on the $\xi^\varkappa$ but also on the η^α. That means that we are now working in the X_{n+r} of the variables $\xi^\varkappa$, η^α from which the L_n can be formed by reduction (cf. II § 4). If a field $v^\varkappa$ or w_λ in L_n depends on the parameters η^α, it may be considered as the reduction of a field $v^\varkappa$, v^α = undetermined or as the section of a field w_λ, $w_\beta = 0$ in the X_{n+r}. Hence the covariant multivectors A^h_λ, $S_{\mu\lambda}^{\cdot\cdot h}$ and $R_{\nu\mu i}^{\cdot\cdot\cdot h}$ may be considered as the sections of the multivectors A^h_B, $S_{CB}^{\cdot\cdot h}$ and $R_{DCi}^{\cdot\cdot\cdot h}$; $A, B, C, D = 1, \ldots, n, \dot{1}, \ldots, \dot{r}$ satisfying the conditions

$$(10.21)\quad \begin{cases} A^h_\beta = 0; \quad S_{\gamma\lambda}^{\cdot\cdot h} = -S_{\lambda\gamma}^{\cdot\cdot h} = 0; \quad S_{\gamma\beta}^{\cdot\cdot h} = 0; \quad R_{\delta\gamma i}^{\cdot\cdot\cdot h} = 0; \\ \qquad R_{\delta\mu i}^{\cdot\cdot\cdot h} = -R_{\mu\delta i}^{\cdot\cdot\cdot h} = 0; \qquad \alpha, \beta, \gamma, \delta = \dot{1}, \ldots, \dot{r}. \end{cases}$$

In the X_{n+r} of $\xi^\varkappa$, η^α the allowable coordinate transformations are $\xi^{\varkappa'} = f^{\varkappa'}(\xi^\varkappa)$; $\eta^{\alpha'} = \varphi^{\alpha'}(\eta^\alpha)$, hence

$$(10.22)\qquad A^{\varkappa'}_\beta = 0; \quad A^\varkappa_{\beta'} = 0; \quad A^{\alpha'}_\lambda = 0; \quad A^\alpha_{\lambda'} = 0.$$

Accordingly the components of some connexion Γ^A_{CB} in X_{n+r} transform as follows

$$(10.23)\qquad \begin{cases} \Gamma^{\varkappa'}_{\mu'\lambda'} = A^{\mu\,\lambda\,\varkappa'}_{\mu'\lambda'\varkappa}\, \Gamma^\varkappa_{\mu\lambda} + A^{\varkappa'}_\lambda\, \partial_{\mu'} A^\lambda_{\lambda'}; \\ \Gamma^{\alpha'}_{\mu'\lambda'} = A^{\mu\,\lambda\,\alpha'}_{\mu'\lambda'\alpha}\, \Gamma^\alpha_{\mu\lambda}; \\ \Gamma^{\varkappa'}_{\gamma'\lambda'} = A^{\gamma\,\lambda\,\varkappa'}_{\gamma'\lambda'\varkappa}\, \Gamma^\varkappa_{\gamma\lambda}; \quad \Gamma^{\varkappa'}_{\mu'\beta'} = A^{\mu\,\beta\,\varkappa'}_{\mu'\beta'\varkappa}\, \Gamma^\varkappa_{\mu\beta}, \\ \Gamma^{\varkappa'}_{\gamma'\beta'} = A^{\gamma\,\beta\,\varkappa'}_{\gamma'\beta'\varkappa}\, \Gamma^\varkappa_{\gamma\beta}; \\ \Gamma^{\alpha'}_{\mu'\beta'} = A^{\mu\,\beta\,\alpha'}_{\mu'\beta'\alpha}\, \Gamma^\alpha_{\mu\beta}; \quad \Gamma^{\alpha'}_{\gamma'\lambda'} = A^{\gamma\,\lambda\,\alpha'}_{\gamma'\lambda'\alpha}\, \Gamma^\alpha_{\gamma\lambda} \\ \Gamma^{\alpha'}_{\gamma'\beta'} = A^{\gamma\,\beta\,\alpha'}_{\gamma'\beta'\alpha}\, \Gamma^\alpha_{\gamma\beta} + A^{\alpha'}_\beta\, \partial_{\gamma'} A^\beta_{\beta'}. \end{cases}$$

[1]) Cartan has dealt with this problem for the case of two V_n's.

[2]) In the case of two V_n's considered by Cartan the local coordinate systems can be taken orthogonal and then $r = \binom{n}{2}$.

We see from (10.23) that the $\Gamma_{\mu\lambda}^{\varkappa}$ transform just like the parameters of a connexion in L_n and that of the other seven sets of indices six undergo a linear homogeneous transformation. Hence in X_{n+r} a connexion can be partially fixed by taking the $\Gamma_{\mu\lambda}^{\varkappa}$ equal to the $\Gamma_{\mu\lambda}^{\varkappa}$ of the connexion in L_n and

$$(10.24)\qquad \Gamma_{\mu\lambda}^{\alpha}=0;\quad \Gamma_{\gamma\lambda}^{\varkappa}=0;\quad \Gamma_{\mu\beta}^{\varkappa}=0;\quad \Gamma_{\gamma\beta}^{\varkappa}=0;\quad \Gamma_{\mu\beta}^{\alpha}=0;\quad \Gamma_{\gamma\lambda}^{\alpha}=0$$

and leaving the $\Gamma_{\gamma\beta}^{\alpha}$ undetermined. This agrees with the fact that there is no connexion given in the space of the parameters η^{α}.

Passing now to the anholonomic system (h, α) in X_{n+r} (the η^{α} remain unchanged) we get

$$(10.25)\begin{cases}\text{a)}\quad \Gamma_{ji}^{h}=A_{j\,i\,\varkappa}^{\mu\lambda h}\,\Gamma_{\mu\lambda}^{\varkappa}-A_{j\,i}^{\mu\lambda}\,\partial_{\mu}A_{\lambda}^{h};\\ \text{b)}\quad \Gamma_{\gamma i}^{h}=-A_{i}^{\lambda}\,\partial_{\gamma}A_{\lambda}^{h};\\ \text{c)}\quad \Gamma_{j\beta}^{h}=0;\quad \text{d)}\ \Gamma_{\gamma\beta}^{h}=0;\quad \text{e)}\ \Gamma_{ji}^{\alpha}=0;\quad \text{f)}\ \Gamma_{\gamma i}^{\alpha}=0;\quad \text{g)}\ \Gamma_{j\beta}^{\alpha}=0\end{cases}$$

and this means that the Γ_{ji}^{h} transform in the ordinary way, that the only new components are the $\Gamma_{\gamma i}^{h}$, and that the $\Gamma_{\gamma\beta}^{\alpha}$ remain undetermined.

Using (10.25) we get besides (10.4)

$$(10.26)\qquad\begin{cases}\text{a)}\qquad \partial_{[\gamma}A_{\lambda]}^{h}=A_{[\gamma}^{i}\,\Gamma_{\lambda]\,i}^{h}+S_{\gamma\lambda}^{\cdot\cdot\,h};\\ \text{b)}\quad 0=\partial_{[\gamma}A_{\beta]}^{h}=A_{[\gamma}^{i}\,\Gamma_{\beta]\,i}^{h}+S_{\gamma\beta}^{\cdot\cdot\,h};\end{cases}$$

or, taking (10.4) and (10.26) together

$$(10.27)\begin{cases}\text{a)}\qquad \partial_{[C}A_{B]}^{h}=A_{[C}^{i}\,\Gamma_{B]i}^{h}+S_{\dot{C}\dot{B}}^{\cdot\cdot\,h};\qquad B, C=1,\ldots,n,\dot{1},\ldots,\dot{r};\\ \text{b)}\ \mathfrak{C}\quad [d\,A^{h}]=[A^{i}\,\Gamma_{i}^{h}]+S^{h},\end{cases}$$

in accordance with CARTAN's general principle that an abridged equation remains in some way valid if the number of independent variables is extended by the introduction of parameters. According to this principle we should now have also next to (10.5)

$$(10.28)\qquad\begin{cases}\text{a)}\qquad \partial_{[D}\Gamma_{C]i}^{h}+\Gamma_{[D\,|l|}^{h}\,\Gamma_{C]i}^{l}=\tfrac{1}{2}R_{DCi}^{\cdot\cdot\cdot\,h};\\ \text{b)}\ \mathfrak{C}\qquad [d\,\Gamma_{i}^{h}]+[\Gamma_{l}^{h}\,\Gamma_{i}^{l}]=\tfrac{1}{2}R_{i}^{\cdot\,h}.\end{cases}$$

In fact the $\Gamma_{\mu\lambda}^{\varkappa}$ are independent of the parameters, hence according to (10.25) after some calculation

$$(10.29)\qquad 0=\partial_{[\delta}\Gamma_{\mu]\lambda}^{\varkappa}=2A_{\lambda}^{i}\,A_{h}^{\varkappa}\,(\partial_{[\delta}\Gamma_{\mu]i}^{h}+\Gamma_{[\delta\,|l|}^{h}\,\Gamma_{\mu]i}^{l})$$

and this proves the validity of (10.28a) for $D=\delta$, $C=\mu$, $R_{\delta\mu i}^{\cdot\cdot\cdot\,h}=0$. From (10.25b) we get

$$(10.30)\qquad \partial_{[\delta}\Gamma_{\gamma]i}^{h}=-(\partial_{[\delta}A_{|i|}^{\lambda})\,\partial_{\gamma]}A_{\lambda}^{h}=-\Gamma_{[\delta\,|l|}^{h}\,\Gamma_{\gamma]i}^{l}.$$

Hence (10.28a) also holds for $D=\delta$, $C=\gamma$, $R_{\delta\gamma i}^{\cdot\cdot\cdot h}=0$. Since the equations (10.4b) and (10.5b) are now valid in the more general sense of (10.27b) and (10.28b) it stands to reason that the same must hold for the second identity (10.7b) and BIANCHI's identity (10.9b), but a direct proof could be given.

In X_{n+r} we now have the $n+r$ vectors (or Pfaffians) A^h, Γ_i^h and their rotations given by (10.27, 28) and in the $'X_{n+r}$ we have in the same way vectors (or Pfaffians) $'A^h$, $'\Gamma_i^h$ whose rotations satisfy equations of the same form. The problem is how to find a transformation $\xi^\varkappa, \eta^\alpha \to \xi^{\varkappa'}, \eta^{\alpha'}$ such that $A^h = 'A^h$, $\Gamma_i^h = '\Gamma_i^h$. If we write (10.27) and (10.28) in the form

$$(10.31)\quad \begin{cases} \text{a)} & \partial_{[C} A_{B]}^h = A_{[C}^i \Gamma_{B]i}^h + S_{ji}^{\cdot\cdot h} A_{[C}^j A_{B]}^i \\ \text{b)} & \complement \quad [d A^h] = [A^i \Gamma_i^h] + S_{ji}^{\cdot\cdot h} [A^j A^i]; \end{cases}$$

$$(10.32)\quad \begin{cases} \text{a)} & \partial_{[D} \Gamma_{C]i}^h = -\Gamma_{[D|l|}^h \Gamma_{C]i}^l + \tfrac{1}{2} R_{kji}^{\cdot\cdot\cdot h} A_{[D}^k A_{C]}^j \\ \text{b)} & \complement \quad [d \Gamma_i^h] = -[\Gamma_l^h \Gamma_i^l] + \tfrac{1}{2} R_{kji}^{\cdot\cdot\cdot h} [A^k A^j] \end{cases}$$

we have got equations of the form (10.16α) and equations of the same form, corresponding to (10.16β) can be derived for $'X_{n+r}$. Comparison with (10.16) shows that many of the coefficients corresponding to the $u_{ji}^{\cdot\cdot h}$ and the $'u_{ji}^{\cdot\cdot h}$ vanish or are simple constants and that the remaining ones are the anholonomic components of S, R, $'S$ and $'R$.

Now the holonomic components of any tensor of X_n ($'X_n$) are independent of the η^α ($\eta^{\alpha'}$). Hence, for instance for a tensor $P_\lambda^{\cdot\varkappa}$ we have

$$(10.33)\quad \begin{cases} \text{a)} & d P_i^{\cdot h} = d\xi^\mu \partial_\mu P_i^{\cdot h} + d\eta^\gamma (\partial_\gamma A_i^\lambda) A_\varkappa^h P_\lambda^{\cdot\varkappa} + d\eta^\gamma A_i^\lambda (\partial_\gamma A_\varkappa^h) P_\lambda^{\cdot\varkappa} \\ & \quad = d\xi^\mu A_\mu^j \nabla_j P_i^{\cdot h} + d\xi^\mu \Gamma_{\mu i}^l P_l^{\cdot h} - d\xi^\mu \Gamma_{\mu l}^h P_i^{\cdot l} + \\ & \qquad + d\eta^\gamma \Gamma_{\gamma i}^l P_l^{\cdot h} - d\eta^\gamma \Gamma_{\gamma l}^h P_i^{\cdot l}; \\ \text{b)} & \complement \quad d P_i^{\cdot h} = A^j \nabla_j P_i^{\cdot h} + \Gamma_i^l P_l^{\cdot h} - \Gamma_l^h P_i^{\cdot l} \,{}^{1)}. \end{cases}$$

Applying this formula to $S_{ji}^{\cdot\cdot h}$, $R_{kji}^{\cdot\cdot\cdot h}$ and their covariant derivatives in L_n we get

$$(10.34)\quad \complement \quad d S_{ji}^{\cdot\cdot h} = A^k \nabla_k S_{ji}^{\cdot\cdot h} + \Gamma_j^l S_{li}^{\cdot\cdot h} + \Gamma_i^l S_{jl}^{\cdot\cdot h} - \Gamma_l^h S_{ji}^{\cdot\cdot l};$$

$$(10.35)\quad \complement \quad d\nabla_k S_{ji}^{\cdot\cdot h} = A^{k_1} \nabla_{k_1} \nabla_k S_{ji}^{\cdot\cdot h} + \Gamma_k^l \nabla_l S_{ji}^{\cdot\cdot h} + \Gamma_j^l \nabla_k S_{li}^{\cdot\cdot h} + \Gamma_i^l \nabla_k S_{jl}^{\cdot\cdot h} - \Gamma_l^h \nabla_k S_{ji}^{\cdot\cdot l}$$

$$\vdots$$

$$(10.36)\quad \complement \quad d R_{kji}^{\cdot\cdot\cdot h} = A^{k_1} \nabla_{k_1} R_{kji}^{\cdot\cdot\cdot h} + \Gamma_k^l R_{lji}^{\cdot\cdot\cdot h} + \Gamma_j^l R_{kli}^{\cdot\cdot\cdot h} + \Gamma_i^l R_{kjl}^{\cdot\cdot\cdot h} - \Gamma_l^h R_{kji}^{\cdot\cdot\cdot l}$$

$$\vdots$$

[1]) E. CARTAN 1946, 1, p. 320 introduced this formula for K_{kjih}, $\nabla_l K_{kjih}$ in V_n as "evident". This optimism we don't share but it has to be granted that the general formula (10.33) after having been proved rigorously gives excellent means to use the abridged calculus for the calculations.

Hence, dealing with the $n+r = n+n^2$ vector fields A^h_B, Γ^h_{Ci} in X_{n+r} in the same way as with the n vector fields w_λ in X_n and using (10.34–36) we get the n.a.s. conditions for the mapping of the L_n and $'L_n$ on each other in the form

$$(10.37)\qquad \begin{cases} \text{a)}\ S_{ji}^{\cdot\cdot h} = 'S_{ji}^{\cdot\cdot h}; \\ \text{b)}\ \nabla_k S_{ji}^{\cdot\cdot h} = '\nabla_k\, 'S_{ji}^{\cdot\cdot h} \\ \qquad\vdots \end{cases}$$

$$(10.38)\qquad \begin{cases} \text{a)}\ R_{kji}^{\cdot\cdot\cdot h} = 'R_{kji}^{\cdot\cdot\cdot h}; \\ \text{b)}\ \nabla_{k_1} R_{kji}^{\cdot\cdot\cdot h} = '\nabla_{k_1}\, 'R_{kji}^{\cdot\cdot\cdot h} \\ \qquad\vdots \end{cases}$$

in which all quantities on the left hand side are expressed as functions of $\xi^\varkappa, \eta^\alpha$ and all on the right hand side as functions of $\xi^{\varkappa'}, \eta^{\alpha'}$. Either after a finite number of steps we get inconsistent equations or after a certain numbers of steps all following steps lead to equations dependent on the equations already derived. Only in the latter case there may exist transformations $\xi^\varkappa, \eta^\alpha \to \xi^{\varkappa'}, \eta^{\alpha'}$ giving the desired mapping and they can be found by solving $\xi^{\varkappa'}, \eta^{\alpha'}$, as functions of $\xi^\varkappa, \eta^\alpha$ and some parameters, from the equations (10.37, 38).[1])

§ 11. Linear connexions depending on a non-symmetric fundamental tensor[2]).

In III § 3 we have seen that a fundamental tensor together with $S_{\mu\lambda}^{\cdot\cdot\varkappa}$ determines uniquely a linear connexion. In order to obtain a unified field theory EINSTEIN started 1945[3]) from a non-symmetric fundamental tensor.[4])

[1]) This example is very instructive, because it reveals at the same time the strength and the weakness of the abridged calculus. The strength lies in the possibility of reading one and the same formula in different aspects, the weakness in the necessity to go back often to non abridged calculus in order to derive the formulae indispensable for further progress or for the formulation of the results.

[2]) This section deals with a fundamental tensor whose symmetric part is not zero. There is another line of thought starting from a fundamental bivector. We mention here only LEE 1943, 1; 1945, 1; 1947, 1; CHERN and WANG 1947, 1; VRANCEANU 1949, 1; GUGGENHEIMER 1951, 1 (relations with the $\tilde{V}_{2m}$).

[3]) EINSTEIN 1945, 1.

[4]) Cf. EINSTEIN and STRAUS 1946, 1; KOSAMBI 1947, 1; 1949, 1; SCHRÖDINGER 1947, 1; 1948, 1; 2; 1950, 1; 1951, 2; EINSTEIN 1948, 1; 1950, 1; JORDAN 1948, 1; 2; 1950, 1; STRAUS 1949, 1; TONNELAT 1949, 1; 1950, 1; 2; 3; 4; 1951, 1; 2; 3; 4; UDESCHINI 1950, 1; 1951, 1; 2; 3; INGRAHAM 1950, 1; 1951, 1; 2; EISENHART 1951, 1; 1952, 1; HLAVATY 1952, 2; 3; 4; 6; 1953, 1; 2; 3; 4; 5; 1954, 1; LICHNEROWICZ 1953, 1. The idea of the deduction given here goes back to SCHRÖDINGER but its very simple form is due to NIJENHUIS (personal communication).

Let this tensor be denoted by $s_{\lambda\varkappa}$, the isomer of its inverse $\overset{-1}{s}{}^{\varkappa\lambda}$ (cf. I § 8) by $S^{\varkappa\lambda}$, hence $s_{\lambda\mu} S^{\varkappa\mu} = A_{\lambda}^{\varkappa}$, and let $\mathfrak{s}$ stand for Det $(s_{\lambda\varkappa})$. Then we introduce the symmetric and alternating part of $s_{\lambda\varkappa}$

$$(11.1) \qquad h_{\lambda\varkappa} \overset{\text{def}}{=} s_{(\lambda\varkappa)}; \qquad k_{\lambda\varkappa} \overset{\text{def}}{=} s_{[\lambda\varkappa]}$$

and use $h_{\lambda\varkappa}$, supposed to be of rank n, and its inverse $h^{\varkappa\lambda}$ for raising and lowering of indices. Let $\mathfrak{h}$ stand for Det $(h_{\lambda\varkappa})$. Of course generally $s^{\varkappa\lambda} \neq S^{\varkappa\lambda}$ and $\mathfrak{h} \neq \mathfrak{s}$.

Let there also be a connexion $\Gamma_{\mu\lambda}^{\varkappa}$ and let $S_{\mu\lambda}^{\cdot\cdot\varkappa}$ not necessarily be zero. Then we establish relations between $s_{\lambda\varkappa}$ and $\Gamma_{\mu\lambda}^{\varkappa}$ by means of a variational equation. *We require that the variation*

$$(11.2) \qquad \overset{v}{d} \int_{\tau_n} R_{\mu\lambda} \mathfrak{S}^{\mu\lambda} d\xi^1 \ldots d\xi^n; \qquad \mathfrak{s}^{\mu\lambda} \overset{\text{def}}{=} \mathfrak{S}^{\frac{1}{2}} S^{\mu\lambda}; \qquad R_{\mu\lambda} = R_{\nu\mu\lambda}^{\cdot\cdot\cdot\nu}$$

should vanish if $s_{\lambda\varkappa}$ *and* $\Gamma_{\mu\lambda}^{\varkappa}$ *are varied independently.* Hence

$$(11.3) \qquad \begin{cases} \text{a)} \int_{\tau_n} \mathfrak{S}^{\mu\lambda} \overset{v}{d} R_{\mu\lambda} d\xi^1 \ldots d\xi^n = 0; \\ \text{b)} \int_{\tau_n} R_{\mu\lambda} \overset{v}{d} \mathfrak{S}^{\mu\lambda} d\xi^1 \ldots d\xi^n = 0. \end{cases}$$

Now we have

$$(11.4) \qquad \begin{cases} \overset{v}{d} R_{\mu\lambda} = \partial_\nu \overset{v}{d} \Gamma_{\mu\lambda}^{\nu} - \partial_\mu \overset{v}{d} \Gamma_{\nu\lambda}^{\nu} + 2 (\overset{v}{d} \Gamma_{[\nu|\varrho|}^{\nu}) \Gamma_{\mu]\lambda}^{\varrho} + 2 \Gamma_{[\nu|\varrho|}^{\nu} \overset{v}{d} \Gamma_{\mu]\lambda}^{\varrho} \\ \qquad = \nabla_\nu \overset{v}{d} \Gamma_{\mu\lambda}^{\nu} - \nabla_\mu \overset{v}{d} \Gamma_{\nu\lambda}^{\nu} + 2 S_{\nu\mu}^{\cdot\cdot\varrho} \overset{v}{d} \Gamma_{\varrho\lambda}^{\nu}; \end{cases}$$

because $\overset{v}{d} \Gamma_{\mu\lambda}^{\varkappa}$ is a tensor. Hence

$$(11.5) \qquad \mathfrak{S}^{\mu\lambda} \overset{v}{d} R_{\mu\lambda} = \partial_\nu (\mathfrak{S}^{\mu\lambda} \overset{v}{d} \Gamma_{\mu\lambda}^{\nu} - \mathfrak{S}^{\nu\mu} \overset{v}{d} \Gamma_{\lambda\mu}^{\lambda}) - (\mathfrak{G}_{\cdot\cdot\varkappa}^{\mu\lambda} - A_{\varkappa}^{\mu} \mathfrak{G}_{\cdot\cdot\sigma}^{\sigma\lambda}) \overset{v}{d} \Gamma_{\mu\lambda}^{\varkappa};$$

where

$$(11.6) \qquad \begin{cases} \mathfrak{G}_{\cdot\cdot\varkappa}^{\mu\lambda} \overset{\text{def}}{=} (\partial_\varkappa \mathfrak{S}^{\mu\lambda} + \Gamma_{\varrho\varkappa}^{\mu} \mathfrak{S}^{\varrho\lambda} + \Gamma_{\varkappa\varrho}^{\lambda} \mathfrak{S}^{\mu\varrho} - \Gamma_{\varrho\varkappa}^{\varrho} \mathfrak{S}^{\mu\lambda}) + \\ \qquad\qquad + \dfrac{2}{n-1} S_{\varrho\nu}^{\cdot\cdot\varrho} \mathfrak{S}^{\nu\lambda} A_{\varkappa}^{\mu}. \end{cases}$$

The terms in brackets in the right hand side of (11.6) do not represent exactly the covariant derivative $\nabla_\varkappa \mathfrak{S}^{\mu\lambda}$ but they differ from this derivative only because of the fact that the term belonging to the first index of $\mathfrak{S}^{\mu\lambda}$ and the term owing its origin to the density-character of $\mathfrak{S}^{\mu\lambda}$ have $\Gamma_{\lambda\mu}^{\varkappa}$ used instead of $\Gamma_{\mu\lambda}^{\varkappa}$. Slightly generalizing a notation of EINSTEIN we therefore write

$$(11.7) \qquad \begin{cases} \nabla_\varkappa \mathfrak{S}^{\underset{-+}{\mu\lambda}} \overset{\text{def}}{=} \partial_\varkappa \mathfrak{S}^{\mu\lambda} + \Gamma_{\varrho\varkappa}^{\mu} \mathfrak{S}^{\varrho\lambda} + \Gamma_{\varkappa\varrho}^{\lambda} \mathfrak{S}^{\mu\varrho} - \Gamma_{\varrho\varkappa}^{\varrho} \mathfrak{S}^{\mu\lambda} \\ \qquad = \nabla_\varkappa \mathfrak{S}^{\mu\lambda} + 2 S_{\varrho\varkappa}^{\cdot\cdot\mu} \mathfrak{S}^{\varrho\lambda} - 2 S_{\varrho\varkappa}^{\cdot\cdot\varrho} \mathfrak{S}^{\mu\lambda}. \end{cases}$$

Making the usual condition that the variations of the $\Gamma_{\mu\lambda}^{\varkappa}$ vanish at the boundary, we get from (11.5)

$$\mathfrak{G}^{\mu\lambda}_{\cdot\cdot\varkappa} - A_{\varkappa}^{\mu}\,\mathfrak{G}^{\sigma\lambda}_{\cdot\cdot\sigma} = 0 \tag{11.8}$$

but this implies that $(1-n)\,\mathfrak{G}^{\sigma\lambda}_{\cdot\cdot\sigma} = 0$ and that (11.8) is consequently equivalent to

$$\mathfrak{G}^{\mu\lambda}_{\cdot\cdot\varkappa} = \nabla_{\varkappa}\,\mathfrak{S}^{\underline{\mu}\,\underline{\lambda}}_{-+} + \frac{2}{n-1}\,S_{\varrho\nu}^{\cdot\cdot\varrho}\,\mathfrak{S}^{\nu\lambda}A_{\varkappa}^{\mu} = 0. \tag{11.9}$$

If the $\mathfrak{S}^{\varkappa\lambda}$ are given there are n^3 algebraic equations for the $\Gamma_{\mu\lambda}^{\varkappa}$. But they cannot be independent because if $\overset{0}{\Gamma}{}_{\mu\lambda}^{\varkappa}$ is a solution $\overset{0}{\Gamma}{}_{\mu\lambda}^{\varkappa} + t_{\mu}A_{\lambda}^{\varkappa}$ is also a solution for every choice of t_{μ}. This is seen very easily if we put (11.9) into the form

$$\partial_{\varkappa}\,\mathfrak{S}^{\mu\lambda} = P_{\varkappa\cdot\varrho}^{\mu\lambda\tau\sigma}\,\Gamma_{\tau\sigma}^{\varrho} \tag{11.10}$$

where

$$P_{\varkappa\cdot\varrho}^{\mu\lambda\tau\sigma} \overset{\text{def}}{=} \mathfrak{S}^{\omega\lambda}\left(-A_{\varkappa\varrho\omega}^{\sigma\mu\tau} + \frac{2}{n-1}A_{\varkappa\ \ \varrho\omega}^{\mu[\sigma\tau]}\right) + \mathfrak{S}^{\mu\omega}\left(-A_{\varrho\varkappa\omega}^{\lambda\tau\sigma} + A_{\varrho\varkappa\omega}^{\tau\sigma\lambda}\right). \tag{11.11}$$

In fact $P_{\varkappa\cdot\sigma}^{\mu\lambda\tau\sigma} = 0$ and this means that the ${}^{\tau\sigma}_{\varrho}$-rank of P is $\leq n^3 - n$. But then the ${}^{\mu\lambda}_{\varkappa}$-rank is also $\leq n^3 - n$ and this is confirmed by the identity $P_{\mu\cdot\ \varrho}^{[\mu\lambda]\tau\sigma} = 0$. This implies that (11.10) can only have solutions $\Gamma_{\tau\sigma}^{\varrho}$ if

$$\boxed{\partial_{\mu}\,\mathfrak{S}^{[\mu\lambda]} = 0}\,. \tag{11.12}$$

Note that this is an *invariant* equation because $\mathfrak{S}^{\varkappa\lambda}$ is a bivector density of weight $+1$.

If a solution $\overset{0}{\Gamma}{}_{\mu\lambda}^{\varkappa}$ is known the vector t_{μ} mentioned above can be chosen in one and only one way so that for ${}^{*}\Gamma_{\mu\lambda}^{\varkappa} \overset{\text{def}}{=} \overset{0}{\Gamma}{}_{\mu\lambda}^{\varkappa} + t_{\mu}A_{\lambda}^{\varkappa}$ the equation (cf. 2.15)

$$\boxed{{}^{*}S_{\mu} = \frac{2}{n-1}\,{}^{*}S_{\mu\lambda}^{\cdot\cdot\lambda} = 0} \tag{11.13}$$

holds. If this identity is introduced as a *condition*, the equations (11.10) have one and only one solution ${}^{*}\Gamma_{\mu\lambda}^{\varkappa}$ if and only if the ${}^{\tau\sigma}_{\varrho}$-rank of P is exactly $n^3 - n$. It is an algebraic problem to find the n.a.s. conditions for this special value of the ${}^{\tau\sigma}_{\varrho}$-rank. But there is a special case where these conditions are not needed. Because of (11.13) we get for ${}^{*}\nabla$ instead of (11.9) the simpler equation

$${}^{*}\nabla_{\mu}\,\mathfrak{S}^{\underline{\varkappa}\,\underline{\lambda}}_{-+} \overset{\text{def}}{=} {}^{*}\nabla_{\mu}\,\mathfrak{S}^{\varkappa\lambda} - 2\,{}^{*}S_{\mu\varrho}^{\cdot\cdot\varkappa}\,\mathfrak{S}^{\varrho\lambda} = 0, \tag{11.14}$$

from which it is easily deduced that

$$*\nabla_\mu \mathfrak{s} = 0 \tag{11.15}$$

and consequently

$$\boxed{\text{a) } *\nabla_\mu \, S^{\underset{-+}{\varkappa\lambda}} = 0; \quad \text{b) } *\nabla_\mu \, s_{\underset{-+}{\lambda\varkappa}} = 0} \tag{11.16}$$

From (11.16b) it follows that

$$\left\{\begin{array}{ll} \text{a)} & \partial_\mu s_{\lambda\varkappa} - *\Gamma^\varrho_{\lambda\mu} s_{\varrho\varkappa} - *\Gamma^\varrho_{\mu\varkappa} s_{\lambda\varrho} = 0; \\ \text{b)} & \partial_\lambda s_{\varkappa\mu} - *\Gamma^\varrho_{\varkappa\lambda} s_{\varrho\mu} - *\Gamma^\varrho_{\lambda\mu} s_{\varkappa\varrho} = 0; \\ \text{c)} & -\partial_\varkappa s_{\mu\lambda} + *\Gamma^\varrho_{\mu\varkappa} s_{\varrho\lambda} + *\Gamma^\varrho_{\varkappa\lambda} s_{\mu\varrho} = 0; \end{array}\right. \tag{11.17}$$

For $s_{[\lambda\varkappa]} = 0$ this gives immediately

$$\text{a) } *\Gamma^\varkappa_{\mu\lambda} = \tfrac{1}{2} S^{\varkappa\varrho}(\partial_\mu s_{\varrho\lambda} + \partial_\lambda s_{\varrho\mu} - \partial_\varrho s_{\mu\lambda}) \qquad \text{b) } *S_{\mu\lambda}{}^{\varkappa} = 0. \tag{11.18}$$

But this proves that the $\overset{\tau\sigma}{\varrho}$-rank of P in (11.11) is exactly $n^3 - n$ if $s_{\lambda\varkappa}$ is symmetric. As this rank can never be $> n^3 - n$ it must be exactly $n^3 - n$ for small values of $s_{[\lambda\varkappa]}$.[1] Hence we see that the equations (11.13,16) determine the $*\Gamma^\varkappa_{\mu\lambda}$ uniquely if the fundamental tensor $s_{\lambda\varkappa}$ differs only little from a symmetric tensor.

If $*\Gamma^\varkappa_{\mu\lambda}$ is known every other solution of (11.9) satisfies the relations

$$\Gamma^\varkappa_{\mu\lambda} = *\Gamma^\varkappa_{\mu\lambda} + S_\mu A^\varkappa_\lambda \tag{11.19}$$

because $*S_\mu = 0$.

The equation (11.3b) leads to

$$\left\{\begin{aligned} R_{\mu\lambda} \overset{v}{d} \mathfrak{S}^{\mu\lambda} &= \mathfrak{s}^{\frac{1}{2}} R_{\mu\lambda} \overset{v}{d} S^{\mu\lambda} + \tfrac{1}{2} R_{\mu\lambda} S^{\mu\lambda} \mathfrak{s}^{-\frac{1}{2}} \frac{\partial \mathfrak{s}}{\partial s_{\varrho\sigma}} \overset{v}{d} s_{\varrho\sigma} \\ &= \mathfrak{s}^{\frac{1}{2}} R_{\mu\lambda} \overset{v}{d} S^{\mu\lambda} - \tfrac{1}{2} R_{\varrho\sigma} S^{\varrho\sigma} \mathfrak{s}^{\frac{1}{2}} s_{\mu\lambda} \overset{v}{d} S^{\mu\lambda} \\ &= \mathfrak{s}^{\frac{1}{2}} (R_{\mu\lambda} - \tfrac{1}{2} R s_{\mu\lambda}) \overset{v}{d} S^{\mu\lambda}; \qquad R = R_{\varrho\sigma} S^{\varrho\sigma}; \end{aligned}\right. \tag{11.20}$$

hence this part of the variation leads to

$$G_{\mu\lambda} \overset{\text{def}}{=} R_{\mu\lambda} - \tfrac{1}{2} R s_{\mu\lambda} = 0. \tag{11.21}$$

Here the general tensor $G_{\mu\lambda}$ plays the role of the symmetric tensor $G_{\mu\lambda}$ defined in (5.29) and it is identical with this latter quantity if $s_{[\lambda\varkappa]} = 0$ and $S_\lambda = 0$. For $n \neq 2$ (11.21) is equivalent to

$$R_{\mu\lambda} = 0 \tag{11.22}$$

[1] This remark was made by Mr. Schrödinger in a letter to Mr. Nijenhuis of 12th November 1951.

and because $R_{\mu\lambda} = {}^*R_{\mu\lambda} - 2\partial_{[\mu} S_{\lambda]}$, as follows from (11.19), also with

(11.23) $$\boxed{{}^*R_{\mu\lambda} - 2\partial_{[\mu} S_{\lambda]} = 0}\,.$$

Now we have for the $4+16+60=80$ unknowns S_λ, $s_{\lambda\varkappa}$ and ${}^*\Gamma^{\varkappa}_{\mu\lambda}$ the $4+60+16=80$ equations (11.12), (11.16b) and (11.23). This number is by four to much because the equations must be invariant for coordinate transformations and a coordinate transformation contains four arbitrary functions. Hence there must be four relations between the left hand sides of these equations and their derivatives, and these relations must hold independent of the vanishing or not vanishing of these left hand sides. They can be obtained by taking in (11.5) and (11.20) only those variations of $\Gamma^{\varkappa}_{\mu\lambda}$ and $S^{\varkappa\lambda}$ that arise from dragging along these fields over some arbitrary $v^\varkappa dt$. Because by dragging along a solution we get another solution, provided that $v^\varkappa$ and its derivatives vanish at the boundary. According to II § 10, III § 5 we get

(11.24) $$\frac{\overset{v}{d}}{dt}\Gamma^{\varkappa}_{\mu\lambda} = -\underset{v}{£}\,\Gamma^{\varkappa}_{\mu\lambda} = -\nabla_\mu \nabla_\lambda v^\varkappa + 2\nabla_\mu S_{\lambda\varrho}^{\cdot\cdot\varkappa} v^\varrho - v^\varrho R_{\varrho\mu\lambda}^{\cdot\cdot\cdot\varkappa}$$

(11.25) $$\frac{\overset{v}{d}}{dt} S^{\varkappa\lambda} = -\underset{v}{£}\, S^{\varkappa\lambda} = -v^\mu \partial_\mu S^{\varkappa\lambda} + S^{\mu\lambda}\partial_\mu v^\varkappa + S^{\varkappa\mu}\partial_\mu v^\lambda .$$

From these variations the four equations can be derived in the ordinary way. All divergences have to be reduced to zero by means of STOKES' theorem and therefore it is profitable to introduce ${}^*\nabla$ instead of ∇ because ${}^*\nabla_\mu S^{\mu\varkappa} = 0$ and ${}^*\nabla_\mu \mathfrak{p}^\mu = \partial_\mu \mathfrak{p}^\mu$ for any vector density $\mathfrak{p}^\varkappa$ of weight $+1$.

If the ${}^*\Gamma^{\varkappa}_{\mu\lambda}$ are solved from (11.16b) as functions of the $s_{\lambda\varkappa}$ and their derivatives these solutions can be substituted in (11.23). Then we get the following $4+10+4=18$ equations

(11.12) $$\partial_\mu \mathfrak{S}^{[\mu\lambda]} = 0$$

(11.26) $${}^*R_{(\mu\lambda)} = 0$$

(11.27) $$\partial_{[\nu}\, {}^*R_{\mu\lambda]} = 0$$

for the 16 unknowns $s_{\lambda\varkappa}$. There must be six relations. Two of them are trivial

(11.28) $$\partial_\lambda \partial_u \mathfrak{S}^{[\mu\lambda]} = 0$$

(11.29) $$\partial_{[\omega}\, \partial_\nu\, {}^*R_{\mu\lambda]} = 0$$

and the other four can be obtained in the way described above. In the symmetric case only the 10 equations (11.26) remain for the 10 unknowns $s_{\lambda\varkappa}$. In this symmetric case $R_{\mu\lambda} \overset{v}{d}\mathfrak{S}^{\mu\lambda}$ vanishes because the

$\Gamma^{\varkappa}_{\mu\lambda}$ are already considered as functions of the $s_{\mu\lambda}$. From (11.20, 25) we get now for $S_\lambda = 0$, neglecting terms that vanish after integration

$$\text{(11.30)}\quad \begin{cases} R_{\mu\lambda}\overset{v}{d}\mathfrak{S}^{\mu\lambda} = \mathfrak{s}^{\frac{1}{2}}\, G_{\mu\lambda}(S^{\varrho\lambda}\nabla_\varrho v^\mu + S^{\mu\varrho}\nabla_\varrho v^\lambda)\, dt \\ \qquad = -2(\nabla_\varrho\, \mathfrak{s}^{\frac{1}{2}}\, G_{\varkappa\lambda}\, S^{\varrho\lambda})\, v^\varkappa\, dt = -2(\nabla_\varrho\, \mathfrak{G}^{\varrho}_{\cdot\varkappa})\, v^\varkappa\, dt \end{cases}$$

from which

$$\text{(11.31)}\qquad \nabla_\mu\, \mathfrak{G}^{\mu\varkappa} = 0.$$

The condition (11.13) that ensures at least in the simplest cases the uniqueness of the connexion ${}^*\Gamma^{\varkappa}_{\mu\lambda}$, means geometrically that this connexion behaves as a symmetric one as far as scalar densities are concerned. But it has yet another meaning as was pointed out by EINSTEIN. If $s_{\lambda\varkappa}$ is changed into $s_{\varkappa\lambda}$, that is, if the alternating part of the fundamental tensor gets another sign while the symmetric part is unchanged, ${}^*\Gamma^{\varkappa}_{\mu\lambda}$ will change into ${}^*\Gamma^{\varkappa}_{\lambda\mu}$ as is seen quite easily from (11.17a). Now let this change induce a change of ${}^*R_{\mu\lambda}$ into ${}^*R'_{\mu\lambda}$. Then we find

$$\text{(11.32)}\quad \begin{cases} \text{a)}\ {}^*R_{(\mu\lambda)} - {}^*R'_{(\mu\lambda)} = -2\partial_{(\mu}\, {}^*S_{|\nu|\lambda)}^{\cdot\,\cdot\,\nu} + 2\,{}^*S_{\nu\varrho}^{\cdot\,\cdot\,\nu}\, {}^*\Gamma^{\varrho}_{(\mu\lambda)}; \\ \text{b)}\ {}^*R_{[\mu\lambda]} + {}^*R'_{[\mu\lambda]} = -2\partial_{[\mu}\, {}^*S_{|\nu|\lambda]}^{\cdot\,\cdot\,\nu} + 2\,{}^*S_{\nu\varrho}^{\cdot\,\cdot\,\nu}\, {}^*S_{\mu\lambda}^{\cdot\,\cdot\,\varrho}; \end{cases}$$

hence, ${}^*S_\lambda = 0$ is sufficient (but not necessary) for the tensor ${}^*R_{\mu\lambda}$ to change into ${}^*R_{\lambda\mu}$ if $s_{\lambda\varkappa}$ is changed into $s_{\varkappa\lambda}$. In the original form of the theory given by EINSTEIN in 1945, $s_{(\lambda\varkappa)}$ was taken real and $s_{[\lambda\varkappa]}$ imaginary. Then $s_{\lambda\varkappa}$ could be considered in a way as "hermitian" and the condition that ${}^*R_{\lambda\varkappa}$ also should be "hermitian" could be fulfilled by taking ${}^*S_\lambda = 0$. This condition however presents itself in a quite natural way if we wish that ${}^*R_{\lambda\varkappa}$ and $s_{\lambda\varkappa}$ should differ only by a scalar factor just as $K_{\lambda\varkappa}$ and $g_{\lambda\varkappa}$ do in the ordinary theory of relativity.

The connexion derived in this way can be used for $n = 4$ to build a unified field theory and the authors mentioned above have got some remarkable results in this line. But it should be remarked that in the end there is nothing but an X_4 with only two fields $s_{(\lambda\varkappa)}$ and $s_{[\lambda\varkappa]}$ and that according to the second reduction theorem (III § 7) the differential concomitants of these fields are ordinary concomitants of $s_{(\lambda\varkappa)}$, $s_{[\lambda\varkappa]}$, the curvature tensor of the *symmetric connexion belonging to* $s_{(\lambda\varkappa)}$ and the covariant derivatives of these quantities with respect to *this* connexion. So, from a mathematical point of view the new connexion derived from (11.10) can never produce anything that could not be expressed in terms of $s_{(\lambda\varkappa)}$, $s_{[\lambda\varkappa]}$ and the ordinary symmetric connexion belonging to $s_{(\lambda\varkappa)}$. It may be useful to introduce this new connexion in order to get some heuristic principles that may suggest for instance a choice of the Hamiltonian or a desirable form of the field equations. But nothing really new can ever arise from following this course. This remark is not valid if the introduction of a non-symmetric fundamental tensor is combined with a change of dimension. In projective relativity in five coordinates a non-symmetric fundamental tensor does not seem to have been used till yet but in conformal relativity in six coordinates INGRAHAM[1]) has recently tried to generalize SCHRÖDINGER's ideas.

By transvecting (11.17a) with $S^{\lambda\varkappa}$ we get

$$\partial_\mu \log \mathfrak{s} = S^{\lambda\varkappa}\,\partial_\mu\, s_{\lambda\varkappa} = {}^*\Gamma^\lambda_{\lambda\mu} + {}^*\Gamma^\lambda_{\mu\lambda}, \tag{11.33}$$

hence

$$\partial_{[\nu}\, {}^*\Gamma^\lambda_{\mu]\lambda} = \partial_{[\nu}\, {}^*S_{\mu]\lambda}^{\cdot\cdot\;\lambda} \tag{11.34}$$

and according to (4.15)

$$ {}^*V_{\nu\mu} = {}^*R_{\nu\mu\lambda}^{\cdot\cdot\cdot\;\lambda} = 2\,\partial_{[\nu}\, {}^*\Gamma_{\mu]} = 2\,\partial_{[\nu}\, {}^*S_{\mu]\lambda}^{\cdot\cdot\;\lambda}. \tag{11.35}$$

This proves that the condition (11.13) implies that ${}^*V_{\nu\mu}$ vanishes, that is, that for the connexion ${}^*\Gamma^\varkappa_{\mu\lambda}$ a covariant constant scalar density field is possible (cf. III § 6).

From (5.6) we get according to the condition (11.13)

$$ {}^*R_{[\mu\lambda]} = {}^*\nabla_\sigma\, {}^*S_{\mu\lambda}^{\cdot\cdot\;\sigma} \tag{11.36}$$

and it is easy to show that this is in accordance with (11.32b).

IV. LIE groups and linear connexions.

§ 1. Finite continuous groups[2]).

An r-parameter finite continuous group in the sense of LIE is a set of elements between which a "multiplication" is defined and that satisfies the following conditions:

a) The elements are in one to one correspondence with the points of an $\mathfrak{N}(\eta^\alpha)$ in an X_r with the coordinates η^α; $\alpha = \dot{1}, \ldots, \dot{r}$.

b) If the element T_η belongs to η^α and T_ζ to ζ^α, the product $T_\theta = T_\eta T_\zeta$ belongs to the set and the θ^α are analytic functions[3]) of the η^α and ζ^α.

c) There is an element I called "unity" corresponding to $\underset{0}{\eta}^\alpha$ such that $I\,T_\eta = T_\eta\, I = T_\eta$ for every choice of T_η.

---

[1]) L. c. p. 179 footnote 4.

[2]) General references also for literature: LIE and ENGEL 1888, 1; BIANCHI 1918, 1; EISENHART 1933, 1; VRANCEANU 1947, 1. According to the general character of this book we only consider the "group germ", that means we only are interested in local properties. There is a vast literature on properties in the large. We mention here only also for literature CARTAN 1927, 2; 1930, 2; 1936, 1; MAYER and T. Y. THOMAS 1935, 1; CHEVALLEY 1946, 1.

[3]) See footnote 1 on page 186.

d) To every T_η there belongs an element T_η^{-1} such that $T_\eta T_\eta^{-1} = T_\eta^{-1} T_\eta = I$. Its coordinates are analytic functions[1]) of the η^α.

e) $(T_\eta T_\zeta) T_\theta = T_\eta (T_\zeta T_\theta)$.

The elements are often transformations in n variables but this is not necessary. From the definition we see that only elements of the group are considered in a neighbourhood of unity. This neighbourhood is called the *group germ* (Gruppenkeim).

We identify each element with its corresponding point in X_r. This X_r is called *group space*.

To every pair of elements S, T there belong two elements $T S^{-1}$ and $S^{-1} T$. Two pairs S, T and S_1, T_1 are called

$$(+)\text{-}\textit{equipollent} \text{ if } T S^{-1} = T_1 S_1^{-1}$$
$$(-)\text{-}\textit{equipollent} \text{ if } S^{-1} T = S_1^{-1} T_1.$$

From this we get the following deductions:

1. If S_1, T_1; S_2, T_2 are $(\pm)$-equipollent, three of these elements determine the fourth uniquely.

2. If S_1, T_1; S_2, T_2 and S_2, T_2; S_3, T_3 are both $(\pm)$-equipollent, the same holds for S_1, T_1; S_3, T_3 (Fig. 12).

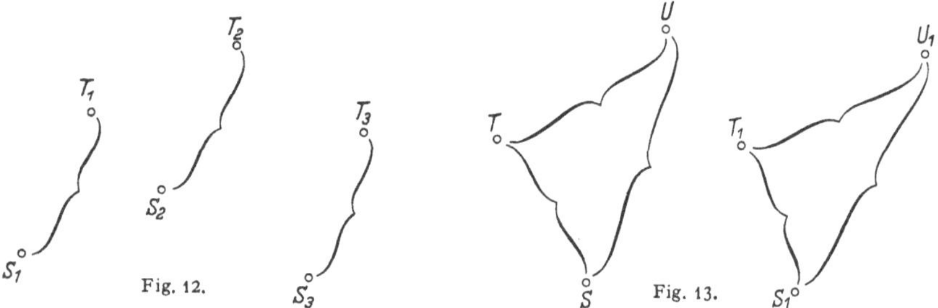

Fig. 12. Fig. 13.

3. If S, T; S_1, T_1 and T, U; T_1, U_1 are $(\pm)$-equipollent, the same holds for S, U; S_1, U_1 (Fig. 13).

4. If S_1, T_1; S_2, T_2 are $(\pm)$-equipollent, then S_1, S_2; T_1, T_2 are $(\mp)$-equipollent (Fig. 14).

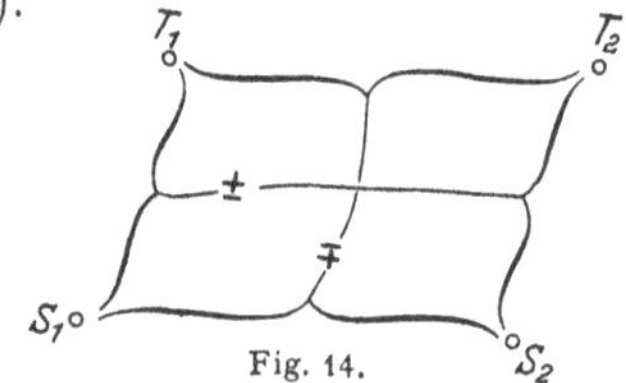

Fig. 14.

[1]) Though in the following all functions will be supposed to be analytic, many theorems remain valid if only it is assumed that all functions are of class u.

5. An $\mathfrak{R}(T_1)$ can always be mapped continuously on an $\mathfrak{R}(T_2)$ such that T_2 is the image of T_1 and all pointpairs of $\mathfrak{R}(T_1)$ are mapped $(\pm)$-equipollent on the pointpairs of $\mathfrak{R}(T_2)$.

6. If $\mathfrak{R}(T_1)$ is mapped in this way on $\mathfrak{R}(T_2)$ and $\mathfrak{R}(T_2)$ in the same way on $\mathfrak{R}(T_3)$ the result is the same as if $\mathfrak{R}(T_1)$ were mapped directly on $\mathfrak{R}(T_3)$.

If a pointpair $T_{\underset{0}{\eta}}, T_{\underset{0}{\eta}+d\underset{0}{\eta}}$ is given, there exists for every choice of T_η in an $\mathfrak{R}(\underset{0}{\eta}^\alpha)$ a $(\pm)$-equipollent pointpair $T_\eta, T_{\eta+d\eta}$. Hence, using for instance the $(+)$-equipollence we get for every choice of an infinitesimal vector $v^\alpha dt$ at $\underset{0}{\eta}^\alpha$ a vector field $v^\alpha dt$ in $\mathfrak{R}(\underset{0}{\eta}^\alpha)$. That means that the $(+)$-equipollence fixes a parallel displacement (cf. III § 2) for contravariant vectors in $\mathfrak{R}(\underset{0}{\eta}^\alpha)$. According to the conditions 1, 2, 3, 5 and 6 the parameters of this displacement establish a connexion that is linear in the sense defined in III § 2. Let $\underset{b}{e}^\alpha, \overset{a}{e}_\beta;\ a, b = 1, \ldots, \mathrm{r}$[1]) be fields of basis vectors covariant constant for this connexion or shortly $(+)$-*constant*. Let the covariant differentiation symbol of the connexion be $\overset{+}{\nabla}$ and the parameters $\overset{+}{\Gamma}{}^\alpha_{\gamma\beta}$, then we have $\overset{+}{\nabla}_\gamma \underset{b}{e}^\alpha = 0$, hence

$$\overset{+}{\Gamma}{}^\alpha_{\gamma\beta} = -\overset{b}{e}_\beta\, \partial_\gamma \underset{b}{e}^\alpha = +\underset{b}{e}^\alpha\, \partial_\gamma \overset{b}{e}_\beta . \tag{1.1}$$

Let this connexion be called the $(+)$-*connexion*. The $(-)$-equipollence leads in the same way to the $(-)$-*connexion* with the symbol $\overset{-}{\nabla}$ and the parameters $\overset{-}{\Gamma}{}^\alpha_{\gamma\beta}$

$$\overset{-}{\Gamma}{}^\alpha_{\gamma\beta} = -\overset{B}{e}_\beta\, \partial_\gamma \underset{B}{e}^\alpha = +\underset{B}{e}^\alpha\, \partial_\gamma \overset{B}{e}_\beta \tag{1.2}$$

where $\underset{B}{e}^\alpha, \overset{A}{e}_\beta;\ A, B = \dot{1}, \ldots, \dot{\mathrm{r}}$ are fields of basis vectors covariant constant for this connexion or shortly $(-)$-*constant*.[2])

Vectors at different points that can be transformed into each other by $(\pm)$-equipollence are said to be $(\pm)$-*parallel*.

As both connexions are integrable (cf. III §4 and deduction 2) their curvature tensors vanish

$$\overset{\pm}{R}{}_{\delta\gamma\beta}{}^{\cdot\cdot\cdot\alpha} \overset{\text{def}}{=} 2\partial_{[\delta} \overset{\pm}{\Gamma}{}^\alpha_{\gamma]\beta} + 2\overset{\pm}{\Gamma}{}^\alpha_{[\delta|\varepsilon|} \overset{\pm}{\Gamma}{}^\varepsilon_{\gamma]\beta} = 0 \tag{1.3}$$

[1]) Or $\dot{1}, \ldots, \dot{\mathrm{r}}$ if the indices 1, 2, ... are already used for another purpose, cf. for instance VII § 4 and 5.

[2]) These connexions were introduced by CARTAN and SCHOUTEN 1926, 3; 4, cf. CARTAN 1927, 1; SCHOUTEN 1927, 1; 1929, 4.

but they are in general not symmetric (cf. III, p. 142, footnote 3)

$$\overset{\pm}{S}_{\gamma\beta}^{\cdot\cdot\alpha} \overset{\text{def}}{=} \overset{\pm}{\Gamma}_{[\gamma\beta]}^{\alpha} \neq 0. \tag{1.4}$$

If a curve contains a pointpair R, S and the first point of a $(+)$-equipollent pointpair T, U, it may happen that it also contains U. If this is the case for every choice of R, S and T the curve is said to be *geodesic*. Now R, T and S, U are $(-)$-equipollent pointpairs, hence, if we use in the definition $(-)$-equipollence instead of $(+)$-equipollence, we get the same geodesics. If a geodesic contains T and VT it also contains V^2T, V^3T, etc. Taking V infinitesimal different from I we see that a geodesic as defined here is identical with a geodesic of the $(+)$-connexion and that for every choice of V the points T, VT; VT, V^2T etc. have the same distance if measured by means of an affine parameter on the geodesic (cf. III § 7). The same reasoning holds for the geodesics of the $(-)$-connexion. Moreover, if $VT = TW$ it follows that $V^2T = TW^2$ etc. Hence the $(+)$- and $(-)$-geodesics are not only identical but the affine parameter on them is also the same for both connexions. Using this parameter, the equation of the geodesic must have the form

$$\frac{d^2\eta^\alpha}{dt^2} + \overset{\pm}{\Gamma}_{\gamma\beta}^{\alpha} \frac{d\eta^\gamma}{dt} \frac{d\eta^\beta}{dt} = 0 \tag{1.5}$$

and from this it follows that

$$\Gamma_{\gamma\beta}^{\alpha} \overset{\text{def}}{=} \overset{+}{\Gamma}_{(\gamma\beta)}^{\alpha} = \overset{-}{\Gamma}_{(\gamma\beta)}^{\alpha} \tag{1.6}$$

are the parameters of a symmetric connexion that can be obtained by symmetrizing the $(+)$-connexion and the $(-)$-connexion. This *(0)-connexion* is in general not integrable (cf. III § 4) and its curvature tensor

$$R_{\delta\gamma\beta}^{\cdot\cdot\cdot\alpha} \overset{\text{def}}{=} 2\partial_{[\delta}\Gamma_{\gamma]\beta}^{\alpha} + 2\Gamma_{[\delta|\varepsilon|}^{\alpha}\Gamma_{\gamma]\beta}^{\varepsilon} \tag{1.7}$$

plays an important role.

Besides the operators $\overset{+}{\nabla}$, $\overset{-}{\nabla}$ and ∇ we also need the LIE derivative (cf. II § 10) with respect to the fields $\underset{b}{e}^\alpha$ and $\underset{B}{e}^\alpha$, for instance

$$\underset{b}{£}\, P_{\cdot\beta}^{\alpha} \overset{\text{def}}{=} \underset{b}{e}^\gamma \partial_\gamma P_{\cdot\beta}^{\alpha} - P_{\cdot\beta}^{\gamma} \partial_\gamma \underset{b}{e}^\alpha + P_{\cdot\gamma}^{\alpha} \partial_\beta \underset{b}{e}^\gamma. \tag{1.8}$$

The LIE derivative $\underset{v}{£}\, P_{\cdot\beta}^{\alpha}$ with respect to the $(+)$-constant field $v^\alpha = v^a A_a^\alpha$ with *constant* v^a is then given by

$$\underset{v}{£}\, P_{\cdot\beta}^{\alpha} = \overset{a}{v}\, \underset{a}{£}\, P_{\cdot\beta}^{\alpha}. \tag{1.9}$$

If in deduction 4 for $S_1, T_1; S_2, T_2$ we take elements that differ from J only in infinitesimals as illustrated in Fig. 15, the quadrangle is built up by means of the vector fields $\underset{b}{e}^{\alpha}\,dt$ and $\underset{B}{e}^{\alpha}\,dt'$. This means that the

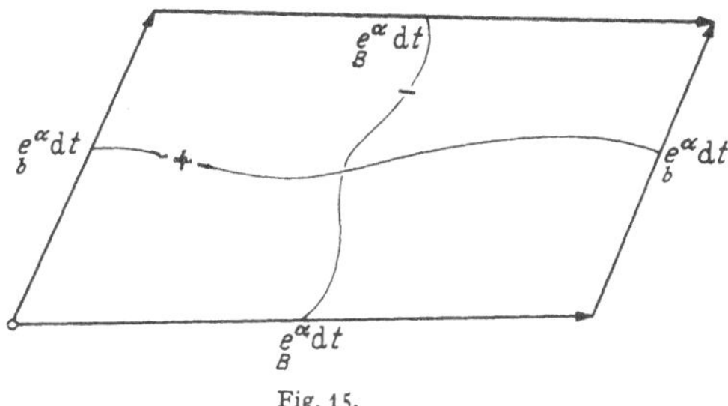

Fig. 15.

field $\underset{b}{e}^{\alpha}$ is absolutely invariant (cf. II § 10) with respect to the field $\underset{B}{e}^{\alpha}$ and vice versa for every choice of B and b:

$$\text{(1.10)}\qquad \text{a)}\ \underset{b}{\pounds}\,\underset{B}{e}^{\alpha} = \underset{b}{e}^{\gamma}\,\partial_{\gamma}\,\underset{B}{e}^{\alpha} - \underset{B}{e}^{\gamma}\,\partial_{\gamma}\,\underset{b}{e}^{\alpha} = 0; \qquad \text{b)}\ \underset{B}{\pounds}\,\underset{b}{e}^{\alpha} = 0.$$

From this it follows that

$$\text{(1.11)}\qquad 0 = \underset{b}{e}^{\gamma}\,\overset{-}{\nabla}_{\gamma}\,\underset{B}{e}^{\alpha} - \underset{B}{e}^{\gamma}\,\overset{+}{\nabla}_{\gamma}\,\underset{b}{e}^{\alpha} = \underset{b}{\pounds}\,\underset{B}{e}^{\alpha} + \left(\overset{-}{\Gamma}{}^{\alpha}_{\gamma\beta} - \overset{+}{\Gamma}{}^{\alpha}_{\beta\gamma}\right)\underset{b}{e}^{\gamma}\,\underset{B}{e}^{\beta};$$

hence

$$\text{(1.12)}\qquad \overset{+}{\Gamma}{}^{\alpha}_{\gamma\beta} = \overset{-}{\Gamma}{}^{\alpha}_{\beta\gamma}.$$

The symmetric part of (1.12) leads back to (1.6) and the alternating part is

$$\text{(1.13)}\qquad \overset{+}{S}_{\gamma\beta}{}^{\cdot\cdot\alpha} = -\overset{-}{S}_{\gamma\beta}{}^{\cdot\cdot\alpha}$$

hence, for every quantity Φ (indices suppressed)

$$\text{(1.14)}\qquad 2\nabla_{\gamma}\Phi = \overset{|}{\nabla}_{\gamma}\Phi + \overset{-}{\nabla}_{\gamma}\Phi.$$

It follows from (1.1, 2, 10, 12) that for any fields v^{α}, w_{β}

$$\text{(1.15)}\qquad \text{a)}\ \begin{cases} \underset{b}{\pounds}\,v^{\alpha} = \underset{b}{e}^{\gamma}\,\overset{-}{\nabla}_{\gamma}\,v^{\alpha}; \\ \underset{b}{\pounds}\,w_{\beta} = \underset{b}{e}^{\gamma}\,\overset{-}{\nabla}_{\gamma}\,w_{\beta}; \end{cases} \qquad \text{b)}\ \begin{cases} \underset{B}{\pounds}\,v^{\alpha} = \underset{B}{e}^{\gamma}\,\overset{+}{\nabla}_{\gamma}\,v^{\alpha}; \\ \underset{B}{\pounds}\,w_{\beta} = \underset{B}{e}^{\gamma}\,\overset{+}{\nabla}_{\gamma}\,w_{\beta}; \end{cases}$$

and this implies that for every quantity Φ (indices suppressed)

$$\text{(1.16)}\qquad \text{a)}\ \underset{b}{\pounds}\,\Phi = \underset{b}{e}^{\gamma}\,\overset{-}{\nabla}_{\gamma}\Phi; \qquad \text{b)}\ \underset{B}{\pounds}\,\Phi = \underset{B}{e}^{\gamma}\,\overset{+}{\nabla}_{\gamma}\Phi$$

expressing the fact that dragging along over a $(\pm)$-constant infinitesimal field is the same as a $(\mp)$-displacement over the same field. Note that this is just what we see in Fig. 15 for an infinitesimal contravariant vector.

Because the fields $\overset{a}{e}_\beta$ are (+)-constant, we have

(1.17) $$0 = \overset{+}{\nabla}_\gamma \overset{a}{e}_\beta = \partial_\gamma \overset{a}{e}_\beta - \overset{+}{\Gamma}{}^\alpha_{\gamma\beta} \overset{a}{e}_\alpha;$$

hence

(1.18) $$\boxed{\partial_{[\gamma} \overset{a}{e}_{\beta]} = -\tfrac{1}{2} c_{\gamma\beta}^{\cdot\cdot\alpha} \overset{a}{e}_\alpha}; \qquad c_{\gamma\beta}^{\cdot\cdot\alpha} \overset{\text{def}}{=} -2\overset{+}{S}{}_{\gamma\beta}^{\cdot\cdot\alpha};$$

or, introducing n^3 *scalar* fields $c^a_{cb} \underset{\text{def}}{\overset{*}{=}} c_{cb}^{\cdot\cdot a}$:

(1.19) $$\boxed{\partial_{[\gamma} \overset{a}{e}_{\beta]} = -\tfrac{1}{2} c^a_{cb} \overset{c}{e}_\gamma \overset{b}{e}_\beta}.$$

Now since the field $\overset{a}{e}_\beta$ is invariant for dragging along over $\underset{B}{e}{}^\alpha dt$, the rotation $\partial_{[\gamma} \overset{a}{e}_{\beta]}$ must have the same property. Hence, according to (1.16b)[1])

(1.20) $$0 = \underset{B}{£}\, c^a_{cb} \overset{c}{e}_\gamma \overset{b}{e}_\beta = \big(\underset{B}{£}\, c^a_{cb}\big) \overset{c}{e}_\gamma \overset{b}{e}_\beta = \big(\underset{B}{e}{}^\delta \partial_\delta c^a_{cb}\big) \overset{c}{e}_\gamma \overset{b}{e}_\beta$$

for every value of b, c and B. This is only possible if

(1.21) $$\boxed{c^a_{cb} = \text{const}}.$$

Since the fields $\underset{b}{e}{}^\alpha$, $\overset{a}{e}_\beta$ are (+)-constant it follows from (1.21) that

(1.22) $$\overset{+}{\nabla}_\delta \overset{+}{S}{}_{\gamma\beta}^{\cdot\cdot\alpha} = -\tfrac{1}{2} \overset{+}{\nabla}_\delta c_{\gamma\beta}^{\cdot\cdot\alpha} = -\tfrac{1}{2} \overset{+}{\nabla}_\delta c^a_{cb} \overset{c}{e}_\gamma \overset{b}{e}_\beta \underset{a}{e}{}^\alpha = 0.$$

In the same way we get

(1.23) $$\boxed{\partial_{[\gamma} \overset{A}{e}_{\beta]} = +\tfrac{1}{2} c_{\gamma\beta}^{\cdot\cdot\alpha} \overset{A}{e}_\alpha}; \qquad c_{\gamma\beta}^{\cdot\cdot\alpha} \overset{\text{def}}{=} +2\overset{-}{S}{}_{\gamma\beta}^{\cdot\cdot\alpha} = -2\overset{+}{S}{}_{\gamma\beta}^{\cdot\cdot\alpha}.$$

(1.24) $$\boxed{c^A_{CB} = \text{const.}}; \qquad c^A_{CB} \underset{\text{def}}{\overset{*}{=}} c_{CB}^{\cdot\cdot A}.$$

For convenience we always choose the basis vectors $\underset{b}{e}{}^\alpha$ and $\underset{B}{e}{}^\alpha$ in such a way that they are identical at $\underset{0}{\eta}{}^\alpha$: $\underset{B}{e}{}^\alpha = \delta^b_B \underset{b}{e}{}^\alpha$. Then at *all* points of X_r we have

(1.25) $$c^A_{CB} \overset{*}{=} \delta^{A\,c\,b}_{a\,C\,B}\, c^a_{cb}.$$

The $\overset{+}{\Gamma}{}^a_{cb}$ and $\overset{-}{\Gamma}{}^A_{CB}$ are all zero because the fields $\underset{b}{e}{}^\alpha$ and $\underset{B}{e}{}^\alpha$ are (+)- and (—)-constant respectively. Hence it follows from (1.21) and also from (1.24) that $c_{\gamma\beta}^{\cdot\cdot\alpha}$ is (+)- and (—)-constant and (*0*)-constant

(1.26) a) $\overset{\pm}{\nabla}_\delta c_{\gamma\beta}^{\cdot\cdot\alpha} = -2\overset{\pm}{\nabla}_\delta \overset{+}{S}{}_{\gamma\beta}^{\cdot\cdot\alpha} = 2\overset{\pm}{\nabla}_\delta \overset{-}{S}{}_{\gamma\beta}^{\cdot\cdot\alpha} = 0$; b) $\overset{0}{\nabla}_\delta c_{\gamma\beta}^{\cdot\cdot\alpha} = 0$.

[1]) The difference between the n^3 scalars c^a_{cb} and the tensor $c_{cb}^{\cdot\cdot a}$ is important here.

From this and (1.3) and the second identity for $\overset{+}{R}_{\delta\gamma\beta}^{\cdot\cdot\cdot\alpha}$ (cf. III 5.2) we get

$$(1.27)\qquad 0=\overset{+}{R}_{[\delta\gamma\beta]}^{\cdot\cdot\cdot\alpha}=-\overset{+}{\nabla}_{[\delta}c_{\gamma\beta]}^{\cdot\cdot\alpha}-c_{[\delta\gamma}^{\cdot\cdot\varepsilon}c_{\beta]\varepsilon}^{\cdot\cdot\alpha}=-c_{[\delta\gamma}^{\cdot\cdot\varepsilon}c_{\beta]\varepsilon}^{\cdot\cdot\alpha};$$

hence

$$(1.28)\qquad \text{a)}\ c_{[\delta\gamma}^{\cdot\cdot\varepsilon}c_{\beta]\varepsilon}^{\cdot\cdot\alpha}=0;\qquad \text{b)}\ c_{[dc}^{\cdot\cdot e}c_{b]e}^{\cdot\cdot a}=0;\qquad \text{c)}\ c_{[DC}^{\cdot\cdot E}c_{B]E}^{\cdot\cdot A}=0.$$

By (1.26) and (1.28) we are now able to compute the curvature tensor $R_{\delta\gamma\beta}^{\cdot\cdot\cdot\alpha}$ of the symmetric connexion $\Gamma_{\gamma\beta}^{\alpha}$ (cf. III, 4.25)

$$(1.29)\qquad \left\{\begin{aligned} R_{\delta\gamma\beta}^{\cdot\cdot\cdot\alpha} &= 2\partial_{[\delta}\Gamma_{\gamma]\beta}^{\alpha}+2\Gamma_{[\delta|\varepsilon|}^{\alpha}\Gamma_{\gamma]\beta}^{\varepsilon}=\overset{+}{R}_{\delta\gamma\beta}^{\cdot\cdot\cdot\alpha}-2\partial_{[\delta}\overset{+}{S}_{\gamma]\beta}^{\cdot\cdot\alpha}-\\ &\quad -2\overset{+}{S}_{[\delta|\varepsilon|}^{\cdot\cdot\alpha}\overset{+}{\Gamma}_{\gamma]\beta}^{\varepsilon}-2\overset{+}{\Gamma}_{[\delta|\varepsilon|}^{\alpha}\overset{+}{S}_{\gamma]\beta}^{\cdot\cdot\varepsilon}-2\overset{+}{S}_{[\delta|\varepsilon|}^{\cdot\cdot\alpha}\overset{+}{S}_{\gamma]\beta}^{\cdot\cdot\varepsilon}\\ &= -2\overset{+}{\nabla}_{[\delta}\overset{+}{S}_{\gamma]\beta}^{\cdot\cdot\alpha}-2\overset{+}{S}_{\delta\gamma}^{\cdot\cdot\varepsilon}\overset{+}{S}_{\varepsilon\beta}^{\cdot\cdot\alpha}-2\overset{+}{S}_{[\delta|\varepsilon|}^{\cdot\cdot\alpha}\overset{+}{S}_{\gamma]\beta}^{\cdot\cdot\varepsilon}\\ &= -\tfrac{1}{4}c_{\delta\gamma}^{\cdot\cdot\varepsilon}c_{\varepsilon\beta}^{\cdot\cdot\alpha}. \end{aligned}\right.$$

Hence $R_{\delta\gamma\beta}^{\cdot\cdot\cdot\alpha}$ is an algebraical concomitant of $c_{\gamma\beta}^{\cdot\cdot\alpha}$ and according to (1.26) this implies

$$(1.30)\qquad \overset{\pm}{\nabla}_{\varepsilon}R_{\delta\gamma\beta}^{\cdot\cdot\cdot\alpha}=\nabla_{\varepsilon}R_{\delta\gamma\beta}^{\cdot\cdot\cdot\alpha}=0$$

from which we see *that every group space is a symmetric A_r with respect to its (0)-connexion* (cf. III § 7). Moreover we see that *all differential concomitants of group space are ordinary concomitants of $c_{\gamma\beta}^{\cdot\cdot\alpha}$* (cf. III § 7). This has as a consequence that all these differential concomitants have constant components with respect to the anholonomic systems (a) and (A).

Exercise.

IV 1,1. Prove (1.3) using (1.1, 2).

§ 2. The parameter-groups and the adjoint group of a finite continuous group.

The equation

$$(2.1)\qquad 'T=UT$$

represents a transformation of the elements of a group by which a one to one correspondence between the elements and their transforms is established. If for U all the elements are taken we get a *group* of element transformations, called the *first parameter-group* of the given group, and the transformations of this new group are in one to one correspondence to the elements of the original group. The following propositions can easily be proved for the transformations of the first parameter-group:

a 1) If S, T is transformed into $'S, 'T$, the pairs S, T; $'S, 'T$ are $(-)$-equipollent and the pairs $S, 'S$; $T, 'T$ are $(+)$-equipollent.

a 2) If a transformation of elements transforms every pair of elements into a $(-)$-equipollent pair, the transformation belongs to the first parameter-group.

b) If S_1, T_1; S_2, T_2 are $(\pm)$-equipollent, their transforms are also $(\pm)$-equipollent.

In the same way, starting from the transformations $'T = TU$ the *second parameter-group* can be defined. For this group the same propositions hold as did for the first group if $(\pm)$ is changed into $(\mp)$ (see Fig. 16).

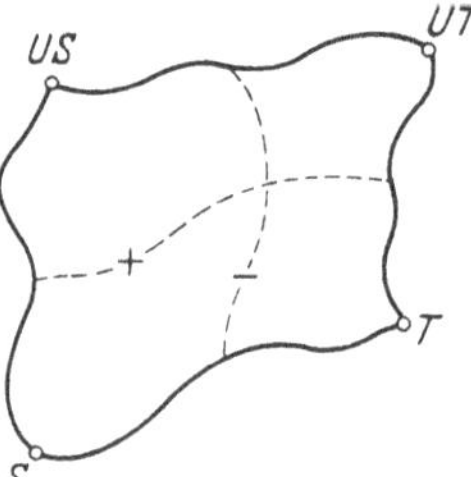

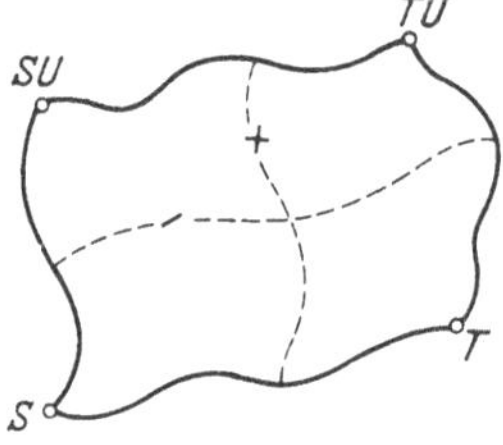

Fig. 16.

Given a pair of elements η^α, ζ^α there is exactly *one* transformation of the first and also exactly *one* transformation of the second parameter-group transforming η^α into ζ^α, namely $T_\zeta T_\eta^{-1}$ and $T_\eta^{-1} T_\zeta$ respectively. Now let there be a transformation of the first parameter-group carrying the element $\underset{1}{\eta}^\alpha$ into $\underset{1}{\eta}^\alpha + \underset{1}{d}\eta^\alpha$ and the element $\underset{2}{\eta}^\alpha$ into $\underset{2}{\eta}^\alpha + \underset{2}{d}\eta^\alpha$. Then the point pairs $\underset{1}{\eta}^\alpha, \underset{1}{\eta}^\alpha + \underset{1}{d}\eta^\alpha$ and $\underset{2}{\eta}^\alpha, \underset{2}{\eta}^\alpha + \underset{2}{d}\eta^\alpha$ are $(+)$-equipollent. Hence, if $\underset{1}{d}\eta^\alpha = e^a A_a^\alpha \{\underset{1}{\eta}^\alpha\}\, dt$ it follows that $\underset{2}{d}\eta^\alpha = e^a A_a^\alpha \{\underset{2}{\eta}^\alpha\}\, dt$ and the constants $e^a dt$ are the components of $\underset{1}{d}\eta^\alpha$ at $\underset{1}{\eta}^\alpha$ and $\underset{2}{d}\eta^\alpha$ at $\underset{2}{\eta}^\alpha$ with respect to the anholonomic system (a). This proves that the transformations

$$\eta^\alpha \to \eta^\alpha + \underset{b}{e}^\alpha\, dt \tag{2.2}$$

are r linearly independent infinitesimal transformations of the first parameter-group. The same reasoning holds for the second parameter-group and the system (A). Hence

$$\eta^\alpha \to \eta^\alpha + \underset{B}{e}^\alpha\, dt \tag{2.3}$$

are r linearly independent infinitesimal transformations of the second parameter-group. Taking $S = \eta^\alpha$, $T = \eta^\alpha + \underset{B}{e^\alpha} dt$, $US = \eta^\alpha + \underset{b}{e^\alpha} dt$ in Fig. 16 we get back the situation of Fig. 15 and again it is proved that the fields $\underset{B}{e^\alpha}$ $(\underset{b}{e^\alpha})$ remain invariant for the infinitesimal transformations (2.2) ((2.3)) of the first (second) parameter-group.

Apart from any considerations about groups we now ask whether it is possible to have in an X_r an anholonomic system (a) satisfying the condition

$$(2.4) \qquad 2\partial_{[\gamma} \overset{a}{e}_{\beta]} = - c_{cb}^{a} \overset{c}{e}_\gamma \overset{b}{e}_\beta; \qquad c_{(cb)}^{a} = 0$$

with $\frac{1}{2} r^2 (r-1)$ *given* (not necessarily constant) coefficients c_{cb}^{a}, and the condition that the fields $\overset{a}{e}_\beta$ are invariant for r linearly independent infinitesimal transformations. (2.4) expresses the fact that (cf. II 9.2)

$$(2.5) \qquad \Omega_{cb}^{a} \overset{*}{=} -\tfrac{1}{2} c_{cb}^{a}.$$

But from this equation and Exerc. II 9,2 we see that the first integrability conditions of (2.4) can be written in the form

$$(2.6) \qquad \partial_{[d} c_{cb]}^{a} + c_{[dc}^{e} c_{b]e}^{a} = 0; \qquad \partial_d \overset{\text{def}}{=} A_d^\delta \partial_\delta.$$

The invariance of the $\overset{a}{e}_\beta$ for an infinitesimal transformation $\eta^\alpha \to \eta^\alpha + v^\alpha dt$ is expressed by [cf. Exerc. II 10,6 and (2.5)]

$$(2.7) \qquad 0 = \underset{v}{\pounds} \underset{b}{e^a} \overset{*}{=} v^c \partial_c \underset{b}{e^a} - (\partial_c v^a - v^d c_{dc}^{a}) \underset{b}{e^c} = -(\partial_c v^a - v^d c_{dc}^{a}) \underset{b}{e^c};$$

and the first integrability condition of this equation is (cf. II 9.4)

$$(2.8) \qquad \Omega_{ec}^{b} \partial_b v^a \overset{*}{=} -\partial_{[e} v^d c_{|d|c]}^{a};$$

or

$$(2.9) \qquad \tfrac{1}{2} c_{ec}^{b} c_{bd}^{a} v^d \overset{*}{=} c_{[e|f|}^{d} c_{|d|c]}^{a} v^f - v^d \partial_{[e} c_{|d|c]}^{a}.$$

Because this must hold for r linearly independent choices of v^a we get

$$(2.10) \qquad 2\partial_{[e} c_{c]d}^{a} = 3 c_{[ed}^{b} c_{c]b}^{a}.$$

Now (2.6) and (2.10) together are equivalent to

$$(2.11) \qquad \text{a)}\ c_{cb}^{a} = \text{const.}; \qquad \text{b)}\ c_{[dc}^{e} c_{b]e}^{a} = 0.$$

If the conditions are satisfied, to every set of values $\overset{a}{e}_\beta$, v^α chosen at a given point $\overset{*}{\eta}{}^\alpha$ of X_r there exist fields $\overset{a}{e}_\beta$, v^α in an $\mathfrak{N}(\overset{*}{\eta}{}^\alpha)$ satisfying (2.4) and (2.7). If r linearly independent vectors v^α are chosen as the r contravariant basis vectors $\underset{B}{e^\alpha}$ of an anholonomic coordinate system (A)

we have

(2.12) $$0 = \underset{B}{\pounds}\, \overset{a}{e}_\beta = \underset{B}{e}^\gamma\, \partial_\gamma\, \overset{a}{e}_\beta + \overset{a}{e}_\gamma\, \partial_\beta\, \underset{B}{e}^\gamma;$$

and accordingly

(2.13) $$(\partial_\beta\, \underset{B}{e}^\alpha)\, \overset{A}{e}_\alpha = -\underset{B}{e}^\gamma\, (\partial_\gamma\, \overset{a}{e}_\beta)\, \underset{a}{e}^\alpha\, \overset{A}{e}_\alpha;$$

hence

(2.14) $$2\,\partial_{[\gamma}\, \overset{A}{e}_{\beta]} = -\,{}'c_{C\,B}^{A}\, \overset{C}{e}_\gamma\, \overset{B}{e}_\beta;$$

where

(2.15) $${}'c_{C\,B}^{A} \overset{\text{def}}{=} -\, c_{cb}^{a}\, \overset{c}{e}_\gamma\, \overset{b}{e}_\beta\, \underset{C}{e}^\gamma\, \underset{B}{e}^\beta\, \underset{a}{e}^\delta\, \overset{A}{e}_\delta.$$

As the $\underset{b}{e}^\alpha$ are invariant for the transformations $\underset{B}{e}^\alpha dt$, the fields $\underset{B}{e}^\alpha$ and $\overset{A}{e}_\beta$ are also invariant for the transformations $\underset{b}{e}^\alpha dt$. But we have already proved that r linearly independent fields $\overset{A}{e}_\beta$ satisfying equations of the form (2.14) and invariant for r linearly independent infinitesimal transformations can only exist if the ${}'c_{C\,B}^{A}$ are constants satisfying the equations ${}'c_{[D\,C}^{E}\, {}'c_{B]\,E}^{A} = 0$.

Now if at some arbitrary point $\underset{0}{\eta}^\alpha$ we choose the $\underset{B}{e}^\alpha$ such that they coincide with the $\underset{b}{e}^\alpha$, we have at that point according to (2.15)

(2.16) $${}'c_{C\,B}^{A} = -\,\delta_{C\,B\,a}^{c\,b\,A}\, c_{cb}^{a}$$

and because the c_{cb}^{a} and ${}'c_{C\,B}^{A}$ are constants, this holds at all points.

To the fields $\underset{b}{e}^\alpha$, $\overset{a}{e}_\beta$ belongs a connexion, say $\overset{+}{\Gamma}{}_{\gamma\beta}^{\alpha}$, for which these fields are covariant constant and in the same way a connexion $\overset{-}{\Gamma}{}_{\gamma\beta}^{\alpha}$ can be constructed with respect to the fields $\underset{B}{e}^\alpha$, $\overset{B}{e}_\beta$. Then it is easily proved from (2.4) and (2.14) that

(2.17) $$\overset{+}{S}{}_{cb}^{\cdot\cdot a} \overset{*}{=} -\tfrac{1}{2}\, c_{cb}^{a}; \qquad \overset{-}{S}{}_{C\,B}^{\cdot\cdot A} \overset{*}{=} -\tfrac{1}{2}\, {}'c_{C\,B}^{A}$$

and from the invariance of the fields $\underset{b}{e}^\alpha$ and $\underset{B}{e}^\alpha$ with respect to each other that

(2.18) $$\overset{+}{S}{}_{\gamma\beta}^{\cdot\cdot\alpha} = -\,\overset{-}{S}{}_{\gamma\beta}^{\cdot\cdot\alpha}.$$

This means that the geometry in X_r is just the same as in a group space of an r-parameter group. But the X_r is not yet a group space; the r-parameter group itself is still missing.

Now we have in fact two groups of point transformations in X_r, viz. the invertible transformations leaving invariant the fields $\underset{B}{e}^\alpha$ and those that leave invariant the fields $\underset{b}{e}^\alpha$. Taking the first group we see that the transformation $\eta^\alpha \to \eta^\alpha + e^b A_b^\alpha\, dt$ represents for all values of the constants e^b one of its infinitesimal transformations (cf. II § 10).

We also see that the set of finite transformations

$$(2.19) \qquad {}'\eta^\alpha = e^{t e^b \partial_b} \eta^\alpha; \qquad \partial_b \overset{\text{def}}{=} A_b^\beta \partial_\beta;$$

depending on r parameters $t e^b$ and consisting of the transformations of all one-parameter subgroups each generated by the infinitesimal transformation for some values of e^b (cf. II § 10), is contained in the set. But we do not yet know whether the set contains all the transformations of the group.

If we take the η^α in (2.19) as fixed, say $\eta^\alpha = \underset{1}{\eta^\alpha}$, the point $'\eta^\alpha$ describes a curve with the differential equation

$$(2.20) \qquad \frac{d\eta^\alpha}{dt} = e^b A_b^\alpha$$

and this is a geodesic in X_r for the connexions $\overset{+}{\Gamma}{}^\alpha_{\gamma\beta}$, $\overset{-}{\Gamma}{}^\alpha_{\gamma\beta}$ and $\Gamma^\alpha_{\gamma\beta} = \frac{1}{2}(\overset{+}{\Gamma}{}^\alpha_{\gamma\beta} + \overset{-}{\Gamma}{}^\alpha_{\gamma\beta})$ because

$$(2.21) \qquad \delta \frac{d\eta^\alpha}{dt} = \overset{+}{\delta} \frac{d\eta^\alpha}{dt} - \overset{+}{S}{}_{\gamma\beta}^{\cdot\cdot\alpha} d\eta^\gamma \frac{d\eta^\beta}{dt} = \overset{+}{\delta} \frac{d\eta^\alpha}{dt} = 0.$$

The $t e^b$ are normal coordinates in X_r with respect to the point $\underset{1}{\eta^\alpha}$ and the connexion $\Gamma^\alpha_{\gamma\beta}$ (cf. III § 7). Hence there is always one and only one choice of $t e^b$ for which $'\eta^\alpha = \underset{2}{\eta^\alpha}$ if $\underset{2}{\eta^\alpha}$ is some arbitrarily given point in $\mathfrak{N}(\underset{1}{\eta^\alpha})$. Moreover it follows that there is one and only one transformation in the set that transforms $\underset{1}{\eta^\alpha}$ into an arbitrary point $\underset{2}{\eta^\alpha}$ of $\mathfrak{N}(\underset{1}{\eta^\alpha})$. This transformation belongs to the group but we have not yet proved that the group does not contain another transformation carrying $\underset{1}{\eta^\alpha}$ into $\underset{2}{\eta^\alpha}$. Now every transformation of the group leaves the fields $\underset{B}{e^\alpha}$ invariant. Hence it leaves invariant the set of all geodesics and also the natural parameters on them. Hence if $\underset{1}{\eta^\alpha}$ is transformed into $\underset{2}{\eta^\alpha}$ the set of all geodesics through $\underset{1}{\eta^\alpha}$ is transformed into the set of all geodesics through $\underset{2}{\eta^\alpha}$ and this transformation is fixed completely by the condition that the direction of a geodesic at $\underset{1}{\eta^\alpha}$ is (+)-parallel to the direction at $\underset{2}{\eta^\alpha}$ of its transform. Moreover the normal coordinates of a point with respect to $\underset{1}{\eta^\alpha}$ must be the same as those of its transform with respect to $\underset{2}{\eta^\alpha}$ and this condition not only fixes the transformation of the geodesics but also the transformation of all points in $\mathfrak{N}(\underset{1}{\eta^\alpha})$.

This proves that every transformation in the group is also a transformation of the set and that accordingly the group is an r-parameter one with the infinitesimal transformations $\eta^\alpha \to \eta^\alpha + e^b A_b^\alpha \, dt$.

If now we associate with every point η^α of X_r the transformation of the group carrying the already chosen point $\underset{1}{\eta^\alpha}$ into η^α, X_r is the group space of this group. The group is now a group of point transformations in its own group space and accordingly it coincides with its first parameter-group.

After this digression on sets of vector fields in an X_r we return to group space and the first and second parameter-group and ask whether the transformation $\eta^\alpha \to \eta^\alpha + \overset{b}{\lambda} \underset{b}{e^\alpha} \, dt$ is an infinitesimal transformation of the first parameter-group. Necessary and sufficient is the invariance of the fields $\underset{B}{e^\alpha}$:

$$\overset{b}{\lambda} \underset{b}{e^\beta} \partial_\beta \underset{B}{e^\alpha} - \underset{B}{e^\beta} (\partial_\beta \overset{b}{\lambda}) \underset{b}{e^\alpha} - \overset{b}{\lambda} \underset{B}{e^\beta} \partial_\beta \underset{b}{e^\alpha} = 0 \tag{2.22}$$

and this is only possible if the $\overset{b}{\lambda}$ are constants.

In group theory it is usual to denote an infinitesimal transformation in some X_n, $\xi^\varkappa \to \xi^\varkappa + X^\varkappa dt = \xi^\varkappa + X^\mu \partial_\mu \xi^\varkappa dt$ by a symbol, for instance X, standing for the operator $X^\mu \partial_\mu$. Then we have for the transformation of $\xi^\varkappa$ and of any function $f(\xi^\varkappa)$

$$d\xi^\varkappa = X \xi^\varkappa \, dt; \qquad df = X f \, dt. \tag{2.23}$$

If X and Y are two different symbols of this kind, we write (XY) for the operator $XY - YX$. This notation was introduced by LIE.

Using $\partial_b \overset{\text{def}}{=} A_b^\alpha \partial_\alpha$ in this way as a symbol for the infinitesimal transformation $\eta^\alpha \to \eta^\alpha + \underset{b}{e^\alpha} dt$ of the first parameter-group we get from (1.19)

$$c_{cb}^{a} = -2 \underset{c}{e^\gamma} \underset{b}{e^\beta} \partial_{[\gamma} \overset{a}{e}_{\beta]} = 2 \underset{[c}{e^\gamma} \overset{a}{e}_{|\beta} \partial_{\gamma|} \underset{b]}{e^\beta} \overset{*}{=} 2 A_\beta^a A_{[c}^\gamma \partial_{|\gamma|} A_{b]}^\beta \tag{2.24}$$

hence

$$\boxed{(\partial_c \partial_b) = c_{cb}^{\cdot\cdot a} \partial_a} \,. \tag{2.25}$$

In the same way it is proved for the second parameter-group

$$\boxed{(\partial_C \partial_B) = -c_{CB}^{\cdot\cdot A} \partial_A}\,; \qquad \partial_B \overset{\text{def}}{=} A_B^\alpha \partial_\alpha\,. \tag{2.26}$$

The equations (2.25, 26) are called the LIE *structural formulae* of the first and second parameter-group and the $c_{cb}^{\cdot\cdot a}$ its *structural constants*. On the other hand (1.19, 23) are called the CARTAN *structural formulae* of these same groups. These latter equations were established first by MAURER.[1]) LIE started from infinitesimal transformations and this is the reason why he looked at the matter from the contravariant side. But CARTAN started from systems of Pfaffians and his point of view was therefore entirely covariant. The rather fine dualism between the two points of view was pointed out by CARTAN.

Comparing the equations (1.19) and (2.25) it is obvious that (1.19) contains an invariant operation, viz. Rot, but that the operator ∂_c is only invariant if it is applied to scalars. Now we have according to (1.16)

$$\underset{c}{\pounds}\,\underset{b}{e}^{\alpha} = \underset{c}{e}^{\gamma}\,\overline{\nabla}_{\gamma}\,\underset{b}{e}^{\alpha} = -\,2\,\underset{c}{e}^{\gamma}\,\overset{+}{S}_{\gamma\beta}{}^{\cdot\cdot\alpha}\,\underset{b}{e}^{\beta} = c_{cb}^{a}\,\underset{a}{e}^{\alpha} \tag{2.27}$$

and therefore, considering Exerc. II 10,9

$$(\underset{c}{\pounds}\,\underset{b}{\pounds})\,\Phi = c_{cb}^{a}\,\underset{a}{\pounds}\,\Phi; \qquad (\underset{c}{\pounds}\,\underset{b}{\pounds}) \overset{\text{def}}{=} \underset{c}{\pounds}\,\underset{b}{\pounds} - \underset{b}{\pounds}\,\underset{c}{\pounds} \tag{2.28}$$

for every quantity Φ (indices suppressed). Hence

$$\boxed{(\underset{c}{\pounds}\,\underset{b}{\pounds}) = c_{cb}^{a}\,\underset{a}{\pounds}} \;\,^{2)} \tag{2.29}$$

and in the same way

$$\boxed{(\underset{C}{\pounds}\,\underset{B}{\pounds}) = -\,c_{CB}^{A}\,\underset{A}{\pounds}}\,. \tag{2.30}$$

(2.29) and (2.30) are equivalent to (2.25, 26) if the operators are applied to a scalar. But in contradistinction to (2.25, 26), in (2.29, 30) only those operators occur that can be applied to *all* quantities.

The finite transformations of the first and second parameter-group can be written in the form (cf. II § 10)

$$'\eta^{\alpha} = e^{t\,e^{b}\partial_{b}}\,\eta^{\alpha}; \qquad '\eta^{\alpha} = e^{t\,e^{B}\partial_{B}}\,\eta^{\alpha} \tag{2.31}$$

and also in the more general form valid for any quantity Φ (indices suppressed) dragged along (cf. II § 10)

$$\overset{m}{\Phi} = e^{-t\overset{b}{e}\underset{b}{\pounds}}\,\Phi; \quad \overset{m}{\Phi} = e^{-t\overset{B}{e}\underset{B}{\pounds}}\,\Phi; \quad \overset{b}{e} \overset{\text{def}}{=} \overset{b}{e}_{c}\,e^{c}; \quad \overset{B}{e} \overset{\text{def}}{=} \overset{B}{e}_{c}\,e^{c}. \tag{2.32}$$

We gather together the following results:

An r-parameter group being given, there exist in group space anholonomic coordinate systems (a) and (A), satisfying the equations

[1]) MAURER 1888, 1, p. 117; cf. for the solution of these equations WHITEHEAD 1932, 2 also for literature.

[2]) The identities (2.29, 30) are proved here only for quantities. But they are valid for all geometric objects having second LIE derivatives (cf. II § 10).

(1.19), (1.23), (1.21), (1.24), (2.25), (2.26). There are two groups of point transformations in X_r, the two parameter-groups, whose transformations are in one to one correspondence with the transformations of the given group. The first leaves invariant the fields of basis vectors of (A) and the second those of (a). Conversely if $\frac{1}{2} r^2(r-1)$ constants $c^a_{cb} = c^a_{[cb]}$ be given that satisfy (1.28b), it is always possible to construct in an arbitrary X_r an anholonomic system (a) satisfying (1.19) and (2.25) and another anholonomic system (A) satisfying (1.23) and (2.26). The ∂_b are the symbols of infinitesimal transformations of a group of point transformations in X_r that leave invariant the fields of the basis vectors of (A) and the same holds for the ∂_B with respect to the basis vectors of (a). The transformations of these groups can be brought into one to one correspondence to the points of the X_r, thus making it a group space.

Besides the two parameter-groups there is a third transformation group connected with every group. The equation

$$'T = U T U^{-1} \tag{2.33}$$

represents a transformation of the elements T of this group. If for U all the elements are taken we get the *adjoint group* of the given group. Because of

$$V U T U^{-1} V^{-1} = (V U) T (V U)^{-1} \tag{2.34}$$

this group and the given group are homomorphic but the homomorphism need not be an isomorphism (that is one to one). In fact, two elements U_1 and U_2 may exist such that

$$U_1 T U_1^{-1} = U_2 T U_2^{-1} \tag{2.35}$$

for every T. This is the case if and only if there exists besides I another element that is commutative with respect to every element of the group. All elements of this kind form an invariant subgroup called the *centre* of the group. Let C be an element of the centre, then the transformation $U T U^{-1}$ of the adjoint group corresponds not only to U but also to $C U$.

An element T and its transform $U T U^{-1}$ are called *homologous*. A subgroup and its transform are also said to be *homologous*. If a subgroup is identical with its transform for every choice of U it is called an *invariant* or *normal subgroup*.

From the definition (2.33) it follows that a transformation of the adjoint group arises if a transformation of the first parameter-group carrying I into U is followed by a transformation of the second group carrying I into U^{-1}. If s^α is a vector at $\underset{0}{\eta}^\alpha$ and $\overset{+}{s}{}^\alpha$ and $\overset{-}{s}{}^\alpha$ are the fields we can get from s^α by $(\pm)$-displacement, the infinitesimal transformation of the adjoint group corresponding to the point $\underset{0}{\eta}^\alpha + s^\alpha du$ arises if the $(-)$-parallel displacement over $\overset{+}{s}{}^\alpha du$ is followed by the $(+)$-parallel displacement over $-\overset{-}{s}{}^\alpha du$. A vector e^a at $\underset{0}{\eta}^\alpha$ is transformed by these transformations first into the vector $e^a - \overset{-}{\Gamma}{}^a_{cb}\, e^b s^c du$ and afterwards into the vector $e^a - c_{cb}^{\cdot\cdot a}\, e^b s^c du$. Hence the infinitesimal transformation corresponding to the point $\underset{0}{\eta}^\alpha + e^\alpha dt$ in group space, with the symbol $e^a A^\beta_a \partial_\beta$, is transformed by an infinitesimal transformation of the adjoint group in the following way

$$(2.36)\qquad d e^a = -\, c_{cb}^{\cdot\cdot a}\, e^b s^c du = s^b E_b^{\cdot a} du; \qquad E_b^{\cdot a} \overset{\text{def}}{=} e^c c_{cb}^{\cdot\cdot a}.$$

This is an infinitesimal transformation of the e^a with the symbol

$$(2.37)\qquad s^b E_b; \quad E_b \overset{\text{def}}{=} E_b^{\cdot a} \frac{\partial}{\partial e^a} = e^c c_{cb}^{\cdot\cdot a} \frac{\partial}{\partial e^a}.$$

Because

$$(2.38)\qquad (E_c E_b) = 2 e^f c_{f[c}^{\cdot\cdot d}\, c_{|d|b]}^{\cdot\cdot e}\, \partial_e = c_{cb}^{\cdot\cdot a} E_a;$$

the E_b and ∂_b satisfy equations with the same structural constants. But the E_b are independent if and only if the c-rank of $c_{cb}^{\cdot\cdot a}$ is equal to r.

Writing $S_b^{\cdot a} \overset{\text{def}}{=} s^c c_{cb}^{\cdot\cdot a}$ we get for a finite transformation of the one-parameter group generated by $s^b E_b$ for some constant values of the s^b

$$(2.39)\qquad {}'e^a = (e^{-uS})^a_b\, e^b;$$

where S symbolizes the matrix of $S_b^{\cdot a}$. The equation (2.39) represents, for variable values of the s^a, a linear homogeneous group of transformations of the e^a, called the *linear adjoint group*. It is the adjoint group but only if considered as a group of transformations working on the infinitesimal transformations of the given group.

If we introduce the $t\, e^a$ as normal coordinates $\eta^{\mathfrak{a}}$ of X_r with respect to the point $\underset{0}{\eta}^\alpha$

$$(2.40)\qquad \eta^{\mathfrak{a}} \overset{\text{def}}{=} \delta^{\mathfrak{a}}_a\, t\, e^a; \qquad \mathfrak{a} = \bar{1}, \ldots, \bar{r}$$

the three coordinate systems (a), (A) and $(\mathfrak{a})$ have the same basis vectors at $\underset{0}{\eta}^\alpha$ and we see that the transformation of these new coordinates is also linear homogeneous. That proves the statement:

The transformations of the adjoint group are linear homogeneous if they are expressed in the normal coordinates $\eta^{\mathfrak{a}}$ with respect to $\underset{0}{\eta}^\alpha$.

We write these transformations

(2.41) $\left\{\begin{array}{ll} \text{a)}\ d\eta^a = -\underset{0}{S}{}_b^{\cdot a}\eta^b\,du; & \underset{0}{S}{}_b^{\cdot a} \overset{\text{def}}{=} S_b^{\cdot a}\{\underset{0}{\eta}{}^\alpha\}\,^{1)} \\ \text{b)}\ {}'\eta^a = \left(e^{-u\underset{0}{S}}\right)_b^a \eta^b; & \underset{0}{S} = \text{matrix of } \underset{0}{S}{}_b^{\cdot a}. \end{array}\right.$

(2.41 b) is the transformation $U\ldots U^{-1}$ if U corresponds to the point $\eta^a = \delta^a_a u s^a$ of group space. This transformation arises if the (+)-equipollence that transforms U into I is followed by the (—)-equipollence that transforms I into U. The first transforms the vectors $\underset{b}{e}{}^\alpha$ at $\delta^a_a u s^a$ into the $\underset{b}{e}{}^\alpha$ at $\underset{0}{\eta}{}^\alpha$ coinciding with the $\underset{B}{e}{}^\alpha$ at that point, and the second transforms these vectors into the $\underset{B}{e}{}^\alpha$ at $\delta^a_a u s^a$. Hence

(2.42) a) $\underset{B}{e}{}^\alpha = \delta^b_B \left(e^{-u\underset{0}{S}}\right)^\alpha_\beta \underset{b}{e}{}^\beta$ or b) $A^a_B = \left(e^{-u\underset{0}{S}}\right)^a_B$;

or in words:

The vectors $\underset{b}{e}{}^\alpha$ at a point of group space that corresponds to the transformation U of the group, transform into the vectors $\underset{B}{e}{}^\alpha$ at that point by the same linear homogeneous transformation that represents the transformation UTU^{-1} of the adjoint group if applied to the normal coordinates η^a.

Exercises.

IV 2,1. If operators $\pounds_b$ are defined by $\pounds_b \overset{\text{def}}{=} \overset{a}{e}_b \underset{a}{\pounds}$, prove that

IV 2,1 α) $$(\pounds_c\, \pounds_b) = 0.$$

This is the reason why we do not use the operators $\pounds_b$. The same difficulty does not arise with the operators ∂_b because $\partial_b\, \overset{a}{e}_c = 0$ whereas $\pounds_b\, \overset{a}{e}_c \neq 0$.

IV 2,2. Prove that

IV 2,2 α) a) $\overset{+}{\Gamma}{}^a_{cb} \overset{*}{=} 0$; b) $\Omega^a_{cb} \overset{*}{=} \overset{+}{S}{}_{cb}^{\cdot\cdot\, a}$.

IV 2,3. Prove that the transformation $\eta^\alpha \to \eta^\alpha + v^\alpha\, dt$ represents an infinitesimal transformation of the first parameter-group if and only if $v^a = \text{const}$.

IV 2,4. Prove that

IV 2,4 α) $$\underset{c}{\pounds}\,\underset{B}{\pounds} = \underset{B}{\pounds}\,\underset{c}{\pounds}$$

and

IV 2,4 β) $$\overset{+}{\nabla}_\gamma \overset{-}{\nabla}_\beta = \overset{-}{\nabla}_\beta \overset{+}{\nabla}_\gamma.$$

---

1) Note that the components $\underset{0}{S}{}_b^{\cdot a}$ are constants but that the $\underset{0}{S}{}_b^{\cdot a}$ equal the $S_b^{\cdot a}$ at $\underset{0}{\eta}{}^\alpha$ only.

IV 2,5. Prove that a quantity is $(+)$-constant if and only if its components with respect to (a) are constants.

IV 2,6[1]). Prove that the (0)-connexion in group space is volume preserving.

§ 3. Finite continuous transformation groups.

So far we have not supposed that the group originally given was in any way a group of transformations.[2])

Let us now take an r-parameter group of transformations in n variables $\xi^\varkappa$; $\varkappa = 1, \ldots, n$

$$(3.1) \qquad '\xi^\varkappa = f^\varkappa(\xi^\nu, \eta^\alpha); \qquad \alpha = \dot{1}, \ldots, \dot{r}.$$

First we have to make sure that the $\xi^\varkappa$ can be solved from (3.1). The necessary and sufficient condition is that the matrix of the $\partial_\lambda f^\varkappa$ has rank n. Secondly it is necessary that the η^α should be *essential*, that is, that there should be no equations of the form

$$(3.2) \qquad f^\varkappa(\xi^\nu, \eta^\alpha) = F^\varkappa(\xi^\nu, \zeta^\mathfrak{a}(\eta^\alpha)); \qquad \mathfrak{a} = \dot{\mathfrak{1}}, \ldots, \dot{\mathfrak{r}}'; \qquad r' < r.$$

From (3.2) it follows that

$$(3.3) \qquad \partial_\beta f^\varkappa = (\partial_\beta \zeta^\mathfrak{a})\, \partial_\mathfrak{a} F^\varkappa; \qquad \partial_\mathfrak{a} = \frac{\partial}{\partial \zeta^\mathfrak{a}}$$

and because the rank of $\partial_\beta \zeta^\mathfrak{a}$ is $\leq r'$ this implies that there exists at least one set of functions $U^\beta(\eta^\alpha)$ such that

$$(3.4) \qquad U^\beta(\eta^\alpha)\, \partial_\beta f^\varkappa(\xi^\nu, \eta^\alpha) = 0.$$

Conversely, if an equation of the form (3.4) holds, the $f^\varkappa(\xi^\nu, \eta^\alpha)$ must be solutions of the equation

$$(3.5) \qquad U^\beta(\eta^\alpha)\, \partial_\beta \varphi(\eta^\alpha) = 0$$

containing parameters $\xi^\varkappa$. But because an equation of this form has at most $r-1$ independent solutions (cf. II § 5) the η^α cannot be essential. Hence the η^α in (3.1) are essential if and only if no equation of the form (3.4) exists.

From now on we suppose that the η^α are essential and that $\underset{0}{\eta}{}^\alpha$ corresponds to the identical transformation. Then there is a one to one correspondence between the transformations of the group in a neighbourhood of the identical transformation and the points of the X_r of the η^α in an $\mathfrak{N}(\underset{0}{\eta}{}^\alpha)$.

---

1) CARTAN 1927, 1, p. 67.

2) HAIMOVICI has proved 1945, 1; 1946, 1; 2; 1947, 1; 1948, 1; 2 that a connexion can also be established in the manifold of a set of transformations. Certain concomitants of this connexion vanish if and only if the set is a group.

We only consider transformations of this group germ. For every definite choice $\eta^\alpha = \underset{1}{\eta}^\alpha$ the equation (3.1)

$$'\xi^\varkappa = f^\varkappa(\xi^\nu, \underset{1}{\eta}^\alpha) \tag{3.6}$$

represents a definite transformation $T_{\underset{1}{\eta}}$ operating on the $\xi^\varkappa$. Other values $\eta^\alpha = \underset{2}{\eta}^\alpha$ give another transformation $T_{\underset{2}{\eta}}$

$$''\xi^\varkappa = f^\varkappa(\xi^\nu, \underset{2}{\eta}^\alpha) \tag{3.7}$$

of the $\xi^\varkappa$. If $\underset{1}{\eta}^\alpha$ is changed into $\underset{1}{\eta}^\alpha + \underset{1}{d\eta}^\alpha$ the change of $'\xi^\varkappa$ is

$$d\,'\xi^\varkappa = (\partial_\beta f^\varkappa)_{\eta^\alpha = \underset{1}{\eta}^\alpha} \underset{1}{d\eta}^\beta \tag{3.8}$$

and if in (3.8) $\xi^\varkappa$ is eliminated from the right hand side by means of (3.6), the resulting equation represents the transformation of $'\xi^\varkappa = T_{\underset{1}{\eta}} \xi^\varkappa$ into $T_{\underset{1}{\eta}+\underset{1}{d\eta}} \xi^\varkappa = T_{\underset{1}{\eta}+\underset{1}{d\eta}} T_{\underset{1}{\eta}}^{-1}\, '\xi^\varkappa$, that is the infinitesimal transformation $T_{\underset{1}{\eta}+\underset{1}{d\eta}} T_{\underset{1}{\eta}}^{-1}$ operating on the $'\xi^\varkappa$. This transformation corresponds to the infinitesimal vector $\underset{1}{d\eta}^\alpha$ at the point $\underset{1}{\eta}^\alpha$ in group space. In the same way

$$d\,''\xi^\varkappa = (\partial_\beta f^\varkappa)_{\eta^\alpha = \underset{2}{\eta}^\alpha} \underset{2}{d\eta}^\beta \tag{3.9}$$

after elimination of the $\xi^\varkappa$ by means of (3.7) represents the transformation $T_{\underset{2}{\eta}+\underset{2}{d\eta}} T_{\underset{2}{\eta}}^{-1}$ operating on the variables $''\xi^\varkappa$ and it corresponds to the infinitesimal vector $\underset{2}{d\eta}^\alpha$ at the point $\underset{2}{\eta}^\alpha$ in group space. If $\underset{2}{d\eta}^\alpha$ is now chosen in such a way that

$$T_{\underset{1}{\eta}+\underset{1}{d\eta}} T_{\underset{1}{\eta}}^{-1} = T_{\underset{2}{\eta}+\underset{2}{d\eta}} T_{\underset{2}{\eta}}^{-1} \tag{3.10}$$

then $\underset{1}{d\eta}^\alpha$ at $\underset{1}{\eta}^\alpha$ is $(+)$-equipollent to $\underset{2}{d\eta}^\alpha$ at $\underset{2}{\eta}^\alpha$. Hence, introducing the coordinate system (a) we have $(\underset{1}{d\eta})^a = (\underset{2}{d\eta})^a$, and the two equations

$$d\,'\xi^\varkappa = (\partial_\beta\, '\xi^\varkappa)_{\eta^\alpha = \underset{1}{\eta}^\alpha} A_b^\beta(\underset{1}{\eta}^\alpha)\,(\underset{1}{d\eta})^b\,[1] \tag{3.11}$$

$$d\,''\xi^\varkappa = (\partial_\beta\, ''\xi^\varkappa)_{\eta^\alpha = \underset{2}{\eta}^\alpha} A_b^\beta(\underset{2}{\eta}^\alpha)\,(\underset{1}{d\eta})^b \tag{3.12}$$

after elimination of the $\xi^\varkappa$ represent one and the same transformation, viz. the transformation corresponding to the point $\underset{0}{\eta}^\alpha + A_b^\alpha(\underset{0}{\eta}^\alpha)\,(\underset{1}{d\eta})^b$ in group space, operating on the $'\xi^\varkappa$ in (3.11) and on the $''\xi^\varkappa$ in (3.12). But *this is only possible if the expressions*

$$\Xi_b^\varkappa \overset{\text{def}}{=} (\partial_\beta\, '\xi^\varkappa)\, A_b^\beta(\underset{1}{\eta}^\alpha) \tag{3.13}$$

---

[1]) Cf. for the use of A_b^β as a function symbol, II § 3.

after elimination of the $\xi^\varkappa$ *by means of* (3.6), *depend only on the* $'\xi^\varkappa$ *and are independent of the* $\underset{1}{\eta}^\alpha$.

The infinitesimal transformation corresponding to the point $\underset{0}{\eta}^\alpha + e^a A^\alpha_a\, dt$ of group space is given by

$$d\xi^\varkappa = \Xi^\varkappa_a(\xi^\nu)\, e^a\, dt; \qquad e^a = \text{const.} \tag{3.14}$$

and to every choice of $e^a dt$ there belongs a definite infinitesimal transformation with $d\xi^\varkappa \neq 0$. In fact no relation of the form

$$c^b\, \Xi^\varkappa_b = 0 \tag{3.15}$$

can exist with coefficients c^b independent of the $\xi^\varkappa$ because in that case there would be a relation

$$c^b A^\beta_b\, \partial_\beta f^\varkappa = 0 \tag{3.16}$$

and as we have proved already this would imply that the parameters were not essential [cf. (3.4)].

According to CAMPBELL[1]) we call a set of infinitesimal transformations *connected* if they satisfy a homogeneous linear equation, and *dependent* if in addition the coefficients in this relation are constants. It frequently happens that a set is independent but connected.

Collecting results we have the first part of the first fundamental theorem of LIE[2]):

I.1. *If the invertible transformations*

$$'\xi^\varkappa = f^\varkappa(\xi^\nu, \eta^\alpha); \qquad \alpha = \dot{1}, \ldots, \dot{r} \tag{3.17}$$

with r essential parameters η^α *form a group, there exist* r^2 *functions* $A^\alpha_b(\eta^\beta)$; $\operatorname{Det}(A^\alpha_b) \neq 0$; $a, b = 1, \ldots, r$ *and nr functions* $\Xi^\varkappa_b(\xi^\lambda)$ *for which no equations exist of the form* $c^b\, \Xi^\varkappa_b = 0$ *with coefficients* c^b *independent of the* $\xi^\varkappa$, *such that*

$$\boxed{\partial_\beta\, '\xi^\varkappa = A^b_\beta(\eta^\alpha)\, \Xi^\varkappa_b('\xi^\nu)}\,. \tag{3.18}$$

Starting from the inverse of (3.17)

$$\xi^\varkappa = F^\varkappa('\xi^\nu, \eta^\alpha) \tag{3.19}$$

it can be proved in the same way that

$$\boxed{\partial_\beta\, \xi^\varkappa = A^B_\beta(\eta^\alpha)\, '\Xi^\varkappa_B(\xi^\lambda)}\,; \qquad '\Xi^\varkappa_B \overset{\text{def}}{=} (\partial_\beta \xi^\varkappa)\, A^\beta_B(\eta^\alpha)\,. \tag{3.20}$$

---

[1]) CAMPBELL 1903, 1.

[2]) In this and the following fundamental theorems all functions are supposed to be analytic in the region considered.

The finite transformations of the group are found as follows. The equation of a geodesic through $\underset{0}{\eta}^\alpha$ is

$$\frac{d\eta^\alpha}{dt} = A_a^\alpha e^a; \quad \eta^\alpha = \underset{0}{\eta}^\alpha \text{ for } t = 0. \tag{3.21}$$

Hence at every point of this curve we have according to (3.18)

$$\frac{d'\xi^\varkappa}{dt} = \Xi_b^\varkappa('\xi^\lambda)\, e^b \tag{3.22}$$

and the solution of this equation with the condition $'\xi^\varkappa = \xi^\varkappa$ for $t=0$ is (cf. II § 10)

$$'\xi^\varkappa = e^{tX}\xi^\varkappa; \quad X \stackrel{\text{def}}{=} e^b X_b; \quad X_b \stackrel{\text{def}}{=} \Xi_b^\mu(\xi^\varkappa)\,\partial_\mu. \tag{3.23}$$

From this, for any function $f(\xi^\varkappa)$ we get

$$f('\xi^\varkappa) = e^{tX} f(\xi^\varkappa) \tag{3.24}$$

representing the transformation of the value at the point $\xi^\varkappa$ into the value at $'\xi^\varkappa$. But if we have a scalar field $p = f(\xi^\varkappa)$ and if this field is dragged along by the point transformation (3.23) the new value at $\xi^\varkappa$ is the original value at $e^{-tX}\xi^\varkappa$ (cf. II § 10), hence this new value is

$$\overset{m}{p} = e^{-tX} p \tag{3.25}$$

and also

$$\overset{m}{p} = e^{-t\pounds} p \tag{3.26}$$

where $\pounds$ is the Lie operator belonging to the field $e^b\Xi_b^\varkappa$.

Unlike (3.25) the formula (3.26) can be generalized for any quantity Φ (indices suppressed) as defined in II § 3 (cf. II 10.38)

$$\overset{m}{\Phi} = e^{-t\pounds}\Phi. \tag{3.27}$$

If we introduce in X_r the normal coordinates $\eta^{\mathfrak{a}}$; $\mathfrak{a} = \bar{1}, \ldots, \bar{r}$, with respect to $\underset{0}{\eta}^\alpha$ and to the coordinate system (a), then the transformation (3.23) corresponds to the point in group space with the normal coordinates

$$\eta^{\mathfrak{a}} \stackrel{*}{=} \delta_a^{\mathfrak{a}}\, t\, e^a. \tag{3.28}$$

The system of equations (3.18) has the solutions

$$'\xi^\varkappa = f^\varkappa(\xi^\nu, \eta^\alpha); \quad f^\varkappa(\xi^\nu, \underset{0}{\eta}^\alpha) = \xi^\varkappa \tag{3.29}$$

depending on the n parameters $\xi^\varkappa$. Hence the system is totally integrable (cf. II § 5) and its integrability conditions must be satisfied identically.

These conditions are [cf. (2.4)]

$$(3.30)\qquad \begin{cases} 0 = A^c_{[\gamma} A^b_{\beta]}\, \Xi^\mu_c('\xi^\lambda)\, \dfrac{\partial\, \Xi^\varkappa_b('\xi^\varrho)}{d\,'\xi^\mu} + \Xi^\varkappa_b('\xi^\lambda)\, \partial_{[\gamma} A^b_{\beta]} \\ \quad = A^c_{[\gamma} A^b_{\beta]} \left(\Xi^\mu_c('\xi^\lambda)\, \dfrac{\partial\, \Xi^\varkappa_b('\xi^\varrho)}{d\,'\xi^\mu} - \dfrac{1}{2}\, c_{cb}^{\cdot\cdot a}\, \Xi^\varkappa_a('\xi^\lambda) \right) \end{cases}$$

or, because the choice of the variables $'\xi^\lambda$ is free and consequently $'\xi^\varkappa$ can be replaced by $\xi^\varkappa$

$$(3.31)\qquad \boxed{(X_c X_b) = c_{cb}^{\cdot\cdot a} X_a}\,.$$

This identity only contains operators operating on scalar functions. But it can be generalized by generalizing the operator $\underset{b}{\pounds}$ defined in (1.8). Let from now on in this section $\underset{b}{\pounds}$ be the LIE operator in the $(n+r)$-dimensional space of the $\xi^\varkappa$ and η^α belonging to the field $\underset{b}{e^a}\, \Xi^\varkappa_a,\ \underset{b}{e^\alpha}$ for every value of b.[1] Then

$$(3.32)\qquad \boxed{(\underset{c}{\pounds}\,\underset{b}{\pounds}) = c^{a}_{cb}\, \underset{a}{\pounds}}$$

is true and equivalent to the combination of (2.25) and (3.31) so long as scalar functions only are operated upon. According to (2.28) it is also true for quantities in X_r. In order to prove that it also holds for quantities in X_n it can first be proved for co- and contravariant vectors and after that the rule of LEIBNIZ can be applied.[2]

Of course besides (3.31, 32) we also have

$$(3.33)\qquad \boxed{(X_C X_B) = -\, c_{CB}^{\cdot\cdot A} X_A}$$

$$(3.34)\qquad \boxed{(\underset{C}{\pounds}\,\underset{B}{\pounds}) = -\, c^{A}_{CB}\, \underset{A}{\pounds}}\,.$$

(3.31, 33) and also (3.32, 34) are called the LIE *structural formulae* of the group and the $c_{cb}^{\cdot\cdot a}$ its *structural constants*. So a group of transformations and both its parameter-groups have the same structural formulae and the same structural constants. This is of course trivial because these groups are isomorphic.

---

[1]) In the classical proof of (3.31) (cf. LIE-ENGEL 1888, 1, p. 158ff.) the equations (3.18) are first brought back to the linear homogeneous system $\underset{b}{\pounds}\,\varphi(\xi^\varkappa, \eta^\alpha) = 0$, so the operator $\underset{b}{\pounds}$ really occurs there in its most general form, except that it is connected with scalar functions only.

[2]) It is also valid for all geometric objects having second LIE derivatives (cf. II § 10).

Collecting results we may now state the first part of the second fundamental theorem:

II.1. *If* X_b *are the symbols of r independent infinitesimal transformations of the group mentioned in* I.1

$$(3.35)\qquad X_b = \Xi^{\mu}_b(\xi^{\varkappa})\,\partial_{\mu}$$

equations of the form (3.31) *with constant coefficients* $c_{cb}^{\cdot\cdot a}$, $c_{(cb)}^{\cdot\cdot a}=0$ *hold* and the first part of the third fundamental theorem:

III.1. *The constants* $c_{cb}^{\cdot\cdot a}$ *satisfy the equations* (1.28b).

The second part of the first fundamental theorem is the converse of the first part:

I.2. *If a set of transformations of the* $\xi^{\varkappa}$ *is given*

$$(3.36)\qquad '\xi^{\varkappa} = f^{\varkappa}(\xi^{\nu}, \eta^{\alpha});\qquad \alpha = \dot{1}, \ldots, \dot{r};$$

with $\mathrm{Det}(\partial_{\lambda}{}'\xi^{\varkappa}) \neq 0$ *in an* $\mathfrak{N}(\underset{0}{\xi^{\varkappa}}, \underset{0}{\eta^{\alpha}})$, *depending on r essential parameters* η^{α} *and containing the identical transformation* $(\eta^{\alpha} = \underset{0}{\eta^{\alpha}})$ *and if there exist* r^2 *functions* $A^{\beta}_b(\eta^{\alpha})$; $\mathrm{Det}(A^{\beta}_b) \neq 0$ *and nr functions* $\Xi^{\varkappa}_b(\xi^{\lambda})$ *not satisfying an equation* $c^b\,\Xi^{\varkappa}_b = 0$ *with coefficients* c^b *independent of* $\xi^{\varkappa}$, *such that*

$$(3.37)\qquad \partial_{\beta}{}'\xi^{\varkappa} = \Xi^{\varkappa}_b('\xi^{\lambda})\,A^b_{\beta}(\eta^{\alpha}),$$

these transformations form an r-parameter group germ.

Proof: though the X_r is not yet proved to be a group space, the A^a_{β} and their inverses A^{α}_b fix an anholonomic coordinate system (a) and a connexion $\overset{+}{\Gamma}{}^{\alpha}_{\gamma\beta}$ satisfying the condition that the $\underset{b}{e^{\alpha}}$, $\overset{a}{e}_{\beta}$ are covariant constant. The geodesics of this connexion satisfy the equations

$$(3.38)\qquad \frac{d\eta^{\alpha}}{dt} = e^b A^{\alpha}_b$$

with constants e^a.

Now let $\underset{1}{\eta^{\alpha}}$ be a point of X_r in $\mathfrak{N}(\underset{0}{\eta^{\alpha}})$ corresponding to the arbitrary transformation $T_{\underset{1}{\eta}}$ of the set. Every point $'\eta^{\alpha}$ of $\mathfrak{N}(\underset{0}{\eta^{\alpha}})$ on a geodesic through $\underset{1}{\eta^{\alpha}}$ is given by

$$(3.39)\qquad '\eta^{\alpha} = (e^{t e^b \partial_b}\,\eta^{\alpha})_{\eta^{\alpha} = \underset{1}{\eta^{\alpha}}};\qquad \partial_b = A^{\alpha}_b\,\partial_{\alpha}$$

for suitable values of t and e^a. If $'\xi^{\varkappa} = f^{\varkappa}(\xi^{\nu}, {}'\eta^{\alpha})$, then, changing $'\eta^{\alpha}$ into the point of the same geodesic $'\eta^{\alpha} + d\eta^{\alpha} = {}'\eta^{\alpha} + e^a A^{\alpha}_a\,dt$ we get

$$(3.40)\qquad d'\xi^{\varkappa} = e^b\,\Xi^{\varkappa}_b('\xi^{\lambda})\,dt$$

and accordingly the transformations of the set belonging to an arbitrary point $\underset{2}{\eta}^{\alpha}$ of this geodesic is given by

$$(3.41)\qquad '\xi^{\varkappa} = (e^{t e^b X_b}\xi^{\varkappa})_{\xi^{\varkappa}=\underset{1}{\xi}^{\varkappa}};\qquad \underset{1}{\xi}^{\varkappa} = f^{\varkappa}(\xi^{\nu}, \underset{1}{\eta}^{\alpha});\qquad X_b \overset{\text{def}}{=} \Xi_b^{\mu}\,\partial_{\mu}$$

for some suitable value of t.

For fixed values of the e^b the transformations $\underset{1}{\xi}^{\varkappa} \to {}'\xi^{\varkappa}$ form a one-parameter group with the parameter t. Every point η^{α} of $\mathfrak{N}(\underset{0}{\eta}^{\alpha})$ lies on one of the geodesics through $\underset{1}{\eta}^{\alpha}$. Hence every transformation of the set can be obtained by the transformation $T_{\underset{1}{\eta}}$ followed by a transformation of one of these one-parameter groups and conversely every transformation constructed in this way belongs to the set. As $\underset{1}{\eta}^{\alpha}$ is entirely arbitrary we may take $\underset{1}{\eta}^{\alpha} = \underset{0}{\eta}^{\alpha}$ and this proves that every transformation of the set belongs to one of these groups. But that implies that two transformations of the set applied one after the other give rise to a new transformation of the set. As the set also contains the identical transformation and the inverses, it is proved that the set is a group and the X_r its group space.

The second part of the second fundamental theorem is the converse of the first part:

II.2. *If $X_b = \Xi_b^{\mu}(\xi^{\varkappa})\,\partial_{\mu}$ are the symbols of r independent infinitesimal transformations of the $\xi^{\varkappa}$ satisfying the equations (3.31) with constant coefficients c_{cb}^{a}, $c_{(cb)}^{a} = 0$; these transformations are the infinitesimal transformations of an r-parameter group germ.*

In order to prove this we remark first that the identities (cf. Exerc. IV 3,3)

$$(3.42)\qquad (X_{[d}\,(X_c\,X_{b]})) = 0$$

can easily be verified and that they imply that the c_{cb}^{a} satisfy the identities (2.11b). Now we have proved in IV §2 that according to these identities and the constancy of the c_{cb}^{a} there exist in the X_r of the auxiliary variables η^{α}, r linearly independent vector fields $\overset{a}{e}_{\beta}$ satisfying the equations (2.4). But then it follows from (3.31, 42) that the integrability conditions of (3.37) are satisfied identically. Hence these equations are totally integrable and they have solutions of the form

$$(3.43)\qquad '\xi^{\varkappa} = f^{\varkappa}(\xi^{\nu}, \eta^{\alpha});\qquad \xi^{\varkappa} = f^{\varkappa}(\xi^{\nu}, \underset{0}{\eta}^{\alpha});\qquad \operatorname{Det}(\partial_{\lambda}\,'\xi^{\varkappa}) \neq 0 \text{ for } \eta^{\alpha} = \underset{0}{\eta}^{\alpha}$$

valid in an $\mathfrak{N}(\underset{0}{\eta}^{\alpha})$ and depending on n parameters $\xi^{\varkappa}$. That finishes the proof because according to the fundamental theorem I.2, (3.43) represents an r-parameter group and the X_b are infinitesimal transformations of this group. As we see an r-parameter group germ is wholly determined by r linearly independent infinitesimal transformations satisfying (3.31) with constants c_{cb}^{a}. It is said to be *generated* by them.

The second part of the third fundamental theorem is the converse of the first part:

III.2. *If* $\frac{1}{2}r^2(r-1)$ *constants* c_{cb}^{a}; $c_{(cb)}^{a}=0$ *satisfy equations of the form* (2.11b) *there always exists an r-parameter group germ with r linearly independent infinitesimal transformations X_b satisfying* (3.31).

In order to prove this we construct in the X_r of the auxiliary variables η^α the fields $\overset{a}{e}_\beta$ satisfying (2.4). In IV § 2 we have already proved that it is possible to construct in X_r an r-parameter group of point transformations with the infinitesimal transformations $\eta^\alpha \rightarrow \eta^\alpha + e^b A_b^\alpha\, dt$ whose symbols X_b satisfy (3.31). But as the fundamental theorem II.2 is already known, it is only necessary to remark that these X_b satisfy (3.31) because then according to II.2 they generate an r-parameter group germ.[1], [2]

The results obtained so far show that the properties of an r-parameter finite continuous group germ of an abstract group correspond to and are in fact identical with the algebraic properties of a tensor $c_{cb}^{\cdot\cdot a}$, which is alternating in cb and satisfies (1.28b) and also identical with the geometric properties of a very special A_r which is symmetric with respect to all its points (cf. III § 7).

As soon as the elements are supposed to be transformations in n variables we also get properties in the space of these variables. But it makes a difference whether the elements are interpreted as point transformations or as coordinate transformations as the following example will show.

Let us assume first that the elements are point transformations in the X_n of the $\xi^\varkappa$. Let T and U be two of these transformations

$$\text{(3.44)} \qquad \begin{cases} T: \xi^\varkappa \rightarrow {}'\xi^\varkappa \\ U: \xi^\varkappa \rightarrow \bar{\xi}^\varkappa; \qquad {}'\xi^\varkappa \rightarrow {}'\bar{\xi}^\varkappa. \end{cases}$$

Then we have

$$\text{(3.45)} \qquad {}'\bar{\xi}^\varkappa = U\,{}'\xi^\varkappa = U\,T\,\xi^\varkappa = U\,T\,U^{-1}\bar{\xi}^\varkappa.$$

[1]) In the proofs of the three fundamental theorems we made use of the existence theorems of partial differential equations. For our purposes this is the natural way because sets of these equations are just what we are interested in. Freudenthal proved in 1933, 1 that it is possible to use only the existence theorem for ordinary differential equations.

[2]) E. Cartan has shown 1930, 1; 1937, 2, p. 189 (with Leray) that the original proofs of the fundamental theorems given by Lie fail if one does not confine oneself to the group germ. Of course the same holds for the proofs given here.

Hence, if T transforms $\xi^\varkappa$ into $'\xi^\varkappa$ we get the transformation of $U\xi^\varkappa$ into $U\,'\xi^\varkappa$ by applying the transformation $U\ldots U^{-1}$ of the adjoint group to T. Now let us assume that T is a point transformation and U a coordinate transformation

$$\left\{\begin{array}{l} T:\xi^\varkappa \to {}'\xi^\varkappa \\ U:\xi^\varkappa \to \xi^{\varkappa'}; \quad {}'\xi^\varkappa \to {}'\xi^{\varkappa'}. \end{array}\right. \tag{3.46}$$

Then we have

$$'\xi^{\varkappa'} = U\,'\xi^\varkappa = U\,T\,\xi^\varkappa = U\,T\,U^{-1}\,\xi^{\varkappa'} \tag{3.47}$$

and this means that T describes a point transformation with respect to $(\varkappa)$ and $U\,T\,U^{-1}$ describes the same transformation with respect to $(\varkappa')$.

In group space the transformations of the adjoint group are of course always point transformations as we have already seen in IV § 2.

Exercises.

IV 3,1. Prove that

IV 3,1 α) $$'\Xi^\varkappa_B\,(\xi^\nu) \overset{*}{=} -\,\delta^b_B\,\Xi^\varkappa_b\,(\xi^\lambda)\,.$$

IV 3,2. Prove that the functions $\Xi^\varkappa_b(\xi^\lambda)$ and $A^b_\beta(\eta^\alpha)$ can be obtained by algebraic operations and differentiations from the equations (3.17). (Use the transformations $'\xi^\varkappa = f^\varkappa(\xi^\nu, \eta^\alpha)$; $''\xi^\varkappa = f^\varkappa('\xi^\nu, \zeta^\alpha) = f^\varkappa(\xi^\nu, \theta^\alpha)$).

IV 3,3. Prove that

IV 3,3 α) $$(\underset{[c}{£}\,(\underset{b}{£}\,\underset{a]}{£})) = 0\,.$$

§ 4. The geometry of group space.

In group space we now have the general coordinates η^α, the integrable but not symmetric (+)- and (−)-connexion and the anholonomic coordinate systems (a) and (A) belonging to them, the symmetric (0)-connexion and the normal coordinates η^a with respect to $\underset{0}{\eta}^\alpha$, the first and second parameter-group with the transformations $T \to U\,T$ and $T \to T\,U$ consisting of all (−)-equipollences and (+)-equipollences, and the adjoint group with the transformations $T \to U\,T\,U^{-1}$ linear homogeneous if expressed in the η^a. In this section a number of important relations will be established.

In III § 7 the relations between general coordinates and normal coordinates were derived. If this time we take $\alpha, \beta, \gamma = 1, \ldots, r$ instead of $\varkappa, \lambda, \mu = 1, \ldots, n$ and $\mathfrak{a}, \mathfrak{b}, \mathfrak{c} = \bar{1}, \ldots, \bar{r}$ instead of $h, i, j = 1, \ldots, n$ we get from (III 7.18, 19)

$$(4.1)\quad \eta^\alpha = \delta^\alpha_{\mathfrak{a}}\eta^{\mathfrak{a}} - \frac{1}{2!}\Gamma^\alpha_{\gamma\beta}\{\underset{0}{\eta}\}\delta^{\gamma\beta}_{\mathfrak{c}\mathfrak{b}}\eta^{\mathfrak{c}}\eta^{\mathfrak{b}} - \frac{1}{3!}\Gamma^\alpha_{\gamma_2\gamma_1\beta}\{\underset{0}{\eta}\}\delta^{\gamma_2\gamma_1\beta}_{\mathfrak{c}_2\mathfrak{c}_1\mathfrak{b}}\eta^{\mathfrak{c}_2}\eta^{\mathfrak{c}_1}\eta^{\mathfrak{b}} - \cdots$$

$$(4.2)\quad \eta^{\mathfrak{a}} = \delta^{\mathfrak{a}}_\alpha\eta^\alpha + \frac{1}{2!}\delta^{\mathfrak{a}}_\alpha\Lambda^\alpha_{\gamma\beta}\{\underset{0}{\eta}\}\eta^\gamma\eta^\beta + \frac{1}{3!}\delta^{\mathfrak{a}}_\alpha\Lambda^\alpha_{\gamma_2\gamma_1\beta}\{\underset{0}{\eta}\}\eta^{\gamma_2}\eta^{\gamma_1}\eta^\beta + \cdots$$

where

$$(4.3)\quad \Gamma^\alpha_{\gamma_p\ldots\gamma_1\beta} \overset{\text{def}}{=} \partial_{(\gamma_p}\Gamma^\alpha_{\gamma_{p-1}\ldots\gamma_1\beta)} - p\,\Gamma^\delta_{(\gamma_p\gamma_{p-1}}\Gamma^\alpha_{\gamma_{p-2}\ldots\gamma_1\beta)\delta}$$

$$(4.4)\quad \begin{cases} \alpha) & \Lambda^\alpha_{\gamma\beta} \overset{\text{def}}{=} \Gamma^\alpha_{\gamma\beta} \\ \beta) & \Lambda^\alpha_{\gamma_2\gamma_1\beta} \overset{\text{def}}{=} \Gamma^\alpha_{\gamma_2\gamma_1\beta} + 3\,\Gamma^\delta_{(\gamma_2\gamma_1}\Gamma^\alpha_{\beta)\delta} \\ \gamma) & \Lambda^\alpha_{\gamma_3\gamma_2\gamma_1\beta} \overset{\text{def}}{=} \Gamma^\alpha_{\gamma_3\gamma_2\gamma_1\beta} + 4\Gamma^\delta_{(\gamma_3\gamma_2\gamma_1}\Gamma^\alpha_{\beta)\delta} + 6\,\Gamma^\delta_{(\gamma_3\gamma_2}\Lambda^\alpha_{\gamma_1\beta)\delta} - 3\Gamma^\alpha_{\delta\varepsilon}\Gamma^\delta_{(\gamma_3\gamma_2}\Gamma^\varepsilon_{\gamma_1\beta)}. \end{cases}$$

Instead of (III 7.23, 24, 25) we have here

$$(4.5)\quad \Gamma^{\mathfrak{a}}_{\mathfrak{c}_p\ldots\mathfrak{c}_1\mathfrak{b}}\{\underset{0}{\eta}\} = 0$$

$$(4.6)\quad \partial_{(\mathfrak{c}_{p+1}}\Gamma^{\mathfrak{a}}_{\mathfrak{c}_p\ldots\mathfrak{c}_1\mathfrak{b})}\{\underset{0}{\eta}\} = 0$$

$$(4.7)\quad \partial_{(\mathfrak{c}_p}\cdots\partial_{\mathfrak{c}_1}\Gamma^{\mathfrak{a}}_{\mathfrak{c}\mathfrak{b})}\{\underset{0}{\eta}\} = 0.$$

Now $A^\alpha_{\mathfrak{b}}$ and $A^{\mathfrak{a}}_\beta$ could be derived from (4.1, 2). But in group space we already have A^α_b and A^a_β and since it is very easy to find expressions for $A^{\mathfrak{a}}_b$ and $A^a_{\mathfrak{b}}$ the indirect way is better. We consider the geodesics through $\underset{0}{\eta}^\alpha$. At every point of such a geodesic we have on one hand

$$(4.8)\quad (d\eta)^a = e^a\,dt = \delta^a_{\mathfrak{a}}\,d\eta^{\mathfrak{a}}$$

and on the other hand

$$(4.9)\quad (d\eta)^a = A^a_{\mathfrak{a}}\,d\eta^{\mathfrak{a}},$$

and (a) and $(\mathfrak{a})$ have the same basis vectors at $\underset{0}{\eta}^\alpha$. Hence, because $d\eta^{\mathfrak{a}} \propto \eta^{\mathfrak{a}}$ we have

$$(4.10)\quad \eta^{\mathfrak{a}}A^a_{\mathfrak{a}} = \eta^{\mathfrak{a}}\delta^a_{\mathfrak{a}}$$

though $A^a_{\mathfrak{a}} = \delta^a_{\mathfrak{a}}$ at $\underset{0}{\eta}^\alpha$ only. By differentiation of (4.10) we get

$$(4.11)\quad \eta^{\mathfrak{b}}\partial_{\mathfrak{c}}A^a_{\mathfrak{b}} + A^a_{\mathfrak{c}} = \delta^a_{\mathfrak{c}}.$$

The identities (1.19) are true for every choice of the coordinate system (α), hence

$$(4.12)\quad \partial_{[\mathfrak{c}}A^a_{\mathfrak{b}]} = -\tfrac{1}{2}\,c_{cb}^{\cdot\cdot a}A^{cb}_{\mathfrak{c}\mathfrak{b}}$$

and if this is substituted in (4.11) we get

$$-\eta^{\mathfrak{b}}\, c_{c b}^{\cdot\cdot a}\, A^{c b}_{c \mathfrak{b}} + \eta^{\mathfrak{b}}\, \partial_{\mathfrak{b}} A^{a}_{\mathfrak{c}} + A^{a}_{\mathfrak{c}} = \delta^{a}_{\mathfrak{c}}. \tag{4.13}$$

For differentiation along the geodesic we have

$$\frac{d}{dt} = \frac{d\eta^{\beta}}{dt}\,\partial_{\beta} = \frac{d\eta^{\mathfrak{b}}}{dt}\,\partial_{\mathfrak{b}} = \delta^{\mathfrak{b}}_{b}\, e^{b}\, \partial_{\mathfrak{b}} = \frac{1}{t}\,\eta^{\mathfrak{b}}\,\partial_{\mathfrak{b}}; \tag{4.14}$$

hence, according to (2.36, 40) and (4.13, 14)

$$\frac{d}{dt} A^{a}_{\mathfrak{c}} + \frac{1}{t}\,(A^{a}_{\mathfrak{c}} - \delta^{a}_{\mathfrak{c}}) = c_{c b}^{\cdot\cdot a}\, A^{c}_{\mathfrak{c}}\, e^{b} = -E_{c}^{\cdot a}\, A^{c}_{\mathfrak{c}}; \qquad E_{b}^{\cdot a} \overset{\text{def}}{=} e^{c}\, c_{c b}^{\cdot\cdot a}. \tag{4.15}$$

Because $A^{a}_{\mathfrak{c}}\{\underset{0}{\eta}\} = \delta^{a}_{\mathfrak{c}}$ this equation can be compared with the ordinary differential equation

$$\frac{dx}{dt} + \frac{1}{t}\,(x - \underset{0}{x}) = \alpha\, x; \qquad x\{0\} = \underset{0}{x}; \tag{4.16}$$

whose well known solution can be written as a series

$$x = \underset{0}{x}\,\frac{e^{\alpha t} - 1}{\alpha t} = \underset{0}{x}\left(1 + \frac{1}{2!}\,\alpha t + \frac{1}{3!}\,\alpha^{2} t^{2} + \cdots\right) \tag{4.17}$$

and from this we derive the solution of (4.15) in the form

$$A^{a}_{\mathfrak{b}} = \delta^{b}_{\mathfrak{b}}\left(\frac{e^{-Et} - A}{-Et}\right)^{a}_{b} = \delta^{b}_{\mathfrak{b}}\left(A - \frac{1}{2!}\,E t + \frac{1}{3!}\,E^{2} t^{2} - \cdots\right)^{a}_{b}. \tag{4.18}$$

The inversion of (4.17)

$$\underset{0}{x} = x\,\frac{\alpha t}{e^{\alpha t} - 1} = x\,(1 + \lambda_{1}\,\alpha t + \lambda_{2}\,\alpha^{2} t^{2} + \cdots) \tag{4.19}$$

with

$$\lambda_{q} = (-1)^{q}\,\frac{B_{q}}{q!}, \tag{4.20}$$

where B_q are the BERNOULLI numbers[1]) $B_0 = 1$; $B_1 = \frac{1}{2}$; $B_2 = \frac{1}{6}$; $B_3 = 0$; $B_4 = -\frac{1}{30}$; $B_5 = 0$; $B_6 = \frac{1}{42}$; $B_7 = 0$; $B_8 = -\frac{1}{30}$, gives the inversion of (4.18)

$$A^{\mathfrak{a}}_{b} = \delta^{\mathfrak{a}}_{a}\left(\frac{-Et}{e^{-Et} - A}\right)^{a}_{b} = \delta^{\mathfrak{a}}_{a}\,(A - \lambda_{1}\, E t + \lambda_{2}\, E^{2} t^{2} - \cdots)^{a}_{b}. \tag{4.21}$$

In the same way we find

$$A^{A}_{\mathfrak{b}} = \delta^{b A}_{\mathfrak{b} a}\left(\frac{e^{Et} - A}{Et}\right)^{a}_{b} = \delta^{b A}_{\mathfrak{b} a}\left(A + \frac{1}{2!}\,E t + \frac{1}{3!}\,E^{2} t^{2} + \cdots\right)^{a}_{b}; \tag{4.22}$$

$$A^{\mathfrak{a}}_{B} = \delta^{\mathfrak{a}\, b}_{a B}\left(\frac{Et}{e^{Et} - A}\right)^{a}_{b} = \delta^{\mathfrak{a}\, b}_{a B}\,(A + \lambda_{1}\, E t + \lambda_{2}\, E^{2} t^{2} + \cdots)^{a}_{b}. \tag{4.23}$$

[1]) Cf. for instance Pascal Repertorium I, p. 437.

From (4.18, 23) and (4.21, 22) we find by transvection once more that

$$A_B^a = \delta_B^b \left(\frac{e^{-Et} - A}{-Et} \cdot \frac{Et}{e^{Et} - A} \right)_b^a = \delta_B^b \, (e^{-Et})_b^a \tag{4.24}$$

$$A_b^A = \delta_a^A \, (e^{Et})_b^a \tag{4.25}$$

in agreement with (2.42b).

The matrix notation can be used to compute the Γ_{cb}^a. From (III 7.28) we get in group space

$$\Gamma_{cb}^a = \eta^b \underset{0}{N}_{bcb}^{\cdot\cdot\cdot a} + \tfrac{1}{2} \eta^{b_2} \eta^{b_1} \underset{0}{N}_{b_2 b_1 cb}^{\cdot\cdot\cdot\cdot a} + \cdots. \tag{4.26}$$

According to (III 7.36) and (1.29) the first term of the right hand side takes the form

$$\tfrac{2}{3} \eta^b \underset{0}{R}_{b(cb)}^{\cdot\cdot\cdot a} = \tfrac{1}{6} \eta^b \underset{0}{c}_{b(c}^{\cdot\cdot e} \underset{0}{c}_{b)e}^{\cdot\cdot a}. \tag{4.27}$$

The normal tensor of valence 5 is zero and the normal tensor of valence 6 can be written as the transvection of two tensors $R_{bcb}^{\cdot\cdot\cdot a}$ (cf. III § 7) from which four factors $c_{cb}^{\cdot\cdot a}$ result. The result is a series with an even number of factors $\underset{0}{c}_{cb}^{\cdot\cdot a}$ in each term. But this result can be obtained much more easily in the following way.

Because the $\overset{a}{e}_\beta$ are (+)-constant we have

$$\text{a)}\ \Gamma_{cb}^a = \overset{+}{\Gamma}_{(cb)}^a = A_a^a \, \partial_{(c} A_{b)}^a; \quad \text{b)}\ c_{cb}^{\cdot\cdot a} = -2 \overset{+}{S}_{cb}^{\cdot\cdot a} = -2 A_a^a \, \partial_{[c} A_{b]}^a. \tag{4.28}$$

Hence, writing P for the matrix $e^c \, c_{cb}^{\cdot\cdot a} \, t = \eta^c \, \delta_c^c \, c_{cb}^{\cdot\cdot a}$ and C_c for the matrix $c_{cb}^{\cdot\cdot a}$, we have

$$\partial_c P = \delta_c^c \, C_c \tag{4.29}$$

and (4.18) can be written

$$A_b^a = \delta_b^b \left(A - \frac{1}{2!} P + \frac{1}{3!} P^2 - \frac{1}{4!} P^3 + \cdots \right)_b^a. \tag{4.30}$$

From this we get, taking into account the non-commutativity of C and P

$$\left\{ \begin{aligned} \partial_c A_b^a = \delta_{cb}^{cb} \Big(& -\frac{1}{2!} C_c + \frac{1}{3!} C_c P + \frac{1}{3!} P C_c - \\ & - \frac{1}{4!} C_c P^2 - \frac{1}{4!} P C_c P - \frac{1}{4!} P^2 C_c + \cdots \Big) \end{aligned} \right. \tag{4.31}$$

and after some calculation (note that the factor A_a^a in (4.28a) comes in matrix notation at the right hand side)

$$\left\{ \begin{aligned} (\partial_c A_b^a) \, A_a^a &= \delta_{bac}^{bac} \big(-\tfrac{1}{2} C_c - \tfrac{1}{12} C_c P + \tfrac{1}{6} P C_c - \tfrac{1}{24} C_c P^2 + \\ &\qquad + \tfrac{1}{24} P C_c P - \tfrac{1}{24} P^2 C_c + \cdots \big)_b^a \\ &= \delta_{bac}^{bac} \big(-\tfrac{1}{2} c_{cb}^{\cdot\cdot a} - \tfrac{1}{12} c_{cb}^{\cdot\cdot d} \eta^e \delta_e^e c_{ed}^{\cdot\cdot a} + \tfrac{1}{6} \eta^e \delta_e^e c_{eb}^{\cdot\cdot d} c_{cd}^{\cdot\cdot a} - \\ &\qquad - \tfrac{1}{24} c_{cb}^{\cdot\cdot d_1} \eta^{e_1} \delta_{e_1}^{e_1} c_{e_1 d_1}^{\cdot\cdot d_2} \eta^{e_2} \delta_{e_2}^{e_2} c_{e_2 d_2}^{\cdot\cdot a} + \\ &\qquad + \tfrac{1}{24} \eta^{e_1} \delta_{e_1}^{e_1} c_{e_1 b}^{\cdot\cdot d_1} c_{cd_1}^{\cdot\cdot d_2} \eta^{e_2} \delta_{e_2}^{e_2} c_{e_2 d_2}^{\cdot\cdot a} - \\ &\qquad - \tfrac{1}{24} \eta^{e_1} \delta_{e_1}^{e_1} c_{e_1 b}^{\cdot\cdot d_1} \eta^{e_2} \delta_{e_2}^{e_2} c_{e_2 d_1}^{\cdot\cdot d_2} c_{cd_2}^{\cdot\cdot a} + \cdots \big). \end{aligned} \right. \tag{4.32}$$

Hence, making use of the fact that (a) and $(\mathfrak{a})$ have the same basis vectors at $\underset{0}{\eta}{}^{\alpha}$ we get

$$
(4.33)\quad \begin{cases} \overset{+}{\Gamma}{}^{\mathfrak{a}}_{\mathfrak{cb}} = -\frac{1}{2}\underset{0}{C}{}_{\mathfrak{cb}}^{\cdot\cdot\mathfrak{a}} - \frac{1}{12}\eta^{\mathfrak{e}}\underset{0}{C}{}_{\mathfrak{cb}}^{\cdot\cdot\mathfrak{d}}\underset{0}{C}{}_{\mathfrak{ed}}^{\cdot\cdot\mathfrak{a}} + \frac{1}{6}\eta^{\mathfrak{e}}\underset{0}{C}{}_{\mathfrak{eb}}^{\cdot\cdot\mathfrak{d}}\underset{0}{C}{}_{\mathfrak{cd}}^{\cdot\cdot\mathfrak{a}} - \\ \quad - \frac{1}{24}\eta^{\mathfrak{e}_1}\eta^{\mathfrak{e}_2}\underset{0}{C}{}_{\mathfrak{cb}}^{\cdot\cdot\mathfrak{d}_1}\underset{0}{C}{}_{\mathfrak{e}_1\mathfrak{d}_1}^{\cdot\cdot\mathfrak{d}_2}\underset{0}{C}{}_{\mathfrak{e}_2\mathfrak{d}_2}^{\cdot\cdot\mathfrak{a}} + \frac{1}{24}\eta^{\mathfrak{e}_1}\eta^{\mathfrak{e}_2}\underset{0}{C}{}_{\mathfrak{e}_1\mathfrak{b}}^{\cdot\cdot\mathfrak{d}_1}\underset{0}{C}{}_{\mathfrak{cd}_1}^{\cdot\cdot\mathfrak{d}_2}\underset{0}{C}{}_{\mathfrak{e}_2\mathfrak{d}_2}^{\cdot\cdot\mathfrak{a}} - \\ \quad - \frac{1}{24}\eta^{\mathfrak{e}_1}\eta^{\mathfrak{e}_2}\underset{0}{C}{}_{\mathfrak{e}_1\mathfrak{b}}^{\cdot\cdot\mathfrak{d}_1}\underset{0}{C}{}_{\mathfrak{e}_2\mathfrak{d}_1}^{\cdot\cdot\mathfrak{d}_2}\underset{0}{C}{}_{\mathfrak{cd}_2}^{\cdot\cdot\mathfrak{a}} + \cdots \end{cases}
$$

and by mixing and alternating over $\mathfrak{cb}$

$$
(4.34)\quad \Gamma^{\mathfrak{a}}_{\mathfrak{cb}} = \overset{+}{\Gamma}{}^{\mathfrak{a}}_{(\mathfrak{cb})} = \frac{1}{6}\eta^{\mathfrak{e}}\underset{0}{C}{}_{\mathfrak{e}(\mathfrak{b}}^{\cdot\cdot\mathfrak{d}}\underset{0}{C}{}_{\mathfrak{c})\mathfrak{d}}^{\cdot\cdot\mathfrak{a}} + \text{terms of an odd degree} \geqq 3 \text{ in } \eta^{\mathfrak{a}}
$$

in accordance with (4.27) and

$$
(4.35)\quad \begin{cases} C_{\mathfrak{cb}}^{\cdot\cdot\mathfrak{a}} = -2\overset{+}{\Gamma}{}^{\mathfrak{a}}_{[\mathfrak{cb}]} = \underset{0}{C}{}_{\mathfrak{cb}}^{\cdot\cdot\mathfrak{a}} + \frac{1}{12}\eta^{\mathfrak{e}_1}\eta^{\mathfrak{e}_2}\underset{0}{C}{}_{\mathfrak{cb}}^{\cdot\cdot\mathfrak{d}_1}\underset{0}{C}{}_{\mathfrak{e}_1\mathfrak{d}_1}^{\cdot\cdot\mathfrak{d}_2}\underset{0}{C}{}_{\mathfrak{e}_2\mathfrak{d}_2}^{\cdot\cdot\mathfrak{a}} - \\ \quad - \frac{1}{12}\eta^{\mathfrak{e}_1}\eta^{\mathfrak{e}_2}\underset{0}{C}{}_{\mathfrak{e}_1[\mathfrak{c}}^{\cdot\cdot\mathfrak{d}_1}\underset{0}{C}{}_{|\mathfrak{e}_2|\mathfrak{b}]}^{\cdot\cdot\mathfrak{d}_2}\underset{0}{C}{}_{\mathfrak{d}_1\mathfrak{d}_2}^{\cdot\cdot\mathfrak{a}} \\ \quad + \text{terms of an even degree} \geqq 4 \text{ in } \eta^{\mathfrak{a}}. \end{cases}
$$

By a reflexion at the point $\underset{0}{\eta}{}^{\mathfrak{a}} = 0$ a field with the values $v^{\mathfrak{a}}$ at $\eta^{\mathfrak{a}}$ is changed into a field with the value $u^{\mathfrak{a}} = -v^{\mathfrak{a}}$ at $-\eta^{\mathfrak{a}}$. If v^a are the (a)-components of the first field at $\eta^{\mathfrak{a}}$, it follows from (4.18, 22) that the u^a at $-\eta^{\mathfrak{a}}$ are not equal to $-v^a$ but that $u^A = -\delta^A_a v^a$. The reflection of a $(+)$-parallel displacement of $v^{\mathfrak{a}}$ over $d\eta^{\mathfrak{a}}$ at $\eta^{\mathfrak{a}}$ is a $(-)$-parallel displacement of $u^{\mathfrak{a}}$ over $-d\eta^{\mathfrak{a}}$ at $-\eta^{\mathfrak{a}}$ because $(+)$-equipollency transforms into $(-)$-equipollency by a reflexion at the origin. Hence the reflexion of a (0)-parallel displacement of $v^{\mathfrak{a}}$ over $d\eta^{\mathfrak{a}}$ at $\eta^{\mathfrak{a}}$ is a (0)-parallel displacement of $u^{\mathfrak{a}}$ over $-d\eta^{\mathfrak{a}}$ at $-\eta^{\mathfrak{a}}$. Accordingly, taking these (0)-displacements we get

$$
(4.36)\quad dv^{\mathfrak{a}} = -d\eta^{\mathfrak{c}}\Gamma^{\mathfrak{a}}_{\mathfrak{cb}}\{\eta\}v^{\mathfrak{b}} = -du^{\mathfrak{a}} = +d\eta^{\mathfrak{c}}\Gamma^{\mathfrak{a}}_{\mathfrak{cb}}\{-\eta\}v^{\mathfrak{b}}
$$

or

$$
(4.37)\quad \Gamma^{\mathfrak{a}}_{\mathfrak{cb}}\{\eta\} = -\Gamma^{\mathfrak{a}}_{\mathfrak{cb}}\{-\eta\};
$$

and this proves that only terms of an *odd* degree in $\eta^{\mathfrak{a}}$ occur in the series of $\Gamma^{\mathfrak{a}}_{\mathfrak{cb}}$ in accordance with (4.34).

Taking a $(+)$-displacement at $\eta^{\mathfrak{a}}$ and its reflexion at $-\eta^{\mathfrak{a}}$ we get

$$
dv^{\mathfrak{a}} = -d\eta^{\mathfrak{c}}\overset{+}{\Gamma}{}^{\mathfrak{a}}_{\mathfrak{cb}}\{\eta\}v^{\mathfrak{b}} = -du^{\mathfrak{a}} = +d\eta^{\mathfrak{c}}\overset{-}{\Gamma}{}^{\mathfrak{a}}_{\mathfrak{cb}}\{-\eta\}v^{\mathfrak{b}}
$$

from which

$$
C_{\mathfrak{cb}}^{\cdot\cdot\mathfrak{a}}\{\eta\} = -2\overset{+}{\Gamma}{}^{\mathfrak{a}}_{[\mathfrak{cb}]}\{\eta\} = 2\overset{-}{\Gamma}{}^{\mathfrak{a}}_{[\mathfrak{cb}]}\{-\eta\} = C_{\mathfrak{cb}}^{\cdot\cdot\mathfrak{a}}\{-\eta\};
$$

and this proves that only terms of an *even* degree in $\eta^{\mathfrak{a}}$ occur in the series of $C_{\mathfrak{cb}}^{\cdot\cdot\mathfrak{a}}$ in accordance with (4.35).

§ 5. Invariants of a transformation group in the X_n of the $\xi^\varkappa$.

In the X_n of the $\xi^\varkappa$ we may have invariant functions of the $\xi^\varkappa$, invariant subspaces or sets of subspaces, invariant fields of quantities or geometric objects, and invariant systems of partial differential equations.

A *subspace* is called *invariant* if its points are interchanged by the transformations of the group; a set of subspaces if the subspaces are interchanged. A *field* is called *invariant* if it does not change when dragged along by the transformations of the group. A *system of differential equations* is called *invariant* if by dragging along all solutions pass into solutions.

We shall deal here only with the most important cases of invariance.

1. *Invariant functions*[1]). A function $\varphi(\xi^\varkappa)$ is invariant if and only if

$$(5.1)\qquad X_b\,\varphi = \Xi^\mu_b\,\partial_\mu\,\varphi = 0;\qquad b = 1,\ldots,r.^{2)}$$

The infinitesimal transformations X_b are independent but they may be connected. The number of linearly independent equations among (5.1) is equal to the b-rank of $\Xi^\varkappa_b$. If this b-rank is $q \leq r$, (5.1) is equivalent to a system of q linearly independent equations and because of

$$(5.2)\qquad 2\,\Xi^\nu_{[c}\,\partial_{|\nu|}\,\Xi^\mu_{b]}\,\partial_\mu = (X_c\,X_b) = c_{cb}^{\cdot\cdot a}\,X_a = c_{cb}^{\cdot\cdot a}\,\Xi^\mu_a\,\partial_\mu;$$

this system is complete (cf. II § 5) and has $n-q$ independent solutions $\overset{1}{\varphi},\ldots,\overset{n-q}{\varphi}$ of which every other solution is a function. The equations $\overset{1}{\varphi} = \text{const.},\ldots,\overset{n-q}{\varphi} = \text{const.}$ represent a set of ∞^{n-q} X_q's and each X_q of this set is invariant for transformations of the group.

If the finite equations of the group

$$(5.3)\qquad {}'\xi^\varkappa = f^\varkappa(\xi^\nu, \eta^\alpha)$$

are given we can see that there must be invariant X_q's by giving the $\xi^\varkappa$ fixed values $\underset{0}{\xi^\varkappa}$ and letting the η^α vary. Then an X_q is described because $\partial_\beta f^\varkappa$ has the rank q. (5.3) represents a system of n equations in $2n+r$ variables, which is regular of dimension $n+r$ and whose rank with respect to the variables η^α is q. Hence according to the well-known theorem of elimination[3]) exactly $2n+r-n-r-q = n-q$ equations not containing the η^α can be derived from (5.3) and from these

---

[1]) Lie-Engel 1888, 1, p. 211 ff.

[2]) The invariance for finite transformations follows from the series (2.31) and also from the series (2.32) because (5.1) can also be written in the form $\underset{b}{\pounds}\,\varphi = 0$.

[3]) P. P. 1949, 1, p. 42.

equations $n-q$ of the $'\xi^\varkappa$ can be solved

$$(5.4)\quad '\xi^{\mathfrak{a}} = \psi^{\mathfrak{a}}('\xi^{\mathfrak{x}}, \xi^\varkappa); \qquad \mathfrak{a} = 1, \ldots, n-q; \qquad \mathfrak{x} = n-q+1, \ldots, n.$$

Now we know that there exist $n-q$ invariant functions $\overset{\mathfrak{a}}{\varphi}$; $\mathfrak{a} = 1, \ldots, n-q$; and this proves that the set of $n-q$ equations

$$(5.5)\qquad \overset{\mathfrak{a}}{\varphi}(\xi^\varkappa) = \overset{\mathfrak{a}}{\varphi}('\xi^\varkappa)$$

is equivalent to the set (5.4). Hence, if we give the $'\xi^\varkappa$ in (5.4) constant values $c^1, \ldots, c^n$ the equations

$$(5.6)\qquad \overset{\mathfrak{a}}{\psi}(c^{\mathfrak{x}}, \xi^\varkappa) = \text{const.}$$

represent the ∞^{n-q} invariant X_q's and the left hand side gives a set of $n-q$ independent invariant functions for each choice of the q constants $c^{\mathfrak{x}}$.

2. *Invariant subspaces*[1]). A set of independent functions $\overset{\mathfrak{a}}{\varphi}(\xi^\varkappa)$ is called *relative invariant*[2]) if the equations

$$(5.7)\qquad X_b \overset{\mathfrak{a}}{\varphi} = \Xi_b^\mu \partial_\mu \overset{\mathfrak{a}}{\varphi} = 0; \qquad \mathfrak{a} = 1, \ldots, n-q;$$

are a consequence of $\overset{\mathfrak{a}}{\varphi} = 0$. This means that the X_q represented by $\overset{\mathfrak{a}}{\varphi} = 0$ is invariant. For every choice of the constants e^b the X_q is built up by streamlines of the field $e^b \Xi_b^\varkappa$. Therefore the dimension q is minimal if and only if q is the rank of $\Xi_b^\varkappa$. Such an X_q through a point $\underset{0}{\xi^\varkappa}$ is called the *smallest invariant subspace* through $\underset{0}{\xi^\varkappa}$. Hence, in order to find all invariant subspaces of a group we have only to construct all points where $\Xi_b^\varkappa$ has a rank q for all values $q \leq r$.[3])

The group interchanges the points of X_q and we may require the group of these transformations. Let

$$(5.8)\qquad \xi^\varkappa = \Phi^\varkappa(y^h); \qquad h = 1, \ldots, q;$$

be the parametric equations of the X_q and let $Y_b \overset{\text{def}}{=} Y_b^j \partial_j$ be the infinitesimal point transformations in X_q corresponding to X_b. Then, if $f(\xi^\varkappa) = f(\Phi^\varkappa(y^h))$ is given at all points of X_q we have for the subgroup generated by $e^b X_b$

$$(5.9)\qquad \frac{df}{dt} = e^b \Xi_b^\mu \partial_\mu f$$

[1]) LIE-ENGEL 1888, 1, p. 222ff.
[2]) EISENHART 1933, 1, p. 63.
[3]) LIE-ENGEL 1888, 1, p. 228, 237.

and for the subgroup generated by $e^b Y_b$

$$\frac{df}{dt} = e^b \mathsf{Y}_b^j B_j^\mu \partial_\mu f; \qquad B_j^\varkappa \stackrel{\text{def}}{=} \partial_j \xi^\varkappa; \tag{5.10}$$

hence

$$\Xi_b^\varkappa = B_h^\varkappa \mathsf{Y}_b^h. \tag{5.11}$$

In general such an equation can not exist because the $\varkappa$-domain of $\Xi_b^\varkappa$ does not coincide with the $\varkappa$-domain of $B_h^\varkappa$. But here the X_q is invariant and this means that its tangent E_q is the support of both domains. This implies that the Y_b^h can be solved from (5.11). The easiest way to find this solution is to choose the coordinate system such that $\xi^{q+1}=0$, $\ldots, \xi^n = 0$ are the equations of the X_q and to take $\xi^1, \ldots, \xi^q$ (suitably chosen) as parameters on the X_q. Then $h=1, \ldots, q$ and

$$B_h^\varkappa = \delta_h^\varkappa; \quad \mathsf{Y}_b^h = \Xi_b^h; \quad \Xi_b^{q+1} = 0; \ldots; \quad \Xi_b^n = 0. \tag{5.12}$$

Hence we get the $\mathsf{Y}_b^h \partial_h$ by dropping in $\Xi_b^\mu \partial_\mu$ all differentiations with respect to $\xi^{q+1}, \ldots, \xi^n$. From this it follows immediately that the group of point transformations in X_q found in this way, and called the group *induced* in X_q, has r infinitesimal transformations Y_b satisfying the equations

$$(Y_c Y_b) = c_{cb}^{\cdot\cdot a} Y_a. \tag{5.13}$$

These r transformations *need not be independent*. Take for instance the case where the X_b are independent but connected. Then there exists at least one equation of the form $\psi^b(\xi^\varkappa) X_b = 0$. It may happen that the $\psi^b(\xi^\varkappa)$ are constant over X_q and in that case the Y_b are not independent. Let the induced group be r'-parametrical, or in other words, let there exist exactly $r-r'$ relations

$$\underset{p}{c^b} Y_b = 0; \quad p = 1, \ldots, r-r'; \quad \underset{p}{c^b} = \text{const}. \tag{5.14}$$

Then there must also exist exactly $r-r'$ relations

$$\underset{p}{c^b} X_b = 0 \tag{5.15}$$

valid at all points of X_q. But that implies that there exist $r-r'$ independent infinitesimal transformations of the original group that leave *every point* of X_q invariant. Of course these transformations form an $(r-r')$-parameter subgroup:

The group induced in an invariant X_q is an r'-parameter group if and only if there exists an $(r-r')$-parameter subgroup of the original group, leaving each point of X_q individually invariant.

A group is called *transitive* if the rank of $\Xi_b^\varkappa$ is n at all points of an $\mathfrak{N}(\underset{0}{\xi^\varkappa})$. If the group is transitive, to every point pair P and Q in a sufficiently small $\mathfrak{N}(\overset{}{\underset{0}{\xi^\varkappa}})$ there is at least one transformation of the group transforming P into Q. Of course for a transitive group $r \geq n$. The group is called *simply transitive* if $r = n$; then there exists exactly one transformation transforming P into Q.

3. *Invariant contravariant vector fields.* A contravariant vector field $Z^\varkappa$ can be looked upon as an infinitesimal transformation $\xi^\varkappa \to \xi^\varkappa + Z^\varkappa dt$ with the symbol $Z \overset{\text{def}}{=} Z^\mu \partial_\mu$. Hence invariance of infinitesimal transformations and invariance of contravariant vector fields is essentially the same problem. Besides the complete invariance defined in the beginning of this section, there are less restricted forms of invariance. We distinguish:

a) invariance to within an arbitrary factor;
b) invariance to within a constant factor;
c) complete invariance.

a) If a field $Z^\varkappa$ is subjected to the infinitesimal transformation X_b of a group, the increase of the field is equal to the negative LIE differential with respect to the field $\underset{b}{e^a}\, \Xi_a^\varkappa$:

$$dZ^\varkappa = -dt \underset{b}{\pounds} Z^\varkappa = -\underset{b}{e^a}\left(\Xi_a^\mu \partial_\mu Z^\varkappa dt - Z^\mu \partial_\mu \Xi_a^\varkappa dt\right) \tag{5.16}$$

or in another form, writing Z for the operator $Z^\mu \partial_\mu$

$$dZ = \underset{b}{e^a} (Z X_a)\, dt. \tag{5.17}$$

Instead of one vector field we now take a set of p independent vector fields $\underset{y}{Z^\varkappa}$; $y = 1, \ldots, p$ and consider the case where the $\underset{b}{\pounds} \underset{y}{Z^\varkappa}$ depend linearly on the $\underset{y}{Z^\varkappa}$. This is a generalization of invariance to within an arbitrary factor:

$$\underset{b}{\pounds} \underset{y}{Z^\varkappa} = \alpha_{b\,y}^{x} (\xi^\varkappa) \underset{x}{Z^\varkappa} \tag{5.18}$$

or in another form, introducing the LIE operators $\underset{y}{\pounds}$ belonging to the fields $\underset{y}{Z^\varkappa}$

$$\left(\underset{b}{\pounds} \underset{y}{\pounds}\right) s = \alpha_{b\,y}^{x} (\xi^\varkappa) \underset{x}{\pounds} s \text{ [1]} \tag{5.19}$$

for every scalar s. The equations (5.18, 19) express the fact that the E_p-field spanned by the p vectors $\underset{y}{Z^\varkappa}$ is invariant. This E_p-field is X_p-forming if and only if the equations $\underset{y}{Z^\mu} \partial_\mu f = 0$ form a complete system (cf. II § 5):

$$\left(\underset{y}{Z} \underset{z}{Z}\right) = \beta_{y z}^{x} (\xi^\varkappa) \underset{x}{Z} \tag{5.20}$$

[1] Note that (5.19) in general does not hold if s is replaced by an arbitrary non-scalar because the $\alpha_{b\,y}^{x}$ are not constant.

or in another form

$$(5.21)\qquad (\underset{y}{\pounds}\,\underset{z}{\pounds})\, s = \beta^{x}_{y z}(\xi^{\varkappa})\,\underset{x}{\pounds}\, s$$

and the equations of the X_p's are obtained by equating the solutions of $\underset{y}{Z}^{\mu}\,\partial_{\mu} f = 0$ to constants. If f is such a solution, then we get from (5.19)

$$(5.22)\qquad \underset{y}{\pounds}\,\underset{b}{\pounds} f = \underset{b}{\pounds}\,\underset{y}{\pounds} f - \alpha^{x}_{b\,y}(\xi^{\varkappa})\,\underset{x}{\pounds} f$$

and this proves that $\underset{b}{\pounds} f$ is also a solution. Every solution of the complete system is a function of $n-p$ independent solutions, hence (cf. II 10.38) the group transforms the set of all solutions into itself and also the set of all ∞^{n-p} X_p's (but not every X_p individually!).

If a group has an invariant complete system (that is an invariant set of ∞^{n-p} X_p's forming a normal system) it is called *imprimitive* and if not it is said to be *primitive*. If a group has an invariant X_p, $p<n$, it can not be transitive, but an imprimitive group can be transitive.

b)[1]) If in (5.18, 19) all coefficients are constants we have a generalization of invariance to within a constant factor:

$$(5.23)\qquad \text{a)}\ \underset{b}{\pounds}\,\underset{y}{Z}^{\varkappa} = k^{x}_{b\,y}\,\underset{x}{Z}^{\varkappa} \quad\text{or}\quad \text{b)}\ (\underset{b}{\pounds}\,\underset{y}{\pounds}) = k^{x}_{b\,y}\,\underset{x}{\pounds}; \qquad k^{x}_{b\,y} = \text{const}.$$

For $p=1$ not only the streamlines of $Z^{\varkappa}$ are invariant but the field $Z^{\varkappa}$ gets a constant factor. For general values of p the infinitesimal transformation $\overset{y}{s}\underset{y}{Z}$ with constant coefficients $\overset{y}{s}$ generates a one-parameter group. These transformations together form a linear set (Schar) of infinitesimal transformations. If the $\underset{y}{Z}$ are independent (though possibly connected) it can be proved[2]) that this linear set always generates ∞^{p} finite transformations. The statement is evidently true if the $\underset{y}{Z}$ belong to one group (though not generating this group). In fact in this case each one-parameter group is represented by a geodesic through $\underset{0}{\eta}^{\alpha}$ and these geodesics form an X_p in $\mathfrak{N}(\underset{0}{\eta}^{\alpha})$. If the $\underset{y}{Z}$ are arbitrary we have only to prove that in the transformations $'\xi^{\varkappa} = e^{t\overset{y}{s}\underset{y}{Z}}\,\xi^{\varkappa}$ the $t\overset{y}{s}$ are essential parameters (cf. IV § 3). Now if these parameters were not essential there would exist equations of the form

$$(5.24)\qquad 0 = \overset{x}{\varphi}(t\overset{z}{s})\,\frac{\partial}{\partial t\overset{x}{s}}\,\overset{x}{s}\,e^{t\overset{y}{s}\underset{y}{Z}}\,\xi^{\varkappa} = \overset{x}{\varphi}\,\underset{x}{Z}\,e^{t\overset{y}{s}\underset{y}{Z}}\,\xi^{\varkappa} = \overset{y}{\varphi}\,\underset{y}{Z}\,'\xi^{\varkappa}$$

and this is only possible if $\overset{x}{\varphi}=0$ because the $\underset{y}{Z}$ are supposed to be independent.

---

[1]) LIE-ENGEL 1888, 1, p. 246ff.

[2]) LIE-ENGEL 1888, 1, p. 62.

If the infinitesimal transformations $\overset{y}{s}\underset{y}{Z}$ are subjected to the infinitesimal transformation $e^b X_b$ of the group, we have

$$\frac{d}{dt}\overset{y}{s}\underset{y}{Z}^\varkappa = -\overset{y}{s}\overset{b}{e}\underset{b}{£}\underset{y}{Z}^\varkappa = -\overset{y}{s}\overset{b}{e}k_{by}^{x}\underset{x}{Z}^\varkappa . \tag{5.25}$$

This equation expresses that the infinitesimal transformation characterized by the constants $\overset{y}{s}$ is transformed into the transformation characterized by the constants $\overset{y}{s} - \overset{b}{e}\overset{x}{s}k_{bx}^{y}dt$. Hence, according to (II 10.38) we have for the transformation of the $\overset{y}{s}$ under a finite transformation of the group

$$\overset{y}{s}(t) = (e^{-tK})_x^y \overset{x}{s}(0) \tag{5.26}$$

where K stands for the matrix of $\overset{b}{e}k_{by}^{x}$. From this we see that the linear set of infinitesimal transformations $\overset{y}{s}\underset{y}{Z}$ is invariant for all finite transformations of the group germ. A finite transformation of the one-parameter group generated by $\overset{y}{s}(0)\underset{y}{Z}$ has the form

$$'\xi^\varkappa = e^{u\overset{y}{s}(0)\underset{y}{Z}}\xi^\varkappa \tag{5.27}$$

with some parameter u. Now if a finite transformation of the group is applied, the transformation (5.27) is transformed into

$$'\xi^\varkappa = e^{u\overset{y}{s}(t)\underset{y}{Z}}\xi^\varkappa \tag{5.28}$$

and this is a finite transformation of the group generated by $\overset{y}{s}(t)\underset{y}{Z}$. Hence the one-parameter groups generated by $\overset{y}{s}\underset{y}{Z}$ for all different values of $\overset{y}{s}$ are interchanged by the transformations of the group. This proves *that (5.23) is the n.a.s. condition for the invariance of the linear set $\overset{y}{s}\underset{y}{Z}$ for all transformations of the group and also for the invariance of the set of all one-parameter subgroups generated by $\overset{y}{s}\underset{y}{Z}$ for different values of the $\overset{y}{s}$.*

If the $\underset{y}{Z}$ belong to the given group and if they generate a group, (5.23) is the n.a.s. condition for this group to be an invariant subgroup. We get as a corollary that a linear set generates a group if and only if it is invariant for all its own infinitesimal transformations.

c) In the case of complete invariance there is no need to consider $p > 1$ because each vector field is dealt with separately. The equations are

$$\text{a) } \underset{b}{£}Z^\varkappa = 0 \quad \text{or} \quad \text{b) } (\underset{b}{£}\underset{Z}{£}) = 0 \tag{5.29}$$

and this proves that the one-parameter group generated by Z is an invariant subgroup of the group generated by the X_b and Z.

§ 6. Invariants of a group in group space.[1]

At the end of IV § 1 we have seen that the differential concomitants of group space are the ordinary concomitants of $c_{\gamma\beta}^{\cdot\cdot\alpha}$ and that accordingly they have constant components with respect to the anholonomic coordinate systems (a) and (A). In IV § 2 we have seen that the infinitesimal transformation of the linear adjoint group corresponding to the transformation $s^b X_b$ ($s^b =$ const.) of the given group is

$$(6.1)\qquad s^b E_b = s^b E_b^{\cdot a} \frac{\partial}{\partial e^a} = s^b e^c c_{cb}^{\cdot\cdot a} \frac{\partial}{\partial e^a}; \qquad E_b^{\cdot a} \stackrel{\text{def}}{=} e^c c_{cb}^{\cdot\cdot a}.$$

This is a linear transformation in the variables e^a and these variables may be considered as affine coordinates in the tangent E_r of $\underset{0}{\eta}^\alpha$.

We have mentioned already (cf. 2.42) the transformation of the $\underset{b}{e}^\alpha$ into the $\underset{B}{e}^\alpha$ in a tangent E_r of a point of group space as an example of a finite transformation of the linear adjoint group. Now the invariance of a quantity for a linear homogeneous vector transformation is the same as the invariance of its components for the corresponding coordinate transformation. Moreover we know from (2.42) that the components $c_{cb}^{\cdot\cdot a}$ are invariant for every coordinate transformation belonging to the linear adjoint group. Hence we know already that the quantity $c_{cb}^{\cdot\cdot a}$ is invariant for vector transformations belonging to this group. This can be checked in the following way. The transformation of a contra- and a covariant vector by (6.1) is

$$(6.2)\qquad \begin{cases} \text{a)} & v^a \to (A_b^a - S_b^{\cdot a}\, dt)\, v^b \\ \text{b)} & w_b \to (A_b^a + S_b^{\cdot a}\, dt)\, w_a \end{cases} \qquad S_b^{\cdot a} \stackrel{\text{def}}{=} s^c c_{cb}^{\cdot\cdot a};$$

hence (cf. 1.28)

$$(6.3)\qquad \begin{cases} c_{cb}^{\cdot\cdot a} \to c_{cb}^{\cdot\cdot a} + (S_c^{\cdot d} c_{db}^{\cdot\cdot a} + S_b^{\cdot d} c_{cd}^{\cdot\cdot a} - S_d^{\cdot a} c_{cb}^{\cdot\cdot d})\, dt \\ \qquad = c_{cb}^{\cdot\cdot a} + s^e (c_{ec}^{\cdot\cdot d} c_{db}^{\cdot\cdot a} + c_{eb}^{\cdot\cdot d} c_{cd}^{\cdot\cdot a} - c_{ed}^{\cdot\cdot a} c_{cb}^{\cdot\cdot d})\, dt = c_{cb}^{\cdot\cdot a} \end{cases}$$

or in other words:

The concomitants of $c_{cb}^{\cdot\cdot a}$ are invariant for the transformations of the linear adjoint group.

There is a set of concomitants, that is of great importance for the theory:

$$(6.4)\qquad \begin{cases} \text{a)} & g_b \stackrel{\text{def}}{=} c_{ab}^{\cdot\cdot a} \\ \text{b)} & g_{ba} \stackrel{\text{def}}{=} c_{cb}^{\cdot\cdot d} c_{da}^{\cdot\cdot c} \\ \text{c)} & g_{cba} \stackrel{\text{def}}{=} c_{dc}^{\cdot\cdot e} c_{eb}^{\cdot\cdot f} c_{fa}^{\cdot\cdot d} \\ & \vdots \end{cases}$$

[1]) Lie-Engel 1888, 1, p. 270ff.

They are all invariant for *cyclic* permutations of the free indices. This implies that g_{ba} is a symmetric tensor and that g_{cba} is the sum of a symmetric tensor and a trivector.

If a concomitant of $c_{cb}^{\cdot\cdot a}$ is transvected with one or more factors e^a and if this expression is equated to zero we get an equation that is invariant for the linear adjoint group. This was proved by CARTAN. We prove this here for the transvection $P^{ab}_{\cdot\cdot cd}\, e^c e^d$ where $P^{ab}_{\cdot\cdot cd}$ is a concomitant. According to (6.2) we have

$$\frac{d}{dt}(P^{ab}_{\cdot\cdot cd}\, e^c e^d) = -S_e^{\cdot a} P^{eb}_{\cdot\cdot cd}\, e^c e^d - S_e^{\cdot b} P^{ae}_{\cdot\cdot cd}\, e^c e^d. \tag{6.5}$$

Hence, because $\frac{d}{dt} P^{ab}_{\cdot\cdot cd} = 0$ we have

$$0 = -P^{ab}_{\cdot\cdot cd}\, S_e^{\cdot c} e^e e^d - P^{ab}_{\cdot\cdot cd}\, S_e^{\cdot d} e^c e^e \tag{6.6}$$

and this shows that if e^a is a solution of the equation $P^{ab}_{\cdot\cdot cd}\, e^c e^d = 0$, then $e^a - S_b^{\cdot a} e^b\, dt$ is also a solution. This proves that the vectors e^a satisfying this equation are *interchanged* by the transformations of the adjoint group (6.2) and that consequently an equation of this kind represents a subspace in the tangent E_r of $\underset{0}{\eta}^\alpha$ that consists of straight lines through $\underset{0}{\eta}^\alpha$ and is invariant for transformations of this group.

To every point e^a in the tangent E_r of $\underset{0}{\eta}^\alpha$ there corresponds the point of group space with the normal coordinates $\eta^{\mathfrak{a}} \overset{*}{=} \delta^{\mathfrak{a}}_a e^a$. Because the e^a and the $\eta^{\mathfrak{a}}$ transform in the same way, to an X_p in E_r consisting of straight lines through $\underset{0}{\eta}^\alpha$ there corresponds an X_p in group space consisting of geodesics through $\underset{0}{\eta}^\alpha$. Hence equations of the form considered represent in group space subspaces that consist of geodesics through $\underset{0}{\eta}^\alpha$ and that are invariant for the adjoint group.

Now consider an E_p through the origin in E_r. To this E_p corresponds an X_p in group space consisting of ∞^{p-1} geodesics through $\underset{0}{\eta}^\alpha$, each of them representing a one-parameter subgroup. In normal coordinates this X_p is represented by $r-p$ homogeneous linear equations. Let the normal coordinates be chosen such that these equations are

$$\eta^{\mathfrak{x}} = 0; \qquad \mathfrak{x}, \mathfrak{y}, \mathfrak{z} = \overline{p+1}, \ldots, \bar{r} \tag{6.7}$$

so that the $\eta^{\mathfrak{m}}$; $\mathfrak{m}, \mathfrak{n}, \mathfrak{p} = \bar{1}, \ldots, \bar{p}$ can be used as coordinates in the X_p. An X_p is said to be *geodesic* in X_r if and only if every contravariant vector in X_p remains in X_p if it is displaced parallel in a direction lying in X_p. Let $v^{\mathfrak{a}}$ be the vector and $d\eta^{\mathfrak{a}}$ the displacement, $\mathfrak{a} = \bar{1}, \ldots, \bar{r}$. Then we have $v^{\mathfrak{x}} = 0$; $d\eta^{\mathfrak{x}} = 0$; $\mathfrak{x} = \overline{p+1}, \ldots, \bar{r}$ and

$$dv^{\mathfrak{a}} = -\Gamma^{\mathfrak{a}}_{\mathfrak{cb}} v^{\mathfrak{b}} d\eta^{\mathfrak{c}} = -\Gamma^{\mathfrak{a}}_{\mathfrak{pn}} v^{\mathfrak{n}} d\eta^{\mathfrak{p}} \tag{6.8}$$

and $dv^{\mathfrak{x}}$ must be zero. Hence

$$\Gamma_{\mathfrak{p}\mathfrak{n}}^{\mathfrak{x}} = 0; \quad \mathfrak{p}, \mathfrak{n} = \bar{1}, \ldots, \bar{p}; \quad \mathfrak{x} = \overline{p+1}, \ldots, \bar{r}; \tag{6.9}$$

at every point of X_p and this condition is n.a.s. Now making use of the expansion (4.34) we see that

$$\underset{0}{C}{}_{\mathfrak{m}(\mathfrak{p}}^{\cdot\,\cdot\,\mathfrak{b}} \underset{0}{C}{}_{\mathfrak{n})\mathfrak{b}}^{\cdot\,\cdot\,\mathfrak{x}} = 0 \tag{6.10}$$

is a *necessary* condition. But (6.10) expresses the fact that $\underset{0}{C}{}_{\mathfrak{m}\mathfrak{p}}^{\cdot\,\cdot\,\mathfrak{b}} \underset{0}{C}{}_{\mathfrak{n}\mathfrak{b}}^{\cdot\,\cdot\,\mathfrak{x}}$ is alternating in $\mathfrak{p}, \mathfrak{n}$ and because the same expression is also alternating in $\mathfrak{m}, \mathfrak{p}$, (6.10) expresses the fact that it is alternating in $\mathfrak{m}, \mathfrak{p}, \mathfrak{n}$. According to (1.28) this is only possible if it is zero. Hence, (6.10) is equivalent to

$$\underset{0}{C}{}_{\mathfrak{m}\mathfrak{p}}^{\cdot\,\cdot\,\mathfrak{b}} \underset{0}{C}{}_{\mathfrak{n}\mathfrak{b}}^{\cdot\,\cdot\,\mathfrak{x}} = 0. \tag{6.11}$$

Now we will prove that (6.11) is not only necessary but also sufficient. Every term of the expression (4.34) [cf. (4.32)] contains one factor C_c and an odd number of factors P, for instance $P^2 C_c P^3$. If this term is written out we get for its contribution to $\Gamma_{\mathfrak{p}\mathfrak{n}}^{\mathfrak{x}}$:

$$\left\{\begin{array}{l} \eta^{\mathfrak{q}_1} \underset{0}{C}{}_{\mathfrak{q}_1(\mathfrak{n}}^{\cdot\,\cdot\,\mathfrak{a}_1} \eta^{\mathfrak{q}_2} \underset{0}{C}{}_{|\mathfrak{q}_2\mathfrak{a}_1|}^{\cdot\,\cdot\,\mathfrak{a}_2} \underset{0}{C}{}_{\mathfrak{p})\mathfrak{a}_2}^{\cdot\,\cdot\,\mathfrak{a}_3} \eta^{\mathfrak{q}_3} \underset{0}{C}{}_{\mathfrak{q}_3\mathfrak{a}_3}^{\cdot\,\cdot\,\mathfrak{a}_4} \eta^{\mathfrak{q}_4} \underset{0}{C}{}_{\mathfrak{q}_4\mathfrak{a}_4}^{\cdot\,\cdot\,\mathfrak{a}_5} \eta^{\mathfrak{q}_5} \underset{0}{C}{}_{\mathfrak{q}_5\mathfrak{a}_5}^{\cdot\,\cdot\,\mathfrak{x}}; \\ \mathfrak{a}_1, \ldots, \mathfrak{a}_5 = \bar{1}, \ldots, \bar{r}; \\ \mathfrak{n}, \mathfrak{p}, \mathfrak{q}_1, \ldots, \mathfrak{q}_5 = \bar{1}, \ldots, \bar{p}; \\ \mathfrak{x} = \overline{p+1}, \ldots, \bar{r}. \end{array}\right. \tag{6.12}$$

But in this expression $\mathfrak{p}_2$ can be written instead of $\mathfrak{a}_2$ according to (6.11). Then by the same argument $\mathfrak{p}_4$ can be written instead of $\mathfrak{a}_4$ and if this is done we see that the whole expression is zero because of the index $\mathfrak{x}$ in the last factor. The reasoning is independent of the number of factors P (if odd) and of the place of the factor C_c in the product. Hence (6.10) has as a consequence that not only the first term in the expansion of $\Gamma_{\mathfrak{p}\mathfrak{n}}^{\mathfrak{x}}$ vanishes but also all other terms.

The E_p at $\underset{0}{\eta}{}^{\alpha}$ corresponding to the X_p (6.7) is spanned by the vectors $\underset{\mathfrak{n}}{e}{}^{\alpha}$ at $\underset{0}{\eta}{}^{\alpha}$. Taking the basis vectors of the system (a) at $\underset{0}{\eta}{}^{\alpha}$ such that they coincide with the basis vectors of $(\mathfrak{a})$, the one-parameter subgroups (geodesics) constituting the X_p are generated by the linear set of infinitesimal transformations $e^m \partial_m$; $m = 1, \ldots, \mathrm{p}$. Writing out the expression $(\partial_n(\partial_m \partial_p))$; $m, n, p = 1, \ldots, \mathrm{p}$; we get

$$(\partial_n(\partial_m \partial_p)) = C_{mp}^{\cdot\,\cdot\,b} C_{nb}^{\cdot\,\cdot\,a} \partial_a = \delta_{m\,p\,n\,a}^{\mathfrak{m}\mathfrak{p}\mathfrak{n}a} \underset{0}{C}{}_{\mathfrak{m}\mathfrak{p}}^{\cdot\,\cdot\,\mathfrak{b}} \underset{0}{C}{}_{\mathfrak{n}\mathfrak{b}}^{\cdot\,\cdot\,a} \partial_a. \tag{6.13}$$

Hence (6.11) is the n.a.s. condition that $(\partial_n(\partial_m \partial_p))$ depends only on $\partial_1, \ldots, \partial_{\mathrm{p}}$.[1])

[1]) Cartan 1927, 1, p. 72ff.

It may happen that $(\partial_n \partial_m)$ already depends on $\partial_1, \ldots, \partial_\mathfrak{p}$ only. The n.a.s. condition is that $c_{nm}^{\cdot\cdot x} = 0$; $m, n = 1, \ldots, \mathfrak{p}$; $x = \mathfrak{p}+1, \ldots, \mathfrak{r}$ or in another form that $\underset{0}{c}_{\mathfrak{nm}}^{\cdot\cdot\mathfrak{x}} = 0$. That means that the X_p represents a subgroup with the structural constants $c_{nm}^{\cdot\cdot p}$; $m, n, p = 1, \ldots, \mathfrak{p}$. Of course the X_p remains geodesic, but in this special case it has another remarkable geometric property. If the vector $v^\mathfrak{a}, v^\mathfrak{x} = 0$ is displaced $(+)$- or $(-)$-parallel along the X_p we have

$$(6.14)\quad d v^\mathfrak{a} = - \Gamma_{\mathfrak{cb}}^{\mathfrak{a}} v^\mathfrak{b} d\eta^\mathfrak{c} \pm \tfrac{1}{2} c_{\mathfrak{cb}}^{\cdot\cdot\mathfrak{a}} v^\mathfrak{b} d\eta^\mathfrak{c} = - \Gamma_{\mathfrak{pn}}^{\mathfrak{a}} v^\mathfrak{n} d\eta^\mathfrak{p} \pm \tfrac{1}{2} c_{\mathfrak{pn}}^{\cdot\cdot\mathfrak{a}} v^\mathfrak{n} d\eta^\mathfrak{p};$$

hence, according to (6.9)

$$(6.15)\qquad d v^\mathfrak{x} = \pm \tfrac{1}{2} c_{\mathfrak{pn}}^{\cdot\cdot\mathfrak{x}} v^\mathfrak{n} d\eta^\mathfrak{p}.$$

Now we use the expansion (4.35) [cf. (4.32)] and remark that all terms contain a factor C_c and an *even* number of factors P, for instance $P C_c P$. If this term is written out, its contribution to $c_{\mathfrak{pn}}^{\cdot\cdot\mathfrak{x}}$ is

$$(6.16)\quad \begin{cases} \eta^{\mathfrak{q}_1} \underset{0}{c}_{\mathfrak{q}_1 [\mathfrak{n}}^{\cdot\cdot\mathfrak{a}_1} \underset{0}{c}_{\mathfrak{p}] \mathfrak{a}_1}^{\cdot\cdot\mathfrak{a}_2} \eta^{\mathfrak{q}_2} \underset{0}{c}_{\mathfrak{q}_2 \mathfrak{a}_2}^{\cdot\cdot\mathfrak{x}}; & \mathfrak{a}_1, \mathfrak{a}_2 = \bar{1}, \ldots, \bar{r}; \\ \mathfrak{n}, \mathfrak{p}, \mathfrak{q}_1, \mathfrak{q}_2 = \bar{1}, \ldots, \bar{p}; & \mathfrak{x} = \overline{p+1}, \ldots, \bar{r}. \end{cases}$$

In this expression $\mathfrak{a}_1$ can be replaced by $\mathfrak{p}_1$ according to $\underset{0}{c}_{\mathfrak{nm}}^{\cdot\cdot\mathfrak{x}} = 0$ and by the same argument $\mathfrak{a}_2$ can be replaced by $\mathfrak{p}_2$ and $\mathfrak{a}_3$ by $\mathfrak{p}_3$. But then the whole expression is zero because of the index $\mathfrak{x}$ in the last factor. The reasoning is independent of the number of factors P (if even) and of the place of the factor C_c. Hence *the X_p represents a subgroup if and only if it is not only geodesic, that is (0)-parallel in itself, but also* $(+)$-*parallel and* $(-)$-*parallel in itself.*

If the vectors $\underset{m}{e}^\alpha$ spanning the E_p at $\underset{0}{\eta}^\alpha$ are subjected to the transformations $s^b E_b$ of the linear adjoint group corresponding to the transformation $s^b X_b$ of the group, we get

$$(6.17)\qquad d \underset{m}{e}^a = - S_b^{\cdot a} \underset{m}{e}^b dt = - s^c c_{cb}^{\cdot\cdot a} \underset{m}{e}^b dt.$$

Hence $c_{nm}^{\cdot\cdot x} = 0$ or $\underset{0}{c}_{\mathfrak{nm}}^{\cdot\cdot\mathfrak{x}} = 0$ is the n.a.s. condition that the E_p is invariant for *those* transformations of the linear adjoint group that correspond to the transformation of the subgroup represented by E_p.

It may happen that not only $(\partial_n \partial_m)$ but also $(\partial_b \partial_m)$; $b = 1, \ldots, \mathfrak{r}$, depends on $\partial_1, \ldots, \partial_\mathfrak{p}$ only. That means that the E_p is invariant for all transformations of the linear adjoint group. But we have already seen in IV § 5 that this means that $\partial_1, \ldots, \partial_\mathfrak{p}$ generate an invariant subgroup. In this case the X_p in X_r has yet another remarkable property. In X_r we form a $(+)$-constant field $v^\mathfrak{a}, v^\mathfrak{x} = 0$. This is always possible because the X_p is (+)-parallel in itself. Then $\overset{+}{\delta} v^{\mathfrak{a}} = 0$ and

$$\overset{-}{\delta} v^{\mathfrak{a}} = \overset{+}{\delta} v^{\mathfrak{a}} - (\overset{+}{\Gamma}{}^{\mathfrak{a}}_{\mathfrak{c}\mathfrak{b}} - \overset{-}{\Gamma}{}^{\mathfrak{a}}_{\mathfrak{c}\mathfrak{b}})\, v^{\mathfrak{b}}\, d\eta^{\mathfrak{c}} = c_{\mathfrak{c}\mathfrak{b}}^{\cdot\cdot\mathfrak{a}}\, v^{\mathfrak{b}}\, d\eta^{\mathfrak{c}}, \tag{6.18}$$

hence

$$\begin{cases} \overset{-}{\delta} v^{\mathfrak{m}} = c_{\mathfrak{c}\mathfrak{n}}^{\cdot\cdot\mathfrak{m}}\, v^{\mathfrak{n}}\, d\eta^{\mathfrak{c}} \\ \overset{-}{\delta} v^{\mathfrak{x}} = c_{\mathfrak{c}\mathfrak{n}}^{\cdot\cdot\mathfrak{x}}\, v^{\mathfrak{n}}\, d\eta^{\mathfrak{c}}. \end{cases} \tag{6.19}$$

Now in the same way as above it can be proved that $\underset{0}{c}{}_{\mathfrak{c}\mathfrak{n}}^{\cdot\cdot\mathfrak{x}} = 0$ implies that also $c_{\mathfrak{c}\mathfrak{n}}^{\cdot\cdot\mathfrak{x}} = 0$. Hence $\overset{-}{\delta} v^{\mathfrak{x}} = 0$ and this proves that *an X_p through $\underset{0}{\eta}{}^{\alpha}$ represents an invariant subgroup if and only if it is geodesic and (+)-parallel and (—)-parallel in itself and if moreover every X_p that is (+)-parallel to it is also (—)-parallel to it.*

This property is clear from the point of view of group theory. A (+)-parallel displacement drags along a (—)-constant field (cf. IV § 1) and this is a transformation of the second parameter-group. Hence the X_p's (+)- or (—)-parallel to the one through $\underset{0}{\eta}{}^{\alpha}$ represent the left hand and right hand co-sets (Nebenklassen) of the invariant subgroup.

Every set of equations linear homogeneous in the e^a, formed by means of concomitants of $c_{cb}^{\cdot\cdot a}$ represents a flat manifold in E_r that is invariant for all transformations of the linear adjoint group. Hence this manifold and the corresponding manifold in X_r represent an invariant subgroup. The following invariant subgroups are most important (cf. 6.4):

a) *the group of g_b*

$$g_b\, e^b = 0. \tag{6.20}$$

If $g_b \neq 0$ this group is an $(r-1)$-parameter group.

b) *The centre*

$$e^c\, c_{cb}^{\cdot\cdot a} = 0. \tag{6.21}$$

This is the group of those transformations of the group that commute with all transformations of the group (cf. IV § 2). The centre is always abelian and it vanishes if the c-rank of $c_{cb}^{\cdot\cdot a}$ is equal to r, that is if the linear adjoint group is an r-parameter group (cf. IV § 2).

c) *The first derived group*, represented by the support of the a-domain of $c_{cb}^{\cdot\cdot a}$. If q is the a-rank of $c_{cb}^{\cdot\cdot a}$ the equation is

$$c_{c_1 b_1}^{\cdot\cdot\,[a_1} \ldots c_{c_q b_q}^{\cdot\cdot\,a_q}\, e^{a]} = 0. \tag{6.22}$$

Its infinitesimal transformations are those transformations of the group that can be written as LIE brackets. This group is contained in the group of g_b. In fact

(6.23) $$c_{cb}^{\cdot\cdot a} g_a = c_{cb}^{\cdot\cdot a} c_{da}^{\cdot\cdot d} = - c_{bd}^{\cdot\cdot a} c_{ca}^{\cdot\cdot d} - c_{dc}^{\cdot\cdot a} c_{ba}^{\cdot\cdot d} = 0 .$$

It may happen that a group is identical with its first derived group. This is the case if and only if the a-rank of $c_{cb}^{\cdot\cdot a}$ is equal to r. *The first derived group of a subgroup is a subgroup of the first derived group of the original group*, as follows from the definition. Let $X_1, \ldots, X_{\mathrm{m}}$ be independent infinitesimal transformations of an m-parameter invariant subgroup. Then $(X_b X_p)$; $p = 1, \ldots, \mathrm{m}$ belongs to this subgroup and $(X_p X_q)$; $p, q = 1, \ldots, \mathrm{m}$ to its first derived group. Now in the identity

(6.24) $$((X_p X_q) X_b) + ((X_q X_b) X_p) + ((X_b X_p) X_q) = 0$$

the last two terms belong to the first derived group of the subgroup and this proves that *the first derived group of an invariant subgroup is an invariant subgroup of the group.*

The first derived group of the first derived group is called the *second derived group*, etc. All derived groups are *invariant* subgroups of the group under consideration.

d) *The group of* g_{ba}

(6.25) $$g_{ba} e^a = 0$$

is an $(r-q)$-parameter group if q is the rank of g_{ba} and vanishes if g_{ba} has rank r.

e) The group of $c_{cb}^{\cdot\cdot d} g_{da}$ viz. $c_{cb}^{\cdot\cdot d} g_{da} e^a = 0$ obviously contains the group of g_{ba}. Notice that $c_{cb}^{\cdot\cdot d} g_{da}$ is always a *trivector*. In fact

(6.26) $$\begin{cases} c_{cb}^{\cdot\cdot d} g_{da} = c_{cb}^{\cdot\cdot d} c_{ed}^{\cdot\cdot f} c_{fa}^{\cdot\cdot e} = - c_{be}^{\cdot\cdot d} c_{cd}^{\cdot\cdot f} c_{fa}^{\cdot\cdot e} - c_{ec}^{\cdot\cdot d} c_{bd}^{\cdot\cdot f} c_{fa}^{\cdot\cdot e} \\ \qquad = -2 g_{[bc]a} = -2 g_{[abc]} \end{cases}$$

because g_{abc} is the sum of a trivector and a symmetric tensor.

§ 7. Properties of integrable groups.

If we form the successive derived groups of a given group it may happen that the process stops because the derived group of the i-th derived group is identical with this latter group. If this is not the case the last derived group is zero and the last but one is abelian. The group is then said to be *integrable*. Every abelian group is integrable and the same holds for every 2-parameter group.

We prove the well known theorem:

An r-parameter group $\mathfrak{G}_r$ is integrable if and only if it is possible to find a sequence of groups $\mathfrak{G}_1, \ldots, \mathfrak{G}_{r-1}, \mathfrak{G}_r$ such that $\mathfrak{G}_p$ is a p-parameter group and an invariant subgroup of $\mathfrak{G}_{p+1}$; $p = 1, \ldots, r-1$.

Proof. Let $\mathfrak{G}_r$ be integrable and let its q-th derived group $\mathfrak{G}^{(q)}$ be an s-parameter group and its $(q-1)$-th derived group $\mathfrak{G}^{(q-1)}$ be a t-parameter group, $t > s$. Let $X_1, \ldots, X_s$ be independent infinitesimal transformations of $\mathfrak{G}^{(q)}$ and $X_1, \ldots, X_t$ independent infinitesimal transformations of $\mathfrak{G}^{(q-1)}$. Then $\mathfrak{G}^{(q)}$ is the derived group of $\mathfrak{G}^{(q-1)}$ and accordingly $c_{cb}^{\cdot\cdot a} = 0$ for $c, b = 1, \ldots, t$; $a = s+1, \ldots, t$. Hence $X_1, \ldots, X_{s+1}$ generate an $(s+1)$-parameter invariant subgroup of $\mathfrak{G}^{(q-1)}$ and this subgroup contains $\mathfrak{G}^{(q)}$ as an *invariant* subgroup. In the same way $X_1, \ldots, X_{s+2}$ generate an $(s+2)$-parameter invariant subgroup of $\mathfrak{G}^{(q-1)}$ and this subgroup contains the $(s+1)$-parameter subgroup as an *invariant subgroup*. Proceeding in this way for all values of q we get at last a sequence of groups $\mathfrak{G}_1, \ldots, \mathfrak{G}_r$ such that $\mathfrak{G}_p$ is always an invariant subgroup of $\mathfrak{G}_{p+t}$ provided that no derived groups occur *between* $\mathfrak{G}_p$ and $\mathfrak{G}_{p+t}$.

Conversely let there be a sequence $\mathfrak{G}_1, \ldots, \mathfrak{G}_r$, such that $\mathfrak{G}_p$ is always an invariant subgroup of $\mathfrak{G}_{p+1}$. If $X_1, \ldots, X_p$ are independent infinitesimal transformations of $\mathfrak{G}_p$ and $X_1, \ldots, X_p, X_{p+1}$ such transformations of $\mathfrak{G}_{p+1}$ we have

$$c_{cb}^{\cdot\cdot p+1} = 0; \qquad c_{p+1\,b}^{\cdot\cdot\, p+1} = 0; \qquad c, b = 1, \ldots, p \tag{7.1}$$

because $\mathfrak{G}_p$ is an invariant subgroup of $\mathfrak{G}_{p+1}$. But it follows from these equations that the derived group of $\mathfrak{G}_{p+1}$ is contained in $\mathfrak{G}_p$. Hence the first derived group of $\mathfrak{G}_r$ is a subgroup of $\mathfrak{G}_{r-1}$, the second is a subgroup of $\mathfrak{G}_{r-2}$, and so on. This proves that it can never happen that one of the derived groups is its own derived group and this implies that the group is integrable.

If the sequence $\mathfrak{G}_1, \ldots, \mathfrak{G}_r$ is known the coordinate system can be chosen such that X_1 belongs to $\mathfrak{G}_1$, X_1 and X_2 to $\mathfrak{G}_2$ and so on. Then we have the following simple equations for the integrable group

$$\text{(7.2)} \quad \begin{cases} \text{a)} & (X_1 X_2) = c_{12}^{\cdot\cdot 1} X_1 \\ \text{b)} & (X_1 X_3) = c_{13}^{\cdot\cdot 1} X_1 + c_{13}^{\cdot\cdot 2} X_2; \quad (X_2 X_3) = c_{23}^{\cdot\cdot 1} X_1 + c_{23}^{\cdot\cdot 2} X_2 \end{cases}$$

or in a general formula

$$\text{(7.3)} \quad \begin{cases} (X_p X_{p+t}) = c_{p\,p+t}^{\cdot\cdot\cdot q} X_q; \quad t > 0; \\ p = 1, \ldots, r-1 \\ q = 1, \ldots, p+t-1 < r. \end{cases}$$

Notice that the centre, the group of g_{ba} and the group of $c_{cb}^{\cdot\cdot d} g_{da}$ are always integrable.[1]) The proof of these propositions is rather long and makes use of the theory of elementary divisors applied to the matrix $e^c c_{cb}^{\cdot\cdot a}$. So we can not show this proof here though we need the integrability of the group of g_{ba} in the next section.

[1]) Cf. Eisenhart 1933, 1, p. 173.

§ 8. Simple and semi-simple groups.

A group is called *simple* if it does not contain any invariant subgroup. It is called *semi-simple* if it does not contain any integrable invariant subgroup. Hence a simple group is also semi-simple. As the group of g_{ba} is an integrable invariant subgroup (cf. IV § 7) it follows that for a semi-simple group g_{ba} has rank r. *As a corollary we get that the group space of a semi-simple group has a riemannian geometry.* But this condition is not only necessary but also sufficient for the group to be semi-simple. We prove that the rank of g_{ba} is always $<r$ if the group contains an integrable invariant subgroup. If such a subgroup exists, either it is abelian itself or one of its derived groups is abelian. This latter abelian group is an invariant subgroup of the group in question. Let $X_1, \ldots, X_r$ be chosen in such a way that X_p; $p=1, \ldots, \mathrm{m}$ generate this abelian group. Then we have

$$(8.1)\qquad c_{pq}^{\cdot\cdot a}=0; \quad c_{pb}^{\cdot\cdot x}=0; \quad p, q=1, \ldots, \mathrm{m}; \quad x=\mathrm{m}+1, \ldots, \mathrm{r}$$

and consequently

$$(8.2)\qquad g_{pb}=c_{cp}^{\cdot\cdot a}\, c_{ab}^{\cdot\cdot c}=c_{cp}^{\cdot\cdot q}\, c_{qb}^{\cdot\cdot c}=c_{sp}^{\cdot\cdot q}\, c_{qb}^{\cdot\cdot s}=0; \quad p, q, s=1, \ldots, \mathrm{m}$$

and this proves that the rank of g_{ba} is $<r$.

V. Imbedding and Curvature.

§ 1. The X_1 in V_n.

In a V_n the length ds of $d\xi^\varkappa$, called the *linear element* of the V_n, is given by

$$(1.1)\qquad d s^2=g_{\lambda\varkappa}\, d\xi^\lambda\, d\xi^\varkappa .$$

Hence a curve in V_n need not be a V_1 because its linear element may have a length zero at some point or at all points. But a real curve in an ordinary V_n is always a V_1.[1])

[1]) In this chapter a V_n is always supposed to be *ordinary* and only *real* subspaces are considered. This means that multi-directions with a singular position with respect to the nullcone do not occur. Cf. for the more general case EISENHART 1926, 1; E I 1935, 1; E II 1938, 2. Also GUICHARD 1928, 1; LENSE 1936, 1. There is a vast literature on isotropic subspaces, that are subspaces with isotropic tangent spaces. We mention here only LENSE 1926, 1; 2; 3; 1928, 1; 1932, 1; 1936, 1; 2; 1939, 1; 2; 1940, 1; 2; SYNGE and MC CONNELL 1928, 2; SCHRENZEL 1929, 1; PINL 1932, 1; 2; 1935, 1; 1936, 1; 1937, 1; 2; 1940, 1; 2; 3; 1947, 1; 1948, 1; 1949, 1; ANDREOLI 1935, 1; BOMPIANI 1940, 1; 1941, 1; MOISIL 1940, 1; STRUBECKER

If $\underset{0}{\xi}^\varkappa$ is a definitely chosen point of the curve and $\xi^\varkappa$ an arbitrary point of it, the length of the curve from $\underset{0}{\xi}^\varkappa$ to $\xi^\varkappa$ is $s = \int ds$ and the parametric equations of the curve can be written in the form

$$\xi^\varkappa = f^\varkappa(s)\,. \tag{1.2}$$

The vector

$$i^\varkappa = \underset{1}{i}^\varkappa \overset{\text{def}}{=} \frac{d\xi^\varkappa}{ds} \tag{1.3}$$

is a unitvector because ds is the length of $d\xi^\varkappa$. The sense of $i^\varkappa$ depends on the (arbitrary) choice of $\underset{0}{\xi}^\varkappa$. The unitvector $i^\varkappa$ is only given at the points of the curve but by the process of prolongation (cf. II § 4) we may make it possible for the operator ∇_μ to be applied to fields of the V_1. Because $i^\varkappa$ is a unitvector, the vector $i^\mu \nabla_\mu i^\varkappa$ is either non zero and perpendicular to $i^\varkappa$ or zero. If it is non zero and if $\underset{2}{i}^\varkappa$ is a unitvector with the same direction *and* the same sense, we have

$$i^\mu \nabla_\mu \underset{1}{i}^\varkappa = \underset{1}{k}\,\underset{2}{i}^\varkappa; \qquad \underset{1}{k} > 0\,. \tag{1.4}$$

If it is zero the same equation is valid with $\underset{1}{k} = 0$. $\underset{1}{k}$ is the *first curvature* and $\underset{1}{k}\,\underset{2}{i}^\varkappa$ is the *curvature vector*[1]) of the curve. The direction of $\underset{2}{i}^\varkappa$ is the *first normal* of the curve and $\underset{1}{i}^\varkappa$ and $\underset{2}{i}^\varkappa$ span its *osculating* R_2. The sense of $\underset{2}{i}^\varkappa$ is independent of the choice of $\underset{0}{\xi}^\varkappa$. At a point where $\underset{1}{k} = 0$ the tangent is said to be *stationary* (cf. III § 2) and $\underset{2}{i}^\varkappa$ and the first normal can not be found by an equation of the form (1.4). If $\underset{1}{k} \neq 0$ in the neighbourhood of the point, a normal may be fixed by a limiting process from one side or from the other side of the curve. If $\underset{1}{k} = 0$ at all points, (1.4) passes into

$$i^\mu \nabla_\mu \frac{d\xi^\varkappa}{ds} = \frac{d^2\xi^\varkappa}{ds^2} + \Gamma^\varkappa_{\mu\lambda} \frac{d\xi^\mu}{ds}\frac{d\xi^\lambda}{ds} = 0 \tag{1.5}$$

and this is the equation of a geodesic with the affine parameter s (cf. III § 7). If $\underset{1}{k}$ is not zero at all points, the vector $i^\mu \nabla_\mu \underset{2}{i}^\varkappa$ is perpendicular to $\underset{2}{i}^\varkappa$, hence it has the direction of $\underset{1}{i}^\varkappa$ or there exists exactly one direc-

---

1941, 1; 1942, 1; 2; 1944, 1; THALER 1941, 1; CHERN 1942, 1; MOGHE 1943, 1; WONG 1945, 2; PETKANTSCHIN 1948, 1; WALKER 1950, 2; 3; 4; PATTERSON and WALKER 1952, 3. Cf. for spaces with a fundamental tensor of rank $< n$ Ch. III p. 133, footnote 1.

[1]) RICCI 1895, 1, p. 298 used the expression "curvatura geodetica".

tion perpendicular to $\underset{1}{j}^\varkappa$ and $\underset{2}{j}^\varkappa$ such that it lies in the R_2 spanned by $\underset{1}{j}^\varkappa$ and this direction. Because

$$(1.6) \qquad 0 = j^\mu \nabla_\mu (\underset{1}{j}^\varkappa \underset{2}{j}_\varkappa) = \underset{1}{k} + \underset{1}{j}_\varkappa j^\mu \nabla_\mu \underset{2}{j}^\varkappa$$

there exists a unitvector $\underset{3}{j}^\varkappa$ perpendicular to $\underset{1}{j}^\varkappa$ and $\underset{2}{j}^\varkappa$ such that

$$(1.7) \qquad j^\mu \nabla_\mu \underset{2}{j}^\varkappa = -\underset{1}{k}\,\underset{1}{j}^\varkappa + \underset{2}{k}\,\underset{3}{j}^\varkappa; \qquad \underset{2}{k} \geq 0.$$

$\underset{2}{k}$ is the *second curvature*, the direction of $\underset{3}{j}^\varkappa$ is the *second normal* and $\underset{1}{j}^\varkappa$, $\underset{2}{j}^\varkappa$ and $\underset{3}{j}^\varkappa$ span the *osculating* R_3. The sense of $\underset{3}{j}^\varkappa$ depends on the choice of $\underset{0}{\xi}^\varkappa$ on the curve. If $\underset{2}{k}=0$ at a point, the osculating R_2 is said to be *stationary* (cf. III § 2) at that point. In this case $\underset{3}{j}^\varkappa$ and the second normal can not be found by an equation of the form (1.7). If $\underset{2}{k}=0$ at only one point, an osculating plane may be fixed at this point by a limiting process from one side or the other. But if $\underset{2}{k}=0$ at all points the process stops.[1]) If $\underset{2}{k}$ is not zero at all points we can go on. Let us suppose that the process stops after the construction of the $(m-1)$-th normal $\underset{m}{j}^\varkappa$. Then we have the equations

$$(1.8) \qquad j^\mu \nabla_\mu \underset{s}{j}^\varkappa = -\underset{s-1}{k}\,\underset{s-1}{j}^\varkappa + \underset{s}{k}\,\underset{s+1}{j}^\varkappa; \qquad s = 1, \ldots, m; \qquad \underset{0}{k} = 0; \qquad \underset{m}{k} = 0;$$

in which the curvatures $\underset{1}{k}, \ldots, \underset{m-1}{k}$ are all > 0 at ordinary points.[2]) It follows immediately from them that

$$(1.9) \qquad j^\mu \nabla_\mu \underset{1}{j}^{[\varkappa_1} \cdots \underset{m}{j}^{\varkappa_m]} = 0$$

and this means that the osculating R_m is displaced parallel along the curve. For a curve in an R_n it means that the curve lies in its osculating R_m. According to (1.9) it is possible to construct along the curve $n-m$ fields of unitvectors $\underset{m+1}{j}^\varkappa, \ldots, \underset{n}{j}^\varkappa$ so that these vectors are perpendicular at each point to the osculating R_m and to each other, and that $\delta \underset{m+1}{j}^\varkappa = 0, \ldots, \delta \underset{n}{j}^\varkappa = 0$. Now taking $\underset{m+1}{k} = 0, \ldots, \underset{n}{k} = 0$, we get the general equations, valid in all cases

$$(1.10) \qquad \begin{cases} \dfrac{\delta}{ds} \underset{s}{j}^\varkappa = j^\mu \nabla_\mu \underset{s}{j}^\varkappa = -\underset{s-1}{k}\,\underset{s-1}{j}^\varkappa + \underset{s}{k}\,\underset{s+1}{j}^\varkappa; \qquad s = 1, \ldots, n; \\ \underset{0}{k} = \underset{m}{k} = \underset{m+1}{k} = \cdots = \underset{n}{k} = 0. \end{cases}$$

---

1) Cartan 1927, 3 has investigated this case.

2) We always take the signs of all curvatures positive. But it may be done in another way. Cf. v. d. Woude 1921, 1; Blaschke 1921, 1, p. 9; Synge 1924, 2; Eisenhart 1926, 1, p. 61.

These are known as the equations of FRENET.[1]) In these formulae the curvatures are all ≥ 0 and the senses of $\underset{2}{j}^{\varkappa}, \underset{4}{j}^{\varkappa}, \ldots$ are all fixed. But the senses of $\underset{1}{j}^{\varkappa}, \underset{3}{j}^{\varkappa}, \ldots$ are only fixed if $\underset{0}{\xi}^{\varkappa}$ is chosen and the senses of $\underset{3}{j}^{\varkappa}, \underset{5}{j}^{\varkappa}, \ldots$ change if the sense of $\underset{1}{j}^{\varkappa}$ changes. From this it follows that the screwsense of $\underset{1}{j}^{\varkappa}, \ldots, \underset{n}{j}^{\varkappa}$ is fixed by the curve and the sense of $\underset{1}{j}^{\varkappa}$, and when $\frac{1}{2} n (n+1)$ is even, by the curve alone (cf. I § 1). $\underset{1}{k}, \ldots, \underset{n-1}{k}$ can be computed as functions of s from the FRENET formulae if the curve is given in the form (1.2). Conversely *the curve is fixed if* $\underset{1}{k}, \ldots, \underset{n-1}{k}$ *are given as functions of* s *and if one point of the curve and the vectors* $\underset{1}{j}^{\varkappa}, \ldots, \underset{n}{j}^{\varkappa}$ *at that point are given.* Accordingly the equations expressing $\underset{1}{k}, \ldots, \underset{n-1}{k}$ as functions of s are called the *natural equations* of the curve.[2]) The proof can be given as follows.[3]) The equations of FRENET can be considered as a system of simultaneous ordinary differential equations of the first order with the unknowns $\xi^{\varkappa}, \underset{1}{j}^{\varkappa}, \ldots, \underset{n}{j}^{\varkappa}$ and the independent variable s, in which $\underset{1}{k}, \ldots, \underset{n}{k}$ are known functions of s and the $\Gamma_{\mu\lambda}^{\varkappa}$ known functions of the $\xi^{\varkappa}$. If these functions are continuous and have continuous derivatives, these equations have one and only one solution such that for $s = \underset{0}{s}$ the $\xi^{\varkappa}$ take some prescribed values $\underset{0}{\xi}^{\varkappa}$ and the $\underset{s}{j}^{\varkappa}$ take some prescribed values satisfying the conditions $\underset{s}{j}^{\varkappa} \underset{t}{j}_{\varkappa} = \delta_{st}$. Then these conditions are satisfied at all points of the curve because $\frac{d}{ds} \underset{s}{j}^{\varkappa} \underset{t}{j}_{\varkappa} = 0$ is a consequence of the FRENET formulae. As a corollary we find that for $V_n = R_n$ a curve is determined by its natural equations to within translations, rotations and reflexions.

If in an R_n the fundamental tensor $g_{\lambda\varkappa}$ is transformed by multiplication with a *constant* real factor $1/c^2$; $c > 0$, the new unit tangent vector at the curve is $'\underset{1}{j}^{\varkappa} = c \underset{1}{j}^{\varkappa}$. From (1.10) it follows that $'\underset{s}{j}^{\varkappa} = c \underset{s}{j}^{\varkappa}$ and $'\underset{s}{k} = c \underset{s}{k}$. Hence the proportions of the $n-1$ curvatures are invariant for this transformation of the fundamental tensor. As a corollary we

[1]) JORDAN 1874, 1; KÜHNE 1903, 1; BLASCHKE 1920, 1; JUVET 1921, 1; 1925, 1; STRUIK 1922, 1, p. 76; SYNGE 1924, 2; DUSCHEK and MAYER 1930, 1, II, p. 62; HLAVATY 1934, 2, p. 12; DAVIES 1938, 1. GOLAB gave 1945, 1 another form, generalizing a formula of LANCRET. If instead of $j^{\varkappa}$ an arbitrary vector field is taken along the curve we get a generalization of the FRENET formulae considered first by MEYER 1911, 1. Cf. STRUIK 1922, 1, p. 77; BIANCHI 1922, 1; EISENHART 1926, 1, p. 105; MC CONNELL 1928, 4; 5; BORTOLOTTI 1930, 4; KRAUS 1934, 1; WONG 1940, 3.

[2]) There are many publications on curves in R_n satisfying relations between the $n-1$ curvatures. Cf. for instance HAYDEN 1930, 1; SYPTÁK 1932, 1; 1934, 1; E II 1938, 2, p. 21; WONG 1940, 3; 1941, 1; 2.

[3]) HLAVATY 1934, 2, p. 16; E II 1938, 2, p. 16.

get that a curve in R_n is fixed to within translations, rotations, reflexions and similarity transformations if the proportions of its curvatures are given as functions of s.

If a hypersurface

$$F(\xi^\varkappa) = 0 \tag{1.11}$$

is given, the points of intersection with the curve (1.2) are determined by the solutions of

$$0 = \Phi(s) \overset{\text{def}}{=} F\left(f^\varkappa(s)\right). \tag{1.12}$$

If $s = \underset{0}{s}$ is a solution of (1.12) and if moreover

$$\frac{d\Phi}{ds}\left\{\underset{0}{s}\right\} = 0;\ \ldots;\ \frac{d^u\Phi}{ds^u}\left\{\underset{0}{s}\right\} = 0;\ \frac{d^{u+1}\Phi}{ds^{u+1}}\left\{\underset{0}{s}\right\} \neq 0 \tag{1.13}$$

there is said to be a *contact of order u* at the point of intersection $\underset{0}{\xi}^\varkappa = f^\varkappa\left(\underset{0}{s}\right)$. If (1.12) is satisfied identically in s the curve lies on the hypersurface. If $u = 1$, the direction of the curve at $\underset{0}{\xi}^\varkappa$ lies in the $(n-1)$-direction of the hypersurface at that point. But if $u = 2$ the osculating R_2 does not necessarily lie in this $(n-1)$-direction and a similar remark can be made for all higher values of u.

§ 2. The X_1 in W_n and L_n.[1]

Curves in a space with a general linear connexion can only be dealt with in a satisfactory manner if it is possible to define in some invariant way a parameter and a tangent vector. If no connexion is given in a space it is not possible to define an invariant parameter and this is the reason why for instance an X_1 in X_n has no osculating 2-direction.

For an X_1 in W_n [cf. III § 3][2]) the factor in $g_{\lambda\varkappa}$ can be fixed at a point $\underset{0}{\xi}^\varkappa$ of the X_1. Then this fixed $g_{\lambda\varkappa}$ can be displaced parallel along the curve and at each point a tangent vector $j^\varkappa$ can be constructed having unit length with respect to this tensor field. The field $j^\varkappa$ is then fixed to within a *constant* scalar factor by the connexion in W_n and the curve. Taking $j^\varkappa = d\xi^\varkappa/ds$ and $s = 0$ for $\xi^\varkappa = \underset{0}{\xi}^\varkappa$ we get a parameter s on the curve which is also fixed to within a *constant* scalar factor. This parameter s can now be used to establish FRENET formulae in exactly the same way as was described in the preceding section. From these formulae we see then that all vectors $\underset{s}{j}^\varkappa$ and all curvatures $\underset{s}{k}$ are fixed to within the same constant scalar factor and that consequently the proportions

[1]) Cf. HLAVATY 1935, 2, I.

[2]) HLAVATY 1928, 3.

of the $\underset{s}{k}$ and also the vectors $\underset{1}{k}\,\underset{s}{j}_\lambda$ and $\underset{1}{k}^{-1}\,\underset{s}{j}^\varkappa$ are invariants of the curve and the connexion.[1])

The case of an X_1 in L_n is much more complicated.[2]) Let such a curve be given by its parametric equation

$$\xi^\varkappa = f^\varkappa(t). \tag{2.1}$$

Then we may form the vectors

$$\underset{1}{v}^\varkappa \overset{\text{def}}{=} \frac{d\,\xi^\varkappa}{d\,t}; \qquad \underset{2}{v}^\varkappa \overset{\text{def}}{=} \frac{\delta}{d\,t}\,\frac{d\,\xi^\varkappa}{d\,t}; \qquad \text{etc.} \tag{2.2}$$

Let $\underset{p+1}{v}^\varkappa$ be the first vector that depends linearly on the preceding vectors. Then it follows from (2.2) that

$$\frac{\delta}{d\,t}\,\underset{1}{v}^{[\varkappa_1}\dots\underset{p}{v}^{\varkappa_p]} = \alpha\,\underset{1}{v}^{[\varkappa_1}\dots\underset{p}{v}^{\varkappa_p]} \tag{2.3}$$

where α is some determinable function of t. This equation expresses the fact that the p-direction of $\underset{1}{v}^{[\varkappa_1}\dots\underset{p}{v}^{\varkappa_p]}$ is covariant constant along the curve. This p-direction is called the *osculating p-direction* of the curve. If the L_n happens to be an E_n it follows that the curve lies in an E_p.

Now let s be another parameter on the curve

$$t = t(s); \qquad t' \overset{\text{def}}{=} \frac{d\,t}{d\,s}; \qquad t'' \overset{\text{def}}{=} \frac{d^2 t}{d\,s^2}. \tag{2.4}$$

Then we have

$$\begin{cases} \underset{1}{u}^\varkappa \overset{\text{def}}{=} \dfrac{d\,\xi^\varkappa}{d\,s} = t'\,\underset{1}{v}^\varkappa \\ \underset{2}{u}^\varkappa \overset{\text{def}}{=} \dfrac{\delta}{d\,s}\,\dfrac{d\,\xi^\varkappa}{d\,s} = (t')^2\,\underset{2}{v}^\varkappa + t''\,\underset{1}{v}^\varkappa \\ \vdots \\ \underset{p}{u}^\varkappa \overset{\text{def}}{=} \left(\dfrac{\delta}{d\,s}\right)^{p-1} \dfrac{d\,\xi^\varkappa}{d\,s} = (t')^p\,\underset{p}{v}^\varkappa + \cdots \end{cases} \tag{2.5}$$

---

[1]) The centre of the osculating circle has the radiusvector $\underset{1}{k}^{-1}\,\underset{2}{j}^\varkappa$ in the tangent space.

[2]) Imbedding in L_n is a generalization of the imbedding in E_n dealt with in the so called "affine geometry" (Affingeometrie). There is a vast literature on this subject. We mention here only some papers on the imbedding in A_n and L_n (cf. Ch. VI § 3 for projective imbedding): WEYL 1922, 3; R. K. 1924, 1, Ch. IV (also for liter.); HLAVATY 1926, 4; 1934, 2 (also for literature); JÄRNEFELT 1928, 1 (L_n); MAYER 1928, 1 (L_n); 1935, 2 (L_n); BORTOLOTTI 1929, 3; 1931, 2; DIENES 1932, 1 (L_n); 1934, 1; GOLAB 1934, 1 (L_n); 2 (curves in X_m in L_n); WEISE 1935, 1; E II 1938, 2, p. 21ff.; T. Y. THOMAS 1938, 3; MAEDA 1939, 1; WAGNER 1946, 1 (variation X_m in X_n); COSSU 1947, 1; COTTON 1948, 1; CASTOLDI 1948, 1; RŽEHNINA 1950, 1; KLINGENBERG 1951, 1; 2; 1952, 1.

hence

$$(2.6)\qquad \underset{1}{u}^{[\varkappa_1} \dots \underset{p}{u}^{\varkappa_p]} = (t')^{\binom{p+1}{2}} \underset{1}{v}^{[\varkappa_1} \dots \underset{p}{v}^{\varkappa_p]}$$

and

$$(2.7)\qquad \frac{\delta}{ds} \underset{1}{u}^{[\varkappa_1} \dots \underset{p}{u}^{\varkappa_p]} = \left(\binom{p+1}{2} (t')^{\binom{p+1}{2}-1} t'' + (t')^{\binom{p+1}{2}+1} \alpha\right) \underset{1}{v}^{[\varkappa_1} \dots \underset{p}{v}^{\varkappa_p]}.$$

This proves that $\underset{1}{u}^{[\varkappa_1} \dots \underset{p}{u}^{\varkappa_p]}$ is covariant constant along the curve if and only if

$$(2.8)\qquad \binom{p+1}{2} t'' + \alpha (t')^2 = 0$$

or

$$(2.9)\qquad \binom{p+1}{2} \frac{d^2 s}{dt^2} - \alpha \frac{ds}{dt} = 0$$

from which s can be solved in the form

$$(2.10)\qquad s = C_1 \varphi(t) + C_2; \qquad C_1, C_2 = \text{constants}.$$

containing two arbitrary constants. A parameter satisfying (2.10) is called an *affine parameter* of the curve. It is determined to within arbitrary linear transformations with constant coefficients.[1]) If $p=1$, the curve is a geodesic and the affine parameter is the same as that defined in III § 7. If $p=n$, the two n-vectors in (2.6) can be replaced by their corresponding Δ-densities $\tilde{\mathfrak{u}}$ and $\tilde{\mathfrak{v}}$:

$$(2.11)\qquad \tilde{\mathfrak{u}} = (t')^{\binom{p+1}{2}} \tilde{\mathfrak{v}}$$

hence

$$(2.12)\qquad s - \underset{0}{s} = \int_{t_0}^{t} \left(\frac{\tilde{\mathfrak{v}}}{\tilde{\mathfrak{u}}}\right)^{\frac{2}{n(n+1)}} dt$$

where $\tilde{\mathfrak{v}}$ is the Δ-density belonging to t and $\tilde{\mathfrak{u}}$ an arbitrary Δ-density of weight -1, that is covariant constant along the curve. In general $\tilde{\mathfrak{u}}$ is determined to within a *constant* scalar factor. But in an L_n with a volume preserving connexion, $V_{\nu\mu}=0$, scalar density fields exist which are covariant constant over the whole space (cf. III § 6). Hence if in this case (f. i. in an E_n) such a field is given, this field can be taken for $\tilde{\mathfrak{u}}$ and then s is determined to within an additive constant $\underset{0}{s}$. The difference of the values of s for two different points is the "affine length" (Affinlänge) known from affine geometry (in E_n).[2]) If the coordinate system

[1]) Hlavaty 1934, 2.

[2]) Cf. footnote 2 on p. 232.

is chosen in such a way that $E^{\varkappa_1 \dots \varkappa_n}$ corresponds to the covariant constant field $\tilde{\mathfrak{u}}$, the affine length between the points s and $\underset{0}{s}$ takes the form

$$s - \underset{0}{s} = \int_{t_0}^{t} \left(\underset{1}{v}^{[1} \dots \underset{n}{v}^{n]} \right)^{\frac{2}{n(n+1)}} dt \tag{2.13}$$

and it is invariant for transformations of coordinates with $\Delta = +1$.[1])

We return now to the general L_n. Once the affine parameter is known, other invariants may be deduced. Let $\underset{0}{\xi}^{\varkappa}$ be a point of the curve with $s = 0$. Then s is determined to within a constant scalar factor. For simplicity we take $p = n$. Then

$$\frac{\delta}{ds} \underset{1}{u}^{[\varkappa_1} \dots \underset{n}{u}^{\varkappa_n]} = 0 \tag{2.14}$$

and from this equation and (2.5) it follows that

$$\frac{\delta}{ds} \underset{n}{u}^{\varkappa} + \underset{1}{\varkappa} \underset{n-1}{u}^{\varkappa} + \dots + \underset{n-1}{\varkappa} \underset{1}{u}^{\varkappa} = 0 \tag{2.15}$$

where $\underset{1}{\varkappa}, \dots, \underset{n-1}{\varkappa}$ are functions of s. If s acquires a constant scalar factor c, $\underset{j}{u}^{\varkappa}$ gets a factor c^{-j} and $\underset{j}{\varkappa}$ a factor $c^{-(j+1)}$. In an E_n the vectors $\underset{j}{u}^{\varkappa}$ and the $\underset{j}{\varkappa}$ are invariants for the special-affine group ($\Delta = +1$, cf. I § 2). The $\underset{j}{\varkappa}$'s could be called special-affine curvatures of the curve. In fact it could be proved that a curve in E_n is determined to within a special-affine transformation if $\underset{1}{\varkappa}, \dots, \underset{n-1}{\varkappa}$ are given as functions of s. But there is one serious difficulty. To each point of the curve there belongs the invariant set of n vectors $\underset{1}{u}^{\varkappa}, \dots, \underset{n}{u}^{\varkappa}$. Now if these vectors are expressed in terms of the derivatives of $\xi^{\varkappa}$ with respect to an *arbitrary* parameter t, it follows that these vectors may contain derivatives up to the orders $n, n+1, \dots, 2n-1$ and that in the $\underset{1}{\varkappa}, \dots, \underset{n-1}{\varkappa}$ derivatives up to the orders $n+2, \dots, 2n$ may occur. But it is possible to find sets of n vectors and a set of "curvatures" belonging to them, that depend only on derivatives of lower order as was proved by WINTERNITZ for $n = 3$.[2])

Of course for a curve in L_n or E_n there are no equations similar to the equations of FRENET in a V_n, because perpendicularity does not exist in an E_n.

---

[1]) Cf. for E_n BERWALD 1923, 1, p. 95; BLASCHKE 1923, 1, p. 73.

[2]) Cf. BLASCHKE 1923, 1, p. 76ff.; HLAVATY 1929, 1; 1934, 2, p. 21ff.; E II 1938, 2, p. 26.

Exercise.

V 2,1. Let the tangent vector $t^\varkappa$ and a normal vector n_λ of a real curve in an ordinary W_n satisfy the conditions

$$\text{V 2,1 }\alpha)\qquad t^\mu n_\mu = 1; \quad n_\lambda t^\mu \nabla_\mu t^\lambda = 0.$$

Prove that the vector $u_\lambda = t^\mu \nabla_\mu n_\lambda$ is perpendicular to $t^\varkappa$ and invariant if $t^\varkappa$ and n_λ are multiplied with a scalar as long as (V 2,1 α) remains valid. u_λ is the curvature vector of the curve in W_n.

§ 3. The X_{n-1} in A_n.[1])

Let an X_{n-1} in A_n be given by its nullform

$$(3.1)\qquad C^{\mathfrak{n}}(\xi^\varkappa) = 0; \quad C^{\mathfrak{n}}_\lambda \overset{\text{def}}{=} \partial_\lambda C^{\mathfrak{n}};$$

or by its parametric form

$$(3.2)\qquad \xi^\varkappa = B^\varkappa(\eta^a); \quad B^\varkappa_b \overset{\text{def}}{=} \partial_b \xi^\varkappa; \quad a, b = 1, \ldots, \mathfrak{n}-1$$

(cf. II 4.4, 12) and let $\overset{0}{\xi}{}^\varkappa$ be a point of X_{n-1}. Then at each point of the X_{n-1} in an $\mathfrak{N}(\underset{0}{\xi}{}^\varkappa)$ we have the connecting quantities $B^\varkappa_b$ and $C^{\mathfrak{n}}_\lambda$ (cf. II § 4). $C^{\mathfrak{n}}_\lambda$ is a covariant vector tangent to the X_{n-1} and determined to within a scalar factor, because here the upper index takes only one value n.

Let the X_{n-1} be rigged (cf. II § 5). To this end we introduce at each point of X_{n-1} an E_1 (in the tangent E_n) not lying in the tangent E_{n-1}. Then in the way described in I § 4 and II § 5 we have at each point of the X_{n-1} in $\mathfrak{N}(\underset{0}{\xi}{}^\varkappa)$ the connecting quantities $C^\varkappa_{\mathfrak{n}}$, B^a_λ; $a = 1, \ldots, \mathfrak{n}-1$. $C^\varkappa_{\mathfrak{n}}$ is a contravariant vector in the direction of the rigging, determined to within a scalar factor. As follows from I § 4 and II § 5 we have in this special case, writing $n^\varkappa$ for $C^\varkappa_{\mathfrak{n}}$ and t_λ for $C^{\mathfrak{n}}_\lambda$,

$$(3.3)\qquad a)\quad B^\lambda_b t_\lambda = 0 \qquad b)\quad B^a_\varkappa n^\varkappa = 0;$$

$$(3.4)\qquad a)\quad B^\mu_b B^a_\mu = B^a_b \qquad b)\quad \boxed{n^\mu t_\mu = 1};$$

$$(3.5)\qquad a)\quad B^\varkappa_b B^b_\lambda = B^\varkappa_\lambda \qquad b)\quad n^\varkappa t_\lambda = C^\varkappa_\lambda;$$

$$(3.6)\qquad A^\varkappa_\lambda = B^\varkappa_\lambda + n^\varkappa t_\lambda.$$

We call $n^\varkappa$ the *pseudo-normal vector* and t_λ the *tangent vector* of X_{n-1} and (3.4b) the *first normalizing condition*. If t_λ acquires a factor σ, $n^\varkappa$ gets a factor σ^{-1} but the quantities $B^\varkappa_b$, B^a_λ, $B^\varkappa_\lambda$, B^a_b and $C^\varkappa_\lambda$ do not change.

---

1) General references on X_{n-1} in A_n: SCHOUTEN 1923, 1; R. K. 1924, 1, p. 133ff.; NOŽISKA 1950, 1 and many papers dealing also with X_m in A_n and L_n.

After the rigging every tensor with valence ≥ 1 of X_{n-1} has components with greek indices and also components with latin indices and every tensor with valence ≥ 1 of A_n at a point of X_{n-1} has an X_{n-1}-part and a normal part (cf. I § 4 and II § 5). If the valence is 1, the vector is the sum of its two parts:

$$(3.7)\qquad \begin{cases} \text{a)} & v^\varkappa = B^\varkappa_\lambda v^\lambda + C^\varkappa_\lambda v^\lambda = {}'v^\varkappa + {}''v^\varkappa \\ \text{b)} & w_\lambda = B^\varkappa_\lambda w_\varkappa + C^\varkappa_\lambda w_\varkappa = {}'w_\lambda + {}''w_\lambda . \end{cases}$$

If p is a scalar field of X_{n-1}, the expression $\nabla_\lambda p = \partial_\lambda p$ has no meaning if the field is not prolonged (cf. II § 4). But $B^\lambda_b \nabla_\lambda p = B^\lambda_b \partial_\lambda p = \partial_b p$ has a meaning independent of any prolongation. It is the gradient of p as a field of X_{n-1} and $B^b_\mu \partial_b p$ are its A_n-components.

If a vector field v^a of X_{n-1} is prolonged we may define a covariant differentiation in X_{n-1} by means of the formula

$$(3.8)\qquad {}'\nabla_c v^a \overset{\text{def}}{=} B^{\mu a}_{c\varkappa} \nabla_\mu v^\varkappa .$$

This means that ${}'\nabla_c v^a$ is the X_{n-1}-part of $\nabla_\mu v^\varkappa$. But ${}'\nabla_c v^a$ is independent of the manner of prolongation as will be seen if we write out the right hand side of (3.8):

$$(3.9)\qquad {}'\nabla_c v^a = B^\mu_c B^a_\varkappa (\partial_\mu v^\varkappa + \Gamma^\varkappa_{\mu\lambda} v^\lambda) = \partial_c v^a + B^{\mu\lambda a}_{c b \varkappa} \Gamma^\varkappa_{\mu\lambda} v^b - B^\varkappa_b (\partial_c B^a_\varkappa) v^b .$$

Moreover this equation proves that the connexion in A_n induces in the rigged X_{n-1} a connexion with the parameters

$$(3.10)\qquad \begin{cases} {}'\Gamma^a_{cb} &= B^{\mu\lambda a}_{c b \varkappa} \Gamma^\varkappa_{\mu\lambda} - B^\varkappa_b \partial_c B^a_\varkappa \\ &= B^{\mu\lambda a}_{c b \varkappa} \Gamma^\varkappa_{\mu\lambda} + B^a_\varkappa \partial_c B^\varkappa_b \\ &= B^{\mu\lambda a}_{c b \varkappa} \Gamma^\varkappa_{\mu\lambda} + B^a_\varkappa \partial_c \partial_b \xi^\varkappa \end{cases}$$

which are independent of the normalization of t_λ. Of course for this connexion we have ${}'\nabla_c p = \partial_c p$. From the third line of (3.10) we see that ${}'\Gamma^a_{[cb]} = 0$, hence:

A rigged X_{n-1} in A_n is an A_{n-1} whose connexion is uniquely determined by the connexion of the A_n and the rigging.

Besides ${}'\Gamma^a_{cb}$ other geometric objects are also induced in the A_{n-1}. The two tensors of A_{n-1}

$$(3.11)\qquad \boxed{h_{cb} \overset{\text{def}}{=} - B^{\mu\lambda}_{cb} \nabla_\mu t_\lambda} = - B^{\mu\lambda}_{cb} \partial_\mu \partial_\lambda C^n - B^{\mu\lambda}_{cb} \Gamma^\varkappa_{\mu\lambda} t_\varkappa ;$$

$$(3.12)\qquad \boxed{l_c^{\cdot a} \overset{\text{def}}{=} - B^{\mu a}_{c\varkappa} \nabla_\mu n^\varkappa}$$

get a factor σ and σ^{-1} respectively if t_λ acquires a factor σ. It follows from (3.11) that h_{cb} is a *symmetric tensor.* The directions in the nullcone of h_{cb} are called the *principal tangents* of the rigged or non rigged X_{n-1}, and curves of X_{n-1} whose tangents are everywhere principal are called *asymptotic lines*. The directions of the eigenvectors of $l_c^{\cdot a}$ are called the *directions of the principal curvature* of the rigged X_{n-1}, and curves of X_{n-1} whose tangent lies everywhere in a principal direction are called *lines of curvature*.

In order to derive $'\nabla_c{}'\nabla_b v^a$ we have first to take the A_n-components $B^{\varrho\,\varkappa}_{\mu\,\sigma}\nabla_\varrho v^\sigma$ of $'\nabla_c v^a$ and to apply a process analogous to (3.8) to this quantity of valence 2. Then we get

$$(3.13)\quad '\nabla_d{}'\nabla_c v^a = B^{\nu\mu a}_{dc\varkappa} h_{\nu\mu} n^\varrho \nabla_\varrho v^\varkappa + B^{\nu\mu a}_{dc\varkappa} l_\nu^{\cdot\varkappa} h_{\mu\sigma} v^\sigma + B^{\nu\mu a}_{dc\varkappa} \nabla_\nu \nabla_\mu v^\varkappa$$

hence

$$(3.14)\quad 'R_{dcb}^{\cdot\cdot\cdot a} v^b = B^{\nu\mu\lambda a}_{dcb\varkappa} R_{\nu\mu\lambda}^{\cdot\cdot\cdot\varkappa} v^b + 2 l_{[d}^{\cdot a} h_{c]b} v^b$$

or

$$(3.15)\quad \boxed{'R_{dcb}^{\cdot\cdot\cdot a} = B^{\nu\mu\lambda a}_{dcb\varkappa} R_{\nu\mu\lambda}^{\cdot\cdot\cdot\varkappa} + 2 l_{[d}^{\cdot a} h_{c]b}}\,.$$

This is the *generalized equation of* GAUSS for a rigged A_{n-1} in A_n.

From (3.11) and (3.12) we get by differentiation and alternation

$$(3.16)\quad \begin{cases} 2\,'\nabla_{[d} h_{c]b} = -\,2 B^{\nu\mu\lambda}_{[dc]b} \nabla_\nu B^{\varrho\sigma}_{\mu\lambda} \nabla_\varrho t_\sigma \\ \qquad = 2 B^{\nu\mu\lambda}_{[dc]b} (\nabla_\nu t_\lambda n^\sigma) \nabla_\mu t_\sigma - 2 B^{\nu\mu\lambda}_{[dc]b} \nabla_\nu \nabla_\mu t_\lambda \\ \qquad = -\,2 B^\mu_{[d} h_{c]b} n^\sigma \nabla_\mu t_\sigma + B^{\nu\mu\lambda}_{dcb} R_{\nu\mu\lambda}^{\cdot\cdot\cdot\varkappa} t_\varkappa; \end{cases}$$

$$(3.17)\quad \begin{cases} 2\,'\nabla_{[d} l_{c]}^{\cdot a} = -\,2 B^{\nu\mu a}_{[dc]\varkappa} \nabla_\nu B^{\varrho\varkappa}_{\mu\sigma} \nabla_\varrho n^\sigma \\ \qquad = 2 B^{\nu\mu a}_{[dc]\varkappa} (\nabla_\nu t_\sigma n^\varkappa) \nabla_\mu n^\sigma - 2 B^{\nu\mu a}_{[dc]\varkappa} \nabla_\nu \nabla_\mu n^\varkappa \\ \qquad = -\,2 B^\mu_{[d} l_{c]}^{\cdot a} t_\sigma \nabla_\mu n^\sigma - B^{\nu\mu a}_{dc\varkappa} R_{\nu\mu\lambda}^{\cdot\cdot\cdot\varkappa} n^\lambda. \end{cases}$$

Now in (3.15) all terms are invariant if the normalization of t_λ is changed. But in (3.16, 17) only the last terms of each of the right hand sides acquire a factor σ and σ^{-1} respectively and the other terms change in a more complicated way because σ need not be a constant. This is very inconvenient and we therefore introduce as a *second condition of normalization* the equations

$$(3.18)\quad \boxed{\text{a)}\ B^\mu_b (\nabla_\mu t_\lambda) n^\lambda = 0; \qquad \text{b)}\ B^\mu_b (\nabla_\mu n^\varkappa) t_\varkappa = 0}$$

which are consequences of each other.[1] If (3.18) is satisfied, t_λ and $n^\varkappa$ are fixed to within a *constant* scalar factor and in (3.16, 17) the first

[1] BERWALD 1922, 1, p. 164 for $A_n = E_n$; SCHOUTEN 1923, 1, p. 169.

term of each right hand side vanishes:

$$\boxed{2\,'\nabla_{[d}\,h_{c]\,b} = B^{\nu\mu\lambda}_{d\,c\,b}\,R_{\nu\mu\lambda}{}^{\cdot\cdot\cdot\varkappa}\,t_\varkappa} \tag{3.19}$$

$$\boxed{2\,'\nabla_{[d}\,l_{c]}^{\cdot\,a} = -\,B^{\nu\mu a}_{d\,c\,\varkappa}\,R_{\nu\mu\lambda}{}^{\cdot\cdot\cdot\varkappa}\,n^\lambda}\;. \tag{3.20}$$

These are the *generalized equations of* CODAZZI for a rigged A_{n-1} in A_n. Moreover we have now from (3.11, 12, 18).

$$2 l_{[d}^{\cdot\,a} h_{c]\,a} = 2\,B^{\nu\;a\;\mu\;\lambda}_{[d\,|\varkappa|\,c]\,a}(\nabla_\nu n^\varkappa)\,\nabla_\mu t_\lambda = 2\,B^{\nu\mu}_{[d\,c]}\,(\nabla_\nu n^\lambda)\,\nabla_\mu t_\lambda = B^{\nu\mu}_{d\,c} R_{\nu\mu\lambda}{}^{\cdot\cdot\cdot\varkappa} t_\varkappa n^\lambda \tag{3.21}$$

and by substituting this value in (3.15) and making use of (3.6)

$$'R_{d\,c\,a}{}^{\cdot\cdot\cdot a} = B^{\nu\mu}_{d\,c}\,R_{\nu\mu\lambda}{}^{\cdot\cdot\cdot\lambda}. \tag{3.22}$$

Hence, if both normalizing conditions are satisfied, and if the connexion in A_n is volume preserving, the induced connexion in A_{n-1} has the same property.[1])

We now have to investigate first whether it is possible to satisfy (3.18) *if the direction of* $n^\varkappa$ *is given.* Let t_λ and $n^\varkappa$ be normalized in any way so as to satisfy (3.4b). Then, if $B^\mu_b\,(\nabla_\mu\,t_\lambda)\,n^\lambda \neq 0$ we try to choose σ such that

$$B^\mu_b\,(\nabla_\mu\,\sigma\,t_\lambda)\,\sigma^{-1}\,n^\lambda = 0\,. \tag{3.23}$$

This gives the following differential equation for σ

$$'\nabla_b \log\sigma = B^\mu_b\,(\nabla_\mu\,t_\lambda)\,n^\lambda \tag{3.24}$$

and after some calculation we find for the integrability conditions of this equation

$$-\tfrac{1}{2}\,B^{\nu\mu}_{d\,c}\,R_{\nu\mu\lambda}{}^{\cdot\cdot\cdot\varkappa}\,t_\varkappa\,n^\lambda + B^{\nu\mu}_{d\,c}(\nabla_{[\mu}\,t_{|\lambda|})\,\nabla_{\nu]}\,n^\lambda = 0. \tag{3.25}$$ [2])

This is equivalent to (3.22) because of (3.6) and (3.15). Hence:

If the direction of rigging be given, t_λ *and* $n^\varkappa$ *can be normalized according to* (3.4b) *and* (3.18) *if and only if* (3.22) *holds*

and as a corollary

If the connexion in A_n *is volume preserving and the direction of* $n^\varkappa$ *given,* t_λ *and* $n^\varkappa$ *can be normalized according to* (3.4b) *and* (3.18) *if and only if the connexion in* A_{n-1} *is volume preserving as well.*

We ask now whether it is always possible to find for a given normalization of t_λ a vector $n^\varkappa$ such that (3.4b) and (3.18) both hold. It

---

[1]) SCHOUTEN 1923, 1, p. 172. On p. 187 this is generalized for A_m in A_n.

[2]) Cf. EISENHART 1927, 1, p. 153 (55.9).

follows from (3.18) that the direction of $n^\varkappa$ must lie in the support of the λ-domain of $B_c^\mu \nabla_\mu t_\lambda$, hence the n.a.s. condition is that this support is not wholly contained in the tangent E_{n-1} or in other words, that t_λ does not lie in the λ-domain of $B_c^\mu \nabla_\mu t_\lambda$. Now if $'r$ is the rank of h_{cb}, the rank r of $B_c^\mu \nabla_\mu t_\lambda$ is $'r$ or $'r+1$ and it is $'r+1$ if and only if t_λ lies in the λ-domain of $B_b^\mu \nabla_\mu t_\lambda$. Hence for $'r = n-1$ the vector t_λ can never lie in this domain, because r can never be greater than $n-1$. Collecting results we get:

If the normalization of t_λ be given, there exists a vector $n^\varkappa$ that satisfies (3.4b) and (3.18) if and only if t_λ does not lie in the λ-domain of $B_c^\mu \nabla_\mu t_\lambda$. If the rank of h_{cb} is equal to $n-1$, t_λ never lies in this domain and $n^\varkappa$ is uniquely determined and has the direction of the support of this domain.[1])

If a vector $n^\varkappa$ satisfying (3.4b) and (3.18) has been found for a given normalization of t_λ and if this normalization is changed, $t_\lambda \to \sigma t_\lambda$, the new value of the normal vector may be written in the form $\sigma^{-1}(n^\varkappa + p^\varkappa)$; $p^\varkappa t_\varkappa = 0$. Then we have

$$0 = B_b^\mu (\nabla_\mu \sigma t_\lambda)(n^\lambda + p^\lambda) = B_b^\mu \nabla_\mu \sigma + \sigma B_b^\mu (\nabla_\mu t_\lambda) p^\lambda \tag{3.26}$$

or

$$h_{ba} p^a = \partial_b \log \sigma . \tag{3.27}$$

Hence, if h_{ba} has rank $n-1$, the vector p^a can be solved

$$p^a = \overset{-1}{h}{}^{ab} \partial_b \log \sigma \tag{3.28}$$

and, as was to be foreseen, the new normal vector is uniquely determined. But if the rank of h_{ba} is $< n-1$, it may happen that the vector $\partial_b \log \sigma$ does not lie in the b-domain of h_{ba} and in this case there is no normal vector for the new normalization of t_λ.

The quantity

$$\tilde{\mathfrak{t}}_\lambda \overset{\text{def}}{=} \frac{1}{(n-1)!} \tilde{\mathfrak{e}}_{\lambda \lambda_2 \ldots \lambda_n} B_{a_2 \ldots a_n}^{\lambda_2 \ldots \lambda_n} \tilde{\mathfrak{E}}^{a_2 \ldots a_n} \tag{3.29}$$

is a covariant vector Δ-density of weight -1 in A_n and weight $+1$ in X_{n-1}. It is a concomitant of the X_{n-1}, independent of any rigging and its $(n-1)$-direction is tangent to the X_{n-1}. Now if h_{cb} has rank $n-1$, the quantity

$$\mathfrak{h} \overset{\text{def}}{=} |\mathrm{Det}\,(h_{cb})| \tag{3.30}$$

is a scalar density of weight $+2$ in X_{n-1}. Hence, if a screwsense is given in X_{n-1} and a scalar Δ-density $\tilde{\mathfrak{q}}$ of weight $+1$ in A_n, at least at all points of X_{n-1}, then densities and Δ-densities in X_{n-1} can be

[1]) Cf. SCHOUTEN 1923, 1, p. 173ff.

identified and a normalization of t_λ can be fixed by the equation

$$t_\lambda = \tilde{\mathfrak{q}}\, \mathfrak{h}^{-\frac{1}{2}} \tilde{\mathfrak{t}}_\lambda. \tag{3.31}$$

If the connexion in A_n happens to be volume preserving, the condition $\nabla_\lambda \tilde{\mathfrak{q}} = 0$ determines a field $\tilde{\mathfrak{q}}$ and therefore also the field t_λ to within a *constant* scalar factor, hence

For an X_{n-1} in a volume preserving A_n for which h_{cb} has rank $n-1$ at all points, special fields t_λ and $n^\varkappa$ satisfying both normalization conditions are determined to within a constant scalar factor and its reciprocal.

In an E_n the equations of GAUSS and CODAZZI take the very simple form

$$\boxed{\begin{aligned} 2 l_{[d}^{\cdot\, a} h_{c]b} &= {}'R_{dcb}^{\cdot\cdot\cdot\, a} && (3.32)\\ {}'\nabla_{[d} h_{c]b} &= 0 && (3.33)\\ {}'\nabla_{[d} l_{c]}^{\cdot\, a} &= 0. && (3.34)\end{aligned}}$$

These equations are the n.a.s. integrability conditions for the construction of an E_n in which the A_{n-1} with a given volume preserving connexion ${}'\Gamma_{cb}^a$ and given fields h_{cb} (of rank $n-1$) and $l_c^{\cdot\, a}$ can be imbedded. Let the coordinate system $\xi^\varkappa$ in the E_n to be constructed be rectilinear, $\Gamma_{\mu\lambda}^\varkappa = 0$. Then the unknowns are $\xi^\varkappa$, t_λ and $n^\varkappa$ as functions of the η^a and the equations for these unknowns are (cf. 3.10, 11, 12)

$$\text{(3.35)}\quad \text{a) } \partial_b \xi^\varkappa = B_b^\varkappa \quad \text{b) } t_\lambda n^\lambda = 1 \quad \text{c) } (\partial_c t_\lambda) n^\lambda = 0 \quad \text{d) } B_b^\lambda t_\lambda = 0$$

$$\text{(3.36)}\quad \text{a) } B_b^\lambda \partial_c t_\lambda = -h_{cb} \quad \text{b) } \partial_c n^\varkappa = -B_a^\varkappa l_c^{\cdot\, a} \quad \text{c) } \partial_c B_b^\varkappa = {}'\Gamma_{cb}^a B_a^\varkappa + h_{cb} n^\varkappa .$$

Taking into account that the given connexion ${}'\Gamma_{cb}^a$ is volume preserving we get the following scheme for the integrability conditions of (3.35, 36):

(3.37)

3.35a	3.36c
3.35b	3.35c, d; 3.36b
3.35c	3.36a; 3.32
3.35d	3.35b; 3.36a, c
3.36a	3.33; 3.35c
3.36b	3.34; 3.35c; 3.36a
3.36c	3.32; 3.33.

In this scheme the integrability conditions of the equation to the left are identically satisfied as a consequence of the equation or equations to the right on the same line. Hence the conditions are all identically satisfied and this implies that there is one and only one solution of (3.35, 36) satisfying the initial conditions for $\eta^a = \underset{0}{\eta}{}^a$, $h_{cb} = \underset{0}{h}{}_{cb}$, $l_c^{\cdot\, a} = \underset{0}{l}{}_c^{\cdot\, a}$, ${}'\Gamma_{cb}^a = {}'\underset{0}{\Gamma}{}_{cb}^a$:

$$
(3.38)\quad \left\{\begin{array}{ll}
\text{a)}\ \xi^\varkappa = \underset{0}{\xi}{}^\varkappa & \text{f)}\ \underset{0}{n}{}^\lambda \partial_c t_\lambda = 0 \\
\text{b)}\ t_\lambda = \underset{0}{t}{}_\lambda & \text{g)}\ \partial_c n^\varkappa = -\underset{0}{B}{}^\varkappa_a \underset{0}{l}{}_c^{\cdot a} \\
\text{c)}\ n^\varkappa = \underset{0}{n}{}^\varkappa & \text{h)}\ \partial_c B^\varkappa_b = \underset{0}{'\Gamma}{}^a_{cb} \underset{0}{B}{}^\varkappa_a + \underset{0}{h}{}_{cb} \underset{0}{n}{}^\varkappa \\
\text{d)}\ B^\varkappa_b = \underset{0}{B}{}^\varkappa_b & \text{i)}\ \underset{0}{t}{}_\lambda \underset{0}{n}{}^\lambda = 1 \\
\text{e)}\ \underset{0}{B}{}^\lambda_b \partial_c t_\lambda = -\underset{0}{h}{}_{cb} & \text{j)}\ \underset{0}{B}{}^\lambda_b \underset{0}{t}{}_\lambda = 0.
\end{array}\right.
$$

This means that the imbedding and rigging of an $\mathfrak{N}(\underset{0}{\eta}{}^a)$ of an A_{n-1} with given fields $'\Gamma^a_{cb}$, h_{cb} (of rank $n-1$) and $l_c^{\cdot a}$, satisfying (3.32, 33, 34) and $'R_{dca}^{\cdot\cdot\cdot a}=0$, in an E_n is determined if we give 1. the point $\underset{0}{\xi}{}^\varkappa$ corresponding to $\underset{0}{\eta}{}^\varkappa$; 2. the two vectors $\underset{0}{t}{}_\lambda$, $\underset{0}{n}{}^\varkappa$ at $\underset{0}{\xi}{}^\varkappa$, satisfying (3.38i); 3. $n-1$ linearly independent vectors $\underset{0}{B}{}^\varkappa_b$ in the E_{n-1} of $\underset{0}{t}{}_\lambda$. If instead we give in an arbitrary way the point $*\underset{0}{\xi}{}^\varkappa$ and the vectors $*\underset{0}{t}{}_\lambda$, $*\underset{0}{n}{}^\varkappa$ and $*\underset{0}{B}{}^\varkappa_b$ satisfying $*\underset{0}{t}{}_\lambda *\underset{0}{n}{}^\lambda = 1$ and $*\underset{0}{B}{}^\lambda_b *\underset{0}{t}{}_\lambda = 0$, obviously there always exist n^2+n constants $C^\varkappa_\lambda$, $c^\varkappa$ such that

$$
(3.39)\quad *\underset{0}{\xi}{}^\varkappa = C^\varkappa_\lambda \underset{0}{\xi}{}^\lambda + c^\varkappa; \quad *\underset{0}{t}{}_\lambda = \overset{-1}{C}{}^\varkappa_\lambda \underset{0}{t}{}_\varkappa; \quad *\underset{0}{n}{}^\varkappa = C^\varkappa_\lambda \underset{0}{n}{}^\lambda; \quad *\underset{0}{B}{}^\varkappa_b = C^\varkappa_\lambda \underset{0}{B}{}^\lambda_b.
$$

If now $\xi^\varkappa$, t_λ, $n^\varkappa$ are the solutions satisfying the old initial conditions it is clear that $C^\varkappa_\lambda \xi^\lambda + c^\varkappa$, $\overset{-1}{C}{}^\varkappa_\lambda t_\varkappa$ and $C^\varkappa_\lambda n^\lambda$ are the solutions satisfying the new conditions. Hence the imbedding and rigging of the given A_{n-1} is determined to within a non homogeneous affine transformation in E_n.

In an X_{n-1} in a general A_n the field h_{ba} is fixed to within an arbitrary scalar factor. This means that, if h_{ba} has rank $n-1$, a conformal geometry with a fundamental pseudotensor $\lfloor h_{ba} \rfloor$ is induced in the X_{n-1}. It also means that the asymptotic lines of an X_{n-1} in A_n are independent of the rigging.

Exercise.

V 3,1. Prove for a rigged X_{n-1} in A_n satisfying (3.4b) and (3.18):

a) The covariant differential of t_λ in an asymptotic direction contains this direction.

b) The covariant differential of $n^\varkappa$ in a direction of principal curvature lies in this direction.

c) For every line of curvature on X_{n-1} the covariant differential of the bivector $n^{[\varkappa} d\xi^{\lambda]}/dt$ in the direction of the curve is simple and contains the direction of $n^\varkappa$.

§ 4. The V_{n-1} in V_n.[1), 2)]

Let $\underset{\mathrm{n}}{i}_\lambda$ be the tangent unitvector of a real X_{n-1} in an ordinary V_n. Then the X_{n-1} is a V_{n-1} and the contravariant vector $\underset{\mathrm{n}}{i}^\varkappa$ determines a rigging. Because of

$$\underset{\mathrm{n}}{i}_\lambda \underset{\mathrm{n}}{i}^\lambda = 1, \tag{4.1}$$

$$B_c^\mu (\nabla_\mu \underset{\mathrm{n}}{i}^\lambda) \underset{\mathrm{n}}{i}_\lambda = 0. \tag{4.2}$$

$\underset{\mathrm{n}}{i}_\lambda$ and $\underset{\mathrm{n}}{i}^\varkappa$ satisfy the normalization conditions of t_λ and $n^\varkappa$ in the preceding section. From now on we write $n^\varkappa$ for $\underset{\mathrm{n}}{i}^\varkappa$. Note that the normalization in A_n is quite different from that in V_n because in V_n it depends entirely on the fundamental tensor $g_{\lambda\varkappa}$. In a V_n the difference between h_{cb} and l_{cb} vanishes because

$$h_{cb} = -B_{c\,b}^{\mu\lambda} \nabla_\mu n_\lambda = -B_{c\,b}^{\mu\lambda} g_{\lambda\varkappa} \nabla_\mu n^\varkappa = -{}'g_{ba} B_{c\varkappa}^{\mu a} \nabla_\mu n^\varkappa = l_c^{\cdot a}\, {}'g_{ba}. \tag{4.3}$$

${}'g_{ba}$ is the section of $g_{\lambda\varkappa}$ and hence ${}'g_{\lambda\varkappa} = g_{\lambda\varkappa} - n_\lambda n_\varkappa$. The induced connexion is Riemannian because

$${}'\nabla_c\, {}'g_{ba} = B_{c\,b\,a}^{\mu\lambda\varkappa} \nabla_\mu {}'g_{\lambda\varkappa} = -B_{c\,b\,a}^{\mu\lambda\varkappa} \nabla_\mu n_\lambda n_\varkappa = 0. \tag{4.4}$$

The equations of GAUSS and CODAZZI take the form

$$\boxed{{}'K_{dcba} = B_{d\,c\,b\,a}^{\nu\mu\lambda\varkappa} K_{\nu\mu\lambda\varkappa} - 2h_{[d[b} h_{c]a]}}; \tag{4.5}$$

$$\boxed{2\,{}'\nabla_{[d} h_{c]b} = B_{d\,c\,b}^{\nu\mu\lambda} K_{\nu\mu\lambda\varkappa} n^\varkappa} \tag{4.6}$$

and for $V_n = R_n$

$$\boxed{{}'K_{dcba} = -2h_{[d[b} h_{c]a]}}; \tag{4.7}$$

$$\boxed{{}'\nabla_{[d} h_{c]b} = 0}. \tag{4.8}$$

---

1) Cf. for a more elaborate treatment of the properties of a V_{n-1} in V_n dealt with in V § 4, 5: EISENHART 1926, 1, Ch. IV; DUSCHEK and MAYER 1930, 1; E II 1938, 2, Ch. II.

2) General references on V_{n-1} in V_n: BERWALD 1922, 1; 2; WEYL 1922, 3 FIALKOW 1938, 2; 3 (R_n and S_n); GHOSH 1938, 1; E II 1938, 2, p. 57ff.; CARTAN 1939, 1; 2 (V_{n-1} isopar. in S_n); COBURN 1940, 1 (EINSTEIN V_{n-1}); COTTON 1940, 1 MUTÔ 1940, 2; MAEDA 1940, 1; URBAN 1947, 1 and 1948, 1 (V_1 on umb. V_{n-1}) VERBICKIĬ 1949, 1 (hyperquadrics); GOLAB 1949, 1 and GOLAB and WRÓBEL 1951, 1 (curves in V_{n-1} in R_n).

h_{ba} is often called the *second fundamental tensor* of the V_{n-1} in V_n. If a V_{n-1} is given in which the curvature tensor can be written in the form (4.7), where h_{cb} is a symmetric tensor satisfying (4.8) it is always possible to construct an R_n in which a sufficiently small part of this V_{n-1} can be imbedded. For $n=3$ this is a fundamental theorem of classical differential geometry. The proof for an arbitrary n gives another example of the significance of the GAUSS-CODAZZI equations as integrability conditions. The unknowns are now $\xi^\varkappa$ and n_λ and the system $(\varkappa)$ in R_n is supposed to be cartesian. Then the equations are

$$(4.9)\quad \begin{cases} \text{a)} & \partial_b \xi^\varkappa = B_b^\varkappa \\ \text{b)} & n_\lambda n^\lambda = 1; \qquad n^\varkappa \overset{\text{def}}{=} g^{\varkappa\lambda} n_\lambda \\ \text{c)} & (\partial_c n_\lambda)\, n^\lambda = 0 \\ \text{d)} & B_b^\lambda n_\lambda = 0 \\ \text{e)} & B_{ba}^{\lambda\varkappa} g_{\lambda\varkappa} = {}'g_{ba}; \qquad g_{\lambda\varkappa} \overset{\text{def}}{=} \delta_{\lambda\varkappa} \end{cases}$$

$$(4.10)\quad \begin{cases} \text{a)} & B_b^\lambda \partial_c n_\lambda = -h_{cb} \\ \text{b)} & \partial_c B_b^\varkappa = {}'\{^{\,a}_{cb}\}\, B_a^\varkappa + h_{cb}\, n^\varkappa \end{cases}$$

and the integrability conditions are identically satisfied according to the scheme (cf. scheme 3.37)

(4.11)

4.9a	4.10b
4.9b	4.9c, d
4.9c	4.10a, 4.7
4.9d	4.9b; 4.10a, b
4.9e	4.9d; 4.10b
4.10a	4.8; 4.9c
4.10b	4.7; 4.8.

Hence the imbedding is fixed if in R_n we give the point $\underset{0}{\xi^\varkappa}$ corresponding to $\underset{0}{\eta^a}$, at this point a unitvector $\underset{0}{n^\varkappa}$ and in the R_{n-1} of $\underset{0}{n_\lambda}$ a set of vectors $\underset{0}{B_1^\varkappa}, \ldots, \underset{0}{B_{n-1}^\varkappa}$ having the same lengths and the same angles between them as the basis vectors $\underset{1}{e^a}, \ldots, \underset{n-1}{e^a}$ at $\underset{0}{\eta^a}$ in V_{n-1}. This proves that the imbedding in R_n is fixed to within translations, rotations and reflexions.

If ${}'g_{ba}$ is given we know ${}'K_{dcba}$ and according to a wellknown theorem of tensor calculus (cf. Exerc. I 8,9) the symmetric tensor h_{cb} is determined by (4.7) to within the sign if such an equation holds and the rank of ${}'K_{dcba}$ is >2. Hence a sufficiently small part of a V_{n-1}, $n \geq 4$, whose curvature tensor can be written in the form (4.7) with a symmetric field h_{cb} satisfying (4.8), can always be imbedded in an R_n and if the

16*

rank of $'K_{dcba}$ is > 2 the imbedding is determined to within translations, rotations and reflexions. For $n = 3$ the curvature tensor has rank 0 or 2 and can always be written in the form (4.7) but in this case this equation determines $\mathrm{Det}(h_{ba})$ only and still leaves two degrees of freedom for the symmetric tensor h_{ba}. A transformation of a V_{n-1} in R_n preserving the linear element is called a *bending* and the bending is said to be *genuine* if it is not a combination of a translation and a rotation or reflexotation. Using this term we are now able to state the theorem of SCHUR:

A V_{n-1} in R_n can for $n = 3$ always suffer a genuine bending and for $n > 3$ can do so if and only if h_{ba} has rank 0, 1 or 2.[1])

In order to prove that the condition is sufficient it is necessary to prove that for a rank < 3 the tensor h_{ba} can always be changed continuously in such a way that the equations of GAUSS and CODAZZI remain valid.

Exercise.

V 4,1. Let a V_{n-1} be given in a V_n. We choose the coordinate system in such a way that $\xi^n = 0$ is the equation of the V_{n-1}, that the ξ^α; $\alpha = 1, \ldots, n-1$, are coordinates in the V_{n-1} and that ξ^n is the distance from a point to the V_{n-1} along a geodesic perpendicular to the V_{n-1}. Prove that with respect to this special coordinate system

$$\text{V 4,1 } \alpha) \qquad h_{\gamma\beta} \overset{*}{=} -\tfrac{1}{2}\,\partial_n\, g_{\gamma\beta}; \qquad \beta, \gamma = 1, \ldots, n-1.$$

§ 5. Congruences in V_n.[2])

The symmetric tensor $h_{\mu\lambda}$ of the preceding section is a special case of a quantity occurring in the theory of congruences. Let $i^\varkappa$ be the unitvector tangent to a real congruence in an ordinary V_n. Then, because $(\nabla_\mu i_\lambda)\, i^\lambda = 0$ the covariant derivative of the field i_λ can always be written in the form

$$(5.1) \qquad \nabla_\mu i_\lambda = l_{\mu\lambda} + i_\mu u_\lambda; \qquad i^\mu l_{\mu\lambda} = 0; \qquad l_{\mu\lambda} i^\lambda = 0,$$

where u_λ is the curvature vector of the congruence (cf. V § 1). Now the n.a.s. condition for i_λ to be V_{n-1}-forming is (cf. II § 5 or § 7)

$$(5.2) \qquad 0 = i_{[\nu} \nabla_\mu i_{\lambda]} = i_{[\nu} l_{\mu\lambda]}$$

[1]) BEEZ 1874, 1; 1875, 1 (in R_n); SCHUR 1886, 1; 2; 1887, 1 (in R_4); BOMPIANI 1914, 1 (in R_n); KILLING 1885, 1; 1893, 1. Cf. for V_{n-1} in S_n CARTAN 1916, 1; E II 1938, 2, p. 143ff., also for literature. Special forms of bending of a surface in R_3 leaving the principal axes of h_{ba} invariant are dealt with by CARTAN 1941, 2; 1942, 1; 1943, 1.

[2]) RICCI 1895, 1; RICCI and LEVI CIVITA 1901, 1; SCHOUTEN-STRUIK 1919, 1; 1921, 2; LEVI CIVITA 1925, 1; EISENHART 1926, 1; E II 1938, 2, p. 27ff.

or

(5.3) $$h_{\mu\lambda} \overset{\text{def}}{=} -l_{(\mu\lambda)} = -l_{\mu\lambda}; \quad l_{[\mu\lambda]} = 0.$$

Now for every V_{n-1} of this set we have

(5.4) $$B_\lambda^\varkappa = A_\lambda^\varkappa - i_\lambda i^\varkappa,$$

and accordingly the symmetric tensor

(5.5) $$h_{\mu\lambda} = -B_{\mu\lambda}^{\sigma\varrho} \nabla_\sigma i_\varrho$$

is the second fundamental tensor of these V_{n-1}'s.

Returning to the general case $l_{[\mu\lambda]} \neq 0$ we may require the principal directions of $h_{\mu\lambda}$. One of these is the direction of $i^\varkappa$, belonging to the eigenvalue zero. The others with unitvectors $\underset{1}{i}^\varkappa, \ldots, \underset{n-1}{i}^\varkappa$ form $n-1$ congruences perpendicular to $i^\varkappa$. They are determined uniquely if and only if the eigenvalues of $h_{\mu\lambda}$ are all different. Ricci[1]) called them the *canonical congruences* of $i^\varkappa$.

If one of these congruences, for instance $\underset{1}{i}^\varkappa$, is V_{n-1}-normal (cf. II § 5, III § 9) we have

(5.6) $$\begin{cases} \left(i^\mu \nabla_\mu \underset{1}{i}_\lambda - \underset{1}{i}^\mu \nabla_\mu i_\lambda\right) \underset{\mathfrak{a}}{i}^\lambda = -2\underset{1}{i}^\mu (\nabla_{(\mu} i_{\lambda)}) \underset{\mathfrak{a}}{i}^\lambda + 2 i^\mu \left(\nabla_{[\mu} \underset{1}{i}_{\lambda]}\right) \underset{\mathfrak{a}}{i}^\lambda \\ \qquad = -2 \underset{1}{i}^\mu \underset{\mathfrak{a}}{i}^\lambda \nabla_{(\mu} i_{\lambda)} = 0; \qquad \mathfrak{a} = 2, \ldots, \mathfrak{n}-1 \end{cases}$$

and this proves (cf. II § 5) that

Every V_{n-1}-normal canonical congruence of $i^\varkappa$ is together with $i^\varkappa$ V_2-forming.

In other words: the equations $i^\mu \partial_\mu f = 0$; $\underset{1}{i}^\mu \partial_\mu f = 0$ form a complete system.

The congruence $i^\varkappa$ is *geodesic* if $u^\varkappa = 0$. A geodesic congruence need not be V_{n-1}-normal as is wellknown from congruences of straight lines in ordinary space. But if a congruence is geodesic and normal to *one* V_{n-1} it is *V_{n-1}-normal throughout.*[2]) To prove this we start from $i^\mu \nabla_\mu i^\varkappa = 0$ and $\nabla_\mu i_\lambda = l_{\mu\lambda}$, valid everywhere and $\nabla_\mu i_\lambda = -h_{\mu\lambda}$ valid at all points of a V_{n-1}. Then we have everywhere (cf. III 4.9a)

(5.7) $$i^\nu \nabla_\nu \nabla_\mu i_\lambda = i^\nu \nabla_\mu \nabla_\nu i_\lambda - i^\nu K_{\nu\mu\lambda}^{\cdot\cdot\cdot\varkappa} i_\varkappa = -(\nabla_\mu i^\nu) \nabla_\nu i_\lambda - i^\nu K_{\nu\mu\lambda}^{\cdot\cdot\cdot\varkappa} i_\varkappa,$$

[1]) Ricci 1895, 1; cf. Schouten-Struik 1919, 1.

[2]) This theorem is due to Beltrami 1869, 1. Cf. Darboux 1889, 1; Struik 1922, 1, p. 51; Eisenhart 1926, 1, p. 57; E II 1938, 2, p. 45. The theorem in its simplest form had already been found by Gauss, 1827, 1, §§ 15, 16.

hence putting $k_{\mu\lambda} \stackrel{\text{def}}{=} - l_{[\mu\lambda]}$

$$\frac{\delta}{ds} k_{\mu\lambda} = k_{[\mu}^{\cdot\,\cdot\,\nu} h_{|\nu|\lambda]} + h_{[\mu}^{\cdot\,\cdot\,\nu} k_{|\nu|\lambda]} = 2 k_{[\mu}^{\cdot\,\cdot\,\nu} h_{\lambda]\nu} \tag{5.8}$$

from which it follows that $k_{\mu\lambda}$ is zero at all points if it is zero at the points of one V_{n-1}. This proves that i_λ is a gradient vector $i_\lambda = \partial_\lambda f$ and from this $i^\mu \partial_\mu f = df/ds = 1$. Hence $f = \text{const.}$ is the equation of the normal V_{n-1}'s and any two of these V_{n-1}'s cut off segments of the same length from all curves of the congruence.

A set of n mutually perpendicular real congruences in an ordinary V_n is called an *orthogonal net.*[1]) It is said to be an *orthogonal system* if the congruences are *all* V_{n-1}-normal.[2]) Let $\underset{j}{i}^\varkappa = \underset{i}{j}_\varkappa;\ j = 1, \ldots, \mathrm{n}$ be the tangent unitvectors. Hence we have for the n-th congruence of an orthogonal system an equation of the form (cf. 5.1, 3)

$$V_\mu \overset{\mathrm{n}}{i}_\lambda = - h_{\mu\lambda} + \overset{\mathrm{n}}{i}_\mu u_\lambda; \quad \underset{\mathrm{n}}{i}^\mu h_{\mu\lambda} = 0; \quad h_{[\mu\lambda]} = 0. \tag{5.9}$$

The orthogonal net is an orthogonal system if and only if each set of two congruences is V_2-forming. The n.a.s. condition is (cf. II § 5)

$$\underset{i}{i}^\mu V_\mu \underset{j}{i}^\varkappa - \underset{j}{i}^\mu V_\mu \underset{i}{i}^\varkappa = \alpha \underset{i}{i}^\varkappa + \beta \underset{j}{i}^\varkappa \tag{5.10}$$

or (cf. III § 9)

$$\Gamma^h_{[j\,i]} \stackrel{*}{=} - \underset{j}{i}^\mu \underset{i}{i}^\lambda V_{[\mu} \overset{h}{i}_{\lambda]} = 0; \quad h, i, j \neq . \tag{5.11}$$

But because $\Gamma^h_{ji} \stackrel{*}{=} - \Gamma^i_{jh}$ this condition is satisfied if and only if

$$- \Gamma^h_{ij} \stackrel{*}{=} \underset{j}{i}^\mu \underset{i}{i}^\lambda V_\mu \overset{h}{i}_\lambda = 0; \quad h, i, j \neq . \tag{5.12}$$

This equation expresses the fact that every set of $n-1$ of the congruences is canonical with respect to the remaining congruence.

Let us now suppose that the principal directions of $h_{\mu\lambda}$ are uniquely determined and that the congruences $\underset{b}{i}^\varkappa;\ b = 1, \ldots, \mathrm{n}-1$ are chosen in these principal directions. Then we have $h_{\mu\lambda} \stackrel{*}{=} \sum_a h_{aa} \overset{a}{i}_\mu \overset{a}{i}_\lambda$. The n con-

[1]) The case of a V_n with an indefinite fundamental tensor is dealt with by many authors. Cf. for instance EISENHART 1926, 1; 1949, 1; E I 1935, 1; E II 1938, 2; WONG 1945, 2.

[2]) Cf. also for literature SCHOUTEN and STRUIK 1919, 1; R. K. 1924, 1, p. 190ff.; EISENHART 1923, 2; 1926, 1, p. 117ff.; SCHOUTEN 1927, 3; WALBERER 1934, 1; E II 1938, 2, p. 27ff.; WONG 1945, 2.

gruences form an orthogonal system if and only if

$$(5.13)\quad \begin{cases} \text{a)} & \underset{\mathrm{n}}{i^\mu}\underset{b}{i^\lambda} \nabla_\mu \overset{a}{i}_\lambda = 0 \\ \text{b)} & \underset{c}{i^\mu}\underset{b}{i^\lambda} \nabla_\mu \overset{a}{i}_\lambda = 0; \qquad a, b, c \neq; \qquad a, b, c = 1, \ldots, \mathrm{n}-1 \end{cases}$$

since the remaining equation $\underset{c}{i^\mu}\underset{b}{i^\lambda} \nabla_\mu \overset{\mathrm{n}}{i}_\lambda = 0$ of (5.12) is already satisfied because of the special choice of the $\underset{b}{i^\varkappa}$. Now by differentiation of

$$(5.14)\qquad \underset{c}{i^\mu}\underset{b}{i^\lambda} h_{\mu\lambda} = 0; \qquad b \neq c$$

we get

$$(5.15)\quad \begin{cases} \text{a)} & (h_{bb} - h_{cc})\underset{\mathrm{n}}{i^\nu}\underset{b}{i^\lambda} \nabla_\nu \overset{c}{i}_\lambda + \underset{\mathrm{n}}{i^\nu}\underset{c}{i^\mu}\underset{b}{i^\lambda} \nabla_\nu h_{\mu\lambda} \overset{*}{=} 0 \\ \text{b)} & (h_{bb} - h_{cc})\underset{d}{i^\nu}\underset{b}{i^\lambda} \nabla_\nu \overset{c}{i}_\lambda + \underset{d}{i^\nu}\underset{c}{i^\mu}\underset{b}{i^\lambda} \nabla_\nu h_{\mu\lambda} \overset{*}{=} 0; \qquad b, c, d \neq; \end{cases}$$

and in these equations $h_{bb} \neq h_{cc}$. This proves the proposition:

A real congruence in an ordinary V_n whose canonical congruences are uniquely determined, belongs to an orthogonal system of n congruences if and only if

$$(5.16)\quad \begin{cases} \text{a)} & \boxed{\underset{\mathrm{n}}{i^\nu}\underset{c}{i^\mu}\underset{b}{i^\lambda} \nabla_\nu h_{\mu\lambda} = 0} \\ \text{b)} & \boxed{\underset{d}{i^\nu}\underset{c}{i^\mu}\underset{b}{i^\lambda} \nabla_\nu h_{\mu\lambda} = 0} \end{cases}; \qquad \begin{matrix} b, c, d = 1, \ldots, \mathrm{n}-1; \\ b, c, d \neq. \end{matrix}$$

The geometrical meaning of these conditions is that the covariant differential of $h_{\mu\lambda}$ for a direction perpendicular to any set of m principal directions of $h_{\mu\lambda}$ has a component in the R_m of these directions whose principal multidirections (cf. I § 9) contain these principal directions.[1])

If the principal directions of $h_{\mu\lambda}$ are not uniquely determined results can be obtained but they are rather complicated.[2]) It can be proved in the same way that the principal directions of a symmetric tensorfield $T_{\mu\lambda}$ in V_n in the special case that they are uniquely determined are V_{n-1}-normal if and only if

$$(5.17)\qquad \underset{j}{i^\nu}\underset{i}{i^\mu}\underset{h}{i^\lambda} \nabla_\nu T_{\mu\lambda} = 0; \qquad j, i, h = 1, \ldots, \mathrm{n}; \qquad j, i, h \neq.^{3)}$$

[1]) LÉVY 1870, 1 for V_2 in R_3.

[2]) SCHOUTEN 1927, 3.

[3]) SCHOUTEN 1927, 3, p. 723ff. also for the general case.

The conditions (5.16, 17) can only be used if the principal directions are already determined. TONOLO[1]) gave for $n=3$ a condition for $T_{\mu\lambda}$ that only involves $T_{\mu\lambda}$ and its adjoint (cf. Exerc. I 8,2) and SCHOUTEN[2]) gave the n.a.s. conditions

$$(5.18)\qquad \text{a)}\ T_{[\nu}^{\cdot\cdot\varkappa}\nabla_\mu T_{\lambda]\varkappa}=0 \qquad \text{b)}\ \overset{-1}{T}{}_{[\nu}^{\cdot\cdot\varkappa}\nabla_\mu T_{\lambda]\varkappa}=0 \qquad \text{c)}\ \bar{T}_{[\nu}^{\cdot\cdot\varkappa}\nabla_\mu \overset{-1}{T}{}_{\lambda]\varkappa}=0$$

for the case when $T_{\mu\lambda}$ has n different eigenvalues none of which is zero. The proof runs as follows. Multiplying (5.17) with the eigenvalue $\underset{h}{\lambda}$ and using the fact that $\underset{h}{i}{}^\varkappa$ is an eigenvector we get

$$(5.19)\qquad \underset{j}{i}{}^\mu\,\underset{i}{i}{}^\lambda\,\underset{h}{i}{}^\nu\,T_\nu^{\cdot\cdot\varkappa}\nabla_\mu T_{\lambda\varkappa}=0;\qquad h,i,j=1,\ldots,n;\qquad h,i,j\neq.$$

Alternation over j, i, h gives (5.18a). The other equations (5.18) are derived in a similar way. Hence (5.18) is necessary. Conversely, using the anholonomic system (h), (5.18) can be replaced by

$$(5.20)\qquad \begin{cases} T_{[k}^{\cdot\cdot h}\nabla_j T_{i]h}=0 \\ \overset{-1}{T}{}_{[k}^{\cdot\cdot h}\nabla_j T_{i]h}=0; \qquad \overset{-2}{T}{}^h_{\cdot i}\overset{\text{def}}{=}\overset{-1}{T}{}^h_{\cdot j}\overset{-1}{T}{}^j_{\cdot i} \\ \overset{-2}{T}{}_{[k}^{\cdot\cdot h}\overset{-1}{T}{}_j^{\cdot\cdot l}\nabla_{i]}T_{hl}=0 \end{cases}$$

and if these equations are written out using the special values of the components of $h_{\mu\lambda}$ with respect to (h) we get

$$(5.21)\qquad \nabla_j T_{ih}=0;\qquad h,i,j\neq$$

which is equivalent to (5.17).

NIJENHUIS[3]) has given yet another form connected with the question of whether a tensor $T_{\cdot\lambda}^{\varkappa}$ in X_n with n different eigenvalues has covariant eigenvectors which are X_{n-1}-forming. For this problem without a metric he introduces the new concomitant (cf. Exerc. II 2,3)

$$(5.22)\qquad H_{\mu\lambda}^{\cdot\cdot\varkappa}\overset{\text{def}}{=}2T_{[\mu}^{\cdot\cdot\varrho}\partial_{|\varrho|}T_{\lambda]}^{\cdot\varkappa}-2T_\varrho^{\cdot\varkappa}\partial_{[\mu}T_{\lambda]}^{\cdot\varrho}$$

of the field $T_\lambda^{\cdot\varkappa}$ and gives several forms of n.a.s. conditions. For instance it is n.a.s. that $H_{\mu\lambda}^{\cdot\cdot\varkappa}$ can be written in the form

$$(5.23)\qquad H_{\mu\lambda}^{\cdot\cdot\varkappa}=\underset{0}{p}_{[\mu}A_{\lambda]}^{\varkappa}+\underset{1}{p}_{[\mu}T_{\lambda]}^{\cdot\varkappa}+\cdots+\underset{n-2}{p}_{[\mu}\overset{n-2}{T}{}_{\lambda]}^{\cdot\varkappa}$$

[1]) TONOLO 1941, 2; 1949, 1; 2; 3; 1953, 1; cf. TONOLO 1941, 1. A V_3 whose principal directions are V_2-normal was already discussed by SLEBODZINSKI 1927, 1.

[2]) SCHOUTEN 1951, 2.

[3]) NIJENHUIS 1951, 1.

where $\underset{0}{p}_\lambda, \ldots, \underset{n-2}{p}_\lambda$ are suitably chosen and where $\overset{q}{T}{}_\lambda^{\cdot\varkappa}$ represents the q-th power of $T_\lambda^{\cdot\varkappa}$. Going back to the metric case he found other forms for the conditions (5.18). The most elegant form is

$$\text{(5.24)} \qquad \left\{ \begin{array}{ll} \text{a)} & H_{[\mu\lambda}^{\cdot\cdot\cdot\varkappa}\, g_{\nu]\varkappa} = 0 \\ \text{b)} & H_{[\mu\lambda}^{\cdot\cdot\cdot\varkappa}\, T_{\nu]\varkappa} = 0 \\ \text{c)} & H_{[\mu\lambda}^{\cdot\cdot\cdot\varkappa}\, \overset{2}{T}_{\nu]\varkappa} = 0 \end{array} \right.$$

also suitable if one of the eigenvalues of $T_{\mu\lambda}$ is zero.

Exercises.

V 5,1 [1]). An ordinary V_n is an R_n if and only if there exists an orthogonal system of real geodesic congruences.

V 5,2 [2]). A real geodesic congruence with unit tangent vector $i^\varkappa$ is V_{n-1}-normal if and only if the integral $\int i_\mu \, d\xi^\mu$ is zero for every closed curve in some non singular V_{n-1} that has a point in common with each curve of the congruence.

V 5,3 [3]). A V_{n-1} in R_n belongs to an orthogonal system in R_n if and only if there exist n different principal directions of h_{cb} that form an orthogonal system in V_{n-1}.

V 5,4 [4]). If the congruences with the unit tangent vectors $\underset{j}{i}^\varkappa$; $j = 1, \ldots, n$, in V_n form an orthogonal system and if the field $\alpha \underset{1}{i}^\varkappa + \beta \underset{2}{i}^\varkappa$ is V_{n-1}-normal, then the field $-\alpha \underset{1}{i}^\varkappa + \beta \underset{2}{i}^\varkappa$ is also V_{n-1}-normal. α and β need not be constants.

V 5,5 [5]). The curvature vector of the congruence of a field $v^\varkappa$ in V_n that satisfies KILLING's equation $\nabla_{(\mu} v_{\lambda)} = 0$, is a gradient vector.

V 5,6 [6]). Every congruence in V_n perpendicular to a field $v^\varkappa$ that satisfies KILLING's equation, has a curvature vector perpendicular to $v^\varkappa$.

V 5,7 [7]). If a field $v^\varkappa$ of V_n is V_{n-1}-normal and satisfies KILLING's equation and if the length of $v^\varkappa$ is constant, its congruence is geodesic and normal to a set of ∞^1 geodesic V_{n-1}'s.

---

[1]) E II 1938, 2, p. 31.

[2]) SCHOUTEN-STRUIK 1921, 2.

[3]) BRAUNER 1951, 1.

[4]) DEMOULIN 1913, 1 for R_3.

[5]) KILLING 1892, 1, p. 167; RICCI 1898, 1; 2; 1901, 1, p. 608.

[6]) RICCI 1898, 1; 2.

[7]) STRUIK 1922, 1, p. 157.

§ 6. Properties of curvature of a V_{n-1} in V_n.

Let $i^\varkappa$ be the tangent unitvector of a real congruence of V_{n-1}. Every curve of this congruence may be considered as a curve of V_n and as a curve of V_{n-1}. Writing $'u^a$ for the curvature vector (cf. V § 1) in V_{n-1} we get

$$(6.1)\quad u^\varkappa \overset{\text{def}}{=} i^\mu \nabla_\mu i^\varkappa = B_a^\varkappa i^c \,'\nabla_c i^a + i^\mu (\nabla_\mu i^\varrho) n_\varrho n^\varkappa = B_a^\varkappa \,'u^a + i^c i^b h_{cb} n^\varkappa .$$

We call $u^\varkappa$ the *absolute curvature vector* of the curve, $'u^a$ the *relative curvature vector,*

$$(6.2)\qquad ''u^\varkappa \overset{\text{def}}{=} i^c i^b h_{cb} n^\varkappa = u^\varkappa - 'u^\varkappa \text{ [1]}$$

the *enforced curvature vector* and $i^c i^b h_{cb}$ the *enforced curvature.* The sign of this latter scalar depends of course on the choice of the sign of $n^\varkappa$. The vector $''u^\varkappa$ is the same for all curves that have the same tangent at the point considered.[2] It is equal to $u^\varkappa$ if and only if the tangent of the curve, considered as a curve of the V_{n-1}, is stationary (cf. VI § 1) at the point and it is equal to $u^\varkappa$ all along the curve if and only if the curve is a geodesic of V_{n-1}. This proves that the curves of V_{n-1} through one of its points, looked upon as curves of the V_n, have extreme values of the enforced curvature if and only if the tangent lies in a principal direction of h_{cb} at that point. These directions are therefore called the *directions of principal curvature* and the corresponding values of $i^c i^b h_{cb}$ are called the *principal curvatures* of the V_{n-1}. The scalar

$$(6.3)\qquad h \overset{\text{def}}{=} \frac{1}{n-1}\, 'g^{ba} h_{ba}$$

is called the *mean curvature.* Principal curvatures and mean curvature are differential invariants of the V_{n-1}.

A curve of V_{n-1} whose tangent lies everywhere in a direction of principal curvature is said to be a *line of curvature.* Hence the lines of curvature belong to canonical congruences of the congruences normal to V_{n-1}. As a result of this *the V_{n-1}'s of an orthogonal system intersect in lines of curvature.* This is the theorem of DUPIN.

If $i^c i^b h_{cb} = 0$ at some point, the direction of i^a is called a *principal tangent* and the curves of V_{n-1} whose tangents are everywhere principal are called *asymptotic lines.*[3]

[1]) If instead of the direction of $n^\varkappa$ the direction of an arbitrary congruence given over V_{n-1} is taken, we get instead of the relative curvature the *union curvature* of SPRINGER 1950, 1 and instead of the geodesics in V_{n-1} the *union curves* with respect to this congruence. Cf. YANO 1948, 1 also for a generalization in A_n; MISHRA 1952, 1.

[2]) This is the quintessence of the theorem of MEUSNIER ($n = 3$).

[3]) KRONECKER 1869, 1; VOSS 1880, 1. Cf. for asymptotic lines of higher order STRUIK 1922, 1, p. 80; E II 1938, 2, p. 73ff.; GOLAB 1949, 1.

If h_{cb} has rank r, $n-1-r$ principal curvatures are zero and the cone of principal tangents consists of ∞^r R_{n-r-1}'s all containing the $(n-1-r)$-direction belonging to these curvatures. If $h_{cb}=0$ at a point, every curve of V_{n-1} through this point has an enforced curvature zero at that point. The V_{n-1} is then called *geodesic at that point.* A *geodesic* V_{n-1} is a V_{n-1} that is geodesic at all its points. Every geodesic of a geodesic V_{n-1} is also a geodesic of V_n and this property is characteristic for a geodesic V_{n-1}.

If h_{cb} equals $'g_{cb}$ to within a scalar factor

$$h_{cb} = h\,'g_{cb} \tag{6.4}$$

the point is called *umbilical.* At an umbilical point all principal curvatures are equal and all directions are directions of principal curvature.[1] If all points of V_{n-1} are umbilical it is called *umbilical.* It follows from (4.6) that for an umbilical V_{n-1}

$$2('\nabla_{[d}\, h)\, 'g_{c]b} = B^{\nu\mu\lambda}_{dcb}\, K_{\nu\mu\lambda\varkappa}\, n^{\varkappa}. \tag{6.5}$$

Hence, if umbilical V_{n-1}'s are possible through every point of V_n and with every $(n-1)$-direction at that point, all components K_{kjih} with respect to any orthogonal (in general anholonomic) coordinate system (h) with *four* different indices are zero. In the next chapter we shall prove that this condition is also sufficient.[2] If the V_{n-1} is not only umbilical but if moreover the mean curvature h is constant, the left hand side of (6.5) vanishes. Now $B^{\varkappa}_{\lambda} = A^{\varkappa}_{\lambda} - n_{\lambda} n^{\varkappa}$, hence by contraction of (6.5)

$$0 = B^{\varrho}_{\nu}\, K_{\varrho\sigma}\, n^{\sigma} \tag{6.6}$$

and this proves that the normal of every umbilical V_{n-1} with constant mean curvature lies at each point in a principal direction of V_n [cf. Exerc. III 5,3].[3] V_{n-1}'s of this kind through every point and with every $(n-1)$-direction at that point are only possible in an EINSTEIN space because according to (6.6) $K_{\mu\lambda}$ must equal $g_{\mu\lambda}$ to within a scalar factor. But in § 5 of the next chapter it will be proved that this condition is not sufficient but that the space must be an S_n [cf. III § 5].[4] As a corollary we get that a geodesic V_{n-1} is at each point always perpendicular to a principal direction of V_n and that only in an S_n geodesic V_{n-1}'s are

---

[1]) Cf. on umbilical points T. Y. THOMAS 1938, 2; SASAKI 1939, 1; MUTÔ 1940, 1; WONG 1943, 3 (EINSTEIN space); ADATI 1951, 1; 2.

[2]) SCHOUTEN 1921, 1, p. 86 for $n > 3$ (cf. footnote 1, p. 309).

[3]) RICCI 1903, 1, p. 415 for $h=0$; RIMINI 1904, 1, p. 35 for $n=3$; STRUIK 1922, 1, p. 143 for the general case.

[4]) SCHOUTEN 1921, 1, p. 87.

possible through every point and with every $(n-1)$-direction through this point.[1])

Some interesting propositions may be derived for $V_n = S_n$ from the integrability equations (4.5, 6). According to (III 5.31) in an S_n these equations take the form

$$(6.7)\qquad 'K_{dcba} = -2\varkappa\, 'g_{[d[b}\, 'g_{c]a]} - 2h_{[d[b}\, h_{c]a]}$$

$$(6.8)\qquad '\nabla_{[d}\, h_{c]b} = 0$$

where $\varkappa$ is the (constant) scalar curvature of the S_n. If the V_{n-1} is geodesic it follows from (6.7) that it is an S_{n-1} with the *same* constant scalar curvature $\varkappa$. In that case (6.8) is trivial. If the V_{n-1} has only umbilical points we have

$$(6.9)\qquad 'K_{dcba} = -2(\varkappa + h^2)\, 'g_{[d[b}\, 'g_{c]a]},$$

hence the V_{n-1} must be an S_{n-1} with the constant curvature $\varkappa + h^2$ and its mean curvature must be constant. From (6.7) it follows that a V_{n-1} in S_n can be an S_{n-1} if and only if

$$(6.10)\qquad h_{[d[b}\, h_{c]a]} = ('\varkappa - \varkappa)\, 'g_{[d[b}\, 'g_{c]a]}.$$

For $n \geq 4$ this implies that the S_{n-1} is umbilical and that $h_{cb} = \pm\, 'g_{cb} \times \times \sqrt{'\varkappa - \varkappa}$ but for $n = 3$ the equation (6.10) only determines the determinant of h_{cb} and leaves the possibility of bending (cf. V § 4).

From (6.7) we get by contraction

$$(6.11)\qquad 'K_{cb} = \varkappa(n-2)\, 'g_{cb} - h^{a}_{\cdot b}\, h_{ca} + h^{a}_{\cdot a}\, h_{cb}$$

from which it follows that the principal directions of h_{cb} are also principal directions of $'K_{cb}$. Hence *the directions of principal curvature of a V_{n-1} in S_n are principal directions of V_{n-1}.*[2])

The quadric in the tangent R_{n-1}

$$(6.12)\qquad h_{ba}\, x^b x^a = \pm 1; \quad a, b = 1, \ldots, n-1$$

is the *indicatrix of* DUPIN. Let (a) be an anholonomic orthogonal coordinate system in V_{n-1} with unit basis vectors $\underset{b}{i}^{a}$ and let $\underset{b}{\alpha}$ be the angles between an arbitrary congruence i^a and the congruences $\underset{b}{i}^{a}$. Then the enforced curvature of the congruence i^a is

$$(6.13)\qquad i^c i^b h_{cb} \overset{*}{=} \sum_b h_{bb} \cos^2 \underset{b}{\alpha}.$$

[1]) A V_n in which geodesic V_{n-1}'s with every $(n-1)$-direction are possible through one definite point is called a SCHUR *space* and the point is called *centre*. DUSCHEK and MAYER gave 1930, 1, p. 167ff. the form of the linear element in normal coordinates with respect to the centre.

[2]) Cf. COBURN 1940, 1.

This is the *theorem of* EULER ($n=3$). The absolute value of the enforced curvature for the direction of i^a equals the inverse quadrate of the length of the radiusvector x^a of DUPIN'S indicatrix in that direction.

Exercises.

V 6,1[1]). Prove for a real curve in a real V_{n-1} in an ordinary V_n:

If the curve is geodesic in V_{n-1} and a line of curvature, then its second curvature in V_n is zero.

If the curve is geodesic in V_{n-1} and if its second curvature in V_n is zero, then it is a line of curvature.

V 6,2[2]). In a real V_{n-1} in an ordinary V_n there are two real congruences with unit tangent vectors $i^\varkappa$ and $j^\varkappa$. Prove that

$$i^\mu \nabla_\mu j^\varkappa = i^\mu{}' \nabla_\mu j^\varkappa + i^c j^b h_{cb} n^\varkappa. \tag{V 6,2 α}$$

§ 7. The rigged X_n^m in L_n and X_n.[3])

Let a rigged X_n^m be given in an L_n and let the basis vectors $\underset{b}{e}^\varkappa$; $a, b, \ldots = 1, \ldots, \mathrm{m}$ span the E_m of the X_n^m and the $\underset{y}{e}^\varkappa$; $x, y, \ldots = \mathrm{m}+1, \ldots, \mathrm{n}$ span the E_{n-m} of the rigging $X_n^{m'}$; $m' = n - m$ at each point. Then at each point we have the basis vectors $\underset{i}{e}^\varkappa$; $\overset{h}{e}_\lambda$; $h, i = 1, \ldots, \mathrm{n}$ of the anholonomic coordinate system (h) and the connecting quantities

$$\left\{\begin{array}{llll} B_b^\varkappa = \overset{a}{e}_b \underset{a}{e}^\varkappa \overset{*}{=} A_b^\varkappa; & C_\lambda^x = \underset{y}{e}^x \overset{y}{e}_\lambda \overset{*}{=} A_\lambda^x; & B_\lambda^\varkappa = \overset{a}{e}_\lambda \underset{a}{e}^\varkappa; & B_a^b \overset{*}{=} \delta_a^b; \\ C_y^\varkappa = \overset{x}{e}_y \underset{x}{e}^\varkappa \overset{*}{=} A_y^\varkappa; & B_\lambda^a = \underset{b}{e}^a \overset{b}{e}_\lambda \overset{*}{=} A_\lambda^a; & C_\lambda^\varkappa = \overset{x}{e}_\lambda \underset{x}{e}^\varkappa; & C_y^x \overset{*}{=} \delta_y^x. \end{array}\right. \tag{7.1}$$

---

1) E II 1938, 2, p. 72.

2) E II 1938, 2, p. 66; cf. GRAUSTEIN 1934, 1; RICCI 1902, 2; AOUST 1864, 1.

3) General references on the X_n^m (cf. also II § 5): SCHOUTEN 1923, 1 (X_m in A_n); 1928, 1; 1929, 3; VRANCEANU 1926, 1; 2; 1927, 1; 2; 1928, 1; 2; 5; 1929, 1; 2; 1930, 1; 1931, 1; 1932, 1; 1934, 1; 1936, 1; 1938, 1; 1942, 2; HORAK 1927, 1; 2; 1928, 1; SYNGE 1928, 1; HLAVATY 1930, 1; 2; 1934, 3; MOISIL 1930, 1; SCHOUTEN and v. KAMPEN 1930, 2; BORTOLOTTI 1931, 1; 2; 1936, 3 (also for literature); 1937, 2 (X_n^m project.); 1941, 1 (also for literature); DIENES 1932, 1; AGOSTINELLI 1933, 1 (V_n); MORINAGA 1934, 1; VANDERSLICE 1934, 1; FABRICIUS-BJERRE 1936, 1 (torsionfree); WAGNER 1936, 1; 1938, 1; 2; 1940, 1 (non linear in E_3); 1941, 1 (X_n^{n-1}); 1943, 2 (non linear in X_n); PAUC 1937, 1; 1938, 1; 2; YANO 1937, 1; 1938, 1; 1939, 1; YANO and PETRESCU 1940, 2; BOMPIANI 1938, 1; MAXIA 1939, 1; 2; 1940, 1; MIKAN 1939, 1; HAIMOVICI 1940, 1; 1946, 3; SU 1943, 1 (X_n^{n-1} project.); PANTAZZI 1943, 1; 1947, 1; PETRESCU 1943, 1 (V_n^{n-1}); 2; 3 ($V_{2p+1}^{2p}, V_{2p+2}^{2p+1}$); 1944, 1 ($L_n^m$ in E_n); 1945, 1; 1948, 1 (V_n^m); WANG 1943, 1; 2; ROZENFEL'D 1947, 2 (X_n^m in E_n); ROGOVOĬ 1949, 1 and 1950, 1 (X_3^2 in E_3); DAVIES 1953, 1; TAKASU 1953, 1 (in the large).

In order to consider questions of curvature we could proceed in the same way as in the case of an X_{n-1} in A_n (cf. V § 3) but then the formulae would become crowded with factors B and C and would be very unreadable. Therefore we introduce here the so-called D-symbolism due to VAN DER WAERDEN and BORTOLOTTI.[1])

If $p^\varkappa = B^\varkappa_a p^a$ and $q_\lambda = B^b_\lambda q_b$ are vectors of X^m_n; $r^\varkappa = C^\varkappa_x r^x$ and $s_\lambda = C^y_\lambda s_y$ vectors of $X^{m'}_n$ and $v^\varkappa$ and w_λ general vectors of X_n, the working of the operators D_μ, D_c and D_z on these quantities and on scalars is defined by the equations

$$(7.2)\qquad \begin{cases} D_\mu p = \nabla_\mu p = \partial_\mu p \\ D_b p = B^\mu_b \nabla_\mu p = B^\mu_b \partial_\mu p \\ D_y p = C^\mu_y \nabla_\mu p = C^\mu_y \partial_\mu p \end{cases}$$

$$(7.3)\quad \text{a)} \begin{cases} \alpha) & D_\mu p^\varkappa = \nabla_\mu p^\varkappa \\ \beta) & D_c p^\varkappa = B^\mu_c \nabla_\mu p^\varkappa \\ \gamma) & D_z p^\varkappa = C^\mu_z \nabla_\mu p^\varkappa \end{cases} \qquad \text{b)} \begin{cases} \alpha) & D_\mu q_\lambda = \nabla_\mu q_\lambda \\ \beta) & D_c q_\lambda = B^\mu_c \nabla_\mu q_\lambda \\ \gamma) & D_z q_\lambda = C^\mu_z \nabla_\mu q_\lambda \end{cases}$$

$$(7.4)\quad \text{a)} \begin{cases} \alpha) & D_\mu p^a = B^a_\varkappa \nabla_\mu p^\varkappa \\ \beta) & D_c p^a = B^{\mu a}_{c \varkappa} \nabla_\mu p^\varkappa \\ \gamma) & D_z p^a = C^\mu_z B^a_\varkappa \nabla_\mu p^\varkappa \end{cases} \qquad \text{b)} \begin{cases} \alpha) & D_\mu q_b = B^\lambda_b \nabla_\mu q_\lambda \\ \beta) & D_c q_b = B^{\mu \lambda}_{c b} \nabla_\mu q_\lambda \\ \gamma) & D_z q_b = C^\mu_z B^\lambda_b \nabla_\mu q_\lambda \end{cases}$$

$$(7.5)\quad \text{a)} \begin{cases} \alpha) & D_\mu r^\varkappa = \nabla_\mu r^\varkappa \\ \beta) & D_c r^\varkappa = B^\mu_c \nabla_\mu r^\varkappa \\ \gamma) & D_z r^\varkappa = C^\mu_z \nabla_\mu r^\varkappa \end{cases} \qquad \text{b)} \begin{cases} \alpha) & D_\mu s_\lambda = \nabla_\mu s_\lambda \\ \beta) & D_c s_\lambda = B^\mu_c \nabla_\mu s_\lambda \\ \gamma) & D_z s_\lambda = C^\mu_z \nabla_\mu s_\lambda \end{cases}$$

$$(7.6)\quad \text{a)} \begin{cases} \alpha) & D_\mu r^x = C^x_\varkappa \nabla_\mu r^\varkappa \\ \beta) & D_c r^x = B^\mu_c C^x_\varkappa \nabla_\mu r^\varkappa \\ \gamma) & D_z r^x = C^{\mu x}_{z \varkappa} \nabla_\mu r^\varkappa \end{cases} \qquad \text{b)} \begin{cases} \alpha) & D_\mu s_y = C^\lambda_y \nabla_\mu s_\lambda \\ \beta) & D_c s_y = B^\mu_c C^\lambda_y \nabla_\mu s_\lambda \\ \gamma) & D_z s_y = C^{\mu \lambda}_{z y} \nabla_\mu s_\lambda \end{cases}$$

$$(7.7)\quad \text{a)} \begin{cases} \alpha) & D_\mu v^\varkappa = \nabla_\mu v^\varkappa \\ \beta) & D_c v^\varkappa = B^\mu_c \nabla_\mu v^\varkappa \\ \gamma) & D_z v^\varkappa = C^\mu_z \nabla_\mu v^\varkappa \end{cases} \qquad \text{b)} \begin{cases} \alpha) & D_\mu w_\lambda = \nabla_\mu w_\lambda \\ \beta) & D_c w_\lambda = B^\mu_c \nabla_\mu w_\lambda \\ \gamma) & D_z w_\lambda = C^\mu_z \nabla_\mu w_\lambda \end{cases}$$

---

[1]) v. D. WAERDEN 1927, 1; BORTOLOTTI 1927, 3. All formulae containing indices $a, b, \ldots$ or $x, y, \ldots$ should of course be written with $\overset{*}{=}$ because they are meant to hold only for the special choice of the basis vectors $\underset{b}{e}^\varkappa$, $\underset{y}{e}^\varkappa$ with respect to the X^m_n and $X^{m'}_n$. Since this is understood we nearly always write in this section $=$. Cf. for historical notes concerning the D-symbolism, SCHOUTEN and v. KAMPEN 1930, 2, p. 774. Cf. BOMPIANI 1952, 2.

and by the rule of LEIBNIZ. Hence, if a quantity of higher valence is given, for instance $v^{\varkappa\lambda}_{\cdot\cdot\mu}$, lying with the index $\varkappa$ in X_n^m, with the index λ in $X_n^{m'}$, and with the index μ in X_n but neither in X_n^m nor in $X_n^{m'}$, there exist full sets of components of four kinds, c.q. $v^{\varkappa\lambda}_{\cdot\cdot\mu}$, $v^{a\lambda}_{\cdot\cdot\mu}$, $v^{\varkappa x}_{\cdot\cdot\mu}$ and $v^{ax}_{\cdot\cdot\mu}$,[1]) and we have the twelve derivatives

$$
(7.8)\quad
\begin{cases}
\text{a)}\begin{cases}
\alpha)\; D_\nu v^{\varkappa\lambda}_{\cdot\cdot\mu} = \nabla_\nu v^{\varkappa\lambda}_{\cdot\cdot\mu} \\
\beta)\; D_d v^{\varkappa\lambda}_{\cdot\cdot\mu} = B^\nu_d \nabla_\nu v^{\varkappa\lambda}_{\cdot\cdot\mu} \\
\gamma)\; D_u v^{\varkappa\lambda}_{\cdot\cdot\mu} = C^\nu_u \nabla_\nu v^{\varkappa\lambda}_{\cdot\cdot\mu}
\end{cases}
\qquad
\text{b)}\begin{cases}
\alpha)\; D_\nu v^{a\lambda}_{\cdot\cdot\mu} = B^a_\varkappa \nabla_\nu v^{\varkappa\lambda}_{\cdot\cdot\mu} \\
\beta)\; D_d v^{a\lambda}_{\cdot\cdot\mu} = B^{\nu a}_{d\varkappa} \nabla_\nu v^{\varkappa\lambda}_{\cdot\cdot\mu} \\
\gamma)\; D_u v^{a\lambda}_{\cdot\cdot\mu} = C^\nu_u B^a_\varkappa \nabla_\nu v^{\varkappa\lambda}_{\cdot\cdot\mu}
\end{cases}
\\[2ex]
\text{c)}\begin{cases}
\alpha)\; D_\nu v^{\varkappa x}_{\cdot\cdot\mu} = C^x_\lambda \nabla_\nu v^{\varkappa\lambda}_{\cdot\cdot\mu} \\
\beta)\; D_d v^{\varkappa x}_{\cdot\cdot\mu} = B^\nu_d C^x_\lambda \nabla_\nu v^{\varkappa\lambda}_{\cdot\cdot\mu} \\
\gamma)\; D_u v^{\varkappa x}_{\cdot\cdot\mu} = C^{\nu x}_{u\lambda} \nabla_\nu v^{\varkappa\lambda}_{\cdot\cdot\mu}
\end{cases}
\qquad
\text{d)}\begin{cases}
\alpha)\; D_\nu v^{ax}_{\cdot\cdot\mu} = B^a_\varkappa C^x_\lambda \nabla_\nu v^{\varkappa\lambda}_{\cdot\cdot\mu} \\
\beta)\; D_d v^{ax}_{\cdot\cdot\mu} = B^{\nu a}_{d\varkappa} C^x_\lambda \nabla_\nu v^{\varkappa\lambda}_{\cdot\cdot\mu} \\
\gamma)\; D_u v^{ax}_{\cdot\cdot\mu} = B^a_\varkappa C^{\nu x}_{u\lambda} \nabla_\nu v^{\varkappa\lambda}_{\cdot\cdot\mu}.
\end{cases}
\end{cases}
$$

From this we see that the following rules hold:

1. With respect to the index of differentiation, D_μ corresponds to ∇_μ; D_c to $B^\mu_c \nabla_\mu$ and D_z to $C^\mu_z \nabla_\mu$;

2. The only possible other indices that can occur in the derivative are those that appear in *possible* components of the quantity to be differentiated. For instance $D_\nu v^{x\lambda}_{\cdot\cdot\mu}$ or $D_\nu v^{\varkappa\lambda}_{\cdot\cdot z}$ can not appear because there are no full sets of components $v^{x\lambda}_{\cdot\cdot\mu}$ or $v^{\varkappa\lambda}_{\cdot\cdot z}$.

3. For an index $a, b, \ldots$ or $x, y, \ldots$ in the quantity to be differentiated a transvection with B or C respectively occurs and for an index $\varkappa, \lambda, \ldots$ there is no such transvection.

4. In order to get the meaning of a formula with ∇ or δ we have only to look at the *skeleton* (cf. I § 11). But the meaning of a formula with D depends also on the kinds of indices used. For instance $\nabla_\nu v^{\varkappa\lambda}_{\cdot\cdot\mu}$, $\nabla_k v^{hi}_{\cdot\cdot j}$, $\nabla_\mu v^{h\lambda}_{\cdot\cdot j}$ and $\nabla_k v^{\varkappa i}_{\cdot\cdot j}$ are all components of the same quantity, the covariant derivative of $v^{\varkappa\lambda}_{\cdot\cdot\mu}$. But $D_\nu v^{\varkappa\lambda}_{\cdot\cdot\mu}$, $D_d v^{a\lambda}_{\cdot\cdot\mu}$, $D_u v^{\varkappa x}_{\cdot\cdot\mu}$ and $D_u v^{ax}_{\cdot\cdot\mu}$ are four different quantities.

In (7.4a β, γ) and (7.4b β, γ) the covariant derivatives $\nabla_\mu p^\varkappa$ and $\nabla_\mu q_\lambda$ are split up into parts that may be considered as a kind of covariant derivatives themselves. The same holds for $\nabla_\mu r^\varkappa$ and $\nabla_\mu s_\lambda$ in (7.6a β, γ) and (7.6b β, γ). The (two) parts of $\nabla_\mu p^\varkappa$, $\nabla_\mu q_\lambda$, $\nabla_\mu r^\varkappa$ and $\nabla_\mu s_\lambda$ not occuring in these formulae are no derivatives of these fields, because for instance $C^\varkappa_\varrho \nabla_\mu p^\varrho = -p^\varrho \nabla_\mu C^\varkappa_\varrho$. This leads to the investigation of

[1]) By a *full set of components* we mean a set that fix es the quantity completely. For instance the components $v^{\varkappa\lambda}_{\cdot\cdot c}$ exist but they fix the quantity $v^{\varkappa\lambda}_{\cdot\cdot\mu}$ only if combined with the $v^{\varkappa\lambda}_{\cdot\cdot z}$.

the derivative $\nabla_\mu B^\varkappa_\lambda = -\nabla_\mu C^\varkappa_\lambda$. If we introduce the quantities[1])

$$(7.9)\qquad \overset{m}{H}{}_{\mu\lambda}^{\cdot\cdot\varkappa} \overset{\text{def}}{=} B^{\tau\sigma}_{\mu\lambda}\nabla_\tau B^\varkappa_\sigma = -B^{\tau\sigma}_{\mu\lambda}\nabla_\tau C^\varkappa_\sigma = B^\tau_\mu C^\varkappa_\sigma \nabla_\tau B^\sigma_\lambda = -B^\tau_\mu C^\varkappa_\sigma \nabla_\tau C^\sigma_\lambda;$$

$$(7.10)\qquad \overset{m}{L}{}_{\mu\cdot\lambda}^{\cdot\varkappa} \overset{\text{def}}{=} B^{\tau\varkappa}_{\mu\varrho}\nabla_\tau B^\varrho_\lambda = -B^{\tau\varkappa}_{\mu\varrho}\nabla_\tau C^\varrho_\lambda = B^\tau_\mu C^\varrho_\lambda \nabla_\tau B^\varkappa_\varrho = -B^\tau_\mu C^\varrho_\lambda \nabla_\tau C^\varkappa_\varrho;$$

$$(7.11)\qquad \overset{m'}{H}{}_{\mu\lambda}^{\cdot\cdot\varkappa} \overset{\text{def}}{=} C^{\tau\sigma}_{\mu\lambda}\nabla_\tau C^\varkappa_\sigma = -C^{\tau\sigma}_{\mu\lambda}\nabla_\tau B^\varkappa_\sigma = C^\tau_\mu B^\varkappa_\sigma \nabla_\tau C^\sigma_\lambda = -C^\tau_\mu B^\varkappa_\sigma \nabla_\tau B^\sigma_\lambda;$$

$$(7.12)\qquad \overset{m'}{L}{}_{\mu\cdot\lambda}^{\cdot\varkappa} \overset{\text{def}}{=} C^{\tau\varkappa}_{\mu\varrho}\nabla_\tau C^\varrho_\lambda = -C^{\tau\varkappa}_{\mu\varrho}\nabla_\tau B^\varrho_\lambda = C^\tau_\mu B^\varrho_\lambda \nabla_\tau C^\varkappa_\varrho = -C^\tau_\mu B^\varrho_\lambda \nabla_\tau B^\varkappa_\varrho;$$

we see that each of these tensors lies with one index or two indices in X^m_n and with the remaining index or indices in $X^{m'}_n$. The full sets of components

$$(7.13)\qquad \overset{m}{H}{}_{cb}^{\cdot\cdot x};\quad \overset{m}{L}{}_{c\cdot y}^{\cdot a};\quad \overset{m'}{H}{}_{zy}^{\cdot\cdot a};\quad \overset{m'}{L}{}_{z\cdot b}^{\cdot x}$$

exist; this follows from (7.9–12). From these full sets we can see which of the indices lie in X^m_n and which in $X^{m'}_n$. The covariant derivative $\nabla_\mu B^\varkappa_\lambda$ can be expressed in terms of these four quantities

$$(7.14)\qquad \nabla_\mu B^\varkappa_\lambda = \overset{m}{H}{}_{\mu\lambda}^{\cdot\cdot\varkappa} + \overset{m}{L}{}_{\mu\cdot\lambda}^{\cdot\varkappa} - \overset{m'}{H}{}_{\mu\lambda}^{\cdot\cdot\varkappa} - \overset{m'}{L}{}_{\mu\cdot\lambda}^{\cdot\varkappa}.$$

The relationship $A^\varkappa_\lambda = B^\varkappa_\lambda + C^\varkappa_\lambda$ shows that a quantity with valence *3* in general splits up into eight parts, but here four of these parts are zero, for instance $B^{\tau\sigma\varkappa}_{\mu\lambda\varrho}\nabla_\tau B^\varrho_\sigma$.

The quantities $\overset{m}{H}{}_{\mu\lambda}^{\cdot\cdot\varkappa}$ and $\overset{m}{L}{}_{\mu\cdot\lambda}^{\cdot\varkappa}$ are called the *first* and *second curvature tensor of valence 3* of X^m_n.[2]) In the same way to $X^{m'}_n$ belong $\overset{m'}{H}{}_{\mu\lambda}^{\cdot\cdot\varkappa}$ and $\overset{m'}{L}{}_{\mu\cdot\lambda}^{\cdot\varkappa}$. The following formulae are useful

$$(7.15)\quad \text{a)}\ \overset{m}{H}{}_{cb}^{\cdot\cdot x} = C^x_\varkappa D_c B^\varkappa_b = -B^\varkappa_b D_c C^x_\varkappa;\quad \text{b)}\ \overset{m}{H}{}_{cb}^{\cdot\cdot\varkappa} = D_c B^\varkappa_b;\quad \text{c)}\ \overset{m}{H}{}_{c\lambda}^{\cdot\cdot x} = -D_c C^x_\lambda$$

$$(7.16)\quad \text{a)}\ \overset{m}{L}{}_{c\cdot y}^{\cdot a} = C^\lambda_y D_c B^a_\lambda = -B^a_\lambda D_c C^\lambda_y;\quad \text{b)}\ \overset{m}{L}{}_{c\cdot\lambda}^{\cdot a} = D_c B^a_\lambda;\quad \text{c)}\ \overset{m}{L}{}_{c\cdot y}^{\cdot\varkappa} = -D_c C^\varkappa_y$$

$$(7.17)\quad \text{a)}\ \overset{m'}{H}{}_{zy}^{\cdot\cdot a} = B^a_\varkappa D_z C^\varkappa_y = -C^\varkappa_y D_z B^a_\varkappa;\quad \text{b)}\ \overset{m'}{H}{}_{zy}^{\cdot\cdot\varkappa} = D_z C^\varkappa_y;\quad \text{c)}\ \overset{m'}{H}{}_{z\lambda}^{\cdot\cdot a} = -D_z B^a_\lambda$$

$$(7.18)\quad \text{a)}\ \overset{m'}{L}{}_{z\cdot b}^{\cdot x} = B^\lambda_b D_z C^x_\lambda = -C^x_\lambda D_z B^\lambda_b;\quad \text{b)}\ \overset{m'}{L}{}_{z\cdot\lambda}^{\cdot x} = D_z C^x_\lambda;\quad \text{c)}\ \overset{m'}{L}{}_{z\cdot b}^{\cdot\varkappa} = -D_z B^\varkappa_b.$$

---

[1]) SCHOUTEN 1923, 1 for X_m in A_n.

[2]) For V_n the curvature tensor of valence 3 (cf. V § 9) was introduced by SCHOUTEN and STRUIK 1921, 3 and at the same time by BOMPIANI 1921, 1. Its components correspond to the Ω_{rtj} of VOSS 1880, 1, to the $b_{\alpha|rs}$ in RICCI 1903, 1 and to the $k^{(r)}_{fg}$ in KÜHNE 1904, 1. Cf. SCHOUTEN and STRUIK 1922, 1; 2; STRUIK 1922, 1; BOMPIANI 1951, 3. Some authors call it the "EULER-SCHOUTEN" tensor which is not correct from a historical point of view.

Using the anholonomic coordinate system (h) in L_n we have the following formulae for the contravariant vectors $p^\varkappa$ and $r^\varkappa$

$$(7.19)\quad \begin{cases} \text{a)} & D_c p^a \overset{*}{=} \partial_c p^a + \Gamma_{cb}^a p^b; \qquad \partial_c \overset{\text{def}}{=} B_c^\mu \partial_\mu \\ \text{b)} & D_z p^a \overset{*}{=} \partial_z p^a + \Gamma_{zb}^a p^b; \qquad \partial_z \overset{\text{def}}{=} C_z^\mu \partial_\mu \\ \text{c)} & D_c r^x \overset{*}{=} \partial_c r^x + \Gamma_{cy}^x r^y \\ \text{d)} & D_z r^x \overset{*}{=} \partial_z r^x + \Gamma_{zy}^x r^y \end{cases}$$

and corresponding formulae for the covariant vectors q_λ, s_λ. These equations fix a linear connexion for quantities of X_n^m and also a linear connexion for quantities of $X_n^{m'}$. We express this by writing L_n^m, $L_n^{m'}$ for X_n^m, $X_n^{m'}$ from now on, and also A_n^m, $A_n^{m'}$ if the L_n happens to be an A_n. In L_n we know that (cf. III 9.3)

$$(7.20)\qquad \Gamma_{[ji]}^h = S_{ji}^{\cdot\cdot h} - \Omega_{ji}^h,$$

hence the expressions

$$(7.21)\qquad \Gamma_{[cb]}^a \overset{*}{=} S_{cb}^{\cdot\cdot a} - \Omega_{cb}^a; \qquad \Gamma_{[zy]}^x \overset{*}{=} S_{zy}^{\cdot\cdot x} - \Omega_{zy}^x$$

depend only on the rigging and on $S_{\mu\lambda}^{\cdot\cdot\varkappa}$ and not on the symmetric part $\Gamma_{(\mu\lambda)}^\varkappa$ of the connexion in L_n. In an A_n they depend only on the rigging and not on the connexion in A_n.

From (7.15b) we get

$$(7.22)\quad \begin{cases} \overset{m}{H}{}_{[cb]}^{\cdot\cdot x} = -B_{[cb]}^{\mu\lambda} (\nabla_\mu C_\lambda^x) C_\varkappa^x = -B_{[cb]}^{\mu\lambda} \nabla_\mu (\overset{y}{e}_\lambda \underset{y}{e}^\varkappa) C_\varkappa^x \\ \qquad = -B_{cb}^{\mu\lambda} (\nabla_{[\mu} \overset{y}{e}_{\lambda]}) \underset{y}{e}^x = -B_{cb}^{\mu\lambda} (\partial_{[\mu} \overset{y}{e}_{\lambda]} - S_{\mu\lambda}^{\cdot\cdot\varkappa} \overset{y}{e}_\varkappa) \underset{y}{e}^x \\ \qquad = B_{cb}^{ji} C_h^x (-\Omega_{ji}^h + S_{ji}^{\cdot\cdot h}) \\ \qquad = -\overset{m}{Z}{}_{cb}^{\cdot\cdot x} + B_{cb}^{ji} C_h^x S_{ji}^{\cdot\cdot h} \overset{*}{=} \Gamma_{[cb]}^x; \quad \overset{m}{Z}{}_{cb}^{\cdot\cdot x} \overset{\text{def}}{=} B_{cb}^{ji} C_h^x \Omega_{ji}^h \overset{*}{=} \Omega_{cb}^x; \end{cases}$$

and this implies that $\overset{m}{Z}{}_{cb}^{\cdot\cdot x}$ is a tensor belonging to the given L_n^m and that it lies with two indices in L_n^m and with one index in $L_n^{m'}$. In II § 9 we have already proved that the X_n^m is holonomic (that is X_m-forming) if and only if $\overset{m}{Z}{}_{cb}^{\cdot\cdot x} = 0$. This n.a.s. condition can now be expressed in the form

$$(7.23)\qquad \overset{m}{H}{}_{[cb]}^{\cdot\cdot x} - B_{cb}^{ji} C_h^x S_{ji}^{\cdot\cdot h} = 0.$$

In the same way the tensor $\overset{m'}{Z}{}_{zy}^{\cdot\cdot a}$ belongs to $L_n^{m'}$ and this $L_n^{m'}$ is holonomic if and only if

$$(7.24)\qquad -\overset{m'}{Z}{}_{zy}^{\cdot\cdot a} \overset{\text{def}}{=} \overset{m'}{H}{}_{[zy]}^{\cdot\cdot a} - C_{zy}^{ji} B_h^a S_{ji}^{\cdot\cdot h} = 0.$$

In an A_n the two quantities $\overset{m}{Z}{}_{cb}^{\cdot\cdot x}$ and $\overset{m'}{Z}{}_{zy}^{\cdot\cdot a}$ are identical with $-\overset{m}{H}{}_{[cb]}^{\cdot\cdot x}$ and $-\overset{m'}{H}{}_{[zy]}^{\cdot\cdot a}$.

Besides the tensors $\overset{m}{Z}{}_{cb}^{\cdot\cdot x}$ and $\overset{m'}{Z}{}_{zy}^{\cdot\cdot a}$ four geometric objects can be formed from Ω_{ji}^{h}, viz. $'\Omega_{cb}^{a} \overset{\text{def}}{\underset{*}{=}} \Omega_{cb}^{a}$ with the transformation (cf. II 9.3)

$$'\Omega_{c'b'}^{a'} \overset{*}{=} A_{c'\,b'\,a}^{c\,b\,a'}\,{}'\Omega_{cb}^{a} + A_{c'\,b'}^{c\,b}\,\partial_{[c}A_{b]}^{a'} \tag{7.25}$$

and $\overset{m}{\Omega}{}_{cy}^{a} \overset{\text{def}}{\underset{*}{=}} \Omega_{cy}^{a}$ with the transformation

$$\overset{m}{\Omega}{}_{c'y'}^{a'} \overset{*}{=} A_{c'\,y'\,a}^{c\,y\,a'}\,\overset{m}{\Omega}{}_{cy}^{a} - \tfrac{1}{2}A_{c'\,y'}^{c\,y}\,\partial_{y}A_{c}^{a'} \tag{7.26}$$

and the two corresponding objects $''\Omega_{zy}^{x}$ and $\overset{m'}{\Omega}{}_{zb}^{x}$. The reason why Ω_{ji}^{h} splits up into six separate geometric objects is of course that here only transformations of (h) are allowed with $A_{y}^{a'}=0$; $A_{y'}^{a}=0$; $A_{b}^{x'}=0$; $A_{b'}^{x}=0$. From now on we write every one of these separate objects with its own kernel letter.

If only one X_m is given in L_n, only the operator D_c is important as long as only fields in X_m are considered. But the other operators can be used also if the fields are prolonged in the neighbourhoods of all points of the X_m. This is often convenient and the results as far as fields of X_m are concerned are independent of the manner of prolongation (cf. II § 4). The vectors $\underset{a}{e}{}^{\varkappa}$ may be chosen in such a way that they are basis vectors of some holonomic coordinate system (a) in X_m. But this is not necessary and it should only be done if it is convenient for some purpose.

The equations concerning the curvature of the L_n^m in L_n can now be obtained in a very simple way using the operators D. We have first to compute the alternated second derivatives of quantities of L_n, L_n^m and $L_n^{m'}$:

$$\text{(7.27A)}\quad \begin{cases} \text{a)}\ D_{[\nu}D_{\mu]}\,p = -S_{\nu\mu}^{\cdot\cdot\varrho}\,\partial_{\varrho}\,p \\ \text{b)}\ D_{[\nu}D_{\mu]}\,v^{\varkappa} = \tfrac{1}{2}R_{\nu\mu\lambda}^{\cdot\cdot\cdot\varkappa}\,v^{\lambda} - S_{\nu\mu}^{\cdot\cdot\varrho}\,\nabla_{\varrho}\,v^{\varkappa} \\ \text{c)}\ D_{[\nu}D_{\mu]}\,w_{\lambda} = -\tfrac{1}{2}R_{\nu\mu\lambda}^{\cdot\cdot\cdot\varkappa}\,w_{\varkappa} - S_{\nu\mu}^{\cdot\cdot\varrho}\,\nabla_{\varrho}\,w_{\lambda} \\ \text{d)}\ D_{[\nu}D_{\mu]}\,p^{a} = \big(\tfrac{1}{2}B_{b\varkappa}^{\lambda a}R_{\nu\mu\lambda}^{\cdot\cdot\cdot\varkappa} + \overset{m}{L}{}_{[\nu\cdot|x|}^{\cdot a}\,\overset{m}{H}{}_{\mu]b}^{\cdot\cdot x} - \overset{m}{L}{}_{[\nu\cdot|x|}^{\cdot a}\,\overset{m'}{L}{}_{\mu]\cdot b}^{\cdot x} - \\ \qquad - \overset{m'}{H}{}_{[\nu|x|}^{\cdot\cdot a}\,\overset{m}{H}{}_{\mu]b}^{\cdot\cdot x} + \overset{m'}{H}{}_{[\nu|x|}^{\cdot\cdot a}\,\overset{m'}{L}{}_{\mu]\cdot b}^{\cdot x}\big)\,p^{b} - S_{\nu\mu}^{\cdot\cdot\varrho}\,D_{\varrho}\,p^{a} \\ \text{e)}\ D_{[\nu}D_{\mu]}\,q_{b} = -\big(\tfrac{1}{2}B_{b\varkappa}^{\lambda a}R_{\nu\mu\lambda}^{\cdot\cdot\cdot\varkappa} + \overset{m}{L}{}_{[\nu\cdot|x|}^{\cdot a}\,\overset{m}{H}{}_{\mu]b}^{\cdot\cdot x} - \overset{m}{L}{}_{[\nu\cdot|x|}^{\cdot a}\,\overset{m'}{L}{}_{\mu]\cdot b}^{\cdot x} - \\ \qquad - \overset{m'}{H}{}_{[\nu|x|}^{\cdot\cdot a}\,\overset{m}{H}{}_{\mu]b}^{\cdot\cdot x} + \overset{m'}{H}{}_{[\nu|x|}^{\cdot\cdot a}\,\overset{m'}{L}{}_{\mu]\cdot b}^{\cdot x}\big)\,q_{a} - S_{\nu\mu}^{\cdot\cdot\varrho}\,D_{\varrho}\,q_{b} \\ \text{f)}\ D_{[\nu}D_{\mu]}\,r^{x} = \big(\tfrac{1}{2}C_{y\varkappa}^{\lambda x}R_{\nu\mu\lambda}^{\cdot\cdot\cdot\varkappa} - \overset{m}{L}{}_{[\nu\cdot|y|}^{\cdot a}\,\overset{m}{H}{}_{\mu]a}^{\cdot\cdot x} + \overset{m}{L}{}_{[\nu\cdot|y|}^{\cdot a}\,\overset{m}{L}{}_{\mu]\cdot a}^{\cdot x} + \\ \qquad + \overset{m'}{H}{}_{[\nu|y|}^{\cdot\cdot a}\,\overset{m}{H}{}_{\mu]a}^{\cdot\cdot x} - \overset{m'}{H}{}_{[\nu|y|}^{\cdot\cdot a}\,\overset{m'}{L}{}_{\mu]\cdot a}^{\cdot x}\big)\,r^{y} - S_{\nu\mu}^{\cdot\cdot\varrho}\,D_{\varrho}\,r^{x} \\ \text{g)}\ D_{[\nu}D_{\mu]}\,s_{y} = -\big(\tfrac{1}{2}C_{y\varkappa}^{\lambda x}R_{\nu\mu\lambda}^{\cdot\cdot\cdot\varkappa} - \overset{m}{L}{}_{[\nu\cdot|y|}^{\cdot a}\,\overset{m}{H}{}_{\mu]a}^{\cdot\cdot x} + \overset{m}{L}{}_{[\nu\cdot|y|}^{\cdot a}\,\overset{m'}{L}{}_{\mu]\cdot a}^{\cdot x} + \\ \qquad + \overset{m'}{H}{}_{[\nu|y|}^{\cdot\cdot a}\,\overset{m}{H}{}_{\mu]a}^{\cdot\cdot x} - \overset{m'}{H}{}_{[\nu|y|}^{\cdot\cdot a}\,\overset{m'}{L}{}_{\mu]\cdot a}^{\cdot x}\big)\,s_{x} - S_{\nu\mu}^{\cdot\cdot\varrho}\,D_{\varrho}\,s_{y}. \end{cases}$$

$$
(7.27\mathrm{B})\quad
\begin{cases}
\text{a)}\; D_{[d}D_{c]}\,p = -\left(B^{\nu\mu}_{dc}\,S_{\nu\mu}^{\cdot\cdot\varkappa} - \overset{m}{H}{}_{[dc]}^{\cdot\cdot\varkappa}\right)\partial_\varkappa p\\
\text{b)}\; D_{[d}D_{c]}\,v^\varkappa = \tfrac12 B^{\nu\mu}_{dc}\,R_{\nu\mu\lambda}^{\cdot\cdot\cdot\varkappa}\,v^\lambda - \left(B^{\nu\mu}_{dc}\,S_{\nu\mu}^{\cdot\cdot\varrho} - \overset{m}{H}{}_{[dc]}^{\cdot\cdot\varrho}\right)\nabla_\varrho v^\varkappa\\
\text{c)}\; D_{[d}D_{c]}\,w_\lambda = -\tfrac12 B^{\nu\mu}_{dc}\,R_{\nu\mu\lambda}^{\cdot\cdot\cdot\varkappa}\,w_\varkappa - \left(B^{\nu\mu}_{dc}\,S_{\nu\mu}^{\cdot\cdot\varrho} - \overset{m}{H}{}_{[dc]}^{\cdot\cdot\varrho}\right)\nabla_\varrho w_\lambda\\
\text{d)}\; D_{[d}D_{c]}\,p^a = \left(\tfrac12 B^{\nu\mu\lambda a}_{dcb\varkappa}\,R_{\nu\mu\lambda}^{\cdot\cdot\cdot\varkappa} - \overset{m}{H}{}_{[d|b|}^{\cdot\cdot x}\,\overset{m}{L}{}_{c]\cdot x}^{\cdot a}\right)p^b -\\
\qquad\qquad - \left(B^{\nu\mu}_{dc}\,S_{\nu\mu}^{\cdot\cdot\varrho} - \overset{m}{H}{}_{[dc]}^{\cdot\cdot\varrho}\right)D_\varrho p^a\\
\text{e)}\; D_{[d}D_{c]}\,q_b = -\left(\tfrac12 B^{\nu\mu\lambda a}_{dcb\varkappa}\,R_{\nu\mu\lambda}^{\cdot\cdot\cdot\varkappa} - \overset{m}{H}{}_{[d|b|}^{\cdot\cdot x}\,\overset{m}{L}{}_{c]\cdot x}^{\cdot a}\right)q_a -\\
\qquad\qquad - \left(B^{\nu\mu}_{dc}\,S_{\nu\mu}^{\cdot\cdot\varrho} - \overset{m}{H}{}_{[dc]}^{\cdot\cdot\varrho}\right)D_\varrho q_b\\
\text{f)}\; D_{[d}D_{c]}\,r^x = \left(\tfrac12 B^{\nu\mu}_{dc}\,C^{\lambda x}_{y\varkappa}\,R_{\nu\mu\lambda}^{\cdot\cdot\cdot\varkappa} + \overset{m}{H}{}_{[d|a|}^{\cdot\cdot x}\,\overset{m}{L}{}_{c]\cdot y}^{\cdot a}\right)r^y -\\
\qquad\qquad - \left(B^{\nu\mu}_{dc}\,S_{\nu\mu}^{\cdot\cdot\varrho} - \overset{m}{H}{}_{[dc]}^{\cdot\cdot\varrho}\right)D_\varrho r^x\\
\text{g)}\; D_{[d}D_{c]}\,s_y = -\left(\tfrac12 B^{\nu\mu}_{dc}\,C^{\lambda x}_{y\varkappa}\,R_{\nu\mu\lambda}^{\cdot\cdot\cdot\varkappa} + \overset{m}{H}{}_{[d|a|}^{\cdot\cdot x}\,\overset{m}{L}{}_{c]\cdot y}^{\cdot a}\right)s_x -\\
\qquad\qquad - \left(B^{\nu\mu}_{dc}\,S_{\nu\mu}^{\cdot\cdot\varrho} - \overset{m}{H}{}_{[dc]}^{\cdot\cdot\varrho}\right)D_\varrho s_y.
\end{cases}
$$

$$
(7.27\mathrm{C})\quad
\begin{cases}
\text{a)}\; D_{[u}D_{z]}\,p = -\left(C^{\nu\mu}_{uz}\,S_{\nu\mu}^{\cdot\cdot\varkappa} - \overset{m'}{H}{}_{[uz]}^{\cdot\cdot\varkappa}\right)\partial_\varkappa p\\
\text{b)}\; D_{[u}D_{z]}\,v^\varkappa = \tfrac12 C^{\nu\mu}_{uz}\,R_{\nu\mu\lambda}^{\cdot\cdot\cdot\varkappa}\,v^\lambda - \left(C^{\nu\mu}_{uz}\,S_{\nu\mu}^{\cdot\cdot\varrho} - \overset{m'}{H}{}_{[uz]}^{\cdot\cdot\varrho}\right)\nabla_\varrho v^\varkappa\\
\text{c)}\; D_{[u}D_{z]}\,w_\lambda = -\tfrac12 C^{\nu\mu}_{uz}\,R_{\nu\mu\lambda}^{\cdot\cdot\cdot\varkappa}\,w_\varkappa - \left(C^{\nu\mu}_{uz}\,S_{\nu\mu}^{\cdot\cdot\varrho} - \overset{m'}{H}{}_{[uz]}^{\cdot\cdot\varrho}\right)\nabla_\varrho w_\lambda\\
\text{d)}\; D_{[u}D_{z]}\,p^a = \left(\tfrac12 C^{\nu\mu}_{uz}\,B^{\lambda a}_{b\varkappa}\,R_{\nu\mu\lambda}^{\cdot\cdot\cdot\varkappa} + \overset{m'}{H}{}_{[u|x|}^{\cdot\cdot a}\,\overset{m'}{L}{}_{z]\cdot b}^{\cdot x}\right)p^b -\\
\qquad\qquad - \left(C^{\nu\mu}_{uz}\,S_{\nu\mu}^{\cdot\cdot\varrho} - \overset{m'}{H}{}_{[uz]}^{\cdot\cdot\varrho}\right)D_\varrho p^a\\
\text{e)}\; D_{[u}D_{z]}\,q_b = -\left(\tfrac12 C^{\nu\mu}_{uz}\,B^{\lambda a}_{b\varkappa}\,R_{\nu\mu\lambda}^{\cdot\cdot\cdot\varkappa} + \overset{m'}{H}{}_{[u|x|}^{\cdot\cdot a}\,\overset{m'}{L}{}_{z]\cdot b}^{\cdot x}\right)q_a -\\
\qquad\qquad - \left(C^{\nu\mu}_{uz}\,S_{\nu\mu}^{\cdot\cdot\varrho} - \overset{m'}{H}{}_{[uz]}^{\cdot\cdot\varrho}\right)D_\varrho q_b\\
\text{f)}\; D_{[u}D_{z]}\,r^x = \left(\tfrac12 C^{\nu\mu\lambda x}_{uzy\varkappa}\,R_{\nu\mu\lambda}^{\cdot\cdot\cdot\varkappa} - \overset{m'}{H}{}_{[u|y|}^{\cdot\cdot a}\,\overset{m'}{L}{}_{z]\cdot a}^{\cdot x}\right)r^y -\\
\qquad\qquad - \left(C^{\nu\mu}_{uz}\,S_{\nu\mu}^{\cdot\cdot\varrho} \quad \overset{m'}{H}{}_{[uz]}^{\cdot\cdot\varrho}\right)D_\varrho r^x\\
\text{g)}\; D_{[u}D_{z]}\,s_y = -\left(\tfrac12 C^{\nu\mu\lambda x}_{uzy\varkappa}\,R_{\nu\mu\lambda}^{\cdot\cdot\cdot\varkappa} - \overset{m'}{H}{}_{[u|y|}^{\cdot\cdot a}\,\overset{m'}{L}{}_{z]\cdot a}^{\cdot x}\right)s_x -\\
\qquad\qquad - \left(C^{\nu\mu}_{uz}\,S_{\nu\mu}^{\cdot\cdot\varrho} - \overset{m'}{H}{}_{[uz]}^{\cdot\cdot\varrho}\right)D_\varrho s_y.
\end{cases}
$$

Using the anholonomic system (h) and the equations (7.20) and

$$
(7.28)\qquad \partial_{[j}\partial_{i]}\,p = -\Omega_{ji}^{h}\,\partial_h p \qquad \text{(cf. II 9.4)}
$$

we find another expression for $D_{[d}D_{c]}\,p^a$

$$
(7.29)\quad
\begin{cases}
D_{[d}D_{c]}\,p^a = \overset{m}{H}{}_{[dc]}^{\cdot\cdot y}\,D_y p^a +\\
\qquad + \{\partial_{[d}\Gamma^a_{c]b} + \Gamma^a_{[d|e|}\Gamma^e_{c]b} + \Omega^j_{dc}\Gamma^a_{jb}\}\,p^b - S_{\mu\lambda}^{\cdot\cdot\varrho}\,B^{\mu\lambda}_{dc}\,D_\varrho p^a.
\end{cases}
$$

Hence, if we write

$$\overset{m}{r}{}_{dcb}^{\cdot\cdot\cdot a} \overset{\text{def}}{=} 2\partial_{[d}\Gamma^a_{c]b} + 2\Gamma^a_{[d|e|}\Gamma^e_{c]b} + 2\Omega^j_{dc}\Gamma^a_{jb} \tag{7.30}$$

it follows from (7.29) that

$$2D_{[d}D_{c]}p^a = \overset{m}{r}{}_{dcb}^{\cdot\cdot\cdot a}p^b - 2\overset{m}{Z}{}_{dc}^{\cdot\cdot y}D_y p^a - 2S_{dc}^{\cdot\cdot e}D_e p^a \tag{7.31}$$

and from (7.27 B d) that

$$\boxed{\overset{m}{r}{}_{dcb}^{\cdot\cdot\cdot a} = B^{\nu\mu\lambda a}_{dcb\varkappa}R_{\nu\mu\lambda}^{\cdot\cdot\cdot\varkappa} - 2\overset{m}{H}{}_{[d|b|}^{\cdot\cdot x}\overset{m}{L}{}_{c]\cdot x}^{\cdot a}}\,. \tag{7.32}$$

In the same way we get for $L_n^{m'}$

$$2D_{[u}D_{z]}r^x = \overset{m'}{r}{}_{uzy}^{\cdot\cdot\cdot x}r^y - 2\overset{m'}{Z}{}_{uz}^{\cdot\cdot a}D_a r^x - 2S_{uz}^{\cdot\cdot y}D_y r^x \tag{7.33}$$

and

$$\boxed{\overset{m'}{r}{}_{uzy}^{\cdot\cdot\cdot x} = C^{\nu\mu\lambda x}_{uzy\varkappa}R_{\nu\mu\lambda}^{\cdot\cdot\cdot\varkappa} - 2\overset{m'}{H}{}_{[u|y|}^{\cdot\cdot a}\overset{m'}{L}{}_{z]\cdot a}^{\cdot x}}\,. \tag{7.34}$$

(7.32) and (7.34) are a kind of generalized equations of Gauss for L_n^m and $L_n^{m'}$ in L_n. But $\overset{m}{r}$ and $\overset{m'}{r}$ can not be considered as the true curvature tensors of the induced connexion in L_n^m and $L_n^{m'}$. In fact there is something queer in (7.31). The left hand side depends only on the rigging and the Γ^a_{cb}, that is on the displacement of vectors of the L_n^m in directions of the L_n^m. But none of the terms of the right hand side has this property. Now because of (7.26) the equations

(7.35)[1]
$$\begin{cases} \text{a)} & \mathfrak{D}_z p^a \overset{\text{def}}{=} \partial_z p^a + 2\overset{m}{\Omega}{}^a_{bz}p^b \\ \text{b)} & \mathfrak{D}_z q_b \overset{\text{def}}{=} \partial_z q_b - 2\overset{m}{\Omega}{}^a_{bz}q_a \end{cases}$$

define a covariant derivative of p^a and q_b depending only on the rigging of the L_n^m and not on any connexion in L_n. As is to be expected this derivative is intimately connected with the Lie derivative. In fact, if $v^\varkappa$ is a vector field of $L_n^{m'}$ it is easy to prove that

$$v^y\mathfrak{D}_y p^a = B^a_\varkappa \underset{v}{\pounds} p^\varkappa.\text{[2]} \tag{7.36}$$

This means that the covariant differential $(d\xi)^y\mathfrak{D}_y p^a$ for a direction lying in $L_n^{m'}$ is zero if the L_n^m-part of the Lie differential vanishes or in other words if the field $p^\varkappa$ dragged along over $d\xi^\varkappa$ differs from the original value at $\xi^\varkappa + d\xi^\varkappa$ only by a vector in $L_n^{m'}$.

[1]) For (7,35c, d) see p. 261.

[2]) Cf. VII § 3.

An operator $\mathfrak{D}_b$ operating on r^x and s_y may be defined in the same way:

$$(7.35)\qquad \begin{cases} \text{c)}\quad \mathfrak{D}_c r^x = \partial_c r^x + 2\overset{m'}{\Omega}{}_{yc}^{x} r^y \\ \text{d)}\quad \mathfrak{D}_c s_y = \partial_c s_y - 2\overset{m'}{\Omega}{}_{yc}^{x} s_x. \end{cases}$$

Note that the operation of $\mathfrak{D}_y$ on quantities of $L_n^{m'}$ and of $\mathfrak{D}_b$ on quantities of L_n^m is not yet defined. The relation between $\mathfrak{D}_y$ and D_y is the identity

$$(7.37)\qquad \mathfrak{D}_z p^a = D_z p^a + \overset{m}{L}{}_{c\cdot z}^{\cdot a} p^c + 2 S_{cz}^{\cdot\cdot a} p^c$$

and if this is substituted in (7.31) we get

$$(7.38)\qquad 2 D_{[d} D_{c]} p^a = -2\overset{m}{Z}{}_{dc}^{\cdot\cdot y} \mathfrak{D}_y p^a - 2 S_{dc}^{\cdot\cdot e} D_e p^a + \overset{m}{R}{}_{dcb}^{\cdot\cdot\cdot a} p^b$$

where

$$(7.39)\qquad \overset{m}{R}{}_{dcb}^{\cdot\cdot\cdot a} \overset{\text{def}}{=} \overset{m}{r}{}_{dcb}^{\cdot\cdot\cdot a} + 2\overset{m}{Z}{}_{dc}^{\cdot\cdot y} (\overset{m}{L}{}_{b\cdot y}^{\cdot a} + 2 S_{by}^{\cdot\cdot a})$$

is a quantity depending only on the displacement of vectors of L_n^m in directions of L_n^m and on the rigging and not extra on any connexion in L_n. This quantity is the *true* curvature tensor of the induced connexion in the L_n^m and there is just such a quantity $\overset{m'}{R}{}_{uzy}^{\cdot\cdot\cdot x}$ for the $L_n^{m'}$.[1]) If the L_n^m is holonomic, $\overset{m}{R}$ is identical with $\overset{m}{r}$ as follows from (7.39). The *true* generalizations of the equations of GAUSS for L_n^m and $L_n^{m'}$ in L_n are (cf. 3.15)

$$(7.40)\qquad \boxed{\begin{aligned} \overset{m}{R}{}_{dcb}^{\cdot\cdot\cdot a} = B_{dcb\varkappa}^{\nu\mu\lambda a} R_{\nu\mu\lambda}^{\cdot\cdot\cdot\varkappa} - 2\overset{m}{H}{}_{[d|b|}^{\cdot\cdot x} \overset{m}{L}{}_{c]\cdot x}^{\cdot a} + \\ + 2\overset{m}{Z}{}_{dc}^{\cdot\cdot y} (\overset{m}{L}{}_{b\cdot y}^{\cdot a} + 2 S_{by}^{\cdot\cdot a}) \end{aligned}}\qquad \text{(GAUSS)}$$

$$(7.41)\qquad \boxed{\begin{aligned} \overset{m'}{R}{}_{uzy}^{\cdot\cdot\cdot x} = C_{uzy\varkappa}^{\nu\mu\lambda x} R_{\nu\mu\lambda}^{\cdot\cdot\cdot\varkappa} - 2\overset{m'}{H}{}_{[u|y|}^{\cdot\cdot a} \overset{m'}{L}{}_{z]\cdot a}^{\cdot x} + \\ + 2\overset{m'}{Z}{}_{uz}^{\cdot\cdot b} (\overset{m'}{L}{}_{y\cdot b}^{\cdot x} + 2 S_{yb}^{\cdot\cdot x}) \end{aligned}}\qquad \text{(GAUSS)}.$$

The following expression can be deduced for $D_{[d} D_{c]} r^x$

$$(7.42)\qquad 2 D_{[d} D_{c]} r^x = \overset{m\,m'}{R}{}_{dcy}^{\cdot\cdot\cdot x} r^y - 2\overset{m}{Z}{}_{dc}^{\cdot\cdot y} D_y r^x - 2 S_{dc}^{\cdot\cdot e} D_e r^x$$

where

$$(7.43)\qquad \boxed{\overset{m\,m'}{R}{}_{dcy}^{\cdot\cdot\cdot x} \overset{\text{def}}{=} B_{dc}^{\nu\mu} C_{y\varkappa}^{\lambda x} R_{\nu\mu\lambda}^{\cdot\cdot\cdot\varkappa} + 2\overset{m}{H}{}_{[d|b|}^{\cdot\cdot x} \overset{m}{L}{}_{c]\cdot y}^{\cdot b}}\qquad \text{(RICCI)}\ ^{2})$$

---

[1]) SCHOUTEN 1929, 3, p. 163f. also for historical notes; SCHOUTEN and v. KAMPEN 1930, 2, p. 777; E II 1938, 2, p. 161. The case of V_n^m in V_n was treated earlier by VRANCEANU 1926, 1; 2; 1928, 1; 3.

[2]) For V_m in V_n FABRICIUS-BJERRE 1934, 1; 1936, 1 called this tensor the "torsion" of the V_m and discussed the case $\overset{m\,m'}{R}{}_{dcy}^{\cdot\cdot\cdot x} = 0$.

and similarly

$$2 D_{[u} D_{z]} p^a = \overset{m'm}{R}{}_{uzb}^{\cdot\cdot\cdot a} p^b - 2 \overset{m'}{Z}{}_{uz}^{\cdot\cdot b} D_b p^a - 2 S_{uz}^{\cdot\cdot y} D_y p^a \tag{7.44}$$

where

$$\boxed{\overset{m'm}{R}{}_{uzb}^{\cdot\cdot\cdot a} \overset{\text{def}}{=} C_{uz}^{\nu\mu} B_{b\varkappa}^{\lambda a} R_{\nu\mu\lambda}^{\cdot\cdot\cdot\varkappa} + 2 \overset{m'}{H}{}_{[u|y|}^{\cdot\cdot\cdot a} \overset{m'}{L}{}_{z]\cdot b}^{\cdot y}} \quad \text{(Ricci)}. \tag{7.45}$$

(7.43, 45) are the generalized equations of Ricci for L_n^m and $L_n^{m'}$ in L_n.

It is easily proved that the rule of Leibniz holds for the operators $D_{[d}D_{c]}$ and $D_{[u}D_{z]}$ (cf. III § 4). Hence for a product $v^\varkappa q_b$ we get from (7.27 B b, e)

$$\begin{cases} D_{[d} D_{c]} v^\varkappa q_b = -\frac{1}{2} \overset{m}{r}{}_{dcb}^{\cdot\cdot\cdot a} v^\varkappa q_a + \frac{1}{2} B_{dc}^{\nu\mu} R_{\nu\mu\lambda}^{\cdot\cdot\cdot\varkappa} v^\lambda q_b + \\ \qquad + \overset{m}{H}{}_{[dc]}^{\cdot\cdot y} D_y v^\varkappa q_b - B_{dc}^{\nu\mu} S_{\nu\mu}^{\cdot\cdot\varrho} D_\varrho v^\varkappa q_b \end{cases} \tag{7.46}$$

and thus

$$\begin{cases} D_{[d} \overset{m}{H}{}_{c]b}^{\cdot\cdot\varkappa} = D_{[d} D_{c]} B_b^\varkappa = -\frac{1}{2} \overset{m}{r}{}_{dcb}^{\cdot\cdot\cdot a} B_a^\varkappa + \frac{1}{2} B_{dc}^{\nu\mu} R_{\nu\mu\lambda}^{\cdot\cdot\cdot\varkappa} B_b^\lambda + \\ \qquad + \overset{m}{H}{}_{[dc]}^{\cdot\cdot y} D_y B_b^\varkappa - B_{dc}^{\nu\mu} S_{\nu\mu}^{\cdot\cdot\varrho} D_\varrho B_b^\varkappa. \end{cases} \tag{7.47}$$

Transvection of this equation with $B_\varkappa^a$ leads back to (7.32) but a new identity arises from transvection with $C_\varkappa^x$ (cf. 3.19):

$$\boxed{\begin{aligned} 2 D_{[d} \overset{m}{H}{}_{c]b}^{\cdot\cdot x} = B_{dcb}^{\nu\mu\lambda} C_\varkappa^x R_{\nu\mu\lambda}^{\cdot\cdot\cdot\varkappa} - 2 \overset{m}{H}{}_{[dc]}^{\cdot\cdot y} \overset{m'}{L}{}_{y\cdot b}^{\cdot x} - \\ - 2 S_{dc}^{\cdot\cdot a} \overset{m}{H}{}_{ab}^{\cdot\cdot x} + 2 S_{\nu\mu}^{\cdot\cdot\varrho} B_{dc}^{\nu\mu} C_\varrho^y \overset{m'}{L}{}_{y\cdot b}^{\cdot x} \end{aligned}} \quad \text{(Codazzi)} \tag{7.48}$$

and in the same way we get for $L_n^{m'}$:

$$\boxed{\begin{aligned} 2 D_{[u} \overset{m'}{H}{}_{z]y}^{\cdot\cdot a} = C_{uzy}^{\nu\mu\lambda} B_\varkappa^a R_{\nu\mu\lambda}^{\cdot\cdot\cdot\varkappa} - 2 \overset{m'}{H}{}_{[uz]}^{\cdot\cdot b} \overset{m}{L}{}_{b\cdot y}^{\cdot a} - \\ - 2 S_{uz}^{\cdot\cdot x} \overset{m'}{H}{}_{xy}^{\cdot\cdot a} + 2 S_{\nu\mu}^{\cdot\cdot\varrho} C_{uz}^{\nu\mu} B_\varrho^b \overset{m}{L}{}_{b\cdot y}^{\cdot a} \end{aligned}} \quad \text{(Codazzi)}. \tag{7.49}$$

Differentiation of $\overset{m}{L}{}_{c\cdot\lambda}^{\cdot a}$ gives, using (7.27 A c, 27 B c, d)

$$\begin{cases} D_{[d} \overset{m}{L}{}_{c]\cdot\lambda}^{\cdot a} = D_{[d} D_{c]} B_\lambda^a = \frac{1}{2} \overset{m}{r}{}_{dcb}^{\cdot\cdot\cdot a} B_\lambda^b - \frac{1}{2} B_{dc}^{\nu\mu} R_{\nu\mu\lambda}^{\cdot\cdot\cdot\varkappa} B_\varkappa^a + \\ \qquad + \overset{m}{H}{}_{[dc]}^{\cdot\cdot x} D_x B_\lambda^a - B_{dc}^{\nu\mu} S_{\nu\mu}^{\cdot\cdot\varrho} D_\varrho B_\lambda^a. \end{cases} \tag{7.50}$$

Transvection with B_b^λ leads back to (7.32) but by transvection with C_y^λ we get the new identity (cf. 3,20)

$$\boxed{2 D_{[d} \overset{m}{L}{}_{c]\cdot y}^{\cdot a} = - B_{dc\varkappa}^{\nu\mu a} C_y^\lambda R_{\nu\mu\lambda}^{\cdot\cdot\cdot\varkappa} + 2 \overset{m}{Z}{}_{dc}^{\cdot\cdot x} \overset{m'}{H}{}_{xy}^{\cdot\cdot a} - 2 S_{dc}^{\cdot\cdot b} \overset{m}{L}{}_{b\cdot y}^{\cdot a}} \quad \text{(Codazzi)} \tag{7.51}$$

and in the same way for $L_n^{m'}$

$$(7.52)\quad \boxed{2D_{[u}\overset{m'}{L}{}_{z]}^{\cdot}{}_{\cdot b}^{x} = -C_{uz\varkappa}^{\nu\mu x}B_b^\lambda R_{\nu\mu\lambda}^{\cdot\cdot\cdot\varkappa} + 2Z_{uz}^{\cdot\cdot a}\overset{m}{H}{}_{ab}^{\cdot\cdot x} - 2S_{uz}^{\cdot\cdot y}\overset{m'}{L}{}_{y}^{\cdot}{}_{\cdot b}^{x}}\quad \text{(CODAZZI)}.$$

(7.48, 49, 51, 52) are the generalized equations of CODAZZI for L_n^m and $L_n^{m'}$ in L_n.[1])

A curve of L_n^m, that is a curve whose direction at every point lies in L_n^m, is called a geodesic of L_n^m if the covariant differential of its tangent vector $d\xi^\varkappa/dt$ along the curve for the in L_n^m induced connexion has the direction of the tangent. Accordingly the n.a.s. condition that every geodesic of L_n^m is at the same time a geodesic of L_n is that the vector

$$(7.53)\quad \frac{d\xi^\mu}{dt}\left(\nabla_\mu \frac{d\xi^\varkappa}{dt} - {}'\nabla_\mu \frac{d\xi^\varkappa}{dt}\right) = \frac{d\xi^\mu}{dt}\left(\nabla_\mu \frac{d\xi^\lambda}{dt}\right)C_\lambda^\varkappa = \frac{d\xi^\mu}{dt}\frac{d\xi^\lambda}{dt}\overset{m}{H}{}_{\mu\lambda}^{\cdot\cdot\varkappa}$$

has the direction of $d\xi^\varkappa/dt$ for every choice of $d\xi^\varkappa/dt$ in the tangent E_m. But from the definition of $\overset{m}{H}{}_{\mu\lambda}^{\cdot\cdot\varkappa}$ we see that this is only possible if $\overset{m}{H}{}_{(\mu\lambda)}^{\cdot\cdot\varkappa} = 0$. If $\overset{m}{H}{}_{(\mu\lambda)}^{\cdot\cdot\varkappa} = 0$ at a point, the L_n^m is said to be *geodesic at this point*. It is called *geodesic* if it is geodesic at all points of L_n. A geodesic A_m in E_n is of course always an E_m. But a geodesic A_n^m in E_n need not have a curvature zero. Here is a very simple example in an R_3. The null system of a system of forces in R_3 fixes at every point the 2-direction of the vector

$$(7.54)\quad p_\lambda = a_\lambda + r^\varkappa f_{\varkappa\lambda}$$

where $r^\varkappa$ is the radiusvector, a_λ a constant vector and $f_{\varkappa\lambda}$ a constant bivector. If $p_\lambda = p\,i_\lambda$ where i_λ is a unitvector, we have

$$(7.55)\quad \text{a)}\ \overset{2}{H}{}_{\mu\lambda}^{\cdot\cdot\varkappa} = -\frac{1}{p}\,{}'f_{\mu\lambda}\,i^\varkappa;\qquad \text{b)}\ \overset{2}{L}{}_{\mu\cdot\lambda}^{\cdot\varkappa} = \frac{1}{p}\,{}'f_\mu^{\cdot\varkappa}\,i_\lambda;\quad {}'f_{\mu\lambda} = B_{\mu\lambda}^{\tau\sigma}f_{\tau\sigma}$$

and accordingly (cf. 7.32, 39)

$$(7.56)\quad \begin{cases} \text{a)}\ \overset{2}{r}{}_{dcb}^{\cdot\cdot\cdot a} = \dfrac{2}{p^2}\,{}'f_{[d|b|}\,{}'f_{c]}^{\cdot a}, \\ \text{b)}\ \overset{2}{R}{}_{dcb}^{\cdot\cdot\cdot a} = \dfrac{2}{p^2}\,{}'f_{[d|b|}\,{}'f_{c]}^{\cdot a} + \dfrac{2}{p^2}\,{}'f_{dc}\,{}'f_b^{\cdot a}. \end{cases}$$

The straight lines of the null system are also geodesics in the L_3^2 but neither $\overset{2}{r}$ nor $\overset{2}{R}$ vanishes.

If a rigged X_n^m is given in X_n, a connexion for quantities of the X_n^m only, can be fixed by giving the ${}'\Gamma_{cb}^a$ as functions of the coordinates $\xi^\varkappa$.

[1]) The GAUSS-CODAZZI equations for V_m in V_n were first given by VOSS in 1880, 1, the RICCI equations for V_m in R_n by RICCI 1888, 1. KÜHNE 1903, 1 had already all equations for V_m in V_n. Then followed SCHOUTEN 1929, 3 for L_n^m in L_n, SCHOUTEN and v. KAMPEN 1930, 2 for V_n^m in V_n and DIENES 1932, 1 for L_n^m in L_n. Cf. also for literature STRUIK 1922, 1, p. 136; EISENHART 1926. 1, p. 192; JÄRNEFELT 1928, 1; MAYER 1928, 1; 1935, 2.

Then $D_c\, p^a$ (and $D_c\, q_b$) can be defined just as in (7.19a) but with $'\Gamma^a_{cb}$ instead of Γ^a_{cb} and besides these derivatives we have also the derivatives $\mathfrak{D}_z\, p^a$ and $\mathfrak{D}_z\, q_b$ defined already in (7.35a, b). These derivatives *together* fix a covariant differentiation for quantities of X^m_n in every direction:

$$(7.57)\quad \begin{cases} \text{a)} & \mathfrak{D}_c\, p^a \overset{\text{def}}{=} D_c\, p^a \overset{\text{def}}{=} \partial_c\, p^a + {}'\Gamma^a_{cb}\, p^b; \quad \mathfrak{D}_z\, p^a \overset{\text{def}}{=} \partial_z\, p^a + 2\overset{m}{\Omega}{}^a_{bz}\, p^b \\ \text{b)} & \mathfrak{D}_c\, q_b \overset{\text{def}}{=} D_c\, q_b \overset{\text{def}}{=} \partial_c\, q_b - {}'\Gamma^a_{cb}\, q_a; \quad \mathfrak{D}_z\, q_b \overset{\text{def}}{=} \partial_z\, q_b - 2\overset{m}{\Omega}{}^a_{bz}\, q_a\,. \end{cases}$$

In the same way, if the $''\Gamma^x_{zy}$ were given as functions of the $\xi^\varkappa$, the derivatives (cf. 7.35c, d)

$$(7.57)\quad \begin{cases} \text{c)} & \mathfrak{D}_z\, r^x \overset{\text{def}}{=} D_z\, r^x \overset{\text{def}}{=} \partial_z\, r^x + {}''\Gamma^x_{zy}\, r^y; \quad \mathfrak{D}_c\, r^x \overset{\text{def}}{=} \partial_c\, r^x + 2\overset{m'}{\Omega}{}^x_{yc}\, r^y \\ \text{d)} & \mathfrak{D}_z\, s_y \overset{\text{def}}{=} D_z\, s_y \overset{\text{def}}{=} \partial_z\, s_y - {}''\Gamma^x_{zy}\, s_x; \quad \mathfrak{D}_c\, s_y \overset{\text{def}}{=} \partial_c\, s_y - 2\overset{m'}{\Omega}{}^x_{yc}\, s_x \end{cases}$$

together would fix a covariant differentiation for quantities of $X^{m'}_n$ in every direction. The quantities $\overset{m}{H}{}_{\mu\lambda}^{\cdot\cdot\,\varkappa}$ and $\overset{m}{L}{}_{\mu\cdot\lambda}^{\cdot\,\varkappa}$ do not exist now but $\overset{m}{H}{}_{[cb]}^{\cdot\cdot\,x}$ can be defined by (cf. 7.22)

$$(7.58)\qquad \overset{m}{H}{}_{[cb]}^{\cdot\cdot\,x} \overset{\text{def}}{=} -\overset{m}{Z}{}_{cb}^{\cdot\cdot\,x}$$

without the term with $S_{ji}^{\cdot\cdot\,h}$ because in X_n there is no connexion and no $S_{ji}^{\cdot\cdot\,h}$. For this connexion repeated differentiation leads to [cf. (7.38)]

$$(7.59)\quad 2\mathfrak{D}_{[d}\,\mathfrak{D}_{c]}\, p^a = -2\overset{m}{Z}{}_{dc}^{\cdot\cdot\,y}\,\mathfrak{D}_y\, p^a + {}^*\overset{m}{R}{}_{dcb}^{\cdot\cdot\cdot\,a}\, p^b - 2({}'\Gamma^e_{[dc]} + {}'\Omega^e_{dc})\, D_e\, p^a$$

with

$$(7.60)\quad {}^*\overset{m}{R}{}_{dcb}^{\cdot\cdot\cdot\,a} \overset{\text{def}}{=} 2\partial_{[d}\, {}'\Gamma^a_{c]b} + 2\,{}'\Gamma^a_{[d|e|}\,{}'\Gamma^e_{c]b} + 2\,{}'\Omega^e_{dc}\,{}'\Gamma^a_{eb} + 4\overset{m}{Z}{}_{dc}^{\cdot\cdot\,y}\,\overset{m}{\Omega}{}^a_{by}$$

and

$$(7.61)\qquad (\mathfrak{D}_u\,\mathfrak{D}_c - \mathfrak{D}_c\,\mathfrak{D}_u)\, p^a = {}^*\overset{m\,m'}{R}{}_{ucb}^{\cdot\cdot\cdot\,a}\, p^b$$

with

$$(7.62)\quad \begin{cases} {}^*\overset{m\,m'}{R}{}_{ucb}^{\cdot\cdot\cdot\,a} \overset{\text{def}}{=} \partial_u\,{}'\Gamma^a_{cb} + 2\partial_c\,\overset{m}{\Omega}{}^a_{ub} + 2\,{}'\Gamma^e_{cb}\,\overset{m}{\Omega}{}^a_{eu} + 2\,{}'\Gamma^a_{eb}\,\overset{m}{\Omega}{}^e_{uc} + \\ \qquad + 2\,{}'\Gamma^a_{ce}\,\overset{m}{\Omega}{}^e_{ub} + 4\overset{m'}{\Omega}{}^y_{uc}\,\overset{m}{\Omega}{}^a_{by}. \end{cases}$$

A rigged X^m_n in X_n provided in this way with a linear connexion is called an L^m_n in X_n and in the special case when ${}'\Gamma^a_{[cb]} + {}'\Omega^a_{cb} = 0$ it is also called an A^m_n in X_n.

Exercises.

V 7,1. If a rigged X^m_n is given in L_n and if we take for the connexion in the X^m_n the connexion induced from the L_n, prove that ${}^*R_{dcb}^{\cdot\cdot\cdot\,a}$ as defined in (V 7.60) with respect to this induced connexion is identical with $\overset{m}{R}{}_{dcb}^{\cdot\cdot\cdot\,a}$ as defined in (V 7.39).

V 7,2. Prove that for a rigged L_n^m in L_n

V 7,2 α) $$\overset{m}{L}{}_{c\cdot y}^{\cdot a} \overset{*}{=} -\Gamma_{cy}^a.$$

V 7,3 [1]). Prove that for a rigged L_n^m in L_n

V 7,3 α) $$\text{a)}\ -\overset{m'}{H}{}_{\mu\lambda}^{\cdot\cdot a} + \overset{m}{L}{}_{\mu\cdot\lambda}^{\cdot a} = D_\mu B_\lambda^a; \qquad \text{b)}\ -\overset{m}{H}{}_{\mu\lambda}^{\cdot\cdot x} + \overset{m'}{L}{}_{\mu\cdot\lambda}^{\cdot x} = D_\mu C_\lambda^x.$$

V 7,4[1]). For every vector field $v^\varkappa$ of a rigged L_n^m in L_n the following identities hold

V 7,4 α) $$\begin{cases} \text{a)} \quad C_\varkappa^x \nabla_\mu v^\varkappa = (\overset{m}{H}{}_{\mu\lambda}^{\cdot\cdot x} - \overset{m'}{L}{}_{\mu\cdot\lambda}^{\cdot x})\, v^\lambda \\ \text{b)} \quad B_c^\mu C_\varkappa^x \nabla_\mu v^\varkappa = \overset{m}{H}{}_{cb}^{\cdot\cdot x} v^b \\ \text{c)} \quad C_{y\varkappa}^{\mu x} \nabla_\mu v^\varkappa = -\overset{m'}{L}{}_{y\cdot b}^{\cdot x} v^b. \end{cases}$$

V 7,5 [2]). Prove for a rigged L_m in L_n that $\overset{m}{H}{}_{[cb]}^{\cdot\cdot\varkappa}$ is always zero if the connexion in L_n is semi-symmetric.

§ 8. The rigged X_m in A_n. [3])

In A_n $S_{\mu\lambda}^{\cdot\cdot\varkappa}$ is zero and therefore an X_n^m in A_n is holonomic, that is, it consists of ∞^{n-m} X_m's if and only if (cf. II §9, V 7.22) $\overset{m}{H}{}_{[cb]}^{\cdot\cdot x} = -\overset{m}{Z}{}_{cb}^{\cdot\cdot x} = 0$. In this case the $\underset{b}{e}{}^\varkappa$ in A_n can be chosen in such a way that they are contravariant basis vectors of a holonomic coordinate system (a) on every X_m. If this has been done we have

(8.1) $$'\Omega_{cb}^a \overset{*}{=} \underset{c}{e}{}^\mu \underset{b}{e}{}^\lambda \partial_{[\mu} \overset{a}{e}_{\lambda]} = \underset{c}{e}{}^e \underset{b}{e}{}^d \partial_{[e} \overset{a}{e}_{d]} = 0.$$

Of course the $X_n^{m'}$ need not be holonomic.

Now we consider the case when only one rigged X_m is given in A_n. Then this X_m is an A_m (cf. V § 7). Let the coordinate system (a) in A_m be holonomic and let the $\underset{y}{e}{}^\varkappa$ be given over A_m in an arbitrary way in the m'-direction of the rigging. Then the $\underset{i}{e}{}^\varkappa$, $\overset{h}{e}_\lambda$ and the $A_i^\varkappa$, A_λ^h; $h, i = 1, \ldots, n$ are known over A_m and the same holds for the Γ_{cb}^a, Γ_{cb}^x, Γ_{cy}^a and Γ_{cy}^x:

(8.2) $$\Gamma_{ci}^h = A_{cix}^{\mu\lambda h} \Gamma_{\mu\lambda}^\varkappa - A_i^\lambda \partial_c A_\lambda^h.$$

The only derivatives that are interesting are those in a direction of A_m. This means that only D_c will be used and that only $\overset{m}{H}{}_{cb}^{\cdot\cdot x}$ and $\overset{m}{L}{}_{c\cdot y}^{\cdot a}$, defined by (7.15, 16), will occur. The equations (7.31), (7.32), (7.42)

[1]) E II 1938, 2, p. 157.
[2]) E I 1935, 1, p. 97.
[3]) Cf. footnote 2 on p. 232.

and (7.43) take the form

(8.3) $$2 D_{[d} D_{c]} p^a = {}'R_{dcb}^{\cdot\cdot\cdot a} p^b$$

(8.4) $$\boxed{{}'R_{dcb}^{\cdot\cdot\cdot a} = B_{dcb\varkappa}^{\nu\mu\lambda a} R_{\nu\mu\lambda}^{\cdot\cdot\cdot\varkappa} - 2 \overset{m}{H}{}_{[d|b|}^{\cdot\cdot x} \overset{m}{L}{}_{c]\cdot x}^{\cdot a}}$$ (Gauss)

(8.5) $$2 D_{[d} D_{c]} r^x = \overset{m m'}{R}{}_{dcy}^{\cdot\cdot\cdot x} r^y$$

(8.6) $$\boxed{\overset{mm'}{R}{}_{dcy}^{\cdot\cdot\cdot x} = B_{dc}^{\nu\mu} C_{y\varkappa}^{\lambda x} R_{\nu\mu\lambda}^{\cdot\cdot\cdot\varkappa} + 2 \overset{m}{H}{}_{[d|b|}^{\cdot\cdot x} \overset{m}{L}{}_{c]\cdot y}^{\cdot b}}$$ (Ricci)

where ${}'R_{dcb}^{\cdot\cdot\cdot a}$ is the curvature tensor of the A_m which equals $\overset{m}{R}{}_{dcb}^{\cdot\cdot\cdot a}$ and also $\overset{m}{r}{}_{dcb}^{\cdot\cdot\cdot a}$ because of (7.30) and (7.39). These equations are only valid at points of the A_m. The equations (7.48) and (7.51) take the form

(8.7) $$\boxed{2 D_{[d} \overset{m}{H}{}_{c]b}^{\cdot\cdot x} = B_{dcb}^{\nu\mu\lambda} C_{\varkappa}^{x} R_{\nu\mu\lambda}^{\cdot\cdot\cdot\varkappa}}$$

(8.8) $$\boxed{2 D_{[d} \overset{m}{L}{}_{c]\cdot y}^{\cdot a} = - B_{dc\varkappa}^{\nu\mu a} C_{y}^{\lambda} R_{\nu\mu\lambda}^{\cdot\cdot\cdot\varkappa}}$$

(Codazzi)

and also are only valid at points of the A_m.

All these equations can be expressed in terms of the tangent vectors $\overset{x}{e}_\lambda$, the rigging vectors $\underset{y}{e}{}^\varkappa$ and their derivatives. It follows from (7.9, 10) that (cf. 3.11, 12)

(8.9) $$\overset{m}{H}{}_{\mu\lambda}^{\cdot\cdot\varkappa} = - B_{\mu\lambda}^{\tau\sigma} \nabla_\tau \overset{x}{e}_\sigma \underset{x}{e}{}^\varkappa = \overset{x}{h}_{\mu\lambda} \underset{x}{e}{}^\varkappa; \quad \overset{x}{h}_{\mu\lambda} \overset{\text{def}}{=} - B_{\mu\lambda}^{\tau\sigma} \nabla_\tau \overset{x}{e}_\sigma$$

(8.10) $$\overset{m}{L}{}_{\mu\cdot\lambda}^{\cdot\varkappa} = - B_{\mu\varrho}^{\tau\varkappa} \nabla_\tau \overset{x}{e}_\lambda \underset{x}{e}{}^\varrho = \underset{x}{l}{}_{\mu}^{\cdot\varkappa} \overset{x}{e}_\lambda; \quad \underset{x}{l}{}_{\mu}^{\cdot\varkappa} \overset{\text{def}}{=} - B_{\mu\varrho}^{\tau\varkappa} \nabla_\tau \underset{x}{e}{}^\varrho$$

hence the equations (7.40, 43) of Gauss and Ricci can be written in the form

(8.11) $$\boxed{{}'R_{dcb}^{\cdot\cdot\cdot a} = B_{dcb\varkappa}^{\nu\mu\lambda a} R_{\nu\mu\lambda}^{\cdot\cdot\cdot\varkappa} - 2 \overset{x}{h}_{[d|b|} \underset{x}{l}{}_{c]}^{\cdot a}}$$ (cf. 3.15) (Gauss)

(8.12) $$\boxed{\overset{m m'}{R}{}_{dcy}^{\cdot\cdot\cdot x} \overset{*}{=} B_{dc}^{\nu\mu} C_{y\varkappa}^{\lambda x} R_{\nu\mu\lambda}^{\cdot\cdot\cdot\varkappa} + 2 \overset{x}{h}_{[d|b|} \underset{y}{l}{}_{c]}^{\cdot b}}$$ (Ricci).

If we introduce the $(m')^2$ vectors of A_m

(8.13) $$\underset{y}{\overset{x}{v}}{}_b \overset{\text{def}}{=} B_b^\mu (\nabla_\mu \overset{x}{e}_\lambda) \underset{y}{e}{}^\lambda = - B_b^\mu (\nabla_\mu \underset{y}{e}{}^\lambda) \overset{x}{e}_\lambda \overset{*}{=} - \Gamma_{by}^{x}$$

we have

(8.14) $$2 {}'\nabla_{[d} \underset{y}{\overset{x}{v}}{}_{c]} \overset{*}{=} - \overset{m m'}{R}{}_{dcy}^{\cdot\cdot\cdot x} + 2 \underset{z}{\overset{x}{v}}{}_{[d} \underset{y}{\overset{z}{v}}{}_{c]}$$

and accordingly the equations of RICCI can be written in the form

$$(8.15)\qquad \boxed{2\,'\nabla_{[d}\overset{x}{\underset{y}{v}}{}_{c]} = -\,B^{\nu\mu}_{dc}\,R_{\nu\mu\lambda}^{\cdot\cdot\cdot\varkappa}\,\underset{y}{e}^{\lambda}\,\overset{x}{e}_{\varkappa} - 2\,\overset{x}{h}_{[d|b|}\,\underset{y}{l}{}_{c]}^{\cdot b} + 2\,\overset{x}{\underset{z}{v}}{}_{[d}\,\overset{z}{\underset{y}{v}}{}_{c]}}\qquad (\text{RICCI}).$$

If (8.13) is substituted in (8.7, 8) we get another form for the equations of CODAZZI

$$(8.16)\qquad \boxed{2\,'\nabla_{[d}\overset{x}{h}_{c]b} = B^{\nu\mu\lambda}_{dcb}\,R_{\nu\mu\lambda}^{\cdot\cdot\cdot\varkappa}\,\overset{x}{e}_{\varkappa} + 2\,\overset{x}{\underset{y}{v}}{}_{[d}\,\overset{y}{h}_{c]b}}\qquad (\text{CODAZZI})$$

$$(8.17)\qquad \boxed{2\,'\nabla_{[d}\underset{y}{l}{}_{c]}^{\cdot a} = -\,B^{\nu\mu a}_{dc\varkappa}\,R_{\nu\mu\lambda}^{\cdot\cdot\cdot\varkappa}\,\underset{y}{e}^{\lambda} - 2\,\overset{x}{\underset{y}{v}}{}_{[d}\,\underset{x}{l}{}_{c]}^{\cdot a}}\qquad (\text{CODAZZI}).$$

The equations (8.11, 15, 16, 17) are especially interesting for the case of an A_m in E_n. Then they take the simple form

$$(8.18)\qquad \boxed{'R_{dcb}^{\cdot\cdot\cdot a} = -\,2\,\overset{x}{h}_{[d|b|}\,\underset{x}{l}{}_{c]}^{\cdot a}}\qquad (\text{GAUSS})$$

$$(8.19)\qquad \boxed{2\,'\nabla_{[d}\overset{x}{\underset{y}{v}}{}_{c]} = -\,2\,\overset{x}{h}_{[d|b|}\,\underset{y}{l}{}_{c]}^{\cdot b} + 2\,\overset{x}{\underset{z}{v}}{}_{[d}\,\overset{z}{\underset{y}{v}}{}_{c]}}\qquad (\text{RICCI})$$

$$\left.\begin{aligned}(8.20)\qquad & '\nabla_{[d}\overset{x}{h}_{c]b} = \overset{x}{\underset{y}{v}}{}_{[d}\,\overset{y}{h}_{c]b}\\ (8.21)\qquad & '\nabla_{[d}\underset{y}{l}{}_{c]}^{\cdot a} = -\,\overset{x}{\underset{y}{v}}{}_{[d}\,\underset{x}{l}{}_{c]}^{\cdot a}\end{aligned}\right\}\qquad (\text{CODAZZI}).$$

We prove that an A_m can be imbedded in an E_n if and only if there exist, in this A_m, $2m' + m'^2$ fields $\overset{x}{h}_{cb}$, $\underset{y}{l}{}_{c}^{\cdot a}$, $\overset{x}{\underset{y}{v}}{}_b$ satisfying (8.18–21) and $\overset{x}{h}_{[cb]} = 0$. Let $(\varkappa)$ be a rectilinear coordinate system in the E_n to be constructed. Then we have from (8.9, 10, 13)

$$(8.22)\qquad \overset{x}{h}_{cb} \overset{*}{=} (\partial_c C^{x}_{\lambda})\,B^{\lambda}_{b} = -\,(\partial_c B^{\lambda}_{b})\,C^{x}_{\lambda}$$

$$(8.23)\qquad \underset{y}{l}{}_{c}^{\cdot a} \overset{*}{=} (\partial_c C^{\varkappa}_{y})\,B^{a}_{\varkappa} = -\,(\partial_c B^{a}_{\varkappa})\,C^{\varkappa}_{y}$$

$$(8.24)\qquad \overset{x}{\underset{y}{v}}{}_c \overset{*}{=} (\partial_c C^{x}_{\lambda})\,C^{\lambda}_{y} = -\,(\partial_c C^{\lambda}_{y})\,C^{x}_{\lambda}$$

and from $'\nabla_c w_b \overset{*}{=} B^{\mu\lambda}_{cb}\,\partial_\mu w_\lambda$

$$(8.25)\qquad \Gamma^{a}_{cb} \overset{*}{=} (\partial_c B^{\lambda}_{b})\,B^{a}_{\lambda} = -\,(\partial_c B^{a}_{\lambda})\,B^{\lambda}_{b}.$$

From these equations we get the differential equations for the unknowns $B^{\varkappa}_{b}$, $C^{\varkappa}_{y}$

$$(8.26)\qquad \partial_c B^{\varkappa}_{b} \overset{*}{=} \Gamma^{a}_{cb}\,B^{\varkappa}_{a} - \overset{x}{h}_{cb}\,C^{\varkappa}_{x}$$

$$(8.27)\qquad \partial_c C^{\varkappa}_{y} \overset{*}{=} \underset{y}{l}{}_{c}^{\cdot a}\,B^{\varkappa}_{a} - \overset{x}{\underset{y}{v}}{}_c\,C^{\varkappa}_{x}$$

and similar equations for B^a_λ, C^x_λ that do not need to be considered. The integrability conditions of (8.26, 27) are

$$
(8.28)\quad \left\{\begin{aligned} 0 &\overset{*}{=} (\partial_{[d}\, \Gamma^a_{c]b})\, B^\varkappa_a + \Gamma^a_{[c|b|}\, \partial_{d]}\, B^\varkappa_a - (\partial_{[d}\, \overset{x}{h}_{c]b})\, C^\varkappa_x - \overset{x}{h}_{[c|b|}\, \partial_{d]}\, C^\varkappa_x \\ &= \tfrac{1}{2}\, ({}'R_{dcb}^{\cdots a} + 2\overset{x}{h}_{[d|b|}\, \underset{x}{l}_{c]}^{\cdot a})\, B^\varkappa_a - ({}'\nabla_{[d}\, \overset{x}{h}_{c]b} - \underset{y}{\overset{x}{v}}_{[d}\, \overset{y}{h}_{c]b})\, C^\varkappa_x \end{aligned}\right.
$$

$$
(8.29)\quad \left\{\begin{aligned} 0 &\overset{*}{=} (\partial_{[d}\, \underset{y}{l}_{c]}^{\cdot a})\, B^\varkappa_a + \underset{y}{l}_{[c}^{\cdot a}\, \partial_{d]}\, B^\varkappa_a - (\partial_{[d}\, \underset{y}{\overset{x}{v}}_{c]})\, C^\varkappa_x - \underset{y}{\overset{x}{v}}_{[c}\, \partial_{d]}\, C^\varkappa_x \\ &= ({}'\nabla_{[d}\, \underset{y}{l}_{c]}^{\cdot a} + \underset{y}{\overset{x}{v}}_{[d}\, \underset{x}{l}_{c]}^{\cdot a})\, B^\varkappa_a - (\partial_{[d}\, \underset{y}{\overset{x}{v}}_{c]} + \overset{x}{h}_{[d|b|}\, \underset{y}{l}_{c]}^{\cdot b} + \underset{y}{\overset{z}{v}}_{[c}\, \underset{z}{\overset{x}{v}}_{d]})\, C^\varkappa_x \end{aligned}\right.
$$

and these are satisfied identically because of (8.18–21). Hence there exist solutions $B^\varkappa_b$, $C^\varkappa_y$ that take arbitrarily given values $\underset{0}{B^\varkappa_b}$, $\underset{0}{C^\varkappa_y}$ at some point $\underset{0}{\eta^a}$ of A_m. When the $B^\varkappa_b$ have been found in this way the equations

$$
(8.30)\qquad B^\varkappa_b = \partial_b\, \xi^\varkappa
$$

are completely integrable because of the symmetry in bc of Γ^a_{cb} and $\overset{x}{h}_{cb}$ in (8.26). The $\xi^\varkappa$ are then the rectilinear coordinates in the E_n that were to be determined.

In these considerations the number n and the fields $\overset{x}{h}_{cb}$, $\underset{y}{\overset{x}{l}}{}_c^{\cdot a}$ and $\underset{y}{\overset{x}{v}}_c$ are given beforehand. But there is another much deeper problem. We may require a number n such that every A_m can always be imbedded (in the neighbourhood of one of its points) in an E_n. By counting the number of equations and of variables for the case of a V_m in R_n SCHLÄFLI 1871 suggested that the value n would be $\binom{m+1}{2}$. The rigorous proof of SCHLÄFLI's supposition was given by JANET 1926 and CARTAN 1927.[1]) The number $n-m$ is called the *class* of the V_m.[2]) T. Y.

[1]) SCHLÄFLI 1871, 1; JANET 1926, 1; CARTAN 1927, 4; BURSTIN 1931, 1; 1932, 1. Cf. also WEISE 1934, 1; TOMPKINS 1941, 1. LAPTEV proved 1943, 1; 1945, 1 that every A_n can be imbedded in an E_N; $N \geq \frac{1}{2}(n^2+2n-1)$. CHERN proved 1937, 1 that every projectively connected n-dimensional space can be imbedded in an ordinary N-dimensional projective space with $N \geq \frac{1}{2}(n^2-1)+n$ dimensions. Cf. for the projective case KANITANI 1947, 1; 1948, 1; 1949, 1; 1950, 2 and for the conformal case RYŽKOV 1950, 1; ICHINOHE 1951, 1. A general theory for imbedding in an R_N was given by ROZENFEL'D 1945, 1. Cf. for imbedding in the large for instance ALLENDOERFER 1937, 1; 1939, 1.

[2]) There is much literature on subspaces of special classes and their properties. We mention here only SCHOUTEN 1918, 2; VRANCEANU 1930, 2; WEISE 1934, 1; T. Y. THOMAS 1936, 4 (class 1, here is the definition of "type"); ALLENDOERFER 1937, 1; 2 (class 1, EINSTEIN spaces); 1939, 1 (class >1, rigidity); BURSTIN 1937, 1; EISENHART 1937, 1 (class >1); E II 1938, 2, p. 141; KOCZIAN 1939, 1 (class 1); LAURA 1940, 1 (class 1); ROSENSON 1940, 1 (class 1); 1941, 1 (class 1); 1943, 1 (class 1); PALATINI 1941, 1 (class 1); MASINI VENTURELLI 1943, 1; 2 (class 1); SCHWARTZ 1946, 1 (second curvature); BLUM 1946, 2; 1947, 2; 3; MATSUMOTO

THOMAS[1]) has proved that for $n-m \geq 4$ the equations of CODAZZI are consequences of the equations of GAUSS.[2])

Exercise.

V 8,1. If A_r is groupspace of an r-parameter group and if X_p is geodesic in A_r, prove that

$$B_{\nu\mu\lambda}^{dcb}\, C_a^{\varkappa}\, c_{dc}^{\cdot\cdot e}\, c_{eb}^{\cdot\cdot a} = 0 \tag{V 8,1 α}$$

(cf. IV 1.29 and IV 6.11 and use CODAZZI).

§ 9. The V_n^m and V_m in V_n and X_n.[3])

We consider a real X_n^m in an ordinary V_n. Such an X_n^m is always rigged by the m'-direction perpendicular to the tangent m-direction and these m'-directions form an $X_n^{m'}$ rigged by the X_n^m. The fundamental tensor of the X_n^m is, taking the anholonomic system (h) as in V § 7 and § 8,

$$'g_{ba} = B_{ba}^{\lambda\varkappa}\, g_{\lambda\varkappa} \tag{9.1}$$

and from (7.9, 12) we get

$$D_c\, 'g_{ba} = B_{cba}^{\mu\lambda\varkappa}\, \nabla_\mu B_{\lambda\varkappa}^{\sigma\varrho}\, g_{\sigma\varrho} = 2\, B_{cba}^{\mu\lambda\varkappa}\, 'g_{\sigma(\varkappa} \nabla_{|\mu|} B_{\lambda)}^{\sigma} = 2\, B_{(a}^{\varkappa} \overset{m}{H}{}_{|c|\,b)}^{\cdot\;\cdot\;\sigma}\, 'g_{\sigma\varkappa} = 0 \tag{9.2}$$

$$D_y\, 'g_{ba} = C_y^{\mu}\, B_{ba}^{\lambda\varkappa} \nabla_\mu B_{\lambda\varkappa}^{\sigma\varrho}\, g_{\sigma\varrho} = -\,2\, B_{(a}^{\varkappa} \overset{m'}{L}{}_{|y|\;\cdot\, b)}^{\cdot\;\;\sigma}\, 'g_{\sigma\varkappa} = 0. \tag{9.3}$$

1950, 1; 2; 3 (class 2); 1951, 1 (class 2); 2 (affine, class 1); 1952, 1 (project., class 1); VERBICKIĬ 1950, 1 (class 2); BRAUNER 1951, 1 (class 1, orth. syst.).

[1]) T. Y. THOMAS 1936, 4.

[2]) The GAUSS-CODAZZI-RICCI equations are not always independent. Cf. MAYER 1935, 2; ALLENDOERFER 1939, 1; SCHWARTZ 1941, 1; BLUM 1946, 2; 1947, 2.

[3]) General references on imbedding in V_n, R_n and S_n (cf. VI § 5 for conformal imbedding): CARTAN 1919, 1; 1920, 3; 1939, 1; 2 (R_n and S_n); SCHOUTEN and STRUIK 1921, 3; 1922, 1; 2; R. K. 1924, 1, Ch. V; BOMPIANI 1921, 1; 1924, 1 (geod. V_m in V_n); 1950, 1; 2 (V_2 in R_n); 1951, 1; 2 (higher order curvature); 3 (V_n); KÄMMERER 1922, 1; V. D. WAERDEN 1927, 1; BORTOLOTTI 1927, 3; HAYDEN 1932, 1; MAYER 1933, 1 (V_n); 1935, 2 (R_n); FABRICIUS-BJERRE 1934, 1 (S_n); 1936, 1 (R_n); KAPLAN 1934, 1 (R_6); 2 (R_{n+1}); 1936, 1 (R_6); PEREPELKINE 1934, 1 (R_n); 1935, 1 (V_n); 1936, 1 (R_n); DUSCHEK 1935, 1 (V_m in R_n, GAUSS-BONNET); DAVIES 1935, 1 (V_n); FUCHS 1935, 1 (geod. in V_4); ANDERSON and INGOLD 1936, 1 (R_n); TOMPKINS 1939, 1 (flat isometric in R_n); 1941, 1 (R_4); T. Y. THOMAS 1939, 3 (R_n); COBURN 1939, 1 (S_n); 1940, 1 (S_n); WONG 1940, 1 (V_n); LENSE 1940, 1 (R_n); 1944, 1 (R_n); FINSLER 1940, 1; GHOSH 1940, 1 (R_n); 1941, 1 (R_n); WAGNER 1942, 2 (anhol. in S_n); CHERN 1944, 1 (R_n); WONG 1945, 1; CASTOLDI 1946, 1 (V_n); 1949, 1 (V_n); LIBER 1947, 1 (S_m in S_n); PETRESCU 1948, 1 (anhol. in V_n); HLAVATY 1949, 1; 1952, 1 (W_n); YANENKO 1949, 1; 1953, 1 (R_n); LOPSCHITZ 1950, 1 (E_{m+1}); 2 (centr. E_{m+2}); ABRAMOV 1950, 1; 1951, 1; 2 (all topol. inv. V_n); OTSUKI 1950, 2 and 1951, 1 (conformal).

The difference between $\overset{m}{H}{}_{\mu\lambda}^{\cdot\cdot\varkappa}$ and $\overset{m}{L}{}_{\mu\lambda}^{\cdot\cdot\varkappa}$ vanishes because we can show from (7.9, 10) that

$$(9.4)\quad \begin{cases} \overset{m}{L}{}_{\mu\lambda}^{\cdot\cdot\varkappa} = \overset{m}{L}{}_{\mu\cdot\sigma}^{\cdot\varrho} g^{\sigma\varkappa} g_{\lambda\varrho} = B_{\mu\nu}^{\tau\varrho} (\nabla_\tau B_\sigma^\nu) g^{\sigma\varkappa} g_{\lambda\varrho} = B_\mu^\tau (\nabla_\tau {}'g^{\nu\varkappa}) {}'g_{\lambda\nu} \\ \qquad = B_{\mu\lambda}^{\tau\varrho} \nabla_\tau {}'g^{\nu\varkappa} {}'g_{\varrho\nu} = B_{\mu\lambda}^{\tau\varrho} \nabla_\tau B_\varrho^\varkappa = \overset{m}{H}{}_{\mu\lambda}^{\cdot\cdot\varkappa} \end{cases}$$

and it can be proved in the same way that $\overset{m'}{L}{}_{\mu\lambda}^{\cdot\cdot\varkappa} = \overset{m'}{H}{}_{\mu\lambda}^{\cdot\cdot\varkappa}$.

For $\mathfrak{D}_y$ we get from (7.37) and (9.3)

$$(9.5)\qquad \mathfrak{D}_y {}'g_{ba} = D_y {}'g_{ba} - \overset{m}{L}{}_{b\cdot y}^{\cdot c} {}'g_{ca} - \overset{m}{L}{}_{a\cdot y}^{\cdot c} {}'g_{bc} = -2\overset{m}{H}_{(ba)y}$$

and from (7.35 b) we get another form

$$(9.6)\qquad \mathfrak{D}_y {}'g_{ba} = \partial_y {}'g_{ba} - 4\overset{m}{\Omega}{}_{(b|y|}^{c} {}'g_{a)c}.$$

If (h) is an anholonomic orthogonal system with unit basis vectors we have $\partial_y {}'g_{ba} \overset{*}{=} 0$ and consequently in this special case

$$(9.7)\qquad \mathfrak{D}_y {}'g_{ba} \overset{*}{=} 4\overset{m}{\Omega}{}_{y(b}^{c} {}'g_{a)c}.$$

Similar formulae hold for the $X_n^{m'}$. We call an X_n^m rigged in this way a V_n^m in V_n.[1])

Now the question arises whether a displacement of quantities of X_n^m in X_n can be fixed by giving only ${}'g_{ba}$ and a rigging. The parameters ${}'\Gamma_{cb}^a$ of this displacement can be found from (9.2)

$$(9.8)\qquad 0 = D_c {}'g_{ba} = \partial_c {}'g_{ba} - {}'\Gamma_{cb}^d {}'g_{da} - {}'\Gamma_{ca}^d {}'g_{bd}$$

and (7.20)

$$(9.9)\qquad {}'\Gamma_{[cb]}^a = -{}'\Omega_{cb}^a.$$

We may proceed as in V § 7, putting $\mathfrak{D}_c p^a = D_c p^a$ and $\mathfrak{D}_z p^a = \partial_z p^a + 2\overset{m}{\Omega}{}_{bz}^a p^b$. Then we get for the parameters of this covariant differentiation

$$(9.10)\qquad {}^*\Gamma_{cb}^a = {}'\Gamma_{cb}^a; \qquad {}^*\Gamma_{yb}^a = -2\overset{m}{\Omega}{}_{yb}^a$$

hence

$$(9.11)\qquad \mathfrak{D}_y {}'g_{ba} = \partial_y {}'g_{ba} + 4\overset{m}{\Omega}{}_{y(b}^c {}'g_{a)c}$$

in accordance with (9.6).

A rigged X_n^m in X_n provided in this way with a metric linear connexion is called a V_n^m in X_n. Note that in this case $\overset{m}{H}{}_{cb}^{\cdot\cdot\varkappa}$ does not exist because there is no connexion in X_n but that $\overset{m}{H}_{(ba)y}$ is already known because of (9.5). For repeated differentiation see V § 7.

---

1) Cf. Schouten 1929, 3, p. 164, 167ff.; Bortolotti 1930, 2; Moisil 1940, 1; Vranceanu 1942, 1.

We will now deduce some of the most important properties of a V_m in V_n.[1]) If $i^\varkappa$ is the tangent unitvector of a curve of V_m we have

$$(9.12)\qquad i^\mu\, '\nabla_\mu i^\varkappa = i^\mu (\nabla_\mu i^\varrho) B_\varrho^\varkappa = i^\mu \nabla_\mu i^\varkappa + i^\mu i^\lambda \nabla_\mu C_\lambda^\varkappa = i^\mu \nabla_\mu i^\varkappa - i^\mu i^\lambda H_{\mu\lambda}^{\cdot\cdot\varkappa}$$

writing $H_{\mu\lambda}^{\cdot\cdot\varkappa}$ instead of $\overset{m}{H}{}_{\mu\lambda}^{\cdot\cdot\varkappa}$, or

$$(9.13)\qquad u^\varkappa = 'u^\varkappa + ''u^\varkappa$$

where $u^\varkappa$ is the *absolute curvature vector*, $'u^\varkappa$ the *relative curvature vector* and $''u^\varkappa$ the *enforced curvature vector* (cf. V § 6). If $''u^\varkappa = 0$, the direction of $i^\varkappa$ is called a *principal tangent* and a curve whose tangent is principal at all points is said to be an *asymptotic line*.[2]) Now let $\underset{b}{i}{}^a$ be the unit basis vectors of an orthogonal anholonomic system (a) of V_m. Then we have

$$(9.14)\qquad i^\varkappa = \sum_b \underset{b}{i}{}^\varkappa \cos \underset{b}{\alpha}$$

where the $\underset{b}{\alpha}$ are the angles between $i^\varkappa$ and the basis vectors. Hence (cf. 6.13)

$$(9.15)\qquad ''u^\varkappa = \sum_{cb} \underset{c}{i}{}^\mu \underset{b}{i}{}^\lambda \cos \underset{c}{\alpha} \cos \underset{b}{\alpha}\, H_{\mu\lambda}^{\cdot\cdot\varkappa}$$

and this proves that, at every point, the vector $''u^\varkappa$ lies in an R_{m_1}; $m_1 \leq n-m$, $m_1 \leq \frac{1}{2}m(m+1)$ in the tangent R_n, and perpendicular to the tangent R_m. The R_m and R_{m_1} determine an R_{m+m_1} called the *first curvature region* of the V_m at the point considered.[3]) If $m_1 = \frac{1}{2}m(m+1)$ the vector $''u^\varkappa$ lies on a cone in the local R_{m_1} and its endpoint describes a V_{m-1} of degree 2^{m-1} which lies in an R_{m_1-1}. This V_{m-1} is the *curvature figure* (Krümmungsgebilde) and its centre of gravity has the radius-vector in tangent space

$$(9.16\text{a})\qquad M^\varkappa = \frac{1}{m}\, 'g^{cb} H_{cb}^{\cdot\cdot\varkappa}.$$

In general the V_{m-1} does not contain the origin, hence principal tangents only occur in special cases. There is a one to one correspondence between the ∞^{m-1} directions in the V_m at the point considered and the points of the curvature figure. For $m_1 < \frac{1}{2}m(m+1)$ there are several subcases.[4])

---

[1]) Cf. for a more detailed treatment EISENHART 1926, 1, Ch. IV, V; DUSCHEK and MAYER 1930, 1, Ch. VII; MAYER 1935, 2; E II 1938, 2, Ch. III, IV.

[2]) SCHOUTEN and STRUIK 1922, 1; 2; HAYDEN 1930, 2; E II 1938, 2, p. 89.

[3]) An osculating space of another kind was introduced by BERZOLARI 1897, 1 making use of an idea of KILLING 1885, 1. Cf. E II 1938, 2, p. 90ff.; BOMPIANI 1950, 3.

[4]) Cf. for an elaborate treatment that deals also with other geometric figures connected with the curvature and that contains many examples E II 1938, 2, p. 94ff.

If $m_1 = 1$, 2 or 3 at some point of the V_m this point is called *axial, planar* and *spatial* respectively.[1])

If the enforced curvature vector is the same for every direction of the V_m at a point, this point is said to be *umbilical.*[2]) N.a.s. conditions are

$$(9.16\text{b}) \qquad 0 = M_{cb}^{\cdot\cdot\varkappa} \overset{\text{def}}{=} H_{cb}^{\cdot\cdot\varkappa} - \frac{1}{m}\,'g_{cb}\,'g^{da}\,H_{da}^{\cdot\cdot\varkappa}.$$

The V_m is called *umbilical* [cf. V § 6][3]) if all its points are umbilical. Every V_1 is umbilical because $H_{cb}^{\cdot\cdot\varkappa} = \underset{1}{\varkappa}\, j_c\, j_b\, \underset{2}{j}^{\varkappa}$. We prove that the section of an umbilical V_p and an umbilical V_q is always umbilical. This section is a V_s; $s \geq p+q-n$. At a point of this section, to every direction of V_p there belongs a curvature vector and the projection of this vector on the normal R_{n-p} is the enforced curvature vector that is the same for all these directions. The same holds for the directions of V_q. Now let us consider the directions of V_s and the projections of the curvature vectors belonging to them on the normal R_{n-s}. If such a projection is projected on the normal R_{n-p} and on the normal R_{n-q} (both lying in R_{n-s}) we get the enforced curvature vectors belonging to the direction of V_s, considered as direction of V_p and of V_q respectively. Hence the enforced curvature vector of a direction of V_s is a vector of R_{n-s} from which we know that its projection on R_{n-p} and on R_{n-q} is the same for all directions of V_s. R_{n-p} and R_{n-q} have an $R_{n-p-q+s}$ in common and $n-p-q+s \geq 0$. Hence the enforced curvature vector of a direction of V_s is wholly determined by these two projections and this proves that it is the same for all directions of V_s.

$M^{\varkappa}$ is the *mean curvature vector* of the V_m. For $m = 1$ it is identical with the curvature vector of the curve. A V_m in V_n is called *minimal* if for every part of it bounded by a closed V_{m-1}, the m-dimensional volume has an extreme value for variations that do not change the boundary. Now let (a) be a *holonomic* coordinate system of V_m with the coordinates η^a. Then $d\eta^1 \ldots d\eta^{\text{m}}$ is the component of a contravariant m-vector represented by an infinitesimal volume element. This component is a scalar density of V_m of weight -1. If (a') is an anholonomic orthogonal coordinate system in V_m with unit basis vectors, the volume of such an element is $|(d\eta)^{1'} \ldots (d\eta)^{\text{m}'}|$. Now $'\mathfrak{g} \overset{\text{def}}{=} |\text{Det}\,('g_{ba})|$ is a scalar density of V_m of weight $+2$ and its value with respect to (a') is $+1$. Hence the volume is

$$(9.17) \qquad d\tau_m \overset{\text{def}}{=} |(d\eta)^{1'} \ldots (d\eta)^{\text{m}'}| = |d\eta^1 \ldots d\eta^{\text{m}}\,'\mathfrak{g}^{\frac{1}{2}}|.$$

---

[1]) Cf. for V_m's in S_n with planar points COBURN 1939, 1.

[2]) Cf. for V_n's admitting umbilical V_m's ADATI 1951, 1; 2.

[3]) Cf. E II 1938, 2, Ch. III; PEREPELKINE 1936, 1; WONG 1943, 1; 3; 1946, 1.

If the part of the V_m considered undergoes the finite point transformation $\overset{tv}{T}$ belonging to the one-parameter group generated by the infinitesimal point transformation $\xi^\varkappa \to \xi^\varkappa + v^\varkappa dt$ (cf. II § 10) the change of the volume can be computed most easily by performing afterwards the inverse transformation $\overset{-tv}{T}$ and dragging along the coordinate systems ($\varkappa$) and (a) and the field $g_{\lambda\varkappa}$ with this latter transformation. Then the V_m returns to its original position and the changed volume we acquire is the volume in this original position with respect to the field $\overset{m}{g}_{\lambda\varkappa}$ dragged along (cf. II § 10). This is the *method of transposition* that will be used often hereafter. Now we have (cf. III 5.48)

$$\text{(9.18)} \qquad \underset{v}{\pounds}\, g_{\lambda\varkappa} = v^\mu \nabla_\mu g_{\lambda\varkappa} + g_{\mu\varkappa} \nabla_\lambda v^\mu + g_{\lambda\mu} \nabla_\varkappa v^\mu = 2\nabla_{(\lambda} v_{\varkappa)}$$

$$\text{(9.19)} \qquad \underset{v}{\pounds}^2 g_{\lambda\varkappa} = 2 v^\mu \nabla_\mu \nabla_{(\lambda} v_{\varkappa)} + 2(\nabla_{(\mu} v_{\varkappa)}) \nabla_\lambda v^\mu + 2(\nabla_{(\lambda} v_{\mu)}) \nabla_\varkappa v^\mu$$

and hence, according to (II 10.38)

$$\text{(9.20)} \qquad \begin{cases} \overset{m}{g}_{\lambda\varkappa} = g_{\lambda\varkappa} - 2t\nabla_{(\lambda} v_{\varkappa)} + t^2 v^\mu \nabla_\mu \nabla_{(\lambda} v_{\varkappa)} + \\ \qquad + t^2 (\nabla_{(\mu} v_{\varkappa)}) \nabla_\lambda v^\mu + t^2 (\nabla_{(\lambda} v_{\mu)}) \nabla_\varkappa v^\mu - \cdots. \end{cases}$$

Because both ($\varkappa$) and (a) are dragged along, $B_b^\varkappa$ is not changed and consequently

$$\text{(9.21)} \qquad \begin{cases} '\overset{m}{g}_{ba} = 'g_{ba} - 2t\, B_{b\,a}^{\lambda\varkappa} \nabla_{(\lambda} v_{\varkappa)} + t^2 B_{b\,a}^{\lambda\varkappa} v^\mu \nabla_\mu \nabla_{(\lambda} v_{\varkappa)} + \\ \qquad + t^2 B_{b\,a}^{\lambda\varkappa} (\nabla_{(\mu} v_{\varkappa)}) \nabla_\lambda v^\mu + t^2 B_{b\,a}^{\lambda\varkappa} (\nabla_{(\lambda} v_{\mu)}) \nabla_\varkappa v^\mu - \cdots. \end{cases}$$

Hence for that variation of the field $'g_{ba}$ at the point considered, which is a consequence of the dragging along, we get

$$\text{(9.22)} \qquad \frac{d}{dt}\, 'g_{ba} = -2 B_{b\,a}^{\lambda\varkappa} \nabla_{(\lambda} v_{\varkappa)} \qquad \text{for} \qquad t = 0$$

and accordingly (cf. Exerc. I, 1,1)

$$\text{(9.23)} \qquad \begin{cases} \dfrac{d}{dt}\, '\mathfrak{g}^{\frac{1}{2}} = \dfrac{1}{2}\, '\mathfrak{g}^{\frac{1}{2}} \dfrac{d}{dt} \log '\mathfrak{g} = \dfrac{1}{2}\, '\mathfrak{g}^{\frac{1}{2}}\, 'g^{ba} \dfrac{d}{dt}\, 'g_{ba} \\ \qquad = -\, '\mathfrak{g}^{\frac{1}{2}}\, 'g^{ba} B_{b\,a}^{\lambda\varkappa} \nabla_\lambda v_\varkappa = -\, '\mathfrak{g}^{\frac{1}{2}} B_\varkappa^\lambda \nabla_\lambda v^\varkappa. \end{cases}$$

Now $v^\varkappa$ can always be split up into $'v^\varkappa$ perpendicular to V_m and $''v^\varkappa$ in V_m. Then we get

$$\text{(9.24)} \qquad \begin{cases} \dfrac{d}{dt}\, d\tau_m = -\, |d\eta^1 \ldots d\eta^m|\, '\mathfrak{g}^{\frac{1}{2}} ('\nabla_b\, ''v^b - 'g^{cb}\, 'v_\varkappa D_c B_b^\varkappa) \\ \qquad = -\, d\tau_m ('\nabla_b\, ''v^b - m\, 'v_\mu M^\mu) \end{cases}$$

and accordingly

$$\frac{d}{dt}\int_{\tau_m} d\tau_m = -\int_{\tau_m} {}'\nabla_b {}''v^b\, d\tau_m + m\int_{\tau_m} {}'v_\mu M^\mu\, d\tau_m . \tag{9.25}$$

The first integral at the right hand side vanishes if ${}''v^\varkappa = 0$ on the boundary (cf. II § 8). Hence the variation of the volume is zero for all infinitesimal point transformations $v^\varkappa dt$ for which the V_m-part of $v^\varkappa$ vanishes at the boundary if and only if the mean curvature vector is zero at all points of the region considered.[1]) For $m = 1$ this proves that the variation of the length of a part of a curve in V_n vanishes for all infinitesimal transformations for which the component in the tangent vanishes at the endpoints if and only if the curve is a geodesic. We can express this in a less rigorous way: geodesics are not only the straightest but also the shortest (longest) curves. Whether the volume really has an extreme value or not, can only be investigated by means of the higher terms of the series (9.20).[2])

Exercises.

V 9,1[3]). In a real V_m in an ordinary V_n there are two congruences with unit tangent vectors $i^\varkappa$ and $j^\varkappa$. Prove that

$$i^\mu \nabla_\mu j^\varkappa = i^\mu {}'\nabla_\mu j^\varkappa + i^c j^b H_{cb}^{\cdot\cdot\varkappa}$$

(cf. (9.12) and Exerc. V 6,2.)

V 9,2[4]). The vectorial mean-value of the enforced curvature vectors with respect to m mutually perpendicular directions of a real V_m in an ordinary V_n is independent of the choice of this set of directions and equals the mean curvature vector.

V 9,3[5]). A set of ∞^1 real V_{n-1}'s in an ordinary V_n contains only minimal V_{n-1}'s if and only if the orthogonal trajectories make correspond the points of the V_{n-1}'s in such a way that the $(n-1)$-dimensional volume is invariant.

V 9,4 (cf. Exerc. VI 5,2)[6]). According to HAANTJES and WRONA the *scalar curvature of a non singular m-direction* at a point of a V_n is the scalar curvature of a geodesic V_m with this m-direction at this

---

1) LIPSCHITZ 1874, 2; BOMPIANI 1921, 1, p. 1141; STRUIK 1922, 1, p. 98; SCHOUTEN and v. KAMPEN 1933, 1, p. 19 also for V_n^m; E II 1938, 2, p. 165 ff.

2) Minimal surfaces in an R_3 with an indefinite fundamental tensor are dealt with by many authors. We mention here only the elder investigators LIE 1879, 1; BIANCHI 1888, 1; CAYLEY 1890, 1 and WOODS 1895, 1.

3) E II 1938, 2, p. 85.

4) E II 1938, 2, p. 87.

5) BIANCHI 1902, 2, II, p. 577 for V_3 in R_4; BOMPIANI 1921, 1, p. 1141 for V_{n-1} in V_n.

6) HAANTJES and WRONA 1939, 1.

point. If now $n=2m$, the V_{2m} is an EINSTEIN space if and only if at every point two mutually perpendicular m-directions have always the same curvature.

V 9,5[1]). Prove that a geodesic V_m in an $S_n(R_n)$ is an $S_m(R_m)$.

§ 10. Higher curvatures of a V_m in V_n.

If $i^\varkappa, j^\varkappa$ and $k^\varkappa$ are the tangent unitvectors of three real congruences in a real V_m in an ordinary V_n we have (cf. Exerc. V 9,1)

$$(10.1)\quad \begin{cases} j^\mu \nabla_\mu i^\varkappa = j^\mu\, '\nabla_\mu i^\varkappa + j^\mu (\nabla_\mu i^\varrho)\, C_\varrho^\varkappa = j^\mu\, '\nabla_\mu i^\varkappa + j^\mu i^\lambda H_{\mu\lambda}^{\cdot\cdot\,\varkappa} \\ \quad = j^\mu\, '\nabla_\mu i^\varkappa + j^c i^b D_c B_b^\varkappa \end{cases}$$

$$(10.2)\quad \begin{cases} k^\nu \nabla_\nu j^\mu \nabla_\mu i^\varkappa = k^\nu\, '\nabla_\nu j^\mu\, '\nabla_\mu i^\varkappa + k^\nu (\nabla_\nu j^\mu\, '\nabla_\mu i^\varrho)\, C_\varrho^\varkappa + k^\nu \nabla_\nu j^\mu i^\lambda H_{\mu\lambda}^{\cdot\cdot\,\varkappa} \\ \quad = k^\nu\, '\nabla_\nu j^\mu\, '\nabla_\mu i^\varkappa + (k^\mu j^\nu\, '\nabla_\nu i^\lambda + i^\lambda k^\nu\, '\nabla_\nu j^\mu + j^\mu k^\nu\, '\nabla_\nu i^\lambda)\, H_{\mu\lambda}^{\cdot\cdot\,\varkappa} + \\ \quad + k^d j^c i^b D_d D_c B_b^\varkappa \end{cases}$$

and in the same way the p-th covariant derivative of $i^\varkappa$ with respect to p directions of V_m can be proved to be a vector lying in the $\varkappa$-region of the set $B_b^\varkappa, D_{b_1} B_b^\varkappa, \ldots, D_{b_p} \ldots D_{b_1} B_b^\varkappa$. This region is called the p-th *curvature region* of V_m at the point considered.[2]) Calling its dimension $m+m_1+\cdots+m_p$ for all values of p for which $m_p \neq 0$ we have the inequalities

$$(10.3)\qquad m_p \leq \tfrac{1}{2} m^p (m+1); \qquad \Sigma\, m_p \leq n-m.\,^{3)}$$

The V_m is called *asymptotic of order p* if the p-th (and not the $(p+1)$-th) curvature region is contained in the tangent R_m. Then every curve of V_m is an *asymptotic line of order p*, i.e. its osculating R_{p+1} is contained in the tangent R_m.[4])

Since the V_m is real, none of the curvature regions has an extraordinary position with respect to the nullcone. There is a number k such that for each value of $l \geq 1$ the $(k+l)$-th curvature region is contained in the k-th curvature region. Then these regions determine uniquely an $R_{m_0}=R_m, R_{m_1}, \ldots, R_{m_k}$ perpendicular to each other. For the space containing all directions perpendicular to $R_{m_0}, \ldots, R_{m_k}$ we write $R_{m_{k+1}}$; $m_{k+1}=n-m-m_1-\cdots-m_k$. The x-th curvature region $x \leq k$ is spanned by $R_{m_0}, \ldots, R_{m_x}$. If the V_n is an R_n it follows immediately from the foregoing that the V_m is imbedded in a flat $R_{n-m_{k+1}}$ in R_n.

[1]) E II 1938, 2, p. 133. Cf. p. 150 for generalizations.

[2]) Higher curvatures of curves in V_m in V_n are considered by DEL PEZZO 1886, 1; HAYDEN 1930, 2; 1934, 1; PEREPELKINE 1936, 1; ANDERSON and INGOLD 1936, 1 (R_n).

[3]) An improvement was given by SCHWARTZ 1946, 1; cf. ALLENDOERFER 1939, 1.

[4]) Cf. p. 250 footnote 3).

Now we choose an anholonomic coordinate system with unit basis vectors such that for each value of x; $x=0,\ldots,k+1$ the vectors $\overset{\varkappa}{\underset{p_x}{i}}$; p_x, $q_x,\ldots=\overset{x}{1},\ldots,\overset{x}{m}_x$ span R_{m_x}. Then if a quantity lies with one index in R_{m_x} this index can be greek or $p_x, q_x, \ldots$. If we write $\overset{x}{g}_{q_x p_x}$ and $\overset{x}{B}{}^{p_x}_{q_x}$ for the fundamental tensor and the unity tensor in R_{m_x} we have

$$(10.4)\qquad {}'g_{q_0 p_0}=\overset{0}{g}_{q_0 p_0};\qquad B^{p_0}_{q_0}=\overset{0}{B}{}^{p_0}_{q_0};\qquad \overset{x}{g}_{q_x p_x}=\overset{x}{B}{}^{\lambda\ \varkappa}_{q_x p_x}\, g_{\lambda\varkappa}$$

$$(10.5)\qquad \overset{x}{B}{}^{\varkappa}_{q_x}\overset{y}{B}{}^{p_y}_{\varkappa}=\begin{cases}\overset{x}{B}{}^{p_x}_{q_x} & \text{for}\quad x=y\\ 0 & \text{for}\quad x\neq y\end{cases}\quad;\qquad x,y=0,\ldots,k+1$$

and the D-symbolism can be extended with the operators D_{r_x}:

$$(10.6)\qquad D_{r_x}p\overset{\text{def}}{=}\overset{x}{B}{}^{\mu}_{r_x}\nabla_\mu p$$

$$(10.7)\qquad D_{r_x}v^{\varkappa}\overset{\text{def}}{=}\overset{x}{B}{}^{\mu}_{r_x}\nabla_\mu v^{\varkappa};\qquad D_{r_x}w_\lambda\overset{\text{def}}{=}\overset{x}{B}{}^{\mu}_{r_x}\nabla_\mu w_\lambda$$

$$(10.8)\qquad \begin{cases}\text{a)}\ D_\mu r^{p_x}\overset{\text{def}}{=}\overset{x}{B}{}^{p_x}_{\varkappa}\nabla_\mu r^{\varkappa}; & D_\mu s_{q_x}\overset{\text{def}}{=}\overset{x}{B}{}^{\lambda}_{q_x}\nabla_\mu s_\lambda\\ \text{b)}\ D_{r_x}r^{p_y}\overset{\text{def}}{=}\overset{x}{B}{}^{\mu}_{r_x}\overset{y}{B}{}^{p_y}_{\varkappa}\nabla_\mu r^{\varkappa}; & D_{r_x}s_{q_y}\overset{\text{def}}{=}\overset{x}{B}{}^{\mu}_{r_x}\overset{y}{B}{}^{\lambda}_{q_y}\nabla_\mu s_\lambda.\end{cases}$$

The curvature tensor $H_{\mu\lambda}^{\cdot\cdot\varkappa}$ appears in the form

$$(10.9)\qquad \overset{1}{H}{}_{r_0 q_0}^{\cdot\cdot\varkappa}=D_{r_0}\overset{0}{B}{}^{\varkappa}_{q_0};\qquad \overset{1}{H}{}_{r_0\cdot\lambda}^{\cdot p_0}=D_{r_0}\overset{0}{B}{}^{p_0}_{\lambda}$$

and is called now the *first curvature tensor of valence 3*. Its $\varkappa$-region coincides with R_{m_1}. The *second curvature tensor of valence 3* is defined by

$$(10.10)\qquad \overset{2}{H}{}_{r_0 q_1}^{\cdot\cdot\varkappa}\overset{\text{def}}{=}D_{r_0}\overset{1}{B}{}^{\varkappa}_{q_1}+\overset{1}{H}{}_{r_0\cdot q_1}^{\cdot\varkappa}.$$

Because

$$(10.11)\qquad \overset{0}{B}{}^{p_0}_{\varkappa}\overset{2}{H}{}_{r_0 q_1}^{\cdot\cdot\varkappa}=-\overset{1}{H}{}_{r_0\cdot\varkappa}^{\cdot p_0}\overset{1}{B}{}^{\varkappa}_{q_1}+\overset{0}{B}{}^{p_0}_{\varkappa}\overset{1}{H}{}_{r_0\cdot q_1}^{\cdot\varkappa}=0$$

and

$$(10.12)\quad \overset{1}{B}{}^{p_1}_{\varkappa}\overset{2}{H}{}_{r_0 q_1}^{\cdot\cdot\varkappa}=\overset{1}{B}{}^{p_1}_{\varkappa}D_{r_0}\overset{1}{B}{}^{\varkappa}_{q_1}=\overset{1}{B}{}^{p_1\lambda}_{\varkappa\ q_1}D_{r_0}\overset{1}{B}{}^{\varkappa}_{\lambda}=\overset{1}{B}{}^{p_1}_{\varkappa}D_{r_0}\overset{1}{B}{}^{\varkappa}_{q_1}-\overset{1}{B}{}^{p_1}_{\lambda}D_{r_0}\overset{1}{B}{}^{\lambda}_{q_1}=0;$$

the $\varkappa$-region of $\overset{2}{H}{}_{r_0 q_1}^{\cdot\cdot\varkappa}$ is m_2-dimensional and perpendicular to R_{m_0} and R_{m_1} and therefore coincides with R_{m_2}. It can be proved in the same way that the $\varkappa$-region of

$$(10.13)\qquad \overset{x+1}{H}{}_{r_0 q_x}^{\cdot\cdot\varkappa}\overset{\text{def}}{=}D_{r_0}\overset{x}{B}{}^{\varkappa}_{q_x}+\overset{x}{H}{}_{r_0\cdot q_x}^{\cdot\varkappa};\qquad x=1,\ldots,k;$$

coincides with $R_{m_{x+1}}$ and that this quantity, the $(x+1)$-th *curvature tensor of valence 3* of the V_m vanishes for $x\geq k$. The formulae

$$(10.14\text{a})\qquad \begin{cases}D_{r_0}\overset{x}{B}{}^{\varkappa}_{q_x}=-\overset{x}{H}{}_{r_0\cdot q_x}^{\cdot\varkappa}+\overset{x+1}{H}{}_{r_0 q_x}^{\cdot\cdot\varkappa}\\ \overset{0}{H}{}_{r_0\cdot q_0}^{\cdot\varkappa}=0;\quad \overset{k+1}{H}{}_{r_0 q_k}^{\cdot\cdot\varkappa}=0;\quad \overset{k+2}{H}{}_{r_0 q_{k+1}}^{\cdot\cdot\varkappa}=0;\qquad x=0,\ldots,k;\end{cases}$$

are a generalization for V_m in V_n [1]) of the FRENET formulae (cf. V § 1) and become these formulae for $m=1$. They can also be written in the form

$$(10.14\text{b})\quad \begin{cases} D_{r_0} \overset{x}{B}{}^{p_x}_{\lambda} = -\overset{x}{H}{}^{\cdot}_{r_0 \lambda}{}^{\cdot p_x} + \overset{x+1}{H}{}^{\cdot p_x}_{r_0 \cdot \lambda} \\ \overset{0}{H}{}^{\cdot}_{r_0 \lambda}{}^{\cdot p_0} = 0; \quad \overset{k+1}{H}{}^{\cdot p_k}_{r_0 \cdot \lambda} = 0; \quad \overset{k+2}{H}{}^{\cdot p_{k+1}}_{r_0 \cdot \lambda} = 0; \quad x = 0, 1, \ldots, k. \end{cases}$$

Note that $\overset{x}{H}{}^{\cdot}_{\mu \lambda}{}^{\cdot \varkappa}$ lies with the second index in $R_{m_{x-1}}$ and with the third index in R_{m_x}.

The equations (10.14a or b) can not be used to find out whether a given V_m can be imbedded in an R_n or S_n because these equations contain connecting quantities. The method that will be followed here, using equations that only contain quantities of the V_m is due to BURSTIN and MAYER and is developed here in the invariant form given by SCHOUTEN and v. KAMPEN. [2])

The following quantities are formed from the $\overset{x}{H}{}^{\cdot}_{r_0 \lambda}{}^{\cdot \varkappa}$

$$(10.15)\quad \begin{cases} \text{a)}\quad \overset{1}{J}{}^{\cdot\cdot\varkappa}_{cb} \overset{\text{def}}{=} \overset{1}{H}{}^{\cdot\cdot\varkappa}_{cb} \\ \text{b)}\quad \overset{x}{J}{}^{\cdot}_{b_{x+1} \ldots b_1}{}^{\cdot \varkappa} \overset{\text{def}}{=} \overset{x}{H}{}^{\cdot}_{b_{x+1} \lambda}{}^{\cdot \varkappa} \overset{x-1}{J}{}^{\cdot}_{b_x \ldots b_1}{}^{\cdot \lambda}; \qquad x = 2, \ldots, k \end{cases}$$

where the indices $b, c, b_1, \ldots, b_{x+1}$ belong to a coordinate system (a) in V_m. Then

$$(10.16)\quad \overset{x}{L}_{b_{x+1} \ldots b_1 d_{x+1} \ldots d_1} \overset{\text{def}}{=} \overset{x}{J}{}^{\cdot}_{b_{x+1} \ldots b_1}{}^{\cdot \varkappa} \overset{x}{J}{}^{\cdot}_{d_{x+1} \ldots d_1}{}^{\cdot \lambda} g_{\varkappa \lambda}; \qquad x = 1, \ldots, k$$

are k quantities of V_m. It can be proved that a V_m in S_n is determined by the quantities $\overset{x}{L}$ to within rotations and reflexions. But the quantities $\overset{x}{L}$ can not be given at random [3]), because they must satisfy certain conditions. A method for building a possible set of quantities $\overset{x}{L}$ was given by SCHOUTEN and STRUIK. [4])

Of course the theory of imbedding involving only the first curvature, developed in V § 8 for the rigged A_m in A_n can also be used for V_m in V_n. If for this purpose we take quite another anholonomic system (h), namely the system (a) orthogonal and anholonomic with unit basis vectors $\underset{b}{i}{}^{\varkappa}$; $a, b = 1, \ldots$, m, in V_m and for $\underset{y}{i}{}^{\varkappa}$; $x, y = \text{m} + 1, \ldots, \text{n}$

[1]) SCHOUTEN and v. KAMPEN 1931, 2, p. 151; E II 1938, 2, p. 120; cf. also WONG 1940, 2; IWAMOTO 1944, 1; BOMPIANI 1951, 2.

[2]) BURSTIN and MAYER 1926, 1; MAYER 1928, 1; 1931, 1; 1935, 2; BURSTIN 1929, 1; DUSCHEK and MAYER 1930, 1, II, p. 201—232; SCHOUTEN and v. KAMPEN 1931, 2, p. 153.

[3]) MAYER 1928, 1; 1935, 2.

[4]) E II 1938, 2, p. 125ff., also for literature.

unitvectors mutually perpendicular and perpendicular to V_m, the difference between $\overset{m}{H}, \overset{x}{h}$ and $\overset{m}{L}, \underset{x}{l}$ as defined in V §§ 7, 8 vanishes and we get immediately from (8.9, 13)

$$\text{(10.17)}\quad \text{a)}\ \overset{x}{h}_{cb} \overset{\text{def}}{=} -B^{\mu\lambda}_{cb} \nabla_\mu \overset{x}{i}_\lambda;\quad \text{b)}\ \underset{y}{\overset{x}{v}}_b \overset{\text{def}}{=} B^\mu_b (\nabla_\mu \overset{x}{i}_\lambda) \underset{y}{i}^\lambda;\quad x, y = m+1, \ldots, n$$

and from (8.11, 15, 16)

$$\text{(10.18)}\qquad \boxed{'K_{dcba} = B^{\nu\mu\lambda\varkappa}_{dcba} K_{\nu\mu\lambda\varkappa} - 2\sum_x \overset{x}{h}_{[d[b} \overset{x}{h}_{c]a]}} \qquad \text{(GAUSS)}$$

$$\text{(10.19)}\qquad \boxed{2'\nabla_{[d} \underset{y}{\overset{x}{v}}_{c]} = -B^{\nu\mu}_{dc} K_{\nu\mu\lambda}{}^{\varkappa} \underset{y}{i}^\lambda \overset{x}{i}_\varkappa - 2\overset{x}{h}_{[d|b|} \overset{y}{h}_{c]}{}^{b} + 2\underset{z}{\overset{x}{v}}_{[d} \underset{y}{\overset{z}{v}}_{c]}} \qquad \text{(RICCI)}^{1)}$$

$$\text{(10.20)}\qquad \boxed{2'\nabla_{[d} \overset{x}{h}_{c]b} = B^{\nu\mu\lambda}_{dcb} K_{\nu\mu\lambda}{}^{\varkappa} \overset{x}{i}_\varkappa + 2\underset{y}{\overset{x}{v}}_{[d} \overset{y}{h}_{c]b}} \qquad \text{(CODAZZI)}.$$

For R_n the terms with $K_{\nu\mu\lambda}{}^{\varkappa}$ drop out and for S_n the quantity $K_{\nu\mu\lambda\varkappa}$ has to be replaced by $-2\varkappa g_{[\nu[\lambda} g_{\mu]\varkappa]}$. Using the method of V § 8 it can be proved that a V_m in R_n or S_n is determined to within translations, rotations and reflexions by $n-m$ symmetric tensor fields $\overset{x}{h}_{cb}$ and $\binom{n-m}{2}$ vector fields $\underset{y}{\overset{x}{v}}_b$ provided that these quantities satisfy the equations (10.18 to 20).[2)]

Another form of the GAUSS-CODAZZI-RICCI equations, involving also higher curvatures can be derived in the following way.[3)] Let the unit basis vectors $\underset{q_x}{i}^{\varkappa}$; $x = 0, \ldots, k+1$ once more be chosen in R_{m_x}. Then we have for any vector of R_{m_x}

$$\text{(10.21)}\quad \begin{cases} D_{[s_0} D_{r_0]} w_{q_x} = D_{[s_0} D_{r_0]} \underset{p_x}{w} \overset{p_x}{i}_{q_x} \\ \qquad = w_{p_x} \left(-\partial_{[s_0} \Gamma^{p_x}_{r_0]q_x} + \Gamma^{t_0}_{[s_0 r_0]} \Gamma^{p_x}_{t_0 q_x} + \Gamma^{t_x}_{[s_0|q_x|} \Gamma^{p_x}_{r_0]t_x}\right) \\ \qquad = -\tfrac{1}{2} \overset{x}{K}_{s_0 r_0 q_x}{}^{p_x} w_{p_x}; \qquad x = 0, \ldots, k+1; \end{cases}$$

where (cf. III 9.3, 20)

$$\text{(10.22)}\quad \begin{cases} \overset{x}{K}_{s_0 r_0 q_x}{}^{p_x} \overset{\text{def}}{=} 2\left(\partial_{[s_0} \Gamma^{p_x}_{r_0]q_x} - \Gamma^{t_x}_{[s_0|q_x|} \Gamma^{p_x}_{r_0]t_x} + \Omega^{t_0}_{s_0 r_0} \Gamma^{p_x}_{t_0 q_x}\right); \\ \qquad x = 0, \ldots, k+1. \end{cases}$$

[1)] RICCI 1888, 1; KÜHNE gave 1903, 1 the whole set (10.18–20) for V_m in V_n. SCHOUTEN and v. KAMPEN gave 1930, 2 the corresponding equations for V^m_n in V_n. An analogous set for L^m_n in L_n was given by DIENES 1932, 1. Cf. also JÄRNEFELT 1928, 1; MAYER 1928, 1; 1935, 2; E II 1938, 2, p. 131.

[2)] E II 1938, 2, p. 121 ff. Cf. EISENHART 1926, 1, p. 192; LENSE 1940, 1; KAPLAN 1941, 1; YANO and MUTO 1942, 1 and YANO 1942, 2 (conformal).

[3)] SCHOUTEN and v. KAMPEN 1931, 2, § 3; E II 1938, 2, p. 121 ff.

Of course $\overset{0}{K}{}_{s_0 r_0 q_0}^{\cdot\cdot\cdot p_0}$ is identical with the curvature tensor of the V_m, viz. $'K_{s_0 r_0 q_0}^{\cdot\cdot\cdot p_0}$.

Differentiation of the FRENET formulae (10.14a) leads to

$$(10.23) \qquad D_{[s_0} D_{r_0]} \overset{x}{B}{}_{q_x}^{\varkappa} = - D_{[s_0} \overset{x}{H}{}_{r_0]\cdot q_x}^{\cdot\cdot\varkappa} + D_{[s_0} \overset{x+1}{H}{}_{r_0]q_x}^{\cdot\cdot\varkappa}; \qquad x = 0, 1, \ldots, k+1$$

or in another form

$$(10.24) \qquad \overset{0}{B}{}^{\nu\mu}_{s_0 r_0} \overset{x}{B}{}^{\lambda}_{q_x} K_{\nu\mu\lambda}^{\cdot\cdot\cdot\varkappa} - \overset{x}{B}{}^{\varkappa}_{p_x} \overset{x}{K}{}_{s_0 r_0 q_x}^{\cdot\cdot\cdot p_x} = -2 D_{[s_0} \overset{x}{H}{}_{r_0]\cdot q_x}^{\cdot\cdot\varkappa} + 2 D_{[s_0}\overset{x+1}{H}{}_{r_0]q_x}^{\cdot\cdot\varkappa}.$$

According to (10.14) the right hand side of this equation lies with the index $\varkappa$ in the space spanned by $R_{m_{x-2}}$, $R_{m_{x-1}}$, R_{m_x}, $R_{m_{x+1}}$ and $R_{m_{x+2}}$. Hence it can be split up into the following four equations[1])

$$(10.25) \begin{cases} \text{a)} & \overset{0}{B}{}^{\nu\mu}_{s_0 r_0} \overset{x-g}{B}{}^{p_{x-g}}_{\varkappa} \overset{x}{B}{}^{\lambda}_{q_x} K_{\nu\mu\lambda}^{\cdot\cdot\cdot\varkappa} = 0; \quad x = g,\ldots,k+1; \quad g \geq 3 \\ \text{b)} & \overset{0}{B}{}^{\nu\mu}_{s_0 r_0} \overset{x-2}{B}{}^{p_{x-2}}_{\varkappa} \overset{x}{B}{}^{\lambda}_{q_x} K_{\nu\mu\lambda}^{\cdot\cdot\cdot\varkappa} = 2\overset{x-1}{H}{}_{[s_0\cdot\;|\varkappa|}^{\cdot p_{x-2}} \overset{x}{H}{}_{r_0]\cdot q_x}^{\cdot\cdot\varkappa}; \quad x=2,\ldots,k+1 \\ \text{c)} & \overset{0}{B}{}^{\nu\mu}_{s_0 r_0} \overset{x-1}{B}{}^{p_{x-1}}_{\varkappa} \overset{x}{B}{}^{\lambda}_{q_x} K_{\nu\mu\lambda}^{\cdot\cdot\cdot\varkappa} = -2 D_{[s_0} \overset{x}{H}{}_{r_0]\cdot q_x}^{\cdot p_{x-1}}; \quad x=1,\ldots,k+1 \\ \text{d)} & \overset{0}{B}{}^{\nu\mu}_{s_0 r_0}\overset{x}{B}{}^{p_x}_{\varkappa}\overset{x}{B}{}^{\lambda}_{q_x}K_{\nu\mu\lambda}^{\cdot\cdot\cdot\varkappa} = \overset{x}{K}{}_{s_0 r_0 q_x}^{\cdot\cdot\cdot p_x} - 2\overset{x}{H}{}_{[s_0\;\lambda}^{\cdot\;\cdot\; p_x}\overset{x}{H}{}_{r_0]\cdot q_x}^{\cdot\cdot\lambda} - \\ & \qquad - 2\overset{x+1}{H}{}_{[s_0\cdot\;|\varkappa|}^{\cdot p_x}\overset{x+1}{H}{}_{r_0]q_x}^{\cdot\cdot\varkappa}; \quad x=0,\ldots,k+1. \end{cases}$$

(10.25) are the GAUSS-CODAZZI-RICCI equations of a V_m in V_n for higher curvatures. For V_m in R_n only three equations remain

$$(10.26) \begin{cases} \text{a)} & \overset{x-1}{H}{}_{[s_0\cdot\;|t_{x-1}|}^{\cdot p_{x-2}} \overset{x}{H}{}_{r_0]\cdot q_x}^{\cdot t_{x-1}} = 0; \qquad x = 2,\ldots,k \\ \text{b)} & D_{[s_0}\overset{x}{H}{}_{r_0]\cdot q_x}^{\cdot p_{x-1}} = 0; \qquad x = 1,\ldots,k \\ \text{c)} & \overset{x}{K}{}_{s_0 r_0 q_x}^{\cdot\cdot\cdot p_x} = 2\overset{x}{H}{}_{[s_0\,|t_{x-1}|}^{\cdot\;\cdot\;p_x}\overset{x}{H}{}_{r_0]\cdot q_x}^{\cdot t_{x-1}} + 2\overset{x+1}{H}{}_{[s_0\cdot\,|t_{x+1}|}^{\cdot p_x}\overset{x+1}{H}{}_{r_0]q_x}^{\cdot\cdot t_{x+1}}; \\ & x = 0,\ldots,k+1. \end{cases}$$

In the equations (cf. 10.14a) in an R_n with *rectilinear* coordinates $\xi^\varkappa$

$$(10.27) \begin{cases} \text{a)} & \partial_{q_0}\xi^\varkappa = \overset{0}{B}{}^{\varkappa}_{q_0}; \qquad \partial_{q_0} \overset{\text{def}}{=} \overset{0}{B}{}^b_{q_0}\partial_b \\ \text{b)} & D_{r_0}\overset{x}{B}{}^\varkappa_{q_x} = -\overset{x}{H}{}_{r_0\cdot q_x}^{\cdot p_{x-1}}\overset{x-1}{B}{}^\varkappa_{p_{x-1}} + \overset{x+1}{H}{}_{r_0 q_x}^{\cdot\cdot p_{x+1}}\overset{x+1}{B}{}^\varkappa_{p_{x+1}}; \\ & x = 0,\ldots,k+1; \end{cases}$$

---

[1]) The first equation is due to SCHWARTZ 1941, 1; cf. CUTLER 1931, 1; 2.

the $\xi^\varkappa$, $\overset{0}{B}{}^\varkappa_{q_0}, \ldots, \overset{k+1}{B}{}^\varkappa_{q_{k+1}}$ may be considered as the unknowns and the $\Gamma^{p_0}_{r_0 q_0}, \ldots, \Gamma^{p_{k+1}}_{r_0 q_{k+1}}, \overset{1}{H}{}^{\cdot\, p_0}_{r_0 \cdot q_1}, \ldots, \overset{k}{H}{}^{\cdot\, p_{k-1}}_{r_0 \cdot q_k}$ as given functions of the coordinates η^a of V_m. Then (10.26) and $\overset{1}{H}{}_{[r_0 q_0]}^{\cdot\;\cdot\; p_1} = 0$ are the integrability conditions.[1]) If these conditions are satisfied identically by the given functions, (10.27) always has a solution that takes initial values $\underset{0}{\xi}{}^\varkappa, \underset{0}{\overset{0}{B}}{}^\varkappa_{q_0}, \ldots, \underset{0}{\overset{k+1}{B}}{}^\varkappa_{q_{k+1}}$ at an arbitrary point $\underset{0}{\eta}{}^a$ of V_m. But these initial values must satisfy the conditions

$$(10.28)\qquad \begin{cases} \text{a)}\quad g_{\lambda\varkappa} \underset{0}{\overset{x}{B}}{}^\lambda_{q_x} \underset{0}{\overset{x}{B}}{}^\varkappa_{p_x} = \delta_{q_x p_x} \\ \text{b)}\quad g_{\lambda\varkappa} \underset{0}{\overset{x}{B}}{}^\lambda_{q_x} \underset{0}{\overset{y}{B}}{}^\varkappa_{p_y} = 0; \qquad x, y = 0, \ldots, k+1; \qquad x \neq y \end{cases}$$

where the $g_{\lambda\varkappa}$ are some given constants with $g_{\lambda\varkappa} = g_{\varkappa\lambda}$ and $\operatorname{Det}(g_{\lambda\varkappa}) \neq 0$. Then similar equations are satisfied by $\overset{0}{B}{}^\varkappa_{q_0}, \ldots, \overset{k+1}{B}{}^\varkappa_{q_{k+1}}$ at all points in an $\mathfrak{N}(\underset{0}{\eta}{}^a)$. In the R_n of the $\xi^\varkappa$ with the fundamental tensor $g_{\lambda\varkappa}$ the first $m + m_1 + \cdots + m_k$ vectors span a flat $R_{m+\cdots+m_k}$ in which the V_m is imbedded. Hence:

If in a V_m the $\Gamma^{p_0}_{r_0 q_0}$ are the parameters of the connexion with respect to an anholonomic coordinate system with unit basis vectors and if in this V_m the $\Gamma^{p_1}_{r_0 q_1}, \ldots, \Gamma^{p_{k+1}}_{r_0 q_{k+1}}$ and the $\overset{1}{H}{}^{\cdot\, p_0}_{r_0 \cdot q_1}, \ldots, \overset{k}{H}{}^{\cdot\, p_{k-1}}_{r_0 \cdot q_k}$; $p_x, q_x = \overset{x}{1}, \ldots, \overset{x}{\mathrm{m}}_x$; $x = 1, \ldots, k+1$ are given functions of the coordinates η^a of the V_m satisfying the conditions $\overset{1}{H}{}_{[r_0 q_0]}^{\cdot\;\cdot\; p_1} = 0$ and (10.26), in which the operator D_{s_0} is defined by

$$(10.29)\qquad \begin{cases} \text{a)}\quad D_{s_0} v^{p_x} \overset{\text{def}}{=} \partial_{s_0} v^{p_x} + \Gamma^{p_x}_{s_0 q_x} v^{q_x} \\ \text{b)}\quad D_{s_0} w_{q_x} \overset{\text{def}}{=} \partial_{s_0} w_{q_x} - \Gamma^{p_x}_{s_0 q_x} w_{p_x} \end{cases}; \qquad x = 0, \ldots, k+1$$

and $\overset{x}{K}{}_{s_0 r_0 q_x}^{\cdot\;\cdot\;\cdot\; p_x}$ by (10.22), then an R_n; $n = m + m_1 + \cdots + m_{k+1}$ can be constructed and in this R_n a flat $R_{m+\cdots+m_k}$, in which the V_m is imbedded in such a way that its x-th curvature region is an $R_{m+m_1+\cdots+m_x}$; $x = 1, \ldots, k$, can also be constructed.

If for a V_m in R_n we have $m_l = 1$, $l \geq 1$ and if $\overset{l}{n}{}^\varkappa$ is a unitvector in R_{m_l} the quantity $\overset{l}{H}{}_{r_0 p_{l-1}}^{\cdot\;\cdot\;\;\varkappa}$ can be written in the form

$$(10.30)\qquad \overset{l}{H}{}_{r_0 p_{l-1}}^{\cdot\;\cdot\;\;\varkappa} = \overset{l}{H}{}_{r_0 p_{l-1}} \overset{l}{n}{}^\varkappa .$$

Then (10.26a) gives for $x = l+1$

$$(10.31)\qquad \overset{l}{H}{}_{[s_0|p_{l-1}} \overset{l}{n}{}_{\lambda|} \overset{l+1}{H}{}_{r_0]}^{\cdot\;\lambda}{}_{\cdot\, q_{l+1}} = 0 .$$

[1]) Schouten and v. Kampen 1931, 2, § 4.

For $l=k$ this is trivial but for $l<k$ (10.31) expresses that for every value of p_{l-1} and q_{l+1} the vectors $\overset{l}{H}_{s_0 p_{l-1}}$ and $\overset{l}{n}_\lambda \overset{l+1}{H}{}_{r_0 \cdot q_{l+1}}^{\cdot\,\lambda}$ *all* have the same direction. Hence for $m=1$ and $1 \leq l<k$ the curvature tensors of valence 3 from $\overset{l}{H}{}_{r_0 \lambda}^{\cdot\,\cdot\,\varkappa}$ to $\overset{k}{H}{}_{r_0 \lambda}^{\cdot\,\cdot\,\varkappa}$ can be written as follows

$$
(10.32)\qquad \begin{cases} \text{a)} & \overset{l}{H}{}_{r_0 \lambda}^{\cdot\,\cdot\,\varkappa} = H_{r_0} u_\lambda \overset{l}{n}{}^\varkappa \\ \text{b)} & \overset{l+1}{H}{}_{r_0 \lambda}^{\cdot\,\cdot\,\varkappa} = \overset{l+1}{\alpha} H_{r_0} \overset{l}{n}_\lambda \overset{l+1}{n}{}^\varkappa \\ & \vdots \\ \text{c)} & \overset{k}{H}{}_{r_0 \lambda}^{\cdot\,\cdot\,\varkappa} = \overset{k}{\alpha} H_{r_0} \overset{k-1}{n}_\lambda \overset{k}{n}{}^\varkappa \end{cases}
$$

in which H_{r_0} is a definite vector of V_m and u_λ a definite *unitvector* of $R_{m_{l-1}}$ and where $\overset{l}{n}{}^\varkappa, \ldots, \overset{k}{n}{}^\varkappa$ are unitvectors in the one dimensional spaces $R_{m_l}, \ldots, R_{m_k}$. For $k=l$ the vector H_{r_0} does not exist and instead of (10.32) we have (10.30) only.[1]) Note that $D_{r_0} \overset{x}{n}_{q_x}=0$; $x=l, \ldots, k$.

From (10.26b) and (10.32) we get for $1 \leq l<k$ and $x=l, \ldots, k$

$$
(10.33)\qquad \begin{cases} \text{a)} & (D_{[s_0} H_{r_0]}) u^{p_{l-1}} \overset{l}{n}_{q_l} + H_{[r_0} (D_{s_0]} u^{p_{l-1}}) \overset{l}{n}_{q_l} = 0 \\ \text{b)} & (D_{[s_0} \overset{l+1}{\alpha} H_{r_0]}) \overset{l}{n}{}^{p_l} \overset{l+1}{n}_{q_{l+1}} = 0 \\ & \vdots \\ \text{c)} & (D_{[s_0} \overset{k}{\alpha} H_{r_0]}) \overset{k-1}{n}{}^{p_{k-1}} \overset{k}{n}_{q_k} = 0 \end{cases}
$$

from which we see (by transvection of (10.33a) with $u_{p_{l-1}}$) that H_{r_0} is a gradientvector, $H_{r_0}=\partial_{r_0} H$, and that the coefficients $\overset{l+1}{\alpha}, \ldots, \overset{k}{\alpha}$ can all be written as functions of H only. The equation $H=\text{const.}$ represents a normal system of V_{m-1}'s in V_m. If v^{p_0} is a vector in these V_{m-1}'s we have $v^{p_0} H_{p_0}=0$ and from (10.33a) it follows that

$$
(10.34)\qquad v^{q_0} D_{q_0} u^{p_{l-1}}=0;
$$

but this equation is only part of a more important result that can be derived from the equations of FRENET.

In fact, according to (10.14a), (10.32) and (10.34) we have

$$
(10.35)\qquad \begin{cases} v^{q_0} D_{q_0} u^\varkappa = v^{q_0} u^{p_{l-1}} D_{q_0} \overset{l-1}{B}{}_{p_{l-1}}^{\varkappa} \\ \qquad = - v^{q_0} u^{p_{l-1}} \overset{l-1}{H}{}_{q_0 \cdot p_{l-1}}^{\cdot\,\varkappa} + v^{q_0} u^{p_{l-1}} H_{q_0} u_{p_{l-1}} \overset{l}{n}{}^\varkappa . \end{cases}
$$

---

[1]) EISENHART 1937, 1. Although in most parts of this paper EISENHART only considers the case $l=2$, his point of view is more general because his fundamental tensor in V_n is also taken to be indefinite. This gives some very interesting results that could not be included here.

But (10.26a) gives, for $x=l$,

(10.36)
$$\overset{l-1}{H}{}_{[s_0}{}^{\cdot p_{l-2}}_{\cdot}{}_{|\lambda|}\, H_{r_0]}\, u^\lambda \overset{l}{n}_{q_l} = 0,$$

hence

(10.37)
$$v^{q_0} D_{q_0} u^\varkappa = 0$$

and this proves *that for* $m_l=1$ *and* $1 \leq l < k$ *the field* $u^\varkappa$ *is covariant constant over each of the* V_{m-1}*'s of the normal system.*[1])

Returning now to the equations of FRENET (10.14) for $x=l,\ldots,k$ we get according to (10.32) for $m_l=1$; $1\leq l<k$

(10.38)
$$\begin{cases} \text{a)} & (D_{r_0}\overset{l}{n}{}^\varkappa)\,\overset{l}{n}_{q_l} = -H_{r_0} u^\varkappa \overset{l}{n}_{q_l} + \overset{l+1}{\alpha} H_{r_0} \overset{l}{n}_{q_l} \overset{l+1}{n}{}^\varkappa \\ \text{b)} & (D_{r_0}\overset{l+1}{n}{}^\varkappa)\,\overset{l+1}{n}_{q_{l+1}} = -\overset{l+1}{\alpha} H_{r_0} \overset{l}{n}{}^\varkappa \overset{l+1}{n}_{q_{l+1}} + \overset{l+2}{\alpha} H_{r_0} \overset{l+1}{n}_{q_{l+1}} \overset{l+2}{n}{}^\varkappa \\ & \vdots \\ \text{c)} & (D_{r_0}\overset{k}{n}{}^\varkappa)\,\overset{k}{n}_{q_k} = -\overset{k}{\alpha} H_{r_0} \overset{k-1}{n}{}^\varkappa \overset{k}{n}_{q_k} \end{cases}$$

or

(10.39)
$$\begin{cases} \text{a)} & D_{r_0}\overset{l}{n}{}^\varkappa = -H_{r_0}(u^\varkappa - \overset{l+1}{\alpha}\overset{l+1}{n}{}^\varkappa) \\ \text{b)} & D_{r_0}\overset{l+1}{n}{}^\varkappa = -H_{r_0}(\overset{l+1}{\alpha}\overset{l}{n}{}^\varkappa - \overset{l+2}{\alpha}\overset{l+2}{n}{}^\varkappa) \\ & \vdots \\ \text{c)} & D_{r_0}\overset{k}{n}{}^\varkappa = -H_{r_0}\overset{k}{\alpha}\overset{k-1}{n}{}^\varkappa, \end{cases}$$

hence

(10.40)
$$v^{r_0} D_{r_0} \overset{x}{n}{}^\varkappa = 0; \qquad x = l,\ldots,k;$$

and this proves *that for* $m_l=1$, $1\leq l<k$ *the fields* $\overset{l}{n}{}^\varkappa,\ldots,\overset{k}{n}{}^\varkappa$ *also are covariant constant over each of the* V_{m-1}*'s of the normal system* $H=\text{const.}$[1])

Substituting (10.32) in (10.26c) we get for $m_l=1$, $1\leq l<k$

(10.41)
$$\overset{x}{K}{}_{s_0 r_0 q_x}{}^{\cdot\cdot\cdot p_x} = 0; \qquad x = l,\ldots,k$$

and for $m_l=1$, $1\leq l=k$

(10.42)
$$\begin{cases} \text{a)} & \overset{k}{K}{}_{s_0 r_0 q_k}{}^{\cdot\cdot\cdot p_k} = 2\overset{k}{H}{}_{[s_0|t_{k-1}|}\overset{k}{H}{}_{r_0]}^{\cdot\, t_{k-1}}\overset{k}{n}{}^{p_k}\overset{k}{n}_{q_k} = 0 \\ \text{b)} & \overset{k-1}{K}{}_{s_0 r_0 q_{k-1}}{}^{\cdot\cdot\cdot p_{k-1}} = 2\overset{k-1}{H}{}_{[s_0|t_{k-2}|}^{\cdot\qquad p_{k-1}}\overset{k-1}{H}{}_{r_0]}^{\cdot\, t_{k-2}}{}_{\cdot\, q_{k-1}} + 2\overset{k}{H}{}_{[s_0}^{\cdot\, p_{k-1}}\overset{k}{H}{}_{r_0]q_{k-1}} \end{cases}$$

We see from (10.26c, 32a, 42b) that for $l=1$

(10.43)
$$\overset{0}{K}{}_{s_0 r_0 q_0}{}^{\cdot\cdot\cdot p_0} = \begin{cases} 2\overset{1}{H}{}_{[s_0}^{\cdot\, p_0}\overset{1}{H}{}_{r_0]q_0} & \text{for} \quad k=1 \\ 0 & \text{for} \quad k>1. \end{cases}$$

---

[1]) EISENHART 1937, 1 for $l=2$.

Hence if $m_1 = 1$ the V_m lies in a flat R_{m+1} in R_n or it is a bent R_m. This implies that if a V_m is imbedded in an R_n such that $m_1 = 1$, it can always be imbedded in an R_{m+1}.[1]) We prove the following more general theorem:

A V_m in R_n with $m_l = 1$; $2 \leq l < k$ can always be imbedded in an $R_{m+m_1+\cdots+m_{l-1}}$ such that its x-th curvature region is an $R_{m+m_1+\cdots+m_x}$; $x = 1, \ldots, l-1$.[2])

If we take (10.32) into account we have from (10.14a) for $2 \leq l < k$

$$(10.44)\quad \begin{cases} \text{a)} & \overset{0}{B}{}^{\varkappa}_{q_0} = \partial_{q_0} \xi^{\varkappa} \\ \text{b)} & D_{r_0} \overset{0}{B}{}^{\varkappa}_{q_0} = \overset{1}{H}{}_{r_0 q_0}^{\cdot\ \cdot\ p_1} \overset{1}{B}{}^{\varkappa}_{p_1} \\ \text{c)} & D_{r_0} \overset{x}{B}{}^{\varkappa}_{q_x} = -\overset{x}{H}{}_{r_0}^{\cdot\ p_{x-1}}{}_{\cdot\ q_x} \overset{x-1}{B}{}^{\varkappa}_{p_{x-1}} + \overset{x+1}{H}{}_{r_0 q_x}^{\cdot\ \cdot\ p_{x+1}} \overset{x+1}{B}{}^{\varkappa}_{p_{x+1}} \quad x = 1, \ldots, l-2 \\ \text{d)} & D_{r_0} \overset{l-1}{B}{}^{\varkappa}_{q_{l-1}} = -\overset{l-1}{H}{}_{r_0}^{\cdot\ p_{l-2}}{}_{\cdot\ q_{l-1}} \overset{l-2}{B}{}^{\varkappa}_{p_{l-2}} + H_{r_0} u_{q_{l-1}} \overset{l}{n}{}^{\varkappa} \\ \text{e)} & D_{r_0} \overset{l}{n}{}^{\varkappa} = -H_{r_0} u^{p_{l-1}} \overset{l-1}{B}{}^{\varkappa}_{p_{l-1}} + \overset{l+1}{\alpha} H_{r_0} \overset{l+1}{n}{}^{\varkappa} \\ \text{f)} & D_{r_0} \overset{y}{n}{}^{\varkappa} = -\overset{y}{\alpha} H_{r_0} \overset{y-1}{n}{}^{\varkappa} + \overset{y+1}{\alpha} H_{r_0} \overset{y+1}{n}{}^{\varkappa}; \quad y = l+1, \ldots, k-1 \\ \text{g)} & D_{r_0} \overset{k}{n}{}^{\varkappa} = -\overset{k}{\alpha} H_{r_0} \overset{k-1}{n}{}^{\varkappa} \\ \text{h)} & D_{r_0} \overset{k+1}{B}{}^{\varkappa}_{q_{k+1}} = 0 \end{cases}$$

and from (10.26) the integrability conditions of (10.44)

$$(10.45)\quad \begin{cases} \text{a)} & \overset{x}{H}{}_{[s_0}^{\cdot\ p_{x-1}}{}_{\cdot\ |t_x|} \overset{x+1}{H}{}_{r_0]}^{\cdot\ t_x}{}_{\cdot\ q_{x+1}} = 0; \quad x = 1, \ldots, l-2 \\ \text{b)} & \overset{l-1}{H}{}_{[s_0}^{\cdot\ p_{l-2}}{}_{\cdot\ |t_{l-1}|} H_{r_0]} u^{t_{l-1}} = 0; \quad u^{p_{l-1}} u_{p_{l-1}} = 1 \\ \text{c)} & D_{[s_0} \overset{z}{H}{}_{r_0]}^{\cdot\ p_{z-1}}{}_{\cdot\ q_z} = 0; \quad z = 1, \ldots, l-1 \\ \text{d)} & D_{[s_0} H_{r_0]} = 0 \\ \text{e)} & H_{[r_0} \partial_{s_0]} \overset{l+1}{\alpha} = 0; \ldots; \quad H_{[r_0} \partial_{s_0]} \overset{k}{\alpha} = 0 \\ \text{f)} & \overset{0}{K}{}_{s_0 r_0 q_0}^{\cdot\ \cdot\ \cdot\ p_0} = 2 \overset{1}{H}{}_{[s_0}^{\cdot\ p_0}{}_{\cdot\ |t_1|} \overset{1}{H}{}_{r_0] q_0}^{\cdot\ \cdot\ t_1} \\ \text{g)} & \overset{x}{K}{}_{s_0 r_0 q_x}^{\cdot\ \cdot\ \cdot\ p_x} = 2 \overset{x}{H}{}_{[s_0 |t_{x-1}|}^{\cdot\ \cdot\ p_x} \overset{x}{H}{}_{r_0]}^{\cdot\ t_{x-1}}{}_{\cdot\ q_x} + 2 \overset{x+1}{H}{}_{[s_0}^{\cdot\ p_x}{}_{\cdot\ |t_{x+1}|} \overset{x+1}{H}{}_{r_0] q_x}^{\cdot\ \cdot\ t_{x+1}}; \\ & \qquad x = 1, \ldots, l-2 \\ \text{h)} & \overset{l-1}{K}{}_{s_0 r_0 q_{l-1}}^{\cdot\ \cdot\ \cdot\ p_{l-1}} = 2 \overset{l-1}{H}{}_{[s_0 |t_{l-2}|}^{\cdot\ \cdot\ p_{l-1}} \overset{l-1}{H}{}_{r_0]}^{\cdot\ t_{l-2}}{}_{\cdot\ q_{l-1}} \\ \text{i)} & \overset{u}{K}{}_{s_0 r_0 q_u}^{\cdot\ \cdot\ \cdot\ p_u} = 0; \quad u = l, \ldots, k+1 \\ \text{j)} & \overset{1}{H}{}_{[r_0 q_0]}^{\cdot\ \cdot\ p_1} = 0. \end{cases}$$

1) EISENHART 1937, 1.

2) EISENHART 1937, 1 for $l = 2$.

In (10.44) the variables $\xi^\varkappa, \overset{0}{B}{}^\varkappa_{q_0}, \ldots, \overset{l-1}{B}{}^\varkappa_{q_{l-1}}, \overset{l}{n}{}^\varkappa, \ldots, \overset{k}{n}{}^\varkappa, \overset{k+1}{B}{}^\varkappa_{q_{k+1}}$ are the unknowns and the $\Gamma^{p_0}_{r_0 q_0}, \ldots, \Gamma^{p_{l-1}}_{r_0 q_{l-1}}, \overset{1}{H}{}^{\cdot\,\cdot\,p_1}_{r_0 q_0}, \ldots, \overset{l-1}{H}{}^{\cdot\,\cdot\,p_{l-1}}_{r_0 q_{l-2}}, H_{r_0}, u^{p_{l-1}}, \overset{l+1}{\alpha}, \ldots, \overset{k}{\alpha}$ are known functions of the coordinates in V_m, satisfying the conditions (10.45 a–h, j). Now this same V_m can be imbedded in a flat $R_{m+m_1+\cdots+m_{l-1}}$ in R_n with curvature regions of dimension $m+m_1, m+m_1+m_2, \ldots, m+\cdots+m_{l-1}$ if and only if equations of the form

$$
(10.46)\quad
\begin{cases}
\text{a)}\ \ {}'\overset{0}{B}{}^\varkappa_{q_0} = \partial_{q_0}\xi^\varkappa \\
\text{b)}\ \ D_{r_0}\,{}'\overset{0}{B}{}^\varkappa_{q_0} = {}'\overset{1}{H}{}^{\cdot\,\cdot\,p_1}_{r_0 q_0}\,{}'\overset{1}{B}{}^\varkappa_{p_1} \\
\text{c)}\ \ D_{r_0}\,{}'\overset{x}{B}{}^\varkappa_{q_x} = -\,{}'\overset{x}{H}{}^{\cdot\,p_{x-1}}_{r_0\,\cdot\,\ \ q_x}\,{}'\overset{x-1}{B}{}^\varkappa_{p_{x-1}} + {}'\overset{x+1}{H}{}^{\cdot\,\cdot\,p_{x+1}}_{r_0 q_x}\,{}'\overset{x+1}{B}{}^\varkappa_{p_{x+1}}; \\
\qquad\qquad\qquad\qquad\qquad\qquad x = 1, \ldots, l-2 \\
\text{d)}\ \ D_{r_0}\,{}'\overset{l-1}{B}{}^\varkappa_{q_{l-1}} = -\,{}'\overset{l-1}{H}{}^{\cdot\,p_{l-2}}_{r_0\,\cdot\,\ \ q_{l-1}}\,{}'\overset{l-2}{B}{}^\varkappa_{p_{l-2}} \\
\text{e)}\ \ D_{r_0}\,{}'\overset{l}{B}{}^\varkappa_{q_l} = 0
\end{cases}
$$

have solutions. Now the integrability conditions of these equations are

$$
(10.47)\quad
\begin{cases}
\text{a)}\ \ {}'\overset{x}{H}{}^{\cdot\,p_{x-1}}_{[s_0\,\cdot\,|t_x|}\,{}'\overset{x+1}{H}{}^{\cdot\,\ \ t_x}_{r_0]\,\cdot\,q_{x+1}} = 0; \qquad x = 1, \ldots, l-2 \\
\text{b)}\ \ D_{[s_0}\,{}'\overset{z}{H}{}^{\cdot\,p_{z-1}}_{r_0]\,\cdot\,\ \ q_z} = 0; \qquad z = 1, \ldots, l-1 \\
\text{c)}\ \ \overset{0}{K}{}^{\cdot\,\cdot\,\cdot\,p_0}_{s_0 r_0 q_0} = 2\,{}'\overset{1}{H}{}^{\cdot\,p_0}_{[s_0\,\cdot\,|t_1|}\,{}'\overset{1}{H}{}^{\cdot\,\cdot\,t_1}_{r_0] q_0} \\
\text{d)}\ \ {}'\overset{x}{K}{}^{\cdot\,\cdot\,\cdot\,p_x}_{s_0 r_0 q_x} = 2\,{}'\overset{x}{H}{}^{\cdot\,\cdot\,p_x}_{[s_0|t_{x-1}|}\,{}'\overset{x}{H}{}^{\cdot\,t_{x-1}}_{r_0]\,\cdot\,\ \ q_x} + 2\,{}'\overset{x+1}{H}{}^{\cdot\,p_x}_{[s_0\,\cdot\,|t_{x+1}|}\,{}'\overset{x+1}{H}{}^{\cdot\,\cdot\,t_{x+1}}_{r_0] q_x}; \\
\qquad\qquad\qquad\qquad\qquad\qquad x = 1, \ldots, l-2 \\
\text{e)}\ \ {}'\overset{l-1}{K}{}^{\cdot\,\cdot\,\cdot\,p_{l-1}}_{s_0 r_0 q_{l-1}} = 2\,{}'\overset{l-1}{H}{}^{\cdot\,\cdot\,p_{l-1}}_{[s_0|t_{l-2}|}\,{}'\overset{l-1}{H}{}^{\cdot\,t_{l-2}}_{r_0]\,\cdot\,q_{l-1}} \\
\text{f)}\ \ {}'\overset{l}{K}{}^{\cdot\,\cdot\,\cdot\,p_l}_{s_0 r_0 q_l} = 0
\end{cases}
$$

and these equations are identical with (10.45 a, c, f, g, h, i) if we take

$$
(10.48)\quad
\begin{cases}
{}'\overset{x}{H}{}^{\cdot\,p_{x-1}}_{s_0\,\cdot\,\ \ q_x} = \overset{x}{H}{}^{\cdot\,p_{x-1}}_{s_0\,\cdot\,\ \ q_x}; \qquad x = 1, \ldots, l-1 \\
{}'\Gamma^{p_x}_{r_0 q_x} = \Gamma^{p_x}_{r_0 q_x}; \qquad x = 1, \ldots, l-1 \\
H_{r_0} = 0\ ^{1)}
\end{cases}
$$

and zero for all ${}'\Gamma^{p_l}_{r_0 q_l}$, where the indices p_l, q_l now belong to a region of dimension $m_l + \cdots + m_{k+1}$.

---

1) It is not necessary to take $H_{r_0} = 0$.

A theory of the imbedding of a W_m in W_n [1]) and of an A_m in A_n [2]) involving higher curvatures was developed by HLAVATY. Problems of higher curvature for $V_{m'}$ in V_m in V_n were dealt with by LONGO 1948[3]). Umbilical subspaces in umbilical subspaces were considered by YANO[4]).

§ 11. Product spaces. [5])

If two riemannian spaces V_m and V_{n-m} are given with coordinates ξ^α; $\alpha, \beta, \gamma = 1, \ldots, m$, and ξ^ξ; $\xi, \eta, \zeta = m+1, \ldots, n$, and fundamental tensors $g_{\beta\alpha}$ and $g_{\eta\xi}$, the V_n with the coordinates $\xi^\varkappa$; $\varkappa, \lambda, \mu = 1, \ldots, n$, and the fundamental tensor $g_{\lambda\varkappa}$ for which $g_{\eta\alpha} = 0$; $g_{\beta\xi} = 0$, is called the *product* of V_m and V_{n-m}. A V_n that is a product space is said to be *decomposable* (or *separable*). That means that V_n is decomposable if and only if there can be found a coordinate system $(\varkappa)$ such that $g_{\eta\alpha} = 0$; $g_{\beta\xi} = 0$ and the $g_{\beta\alpha}$ and the $g_{\eta\xi}$ are functions of the ξ^α and the ξ^ξ respectively.

If two spaces L_m and L_{n-m} are given with coordinates ξ^α and ξ^ξ and the connexions $\Gamma^\alpha_{\gamma\beta}$ and $\Gamma^\xi_{\zeta\eta}$, the L_n with the coordinates $\xi^\varkappa$ and the connexion $\Gamma^\varkappa_{\mu\lambda}$ for which all components with two indices of different kind vanish, is called the *product* of L_m and L_{n-m}. An L_n that is a product space is also called *decomposable*. Hence an L_n is decomposable if and only if there exists a coordinate system $(\varkappa)$ such that all parameters of the connexion with two indices of different kind vanish and that the others only depend on the set of coordinates to which they belong. Obviously an L_n that is decomposable according to the second definition and happens to be a V_n is also decomposable according to the first definition.

An object in a decomposable L_n is called *breakable* if its components with respect to a special coordinate system as defined above are always zero when they have indices from both ranges. A breakable object is said to be *decomposable* if and only if the components belonging to the subspace L_m (L_{n-m}) depend on the ξ^α (ξ^ξ) only. The objects $g_{\lambda\varkappa}$ and $\Gamma^\varkappa_{\mu\lambda}$ in a decomposable V_n or L_n are examples of decomposable objects.

A decomposable L_n contains a set of ∞^{n-m} X_m's and a set of ∞^m X_{n-m}'s with the equations

$$(11.1) \quad \text{a)}\ \xi^\xi \overset{*}{=} 0;\ \xi = m+1, \ldots, n \quad \text{and} \quad \text{b)}\ \xi^\alpha \overset{*}{=} 0;\ \alpha = 1, \ldots, m$$

[1]) HLAVATY 1949, 1; 1952, 1.

[2]) HLAVATY 1949, 2; 3; 4.

[3]) LONGO 1948, 1.

[4]) YANO 1940, 3.

[5]) General references: EISENHART 1923, 1; LEVY 1925, 2; DUSCHEK and MAYER 1930, 1; T. Y. THOMAS 1939, 1; 2; FICKEN 1939, 1; LICHNEROWICZ 1944, 1.

such that nowhere an X_m has a direction in common with an X_{n-m}. Each of these sets is geodesic and parallel in L_n. In fact, if v^α; $v^\xi = 0$ is a field in the X_m's, we have for each displacement in these spaces $(d\xi^\xi = 0)$

$$\delta v^\xi \overset{*}{=} d\xi^\gamma \Gamma^\xi_{\gamma\beta} v^\beta \overset{*}{=} 0 \tag{11.2}$$

and for each displacement in the other spaces $(d\xi^\alpha = 0)$

$$\delta v^\xi \overset{*}{=} d\xi^\zeta \Gamma^\xi_{\zeta\beta} v^\beta \overset{*}{=} 0. \tag{11.3}$$

Conversely an L_n is decomposable if there exist a set of ∞^{n-m} geodesic and parallel X_m's and a set of ∞^m geodesic and parallel X_{n-m}'s having nowhere a direction in common. In an ordinary V_n it is sufficient that there is a set of ∞^{n-m} real geodesic and parallel V_m's because in this case the set of orthogonal V_{n-m}'s exists and is geodesic and parallel. But if the fundamental tensor of V_n is indefinite it may happen that the tangent m-direction and the $(n-m)$-directions perpendicular to it have a direction in common.

If a V_n is decomposable, $g_{\lambda\varkappa}$ can be split up into two symmetric tensors

$$\begin{cases} g_{\lambda\varkappa} = 'g_{\lambda\varkappa} + ''g_{\lambda\varkappa}; & 'g_{\beta\xi} \overset{*}{=} 0; \quad 'g_{\eta\alpha} \overset{*}{=} 0; \quad 'g_{\eta\xi} \overset{*}{=} 0 \\ & ''g_{\beta\xi} \overset{*}{=} 0; \quad ''g_{\eta\alpha} \overset{*}{=} 0; \quad ''g_{\beta\alpha} \overset{*}{=} 0 \end{cases} \tag{11.4}$$

and each of these tensors is covariant constant.

$$\text{a) } \nabla_\mu 'g_{\lambda\varkappa} = 0; \qquad \text{b) } \nabla_\mu ''g_{\lambda\varkappa} = 0. \tag{11.5}$$

Moreover the $'g_{\beta\alpha}$ ($''g_{\eta\xi}$) depend on the ξ^α (ξ^ξ) only. But this implies that also the symmetric tensor $'c\ 'g_{\lambda\varkappa} + ''c\ ''g_{\lambda\varkappa}$ with two constants $'c$ and $''c$, is covariant constant. Conversely, if there exists in an ordinary V_n a real covariant constant symmetric tensor $h_{\lambda\varkappa}$ that is not a multiple of $g_{\lambda\varkappa}$, the principal multidirections of $h_{\lambda\varkappa}$ form parallel fields. Hence:

An ordinary V_n is decomposable if and only if there exists a real covariant constant symmetric tensor field $h_{\lambda\varkappa}$ that is not a multiple of $g_{\lambda\varkappa}$.[1])

In a decomposable L_n any figure can be "projected" by the X_{n-m}'s on one of the X_m's and by the X_m's on one of the X_{n-m}'s. These projections have many interesting properties. For instance it can be proved that a curve in a decomposable A_n is a geodesic if and only if its projections are both geodesics.

There are several generalizations possible. WONG[2]) considered a V_n whose linear element can be written in the form

$$ds^2 \overset{*}{=} g_{\beta\alpha}\, d\xi^\beta\, d\xi^\alpha + g_{\eta\xi}\, d\xi^\eta\, d\xi^\xi \tag{11.6}$$

---

1) T. Y. THOMAS 1939, 1; 2; LICHNEROWICZ 1944, 1.

2) WONG 1943, 1.

where the $g_{\beta\alpha}$ and $g_{\eta\zeta}$ are general functions of the ξ^α and ξ^ξ both. Such a V_n need not be decomposable. In this case the V_m's : $\xi^\xi = 0$ and the V_{n-m}'s : $\xi^\alpha = 0$ are *called complementary subspaces.*

If in (11.6) $g_{\beta\alpha}$ and $g_{\eta\xi}$ have the form

$$g_{\beta\alpha} \overset{*}{=} \varrho\, 'g_{\beta\alpha}; \quad g_{\eta\xi} \overset{*}{=} \sigma\, ''g_{\eta\xi} \tag{11.7}$$

where the $'g_{\beta\alpha}$ ($''g_{\eta\zeta}$) depend on the ξ^α (ξ^ξ) only but ϱ and σ are general functions of the ξ^α and ξ^ξ both, the V_n is said to be *conformally decomposable* (*separable*). [1])

Many recent investigations deal with the geometry in the large of decomposable and indecomposable compact spaces.

Exercise.

V 11,1 [2]). Every symmetric V_n that is not decomposable is an EINSTEIN space.

VI. Projective and conformal transformations of connexions. [3])

§ 1. Projective transformations of a symmetric connexion.

In Ch. III § 7 we proved that the transformations of the $\Gamma^\varkappa_{\mu\lambda}$ in A_n which preserve the geodesics have the form

$$'\Gamma^\varkappa_{\mu\lambda} = \Gamma^\varkappa_{\mu\lambda} + p_\mu A^\varkappa_\lambda + p_\lambda A^\varkappa_\mu \tag{1.1}$$ [4])

where p_λ is an arbitrary vector. We call this transformation *projective* and p_λ the *vector of the transformation*. From (1.1) and (III 2.3, 7) we get

$$\begin{cases} \text{a)} \quad '\nabla_\mu v^\varkappa = \nabla_\mu v^\varkappa + A^\varkappa_\mu p_\lambda v^\lambda + v^\varkappa p_\mu \\ \text{b)} \quad '\nabla_\mu w_\lambda = \nabla_\mu w_\lambda - 2 w_{(\mu} p_{\lambda)}. \end{cases} \tag{1.2}$$

1) YANO 1940, 1; WONG 1943, 2.

2) LICHNEROWICZ 1944, 1.

3) In this chapter a V_n is always ordinary and imbedded spaces are always real if another supposition is not made explicitly.

4) A transformation of the $\Gamma^\varkappa_{\mu\lambda}$ in L_n of the form $\Gamma^\varkappa_{\mu\lambda} \to \Gamma^\varkappa_{\mu\lambda} + p_\mu A^\varkappa_\lambda$ preserves not only the geodesics but also parallelism of directions (cf. VI § 9). The simplest object that can be formed of a connexion given to within transformations of this kind is $\Gamma^\varkappa_{\mu\lambda} - \frac{1}{n} \Gamma_\mu A^\varkappa_\lambda$ as was proved by J. M. THOMAS 1926, 1. Cf. FRIESECKE 1925, 1; BORTOLOTTI 1930, 3, p. 75ff.; 1931, 4; SCHEIBE 1952, 1 and footnotes 5, p. 322 and 6, p. 334.

If (1.1) is substituted in (III 4.2) it follows that

$$(1.3)\qquad 'R_{\nu\mu\lambda}^{\cdot\cdot\cdot\varkappa} = R_{\nu\mu\lambda}^{\cdot\cdot\cdot\varkappa} - 2p_{[\nu\mu]}A_\lambda^\varkappa + 2A_{[\nu}^\varkappa p_{\mu]\lambda}; \qquad p_{\mu\lambda} \overset{\text{def}}{=} -\nabla_\mu p_\lambda + p_\mu p_\lambda$$

and accordingly

$$(1.4)\qquad \begin{cases} \text{a)}\ \ 'R_{\mu\lambda} = R_{\mu\lambda} + n\,p_{\mu\lambda} - p_{\lambda\mu} \\ \text{b)}\ \ 'V_{\nu\mu} = -2'R_{[\nu\mu]} = V_{\nu\mu} - 2(n+1)\,p_{[\nu\mu]} = V_{\nu\mu} + 2(n+1)\nabla_{[\nu}p_{\mu]} \end{cases}$$

(1.4b) implies that a volume preserving connexion (Exerc. III 4,5) is transformed into a connexion with the same property if and only if p_λ is a gradient. The transformation is then called *restricted projective.*[1]) If we wish to make $'V_{\nu\mu}$ zero, the integrability conditions of (1.4b) are $\nabla_{[\omega}V_{\nu\mu]} = 0$. But these are identically satisfied (cf. III 5.24), hence *every connexion can be transformed into a volume preserving connexion by a projective transformation.*[2])

An A_n and its connexion are said to be *projectively euclidean*[3]) and the A_n is called a D_n if it can be transformed projectively into an E_n.[4]) The n.a.s. condition is that there exists a vector field P_λ such that

$$(1.5)\qquad \begin{cases} \text{a)}\ \ R_{\nu\mu\lambda}^{\cdot\cdot\cdot\varkappa} = 2P_{[\nu\mu]}A_\lambda^\varkappa - 2A_{[\nu}^\varkappa P_{\mu]\lambda} \\ \text{b)}\ \ \ P_{\mu\lambda} \overset{\text{def}}{=} -\nabla_\mu P_\lambda + P_\mu P_\lambda. \end{cases}$$

Now in an A_n the curvature tensor satisfies only the first and the second identity (III 5.1, 4) and therefore the number of its independent components is $n^2\binom{n}{2} - n\binom{n}{3} = \frac{1}{3}n^2(n^2-1)$. This number is $> n^2$ for $n > 2$, hence (1.5a) can always be satisfied for $n = 2$, but for $n > 2$ only if $R_{\nu\mu\lambda}^{\cdot\cdot\cdot\varkappa}$ satisfies certain algebraic conditions. In this case we get from (1.4a) for $p_{\mu\lambda} = P_{\mu\lambda}$ and $'R_{\mu\lambda} = 0$

$$(1.6)\qquad \begin{cases} \text{a)}\ \ P_{\mu\lambda} = -\dfrac{n}{n^2-1}R_{\mu\lambda} - \dfrac{1}{n^2-1}R_{\lambda\mu} \\ \qquad\quad = -\dfrac{1}{n-1}R_{\mu\lambda} - \dfrac{1}{n^2-1}V_{\mu\lambda} = -\dfrac{1}{n-1}R_{(\mu\lambda)} - \dfrac{1}{n+1}R_{[\mu\lambda]} \\ \text{b)}\ \ P_{[\mu\lambda]} = \dfrac{1}{2(n+1)}V_{\mu\lambda}. \end{cases}$$

---

[1]) The term was introduced by COBURN 1941, 1.

[2]) Cf. EISENHART 1922, 2, p. 236.

[3]) Many authors use the term projectively flat. But an A_{n-1} in E_n can be euclidean, that is an E_{n-1}, without being flat in the ordinary sense, for instance a cone has this property. There is no objection to the term *locally projectively flat.*

[4]) In BORTOLOTTI 1931, 4 a D_n is an L_n whose connexion is given to within a transformation of the form $\Gamma_{\mu\lambda}^\varkappa \to \Gamma_{\mu\lambda}^\varkappa + p_\mu A_\lambda^\varkappa$. Cf. footnote 4 on page 287.

The first integrability condition of (1.5b) is

$$R_{\nu\mu\lambda}^{\cdot\cdot\cdot\varkappa} P_\varkappa = 2\nabla_{[\nu} P_{\mu]\lambda} - 2(\nabla_{[\nu} P_{\mu]}) P_\lambda - 2P_{[\mu} \nabla_{\nu]} P_\lambda \tag{1.7}$$

or, according to (1.5a, b)

$$\nabla_{[\nu} P_{\mu]\lambda} = 0. \tag{1.8}$$

Now, if the identity of BIANCHI (III 5.21) is applied to (1.5a) we get

$$0 = \nabla_{[\omega} P_{\nu\mu]} A_\lambda^\varkappa - A_{[\nu}^\varkappa \nabla_\omega P_{\mu]\lambda} \tag{1.9}$$

and from this equation by contraction over $\varkappa\lambda$

$$\nabla_{[\omega} P_{\nu\mu]} = 0 \tag{1.10}$$

and by contraction over $\varkappa\nu$

$$\nabla_{[\omega} P_{\lambda\mu]} = \tfrac{1}{3}(n-2) \nabla_{[\omega} P_{\mu]\lambda}. \tag{1.11}$$

Hence, for $n > 2$ the integrability condition of (1.5b) is a consequence of (1.5a). This proves the theorem[1]):

An A_n is projectively euclidean if and only if

$$P_{\nu\mu\lambda}^{\cdot\cdot\cdot\varkappa} \overset{\text{def}}{=} R_{\nu\mu\lambda}^{\cdot\cdot\cdot\varkappa} - 2P_{[\nu\mu]} A_\lambda^\varkappa + 2A_{[\nu}^\varkappa P_{\mu]\lambda} = 0 \tag{1.12}$$

where

$$P_{\mu\lambda} \overset{\text{def}}{=} -\frac{n}{n^2-1} R_{\mu\lambda} - \frac{1}{n^2-1} R_{\lambda\mu} \tag{1.13}$$

and if for $n=2$ moreover

$$\nabla_{[\nu} P_{\mu]\lambda} = 0. \tag{1.14}$$

For $n=2$, (1.12) is identically satisfied and for $n>2$ (1.14) is a consequence of (1.12).

$P_{\nu\mu\lambda}^{\cdot\cdot\cdot\varkappa}$ is called the *projective curvature tensor.*[2]) From (1.13) and (1.4a) it follows that

$$'P_{\mu\lambda} = P_{\mu\lambda} - p_{\mu\lambda} \tag{1.15}$$

and by substituting this in (1.12) we see that $P_{\nu\mu\lambda}^{\cdot\cdot\cdot\varkappa}$ *is invariant for projective transformations of the connexion.* For $n=2$ it vanishes identically and for $n>2$ it satisfies the identities

$$\begin{cases} \text{a)}\ P_{(\nu\mu)\lambda}^{\cdot\cdot\cdot\varkappa} = 0; & \text{b)}\ P_{[\nu\mu\lambda]}^{\cdot\cdot\cdot\varkappa} = 0; \\ \text{c)}\ P_{\nu\mu\lambda}^{\cdot\cdot\cdot\nu} = 0; & \text{d)}\ P_{\nu\mu\lambda}^{\cdot\cdot\cdot\lambda} = 0. \end{cases} \tag{1.16}$$

---

[1]) WEYL 1921, 2, p. 105. In EISENHART 1927, 1, p. 97 there is a mistake in the condition for a V_n of being a D_n. It is probably caused by a wrong interpretation of WEYL's term "scalar curvature". The same mistake occurs in LOVELL 1934, 1. Cf. footnote 4, Ch. III, p. 148. For $n=2$ there is a mistake in SCHOUTEN 1953, 1, p. 74.

[2]) Cf. for the semi-symmetric case SCHOUTEN 1925, 2.

As a corollary we get from (1.5a) that for a projectively euclidean connexion the curvature tensor has at the most n^2 independent components in the general case and $\frac{1}{2}n(n+1)$ if the connexion is volume preserving.

From (1.12) we see that in a D_n all the components of $R_{\nu\mu\lambda}^{\cdot\cdot\cdot\varkappa}$ vanish for which $\varkappa$ is not equal to at least one of the indices ν, μ or λ. It can be proved that for $n > 2$ this condition, if valid for every coordinate system, is also sufficient:[1])

An A_n, $n > 2$, is a D_n if and only if the equation

$$R_{\nu\mu\lambda}^{\cdot\cdot\cdot\varkappa} = 0 \quad \text{for} \quad \varkappa \neq \nu, \mu, \lambda \tag{1.17}$$

holds for every coordinate system at every point.

Another n.a.s. condition can be derived as follows. The X_{n-1} with the equation $f(\xi^\varkappa) = c$ is geodesic (V § 7) if and only if $v^\mu v^\lambda \nabla_\mu \nabla_\lambda f = 0$ is a consequence of $v^\mu \nabla_\mu f = 0$ and $f = c$. Hence a n.a.s. condition is that (cf. Exerc. I 3,2)

$$\nabla_\mu w_\lambda = 2q_{(\mu} w_{\lambda)}; \qquad w_\lambda \stackrel{\text{def}}{=} \nabla_\lambda f, \tag{1.18}$$

at all points of the X_{n-1}. Now the first integrability conditions of this equation are

$$\begin{cases} -R_{\nu\mu\lambda}^{\cdot\cdot\cdot\varkappa} w_\varkappa = 2(\nabla_{[\nu} q_{\mu]}) w_\lambda + 2q_{[\mu} \nabla_{\nu]} w_\lambda + 2w_{[\mu} \nabla_{\nu]} q_\lambda \\ \qquad = 2w_\lambda \nabla_{[\nu} q_{\mu]} + 2w_{[\nu} (q_{\mu]} q_\lambda - \nabla_{\mu]} q_\lambda). \end{cases} \tag{1.19}$$

If (1.18) is totally integrable, that is if (1.19) can be identically satisfied for every value of w_λ, then for every point of A_n and every $(n-1)$-direction through this point, there exists a geodesic X_{n-1} having at this point this given $(n-1)$-direction. Now taking $w_\lambda = \overset{1}{e}_\lambda$ it follows from (1.19) that $R_{\nu\mu\lambda}^{\cdot\cdot\cdot 1} = 0$ for $\lambda, \mu, \nu \neq 1$ and this holds also mutatis mutandis for $w_\lambda = \overset{2}{e}_\lambda, \ldots, w_\lambda = \overset{n}{e}_\lambda$ and for every choice of the coordinate system. Hence, for $n > 2$, the A_n must be a D_n and because in an E_n and therefore also in a D_n there exist geodesic X_{n-1}'s through every point in every $(n-1)$-direction, the condition is also necessary: [2])

An A_n, $n > 2$, is projectively euclidean if and only if there exist through every point geodesic X_{n-1}'s with every $(n-1)$-direction at that point.

In the case when the A_n is already an E_n we may require all projective transformations of the $\Gamma_{\mu\lambda}^\varkappa$ for which the E_n remains an E_n. From (1.4, 5) it follows that $p_{\mu\lambda} = 0$ and that accordingly p_λ is a solution of the equation

$$\partial_\mu p_\lambda - p_\mu p_\lambda = 0 \tag{1.20}$$

---

[1]) E II 1938, 2, p. 181.

[2]) E II 1938, 2, p. 182.

if the coordinate system is chosen such that $\Gamma_{\mu\lambda}^{\varkappa}=0$. From (1.20) it follows that

$$d p_\lambda = p_\lambda p_\mu d\xi^\mu \tag{1.21}$$

and this means that p_λ is a parallel field (cf. III § 1). Taking the rectilinear coordinate system such that $p_\lambda = p\overset{1}{e}_\lambda$ we get for p the equations

$$\partial_1 p = p p\,; \quad \partial_\beta p = 0; \quad \beta = 2, \ldots, n \tag{1.22}$$

leading to the general solution for p_λ

$$p_\lambda = p\overset{1}{e}_\lambda = -(\xi^1 + C)^{-1}\overset{1}{e}_\lambda; \quad C = \text{const.} \tag{1.23}$$

For every choice of $\overset{1}{e}_\lambda$ and C there is one solution. If for any definitely chosen solution we lay the first E_{n-1} of the p_λ belonging to the point $\xi^\varkappa$ through this point, the second E_{n-1} goes through the point $\xi^1 = -C$ on the 1-axis.[1] Hence this second E_{n-1} is common to the p_λ's of the whole field. In fact it is the improper E_{n-1} for the transformed connexion. To prove this we take the affine parameter z on the geodesic 1-axis for the given connexion equal to ξ^1. Then, according to (III 7.11) we get for the affine parameter belonging to the transformed connexion

$$\left\{\begin{aligned} {}'z &= C_1 \int e^{+2\int p_\mu d\xi^\mu} dz + C_2 \\ &= C_1 \int e^{-2\log(\xi^1 + C)} d\xi^1 + C_2 \\ &= -C_1(\xi^1 + C)^{-1} + C_2 \end{aligned}\right. \tag{1.24}$$

which takes the value ∞ for $\xi^1 = -C$. Hence

If the connexion of an E_n is transformed projectively so that the E_n remains an E_n, the vector field p_λ consists of parallel covariant vectors which all have the same second E_{n-1} and this E_{n-1} is the improper E_{n-1} of the transformed E_n.[2]

Exercises.

VI 1,1[3]). Prove that $V_{\nu\mu}=0$ in a D_n in which a covariant constant field $v^\varkappa$ exists.

VI 1,2. An A_n is a E_n if and only if for every value of $\overset{0}{p}_\lambda$ at an arbitrary point $\overset{0}{\xi^\varkappa}$ there exists a projective transformation that leaves the curvature tensor invariant and whose vector takes the value $\overset{0}{p}_\lambda$ at $\overset{0}{\xi^\varkappa}$.

---

[1]) Cf. for the geometric representation of a covariant vector I § 2.

[2]) R. K. 1924, 1, p. 133; E II 1938, 2, p. 184.

[3]) LEVINE 1949, 1. The term parallel is used there in the sense of covariant constant. Cf. III § 2, VI § 9. Cf. LEVINE 1950, 1 for the C_n.

VI 1,3 [1]). The projective transformations with vectors p_λ, $2p_\lambda$ and $3p_\lambda$ transform $R_{\nu\mu\lambda}^{\cdot\cdot\cdot\varkappa}$ into $\overset{1}{R}{}_{\nu\mu\lambda}^{\cdot\cdot\cdot\varkappa}$, $\overset{2}{R}{}_{\nu\mu\lambda}^{\cdot\cdot\cdot\varkappa}$ and $\overset{3}{R}{}_{\nu\mu\lambda}^{\cdot\cdot\cdot\varkappa}$. Prove that

VI 1,3 α) $$R_{\nu\mu\lambda}^{\cdot\cdot\cdot\varkappa} - 3\overset{1}{R}{}_{\nu\mu\lambda}^{\cdot\cdot\cdot\varkappa} + 3\overset{2}{R}{}_{\nu\mu\lambda}^{\cdot\cdot\cdot\varkappa} - \overset{3}{R}{}_{\nu\mu\lambda}^{\cdot\cdot\cdot\varkappa} = 0.$$

VI 1,4 [2]). Prove the identity for A_n, $n>2$

VI 1,4 α) $$\nabla_{[\omega} P_{\nu\mu]\lambda}^{\cdot\cdot\cdot\varkappa} - \frac{1}{n-2} A^{\varkappa}_{[\omega} \nabla_{|\sigma|} P_{\nu\mu]\lambda}^{\cdot\cdot\cdot\sigma} = 0.$$

§ 2. Projective transformation of the connexion in a V_n. [3])

Every S_n is a D_n because in consequence of (cf. III § 5)

(2.1) $$K_{\nu\mu\lambda}^{\cdot\cdot\cdot\varkappa} = 2\varkappa\, A^{\varkappa}_{[\nu}\, g_{\mu]\lambda}$$

the equations (1.12) are identically satisfied for $P_{\mu\lambda} = -\varkappa g_{\mu\lambda}$ and P_λ is a solution of the equations

(2.2) $$\nabla_\mu P_\lambda - P_\mu P_\lambda = \varkappa\, g_{\mu\lambda}$$

whose integrability conditions are identically satisfied. Because two D_n's can always be mapped on each other so that geodesics remain geodesics, the same holds for two S_n's and for an S_n and an R_n. [4])

If, in a V_n, a new (symmetric) fundamental tensor $'g_{\lambda\varkappa}$ is introduced, a new connexion arises. As follows from (1.1) this change of connexion preserves the geodesics if and only if a vector p_λ exists such that

(2.3) $$\nabla_\mu\, 'g_{\lambda\varkappa} = 2p_\mu\, 'g_{\lambda\varkappa} + p_\lambda\, 'g_{\mu\varkappa} + p_\varkappa\, 'g_{\lambda\mu}.$$

By transvection with $'\overset{-1}{g}{}^{\varkappa\lambda}$ we get (cf. III 3.12)

(2.4) $$\begin{cases} 2(n+1)\, p_\mu = '\overset{-1}{g}{}^{\varkappa\lambda}\, \nabla_\mu\, 'g_{\lambda\varkappa} = '\overset{-1}{g}{}^{\varkappa\lambda}\, \partial_\mu\, 'g_{\lambda\varkappa} - 2\Gamma_\mu = \partial_\mu \log \dfrac{'\mathfrak{g}}{\mathfrak{g}}; \\ '\mathfrak{g} \overset{\text{def}}{=} |\mathrm{Det}\,('g_{\lambda\varkappa})|. \end{cases}$$

As a corollary we get that p_λ is a gradient vector, which could have been foreseen because the connexion remains riemannian and therefore remains volume preserving. From (2.3) and (2.4) it follows that the transformation of the connexion is projective if and only if

(2.5) $$2(n+1)\, \nabla_\mu\, 'g_{\lambda\varkappa} = 'g_{\lambda\varkappa}\, \partial_\mu \log \frac{'\mathfrak{g}}{\mathfrak{g}} + 3\, 'g_{(\lambda\varkappa}\, \partial_{\mu)} \log \frac{'\mathfrak{g}}{\mathfrak{g}}.$$

---

1) Berwald gave a similar proposition for conformal transformations in V_2 (Arch. Math. u. Phys. 1918, 27, p. 81; Jahr. D. M. V. 1923, 22, p. 48).

2) R. K. 1924, 1, p. 132; E II 1938, 2, p. 182.

3) Cf. E II 1938, 2, p. 185ff. also for S_n.

4) Note that we do not consider spaces in the large.

The differential equation of the geodesics of an A_n

$$\frac{\delta}{dz}\frac{d\xi^\varkappa}{dz} = 0 \tag{2.6}$$

is said to have a *first integral of degree* p[1]) if a tensor field $q_{\lambda_1 \ldots \lambda_p}$ exists such that

$$q_{\lambda_1 \ldots \lambda_p}\frac{d\xi^{\lambda_1}}{dz}\cdots\frac{d\xi^{\lambda_p}}{dz} = \text{const.} \tag{2.7}$$

along every geodesic, and this condition is satisfied if and only if

$$\nabla_{(\mu}\, q_{\lambda_1 \ldots \lambda_p)} = 0. \tag{2.8}$$

Now it follows from (2.5) that

$$\nabla_{(\mu}\, \mu^2\, 'g_{\lambda\varkappa)} = 0; \quad \mu^{-(n+1)} = \frac{'\mathfrak{g}}{\mathfrak{g}} \tag{2.9}$$

and this proves that in a V_n which can be transformed projectively into another V_n, the equation of the geodesics admits besides $g_{\lambda\varkappa}\frac{d\xi^\lambda}{ds}\frac{d\xi^\varkappa}{ds} = \text{const.}$ another first integral $\mu^2\, 'g_{\lambda\varkappa}\frac{d\xi^\lambda}{ds}\frac{d\xi^\varkappa}{ds} = \text{const.}$

If the principal directions of $'g_{\lambda\varkappa}$ are uniquely determined, it follows from (2.3) and (V 5.17) that they are all V_{n-1}-normal. The integrability conditions of (2.3) are

$$2K_{\nu\mu(\lambda}^{\cdot\cdot\cdot\varrho}\, 'g_{\varkappa)\varrho} = +4p_{[\nu(\lambda}\, 'g_{\mu]\varkappa)} \tag{2.10}$$

because p_λ is a gradientvector. This equation could also be derived from (1.3) by substituting $K_{\nu\mu\lambda}^{\cdot\cdot\cdot\varkappa}$ and $'K_{\nu\mu\lambda}^{\cdot\cdot\cdot\varkappa}$ for $R_{\nu\mu\lambda}^{\cdot\cdot\cdot\varkappa}$ and $'R_{\nu\mu\lambda}^{\cdot\cdot\cdot\varkappa}$ because $p_{[\nu\mu]} = 0$ and $'K_{\nu\mu(\lambda}^{\cdot\cdot\cdot\varrho}\, 'g_{\varkappa)\varrho} = 0$.

If the V_n is an S_n (2.10) takes the form (cf. III § 5)

$$0 = +4\varkappa\, g_{[\nu(\lambda} A_{\mu]}^{\varrho}\, 'g_{\varkappa)\varrho} + 4p_{[\nu(\lambda}\, 'g_{\mu]\varkappa)} = 4(p_{[\nu(\lambda} + \varkappa\, g_{[\nu(\lambda})\, 'g_{\mu]\varkappa)}; \tag{2.11}$$

but this equation expresses the fact that $p_{\nu\lambda} + \varkappa g_{\nu\lambda}$ and $'g_{\nu\lambda}$ differ only by a scalar factor. Let this factor be α

$$p_{\nu\lambda} + \varkappa\, g_{\nu\lambda} = \alpha\, 'g_{\nu\lambda} \tag{2.12}$$

then we get from (1.3)

$$'K_{\nu\mu\lambda}^{\cdot\cdot\cdot\varkappa} = -2\varkappa\, g_{[\nu|\lambda|} A_{\mu]}^{\varkappa} - 2\varkappa\, A_{[\nu}^{\varkappa}\, g_{\mu]\lambda} + 2\alpha\, A_{[\nu}^{\varkappa}\, 'g_{\mu]\lambda} = 2\alpha\, A_{[\nu}^{\varkappa}\, 'g_{\mu]\lambda} \tag{2.13}$$

and this proves that for $n > 2$ after the transformation once more we have an S_n but with the scalar curvature $+\alpha$. Hence

[1]) Cf. EISENHART 1924, 1; 1927, 1, p. 83–86; LIBER 1941, 1.

If the connexion of an S_n, $n>2$, is transformed projectively and if the transform happens to be the connexion of a V_n, this V_n is an S_n.[1])

As a corollary we get that a V_n is projectively euclidean if and only if it is an S_n [2]).

We return now to the general V_n. By transvecting (2.10) with $g^{\mu\lambda}$ we get

$$(2.14)\quad K_{\nu}^{\cdot\varrho}\,'g_{\varrho\varkappa}+K_{\nu\cdot\varkappa}^{\cdot\sigma\cdot\varrho}\,'g_{\varrho\sigma}=p_{\nu}^{\cdot\varrho}\,'g_{\varrho\varkappa}-p_{\varrho\sigma}g^{\varrho\sigma}\,'g_{\nu\varkappa}+p_{\nu\varkappa}\,'g_{\varrho\sigma}g^{\varrho\sigma}-p^{\varrho}_{\cdot\varkappa}\,'g_{\varrho\nu}$$

hence from (2.14) and (2.10) transvected with $g^{\lambda\varkappa}$

$$(2.15)\qquad K_{[\nu}^{\cdot\varrho}\,'g_{|\varrho|\varkappa]}=0.$$

But this equation expresses the fact that the symmetric tensors $'g_{\varkappa\lambda}$ and $'K_{\varkappa\lambda}$ have at least one set of n mutually perpendicular principal directions in common (cf. Exerc. I 9,3). Transvection of (2.10) with $g^{\lambda\varkappa}$ leads to

$$(2.16)\qquad p_{[\nu}^{\cdot\lambda}\,'g_{\mu]\lambda}=0$$

and this expresses the fact that also the symmetric tensors $'g_{\varkappa\lambda}$ and $p_{\varkappa\lambda}$ have at least one set of n mutually perpendicular principal directions in common. But (2.10) contains still more information about these principal directions. If a local coordinate system (h) with unit basis vectors is laid in principal directions common to $'g_{\varkappa\lambda}$ and $p_{\varkappa\lambda}$, (2.10) can be written as

$$(2.17)\quad K_{kjih}\,('g_{hh}-'g_{ii})\overset{*}{=}p_{ki}\,'g_{jh}-p_{ji}\,'g_{kh}+p_{kh}\,'g_{ji}-p_{jh}\,'g_{ki}.$$

For $k=j$ both sides vanish and the same holds for $i=h$. For $k, j, i \neq$ we get[3])

$$(2.18\text{a})\qquad K_{kjih}\overset{*}{=}0$$

provided that h and i do not belong to the same principal multidirection. For $k=i$, $h=j$ we get

$$(2.18\text{b})\qquad K_{ijij}\,('g_{jj}-'g_{ii})\overset{*}{=}p_{ii}\,'g_{jj}-p_{jj}\,'g_{ii}.$$

Hence, if $'g_{ii}='g_{jj}$, it follows that $p_{ii}=p_{jj}$ and this means that every principal multidirection of $'g_{\lambda\varkappa}$ is contained in a principal multidirection of $p_{\lambda\varkappa}$. But this has as a consequence that every principal direction of $'g_{\lambda\varkappa}$ is also a principal direction of $p_{\lambda\varkappa}$. Hence the mutually perpendicular principal directions common to $'g_{\lambda\varkappa}$ and $K_{\lambda\varkappa}$ are also principal directions of $p_{\lambda\varkappa}$, and thus from (1.4a) in the form

$$(2.19)\qquad 'K_{\mu\lambda}=K_{\mu\lambda}+(n-1)\,p_{\mu\lambda}$$

---

[1]) BELTRAMI 1868, 1; cf. EISENHART 1926, 1, p. 134.
[2]) WEYL 1921, 2, p. 110.
[3]) FUBINI 1905, 1, p. 306.

it follows that they are also principal directions of $'K_{\lambda\varkappa}$. This proves the theorem

The symmetric tensors $'g_{\lambda\varkappa}$, $p_{\lambda\varkappa}$, $K_{\lambda\varkappa}$ *and* $'K_{\lambda\varkappa}$ *have always at least one set of* n *mutually perpendicular principal directions in common.*[1])

We now consider the special case where the principal directions of $'g_{\lambda\varkappa}$ are uniquely determined. Taking the anholonomic coordinate system (h) with unit basis vectors $\underset{i}{i}^{\varkappa}$, $\overset{h}{i}_{\lambda}$ in these directions, (2.3) takes the form

$$(2.20)\quad \left\{\begin{aligned} \partial_j\, 'g_{ih} - 2\,\Gamma^l_{j(i}\, 'g_{h)l} &\overset{*}{=} -\,'g_{ih}\,\partial_j \log p - \tfrac{1}{2}\,'g_{hj}\,\partial_i \log p - \tfrac{1}{2}\,'g_{ij}\,\partial_h \log p;\\ p_i &\overset{\text{def}}{=} -\tfrac{1}{2}\,\partial_i \log p \end{aligned}\right.$$

and from this we derive (cf. III 9.11) for all possible assumptions with respect to h, i and j

$$(2.21)\quad \left\{\begin{array}{lll} \text{a)} & \Gamma_{jih}('g_{hh} - 'g_{ii}) \overset{*}{=} 0; & (h, i, j \neq)\\ \text{b)} & \partial_j\, 'g_{ii} \overset{*}{=} -\,'g_{ii}\,\partial_j \log p; & (h = i \neq j)\\ \text{c)} & \Gamma_{iih}('g_{ii} - 'g_{hh}) \overset{*}{=} -\tfrac{1}{2}\,'g_{ii}\,\partial_h \log p; & (h \neq i = j)\\ \text{d)} & \partial_i\, 'g_{ii} \overset{*}{=} -\,2\,'g_{ii}\,\partial_i \log p; & (h = i = j). \end{array}\right.$$

From (2.21 a) and (V 5.12) it follows once more that the n congruences $\underset{i}{i}^{\varkappa}$ are V_{n-1}-normal. Accordingly the coordinate system $(\varkappa)$ can be chosen such that its contravariant basis vectors are tangent to those congruences. This means that the linear element can be written in the form

$$(2.22)\qquad ds^2 = \sum_h \overset{h}{H}\overset{h}{H}\,\overset{h}{e}_{\mu}\,\overset{h}{e}_{\lambda}\,d\xi^{\mu}\,d\xi^{\lambda}; \qquad \overset{h}{e}_{\lambda} \overset{\text{def}}{=} \delta^h_{\varkappa}\,\overset{\varkappa}{e}_{\lambda}$$

and

$$(2.23)\qquad \overset{h}{i}_{\lambda} = \overset{h}{H}\,\overset{h}{e}_{\lambda}.$$

Then we have for $h \neq i$

$$(2.24)\quad \left\{\begin{aligned} \Gamma^h_{ii} &\overset{*}{=} -\,V_i\,\overset{h}{i}_i \overset{*}{=} -\,\underset{i}{i}^{\mu}\,\underset{i}{i}^{\lambda}\,V_{\mu}\,\overset{h}{i}_{\lambda} = \overset{i}{i}^{\mu}\,\overset{h}{i}^{\lambda}\,V_{\mu}\,\overset{i}{i}_{\lambda} = \overset{i}{i}^{\mu}\,\overset{h}{i}^{\lambda}\,V_{\mu}\overset{i}{H}\,\overset{i}{e}_{\lambda}\\ &= \overset{i}{H}\,\overset{i}{i}^{\mu}\,\overset{h}{i}^{\lambda}\,V_{\mu}\,\overset{i}{e}_{\lambda} = \overset{i}{H}\,\overset{i}{i}^{\mu}\,\overset{h}{i}^{\lambda}\,V_{\mu}\overset{i}{H}{}^{-1}\,\overset{i}{i}_{\lambda} \overset{*}{=} \overset{i}{H}\,\partial_h \overset{i}{H}{}^{-1} \end{aligned}\right.$$

and accordingly from (2.21 b, c, d), writing ϱ_i for the $'g_{ii}$ from (2.21)

$$(2.25)\quad \left\{\begin{array}{lll} \text{a)} & \partial_j\, p\,\varrho_i \overset{*}{=} 0; & i \neq j\\ \text{b)} & (\varrho_h - \varrho_i)\,\partial_h \log \overset{i}{H} \overset{*}{=} \tfrac{1}{2}\,\partial_h\,\varrho_i; & h \neq i\\ \text{c)} & \partial_i\, p\,\varrho_i + \varrho_i\,\partial_i\, p \overset{*}{=} 0 & \end{array}\right.$$

[1]) R. K. 1924, 1, p. 206.

in which ∂_j can be interpreted as $A_j^\mu \partial_\mu$ but also as $\delta_j^\mu \partial_\mu$, because $A_j^\mu \overset{*}{=} \overset{j}{H}{}^{-1} \delta_j^\mu$ (no summation over j). The integrability conditions of (2.25) are the well known equations of LAMÉ[1])

$$(2.26)\quad \partial_j \partial_i \overset{h}{H} \overset{*}{=} \overset{i}{H}{}^{-1} (\partial_j \overset{i}{H})\, \partial_i \overset{h}{H} + \overset{j}{H}{}^{-1} (\partial_i \overset{j}{H})\, (\partial_j \overset{h}{H})\,; \qquad h, i, j \neq;$$

(no summation over i and j)

which are equivalent to

$$(2.27)\qquad K_{kjik} \overset{*}{=} 0; \qquad k, j, i \neq$$

and this is a special case of (2.18a).

Exercises.

VI 2,1[2]). A V_n can be transformed projectively into a V_n if and only if there exist a gradient vector $p_\lambda \neq 0$ and a symmetric tensor $'g_{\lambda\varkappa}$ of rank n satisfying (2.3).

VI 2,2[3]). In a V_n a covariant constant symmetric tensor field $'g_{\lambda\varkappa}$ of rank n can exist if and only if

a) the differential equation of the geodesics admits a first integral $'g_{\lambda\varkappa} \dfrac{d\xi^\lambda}{dz} \dfrac{d\xi^\varkappa}{dz} = \text{const.}$;

b) the connexion of the V_n can be transformed projectively into the riemannian connexion belonging to $'g_{\lambda\varkappa}$.

§ 3. Imbedded spaces in A_n under projective transformations of the connexion.[4])

Let η^a; $a = 1, \ldots, m$ be a set of coordinates in an X_m imbedded and rigged in A_n and let Γ'^a_{cb} be the parameters of the connexion induced in X_m (cf. V § 3). Then we have for each vector field v^a of X_m

$$(3.1)\qquad \nabla'_c v^a = \partial_c v^a + \Gamma'^a_{cb} v^b = B^{\mu a}_{c\varkappa} (\partial_\mu v^\varkappa + \Gamma^\varkappa_{\mu\lambda} v^\lambda)$$

---

[1]) Cf. BIANCHI-LUKAT 1899, 1, p. 485. Cf. for the integration of (2.25) LEVI CIVITA 1896, 1; FUBINI 1905, 1; WRIGHT 1908, 1, p. 80ff.

[2]) E II 1938, 2, p. 186.

[3]) LEVY 1925, 2; EISENHART 1927, 1, p. 80; E II 1938, 2, p. 193. Cf. CARTAN 1927, 3 for another property of V_n's of this kind.

[4]) General references on projective imbedding: SCHOUTEN and HAANTJES 1936, 1; BORTOLOTTI 1941, 2; BOMPIANI 1943, 1; KIMPARA 1943, 1; KANITANI 1943, 1; 2; 1947, 1; 1948, 1; 1949, 1; 1950, 2; NORDEN 1945, 3; 1947, 1; 1948, 1; 1949, 1; BOL 1950, 1.

hence (cf. V 3.10) [1])

$$\Gamma'^{a}_{cb} = B^{\mu\lambda a}_{cb\varkappa}\Gamma^{\varkappa}_{\mu\lambda} - B^{\mu\lambda}_{cb}\partial_\mu B^a_\lambda. \tag{3.2}$$

If now the connexion in A_n suffers the projective transformation (1.1) we get for the new connexion in X_m

$$\begin{cases} {}'\Gamma'^{a}_{cb} = B^{\mu\lambda a}_{cb\varkappa}(\Gamma^{\varkappa}_{\mu\lambda} + p_\mu A^\varkappa_\lambda + p_\lambda A^\varkappa_\mu) - B^{\mu\lambda}_{cb}\partial_\mu B^a_\lambda \\ \quad = \Gamma'^{a}_{cb} + p'_c B^a_b + p'_b B^a_c; \quad p'_b \overset{\text{def}}{=} B^\lambda_b p_\lambda \end{cases} \tag{3.3}$$

and this proves that *the induced connexion is also transformed projectively*. From this it follows at once that a *geodesic X_m in A_n remains geodesic*. This is also one of the consequences of the fact that the quantities $\overset{m}{H}{}^{\cdot\cdot\varkappa}_{\mu\lambda}$ and $\overset{m'}{H}{}^{\cdot\cdot\varkappa}_{\mu\lambda}$ defined in (V 7.9, 11) are invariant for projective transformations of the connexion in A_n. The quantities $\overset{m}{L}{}^{\cdot\,\varkappa}_{\mu\cdot\lambda}$ and $\overset{m'}{L}{}^{\cdot\,\varkappa}_{\mu\cdot\lambda}$ are in general not invariant, for instance

$$\begin{cases} {}'\overset{m}{L}{}^{\cdot\,\varkappa}_{\mu\cdot\lambda} = \overset{m}{L}{}^{\cdot\,\varkappa}_{\mu\cdot\lambda} + B^{\tau\varkappa}_{\mu\varrho}(-p_\tau A^\sigma_\lambda B^\varrho_\sigma - p_\lambda A^\sigma_\tau B^\varrho_\sigma + p_\tau A^\varrho_\sigma B^\sigma_\lambda + p_\sigma A^\varrho_\tau B^\sigma_\lambda) \\ \quad = \overset{m}{L}{}^{\cdot\,\varkappa}_{\mu\cdot\lambda} - B^\varkappa_\mu p''_\lambda; \quad p''_\lambda \overset{\text{def}}{=} C^\varkappa_\lambda p_\varkappa. \end{cases} \tag{3.4}$$

If for an X_{n-1} in A_n the rigging is chosen in the way described in V § 3, it depends on the connexion in A_n. Because it was assumed in that section that the connexion was volume preserving we need only consider here those projective transformations for which p_λ is a gradient, $p_\lambda = \partial_\lambda p$. In V § 3 the vector t_λ was fixed by (V 3.31) and in that formula $\mathfrak{q}$ was a covariant constant scalar density field of weight $+1$. Let now $'\mathfrak{q} = \varrho\,\mathfrak{q}$ be covariant constant for the transformed connexion

$$'\nabla_\lambda\,'\mathfrak{q} = (\nabla_\lambda\varrho)\,\mathfrak{q} + \varrho\,'\nabla_\lambda\mathfrak{q} = \mathfrak{q}\,\nabla_\lambda\varrho - (n+1)\,\varrho\,p_\lambda\,\mathfrak{q}, \tag{3.5}$$

then

$$\nabla_\lambda \log\varrho = (n+1)\,p_\lambda \tag{3.6}$$

and according to (V 3.31), if $'t_\lambda = \sigma t_\lambda$ is the new normalized tangent vector, we have

$$\sigma t_\lambda = {}'\mathfrak{q}\,\sigma^{-\frac{n}{2}}\mathfrak{h}^{-\frac{1}{2}}\,\mathfrak{t}_\lambda = \varrho\,\sigma^{-\frac{n}{2}}\,t_\lambda \tag{3.7}$$

or

$$(n+2)\log\sigma = 2\log\varrho \tag{3.8}$$

at all points of X_{n-1}, hence

$$\partial_b \log\sigma = 2\,\frac{n+1}{n+2}\,p'_b; \quad p'_b \overset{\text{def}}{=} B^\lambda_b p_\lambda. \tag{3.9}$$

[1]) We use an accent on the right here to denote the induced connexion in order to preserve the accent on the left for the transformed objects.

For the new vector $'n^\varkappa$ we find from (V 3.28)

$$(3.10)\quad 'n^\varkappa = \sigma^{-1}\left(n^\varkappa + B_b^\varkappa \overset{-1}{h^{ab}}\,\partial_a \log\sigma\right) = \sigma^{-1}\left(n^\varkappa + 2\,\frac{n+1}{n+2}\,B_b^\varkappa \overset{-1}{h^{ab}}\,'p_a\right).$$

According to (V 3.11, 12) we get for the new values $'h_{cb}$ and $'l_c^{\cdot a}$

$$(3.11)\qquad 'h_{cb} = -\,B_{cb}^{\mu\lambda}\,'\nabla_\mu\,'t_\lambda = \sigma\,h_{cb} - 2\,B_{cb}^{\mu\lambda}\,p_\mu\,t_\lambda = \sigma\,h_{cb}$$

$$(3.12)\quad \begin{cases} 'l_c^{\cdot a} = -\,B_{c\varkappa}^{\mu a}\,'\nabla_\mu\,'n^\varkappa \\ \qquad = \sigma^{-1}\left(l_c^{\cdot a} - B_c^a\,p_\lambda\,n^\lambda + \dfrac{n}{n+2}\,v^a\,p_c' - B_c^a\,p_b'\,v^b - B_{c\varkappa}^{\mu a}\,\nabla_\mu\,v^\varkappa\right)\end{cases}$$

where

$$(3.13)\qquad v^a \overset{\text{def}}{=} 2\,\frac{n+1}{n+2}\,\overset{-1}{h^{ab}}\,p_b'.$$

From (3.11) we see that the asymptotic lines are invariant for the transformations under consideration (cf. V § 3). The lines of curvature have not this independence as can be seen from (3.12).

If a curve is given in A_n with an arbitrary parameter t, the covariant derivative $\dfrac{\delta}{dt}\dfrac{d\xi^\varkappa}{dt}$ depends on the choice of this parameter. If $z = z(t)$ is introduced as a new parameter, we get

$$(3.14)\quad \frac{\delta}{dt}\frac{d\xi^\varkappa}{dt} = z'\,\frac{\delta}{dz}\,z'\,\frac{d\xi^\varkappa}{dz} = z'^2\,\frac{\delta^2\xi^\varkappa}{dz^2} + z''\,\frac{d\xi^\varkappa}{dz};\qquad z' \overset{\text{def}}{=} \frac{dz}{dt};\qquad z'' \overset{\text{def}}{=} \frac{d^2 z}{dt^2}$$

hence the 2-direction spanned by $d\xi^\varkappa/dt$ and $\delta^2\xi^\varkappa/dt^2$ is independent of the choice of the parameter. Now if we write

$$(3.15)\qquad t^{\varkappa\lambda} \overset{\text{def}}{=} \frac{d\xi^{[\varkappa}}{dt}\,\frac{\delta^2\xi^{\lambda]}}{dt^2},$$

this quantity transforms into

$$(3.16)\qquad 't^{\varkappa\lambda} \overset{\text{def}}{=} \frac{d\xi^{[\varkappa}}{dz}\,\frac{\delta^2\xi^{\lambda]}}{dz^2} = z'^{-3}\,t^{\varkappa\lambda}.$$

A curve in A_n will be called *quasi-plane*[1]) if $t^{\varkappa\lambda}$ is parallel along the curve:

$$(3.17)\qquad \frac{\delta}{dt}\,t^{\varkappa\lambda} = \alpha\,t^{\varkappa\lambda}.$$

In this case we have

$$(3.18)\qquad z'^{-1}\,\frac{\delta}{dz}\,z'^3\,'t^{\varkappa\lambda} = 3\,z''\,'t^{\varkappa\lambda} + z'^2\,\frac{\delta}{dz}\,'t^{\varkappa\lambda} = \alpha\,z'\,'t^{\varkappa\lambda}$$

and this proves that

$$(3.19)\qquad \frac{\delta}{dz}\,'t^{\varkappa\lambda} = 0$$

[1]) YANO and TAKANO 1944, 1; 2; cf. YANO, TAKANO and TOMONAGA 1948, 2; YANO 1949, 2.

provided that z is a solution of the equation

$$\frac{z''}{z'} = \frac{1}{3}\alpha. \tag{3.20}$$

The general solution of (3.20) has the form

$$z = C_1 \int e^{\int \frac{1}{3}\alpha\, dt}\, dt + C_2; \tag{3.21}$$

where C_1 and C_2 are constants and $C_1 \neq 0$. z is called the *affine parameter* of the quasi-plane curve. It is fixed to within an affine transformation. C_2 fixes the nullpoint of the scale and C_1 fixes a kind of gauge. If z is chosen in this way we call $n^\varkappa = \delta^2 \xi^\varkappa / d z^2$ the *affine normal* of the curve. It is fixed to within a *constant* factor. According to (3.19) we have along the curve

$$\frac{\delta^3 \xi^\varkappa}{d z^3} + k(z) \frac{d \xi^\varkappa}{d z} = 0 \tag{3.22}$$

where k is a function of z. From (3.17) we get

$$\frac{\delta^3 \xi^\varkappa}{\delta t^3} - \alpha(t) \frac{\delta^2 \xi^\varkappa}{d t^2} - \beta(t) \frac{d \xi^\varkappa}{d t} = 0 \tag{3.23}$$

and after the introduction of z

$$z'^3 \frac{\delta^3 \xi^\varkappa}{d z^3} + (z''' - \alpha z'' - \beta z') \frac{d \xi^\varkappa}{d z} = 0 \tag{3.24}$$

from which

$$k = z'^{-3} (z''' - \alpha z'' - \beta z') = C_1^{-2} e^{-\frac{2}{3}\int \alpha\, dt} \left(\frac{1}{3}\alpha' - \frac{2}{9}\alpha^2 - \beta \right). \tag{3.25}$$

If $k = \text{const.}$, the curve is called an *affine conic*. Hence the n.a.s. conditions for a quasi-plane curve to be an affine conic are

$$\frac{1}{3}\alpha'' - \frac{2}{3}\alpha\alpha' + \frac{4}{27}\alpha^3 + \frac{2}{3}\alpha\beta - \beta' = 0. \tag{3.26}$$

If the connexion is transformed projectively we have for any vector $v^\varkappa$ along the curve (cf. 1.1)

$$\frac{'\delta}{dt} v^\varkappa = \frac{\delta}{dt} v^\varkappa + p_\mu \frac{d\xi^\mu}{dt} v^\varkappa + p_\mu v^\mu \frac{d\xi^\varkappa}{dt} \tag{3.27}$$

hence, for any functions $'\alpha(t)$ and $'\beta(t)$

$$\left\{ \begin{aligned} & \frac{'\delta^3 \xi^\varkappa}{dt^3} - {'\alpha} \frac{'\delta^2 \xi^\varkappa}{dt^2} - {'\beta} \frac{d\xi^\varkappa}{dt} \\ & \quad = \frac{\delta^3 \xi^\varkappa}{dt^3} - ('\alpha - 3p) \frac{\delta^2 \xi^\varkappa}{dt^2} - ('\beta + 3p' - 2\alpha p - 2p^2 - q) \frac{d\xi^\varkappa}{dt}; \\ & p \overset{\text{def}}{=} \frac{d\xi^\mu}{dt} p_\mu; \qquad q \overset{\text{def}}{=} \frac{d\xi^\mu}{dt} \frac{\delta p_\mu}{dt}. \end{aligned} \right. \tag{3.28}$$

Equating this to zero we get on account of (3.23)

$$(3.29)\quad \begin{cases} '\alpha = \alpha + 3p \\ '\beta = \beta + 3p' - q - 2p^2 - 2\alpha p. \end{cases}$$

From (3.23) and (3.29) it follows that (cf. 1.1)

$$(3.30)\quad \begin{cases} \frac{1}{3}{}'\alpha'' - \frac{2}{3}{}'\alpha \cdot {}'\alpha' + \frac{4}{27}{}'\alpha^3 + \frac{2}{3}{}'\alpha \cdot {}'\beta - {}'\beta' = \frac{1}{3}\alpha'' - \frac{2}{3}\alpha\alpha' + \\ \qquad + \frac{4}{27}\alpha^3 + \frac{2}{3}\alpha\beta - \beta' + 2\frac{d\xi^\mu}{dt}\frac{\delta^2\xi^\lambda}{dt^2}p_{\mu\lambda} - \frac{4}{3}\alpha\Pi + \Pi' \\ \Pi \overset{\text{def}}{=} p_{\mu\lambda}\frac{d\xi^\mu}{dt}\frac{d\xi^\lambda}{dt} \quad \text{(cf. 1.3)}. \end{cases}$$

Hence k remains a constant for all conics if and only if $p_{\mu\lambda}=0$. This proves the theorem[1]) (cf. 1.3):

A projective transformation (1.1) *of the connexion in A_n changes all affine conics into affine conics if and only if $p_{\mu\lambda}=0$, that is for $n \geq 2$, if and only if the transformation leaves the curvature tensor invariant.*

That means that an affine conic is in general not invariant for all projective transformations.[2])

Exercises.

VI 3,1[3]). If a connexion in L_n suffers a transformation that preserves parallelism of directions, the transformation of the connexion induced in a rigged X_m has the same property.

VI 3,2[4]). Among all parameters $\Gamma^\varkappa_{\mu\lambda}$ in L_n with the same parallelism of directions there is at most one set that is symmetric in $\mu\lambda$.

§ 4. Projective connexions.[5])

T. Y. Thomas was the first to remark that the expressions

$$(4.1)\quad \overset{p}{\Gamma}{}^\varkappa_{\mu\lambda} \overset{\text{def}}{=} \Gamma^\varkappa_{\mu\lambda} - \frac{2}{n+1} A^\varkappa_{(\mu}\Gamma_{\lambda)}$$

---

1) Yano and Takano 1944, 2, p. 424.

2) Hombu and Mikami 1941, 1 and Mikami 1941, 2 defined projective conics. Cf. Hombu and Mikami 1942, 1; Yano and Takano 1944, 1 and 1949, 3 also for literature. It can be proved that an affine conic is transformed into a projective conic by every projective transformation of the connexion.

3) Bortolotti 1931, 2, p. 19ff.

4) Bortolotti 1931, 4.

5) *References. General:* Veblen 1922, 1; 2; 1928, 1; 1929, 1; T. Y. Thomas 1922, 1; 1925, 2; 3; 1926, 2; 1927, 3; 1934, 2, Ch. III; Veblen and T. Y. Thomas 1923, 1; J. M. Thomas 1925, 2; Veblen and J. M. Thomas 1926, 1; Weyl 1929, 2; Weyl and Robertson 1929, 1; Whitehead 1929, 1; Golab 1930, 1; Schouten

are invariant for the projective transformations (1.1) of a symmetric connexion. They are called the *projective parameters of* T. Y. THOMAS.[1]) Their transformation is

$$\overset{p}{\Gamma}{}^{\varkappa'}_{\mu'\lambda'} = A^{\mu\lambda\varkappa'}_{\mu'\lambda'\varkappa}\,\overset{p}{\Gamma}{}^{\varkappa}_{\mu\lambda} + A^{\varkappa'}_{\varrho}\,\partial_{\mu'} A^{\varrho}_{\lambda'} + \frac{2}{n+1}\,A^{\varkappa'}_{(\mu'}\,\partial_{\lambda')} \log \Delta\,. \tag{4.2}$$

Hence the $\overset{p}{\Gamma}{}^{\varkappa}_{\mu\lambda}$ are components of a geometric object of a more complicated kind than the $\Gamma^{\varkappa}_{\mu\lambda}$. Therefore the $\overset{p}{\Gamma}{}^{\varkappa}_{\mu\lambda}$ do not fix one linear connexion but only an infinity of symmetric linear connexions that transform into each other by projective transformations and that all belong to the same system of geodesics.

If we first prefer one coordinate system and then consider coordinate transformations with $\Delta = \text{const.}$ only, the $\overset{p}{\Gamma}{}^{\varkappa}_{\mu\lambda}$ transform in the same way as the $\Gamma^{\varkappa}_{\mu\lambda}$ and in this case the $\overset{p}{\Gamma}{}^{\varkappa}_{\mu\lambda}$ represent the only connexion that has the same geodesics as the connexion $\Gamma^{\varkappa}_{\mu\lambda}$ and whose parameters give zero if transvected with $A^{\lambda}_{\varkappa}$.

The $\overset{p}{\Gamma}{}^{\varkappa}_{\mu\lambda}$ can be used to investigate an affine geometry that is given to within projective transformations. On every geodesic a projective parameter can be defined that is fixed to within an affine or a projective transformation with constant coefficients. The equation of the geodesics can be written in the form (cf. III 7.2)

$$\frac{d^2\xi^\varkappa}{dt^2} + \overset{p}{\Gamma}{}^{\varkappa}_{\mu\lambda}\frac{d\xi^\mu}{dt}\frac{d\xi^\lambda}{dt} = \alpha(t)\frac{d\xi^\varkappa}{dt} - \frac{2}{n+1}A^{\varkappa}_{\mu}\Gamma_\lambda\frac{d\xi^\mu}{dt}\frac{d\xi^\lambda}{dt} \tag{4.3}$$

with an arbitrary parameter t. If a new parameter ζ is introduced and if we write $\zeta' \overset{\text{def}}{=} d\zeta/dt$; $\zeta'' \overset{\text{def}}{=} d^2\zeta/dt^2$, (4.3) takes the form

$$\frac{d^2\xi^\varkappa}{d\zeta^2} + \overset{p}{\Gamma}{}^{\varkappa}_{\mu\lambda}\frac{d\xi^\mu}{d\zeta}\frac{d\xi^\lambda}{d\zeta} = \frac{1}{\zeta'}\,\alpha\,\frac{d\xi^\varkappa}{d\zeta} - \frac{2}{n+1}\Gamma_\lambda\frac{d\xi^\lambda}{d\zeta}\frac{d\xi^\varkappa}{d\zeta} - \frac{\zeta''}{(\zeta')^2}\frac{d\xi^\varkappa}{d\zeta} \tag{4.4}$$

and the right hand side of this equation is zero if and only if ζ is a solution of

$$\frac{\zeta''}{\zeta'} = \alpha - \frac{2}{n+1}\Gamma_\lambda\frac{d\xi^\lambda}{dt} \tag{4.5}$$

---

and GOLAB 1930, 3; EISENHART 1930, 1; BORTOLOTTI 1931, 6; 1932, 1; 1933, 2; V. DANTZIG 1932, 1; 2; 3; 1934, 1; HLAVATY 1933, 1; SCHOUTEN and V. DANTZIG 1933, 2; CARTAN 1934, 1; CARTAN 1935, 1; 1937, 1; 3; BERWALD 1936, 1; SCHOUTEN and HAANTJES 1936, 1; BORTOLOTTI and HLAVATY 1936, 2 (literature); HAANTJES 1937, 1; KANITANI 1941, 1; YANO 1942, 4; CHI TA YEN 1948, 1. *Imbedding:* BORTOLOTTI 1941, 2 (literature); BOMPIANI 1943, 1; KIMPARA 1943, 1; KANITANI 1943, 1; 2; 1947, 1; 2; 1948, 1; 1949, 1; 1950, 2; NORDEN 1945, 3; 1947, 1; 1948, 1; 1949, 1; BOL 1950, 1. *Mapping:* EISENHART and KNEBELMAN 1927, 2; WHITEHEAD 1929, 1 (D_n); KANITANI 1947, 2; CZECH 1949, 1; 1950, 1; 2.

[1]) T. Y. THOMAS 1925, 3. The special case $\Delta = +1$ was called by him *equiprojective.*

hence

$$\zeta = C_1 \int e^{\int \alpha\, dt}\, e^{-\frac{2}{n+1}\int \Gamma_\lambda d\xi^\lambda}\, dt + C_2. \tag{4.6}$$

Now it follows from (1.1) and (III 7.2) that $\alpha\, dt - \frac{2}{n+1}\Gamma_\lambda\, d\xi^\lambda$ is invariant for projective transformations of $\Gamma_{\mu\lambda}^{\varkappa}$, hence ζ is a projective invariant. It is the *projective parameter* on a geodesic introduced by T. Y. THOMAS.[1]) If $\overset{p}{\Gamma}{}_{\mu\lambda}^{\varkappa}$ is given, ζ is determined on each geodesic to within an affine transformation. It is remarkable that ζ is not a scalar. If the coordinates are transformed we get

$$e^{\frac{2}{n+1}\int \Gamma_{\lambda'} d\xi^{\lambda'}} = e^{\frac{2}{n+1}\int \Gamma_\lambda d\xi^\lambda}\, e^{-\frac{2}{n+1}\int d\log \Delta} \tag{4.7}$$

hence

$$\overset{(\varkappa')}{d\zeta} = \Delta^{\frac{2}{n+1}}\, \overset{(\varkappa)}{d\zeta} \tag{4.8}$$

and this proves that $d\zeta$ is a scalar Δ-density of weight $-\frac{2}{n+1}$. If a projective parameter on a curve is fixed in any way this parameter can be used to build up the projective differential geometry of the curve, generalized FRENET formulae etc. [2])

But the $\overset{p}{\Gamma}{}_{\mu\lambda}^{\varkappa}$ can also be used in a more general way. The most simple method is to construct a scalar density that is a concomitant of the $\overset{p}{\Gamma}{}_{\mu\lambda}^{\varkappa}$. [3]) Such a density is invariant for projective transformations of the connexion. It fixes a set of coordinate systems with respect to which it has the component $+1$. These coordinate systems transform into each other with $\Delta = +1$ and accordingly the $\overset{p}{\Gamma}{}_{\mu\lambda}^{\varkappa}$ now fix a linear connexion that is projectively invariant. The disadvantage of the method is not that it would be difficult to form a projectively invariant density, in fact there are many of them, but there is no one specially preferred and they all depend on higher derivatives. CARTAN[4]) proposed in 1924 to use a flat projective space instead of an affine space as local space. SCHOUTEN[5]) proved in the same year that this amounts to introducing a linear connexion for local projective spaces and that it is intimately connected with the linear connexions of local

[1]) T. Y. THOMAS 1925, 3; cf. VEBLEN and J. M. THOMAS 1926, 1.

[2]) WHITEHEAD 1931, 2 has introduced a scalar projective parameter determined to within a "projective" transformation and BORTOLOTTI 1932, 1 has used another one. Cf. HLAVATY 1931, 1; BERWALD 1936, 1; HAANTJES 1937, 1; SASAKI 1937, 1; YANO 1937, 2; 3; 1944, 3; 4; E II 1938, 2, p. 194.

[3]) WEITZENBÖCK 1932, 1.

[4]) CARTAN 1924, 1.

[5]) SCHOUTEN 1924, 2.

E_N's; $N \neq n$ already introduced by KÖNIG in 1920.[1]) EISENHART, VEBLEN and T. Y. THOMAS first considered the generalized projective geometry as "geometry of paths"[2]), [3]) but soon after the introduction of the projective parameter of T. Y. THOMAS several authors developed theories that could all be considered in some way as theories of linear connexion of local E_{n+1}'s with a restricted transformation group.[4]) Another line of thought found its origin in the idea of v. DANTZIG[5]) of using in X_n a set of $n+1$ homogeneous coordinates subjected to homogeneous transformations of degree one. Then local space is automatically a flat projective space and for the vectors in this space (called *projectors*) a linear displacement can be determined uniquely.[6]), [7])

Exercise.

VI 4,1. For every choice of an object $\overset{*}{\Gamma}_\lambda$ with the transformation

$$\overset{*}{\Gamma}_{\lambda'} = A^{\lambda}_{\lambda'} \overset{*}{\Gamma}_\lambda - \partial_{\lambda'} \log \Delta \tag{VI) 4,1 α)}$$

the $\overset{*}{\Gamma}{}^{\varkappa}_{\mu\lambda}$ defined by

$$\overset{*}{\Gamma}{}^{\varkappa}_{\mu\lambda} = \overset{p}{\Gamma}{}^{\varkappa}_{\mu\lambda} + \frac{2}{n+1} A^{\varkappa}_{(\mu} \overset{*}{\Gamma}_{\lambda)} \tag{VI 4,1 β)}$$

transform like the parameters of a symmetric linear connexion and $\overset{*}{\Gamma}{}^{\lambda}_{\mu\lambda} = \overset{*}{\Gamma}_\mu$ (cf. Exerc. VI 7,1).

---

1) KÖNIG 1920, 1.

2) EISENHART and VEBLEN 1922, 3; VEBLEN 1922, 1; 1928, 1; VEBLEN and T. Y. THOMAS 1923, 1.

3) DOUGLAS gave two generalizations. In 1928, 1 he considered systems of curves such that one and only one curve passes through every point in every direction. A general system of this kind can be defined as solutions of a differential equation of the second order. This theory is intimately connected with FINSLER geometry. Cf. T. Y. THOMAS 1934, 2, p. 7, also for literature. In 1931, 1 DOUGLAS considered systems of X_K's (K-spreads) such that one and only one passes through every point in every K-direction. There are many recent investigations on the geometry in the space of K-spreads especially by chinese and japanese authors. These theories are beyond the realm of this book.

4) T. Y. THOMAS 1925, 3; 1926, 2; 1934, 2; WEYL 1929, 2; WEYL and ROBERTSON 1929, 1; GOLAB 1930, 1; SCHOUTEN and GOLAB 1930, 3; BORTOLOTTI 1932, 1 (literature); HLAVATY 1933, 1 (literature); BERWALD 1936, 1.

5) v. DANTZIG 1932, 1; 2; 3.

6) v. DANTZIG 1932, 3; SCHOUTEN and v. DANTZIG 1932, 1; 2; SCHOUTEN and HAANTJES 1936, 1 (literature).

7) Cf. for the relations between the different points of view also CARTAN 1924, 1; SCHOUTEN 1924, 2; 3; WEYL 1929, 2; WEYL and ROBERTSON 1929, 1; v. DANTZIG 1932, 1; 2; 1934, 1; YANO 1937, 3; 1938, 4; 5; 6; 1940, 4; 1941, 6; 1942, 4; 1944, 3; 8; 9; 10; 11; 1945, 4; 1947, 1; HLAVATY 1939, 1; YANO and SASAKI 1949, 4.

§ 5. Conformal transformation of a connexion in V_n.

If the fundamental tensor $g_{\lambda\varkappa}$ in V_n is transformed into

$$(5.1) \qquad 'g_{\lambda\varkappa} = \sigma\, g_{\lambda\varkappa}; \qquad '\overset{-1}{g^{\varkappa\lambda}} = \sigma^{-1}\, g^{\varkappa\lambda}$$ [1]

where σ is an arbitrary scalar, the parameters $\Gamma_{\mu\lambda}^{\varkappa}$ transform into

$$(5.2) \qquad \begin{cases} '\Gamma_{\mu\lambda}^{\varkappa} = \frac{1}{2}\, '\overset{-1}{g^{\varkappa\sigma}} (\partial_\mu\, 'g_{\lambda\sigma} + \partial_\lambda\, 'g_{\mu\sigma} - \partial_\sigma\, 'g_{\mu\lambda}) \\ \qquad = \Gamma_{\mu\lambda}^{\varkappa} + \frac{1}{2}(A_\lambda^\varkappa\, s_\mu + A_\mu^\varkappa\, s_\lambda - g^{\varkappa\sigma}\, g_{\mu\lambda}\, s_\sigma); \\ \qquad s_\lambda \overset{\text{def}}{=} \partial_\lambda \log \sigma. \end{cases}$$ [2]

We call this transformation *conformal*. From (5.2) and (III 2.3, 7) we get

$$(5.3) \qquad \begin{cases} \text{a)} \quad '\nabla_\mu\, v^\varkappa = \nabla_\mu\, v^\varkappa + \frac{1}{2}\, s_\mu\, v^\varkappa + \frac{1}{2}\, s_\lambda\, v^\lambda\, A_\mu^\varkappa - \frac{1}{2}\, g^{\varkappa\sigma}\, s_\sigma\, g_{\mu\lambda}\, v^\lambda \\ \text{b)} \quad '\nabla_\mu\, w_\lambda = \nabla_\mu\, w_\lambda - w_{(\mu}\, s_{\lambda)} + \frac{1}{2}\, g_{\mu\lambda}\, s_\sigma\, g^{\sigma\tau}\, w_\tau. \end{cases}$$

In dealing with these transformations the process of raising and lowering should be used very carefully because there are now two fundamental tensors $g_{\lambda\varkappa}$ and $'g_{\lambda\varkappa}$ and the process can only be defined with respect to one of them. From (5.3) it follows that

$$(5.4) \qquad '\nabla_\mu\, v^\mu = \nabla_\mu\, v^\mu + \frac{1}{2}\, n\, s_\mu\, v^\mu.$$

If a transformation of the $\Gamma_{\mu\lambda}^{\varkappa}$ in V_n is at the same time projective and conformal we have from (1.1) and (5.2)

$$(5.5) \qquad 2 p_\mu\, A_\lambda^\varkappa + 2 p_\lambda\, A_\mu^\varkappa = A_\lambda^\varkappa\, s_\mu + A_\mu^\varkappa\, s_\lambda - g^{\varkappa\sigma}\, g_{\mu\lambda}\, s_\sigma$$

from which by transvection with $g_{\mu\lambda}$ and by contraction with respect to $\varkappa\lambda$ two equations can be derived which are consistent if and only if $s_\lambda = 0$. Hence

A riemannian connexion is wholly determined if its geodesics are given and the fundamental tensor to within a scalar factor. Then the fundamental tensor is also fixed to within a constant scalar factor [3]

or in other words

The metric of a V_n is determined to within a constant scalar factor by the projective and the conformal properties of the V_n together.

---

[1]) Since raising and lowering of indices is performed by means of $g^{\varkappa\lambda}$ and $g_{\lambda\varkappa}$, $'\overset{-1}{g^{\varkappa\lambda}}$ can not be written as $'g^{\varkappa\lambda}$.

[2]) Fubini 1909, 1, p. 144; Schouten 1918, 1; Weyl 1918, 1.

[3]) Weyl 1921, 2, p. 100.

If (5.2) is substituted in (III 4.2) it follows that

$$'K_{\nu\mu\lambda}^{\cdot\cdot\cdot\varkappa} = K_{\nu\mu\lambda}^{\cdot\cdot\cdot\varkappa} + g_{[\nu[\lambda} s_{\mu]\sigma]} g^{\sigma\varkappa} \quad (5.6)$$

[1]

where

$$s_{\mu\lambda} = s_{\lambda\mu} \stackrel{\text{def}}{=} 2\nabla_\mu s_\lambda - s_\mu s_\lambda + \tfrac{1}{2} g_{\mu\lambda} s_\sigma s^\sigma \quad (5.7)$$

and accordingly

$$'K_{\mu\lambda} = K_{\mu\lambda} - \tfrac{1}{4}(n-2) s_{\mu\lambda} - \tfrac{1}{4} g_{\mu\lambda} s_{\sigma\tau} g^{\sigma\tau} \quad (5.8)$$

$$'K \;= \sigma^{-1} K - \tfrac{1}{2}(n-1)\sigma^{-1} s_{\mu\lambda} g^{\mu\lambda} \quad (5.9)$$

$$'G_{\mu\lambda} = G_{\mu\lambda} - \tfrac{1}{4}(n-2) s_{\mu\lambda} + \tfrac{1}{4}(n-2) g_{\mu\lambda} s_{\sigma\tau} g^{\sigma\tau}. \quad (5.10)$$

A V_n and its connexion are said to be *conformally euclidean* and the V_n is called a C_n if it can be transformed conformally into an R_n. [2] N.a.s. condition is for $n > 2$ that there exists a vector field L_λ such that

$$(5.11)\quad \begin{cases} \text{a)}\quad K_{\nu\mu\lambda\varkappa} = \dfrac{4}{n-2} g_{[\nu[\lambda} L_{\mu]\varkappa]}; \qquad L_{[\mu\lambda]} = 0 \\ \text{b)}\quad -\dfrac{4}{n-2} L_{\mu\lambda} = 2\nabla_\mu L_\lambda - L_\mu L_\lambda + \dfrac{1}{2} g_{\mu\lambda} L_\sigma L^\sigma \end{cases}$$

(the factor $\frac{4}{n-2}$ is introduced only for convenience). The number of the independent components of $K_{\nu\mu\lambda\varkappa}$ is $\frac{1}{12} n^2 (n^2-1)$ and this is $> \binom{n+1}{2}$ for $n > 3$, hence (5.11a) can always be satisfied for $n \leq 3$ but for $n > 3$ only if $K_{\nu\mu\lambda\varkappa}$ satisfies certain algebraical conditions. In this case we get from (5.8) for $s_{\mu\lambda} = -\frac{4}{n-2} L_{\mu\lambda}$ and $'K_{\mu\lambda} = 0$

$$L_{\mu\lambda} = -K_{\mu\lambda} + \frac{1}{2(n-1)} K g_{\mu\lambda}. \quad (5.12)$$

The first integrability condition of (5.11b) is

$$\frac{n-2}{4} K_{\nu\mu\lambda}^{\cdot\cdot\cdot\varkappa} L_\varkappa = + \nabla_{[\nu} L_{\mu]\lambda} + g_{[\nu[\lambda} L_{\mu]\varkappa]} L^\varkappa \quad (5.13)$$

from which we get by substituting (5.11a)

$$\nabla_{[\nu} L_{\mu]\lambda} = 0. \quad (5.14)$$

[1]) Note that $\sigma^{-1}\,'K_{\nu\mu\lambda\varkappa} = \sigma^{-1}\,'g_{\varkappa\varrho}\,'K_{\nu\mu\lambda}^{\cdot\cdot\cdot\varrho} = K_{\nu\mu\lambda\varkappa} + g_{[\nu[\lambda} s_{\mu]\varkappa]}$; cf. Struik 1922 1, p. 151; Finzi 1923, 1; Schouten and Struik 1921, 4; R. K. 1924, 1, p. 168, footnote [3]).

[2]) Cf. for the geometry of a C_n: Haantjes and Wrona 1939, 1; Haantjes 1940, 1; 1941, 1; 1942, 1; 2; 3; 1943, 1; Yano 1945, 4; 5; 6; 1946, 2; 3; Wrona 1948, 1; 2; 1949, 1; 1950, 1; Levine 1950, 1 (parallel fields); Matsumoto 1951, 3 (class one).

For an S_n, $n \geq 2$ we have (cf. III 5.31) a solution of (5.11a)

$$L_{\mu\lambda} = -\tfrac{1}{2}(n-2)\varkappa g_{\mu\lambda} \tag{5.15}$$

and for $n>2$ this is the only solution. Because $\varkappa$ is constant (5.14) is satisfied and this implies that *every S_n is a C_n*. For a V_2 which is not an S_2 (5.15) is also a solution of (5.11a) but it does not satisfy (5.14) because $\varkappa$ is no longer constant. But in this case (5.11a) has an infinite number of solutions and it is well known from elementary differential geometry that every V_2 can be conformally mapped on an R_2.[1]) For a V_n the identity of BIANCHI (III 5.21) can be applied to (5.11a). Then we get

$$(n-3)\nabla_{[\nu} L_{\mu]\lambda} = 0. \tag{5.16}$$

Hence for $n>3$ the integrability condition of (5.11b) is a consequence of (5.11a). That proves[2])

A V_2 is always a C_2. A V_n, $n>2$, is conformally euclidean if and only if

$$C_{\nu\mu\lambda}^{\cdot\cdot\cdot\varkappa} \overset{\text{def}}{=} K_{\nu\mu\lambda}^{\cdot\cdot\cdot\varkappa} - \frac{4}{n-2} g_{[\nu[\lambda} L_{\mu]\varrho]} g^{\varrho\varkappa} = 0 \tag{5.17}$$

where

$$L_{\mu\lambda} \overset{\text{def}}{=} -K_{\mu\lambda} + \frac{1}{2(n-1)} K g_{\mu\lambda} \tag{5.18}$$

and if for $n=3$ moreover

$$\nabla_{[\nu} L_{\mu]\lambda} = 0. \tag{5.19}$$

For $n=3$, (5.17) is identically satisfied and for $n>3$, (5.19) is a consequence of (5.17).

$C_{\nu\mu\lambda}^{\cdot\cdot\cdot\varkappa}$ is called the *conformal curvature tensor*. From (5.8, 12) it follows that

$$'L_{\mu\lambda} = L_{\mu\lambda} + \frac{n-2}{4} s_{\mu\lambda} \tag{5.20}$$

and by substituting this in (5.17) it follows that $C_{\nu\mu\lambda}^{\cdot\cdot\cdot\varkappa}$ (but *not* $C_{\nu\mu\lambda\varkappa}$) *is invariant for conformal transformations of the connexion.*[3]) For $n \leq 3$ it vanishes identically and for $n>3$ it satisfies the identities

$$\text{(5.21)}\quad \begin{cases} \text{a)}\ C_{(\nu\mu)\lambda}^{\cdot\cdot\cdot\varkappa} = 0; & \text{b)}\ C_{[\nu\mu\lambda]}^{\cdot\cdot\cdot\varkappa} = 0; \\ \text{c)}\ C_{\nu\mu\lambda}^{\cdot\cdot\cdot\nu} = 0; & \text{d)}\ C_{\nu\mu\lambda}^{\cdot\cdot\cdot\lambda} = 0; \quad \text{e)}\ C_{\nu\mu\lambda}^{\cdot\cdot\cdot\varkappa} = -C_{\nu\mu\cdot\lambda}^{\cdot\cdot\varkappa}. \end{cases}$$

---

[1]) Cf. for instance WEATHERBURN 1927, 1, p. 168; HAACK 1948, 1.

[2]) SCHOUTEN 1921, 1, p. 82ff.; COTTON 1899, 1, p. 412 and FINZI 1902, 1 for $n=3$. Cf. FINZI 1921, 1. WEYL proved 1918, 1, p. 404 that $C_{\nu\mu\lambda}^{\cdot\cdot\cdot\varkappa}$ is zero in a C_n. FINZI proved 1921, 1 that (5.17) and (5.19) together are n.a.s. for the V_n to be a C_n. Cf. also FINZI 1922, 1; CARTAN 1922, 3 for $n=4$ and LAGRANGE 1923, 1, p. 43 for $n \neq 3$. DOUGLAS gave a synthetic criterion 1925, 1.

[3]) Cf. for the semi-symmetric case SCHOUTEN 1925, 2.

As a corollary we get that for a conformally euclidean connexion the curvature tensor has at most $\frac{1}{2}n(n+1)$ independent components.

From (5.11a) we see that for a C_n all components of $K_{\nu\mu\lambda}{}^{\varkappa}$ with respect to an (holonomic or anholonomic) orthogonal coordinate system, with four different indices vanish. It can be proved[1]) that for $n>3$ this condition is also sufficient:

A V_n, $n>3$, is a C_n if and only if with respect to every holonomic or anholonomic orthogonal coordinate system (h)

$$K_{kjih} \stackrel{*}{=} 0; \quad h, i, j, k \neq \tag{5.22}$$

at every point.[2])

Another form of the n.a.s. conditions is found as follows. We consider here only an ordinary V_n and real subspaces. Let a V_n be such that it contains orthogonal systems (cf. definition V § 5) in every direction, i.e. that for every set of n directions at any arbitrarily chosen point there exists at least one orthogonal system having just these directions at that point. Then writing $\underset{b}{i}{}^{\varkappa}$; $b=1, \ldots, \mathrm{n}-1$ and $\underset{\mathrm{n}}{i}{}^{\varkappa}$ for the unitvectors in the directions of this system we have (cf. V 5.16b)

$$0 = 2\,\underset{d}{i}{}^{\nu}\,\underset{c}{i}{}^{\mu}\,\underset{b}{i}{}^{\lambda}\,\nabla_{[\nu}\,h_{\mu]\lambda} = \underset{d}{i}{}^{\nu}\,\underset{c}{i}{}^{\mu}\,K_{\nu\mu\lambda\varkappa}\,\underset{b}{i}{}^{\lambda}\,\underset{\mathrm{n}}{i}{}^{\varkappa} \stackrel{*}{=} K_{dcb\mathrm{n}}; \quad b, c, d \neq \tag{5.23}$$

from which we see that all orthogonal components of $K_{\nu\mu\lambda\varkappa}$ with four different indices vanish and that accordingly for $n>3$ the V_n is a C_n. Now an R_n contains of course orthogonal systems in every direction and orthogonal systems remain orthogonal under conformal transformations. Hence

A V_n, $n>3$ is a C_n if and only if for every choice of a point and an $(n-1)$-direction at this point, it contains at least one V_{n-1} belonging to an orthogonal system and having just this $(n-1)$-direction at this point.[3])

In V § 6 it was proved that if in a V_n there exist umbilical V_{n-1}'s through every point and with every $(n-1)$-direction at this point, all orthogonal components K_{kjih} with four different indices vanish. But we just proved that for $n>3$ this means that the V_n is a C_n. Now let an anholonomic coordinate system with mutually perpendicular unitvectors $\underset{j}{i}{}^{\varkappa}$ be chosen such that $\underset{\mathrm{n}}{i}{}^{\varkappa}$ is normal to a set of ∞^1 V_{n-1}'s. Then we have for the second fundamental tensor h_{cb} of these V_{n-1}'s

$$h_{cb} = -B_{cb}^{\mu\lambda}\,\nabla_{\mu}\,\underset{\mathrm{n}}{i}_{\lambda}; \quad b, c = 1, \ldots, \mathrm{n}-1 \tag{5.24}$$

[1]) Schouten 1921, 1, p. 84; cf. Eisenhart 1926, 1, p. 122ff.

[2]) Cf. Exerc. III 5,5.

[3]) Schouten 1927, 3, p. 719; E II 1938, 2, p. 204 also for $n=3$ and for literature.

and after the transformation $g_{\lambda\varkappa} \to \sigma g_{\lambda\varkappa}$ according to (5.3b), denoting the components of the new h_{cb} with respect to the old unitvectors $\underset{b}{i}^{\varkappa}$ by $'h_{cb}$

$$(5.25)\quad \begin{cases} 'h_{cb} = -B^{\mu\lambda}_{cb}\, '\nabla_\mu \sigma^{\frac{1}{2}} \underset{n}{i}_\lambda = -\sigma^{\frac{1}{2}} B^{\mu\lambda}_{cb} (\nabla_\mu \underset{n}{i}_\lambda - \underset{n}{i}_{(\mu} s_{\lambda)} + \tfrac{1}{2} g_{\mu\lambda} s_\sigma \underset{n}{i}_\tau g^{\sigma\tau}) \\ \qquad = \sigma^{\frac{1}{2}} (h_{cb} - \tfrac{1}{2} s_\sigma \underset{n}{i}_\tau g^{\sigma\tau}\, 'g_{cb}). \end{cases}$$

This proves that the directions of principal curvature are conformally invariant and that all umbilical points remain umbilical points after the conformal transformation. Now in an R_n, umbilical V_{n-1}'s are possible through every point with every $(n-1)$-direction at that point and this implies that the same holds in a C_n. So we have proved for $n > 3$ that the existence of these V_{n-1}'s is a necessary and sufficient condition for the V_n to be a C_n. STELLMACHER[1]) has proved that the condition is also n.a.s. for $n=3$. We prove this in the following way:

Let η^α; $\alpha = \dot{1}, \dot{2}$ be coordinates in an umbilical V_2 in V_3. Then we have on the one hand from (V 5.9)

$$(5.26)\qquad 2 B^{\nu\mu\lambda}_{\delta\gamma\beta} \nabla_{[\nu} h_{\mu]\lambda} = -2 B^{\nu\mu\lambda}_{\delta\gamma\beta} \nabla_{[\nu} \nabla_{\mu]} \underset{3}{i}_\lambda = B^{\nu\mu\lambda}_{\delta\gamma\beta} K_{\nu\mu\lambda\varkappa} \underset{3}{i}^{\varkappa}$$

if $\underset{3}{i}^{\varkappa}$ is a congruence normal to the V_2. On the other hand we have because all points are umbilical

$$(5.27)\qquad 2 B^{\nu\mu\lambda}_{\delta\gamma\beta} \nabla_{[\nu} h_{\mu]\lambda} = 2 B^{\nu\mu\lambda}_{\delta\gamma\beta} \nabla_{[\nu} h\, (g_{\mu]\lambda} - \underset{3}{i}_{\mu]} \underset{3}{i}_\lambda) = 2\, 'g_{\beta[\gamma} \partial_{\delta]} h.$$

But in a V_3 the equation (5.17) is identically satisfied, hence from (5.26, 27)

$$(5.28)\qquad B^{\nu\mu\lambda}_{\delta\gamma\beta} (g_{\nu\lambda} L_{\mu\varkappa} - g_{\mu\lambda} L_{\nu\varkappa}) \underset{3}{i}^{\varkappa} = 2\, 'g_{\beta[\gamma} \partial_{\delta]} h$$

or

$$(5.29)\qquad 'g_{\beta[\delta} B^{\mu}_{\gamma]} L_{\mu\varkappa} \underset{3}{i}^{\varkappa} = 'g_{\beta[\gamma} \partial_{\delta]} h$$

and this implies that $B^\lambda_\beta L_{\lambda\varkappa} \underset{3}{i}^{\varkappa}$ is a gradient vector in V_2:

$$(5.30)\qquad B^\lambda_\beta L_{\lambda\varkappa} \underset{3}{i}^{\varkappa} = -\partial_\beta h.$$

The rotation of this vector is zero and equal to the V_2-part of the rotation of $L_{\lambda\varkappa} \underset{3}{i}^{\varkappa}$

$$(5.31)\quad \begin{cases} 0 = B^{\mu\lambda}_{\gamma\beta} \nabla_{[\mu} L_{\lambda]\varkappa} \underset{3}{i}^{\varkappa} = B^{\mu\lambda}_{\gamma\beta} \underset{3}{i}^{\varkappa} \nabla_{[\mu} L_{\lambda]\varkappa} - B^{\mu\lambda}_{\gamma\beta} L^{\varkappa}_{\cdot[\lambda} h_{\mu]\varkappa} \\ \quad = B^{\mu\lambda}_{\gamma\beta} \underset{3}{i}^{\varkappa} \nabla_{[\mu} L_{\lambda]\varkappa} - h\, B^{\mu\lambda}_{\gamma\beta} L^{\varkappa}_{\cdot[\lambda} g_{\mu]\varkappa} = B^{\mu\lambda}_{\gamma\beta} \underset{3}{i}^{\varkappa} \nabla_{[\mu} L_{\lambda]\varkappa}. \end{cases}$$

---

[1]) STELLMACHER 1951, 1.

Hence, if in V_3 there exist umbilical V_2's through every point with every 2-direction, *all orthogonal components of $\nabla_{[\mu} L_{\lambda]\kappa}$ with three different indices vanish.* From this we see that from the quantity (cf. I 9.20)

$$N_{\nu\kappa} \overset{\text{def}}{=} I_{\nu}^{\cdot\mu\lambda} \nabla_{[\mu} L_{\lambda]\kappa} \tag{5.32}$$

all orthogonal components with two *equal* indices vanish. But this is only possible with respect to all orthogonal systems if $N_{\nu\kappa} = N_{[\nu\kappa]}$. Returning now to the identity of BIANCHI (III 5.21) we have according to (5.17)

$$0 = g_{\lambda[\nu} \nabla_{\omega} L_{\mu]\kappa} - g_{\kappa[\nu} \nabla_{\omega} L_{\mu]\lambda} \tag{5.33}$$

or, transvecting with $I^{\nu\omega\mu}$

$$N_{\lambda\kappa} = N_{[\lambda\kappa]} = 0 \tag{5.34}$$

which implies that in a V_3 of the kind considered $\nabla_{[\mu} L_{\lambda]\kappa} = 0$.

Gathering results we have[1])

A V_n, $n > 2$, is conformally euclidean if and only if there exist umbilical V_{n-1}'s through every point with every $(n-1)$-direction at that point.

If a V_m in V_n can be transformed into a geodesic V_m by a conformal transformation of the connexion in V_n it is called *conformally geodesic.* If $m = 1$ we have for the tangent unitvector $(\sigma > 0)$

$$'j^{\kappa} = \sigma^{-\frac{1}{2}} j^{\kappa}; \qquad 'j_{\lambda} = \sigma^{\frac{1}{2}} j_{\lambda} \tag{5.35}$$

hence (cf. V § 1)

$$'u^{\kappa} = 'j^{\mu}\, '\nabla_{\mu}\, 'j^{\kappa} = \sigma^{-1} (u^{\kappa} - \tfrac{1}{2} z^{\kappa}), \tag{5.36}$$

where u^{κ} is the curvature vector and z^{κ} is defined by

$$z^{\kappa} \overset{\text{def}}{=} s^{\kappa} - s_{\lambda} j^{\lambda} j^{\kappa} \tag{5.37}$$

is the projection of s^{κ} on the R_{n-1} perpendicular to j^{κ}. If the transformed curve is geodesic it follows that u^{κ} is the projection of a gradient vector. Conversely, if we know that u^{κ} is the projection of a vector $g^{\kappa\lambda} \partial_{\lambda} \log \sigma$, the transformation $g_{\lambda\kappa} \to \sigma g_{\lambda\kappa}$ transforms the curve into a geodesic.

---

[1]) For $n > 3$ this was proved by SCHOUTEN 1921, 1, p. 86. But for $n = 3$ it was stated there erroneously that every V_3 contains umbilical V_2's through every point in every 2-direction. This wrong statement occurred also in the first edition of this book R. K. 1924, 1, p. 180. But it was not republished in E II 1938, 2 § 19 because at that time we smelled a rat without knowing exactly its lare. STELLMACHER gave the solution in a letter of 19. 11.'49 to the author and published this with many other results concerning conformal geometry in 1951, 1; cf. SASAKI 1939, 1; 1940, 1.

Hence[1])

Every single curve is conformally geodesic. A real congruence of curves in an ordinary V_n is conformally geodesic if and only if the curvature vector is the projection of a gradientvector.

Orthogonality being invariant for conformal transformations, the theorem of BELTRAMI (cf. V § 5) leads to the theorem of THOMSON and TAIT[2])

If a real conformally geodesic congruence in an ordinary V_n is normal to one V_{n-1} it is V_{n-1}-normal throughout.

But such a congruence is not in any way special because it can be proved that every V_{n-1}-normal congruence is conformally geodesic.[3])

Because of

$$v^\mu \, '\nabla_\mu v^\varkappa = v^\mu \nabla_\mu v^\varkappa + v^\mu s_\mu v^\varkappa - \tfrac{1}{2} v^\mu v_\mu s^\varkappa \tag{5.38}$$

every null geodesic (geodesic whose tangent lies everywhere in a null direction) remains a null geodesic after conformal transformation of $g_{\lambda\varkappa}$.

For $m>1$ we get [cf. (5.37) and V § 9][4])

$$\left\{ \begin{array}{l} 'H_{cb}^{\cdot\cdot\varkappa} = - B_{cb}^{\mu\lambda} \, ('\nabla_\mu \, '\overset{x}{i}_\lambda) \, '\underset{x}{i}^\varkappa = H_{cb}^{\cdot\cdot\varkappa} - \tfrac{1}{2} \, 'g_{cb} \, z^\varkappa ; \\ b, c = 1, \ldots, \mathrm{m}; \qquad x = \mathrm{m}+1, \ldots, \mathrm{n}; \end{array} \right. \tag{5.39}$$

hence

An umbilical point of a V_m in V_n remains umbilical after conformal transformation of $g_{\lambda\varkappa}$.

This proves that a conformally geodesic V_m is for $m>1$ the same as an umbilical V_m. But this does not imply that a system of ∞^{n-m} umbilical V_m's; $m=1, \ldots, n-1$, can be transformed at the same time by one conformal transformation of the $g_{\lambda\varkappa}$ into ∞^{n-m} geodesic V_m's. According to (5.39) this is possible if and only if the mean curvature vector (cf. V § 9) is the projection of a gradientvector.[5])

According to (5.39) the tensor $M_{cb}^{\cdot\cdot\varkappa}$ defined in (V 9.16b) is conformal invariant. The same holds for the vectors $\underset{y}{\overset{x}{v}}_b$ defined in (V 8.13).

---

[1]) SCHOUTEN 1928, 2.

[2]) THOMSON and TAIT 1879, 1, p. 353 for $n=3$ and SCHOUTEN and STRUIK 1921, 2 for general values of n.

[3]) SCHOUTEN and STRUIK 1921, 2; 1922, 4; E II 1938, 2, p. 51; SCHOUTEN 1928, 2. Cf. for families of curves that may be considered as conformal geodesics PAINLEVÉ 1894, 1; KASNER 1910, 1; 1913, 1; DOUGLAS 1924, 1; SCHOUTEN 1928, 2; BLASCHKE 1928, 1; FIALKOW 1939, 2; 1940, 1; 1942, 2; YANO 1950, 1.

[4]) SCHOUTEN and STRUIK 1923, 4; R. K. 1924, 1, p. 202.

[5]) E II 1938, 2, p. 211.

For a V_m they take the form

$$\underset{y}{\overset{x}{v}}_b = B_b^\mu \, (\nabla_\mu \overset{x}{i}_\lambda) \, \underset{y}{i}^\lambda, \tag{5.40}$$

hence

$$'\underset{y}{\overset{x}{v}}_b = B_b^\mu \, ('\nabla_\mu \, '\overset{x}{i}_\lambda) \, '\underset{y}{i}^\lambda = \underset{y}{\overset{x}{v}}_b . \tag{5.41}$$ [1]

In (V § 8) it was proved that an A_m can be imbedded in an E_n if there exist fields $\overset{x}{h}_{cb}$, $\underset{y}{\overset{x}{l}}{}_c^{\cdot a}$ and $\underset{y}{\overset{x}{v}}_b$ satisfying the equations (V 8.18–21) of GAUSS, CODAZZI and RICCI. For a V_m in R_n the fields $\overset{x}{h}_{cb}$ and $\underset{y}{l}_{cb}$ coincide and accordingly a V_m can be imbedded in an R_n if there exist fields $\overset{x}{h}_{cb}$ and $\underset{y}{\overset{x}{v}}_b$ satisfying the GAUSS-CODAZZI-RICCI equations. YANO and MUTO[2]) proved that a V_m with the fields $M_{cb(x)}$ and $\underset{y}{\overset{x}{v}}_b$ can be conformally imbedded in a C_n provided that these fields satisfy three equations that may be considered as the conformal generalizations of the GAUSS-CODAZZI-RICCI equations. These equations were derived before by YANO.[3])

In V § 6 we have found that an umbilical V_{n-1} with constant mean curvature ($h = \text{const.}$) is at every point perpendicular to a principal direction of V_n and that accordingly V_{n-1}'s of this kind are only possible through every point and with every $(n-1)$-direction at that point if the V_n is an EINSTEIN space, $K_{\mu\lambda} \propto g_{\mu\lambda}$. But then also $L_{\mu\lambda}$ equals $g_{\mu\lambda}$ to within a scalar factor and according to (5.17) the V_n is an S_n [4]) (cf. V § 6):

A V_n, $n>2$, is an S_n if and only if there exist umbilical V_{n-1}'s with constant mean curvature through every point with every $(n-1)$-direction at this point.

A geodesic V_{n-1} is a special case of an umbilical V_{n-1} ($h=0$), hence

A V_n, $n>2$, is an S_n if and only if there exist geodesic V_{n-1}'s through every point with every $(n-1)$-direction at this point.

According to (5.7, 12) an R_n is transformed conformally into another R_n if $s_\lambda = \partial_\lambda \log \sigma$ satisfies the equations

$$2\nabla_\mu s_\lambda - s_\mu s_\lambda + \tfrac{1}{2} g_{\mu\lambda} s_\sigma s_\tau g^{\sigma\tau} = 0. \tag{5.42}$$

This equation is totally integrable[5]).

[1]) SASAKI 1936, 1.
[2]) YANO and MUTO 1942, 1; 1946, 1.
[3]) YANO 1939, 2; 3; 1940, 3; 1943, 2.
[4]) SCHOUTEN 1921, 1, p. 87.
[5]) Cf. for R_n transformed conformally into R_n, E II 1938, 2, p. 205ff.

LIOUVILLE proved the theorem

For $n>2$ every conformal mapping of an R_n on itself is always a similarity transformation with respect to a point or a combination of an inversion at a point and a similarity transformation with respect to that point

for $n=3$ and LIE gave the generalization for $n>3$. [1]) The conformal transformations of an R_2 that leads to another R_2 are well known from the theory of functions. [2])

Exercises.

VI 5,1 [3]). A V_2 is transformed conformally into an R_2. If $'g_{\lambda\varkappa}=\sigma\, g_{\lambda\varkappa}$ and $s_\lambda=\partial_\lambda \log\sigma$, prove that

$$\text{VI 5,1 }\alpha)\qquad \nabla_\lambda s^\lambda = K.$$

VI 5,2 [cf. Exerc. V 9,4] [4]). A V_{2m} is a C_{2m} if and only if the sum of the scalar curvatures of 2 arbitrary mutually perpendicular m-directions is always independent of the choice of these m-directions.

VI 5,3 [5]). A normal system of ∞^{n-m} V_m's in a V_n can be transformed conformally into a system of ∞^{n-m} minimal V_m's if and only if the mean curvature vector is the projection of a gradient vector on the local R_{n-m} perpendicular to the V_m.

VI 5,4 [6]). If a congruence in V_m in V_n is given, there is always a conformal transformation in V_n that transforms the congruence into a congruence of asymptotic lines of the V_m.

§ 6. Conformal transformations of the connexion in an EINSTEIN space. [7])

A V_n, $n>2$, is transformed into an EINSTEIN space by a conformal transformation of the connexion if and only if

$$(6.1)\qquad 'K_{\mu\lambda}=\frac{'K}{n}\,\sigma\, g_{\mu\lambda};\qquad 'K=\sigma^{-1}K-\frac{1}{2}(n-1)\,\sigma^{-1}s_{\mu\lambda}\,g^{\mu\lambda}=\text{const}.$$

[1]) LIE 1872, 1; BIANCHI 1902, 2, p. 375. The proof is not difficult but rather tiresome and uninteresting. A generalization was given by HAANTJES 1937, 2. Cf. E II 1938, 2, p. 209.

[2]) Cf. E II 1938, 2, p. 210.

[3]) E II 1938, 2, p. 205.

[4]) HAANTJES and WRONA 1939, 1.

[5]) SCHOUTEN and STRUIK 1923, 4; E II 1938, 2, p. 212.

[6]) SCHOUTEN and STRUIK 1923, 4; R. K. 1924, 1, p. 202.

[7]) General references: BRINKMAN 1923, 1; 1924, 1; 1925, 1; HAANTJES and WRONA 1939, 1; FIALKOW 1939, 1; 1942, 1; SASAKI 1942, 1; YANO 1943, 1; WONG 1943, 2; 3; WRONA 1948, 2; KUIPER 1950, 1; 2; 1951, 1.

On account of (5.12) this is equivalent to

$$'L_{\mu\lambda} = -\frac{n-2}{2n(n-1)}\,'K\,\sigma\,g_{\mu\lambda}, \tag{6.2}$$

hence, according to (5.20)

$$s_{\mu\lambda} = -\frac{4}{n-2}L_{\mu\lambda} - \frac{2}{n(n-1)}\,'K\,\sigma\,g_{\mu\lambda}. \tag{6.3}$$

This gives for σ the differential equation (cf. 5.7)

$$2\nabla_\mu s_\lambda - s_\mu s_\lambda + \frac{1}{2}g_{\mu\lambda}s_\sigma s^\sigma = -\frac{4}{n-2}L_{\mu\lambda} - \frac{2}{n(n-1)}\,'K\,\sigma\,g_{\mu\lambda} \tag{6.4}$$

with the first integrability conditions (cf. 5.13)

$$\boxed{\nabla_{[\nu}L_{\mu]\lambda} - \frac{n-2}{4}C_{\nu\mu\lambda}{}^{\varkappa}s_\varkappa = 0.} \tag{6.5}$$

These conditions can also be derived immediately from (6.2):

$$'\nabla_{[\nu}\,'L_{\mu]\lambda} = 0 \tag{6.6}$$

because the left hand sides of (6.5) and (6.6) are equal.[1])

Now from (III 5.27) it can be derived that[2])

$$\nabla_\varkappa C_{\nu\mu\lambda}{}^{\varkappa} = -\frac{2(n-3)}{n-2}\nabla_{[\nu}L_{\mu]\lambda} \tag{6.7}$$

and from (6.5) and (6.7) after some calculation

$$\boxed{\nabla^\mu\nabla_{[\nu}L_{\mu]\lambda} + \frac{1}{2}L^{\mu\sigma}C_{\nu\mu\lambda\sigma} = -\frac{n-4}{2}s^\sigma\nabla_{[\lambda}L_{\sigma]\nu}.} \quad ; \quad \nabla^\mu \overset{\text{def}}{=} g^{\mu\nu}\nabla_\nu \tag{6.8}$$

This equation is a part of the second integrability condition of (6.4). It is remarkable that the right hand side is zero for $n=4$. That implies that for $n=4$ the left hand side must be a conformal concomitant to within a scalar factor. In fact it can be proved that for $n=4$ the symmetric tensor density of weight $+\frac{1}{2}$

$$\mathfrak{C}_{\nu\lambda} \overset{\text{def}}{=} \mathfrak{g}^{\frac{1}{4}}\left(\nabla^\mu\nabla_{[\nu}L_{\mu]\lambda} + \tfrac{1}{2}L^{\mu\sigma}C_{\nu\mu\lambda\sigma}\right) \tag{6.9}$$

[1]) E II 1938, 2, p. 203.
[2]) E II 1938, 2, p. 204.

is invariant for conformal transformations of the connexion.[1]) For $n=4$ the vanishing of $\mathfrak{C}_{\nu\lambda}$ is a necessary though not a sufficient condition for the V_4 to be conformally transformable into an EINSTEIN V_4.

Returning to the first integrability conditions (6.5) for $n>2$ we first remark that (6.4) is totally integrable if and only if $C_{\nu\mu\lambda}^{\cdot\cdot\cdot\varkappa}$ and $\nabla_{[\nu} L_{\mu]\lambda}$ vanish, that is if the V_n is conformally euclidean. But in that case the EINSTEIN space is conformally euclidean as well, and from (5.17) it follows that a conformally euclidean EINSTEIN space is always an S_n.[2]) Now let us suppose that the V_n is a *special* EINSTEIN space, $L_{\mu\lambda}=0$. Then (6.5) reduces to

$$K_{\nu\mu\lambda}^{\cdot\cdot\cdot\varkappa} s_\varkappa = 0; \qquad s_\varkappa \neq 0 \tag{6.10}$$

because in this case $C_{\nu\mu\lambda}^{\cdot\cdot\cdot\varkappa}$ is equal to $K_{\nu\mu\lambda}^{\cdot\cdot\cdot\varkappa}$ according to its definition. Now (6.10) implies that $K_{\nu\mu\lambda\varkappa}$ can not have the $\varkappa$-rank n. But the rank of $K_{\nu\mu\lambda\varkappa}$ is the same with respect to all indices, hence at every point $K_{\nu\mu\lambda\varkappa}$ lies in an R_{n-1} perpendicular to $s^\varkappa$. For $n=4$ this has an important consequence. According to its symmetry properties $K_{\nu\mu\lambda\varkappa}$ can always be written in the form $'a_{[\nu[\lambda} Q_{\mu]\varkappa]}$ where $'a_{\lambda\varkappa}$ is the fundamental tensor of the R_3 perpendicular to $s^\varkappa$, $'a_{\lambda\varkappa}=a_{\lambda\varkappa}-\varrho\, s_\lambda s_\varkappa$; $\varrho\, s_\lambda s^\lambda=1$. But because $K_{\nu\mu\lambda}^{\cdot\cdot\cdot\varkappa}=C_{\nu\mu\lambda}^{\cdot\cdot\cdot\varkappa}$, the RICCI tensor $K_{\mu\lambda}$ must vanish and this is only possible if $Q_{\lambda\varkappa}=0$. This proves the theorem of BRINKMAN:[3])

A conformally euclidean special EINSTEIN *space is always an* R_n. *If a special* EINSTEIN *space is not conformally euclidean then for* $n=4$ *it is impossible to map it conformally on another special* EINSTEIN *space*[4]).

This is very important for general relativity. Time-space is a special EINSTEIN space in all regions without matter. In these regions the metric can only be investigated by means of light signals and these signals fix only the null geodesics. That would imply that in empty time-space a metric could only be determined to within conformal transformations. But according to the theorem of BRINKMAN the condition $G_{\mu\lambda}=0$ (usually derived from a variational principle) is sufficient to fix the metric[5]).

---

[1]) This quantity was discovered by SCHOUTEN and HAANTJES 1936, 2. It played an important role in the imbedding of an EINSTEIN V_4 in a five-dimensional space with a general projective geometry and it was rediscovered by HESSELBACH 1949, 1. He proved that $\nabla_\mu \mathfrak{C}^{\mu}_{\cdot\lambda}=0$.

[2]) SCHOUTEN and STRUIK 1921, 4.

[3]) BRINKMAN 1923, 1; 1924, 1; 1925, 1 also for many other properties of conformal mapping of EINSTEIN spaces; cf. E II 1938, 2, p. 213; KUIPER 1950, 1; 2 also for literature on EINSTEIN spaces.

[4]) It is also impossible to map it conformally on a non special EINSTEIN space.

[5]) This remark is due to KASNER 1921, 1; 1922, 1 who proved the theorem of BRINKMAN for a special case.

§ 7. Conformal connexions.[1]

If the (symmetric) fundamental tensor $g_{\lambda\varkappa}$ is given to within a scalar factor the (symmetric) tensor density of weight $-\frac{2}{n}$

$$\mathfrak{G}_{\lambda\varkappa} \overset{\text{def}}{=} \mathfrak{g}^{-\frac{1}{n}} g_{\lambda\varkappa} \tag{7.1}$$

is uniquely determined. Its determinant (or in the indefinite case the absolute value) is $+1$. This has as a consequence that on each curve a *conformal parameter* ζ is defined whose differential

$$\overset{(\varkappa)}{d\zeta} = (\mathfrak{G}_{\lambda\varkappa}\, d\xi^\lambda\, d\xi^\varkappa)^{\frac{1}{2}} \tag{7.2}$$

is a scalar density of weight $-\frac{1}{n}$.[2] If the process that leads from the $g_{\lambda\varkappa}$ to the $\Gamma^\varkappa_{\mu\lambda}$ belonging to $g_{\lambda\varkappa}$, is applied to the $\mathfrak{G}_{\lambda\varkappa}$ we get

$$\left\{\begin{aligned} \overset{c}{\Gamma}{}^\varkappa_{\mu\lambda} &\overset{\text{def}}{=} \frac{1}{2} \overset{-1}{\mathfrak{G}}{}^{\varkappa\sigma} (\partial_\mu \mathfrak{G}_{\lambda\sigma} + \partial_\lambda \mathfrak{G}_{\mu\sigma} - \partial_\sigma \mathfrak{G}_{\mu\lambda}) \\ &= \Gamma^\varkappa_{\mu\lambda} - \frac{2}{n} \Gamma_{(\mu} A^\varkappa_{\lambda)} + \frac{1}{n} g_{\mu\lambda} g^{\varkappa\varrho} \Gamma_\varrho . \end{aligned}\right. \tag{7.3}$$

These are the *conformal parameters* of J. M. Thomas[3]), which are invariant for conformal transformations of $g_{\lambda\varkappa}$. Their transformation is

$$\left\{\begin{aligned} \overset{c}{\Gamma}{}^{\varkappa'}_{\mu'\lambda'} = A^{\mu\,\lambda\,\varkappa'}_{\mu'\lambda'\varkappa} \overset{c}{\Gamma}{}^\varkappa_{\mu\lambda} + A^{\varkappa'}_\varrho \partial_{\mu'} A^\varrho_{\lambda'} + \frac{2}{n} A^{\varkappa'}_{(\mu'} \partial_{\lambda')} \log \Delta - \\ - \frac{1}{n} \mathfrak{G}_{\mu'\lambda'} \overset{-1}{\mathfrak{G}}{}^{\varkappa'\varrho'} \partial_{\varrho'} \log \Delta . \end{aligned}\right. \tag{7.4}$$

---

1) *References. General:* Cartan 1923, 3; 1937, 3; Schouten and Struik 1923, 4; T. Y. Thomas 1926, 3; 1932, 1; 1934, 2, Ch. IV; Veblen 1928, 3; 1930, 1; 1935, 1; Whitehead 1929, 1; Vanderslice 1934, 2; Hlavatý 1935, 2; 1936, 1; Schouten and Haantjes 1936, 2; Yano 1938, 2; 1939, 4; 5; Sasaki 1940, 2; 1941, 2; 1948, 1; Schirokow 1940, 1; Vranceanu 1940, 1; 1943, 1; Yano and Muto 1941, 1; 2; Schmidt 1949, 1. *Curves:* Agostinelli 1933, 3, Hlavatý 1935, 2, III; Modesitt 1938, 1; Yano 1938, 3; 1944, 4; 1950, 1; Fialkow 1939, 2; 1940, 1; 1942, 2; Sasaki 1939, 2; Muto 1939, 1; 1942, 1; Haantjes 1941, 1; 1942, 1; Lagrange 1941, 1; 2; Yano and Muto 1941, 1; 2; 3; 4. *Imbedding:* Brinkman 1923, 2; Perepelkine 1935, 2; Hlavatý 1935, 2, II; Sasaki 1936, 1; 1939, 3; 1940, 2; 3; Haimovici 1938, 1; Fialkow 1939, 1; 1942, 1; 1944, 1; Yano 1939, 2; 3 (Gauss and Codazzi); 1940, 3; 1942, 2; 1943, 2; 4; Muto 1939, 2; 1940, 3; 1942, 1; Yano and Muto 1940, 9; 1941, 5; 1942, 1; 1946, 1; Haantjes 1942, 2; 3; 1943, 1; Beckenbach and Bing 1943, 1; Norden 1945, 3; Petrescu 1946, 1; 1948, 1; Otsuki 1950, 2; Verbickiĭ 1951, 1; 2 (class); Ichilione 1951, 1; Akivis 1952, 1. *Product spaces:* Yano 1940, 1; Wong 1943, 2. *Mapping:* Golab 1935, 1; Haantjes 1937, 2; Zaremba 1937, 1; Coburn 1941, 2.

2) T. Y. Thomas 1926, 3.

3) J. M. Thomas 1925, 1; 1926, 3.

Hence the $\overset{c}{\Gamma}{}^{\varkappa}_{\mu\lambda}$ are not the components of a geometric object because their transformation formula contains not only $\overset{c}{\Gamma}{}^{\varkappa}_{\mu\lambda}$ and $A^{\varkappa'}_{\lambda}$ but also $\mathfrak{G}_{\mu\lambda}$ (or $g_{\mu\lambda}$, because $\mathfrak{G}_{\mu\lambda}\overset{-1}{\mathfrak{G}}{}^{\varkappa\varrho} = g_{\mu\lambda}g^{\varkappa\varrho}$). But $\overset{c}{\Gamma}{}^{\varkappa}_{\mu\lambda}$ together with $\mathfrak{G}_{\mu\lambda}$ form a geometric object. This geometric object does not fix one metric and symmetric connexion but an infinity of such connexions transforming into each other by conformal transformations. If we first prefer one coordinate system and then consider coordinate systems with $\Delta = \text{const.}$ only, the $\overset{c}{\Gamma}{}^{\varkappa}_{\mu\lambda}$ transform in the same way as the $\Gamma^{\varkappa}_{\mu\lambda}$ and in this case the $\overset{c}{\Gamma}{}^{\varkappa}_{\mu\lambda}$ represent the only symmetric connexion that leaves $\mathfrak{G}_{\lambda\varkappa}$ invariant and whose parameters give zero if transvected with $A^{\lambda}_{\varkappa}$.

The $\overset{c}{\Gamma}{}^{\varkappa}_{\mu\lambda}$ can be used in several ways to fix some kind of connexion. The simplest method is, as in the projective case, to construct an invariant scalar density.[1]) In 1924 CARTAN[2]) proposed to use a flat projective $(n+1)$-dimensional space with a quadric instead of an affine space as local space. SCHOUTEN[3]) proved in the same year that this amounts to introducing a linear connexion for local conformal spaces and that it is intimately connected with the linear connexions of local E_N's; $N \neq n$ already introduced by KÖNIG in 1920.[4]) T. Y. THOMAS[5]) succeeded in constructing from the $\overset{c}{\Gamma}{}^{\varkappa}_{\mu\lambda}$ a connexion for certain quantities in an X_{n+1} with a restricted group of coordinate transformations.

It is well known that the conformal geometry of an R_n is identical with the projective geometry on a quadric in a flat projective $(n+1)$-dimensional space. This can be proved for instance by introducing $n+2$ polyspherical coordinates.[6]) This can be generalized for the conformal geometry of a V_n. SCHOUTEN and HAANTJES[7]) have proved that for n odd an X_n with a general conformal geometry can be imbedded in an X_{n+1} with a general projective connexion, and that *this connexion is uniquely determined.* This means that n-dimensional general conformal geometry can for n odd be treated with $n+2$ homogeneous coordinates that satisfy one condition. For n even the theorem holds only if special conditions are satisfied. For $n=4$ the only condition is that $\mathfrak{C}_{\lambda\varkappa}=0$ (cf. 6.9).

1) COTTON 1899, 1 for $n=3$; HLAVATY 1935, 2; LEVINE 1935, 1.

2) CARTAN 1924, 1; cf. 1923, 3.

3) SCHOUTEN 1924, 2; 3.

4) KÖNIG 1920, 1.

5) T. Y. THOMAS 1932, 1; 1934, 2 also for literature; cf. VEBLEN 1928, 2; 3; 1935, 1; VANDERSLICE 1934, 2; HLAVATY 1935, 2; 1936, 1; YANO 1938, 2; 1940, 4.

6) Cf. for instance WOODS 1922, 1, p. 251; KLEIN 1926, 1, p. 193; SCHOUTEN and HAANTJES 1936, 2.

7) SCHOUTEN and HAANTJES 1935, 3; 1936, 2.

Exercise.

VI 7,1. For every choice of an object $\overset{*}{\Gamma}_\lambda$ with the transformation

VI 7,1 α) $$\overset{*}{\Gamma}_{\lambda'} = A^{\lambda}_{\lambda'} \overset{*}{\Gamma}_\lambda - \partial_{\lambda'} \log \Delta$$

the $\overset{*}{\Gamma}{}^{\varkappa}_{\mu\lambda}$ defined by

VI 7,1 β) $$\overset{*}{\Gamma}{}^{\varkappa}_{\mu\lambda} = \overset{c}{\Gamma}{}^{\varkappa}_{\mu\lambda} + \frac{2}{n} \overset{*}{\Gamma}_{(\mu} A^{\varkappa}_{\lambda)} - \frac{1}{n} g_{\mu\lambda} g^{\varkappa\varrho} \overset{*}{\Gamma}_\varrho$$

transform like the parameters of a linear connexion leaving $\mathfrak{G}_{\lambda\varkappa}$ invariant and $\overset{*}{\Gamma}{}^{\lambda}_{\mu\lambda} = \overset{*}{\Gamma}_\mu$ (cf. Exerc. VI 4,1).

§ 8. Subprojective connexions.

An A_n is said to be *k-fold projective*[1]) if there exists a coordinate system with respect to which every geodesic can be given by means of k linear equations and $n-k-1$ equations that need not be linear. Hence a D_n is $(n-1)$-fold projective. For $k=n-2$ it may happen that there exists a coordinate system such that every geodesic is given with respect to this system by $n-2$ *homogeneous* linear equations and one other equation that need not be linear. Then the A_n is called *subprojective*[2]) and the origin of the special coordinate system a *pole*.[3]) At any point the direction of the geodesic through this point and a pole is called a *pole direction*. The special coordinate systems with respect to which $n-2$ equations of any geodesic are homogeneous and linear transform into each other by the transformations of the group

(8.1) $$\xi^{h'} \overset{*}{=} \varrho(\xi^i)\, P^{h'}_{\cdot h} \xi^h; \qquad P^{h'}_{\cdot h} = \text{const.}$$

if they have the same origin, as was proved by RACHEVSKI and will be proved here later on.[4])

$n-2$ homogeneous linear equations in one of these special coordinate systems fix an X_2 that is totally geodesic in A_n because every geodesic having two points in common with such an X_2 lies wholly in it. Hence through the origin totally geodesic X_{n-1}'s can be laid with every $(n-1)$-direction and every one of them is given by one homogeneous linear equation. In the following we do not consider the case $n=2$ because according to the definition every 1-fold projective A_2 is a D_2 and every A_2 is subprojective.

1) KAGAN 1930, 1; 1933, 1; RACHEVSKI 1930, 1; 1933, 1; SCHAPIRO 1930, 1; 1933, 1.

2) KAGAN 1930, 1; 1935, 1.

3) The case $k<n-2$ was recently discussed by VRANCEANU 1947, 2; 1950, 1; 2. A linear element was given for a k-fold projective V_n.

4) RASCHEVSKI 1933, 1, p. 128. Cf. for this and some of the following results also E II 1938, 2, p. 215ff.

Now let (h) be such a special coordinate system and let v^h be an arbitrary vector field with *constant* components v^h. The equations

$$(8.2)\qquad \xi^h \overset{*}{=} c^h \underset{1}{t} + v^h \underset{2}{t}; \quad c^h = \text{const}; \quad v^h = \text{const}.$$

represent the parameter form of a geodesic X_2 at every point of which the vector v^h has a tangent direction. But because the X_2 is geodesic this implies that

$$(8.3)\qquad v^j \nabla_j v^h \overset{*}{=} \Gamma^h_{ji} v^j v^i \overset{*}{=} \lambda v^h + \mu \xi^h$$

for every choice of the direction of v^h. Because the field v^h is arbitrary, we may now look upon the v^h and ξ^h in (8.3) as independent variables. Then by differentiation with respect to v^h we get

$$(8.4)\qquad 2\Gamma^h_{ji} \overset{*}{=} \frac{\partial^2 \lambda}{\partial v^j \partial v^i} v^h + \frac{\partial \lambda}{\partial v^j} A^h_i + \frac{\partial \lambda}{\partial v^i} A^h_j + \xi^h \frac{\partial^2 \mu}{\partial v^j \partial v^i},$$

which is only possible if $\dfrac{\partial^2 \lambda}{\partial v^j \partial v^i} = 0$ because the Γ^h_{ji} depend on the ξ^h only and not on the v^h. Hence $\dfrac{\partial \lambda}{\partial v^j}$ and $\dfrac{\partial^2 \mu}{\partial v^j \partial v^i}$ do not depend on the v^h and we get for Γ^h_{ji} an expression of the form

$$(8.5)\qquad \Gamma^h_{ji} \overset{*}{=} 2\varphi_{(j} A^h_{i)} + \varphi_{ji} \xi^h; \quad \varphi_{[ji]} = 0$$

valid with respect to all special coordinate systems.[1]) If we apply the coordinate transformation (8.1) we have

$$(8.6)\qquad \begin{cases} \text{a)} & A^{h'}_i \overset{*}{=} \varrho P^{h'}_{\cdot i} + \xi^{h'} \partial_i \log \varrho \\ \text{b)} & \partial_j A^{h'}_i \overset{*}{=} (\partial_j \varrho) P^{h'}_{\cdot i} + \varrho P^{h'}_{\cdot j} \partial_i \log \varrho + \xi^{h'} (\partial_j \log \varrho) \partial_i \log \varrho + \xi^{h'} \partial_j \partial_i \log \varrho \end{cases}$$

and

$$(8.7)\qquad \Gamma^{h'}_{j'i'} = 2\varphi_{(j'} A^{h'}_{i')} + \varphi_{j'i'} \xi^{h'}$$

with[2])

$$(8.8)\qquad \begin{cases} \text{a)} & \varphi_{i'} \overset{*}{=} A^i_{i'} \varphi_i - \partial_{i'} \log \varrho \\ \text{b)} & \varphi_{j'i'} \overset{*}{=} (1 + \xi^h \partial_h \log \varrho)(A^{ji}_{j'i'} \varphi_{ji} - \varrho^{-1} \partial_{j'} \partial_{i'} \varrho). \end{cases}$$

From these equations we see that indeed a transformation of the form (8.1) transforms a special coordinate system into another special system with the same pole. The other part of RACHEVSKI's theorem follows from the condition that the transformation must transform every homogeneous linear equation into an equation with the same property. The φ_i and φ_{ji} do not behave like a vector field or a tensor field respectively and each of them forms a geometric object for transformations of the

---

[1]) KAGAN 1933, 1, p. 41.
[2]) Cf. E II 1938, 2, p. 219.

form (8.1) only. More general transformations disturb the form of (8.5). It is often convenient to consider a vector field and a symmetric tensor field defined by $u_i \overset{*}{=} \varphi_i$ and $u_{ji} \overset{*}{=} \varphi_{ji}$ but these fields depend on the choice of (h) and they change if another special coordinate system is introduced instead of (h). From (8.18a) we see that $\varphi_{i'}$ can be made zero if and only if the field u_i is a gradient field. That implies that $2\partial_{[j}\varphi_{i]}$ is an invariant bivector and a concomitant of the connexion. Another consequence of (8.7) is that *a subprojective connexion remains subprojective for every projective transformation* (1.1). From a geometrical point of view this is quite obvious because a projective transformation of the connexion does not change the geodesics.

The form (8.5) leads to the following expression for the curvature tensor

$$R_{kji}^{\cdot\cdot\cdot h} = 2A_{[k}^h T_{j]i} - 2A_i^h T_{[kj]} + 2U_{kji}x^h \text{ [1]}, \tag{8.9}$$

where U_{kji}, T_{ji} and x^h are tensor fields and a vector field, defined by

$$U_{kji} \overset{*}{\underset{\text{def}}{=}} \partial_{[k}\varphi_{j]i} + \xi^l \varphi_{l[k}\varphi_{j]i}, \tag{8.10}$$

$$T_{ji} \overset{*}{\underset{\text{def}}{=}} -\partial_j \varphi_i + \varphi_{ji}(1+\xi^l\varphi_l) + \varphi_j\varphi_i, \tag{8.11}$$

$$x^h \overset{*}{\underset{\text{def}}{=}} \xi^h. \tag{8.12}$$

For R_{ji}, V_{kj}, P_{ji} and $P_{kji}^{\cdot\cdot\cdot h}$ (cf. VI § 1) we have

$$R_{ji} \overset{*}{=} (n-1)\,T_{ji} - 2\,\partial_{[j}\varphi_{i]} + 2\,x^h U_{hji}, \tag{8.13}$$

$$V_{ji} \overset{*}{=} 2\,(n+1)\,\partial_{[j}\varphi_{i]} + 2\,U_{jih}x^h, \tag{8.14}$$

$$P_{ji} = -T_{ji} + U_{ji}, \tag{8.15}$$

$$P_{kji}^{\cdot\cdot\cdot h} = 2\,U_{kji}x^h + 2\,A_{[k}^h U_{j]i} - 2\,A_i^h U_{[kj]}, \tag{8.16}$$

where

$$U_{ji} \overset{\text{def}}{=} -\frac{2}{n-1}\,x^l U_{lji} - \frac{2}{n^2-1}\,U_{jil}x^l, \tag{8.17}$$

$$U_{[ji]} = \frac{1}{n+1}\,U_{jil}x^l. \tag{8.18}$$

From (8.16, 17) it is seen that for $n>2$ the connexion is projectively euclidean if $U_{kji}=0$. In order to prove that this condition is also necessary we consider an expression of the form [2]

$$\begin{cases} A_{[k}^h B_{j]i} + A_i^h C_{kj} + \underset{1}{z}^h \underset{1}{Z}_{ikj} + \cdots + \underset{p}{z}^h \underset{p}{Z}_{ikj}; \\ \qquad C_{(kj)} = 0; \quad \underset{1}{Z}_{i(kj)} = 0; \ldots; \underset{p}{Z}_{i(kj)} = 0 \end{cases} \tag{8.19}$$

---

[1]) In this equation the sign * can be dropped because the equation has the invariant form.

[2]) SCHOUTEN 1953, 1.

with $\underset{1}{z}^h, \ldots, \underset{p}{z}^h$ linearly independent. For $B_{ji} \neq 0$, $C_{kj} \neq 0$ the h-rank of the first term is $\geq n-1$ and the second term has h-rank n [1]). In order to find the h-rank of the first two terms together we try to find a solution w_i of

$$w_{[k} B_{j]i} + w_i C_{kj} \overset{*}{=} 0. \tag{8.20}$$

Such a solution is only possible if B_{ji} has the form $p_j r_i + r_j q_i$ and if $C_{kj} = p_{[k} r_{j]}$. [2]) For $p_i \neq q_i$ the solution is $w_i \propto r_i$. Hence the h-rank of the first two terms together is $n-1$ in this exceptional case and n in the ordinary case. The h-rank of the last p terms of (8.19) together is $\leq p$, hence

Auxiliary theorem:

For $n > p+2$ an expression of the form (8.19) *is zero if and only if B_{kj}, C_{kj} and $\underset{1}{Z}_{ikj}, \ldots, \underset{p}{Z}_{ikj}$ are all zero. For $n = p+1$ the expression can possibly also be zero if B_{ji} has the form $p_j r_i + r_j q_i$ and if $C_{kj} = p_{[k} r_{j]}$.*

If an expression can be written in the form (8.19) *and also in the form*

$$\begin{cases} A^h_{[k} {}'B_{j]i} + A^h_i {}'C_{kj} + {}'\underset{1}{z}^h {}'\underset{1}{Z}_{ikj} + \cdots + {}'\underset{q}{z}^h {}'\underset{q}{Z}_{ikj} \\ {}'C_{(kj)} = 0; \quad {}'\underset{1}{Z}_{i(kj)} = 0; \quad \ldots; \quad {}'\underset{q}{Z}_{i(kj)} = 0 \end{cases} \tag{8.21}$$

and if there are $p+q-s$ linearly independent ones among the z^h and ${}'z^h$, it follows that

$${}'B_{ji} = B_{ji}; \quad {}'C_{kj} = C_{kj} \tag{8.22}$$

for $n > p+q-s+2$. This is also true for $n = p+q-s+1$ but it need not be true if $B_{ji} - {}'B_{ji}$ has the form $p_j r_i + r_j q_i$ and if $C_{kj} - {}'C_{kj} = p_{[k} r_{j]}$.

Now let us suppose that there exists another decomposition of $P_{kji}^{\cdot\cdot\cdot h}$

$$P_{kji}^{\cdot\cdot\cdot h} \overset{*}{=} 2 Z_{kji} s^h + 2 A^h_{[k} Z_{j]i} - 2 A^h_i Z_{[kj]} \qquad Z_{(kj)i} = 0 \tag{8.23}$$

then we have $p=1$, $q=1$ and $s=1$ or $s=0$. Then the exceptional case can for $n>2$ only occur for $s=0$, $n=3$ (necessary conditions for this case are that $U_{ji} - Z_{ji}$ has the form $p_j r_i + r_j q_i$ and that $2 p_{[k} r_{j]} = q_{[k} r_{j]}$). This proves that, for $n>2$, U_{kj} and U_{kji} are uniquely determined by $P_{kji}^{\cdot\cdot\cdot h}$ provided that the direction of x^h is known and that, for $n>3$, the direction of x^h is uniquely determined by $P_{kji}^{\cdot\cdot\cdot h}$ provided that $P_{kji}^{\cdot\cdot\cdot h} \neq 0$. For $n>2$, $U_{kji} = 0$ if $P_{kji}^{\cdot\cdot\cdot h} = 0$.

Gathering results we have

---

[1]) The expressions B_{ij}, C_{kj} etc. need not be tensors. Notwithstanding it is possible to speak of a h-rank as long as the coordinate system is not transformed.

[2]) In SCHOUTEN 1953, 1 only the case $q_i = 0$ was considered. I owe this correction to a personal communication of J. HAANTJES.

A subprojective connexion is for $n>2$ projectively euclidean if and only if $U_{kji}=0$.[1]) *If it is not projectively euclidean the pole direction at every point is for $n>3$ uniquely determined and can be computed from $P_{kji}^{\cdot\cdot\cdot h}$.*

§ 9. ADATI's problem[2]).

ADATI considered the question whether a connexion that is not projectively euclidean and that takes the form

$$\Gamma_{ji}^{h} \overset{*}{=} 2 w_{(j} A_{i)}^{h} + w_{ji} v^{h}; \qquad w_{[ji]} = 0; \tag{9.1}$$

with respect to some coordinate system (h) is subprojective if some suitable conditions are introduced for the field v^h. In (9.1) w_j and the product $w_{ji} v^h$ are supposed to be given as functions of the coordinates. Of course it can not be said of w_j, v^h and w_{ji} that they transform in the way of vectors or of a tensor because a coordinate transformation generally disturbs the form of (9.1). But it is allowed and also convenient to consider a vector field and a symmetric tensor field that have just the components w_j, v^h and w_{ji} with respect to (h). Of course these fields belong then to the coordinate system (h) used in (9.1) and to those systems that arise from (h) by a homogeneous linear transformation with *constant* coefficients.

N.a.s. condition for the connexion (9.1) to be subprojective is that there exists a coordinate system (h') such that (cf. 8.5)

$$\Gamma_{j'i'}^{h'} \overset{*}{=} 2 \varphi_{(j'} A_{i')}^{h'} + \varphi_{j'i'} \xi^{h'}. \tag{9.2}$$

From (9.1) we get for the curvature tensor

$$\left\{ \begin{aligned} R_{kji}^{\cdot\cdot\cdot h} \overset{*}{=} 2 A_i^h \partial_{[k} w_{j]} &+ 2 A_{[k}^h (-\partial_{j]} w_i + w_{j]} w_i + w_{j]i} w_l v^l) + \\ &+ 2 v^h (\partial_{[k} w_{j]i} + 2 w_{[k|l|} w_{j]i} v^l) + 2 w_{i[j} \partial_{k]} v^h \end{aligned} \right. \tag{9.3}$$

and this expression could be equated to (8.9) because (8.9) is an invariant equation valid with respect to every coordinate system. Of course (h') is now the system for which (9.2) holds and for x^h we have now

$$x^{h'} = A_h^{h'} x^h \overset{*}{=} \xi^{h'}; \qquad x^h \neq \xi^h. \tag{9.4}$$

But comparing (9.3) and (8.9) we see that simple results can only be obtained if $\partial_j v^h$ has the form

$$\partial_j v^h \overset{*}{=} \alpha' A_j^h + \gamma_j v^h. \tag{9.5}$$

---

1) ADATI 1951, 8, p. 136.

2) ADATI 1951, 3; 4; 5; 6; 7; 8; SCHOUTEN 1953, 1.

This is equivalent to the invariant condition introduced by ADATI for the vector field $v^\varkappa$

$$\nabla_\mu v^\varkappa = \alpha A_\mu^\varkappa + \beta_\mu v^\varkappa. \tag{9.6}$$

A vector field in L_n satisfying an equation of the form (9.6) is called *torse forming* because in an E_n the lines of the vectors of such a field at the points of any curve form a torse.[1]) By multiplying $v^\varkappa$ with a scalar σ another torse forming field arises and if $\alpha \neq 0$, σ can always be chosen such that the new factor α equals $+1$.

Important special cases are:

the *concircular* field[2]): β_μ = gradient; α arbitrary. If $v^\varkappa$ is concircular, $\sigma v^\varkappa$ has the same property and $\beta_\mu = \partial_\mu \beta$ can always be reduced to zero by taking $\sigma = C e^{-\beta}$; C = const.;

the *special concircular* field: $\beta_\mu = 0$; α arbitrary. The property is only invariant if σ = const.;

the *concurrent* field[3]): $\beta_\mu = 0$; α = const. By multiplying $v^\varkappa$ with α^{-1} the constant α is reduced to $+1$;

the *recurrent* or *parallel* field (cf. III § 2)[4]): $\alpha = 0$; β_μ arbitrary;

the *covariant constant* field (cf. III § 2): $\alpha = 0$; $\beta_\mu = 0$.

It is easily proved that the only transformations of the $\Gamma_{\mu\lambda}^\varkappa$ that leave every torse forming field torse forming are the transformations of the form (cf. Exerc. VI 9,1)

$$'\Gamma_{\mu\lambda}^\varkappa = \Gamma_{\mu\lambda}^\varkappa + p_\mu A_\lambda^\varkappa + q_\lambda A_\mu^\varkappa. \tag{9.7}$$ [5])

These are at the same time those transformations in L_n that preserve all geodesics and transform every semi-symmetric connexion into a connexion with the same property. Here is a survey of the invariance of the six vector fields just defined under the transformations (9.7)

---

[1]) If for a field v^h in an A_n or L_n satisfying (9.5) a CARTAN displacement (cf. III § 2) is effected along a curve the images of the lines of the vectors in the moving E_n form a torse. Fields of this kind were introduced by YANO 1944, 5. The property of being torse forming is in fact not a property of the vector field but of its field of directions. Obviously the field of the pole directions of a subprojective connexion is torse forming.

[2]) YANO 1940, 5; 6; 7; 8; 1943, 3; ADATI 1951, 3, p. 161.

[3]) Concurrent fields were first introduced by MYLLER 1924, 1; 1928, 1 for V_m in R_n, cf. BORTOLOTTI 1931, 5; HAIMOVICI 1938, 1; SCHIROKOW 1939, 1; YANO 1943, 3.

[4]) Cf. for instance EISENHART 1922, 4; 1925, 1; 1926, 1, p. 67ff.; 1938, 1; LOPSCHITZ 1936, 1; YANO 1943, 3; LEVINE 1948, 1; 1949, 1; RUSE 1949, 1; 1950, 1; WALKER 1949, 1; 1950, 1; 3; 4; PATTERSON 1951, 1.

[5]) Transformations of this kind were considered by HLAVATY 1927, 3; 1928, 2; 1933, 1; SCHOUTEN and GOLAB 1930, 3; HOMBU and OKADA 1941, 2; YANO 1944, 12; MIKAMI 1949, 1.

	α	β_μ	Transformation
Torse forming	arbitrary	arbitrary	$\alpha \to \alpha + q_\mu v^\mu;\quad \beta_\mu \to \beta_\mu + p_\mu$
concircular	arbitrary	gradient	invariant for $p_\mu =$ gradient
spec. concircular	arbitrary	0	invariant for $p_\mu = 0$
concurrent	constant	0	invariant for $p_\mu = 0;\quad q_\mu = 0$
recurrent (parallel)	0	arbitrary	invariant for $q_\mu = 0$ [1])
cov. constant	0	0	invariant for $p_\mu = 0;\quad q_\mu = 0$

Torse forming fields are not possible in every A_n.

From (9.1, 5, 6) we get for $n>2$

$$\alpha' \overset{*}{=} \alpha - w_i v^i; \qquad \gamma_j \overset{*}{=} \beta_j - w_j - w_{ji} v^i. \tag{9.8}$$

If (9.5) is satisfied, its integrability conditions

$$(\partial_{[k} \alpha' - \alpha' \gamma_{[k}) A^h_{j]} + \partial_{[k} \gamma_{j]} v^h \overset{*}{=} 0 \tag{9.9}$$

must also hold. But if in the auxiliary theorem (cf. VI § 8) i is strangled and C_{kj} dropped it follows that for $n>2$ (9.9) is equivalent to

$$\begin{cases} \text{a)} & \partial_j \alpha' - \alpha' \gamma_j \overset{*}{=} 0 \\ \text{b)} & \partial_{[j} \gamma_{i]} \overset{*}{=} 0. \end{cases} \tag{9.10}$$

(9.10b) is a consequence of (9.10a), and (9.10a) can also be written in the form

$$\partial_j \alpha - \alpha \beta_j \overset{*}{=} - v^i (\alpha w_{ji} - \partial_j w_i + w_j w_i). \tag{9.11}$$

If (9.5) is substituted in (9.3) it follows that $R_{kji}^{\cdot\cdot\cdot h}$ has the special form (8.9). If then the auxiliary theorem is applied to (8.9) and (9.3) it follows for $n>3$ that

$$\begin{cases} \text{a)} & T_{ji} \overset{*}{=} \alpha w_{ji} - \partial_j w_i + w_j w_i \\ \text{b)} & T_{[ji]} = - \partial_{[j} w_{i]} \\ \text{c)} & v^h \overset{*}{=} \varrho x^h \\ \text{d)} & U_{kji} \overset{*}{=} \varrho (\partial_{[k} w_{j]i} + \beta_{[k} w_{j]i} - w_{[k} w_{j]i}). \end{cases} \tag{9.12}$$

Consequently (9.11) takes for $n>3$ the form

$$\partial_j \alpha - \alpha \beta_j \overset{*}{=} - v^i T_{ji}. \tag{9.13}$$ [2])

---

[1]) Cf. footnote 4 on p. 287.

[2]) In order to derive this equation we have only used (9.1) and (9.5) and these conditions are sufficient to ensure that $R_{kji}^{\cdot\cdot\cdot h}$ has the special form (8.9) without it being known whether the connexion is subprojective or not.

The Γ_{ji}^{h} transform into the $\Gamma_{j'i'}^{h'}$ as follows

$$\Gamma_{j'i'}^{h'} = A_{j'i'h}^{j\,i\,h'} \Gamma_{ji}^{h} - A_{j'i'}^{j\,i} \partial_j A_i^{h'} . \tag{9.14}$$

Hence, in order that $\Gamma_{j'i'}^{h'}$ takes the form (9.2) it is necessary and sufficient that $\partial_j A_i^{h'}$ takes the form

$$\partial_j A_i^{h'} \overset{*}{=} \psi_j A_i^{h'} + \psi_i A_j^{h'} + \psi_{ji} v^{h'}; \qquad \psi_{[ji]} = 0. \tag{9.15}$$

Now we have from (9.4, 5, 12c, 15)

$$\begin{cases} A_j^{h'} \overset{*}{=} \partial_j \xi^{h'} \overset{*}{=} \partial_j \varrho^{-1} A_i^{h'} v^i \overset{*}{=} A_j^{h'} \varrho^{-1} (\alpha' + \psi_i v^i) + \\ \qquad\qquad + \varrho^{-1} v^{h'} (\psi_j + \psi_{ji} v^i + \gamma_j - \partial_j \log \varrho) \end{cases} \tag{9.16}$$

hence for $n > 3$

$$\alpha' \overset{*}{=} \varrho - \psi_i v^i \tag{9.17}$$

$$\gamma_j \overset{*}{=} -\psi_j - \psi_{ji} v^i + \partial_j \log \varrho . \tag{9.18}$$

The integrability conditions of (9.15) are

$$\begin{cases} 0 \overset{*}{=} A_i^{h'} \partial_{[k} \psi_{j]} + A_{[k}^{h'} (\psi_{j]} \psi_i - \partial_{j]} \psi_i + \psi'_{j]i}) + \\ \qquad + \varrho^{-1} v^{h'} (\psi'_{i[k} \psi_{j]} + \partial_{[k} \psi'_{j]i}); \qquad \psi'_{ji} \overset{*}{\underset{\text{def}}{=}} \varrho \psi_{ji}; \end{cases} \tag{9.19}$$

and according to the auxiliary theorem these are for $n > 3$ equivalent to

$$\begin{cases} \text{a)} \quad \partial_j \psi_i - \psi_j \psi_i - \psi'_{ji} \overset{*}{=} 0 \\ \text{b)} \quad \partial_{[k} \psi'_{j]i} - \psi_{[k} \psi'_{j]i} \overset{*}{=} 0 . \end{cases} \tag{9.20}$$

Now the integrability condition of (9.20a) is (9.20b) and the integrability condition of (9.20b) is identically satisfied. Moreover if (9.17, 18) is substituted in the integrability conditions (9.10) these equations are identically satisfied on account of (9.17, 18, 20). Hence the system of differential equations (9.4, 12c, 15, 17, 18, 20a, 20b) for the unknowns $\xi^{h'}$, ϱ, $A_j^{h'}$, ψ_j, ψ_{ji} is totally integrable. Every solution of (9.20a) if substituted in (9.17) gives a value of ϱ that satisfies (9.18) on account of (9.20a). That proves the theorem[1])

If the connexion of a non projectively euclidean A_n can be written in the form

$$\Gamma_{ji}^{h} \overset{*}{=} 2 w_{(j} A_{i)}^{h} + w_{ji} v^h; \qquad w_{[ji]} = 0 \tag{9.21}$$

[1]) SCHOUTEN 1953, 1; ADATI proved this theorem for V_n but for A_n he introduced (9.13), equivalent to (9.10a) as an additional condition and he did not observe that this condition is in fact the integrability condition of (9.5) or (9.6) and that it can be derived using (9.1) and (9.5) only [see footnote 2 on p. 323]. Cf. ADATI 1951, 7, p. 127 where he gives the final form of his theorem for A_n.

with respect to some coordinate system (h) and if the vector field which has the components v^h with respect to (h) is torse forming the connexion is subprojective. The pole direction is the direction of v^h. [1])

The system (9.20a) has some very simple solutions. There are two cases:

1. $\alpha' \neq 0$. Because the product $w_{ji} v^h$ is given, we may fix the unknown factor in v^h in such a way that $\alpha' = 1$. Then

$$\partial_j v^h \overset{*}{=} A_j^h + \gamma_j v^h. \tag{9.22}$$

But the integrability conditions of this equation are

$$(\partial_{[k} \gamma_{j]}) v^h + A^h_{[k} \gamma_{j]} \overset{*}{=} 0 \tag{9.23}$$

and these can only be satisfied if $\gamma_j = 0$. Then from (9.22) we get

$$\partial_j v^h \overset{*}{=} \partial_j \xi^h \tag{9.24}$$

hence

$$v^h \overset{*}{=} \xi^h + c^h; \qquad c^h = \text{const.} \tag{9.25}$$

Now if we take

$$\xi^{h'} \overset{*}{=} \delta_h^{h'} (\xi^h + c^h) \tag{9.26}$$

we have $A_h^{h'} \overset{*}{=} \delta_h^{h'}$ and accordingly

$$\Gamma_{j'i'}^{h'} \overset{*}{=} 2 w_{(j'} A_{i')}^{h'} + w_{j'i'} \xi^{h'} \tag{9.27}$$

as was demanded.

2. $\alpha' = 0$. In this case we have

$$\partial_j v^h \overset{*}{=} \gamma_j v^h \tag{9.28}$$

for any choice of the scalar factor in v^h. The integrability conditions are

$$\partial_{[k} \gamma_{j]} \overset{*}{=} 0, \tag{9.29}$$

hence by changing the scalar factor in v^h we can always obtain

$$\partial_j v^h \overset{*}{=} 0; \qquad v^h \overset{*}{=} C^h = \text{const.} \tag{9.30}$$

Now we take for the ψ_i n arbitrary constants c_i such that $c_i C^i = 1$ and for ψ'_{ji} we take $-c_j c_i$. Then (9.20a, 20b) are satisfied and from (9.17) we get $\varrho = 1$, satisfying also (9.18). That leaves for the $A_i^{h'}$ the equations (9.15) in the form

$$\partial_j A_i^{h'} \overset{*}{=} 2 c_{(j} A_{i)}^{h'} - c_j c_i C^h A_h^{h'} \tag{9.31}$$

[1]) The condition for v^h is sufficient but not necessary. Necessary and sufficient conditions are not yet found. The theorem holds also for $n = 3$ but for that case it is not proved that the direction of v^h is the only pole-direction at every point.

with constants c_i and C^h. Because the c_i and C^h are constants the coordinate system (h) can be transformed by a homogeneous linear transformation with *constant* coefficients such that after the transformation c_i coincides with $\overset{1}{e}_i$ and C^h with $\underset{1}{e}^h$ and this transformation does not disturb the form of (9.1) and (9.30).

Then (9.31) takes the form

$$(9.32)\qquad \begin{cases} \text{a)}\ \ \partial_1 A_1^{h'} = A_1^{h'} \\ \text{b)}\ \ \partial_b A_1^{h'} = \partial_1 A_b^{h'} = A_b^{h'}; \qquad b, c = 2, \ldots, n \\ \text{c)}\ \ \partial_c A_b^{h'} \overset{*}{=} 0 \end{cases}$$

from which by integration

$$(9.33)\quad \begin{cases} \text{a)}\ \partial_1 \xi^{h'} = A_1^{h'} = (C_b^{h'} \xi^b + C_1^{h'})\, e^{\xi^1}; \quad C_h^{h'} = \text{const.}; \quad b = 2, \ldots, n \\ \text{b)}\ \partial_b \xi^{h'} = A_b^{h'} \overset{*}{=} C_b^{h'} e^{\xi^1} \end{cases}$$

with constants $C_h^{h'}$ that can be chosen arbitrarily provided that $\operatorname{Det}(C_h^{h'}) \neq 0$. By a second integration we get

$$(9.34)\qquad \xi^{h'} \overset{*}{=} (C_b^{h'} \xi^b + C_1^{h'})\, e^{\xi^1} = \underset{1}{e}^{h'} = v^{h'}$$

and this leads to the following expression for the $\Gamma_{j'i'}^{h'}$

$$(9.35)\qquad \Gamma_{j'i'}^{h'} \overset{*}{=} (w_{j'} + \overset{1}{e}_{j'})\, A_{i'}^{h'} + (w_{i'} + \overset{1}{e}_{i'})\, A_{j'}^{h'} + (w_{j'i'} - \overset{1}{e}_{j'} \overset{1}{e}_{i'})\, \xi^{h'}$$

as was demanded. (9.35) is independent of the choice of the $C_h^{h'}$. In fact a change of these constants is equivalent to a homogeneous linear transformation of the $\xi^{h'}$ with constant coefficients and such a transformation does not disturb the form of (9.35).

From (9.34) we see that the case $\alpha' = 0$ is not at all abnormal and that it only arises because the coordinates are chosen such that they get infinite at the pole. VRANCEANU[1]) used 1947 this special choice of the coordinates in order to simplify his calculations.

Exercises.

VI 9,1[2]). The transformations $\Gamma_{\mu\lambda}^{\varkappa} \to \Gamma_{\mu\lambda}^{\varkappa} + P_{\mu\lambda}^{\cdot\cdot\varkappa}$ in L_n transforms every torse forming vector field into a field with the same property if and only if $P_{\mu\lambda}^{\cdot\cdot\varkappa}$ has the form

$$\text{VI 9,1 } \alpha)\qquad P_{\mu\lambda}^{\cdot\cdot\varkappa} = p_\mu A_\lambda^\varkappa + A_\mu^\varkappa q_\lambda.$$

VI 9,2. The transformation of Exerc. VI 9,1 leaves the tensor $R_{\nu\mu\lambda}^{\cdot\cdot\cdot\varkappa} + A_\lambda^\varkappa R_{\nu\mu}$ invariant for $q_\lambda = 0$.

---

[1]) VRANCEANU 1947, 2.

[2]) Cf. footnote 5 on p. 322.

VI 9,3[1]). The object

VI 9,3 α) $$\Lambda^{\varkappa}_{\mu\lambda} = \Gamma^{\varkappa}_{\mu\lambda} + \frac{1}{n^2-1}\{A^{\varkappa}_{\mu}(\Gamma^{\varrho}_{\lambda\varrho} - n\,\Gamma^{\varrho}_{\varrho\lambda}) + A^{\varkappa}_{\lambda}(\Gamma^{\varrho}_{\varrho\mu} - n\,\Gamma^{\varrho}_{\mu\varrho})\}.$$

is invariant for the transformations of Exerc. VI 9,1.

§ 10. Subprojective transformations of a connexion in A_n.

A transformation of a connexion of the form

(10.1) $$'\Gamma^h_{ji} = \Gamma^h_{ji} + 2p_{(j}A^h_{i)} + q_{ji}v^h; \qquad q_{[ji]} = 0$$

with arbitrary vectors p_j and v^h and an arbitrary symmetric tensor q_{ji} is called a *subprojective transformation belonging to the direction* $\lfloor v^h \rfloor$.[2]) All subprojective transformations belonging to the same direction $\lfloor v^h \rfloor$ form a group.

The curves satisfying equations of the form

(10.2) $$\frac{d^2\xi^h}{dt^2} + \Gamma^h_{ji}\frac{d\xi^j}{dt}\frac{d\xi^i}{dt} = \varkappa\frac{d\xi^h}{dt} + \lambda v^h$$

are called *subgeodesics with respect to the field* $\lfloor v^h \rfloor$.[3]) The subprojective transformations belonging to $\lfloor v^h \rfloor$ leave the form of these equations invariant. If $\lfloor v^h \rfloor$ is torse forming we know that subprojective connexions exist with the pole-direction $\lfloor v^h \rfloor$ and in this case the subprojective transformations belonging to $\lfloor v^h \rfloor$ are those transformations that transform these subprojective connexions into each other. The set of all subprojective connexions belonging to $\lfloor v^h \rfloor$ and the set of all streamlines of $\lfloor v^h \rfloor$ determine each other uniquely. Hence a transformation $\xi^\varkappa \to \xi^\varkappa + u^\varkappa\,dt$ leaves this set of connexions invariant if and only if it leaves the streamlines invariant, that is if $\underset{u}{\pounds}\, v^h \propto v^h$. This can be proved also from (II 10.34). If this equation is written in the form[4])

(10.3) $$\underset{u}{\pounds}\,\Gamma^h_{ji} = u^l\,\partial_l\Gamma^h_{ji} - \Gamma^l_{ji}\,\partial_l u^h + \Gamma^h_{li}\,\partial_j u^l + \Gamma^h_{jl}\,\partial_i u^l + \partial_j\,\partial_i u^h$$

it must be proved that $\underset{u}{\pounds}\,\Gamma^h_{ji}$ has the form $\omega_j A^h_i + \omega_i A^h_j + \omega_{ji}v^h$[5]) if and only if $\underset{u}{\pounds}\, v^h \propto v^h$. Now if (9.1) is substituted in (10.3) and if all terms

---

1) Hlavaty 1927, 3, p. 85.

2) Yano 1944, 6; Adati 1951, 6, p. 106.

3) Equations of the form (10.1) and (10.2) were first considered by v. Dantzig 1932, 3; cf. Schouten and Haantjes 1936, 1; Haantjes 1937, 1; Yano 1944, 3; 6, he introduced the name "subpaths".

4) Cf. E I 1935, 1, p. 142.

5) Yano 1944, 6 has formulated this condition but he seems to have overlooked that it is satisfied if and only if the streamlines are invariant.

are dropped that have already the right form, the only remaining terms are

$$w_{ji}\,\alpha'\,u^h - w_{ji}\,v^l\,\partial_l\,u^h = w_{ji}\,\underset{u}{\pounds}\,v^h - w_{ji}\,u^l\,\gamma_l\,v^h; \qquad \text{(cf. 9.5)} \tag{10.4}$$

and these take the right form if and only if $\underset{u}{\pounds}\,v^h \propto v^h$.

Exercise.

VI 10,1 [1]). Every subprojective A_n can be transformed into an E_n by a subprojective transformation of its connexion.

§ 11. The subprojective V_n.

If the subprojective A_n is a V_n, (8.9) takes the form

$$K_{kjih} \overset{*}{=} 2g_{[k|h|}\,T_{j]i} - 2g_{ih}\,T_{[kj]} + 2U_{kji}\,x_h \tag{11.1}$$

and this expression must be alternating in ih:

$$g_{[k(h}\,T_{j]i)} - g_{ih}\,T_{[kj]} + U_{kj(i}\,x_{h)} \overset{*}{=} 0. \tag{11.2}$$

For $n>2$ we may take two vectors y^h, z^h, perpendicular to x^h and to each other. Then by transvection of (11.2) with $y^h z^i$ we get

$$y_{[k}\,T_{j]i}\,z^i + z_{[k}\,T_{j]i}\,y^i \overset{*}{=} 0 \tag{11.3}$$

and this means that $T_{ji}\,y^i \propto y_j$ for every vector y^h perpendicular to x^h. But this is only possible if T_{ji} has the form

$$T_{ji} \overset{*}{=} 2\mu\,g_{ji} + l_j\,x_i \tag{11.4}$$

with suitably chosen μ and l_j. If (11.4) is substituted in (11.1) we get

$$K_{kjih} \overset{*}{=} 4\mu\,g_{[k|h|}\,g_{j]i} + 2g_{[k|h|}\,l_{j]}\,x_i - 2g_{ih}\,l_{[k}\,x_{j]} + 2U_{kji}\,x_h. \tag{11.5}$$

As this expression must be alternating in ih we get

$$g_{[k(h}\,l_{j]}\,x_{i)} - g_{ih}\,l_{[k}\,x_{j]} + U_{kj(i}\,x_{h)} \overset{*}{=} 0 \tag{11.6}$$

and by transvection with $y^i\,y^h$ it follows that $l_{[k}\,x_{j]} = 0$. Hence T_{ji} is symmetric

$$T_{ji} \overset{*}{=} 2\mu\,g_{ji} + \lambda\,x_j\,x_i \tag{11.7}$$

and this implies that φ_j is a gradient and that there exist special coordinate systems for which φ_j vanishes (cf. VI § 8). If (11.7) is substituted in (11.2) we get

$$\text{a)}\quad U_{kjh} = \lambda\,x_{[k}\,g_{j]h}; \qquad \text{b)}\quad U_{kjh}\,x^h = 0 \tag{11.8}$$

[1]) Yano 1944, 6, p. 104.

and consequently

$$(11.9)\quad \begin{cases} \text{a)} & K_{kjih} = \frac{4}{n-2} g_{[k[i} L_{j]h]} \\ \text{b)} & L_{jh} \overset{*}{\underset{\text{def}}{=}} -(n-2)(\mu g_{jh} + \lambda x_j x_h). \end{cases}$$

For $n>3$, (11.9) is the n.a.s. condition for the V_n to be conformally euclidean. For $n=3$, K_{kjih} can always be written as the double alternating product of g_{ki} with another symmetric tensor. But in that case $\nabla_{[k} L_{j]h} = 0$ is the n.a.s. condition (cf. VI § 5). Now we can prove that this latter condition [that is a consequence of (11.9) for $n>3$] is always satisfied for a subprojective V_n.[1]) We take the coordinate system such that $\varphi_i = 0$. Then $\alpha = 1$ and from (8.5, 10, 11, 12) it follows that

$$(11.10)\qquad T_{ji} \overset{*}{=} \varphi_{ji} \overset{*}{=} 2\mu g_{ji} + \lambda x_j x_i,$$

$$(11.11)\qquad \Gamma_{ji}^h \overset{*}{=} \varphi_{ji} x^h,$$

$$(11.12)\qquad U_{kji} \overset{*}{=} \nabla_{[k} \varphi_{j]i}.$$

But on account of (11.8) this leads to

$$(11.13)\qquad \nabla_i \mu = \lambda x_i; \quad \nabla_i \lambda = \zeta x_i$$

where ζ is a scalar. If (11.13) is substituted in (11.9), differentiating gives $\nabla_{[k} L_{j]h} = 0$. Moreover RACHEVSKI[1]) proved that the condition is also sufficient and can be put in the following form

A V_n, $n>2$ is subprojective if and only if it is a C_n satisfying the special condition

$$(11.14)\qquad L_{jh} = \sigma g_{jh} + \tau (\nabla_j \sigma) \nabla_h \sigma$$

where σ is an arbitrary scalar and τ a function of σ only.

From

$$(11.15)\qquad \nabla_j x^h \overset{*}{=} A_j^h + \varphi_{ji} x^i x^h \overset{*}{=} A_j^h + (2\mu + \lambda x_i x^i) x_j x^h$$

and (11.13) it follows immediately that β_j is a gradient. Hence x^h is not only torse forming but also concircular.[2]) ADATI gave also the forms of L_{jh} for the cases that x^h is parallel or the unitvector of x^h concurrent and he gave also other forms of the linear element of subprojective V_n's for different cases.[3])

[1]) RACHEVSKI 1933, 1, p. 137; cf. E II 1938, 2, p. 220ff.; cf. ADATI 1951, 5, p. 343ff.

[2]) ADATI 1951, 3, p. 172.

[3]) ADATI 1951, 4; 5, p. 333, 357f.

Exercises.

VI 11,1 [1]). If in a subprojective V_n; $n>2$, the coefficient μ in (11.4) is a constant, the V_n is an S_n (use 11.9 and 11.13).

VI 11,2 [2]). The C_n with the linear element

$$d s^2 \overset{*}{=} f(p) \sum_\lambda d\xi^\lambda d\xi^\lambda; \qquad p \overset{*}{=} \sum_\mu \xi^\mu \xi^\mu$$

is subprojective.

§12. Concircular transformations of a V_n.

From (5.36) we get for the curvature vector $u^\varkappa = j^\mu \nabla_\mu j^\varkappa$ of a curve with tangent unitvector $j^\varkappa$ under conformal transformations of $g_{\lambda\varkappa}$

$$(12.1)\quad 'u^\varkappa = \sigma^{-1}(u^\varkappa - \tfrac{1}{2} z^\varkappa); \qquad z^\varkappa \overset{\text{def}}{=} s^\varkappa - j^\mu s_\mu j^\varkappa; \qquad s_\lambda \overset{\text{def}}{=} \partial_\lambda \log \sigma.$$

Now we consider a *geodesic circle* [3]), that is a curve whose curvature vector has constant length and whose second curvature is zero

$$(12.2)\qquad \text{a)}\ \frac{d}{ds} u^\varkappa u^\lambda g_{\lambda\varkappa} = 0; \qquad \text{b)}\ j^{[\varkappa} \frac{\delta}{ds} u^{\lambda]} = 0.$$

Then we have after the transformation

$$(12.3)\qquad \frac{d}{ds}('u^\varkappa\, 'u^\lambda\, 'g_{\lambda\varkappa}) = \sigma j^\mu\, 'u^\lambda \left(-\nabla_\mu s_\lambda + \frac{1}{2} s_\mu s_\lambda\right)$$

and this proves that a conformal transformation transforms every geodesic circle into a geodesic circle if and only if an equation holds of the form [4])

$$(12.4)\qquad 2\nabla_\mu s_\lambda - s_\mu s_\lambda = 4\psi g_{\mu\lambda}\ ^{5})$$

or (cf. 5.7)

$$(12.5)\qquad s_{\mu\lambda} = 4\varphi g_{\mu\lambda}; \qquad \varphi = \psi + \tfrac{1}{8} s_\lambda s^\lambda.$$

Conformal transformations satisfying this condition are called *concircular*. They leave invariant the equation of all geodesic circles, that

[1]) SCHAPIRO 1933, 1, p. 112; E II 1938, 2, p. 222 and 225.

[2]) Cf. BUCHHOLZ 1899, 1; KAGAN 1933, 1, p. 94; E II 1938, 2, p. 225.

[3]) FIALKOW 1939, 2. YANO introduced 1938, 3 other generalized circles and proved 1940, 6 that these curves are geodesic circles if their tangent direction is everywhere a principal direction of V_n.

[4]) BRINKMAN 1925, 1; FIALKOW 1939, 2, p. 461; cf. also FIALKOW 1940, 1; 1942, 2; YANO 1940, 5; 6; 7; 8; YANO and ADATI 1944, 7.

[5]) The factor 4 is introduced only in order to get the same φ and ψ as in the paper of YANO. His $\varrho_{\mu\lambda}$ is equal to $\frac{1}{4} s_{\mu\lambda}$.

can be proved to be [1])

$$\frac{\delta^3 \xi^\varkappa}{d s^3} + g_{\mu\lambda} \frac{\delta^2 \xi^\mu}{d s^2} \frac{\delta^2 \xi^\lambda}{d s^2} \frac{d \xi^\varkappa}{d s} = 0. \tag{12.6}$$

Transvection of (12.4) with s^μ leads to

$$2 s^\mu \nabla_\mu s^\varkappa = s^\mu s_\mu s^\varkappa + 4\psi s^\varkappa \tag{12.7}$$

hence the streamlines of $s^\varkappa$ are geodesics and the congruence of these geodesics is normal to the V_{n-1}'s tangent to the gradient s_λ. The integrability conditions of (12.4) are

$$- K_{\nu\mu\lambda}^{\cdots\varkappa} s_\varkappa - (2\psi s_{[\mu} - 4 \nabla_{[\mu} \psi) g_{\nu]\lambda} = 0 \tag{12.8}$$

from which by transvection with $g^{\mu\lambda}$

$$K_\nu^{\cdot\varkappa} s_\varkappa = (n-1)(2\psi s_\nu - 4 \nabla_\nu \psi). \tag{12.9}$$

But by transvection of (12.8) with s^λ it follows that $\nabla_\nu \psi$ is a multiple of s_ν. Hence $s^\varkappa$ *lies in a principal direction* (cf. III § 5) *of* V_n [2]) and the function ψ is constant on each of the V_{n-1}'s orthogonal to the congruence $s^\varkappa$.

If $i^\varkappa$ is the unitvector of $s^\varkappa$ and $s^\varkappa = \zeta i^\varkappa$ we have from (12.4)

$$2 (\nabla_\mu \zeta) i_\lambda + 2\zeta \nabla_\mu i_\lambda - \zeta^2 i_\mu i_\lambda = 4\psi g_{\mu\lambda} \tag{12.10}$$

and

$$2 \nabla_\mu \zeta = \zeta^2 i_\mu + 4\psi i_\mu. \tag{12.11}$$

But this implies that $\nabla_\mu i_\lambda$ has the form

$$\nabla_\mu i_\lambda = - h (g_{\mu\lambda} - i_\mu i_\lambda). \tag{12.12}$$

Hence the second fundamental tensor of the V_{n-1}'s normal to $i^\varkappa$ is

$$h_{cb} = - B_{c\,b}^{\mu\lambda} \nabla_\mu i_\lambda = h\, 'g_{cb} \tag{12.13}$$

and this proves that these V_{n-1}'s are umbilical. From the CODAZZI equation (V 6.5)

$$2 ('\nabla_{[d} h)\, 'g_{c]b} = B_{dcb}^{\nu\mu\lambda} K_{\nu\mu\lambda\varkappa} i^\varkappa \tag{12.14}$$

we get by transvection with $'g^{cb}$

$$(n-2)\, '\nabla_d h = B_d^\nu (g^{\mu\lambda} - i^\mu i^\lambda) K_{\nu\mu\lambda\varkappa} i^\varkappa = B_d^\nu K_{\nu\varkappa} i^\varkappa \tag{12.15}$$

---

[1]) YANO 1940, 5, p. 195. He also considers curves that can be transformed conformally into a geodesic circle and calls them conformally geodesic circles. But because every single curve is conformally geodesic (cf. VI § 5), it is also a conformally geodesic circle and the term is useful only in connexion with systems of curves.

[2]) YANO 1940, 6, p. 356.

but, because $i^\varkappa$ lies in a principal direction of $K_{\nu\varkappa}$ this proves that h is constant over V_{n-1}, i.e. that the V_{n-1}'s have each constant mean curvature. So we see that the V_{n-1}'s normal to a geodesic and normal congruence are always umbilical if the tangent direction of the congruence is torse forming[1]) and that the constancy of the mean curvature is due to the fact that in the special case under consideration the tangent lies in a principal direction of the V_n.

Now let us suppose that there exists in a V_n a congruence with unit-vector $i^\varkappa$ satisfying (12.12) with $'\nabla_b h = 0$. Then we prove that there exists a scalar η such that the field defined by $s_\lambda = \eta i_\lambda$ satisfies an equation of the form (12.4). From

$$(12.16) \qquad 2(\nabla_\mu \eta)\, i_\lambda - 2\eta\, h\, g_{\mu\lambda} + 2\eta\, h\, i_\mu i_\lambda - \eta^2 i_\mu i_\lambda = 4\psi\, g_{\mu\lambda}$$

follows

$$(12.17) \qquad \begin{cases} \text{a)}\ \ 2\partial_\mu \eta = (\eta^2 - 2\eta\, h)\, i_\mu \\ \text{b)}\ \ \ \eta\, h = -2\psi. \end{cases}$$

The integrability condition of (12.17a) is

$$(12.18) \qquad (2\eta\, \partial_{[\nu}\eta - 2\eta\, \partial_{[\nu} h - 2h\, \partial_{[\nu}\eta)\, i_{\mu]} = 0$$

and because $i_{[\mu}\, \partial_{\nu]} h = 0$, this implies that η is a function of u only, if we write $i_\mu = \partial_\mu u$. Now h is also a function of u only, hence if $\eta \overset{\text{def}}{=} f(u)$ and $f'(u) \overset{\text{def}}{=} \frac{df}{du}$ the differential equation for η takes the form

$$(12.19) \qquad 2f' = f^2 - 2hf$$

and this is a RICATTI equation. Gathering results we have[2])

A V_n admits a concircular transformation if and only if there exists at least one geodesic congruence, normal to a set of umbilical V_{n-1}'s with constant mean curvature.

The concircular transformations form a subgroup of all conformal transformations. If a V_m in V_n is given, every concircular transformation $g_{\mu\lambda} \to \sigma g_{\mu\lambda}$ in V_n induces a conformal transformation $'g_{cb} \to \sigma' g_{cb}$ in V_m and we may ask[3]) whether the induced transformation is also concircular. Now if $'s_b \overset{\text{def}}{=} B_b^\lambda s_\lambda$ we have

$$(12.20) \qquad \begin{cases} 2'\nabla_c\, 's_b = 2'\nabla_c B_b^\lambda s_\lambda = 2B_{cb}^{\mu\lambda} \nabla_\mu s_\lambda + 2s_\lambda D_c B_b^\lambda \\ \qquad\qquad\qquad\qquad = {'s_c}\, {'s_b} + 4\psi\, 'g_{cb} + 2H_{cb}^{\cdot\cdot\lambda} s_\lambda, \end{cases}$$

---

[1]) YANO 1944, 5.

[2]) YANO 1940, 6, p. 359; cf. also FIALKOW 1939, 2. The proof given here is much simpler than YANO's proof that is based on a special form of the linear element and makes use of a function suggested by A. KAWAGUCHI.

[3]) YANO 1940, 8, p. 506.

hence the induced transformation is concircular if and only if $H_{cb}^{\cdot\cdot\lambda} s$ is a multiple of $'g_{cb}$ or, in another form, if and only if[1]) (cf. V 9.16b)

$$M_{cb}^{\cdot\cdot\lambda} s_\lambda = 0. \tag{12.21}$$

As a corollary we get that a concircular transformation induces a concircular transformation in every umbilical V_m. This implies that *every geodesic and every geodesic circle in an umbilical V_m is a geodesic circle in V_n.* [2])

If a field $u^\varkappa$ is subjected to a parallel displacement over $v^\varkappa\, dt$ we have for the field value of the new field at $\xi^\varkappa + v^\varkappa\, dt$

$$u^\varkappa - v^\mu \Gamma_{\mu\lambda}^\varkappa u^\lambda\, dt. \tag{12.22}$$

In order that this value has the same direction as the field dragged along over $v^\varkappa\, dt$

$$u^\varkappa + u^\mu \partial_\mu v^\varkappa\, dt \tag{12.23}$$

it is necessary and sufficient that

$$u^\mu \nabla_\mu v^\varkappa = \alpha u^\varkappa \tag{12.24}$$

where α is a function of the $\xi^\varkappa$. Hence, a point transformation $v^\varkappa\, dt$ leaves invariant the tangent direction of every curve if and only if $v^\varkappa$ is a special concircular field (cf. VI § 9):

$$\nabla_\mu v^\varkappa = \alpha A_\mu^\varkappa. \tag{12.25}$$

Because $\nabla_\mu v_\lambda = \alpha g_{\mu\lambda}$, the vector v_λ is a gradient. If we write $v_\lambda = \partial_\lambda \varrho^{-1}$ it follows that

$$\nabla_\mu \nabla_\lambda \log \varrho - (\nabla_\mu \log \varrho) \nabla_\lambda \log \varrho = -\alpha \varrho g_{\mu\lambda} \tag{12.26}$$

and this proves that the V_n admits a concircular transformation $g_{\mu\lambda} \to \varrho^2 g_{\mu\lambda}$ with $\psi = -\alpha\varrho$ (cf. 12.4). Conversely, if we have a V_n with a field s_λ satisfying (12.4) and if $s_\lambda = \partial_\lambda \log \sigma$, the field $v^\varkappa = g^{\lambda\varkappa} \partial_\lambda \sigma^{-\frac{1}{2}}$ is a special concircular field. Hence[3])

A V_n admits a concircular transformation if and only if it admits at least one special concircular vector field and this is the case if and only if it admits at least one infinitesimal point transformation that displaces the tangent directions of every curve parallel.

---

[1]) See footnote 3, page 332.

[2]) YANO 1943, 3, p. 193.

[3]) YANO and ADATI 1944, 7. This paper deals also with the case where $v^\varkappa$ is a concurrent field, $\alpha = \text{const.}$

FRENET formulae for curves in concircular geometry can be developed by means of the vector (cf. 12.6)

$$V^{\varkappa} \overset{\text{def}}{=} \frac{\delta^3 \xi^{\varkappa}}{ds^3} + g_{\mu\lambda} \frac{\delta^2 \xi^{\mu}}{ds^2} \frac{\delta^2 \xi^{\lambda}}{ds^2} \frac{d\xi^{\varkappa}}{ds} \tag{12.27}$$

that transforms into $\sigma^{-\frac{3}{2}} V^{\varkappa}$ for a conformal transformation of $g_{\mu\lambda}$ if and only if this transformation is concircular.[1])

From (12.5) and (5.6) it follows that for a concircular transformation

$$'K_{\nu\mu\lambda}^{\cdot\cdot\cdot\varkappa} = K_{\nu\mu\lambda}^{\cdot\cdot\cdot\varkappa} + 4\varphi\, g_{[\nu[\lambda}\, g_{\mu]\sigma]}\, g^{\sigma\varkappa}. \tag{12.28}$$

Hence, a V_n that can be transformed into an R_n by a concircular transformation must be an S_n. Moreover it follows from (12.28) that a concircular transformation transforms every EINSTEIN space into an EINSTEIN space.[2]) Another consequence of (12.28) is that[3])[4])

$$Z_{\nu\mu\lambda}^{\cdot\cdot\cdot\varkappa} \overset{\text{def}}{=} K_{\nu\mu\lambda}^{\cdot\cdot\cdot\varkappa} - \frac{2}{n(n-1)} K A_{[\nu}^{\varkappa}\, g_{\mu]\lambda} \tag{12.29}$$

is invariant for concircular transformations. YANO[3]) proved that every conformal transformation that leaves $Z_{\nu\mu\lambda}^{\cdot\cdot\cdot\varkappa}$ invariant is concircular. $Z_{\nu\mu\lambda}^{\cdot\cdot\cdot\varkappa}$ is called the *concircular curvature tensor*. In the same paper YANO proved that its vanishing is for $n>2$ n.a.s. for the V_n to be transformable into an R_n by a concircular transformation. The V_n could be called *concircular euclidean* in this case but because for $n>2$ there is no difference between an S_n and a concircular euclidean space it is better not to introduce a new name. For $Z_{\nu\mu\lambda}^{\cdot\cdot\cdot\varkappa}$ the following identities hold

$$\begin{cases} \text{a)} & Z_{\varkappa\mu\lambda}^{\cdot\cdot\cdot\varkappa} = K_{\mu\lambda} - \frac{1}{2} K\, K_{\mu\lambda} \\ \text{b)} & g^{\mu\lambda} Z_{\varkappa\mu\lambda}^{\cdot\cdot\cdot\varkappa} = 0. \end{cases} \tag{12.30}$$

In the last of YANO's papers on concircular transformations and in a paper of 1943[5]) he studied conformal and concircular properties of EINSTEIN spaces. Recently TACHIBANA[6]) dealt with the relations between concircular transformations and the investigations of SASAKI[7]) on the holonomy group of conformal spaces.

Exercise.

VI 12,1[8]). If every geodesic circle of a V_{n-1} in V_n is a geodesic circle of V_n, the V_{n-1} is umbilical and its mean curvature is constant.

---

1) YANO 1940, 4; 7.
2) YANO 1940, 5, p. 200.
3) YANO 1940, 5.
4) TACHIBANA 1951, 2, p. 150.
5) YANO 1942, 3; 1943, 1.
6) TACHIBANA 1951, 2.
7) SASAKI 1943, 1; 2; 3.
8) YANO 1940, 7; 1943, 3.

VII. Variations and deformations.

§ 1. General deformation problems. [1]

In this chapter several problems concerning variations and deformations are dealt with. The general form of such a problem is as follows: certain geometrical objects suffer a finite or infinitesimal transformation and the behaviour of some other objects depending on them is required. In deformation problems we deal with the special case where the variation is due to displacements of some kind, for instance a dragging along or a parallel displacement. The case occurring most frequently is that some objects are left at rest, others are dragged along and others are displaced parallel. The theory of variation and deformation is very important because a great number of problems in differential geometry can be treated in a very elegant way by using the methods of this theory.

If a geometric object field is defined in an $\mathfrak{N}(\xi^\varkappa)$ we have at $\xi^\varkappa + v^\varkappa dt$. [2])

1. the *natural value* of the field or the value of the field *at rest*, for instance for a contravariant vector

$$p^\varkappa + d\,p^\varkappa = p^\varkappa + v^\mu\,\partial_\mu\,p^\varkappa\,dt. \tag{1.1}$$

2. the value arising from *dragging along* the field over $v^\varkappa dt$, for instance (cf. II § 10) for vectors

$$\begin{cases} \text{a)} \quad p^\varkappa + \overset{m}{d}\,p^\varkappa = p^\varkappa + p^\mu\,\partial_\mu\,v^\varkappa\,dt \\ \text{b)} \quad q_\lambda + \overset{m}{d}\,q_\lambda = q_\lambda - q_\mu\,\partial_\lambda\,v^\mu\,dt. \end{cases} \tag{1.2}$$

If a parallel displacement in the direction of $v^\varkappa$ is defined for the quantities considered we also have

3. the value arising from parallel displacement of the field over $v^\varkappa\,dt$, for instance (cf. III § 2)

$$\begin{cases} \text{a)} \quad p^\varkappa + \overset{*}{d}p^\varkappa = p^\varkappa - \Gamma^\varkappa_{\mu\lambda}\,p^\lambda\,v^\mu\,dt \\ \text{b)} \quad q_\lambda + \overset{*}{d}q_\lambda = q_\lambda + \Gamma^\varkappa_{\mu\lambda}\,q_\varkappa\,v^\mu\,dt\,. \end{cases} \tag{1.3}$$

---

[1]) Cf. Mc Connell 1929, 1 (curves in V_n); Hayden 1931, 1; Schouten and v. Kampen 1933, 1; Davies 1933, 1 (in U_n); 1937, 1; 1938, 1; Dienes 1933, 3 (in L_n); E I 1935, 1, § 12; E II 1938, 2, § 16; Coburn 1940, 2 (on Schouten and v. Kampen); Shabbar 1942, 1 (finite deform.); Yano 1945, 7; 8; 9; 10 (on Coburn); 1949, 2 (literature); Tonolo 1950, 1 (in V_3 and S_3).

[2]) We follow here the line of thought developed in Schouten and v. Kampen 1933, 1 and E I 1935, 1, p. 140ff.

The three differentials arising in this way are all non invariant[1]) but their differences are invariant:

1. the *covariant differential* δ (cf. III § 2), for instance for vectors

$$(1.4)\quad \begin{cases} \text{a)}\quad \delta p^\varkappa = d p^\varkappa - \overset{*}{d} p^\varkappa = v^\mu \nabla_\mu p^\varkappa\, dt \\ \text{b)}\quad \delta q_\lambda = d q_\lambda - \overset{*}{d} q_\lambda = v^\mu \nabla_\mu q_\lambda\, dt. \end{cases}$$

2. the LIE *differential* $\underset{v}{£}\, dt$ (cf. II § 10 and III 5.48), for instance

$$(1.5)\quad \begin{cases} \text{a)}\quad \underset{v}{£}\, p^\varkappa\, dt = d p^\varkappa - \overset{m}{d} p^\varkappa = (v^\mu \partial_\mu p^\varkappa - p^\mu \partial_\mu v^\varkappa)\, dt \\ \qquad\qquad = v^\mu \nabla_\mu p^\varkappa\, dt - p^\mu v_{\mu}^{\cdot\,\varkappa}\, dt. \\ \text{b)}\quad \underset{v}{£}\, q_\lambda\, dt = d q_\lambda - \overset{m}{d} q_\lambda = (v^\mu \partial_\mu q_\lambda + q_\mu \partial_\lambda v^\mu)\, dt \\ \qquad\qquad = v^\mu \nabla_\mu q_\lambda\, dt + q_\mu v_{\lambda}^{\cdot\,\mu}\, dt. \\ \qquad v_{\lambda}^{\cdot\,\varkappa} \overset{\text{def}}{=} \partial_\lambda v^\varkappa + \Gamma_{\mu\lambda}^{\varkappa} v^\mu. \end{cases}$$

3. the *apparent differential*[2]) $\underset{s}{D}\, dt$, for instance

$$(1.6)\quad \begin{cases} \text{a)}\quad \underset{s}{D}\, p^\varkappa\, dt = \overset{*}{d} p^\varkappa - \overset{m}{d} p^\varkappa = -\, p^\mu v_{\mu}^{\cdot\,\varkappa}\, dt \\ \text{b)}\quad \underset{s}{D}\, q_\lambda\, dt = \overset{*}{d} q_\lambda - \overset{m}{d} q_\lambda = +\, q_\mu v_{\lambda}^{\cdot\,\mu}\, dt. \end{cases}$$

Between these three differentials applied to quantities we have the relation

$$(1.7)\qquad \delta - \underset{v}{£}\, dt + \underset{s}{D}\, dt = 0.$$

The geometrical interpretation is as follows. Let the field $v^\varkappa$ be given in an $\mathfrak{N}(\overset{}{\underset{0}{\xi^\varkappa}})$ and let the field of a quantity Ω (indices suppressed) undergo on the one hand a parallel displacement over $v^\varkappa\, dt$. Then the value $\underset{0}{\Omega}$ at $\underset{0}{\xi^\varkappa}$ changes into $\underset{0}{\Omega} - \delta\Omega$. On the other hand let the field Ω be dragged along over $v^\varkappa\, dt$. Then the value at $\underset{0}{\xi^\varkappa}$ changes into $\underset{0}{\Omega} - \underset{v}{£}\,\Omega dt$. The difference between the two new values at $\underset{0}{\xi^\varkappa}$ is $-\underset{s}{D}\,\Omega dt$, hence we get exactly the field value $\underset{0}{\Omega} + \underset{s}{D}\,\Omega\, dt$ at $\underset{0}{\xi^\varkappa}$ if the field Ω is first dragged along over $-v^\varkappa\, dt$ and afterwards displaced parallel over $+v^\varkappa\, dt$. This illustrates the fact that $\underset{s}{D}\,\Omega\, dt$ is really not a true differential of the field Ω but only a linear transformation of the components of Ω at $\underset{0}{\xi^\varkappa}$

---

[1]) If by applying a differential operator to the components of a geometric object of a certain kind the components of another geometric object of the same kind arise, we call the operator *invariant with respect to objects of this kind.*

[2]) "Scheinbares Differential" in SCHOUTEN and v. KAMPEN 1933, 1 and E I 1935, 1, p. 142.

depending on the values of Ω and of the field $v_{\lambda}^{\cdot\,\varkappa}$ at the point $\underset{0}{\xi}^{\varkappa}$ only, and not on the values of these fields at neighbouring points.

From (II 10.34) or (III 5.47) we get

$$(1.8)\quad \underset{v}{\pounds}\, \Gamma_{\mu\lambda}^{\varkappa} = \nabla_\mu \nabla_\lambda v^\varkappa + 2\nabla_\mu S_{\varrho\lambda}^{\cdot\cdot\,\varkappa} v^\varrho + v^\nu R_{\nu\mu\lambda}^{\cdot\cdot\cdot\,\varkappa} = \nabla_\mu v_{\lambda}^{\cdot\,\varkappa} + v^\nu R_{\nu\mu\lambda}^{\cdot\cdot\cdot\,\varkappa}.$$

In (III 5.50) we obtained the first of the following three identities

$$(1.9)\quad \begin{cases} \text{a)} & (\underset{v}{\pounds}\nabla_\mu - \nabla_\mu \underset{v}{\pounds})\, p^\varkappa = (\underset{v}{\pounds}\,\Gamma_{\mu\lambda}^{\varkappa})\, p^\lambda = (\nabla_\mu v_{\lambda}^{\cdot\,\varkappa} + v^\nu R_{\nu\mu\lambda}^{\cdot\cdot\cdot\,\varkappa})\, p^\lambda \\ \text{b)} & (\underset{v}{\pounds}\nabla_\mu - \nabla_\mu \underset{v}{\pounds})\, q_\lambda = -(\underset{v}{\pounds}\,\Gamma_{\mu\lambda}^{\varkappa})\, q_\varkappa = -(\nabla_\mu v_{\lambda}^{\cdot\,\varkappa} + v^\nu R_{\nu\mu\lambda}^{\cdot\cdot\cdot\,\varkappa})\, q_\varkappa \\ \text{c)} & (\underset{v}{\pounds}\nabla_\mu - \nabla_\mu \underset{v}{\pounds})\, \mathfrak{q} = -w\,(\underset{v}{\pounds}\,\Gamma_\mu)\, \mathfrak{q} = -w\,(\nabla_\mu v_{\lambda}^{\cdot\,\lambda} + v^\nu V_{\nu\mu})\, \mathfrak{q} \end{cases}$$

for the vectors $p^\varkappa$, q_λ and the density $\mathfrak{q}$ of weight w, from which the last two can be derived easily from the first. From (III 4.9a, b) we get for these same quantities

$$(1.10)\quad \begin{cases} \text{a)} & (\delta\nabla_\mu - \nabla_\mu\delta)\, p^\varkappa = (v^\nu R_{\nu\mu\lambda}^{\cdot\cdot\cdot\,\varkappa} p^\lambda - v_{\mu}^{\cdot\,\varrho}\nabla_\varrho p^\varkappa)\, dt \\ \text{b)} & (\delta\nabla_\mu - \nabla_\mu\delta)\, q_\lambda = -(v^\nu R_{\nu\mu\lambda}^{\cdot\cdot\cdot\,\varkappa} q_\varkappa + v_{\mu}^{\cdot\,\varrho}\nabla_\varrho q_\lambda)\, dt \\ \text{c)} & (\delta\nabla_\mu - \nabla_\mu\delta)\, \mathfrak{q} = -w\,(v^\nu V_{\nu\mu}\,\mathfrak{q} + v_{\mu}^{\cdot\,\varrho}\nabla_\varrho \mathfrak{q})\, dt \end{cases}$$

and from these equations and (1.7) it follows that[1])

$$(1.11)\quad \begin{cases} \text{a)} & (\underset{s}{D}\nabla_\mu - \nabla_\mu \underset{s}{D})\, p^\varkappa = p^\lambda \nabla_\mu v_{\lambda}^{\cdot\,\varkappa} + v_{\mu}^{\cdot\,\varrho}\nabla_\varrho p^\varkappa \\ \text{b)} & (\underset{s}{D}\nabla_\mu - \nabla_\mu \underset{s}{D})\, q_\lambda = -q_\varkappa \nabla_\mu v_{\lambda}^{\cdot\,\varkappa} + v_{\mu}^{\cdot\,\varrho}\nabla_\varrho q_\lambda \\ \text{c)} & (\underset{s}{D}\nabla_\mu - \nabla_\mu \underset{s}{D})\, \mathfrak{q} = -w\,\mathfrak{q}\,\nabla_\mu v_{\lambda}^{\cdot\,\lambda} + w\, v_{\mu}^{\cdot\,\varrho}\nabla_\varrho \mathfrak{q}. \end{cases}$$

All these identities are important because they give us an opportunity of changing the order of the operators $\underset{v}{\pounds}$, δ or $\underset{s}{D}$ and the operator ∇_μ in any formula.

If the operators δ, $\underset{v}{\pounds}\,dt$ and $\underset{s}{D}\,dt$ are applied to a number of objects, other objects, depending on these will in general suffer variations of a more complicated kind, that could not arise by the mere dragging along or parallel displacement of these latter objects. Let such a variation transform an object field $\underset{v}{\Omega}$ (indices suppressed) with fieldvalues $\underset{0}{\overset{v}{\Omega}}$ at $\underset{0}{\xi}^\varkappa$ into a field with fieldvalues $\underset{0}{\Omega} + \overset{v}{d}\Omega$ at $\underset{0}{\xi}^\varkappa + v^\varkappa\,dt$. Then $\overset{v}{d}$ is not an invariant operator but the following three invariant operators can be defined

[1]) SCHOUTEN and v. KAMPEN 1933, 1, p. 9.

1. The *natural variation* $\overset{n}{D}\Omega dt$, that is the difference between the new value at $\underset{0}{\xi}^\varkappa + v^\varkappa dt$ and the old value at that same point

(1.12)
$$\overset{n}{D} \overset{\text{def}}{=} \frac{\overset{v}{d}}{dt} - \frac{d}{dt} = \frac{\overset{v}{d}}{dt} - v^\mu \partial_\mu .$$

The natural variation does in general not depend on a connexion.

2. The *absolute variation* $\overset{a}{D}\Omega dt$, that is the difference between the new value at $\underset{0}{\xi}^\varkappa + v^\varkappa dt$ and the value arising at that point from dragging along from $\underset{0}{\xi}^\varkappa$ over $v^\varkappa dt$

(1.13)
$$\overset{a}{D} \overset{\text{def}}{=} \frac{\overset{v}{d}}{dt} - \frac{\overset{m}{d}}{dt} .$$

The absolute variation does in general not depend on a connexion.

3. The *geodesic variation* $\overset{g}{D}\Omega dt$, that is the difference between the new value at $\underset{0}{\xi}^\varkappa + v^\varkappa dt$ and the value arising at that point from parallel displacement from $\underset{0}{\xi}^\varkappa$ to $\underset{0}{\xi}^\varkappa + v^\varkappa dt$

(1.14)
$$\overset{g}{D} \overset{\text{def}}{=} \frac{\overset{v}{d}}{dt} - \frac{\overset{*}{d}}{dt} .$$

Of course these operators do not always exist. Between them and the operators $v^\mu \nabla_\mu$, $\underset{v}{£}$ and $\underset{s}{D}$ the following relations hold

(1.15)
$$\begin{cases} \overset{n}{D} = & = \overset{a}{D} - \underset{v}{£} = \overset{g}{D} - v^\mu \nabla_\mu \\ \overset{a}{D} = \overset{n}{D} + \underset{v}{£} & = \qquad\qquad = \overset{g}{D} + \underset{s}{D} \\ \overset{g}{D} = \overset{n}{D} + v^\mu \nabla_\mu = \overset{a}{D} - \underset{s}{D} & \end{cases}$$

If Φ is a field (indices suppressed) in an $\mathfrak{N}(\underset{0}{\xi}^\varkappa)$ we also have the following geometrical interpretation. After the variation the field value at $\underset{0}{\xi}^\varkappa$ is $\Phi - v^\mu \partial_\mu \Phi dt + \overset{v}{d}\Phi = \Phi + \overset{n}{D}\Phi dt$. So the natural variation $\overset{n}{D}\Phi dt$ is just the change of the field at $\underset{0}{\xi}^\varkappa$ that we observe looking at the field during the variation. The changes $\overset{g}{D}\Phi dt$ and $\overset{a}{D}\Phi dt$ appear if we compare the new fieldvalues at $\underset{0}{\xi}^\varkappa$ not with the original fieldvalues at that point but with the values arising at $\underset{0}{\xi}^\varkappa$ after a parallel displacement or a dragging along respectively of the field over $v^\varkappa dt$.

Accordingly we have for a field *left at rest:*

(1.16)
$$\overset{n}{D} = 0; \qquad \overset{a}{D} = \underset{v}{£}; \qquad \overset{g}{D} = v^\mu \nabla_\mu ,$$

for a *field dragged along*

$$(1.17)\qquad \overset{n}{D} = -\underset{v}{\pounds}; \qquad \overset{a}{D} = 0; \qquad \overset{g}{D} = -\underset{s}{D}$$

and for a field *displaced parallel*

$$(1.18)\qquad \overset{n}{D} = -v^{\mu}\nabla_{\mu}; \quad \overset{a}{D} = +\underset{s}{D}; \quad \overset{g}{D} = 0.$$

As a first example of a deformation problem we try to compute the variation of the expression $u^{\mu}\nabla_{\mu}P^{\varkappa}_{\cdot\lambda}$ if 1) the vector field $u^{\varkappa}$ is left unchanged, 2) the tensor field $P^{\varkappa}_{\cdot\lambda}$ is displaced parallel over $v^{\varkappa}dt$ and 3) the object field $\Gamma^{\varkappa}_{\mu\lambda}$ is dragged along over $v^{\varkappa}dt$. Then we have

$$(1.19)\qquad \begin{cases} \text{a)} & \overset{n}{D}u^{\varkappa} = 0; \\ \text{b)} & \overset{n}{D}P^{\varkappa}_{\cdot\lambda} = -v^{\mu}\nabla_{\mu}P^{\varkappa}_{\cdot\lambda}; \\ \text{c)} & \overset{n}{D}\Gamma^{\varkappa}_{\mu\lambda} = -\underset{v}{\pounds}\Gamma^{\varkappa}_{\mu\lambda} = -\nabla_{\mu}v_{\lambda}^{\cdot\varkappa} - v^{\nu}R_{\nu\mu\lambda}^{\cdot\cdot\cdot\varkappa}, \end{cases}$$

hence after some calculation

$$(1.20)\qquad \begin{cases} \overset{n}{D}u^{\mu}\nabla_{\mu}P^{\varkappa}_{\cdot\lambda} = u^{\mu}\nabla_{\mu}\overset{n}{D}P^{\varkappa}_{\cdot\lambda} - u^{\mu}(\overset{n}{D}\Gamma^{\sigma}_{\mu\lambda})P^{\varkappa}_{\cdot\sigma} + u^{\mu}(\overset{n}{D}\Gamma^{\varkappa}_{\mu\varrho})P^{\varrho}_{\cdot\lambda} \\ \qquad = -v^{\nu}\nabla_{\nu}u^{\mu}\nabla_{\mu}P^{\varkappa}_{\cdot\lambda} + (\underset{v}{\pounds}u^{\mu})\nabla_{\mu}P^{\varkappa}_{\cdot\lambda} + \\ \qquad + (u^{\mu}\nabla_{\mu}v_{\lambda}^{\cdot\sigma})P^{\varkappa}_{\cdot\sigma} - (u^{\mu}\nabla_{\mu}v_{\varrho}^{\cdot\varkappa})P^{\varrho}_{\cdot\lambda}; \end{cases}$$

and, according to (1.15)

$$(1.21)\qquad \overset{g}{D}u^{\mu}\nabla_{\mu}P^{\varkappa}_{\cdot\lambda} = (\underset{v}{\pounds}u^{\mu})\nabla_{\mu}P^{\varkappa}_{\cdot\lambda} + (u^{\mu}\nabla_{\mu}v_{\lambda}^{\cdot\sigma})P^{\varkappa}_{\cdot\sigma} - (u^{\mu}\nabla_{\mu}v_{\varrho}^{\cdot\varkappa})P^{\varrho}_{\cdot\lambda}.$$

$$(1.22)\qquad \begin{cases} \overset{a}{D}u^{\mu}\nabla_{\mu}P^{\varkappa}_{\cdot\lambda} = \overset{g}{D}u^{\mu}\nabla_{\mu}P^{\varkappa}_{\cdot\lambda} + \underset{s}{D}u^{\mu}\nabla_{\mu}P^{\varkappa}_{\cdot\lambda} \\ \qquad = (\underset{v}{\pounds}u^{\mu})\nabla_{\mu}P^{\varkappa}_{\cdot\lambda} + u^{\mu}\nabla_{\mu}(v_{\lambda}^{\cdot\sigma}P^{\varkappa}_{\cdot\sigma} - v_{\varrho}^{\cdot\varkappa}P^{\varrho}_{\cdot\lambda}). \end{cases}$$

Formula (1.22) can also be found by starting from

$$(1.23)\qquad \begin{cases} \overset{a}{D}u^{\varkappa} = \underset{v}{\pounds}u^{\varkappa} \\ \overset{a}{D}P^{\varkappa}_{\cdot\lambda} = \underset{s}{D}P^{\varkappa}_{\cdot\lambda} \\ \overset{a}{D}\Gamma^{\varkappa}_{\mu\lambda} = 0. \end{cases}$$

The formulae (1.15–18) are very convenient for computational purposes because they make it possible to choose always the easiest operators. But it should be remembered that $v^{\mu}\nabla_{\mu}$, $\underset{s}{D}$ and $\overset{g}{D}$ can only be applied if a parallel displacement is given and then only to quantities.

A more important application, connected with SCHLESINGER's product integrals, is the following. Let C be a closed curve $\xi^\varkappa = f^\varkappa(t)$ in L_n through $\underset{0}{\xi}^\varkappa$ and let t be a parameter on this curve, $t_0 \leq t \leq t_1$, such that $f^\varkappa(t_0) = f^\varkappa(t_1) = \underset{0}{\xi}^\varkappa$. Now let the vectors in the tangent E_n at $\underset{0}{\xi}^\varkappa$ be displaced parallel along C till they return to $\underset{0}{\xi}^\varkappa$. Then they have suffered a linear homogeneous transformation, depending on the form of C and on the connexion. This transformation may be written in the form

$$\underset{*}{u}^\varkappa = \left|{}^{t_1}_{t_0}\right|^\varkappa_\lambda \underset{0}{u}^\lambda, \tag{1.24}$$

in which the kernel of the tensor of the transformation is symbolized by $\left|{}^{t_1}_{t_0}\right|$. This transformation can be expressed by means of a product integral

$$\left|{}^{t_1}_{t_0}\right|^\varkappa_\lambda = \overset{\frown t_1}{\underset{t_0}{\int}} \left(A^\varkappa_\lambda - \Gamma^\varkappa_{\mu\lambda} \frac{d\xi^\mu}{dt}\, dt\right). \tag{1.25}$$

A product integral is defined in matrix notation by

$$\overset{\frown t_n}{\underset{t_0}{\int}} (I + P\, dt) \overset{\text{def}}{=} \underset{\substack{n \to \infty \\ t_k - t_{k-1} \to 0}}{\text{Lim}} \prod_{k=1}^{k=n} \{I + P(t_k)(t_k - t_{k-1})\}. \tag{1.26}$$

Product integrals have many properties that can be found in the papers of SCHLESINGER and other authors.[1] Here we need only the generalization of LEIBNIZ' rule for the variation[2]:

$$\overset{v}{d} \overset{\frown t_1}{\underset{t_0}{\int}} (I + P\, dt) = \int_{t_0}^{t_1} \left\{\overset{\frown t}{\underset{t_0}{\int}} (I + P\, du)\right\} \overset{v}{d} P(t) \left\{\overset{\frown t_1}{\underset{t}{\int}} (I + P\, du)\right\} dt \tag{1.27}$$

and the differentiation rule

$$\frac{d}{dt} \overset{\frown t}{\underset{t_0}{\int}} (I + P\, du) = \left\{\overset{\frown t}{\underset{t_0}{\int}} (I + P\, du)\right\} P(t) \tag{1.28}$$

that can be derived easily.

Starting from a vector $\underset{0}{u}^\varkappa$ at $\underset{0}{\xi}^\varkappa$, at an arbitrary point $\xi^\varkappa$ of C we get

$$u^\varkappa = \left|{}^{t}_{t_0}\right|^\varkappa_\lambda \underset{0}{u}^\lambda; \quad t_0 \leq t \leq t_1. \tag{1.29}$$

Here the tensor $\left|{}^{t}_{t_0}\right|^\varkappa_\lambda$ is a connecting quantity lying with the lower index in the tangent E_n of $\underset{0}{\xi}^\varkappa$ and with the upper one in the tangent E_n

---

[1]) SCHLESINGER 1928, 1; 1931, 1; 2; 1932, 1; cf. VOLTERRA 1887, 1; MORINAGA 1934, 2; E I 1935, 1, p. 134; NIJENHUIS 1952, 1, Ch. II § 13.

[2]) Cf. MICHAL 1945, 1; JOHNSON 1948, 1.

of $\xi^{\varkappa}$. Because $u^{\varkappa}$ is covariant constant along C we have along this curve according to (1.28)

$$(1.30)\qquad 0 = \frac{\delta}{dt}\left[{t \atop t_0}\right]^{\varkappa}_{\lambda} = \frac{d}{dt}\left[{t \atop t_0}\right]^{\varkappa}_{\lambda} + \frac{d\xi^{\mu}}{dt}\,\Gamma^{\varkappa}_{\mu\varrho}\left[{t \atop t_0}\right]^{\varrho}_{\lambda}$$

and in the same way

$$(1.31)\qquad 0 = \frac{\delta}{dt}\left[{t_1 \atop t}\right]^{\varkappa}_{\lambda} = \frac{d}{dt}\left[{t_1 \atop t}\right]^{\varkappa}_{\lambda} - \frac{d\xi^{\mu}}{dt}\,\Gamma^{\sigma}_{\mu\lambda}\left[{t_1 \atop t}\right]^{\varkappa}_{\sigma}.$$

Now let C be subjected to the variation $\varepsilon v^{\varkappa}$. Then a point $\xi^{\varkappa}$ of C is transformed into $\xi^{\varkappa}+\varepsilon v^{\varkappa}$ and the variation $\overset{v}{d}\left[{t_1 \atop t_0}\right]^{\varkappa}_{\lambda}$ must be computed from the values $\Gamma^{\varkappa}_{\mu\lambda}+\varepsilon v^{\nu}\partial_{\nu}\Gamma^{\varkappa}_{\mu\lambda}$ at the points of the displaced curve.[1]) Instead we transpose the problem in the way described in V § 9 by dragging back the curve over $-\varepsilon v^{\varkappa}$ to its original position and at the same time dragging along the field $\Gamma^{\varkappa}_{\mu\lambda}$ over $-\varepsilon v^{\varkappa}$. If now we compute the natural variation of $\left[{t_1 \atop t_0}\right]^{\varkappa}_{\lambda}$ for the transposed problem this variation is the difference between the new value and the original value both at $\underset{0}{\xi}^{\varkappa}$. But this new value arises from the new value of the original (not transposed) problem at $\underset{0}{\xi}^{\varkappa}+\varepsilon v^{\varkappa}$ by dragging along over $-\varepsilon v^{\varkappa}$. This means that *the natural variation for the transposed problem equals the absolute variation for the original problem.*

Hence, *if we denote the natural variation for the transposed problem by* $'\overset{n}{D}$, *we have* $'\overset{n}{D}=\overset{a}{D}$. Now we have according to (1.25, 27)

$$(1.32)\qquad \begin{cases} \left[{t_1 \atop t_0}\right]^{\varkappa}_{\lambda} + \varepsilon\,'\overset{n}{D}\left[{t_1 \atop t_0}\right]^{\varkappa}_{\lambda} = \displaystyle\int_{t_0}^{\frown t_1}\left\{A^{\varkappa}_{\lambda} - \left(\Gamma^{\varkappa}_{\mu\lambda} + \varepsilon\,\underset{v}{\pounds}\,\Gamma^{\varkappa}_{\mu\lambda}\right)\frac{d\xi^{\mu}}{dt}dt\right\} \\ \qquad = \left[{t_1 \atop t_0}\right]^{\varkappa}_{\lambda} - \varepsilon\displaystyle\int_{t_0}^{t_1}\left[{t \atop t_0}\right]^{\sigma}_{\lambda}\frac{d\xi^{\mu}}{dt}\left(\underset{v}{\pounds}\,\Gamma^{\varrho}_{\mu\sigma}\right)\left[{t_1 \atop t}\right]^{\varkappa}_{\varrho}dt, \end{cases}$$

hence, using (1.8)

$$(1.33)\qquad '\overset{n}{D}\left[{t_1 \atop t_0}\right]^{\varkappa}_{\lambda} = -\int_{t_0}^{t_1}\left[{t \atop t_0}\right]^{\sigma}_{\lambda}\frac{d\xi^{\mu}}{dt}\left(\nabla_{\mu}v_{\sigma}^{\cdot\varrho} + v^{\nu}R_{\nu\mu\sigma}^{\cdot\cdot\cdot\varrho}\right)\left[{t_1 \atop t}\right]^{\varkappa}_{\varrho}dt.$$

According to (1.30) we get, integrating by parts

$$(1.34)\qquad \int_{t_0}^{t_1}\left[{t \atop t_0}\right]^{\sigma}_{\lambda}\delta v_{\sigma}^{\cdot\varrho}\left[{t_1 \atop t}\right]^{\varkappa}_{\varrho} = \left(\left[{t \atop t_0}\right]^{\sigma}_{\lambda}v_{\sigma}^{\cdot\varrho}\left[{t_1 \atop t}\right]^{\varkappa}_{\varrho}\right)\Bigg|_{t_0}^{t_1} = \left[{t_1 \atop t_0}\right]^{\sigma}_{\lambda}v_{\sigma}^{\cdot\varkappa} - v_{\lambda}^{\cdot\varrho}\left[{t_1 \atop t_0}\right]^{\varkappa}_{\varrho};$$

hence, returning to the original problem

$$(1.35)\qquad \overset{a}{D}\left[{t_1 \atop t_0}\right]^{\varkappa}_{\lambda} = -\int_{t_0}^{t_1}\left[{t \atop t_0}\right]^{\sigma}_{\lambda}v^{\nu}\frac{d\xi^{\mu}}{dt}R_{\nu\mu\sigma}^{\cdot\cdot\cdot\varrho}\left[{t_1 \atop t}\right]^{\varkappa}_{\varrho}dt - \left[{t_1 \atop t_0}\right]^{\sigma}_{\lambda}v_{\sigma}^{\cdot\varkappa} + v_{\lambda}^{\cdot\varrho}\left[{t_1 \atop t_0}\right]^{\varkappa}_{\varrho}.$$

---

[1]) Cf. JOHNSON 1948, 1, who also considers finite shifts of C.

The natural variation of $\left[{}^{t_1}_{t_0}\right]^{\varkappa}_{\lambda}$ for the original problem does not exist because $\left[{}^{t_1}_{t_0}\right]^{\varkappa}_{\lambda}$ is not a field defined in an $\mathfrak{N}(\underset{0}{\xi^{\varkappa}})$ and accordingly has no field value at $\underset{0}{\xi^{\varkappa}} + \varepsilon v^{\varkappa}$. But the geodesic variation $\overset{g}{D} \left[{}^{t_1}_{t_0}\right]^{\varkappa}_{\lambda}$ exists because it is the difference between the new value at $\underset{0}{\xi^{\varkappa}} + \varepsilon v^{\varkappa}$ and the value arising from the parallel displacement of $\left[{}^{t_1}_{t_0}\right]^{\varkappa}_{\lambda}$ from $\underset{0}{\xi^{\varkappa}}$ to $\underset{0}{\xi^{\varkappa}} + \varepsilon v^{\varkappa}$. From (1.15) and (1.35) we get

$$\overset{g}{D} \left[{}^{t_1}_{t_0}\right]^{\varkappa}_{\lambda} = - \int_{t_0}^{t_1} \left[{}^{t}_{t_0}\right]^{\sigma}_{\lambda} v^{\nu} \frac{d\xi^{\mu}}{dt} R_{\nu\mu\sigma}{}^{\cdot\cdot\cdot\varrho} \left[{}^{t_1}_{t}\right]^{\varkappa}_{\varrho} dt; \tag{1.36}$$

and this equation does not contain any derivatives of the field $v^{\varkappa}$. The case where $v^{\varkappa} = 0$ at $\underset{0}{\xi^{\varkappa}}$ is very important. Since $\underset{0}{\xi^{\varkappa}}$ is then fixed, $\overset{*}{d}$ is zero at that point and $\overset{g}{D}\left[{}^{t_1}_{t_0}\right]$ equals the $\overset{n}{D}\left[{}^{t_1}_{t_0}\right]$ that exists in this special case and represents the variation at $\underset{0}{\xi^{\varkappa}}$ that really appears. The formula (1.36) will be used for the discussion of the holonomy group of a connexion in VII § 4.

As another interesting application we consider the generalization of BERTRAND curves for V_n. In R_3 a curve is called a BERTRAND curve if there exists another curve such that the first normals of both curves are the same.[1]) Let a curve in R_n be given by

$$\xi^{\varkappa} = f^{\varkappa}(s) \tag{1.37}$$

as in V § 1 with s as parameter of length but with the restriction that $(\varkappa)$ is a rectilinear coordinate system. Then a point on the first normal of a point $f^{\varkappa}(s)$ has the coordinates

$$\varphi^{\varkappa}(s) = f^{\varkappa}(s) + \psi(s) \underset{2}{j^{\varkappa}}(s). \tag{1.38}$$

We try to choose $\psi(s)$ in such a way that the curve $\xi^{\varkappa} = \varphi^{\varkappa}(s)$ has the same first normals as the curve (1.37). According to the FRENET formulae (V 1.10) we have

$$\frac{d\varphi^{\varkappa}}{ds} = \frac{df^{\varkappa}}{ds} + \frac{d\psi}{ds} \underset{2}{j^{\varkappa}} + \psi \frac{d\underset{2}{j^{\varkappa}}}{ds} = (1 - \underset{1}{k}\psi)\, \underset{1}{j^{\varkappa}} + \frac{d\psi}{ds} \underset{2}{j^{\varkappa}} + \underset{2}{k} \psi \underset{3}{j^{\varkappa}} \tag{1.39}$$

and because $d\varphi^{\varkappa}/ds$ must be perpendicular to $\underset{2}{j^{\varkappa}}$, this means that ψ is constant: $\psi = \underset{1}{c}$. Now s need not be a length on the second curve, hence, if $'s$ is the parameter of length on this second curve we have for its tangent unitvector

$$\underset{1}{k^{\varkappa}} = \frac{ds}{d's} \frac{d\varphi^{\varkappa}}{ds}; \qquad \frac{ds}{d's} \neq 0; \qquad \frac{d's}{ds} \neq 0. \tag{1.40}$$

---

[1]) Cf. for instance WEATHERBURN 1927, 1, p. 34ff.; HAACK 1948, 1, p. 31f.

From this it follows that the vector

$$(1.41)\qquad \frac{d^2 s}{d\,'s^2}\,\frac{d\varphi^\varkappa}{d s} + \left(\frac{d s}{d\,'s}\right)^2 \frac{d^2 \varphi^\varkappa}{d s^2}$$

has the direction of $\underset{2}{j}^\varkappa$. But this is only possible if

$$(1.42)\qquad \begin{cases} \text{a)}\quad \left(1 - \underset{1}{c}\,\underset{1}{k}\right)\dfrac{d^2 s}{d\,'s^2} - \underset{1}{c}\,\dfrac{d\underset{1}{k}}{d s}\left(\dfrac{d s}{d\,'s}\right)^2 = 0 \\ \text{b)}\quad \underset{1}{c}\,\underset{2}{k}\,\dfrac{d^2 s}{d\,'s^2} + \underset{1}{c}\,\dfrac{d\underset{2}{k}}{d s}\left(\dfrac{d s}{d\,'s}\right)^2 = 0 \\ \text{c)}\quad \underset{1}{c}\,\underset{2}{k}\,\underset{3}{k}\left(\dfrac{d s}{d\,'s}\right)^2 = 0. \end{cases}$$

There are two trivial solutions: $\underset{1}{c} = 0$ and $\underset{2}{k} = 0$ (hence $\underset{3}{k} = 0$). If we exclude these, (1.42) leads to the n.a.s. conditions

$$(1.43)\qquad \begin{cases} \text{a)}\quad \underset{1}{c}\,\underset{1}{k} + \underset{2}{c}\,\underset{2}{k} = 1; \quad \underset{1}{c} \text{ and } \underset{2}{c} \text{ constants} \\ \text{b)}\quad \underset{3}{k} = 0 \end{cases}$$

for a curve to be a BERTRAND curve in R_n. If these conditions are satisfied it is easily proved that

$$(1.44)\qquad \frac{d\,'s}{d s} = \underset{2}{k}\sqrt{\underset{1}{c}^2 + \underset{2}{c}^2}$$

and that $\underset{1}{c}/\underset{2}{c}$ is the tangent of the (constant) angle between the tangents of the curves at corresponding points. $\underset{1}{c}$ is the (constant) distance between corresponding points and from this we see that to a BERTRAND curve there belongs in general one and only one corresponding curve. But if $\underset{1}{k}$ and $\underset{2}{k}$ are constants, there is a second curve for every choice of $\underset{1}{c}$ as is well known from the theory of circular helices in R_3.

Now this problem can be generalized for curves in V_n by requiring that a curve passes into another curve by an infinitesimal deformation in the direction of its first normal and such that the first normals of both curves are parallel at corresponding points.[1]) The curve may be given by (1.37) but $(\varkappa)$ is no longer a rectilinear system. Instead of (1.38) we get

$$(1.45)\qquad \varphi^\varkappa(s) = f^\varkappa(s) + \varepsilon\,\underset{2}{j}^\varkappa(s)$$

and it can be proved as before that ε is an infinitesimal constant.[2]) But from here we must proceed in quite another way because the covariant derivatives are no longer ordinary derivatives and the $\Gamma^\varkappa_{\mu\lambda}$

---

[1]) PEARS 1935, 1; YANO, TAKANO and TOMONAGA 1948, 2. The latter paper contains many theorems on deformation of curves.

[2]) It is convenient to use FERMI coordinates, cf. III § 8.

do not have the same values at corresponding points of the two curves. Again we transpose the problem, dragging back the second curve and the field $g_{\lambda\varkappa}$ over $-\varepsilon \underset{2}{j}^\varkappa$. We note that although the curve dragged back coincides with the original curve, s need no longer be a length parameter on it.

For the fields $g_{\lambda\varkappa}$ and $\Gamma_{\mu\lambda}^{\varkappa}$ dragged back over $-\varepsilon v^\varkappa$ we have according to (1.17)

$$(1.46)\qquad '\overset{n}{D}\, g_{\mu\lambda} = \underset{v}{\pounds}\, g_{\mu\lambda} = 2 \nabla_{(\mu} v_{\lambda)}$$

$$(1.47)\qquad '\overset{n}{D}\, \Gamma_{\mu\lambda}^{\varkappa} = \underset{v}{\pounds}\, \Gamma_{\mu\lambda}^{\varkappa} = \nabla_\mu \nabla_\lambda v^\varkappa + v^\nu K_{\nu\mu\lambda}^{\cdot\cdot\cdot\varkappa}$$

but, because $\underset{2}{j}^\varkappa$ is defined over C only, we must put $v^\varkappa = \underset{2}{j}^\varkappa$ on this curve and prolong this field over a neighbourhood of C in order that the derivatives in (1.46, 47) get a meaning. Of course all results will be independent of the manner of prolongation.

From (1.46) it follows that

$$(1.48)\qquad '\overset{n}{D}\,(ds)^2 = 2\, d\xi^\mu\, d\xi^\lambda\, \nabla_\mu v_\lambda = 2\,(ds)^2 j^\mu j^\lambda \nabla_\mu \underset{2}{j}_\lambda = -\,2 \underset{1}{k}\,(ds)^2$$

hence

$$(1.49)\qquad \begin{cases} \text{a)}\quad \dfrac{'\overset{n}{D}\,ds}{ds} = -\underset{1}{k}; \\[2ex] \text{b)}\quad '\overset{n}{D} j^\varkappa = \overset{a}{D} \dfrac{d\xi^\varkappa}{ds} = -\dfrac{d\xi^\varkappa}{(ds)^2} \overset{a}{D}\, ds = \dfrac{d\xi^\varkappa}{(ds)^2} \underset{1}{k}\, ds = \underset{1}{k}\, j^\varkappa. \end{cases}$$

For the first curvature vector we get

$$(1.50)\qquad \begin{cases} '\overset{n}{D} j^\mu \nabla_\mu j^\varkappa = ('\overset{n}{D} j^\mu)\, \nabla_\mu j^\varkappa + j^\mu \nabla_\mu\, '\overset{n}{D} j^\varkappa + j^\mu j^\lambda\, '\overset{n}{D}\, \Gamma_{\mu\lambda}^{\varkappa} \\[1ex] \qquad = \underset{1}{k}\, j^\mu \nabla_\mu j^\varkappa + j^\mu (\nabla_\mu \underset{1}{k})\, j^\varkappa + j^\mu \underset{1}{k}\, \nabla_\mu j^\varkappa + \\[1ex] \qquad\qquad + j^\mu j^\lambda \nabla_\mu \nabla_\lambda v^\varkappa + j^\nu j^\mu j^\lambda K_{\nu\mu\lambda}^{\cdot\cdot\cdot\varkappa} \\[1ex] \qquad = 2 \underset{1}{k}\, j^\mu \nabla_\mu j^\varkappa + j^\varkappa \dfrac{d}{ds} \underset{1}{k} + \dfrac{\delta^2}{ds^2} \underset{2}{j}^\varkappa - \underset{1}{k}\, \underset{2}{j}^\mu \nabla_\mu v^\varkappa + j^\nu j^\mu j^\lambda K_{\nu\mu\lambda}^{\cdot\cdot\cdot\varkappa}; \end{cases}$$

hence, returning to the original untransposed problem

$$(1.51)\qquad (\overset{a}{D} \underset{1}{k})\, \underset{2}{j}^\varkappa + \underset{1}{k}\, \overset{a}{D} \underset{2}{j}^\varkappa = 2 \underset{1}{k}^2 \underset{2}{j}^\varkappa + j^\varkappa \frac{d}{ds} \underset{1}{k} + \frac{\delta^2}{ds^2} \underset{2}{j}^\varkappa - \underset{1}{k}\, \underset{2}{j}^\mu \nabla_\mu v^\varkappa + j^\nu j^\mu j^\lambda K_{\nu\mu\lambda}^{\cdot\cdot\cdot\varkappa};$$

or, according to (1.15) (cf. V 1.10)

$$(1.52)\qquad \begin{cases} \underset{1}{k}\, \overset{g}{D} \underset{2}{j}^\varkappa = (2 \underset{1}{k}^2 - \overset{a}{D} \underset{1}{k})\, \underset{2}{j}^\varkappa + j^\varkappa \dfrac{d}{ds} \underset{1}{k} + \dfrac{\delta^2}{ds^2} \underset{2}{j}^\varkappa + \underset{2}{j}^\nu j^\mu j^\lambda K_{\nu\mu\lambda}^{\cdot\cdot\cdot\varkappa} \\[2ex] \qquad = (\underset{1}{k}^2 - \overset{a}{D} \underset{1}{k} - \underset{2}{k}^2)\, \underset{2}{j}^\varkappa + \underset{3}{j}^\varkappa \dfrac{d}{ds} \underset{2}{k} + \underset{2}{k}\, \underset{3}{k}\, \underset{4}{j}^\varkappa + \underset{2}{j}^\nu j^\mu j^\lambda K_{\nu\mu\lambda}^{\cdot\cdot\cdot\varkappa}. \end{cases}$$

Now $\overset{g}{D}\underset{2}{j}^{\varkappa}=0$ according to our suppositions, hence, taking an anholonomic coordinate system (h) whose unit basis vectors coincide with the $\underset{2}{j}^{\varkappa}, \underset{}{j}^{\varkappa}, \ldots, \underset{n}{j}^{\varkappa}$ at all points of C we have

$$(1.53)\quad \begin{cases} \text{a)}\ \overset{a}{D}\underset{1}{k}=\underset{1}{k}^2-\underset{2}{k}^2+K_{2112} \\ \text{b)}\ \dfrac{d}{ds}\underset{2}{k}=-K_{2113} \\ \text{c)}\ \underset{2}{k}\underset{3}{k}=-K_{2114} \\ \text{d)}\ 0=K_{211p};\qquad p=5,\ldots,\mathrm{n}. \end{cases}$$

This is in accordance with the results (1.42) derived for R_n ($\underset{1}{c}$ infinitesimal). The problem of two curves in a general V_n with common normal geodesics having on both curves the direction of the first normal can not be tackled in this way. In an R_n the components of $K_{\nu\mu\lambda\varkappa}$ occurring in (1.53b–d) are zero, hence either $\underset{3}{k}=0$ and $\underset{2}{k}=\text{const.}$ or $\underset{2}{k}=\underset{3}{k}=0$.

The direction of $j^{\varkappa}$ can be compared with the tangent direction at the corresponding point of the other curve if this tangent is displaced back parallel. This displaced tangent has the direction of the unit-vector $j^{\varkappa}+\varepsilon\overset{g}{D}j^{\varkappa}$. This gives for the angle $d\alpha$ between both directions according to (1.15) and (1.49b)

$$(1.54)\quad \begin{cases} (d\alpha)^2=\varepsilon^2\,(\overset{g}{D}j^{\varkappa})\,(\overset{g}{D}j^{\lambda})\,g_{\lambda\varkappa} \\ \qquad =\varepsilon^2\,(\overset{a}{D}j^{\varkappa}+j^{\mu}\,\nabla_{\mu}\underset{2}{j}^{\varkappa})\,(\overset{a}{D}j^{\lambda}+j^{\mu}\,\nabla_{\mu}\underset{2}{j}^{\lambda})\,g_{\lambda\varkappa}=\varepsilon^2\,\underset{2}{k}^2\,\underset{3}{j}^{\varkappa}\,\underset{3}{j}^{\lambda}\,g_{\lambda\varkappa}=\varepsilon^2\,\underset{2}{k}^2. \end{cases}$$

Hence the angle $d\alpha$ is constant if $\underset{2}{k}$ is constant. This is always the case in an R_n.

Exercises.

VII 1,1. Prove that the rule of LEIBNIZ holds for the operators $O_1O_2-O_2O_1$ if it holds for both operators O_1 and O_2 and if in the formulae considered all operations can be effected.

VII 1,2. If the variations of a quantity Φ (indices suppressed) and of the connexion $\Gamma_{\mu\lambda}^{\varkappa}$ are given, the quantity $(\overset{n}{D}\nabla_{\mu}-\nabla_{\mu}\overset{n}{D})\,\Phi$ is known. Prove that this quantity does not change if the variation of Φ is changed, as long as the variation of $\Gamma_{\mu\lambda}^{\varkappa}$ remains unchanged. Prove the same for the operators $\overset{g}{D}\nabla_{\mu}-\nabla_{\mu}\overset{g}{D}$ and $\overset{a}{D}\nabla_{\mu}-\nabla_{\mu}\overset{a}{D}$.[1])

[1]) SCHOUTEN and v. KAMPEN 1933, 1, p. 8ff.

VII 1,3. Prove that

VII 1,3 α) $$\overset{n}{D}\nabla_\mu - \nabla_\mu \overset{n}{D} = 0$$

if the connexion is not changed.[1])

§ 2. Groups of "motions" in V_n and L_n. [2])

A deformation in V_n is called a *motion* if it leaves the metric invariant. Hence an infinitesimal deformation $\xi^\varkappa \to \xi^\varkappa + v^\varkappa\, dt$ is a motion if and only if $\underset{v}{£}\, g_{\lambda\varkappa} = 0$. This idea can be generalized. A deformation in U_n is a motion if $\underset{v}{£}\, g_{\lambda\varkappa} = 0$; $\underset{v}{£}\, S_{\mu\lambda}^{\cdot\cdot\varkappa} = 0$. In L_n we may call a deformation an *affine motion* if the connexion is left invariant. Hence an infinitesimal affine motion in L_n is characterized by $\underset{v}{£}\, \Gamma_{\mu\lambda}^\varkappa = 0$. Further generalizations are the *conformal motions* in V_n characterized by $\underset{v}{£}\, g_{\lambda\varkappa} = \alpha\, g_{\lambda\varkappa}$, the *homothetic motion* in V_n, being a conformal motion for which α is a constant and the *projective motion* in A_n characterized by $\underset{v}{£}\, \Gamma_{\mu\lambda}^\varkappa = \varphi_\mu A_\lambda^\varkappa + \varphi_\lambda A_\mu^\varkappa$.

The conformal and projective motions are also characterized by $\underset{v}{£}\, \overset{c}{\Gamma}{}_{\mu\lambda}^\varkappa = 0$ and $\underset{v}{£}\, \overset{p}{\Gamma}{}_{\mu\lambda}^\varkappa = 0$ where $\overset{c}{\Gamma}{}_{\mu\lambda}^\varkappa$ and $\overset{p}{\Gamma}{}_{\mu\lambda}^\varkappa$ are the conformal and projective parameters respectively of J. M. Thomas (cf. VI § 7) and of T. Y. Thomas (cf. VI § 4). Hence every motion is characterized by the *absolute* invariance (cf. II § 10) of one or more fields.[3])

The differential equations of the affine motions in L_n are according to (1.8)

$$\text{(2.1)} \qquad \begin{cases} \text{a)} & \nabla_\mu v_\lambda^{\cdot\varkappa} = - v^\nu R_{\nu\mu\lambda}^{\cdot\cdot\cdot\varkappa} \\ \text{b)} & \nabla_\lambda v^\varkappa = v_\lambda^{\cdot\varkappa} + 2 S_{\mu\lambda}^{\cdot\cdot\varkappa} v^\mu \end{cases}$$

with the unknowns $v^\varkappa$ and $v_\lambda^{\cdot\varkappa}$. The first integrability conditions (cf. II § 6, III § 6) can be derived by differentiation, alternation and using

[1]) Schouten and v. Kampen 1933 1. p. 8 ff.

[2]) Slebodzinski 1932, 1. Cf. Eisenhart 1935, 1; 1936, 1 (simply trans. gr. in V_n); Levine 1937, 1 and 1939, 1 (in C_n); 1950, 2 (in A_2 and project.); 1951, 1 (in W_2); 1951, 2 (in A_2); Knebelman 1945, 1 (in A_n and V_n); Yano 1946, 5; Yano and Takano 1946, 6 and Yano 1946, 7; 1947, 3 (in L_n); 1949, 1; 1951, 2 (homoth. in V_n); Yano and Tomonaga 1946, 8; 9 (in L_n); Egorov 1947, 1 (in A_n); 1948, 1 (project.); 1949, 1 (in V_n); 1949, 2 (in L_n); 1950, 1 (in L_n); Su 1949, 1 (conform); Shanks 1950, 1 (homothet. in V_n); Yano and Imai 1950, 2; Muto 1951, 1 (simply trans. gr. in V_n).

[3]) Many results on these different "motions" are to be found in Yano's booklet 1949, 1. Cf. also Kosambi 1951, 1; 1952, 1.

the second identity and the identity of BIANCHI. This gives

$$(2.2)\quad \begin{cases} \text{a)} & \underset{v}{\pounds} R_{\nu\mu\lambda}^{\cdot\cdot\cdot\varkappa} = v^{\omega} \nabla_{\omega} R_{\nu\mu\lambda}^{\cdot\cdot\cdot\varkappa} + v_{\nu}^{\cdot\sigma} R_{\sigma\mu\lambda}^{\cdot\cdot\cdot\varkappa} + v_{\mu}^{\cdot\sigma} R_{\nu\sigma\lambda}^{\cdot\cdot\cdot\varkappa} + \\ & \quad + v_{\lambda}^{\cdot\sigma} R_{\nu\mu\sigma}^{\cdot\cdot\cdot\varkappa} - v_{\varrho}^{\cdot\varkappa} R_{\nu\mu\lambda}^{\cdot\cdot\cdot\varrho} = 0 \\ \text{b)} & \underset{v}{\pounds} S_{\mu\lambda}^{\cdot\cdot\varkappa} = v^{\nu} \nabla_{\nu} S_{\mu\lambda}^{\cdot\cdot\varkappa} + v_{\mu}^{\cdot\sigma} S_{\sigma\lambda}^{\cdot\cdot\varkappa} + v_{\lambda}^{\cdot\sigma} S_{\mu\sigma}^{\cdot\cdot\varkappa} - v_{\varrho}^{\cdot\varkappa} S_{\mu\lambda}^{\cdot\cdot\varrho} = 0. \end{cases}$$

The second and following integrability conditions are found by covariant differentiation of (2.2). As ∇_{μ} and $\underset{v}{\pounds}$ can be interchanged on account of (1.9) and $\underset{v}{\pounds} \Gamma_{\mu\lambda}^{\varkappa} = 0$, we get

$$(2.3)\quad \begin{cases} \text{a)} & \underset{v}{\pounds} \nabla_{\omega_k \ldots \omega_1} R_{\nu\mu\lambda}^{\cdot\cdot\cdot\varkappa} = 0; \\ \text{b)} & \underset{v}{\pounds} \nabla_{\omega_k \ldots \omega_1} S_{\mu\lambda}^{\cdot\cdot\varkappa} = 0; \qquad k = 0, 1, 2, 3, \ldots \end{cases}$$

These equations express that all covariant derivatives of the fields $R_{\nu\mu\lambda}^{\cdot\cdot\cdot\varkappa}$ and $S_{\mu\lambda}^{\cdot\cdot\varkappa}$ are absolutely invariant for the field $v^{\varkappa}$ (cf. II § 10). If the set (2.3) for $k = m+1$ depends algebraically on the sets for $k = 0, \ldots, m$, the process stops because all other sets depend for $k > m$ on the foregoing ones (cf. II § 6). The equations (2.1) have solutions if and only if at some point $\underset{0}{\xi}^{\varkappa}$ there exist values $\underset{0}{v}^{\varkappa}$, $\underset{0}{v}_{\lambda}^{\cdot\varkappa}$ satisfying the algebraic equations (2.3). Then these values can be taken as the initial values at $\underset{0}{\xi}^{\varkappa}$ and there is one and only one solution of (2.1) taking these values at this point. As the number of independent sets of values $\underset{0}{v}^{\varkappa}$, $\underset{0}{v}_{\lambda}^{\cdot\varkappa}$ satisfying (2.3) is finite, the $v^{\varkappa}$ generate a set of one-parameter groups of transformations, depending on a finite number of parameters. Let $u^{\varkappa}$ and $v^{\varkappa}$ be two solutions:

$$(2.4)\qquad \underset{u}{\pounds} \Gamma_{\mu\lambda}^{\varkappa} = 0; \qquad \underset{v}{\pounds} \Gamma_{\mu\lambda}^{\varkappa} = 0$$

then we have (cf. Exerc. II 10,9)

$$(2.5)\qquad 0 = (\underset{u}{\pounds}\underset{v}{\pounds} - \underset{v}{\pounds}\underset{u}{\pounds}) \Gamma_{\mu\lambda}^{\varkappa} = \underset{w}{\pounds} \Gamma_{\mu\lambda}^{\varkappa}; \qquad w^{\varkappa} \overset{\text{def}}{=} \underset{u}{\pounds} v^{\varkappa};$$

and this proves that $\underset{u}{\pounds} v^{\varkappa}$ is also a solution. This shows that the one-parameter groups together form a group.

The other cases mentioned above can be treated in the same way. The equations of motion can always be found by equating the LIE derivative of the fundamental geometric object or objects to zero. The integrability conditions of these equations only express that the LIE derivatives of all differential concomitants vanish.[1])

[1]) KOSAMBI 1951, 1 has hinted at the possibility of a theorem of this kind, valid for every geometry whose fundamental geometric objects possess a LIE derivative. But a proof of such a general theorem has not yet been given. We only know that it is found to be true in all cases so far investigated. The same holds for the supposition that $\underset{v}{\pounds}$ commutes with all invariant differential operators if and only if $v^{\varkappa}$ defines a motion.

The maximal number of parameters of a group of motions in a V_n that is not an S_n was proved to be $\frac{1}{2}n(n+1)-1$ for $n>2$ by FUBINI.[1]) EGOROV proved[2]) that for a V_n that is not an EINSTEIN V_n the maximum is $\frac{1}{2}n(n-1)+1$ and that it is $\frac{1}{2}n(n-1)+2$ for a V_n that is not an S_n. VRANCEANU reduced this latter number in 1951 [3]) to $\frac{1}{2}(n-1)(n-2)+5$. In 1947 EGOROV proved[4]) that for an L_n that is not an A_n the maximum is n^2 and that an A_n with an n^2-parameter group of motions is necessarily projective euclidean (cf. VI § 1). The maximum has been established by EGOROV and VRANCEANU for a great number of other cases.[5]) YANO proved 1953[6]) that the number of parameters for $n>4$, $n\neq 8$ is $\binom{n}{2}+1$ if and only if the V_n is either the product space of an R_1 and an S_{n-1} or an S_n with negative curvature.

For the motions in a V_n we have because of $\underset{v}{\pounds}\, g_{\lambda\varkappa}=0$; $\underset{v}{\pounds}\, g^{\varkappa\lambda}=0$ and the commutativity of $\underset{v}{\pounds}$ and ∂_μ

$$(2.6)\quad \begin{cases} \text{a)} \quad \underset{v}{\pounds}\left\{{}^{\,\varkappa}_{\mu\lambda}\right\} = \frac{1}{2}g^{\varkappa\nu}\left(\partial_\mu \underset{v}{\pounds}\, g_{\lambda\nu} + \partial_\lambda \underset{v}{\pounds}\, g_{\mu\nu} - \partial_\nu \underset{v}{\pounds}\, g_{\mu\lambda}\right) = 0; \\ \qquad\qquad\qquad\qquad\qquad\qquad \text{(also valid with } \nabla \text{ instead of } \partial) \\ \text{b)} \quad \nabla_\nu \underset{v}{\pounds}\, g_{\mu\lambda} = \left(\underset{v}{\pounds}\left\{{}^{\,\sigma}_{\nu\mu}\right\}\right) g_{\sigma\lambda} - \left(\underset{v}{\pounds}\left\{{}^{\,\sigma}_{\nu\lambda}\right\}\right) g_{\mu\sigma} = 0. \end{cases}$$

From (2.6) we see that the group of motions in V_n is a subgroup of the group of affine motions and that the former group is determined by those solutions of (2.1) with $R_{\nu\mu\lambda}^{\cdot\cdot\cdot\,\varkappa}=K_{\nu\mu\lambda}^{\cdot\cdot\cdot\,\varkappa}$, $S_{\mu\lambda}^{\cdot\cdot\,\varkappa}=0$ which also satisfy the equations

$$(2.7)\qquad \underset{v}{\pounds}\, g_{\mu\lambda} = 2\nabla_{(\mu}v_{\lambda)} = 2v_{(\mu\lambda)} = 0.$$

Now from (2.1 a) it follows that $\nabla_\mu v_{(\lambda\varkappa)}=0$ because $K_{\nu\mu\lambda\varkappa}$ is alternating in $\lambda\varkappa$, hence, if $v_{(\lambda\varkappa)}$ is zero at one point of V_n, this quantity vanishes at all points. Therefore an affine motion in V_n is an ordinary motion in that space if and only if $v_{(\lambda\varkappa)}=0$ at one point. Then $v_{(\lambda\varkappa)}$ is also zero (cf. Exerc. V 5, 6, 7) at all other points. The equation $v_{(\lambda\varkappa)}=0$ is known as KILLING's equation.[7])

---

1) FUBINI 1903, 1; 1904, 1.

2) EGOROV 1949, 1.

3) VRANCEANU 1951, 2; 1953, 1.

4) EGOROV 1947, 1.

5) EGOROV 1948, 1; 1949, 1; 2; 1950, 1; 1951, 1; 1952, 1; 2; VRANCEANU 1947, 1, 1949, 1; 2; 1951, 1; 2; 3; this last publication gives a good summary of results in french.

6) YANO 1953, 1.

7) KILLING 1892, 1; cf. BIANCHI 1918, 1; EISENHART 1926, 1 and for the more general case where $S_{\mu\lambda\varkappa}=S_{[\mu\lambda\varkappa]}\neq 0$ HAYDEN 1934, 2; E II 1938, 2, p. 163.

We prove here only a few theorems on motions in V_n.[1]) From $\underset{v}{£}\, g_{\lambda\varkappa}=0$ and $\underset{v}{£}\, v^\varkappa=0$ it follows:

Along the streamlines of a one-parameter group of motions in V_n the length of the displacement vector is constant.

A *translation* is a motion whose streamlines are geodesics. Because of

$$V_\mu v_\lambda v^\lambda = 2v^\lambda V_\mu v_\lambda = -2v^\lambda V_\lambda v_\mu \tag{2.8}$$

we have:

A motion in V_n is a translation if and only if the length of the displacement vector is constant.[2])

We consider two V_n's that can be mapped on each other so that geodesics correspond to geodesics. Then we wish to prove that the groups of motions in both V_n's have the same number of parameters. This is not trivial because the notion of a motion in V_n is based on the metric and the mapping is projective and will in general disturb the metric. Let corresponding points have the same coordinates $\xi^\varkappa$ and let $g_{\lambda\varkappa}, \{{}^{\,\varkappa}_{\mu\lambda}\}, V_\mu$ belong to one V_n and $'g_{\lambda\varkappa}, '\{{}^{\,\varkappa}_{\mu\lambda}\}, 'V_\mu$ to the other V_n, called $'V_n$ hereafter. Because the mapping is projective we have (cf. III § 7)

$$'\{{}^{\,\varkappa}_{\mu\lambda}\} = \{{}^{\,\varkappa}_{\mu\lambda}\} + 2p_{(\mu} A^\varkappa_{\lambda)} \tag{2.9}$$

and (cf. III 3.19)

$$\partial_\mu \log {}'\mathfrak{g} = \partial_\mu \log \mathfrak{g} + 2(n+1)\, p_\mu \tag{2.10}$$

which proves that p_λ is a gradientvector

$$p_\lambda = \frac{1}{2}\partial_\lambda \log\varphi; \qquad \varphi \overset{\text{def}}{=} \left(\frac{'\mathfrak{g}}{\mathfrak{g}}\right)^{\frac{1}{n+1}}. \tag{2.11}$$

Now let $d\xi^\varkappa = v^\varkappa dt$ be an infinitesimal motion in V_n. Then we consider in $'V_n$ the transformation $d\xi^\varkappa = {}'v^\varkappa dt$ with

$$'v_\lambda \overset{\text{def}}{=} \varphi v_\lambda; \qquad 'v^\varkappa = {}'g^{\varkappa\lambda}\, 'v_\lambda = \varphi\, 'g^{\varkappa\mu} g_{\mu\lambda} v^\lambda. \tag{2.12}$$

For this transformation we get

$$\left\{\begin{aligned} \underset{'v}{£}\, 'g_{\lambda\varkappa} &= 2'V_{(\lambda}\, 'v_{\varkappa)} = 2v_{(\varkappa}\partial_{\lambda)}\varphi + 2\varphi\, 'V_{(\lambda} v_{\varkappa)} \\ &= 2v_{(\varkappa}\partial_{\lambda)}\varphi + 2\varphi V_{(\lambda} v_{\varkappa)} - 4\varphi A^\mu_{(\lambda} p_{\varkappa)} v_\mu \\ &= 2\varphi V_{(\lambda} v_{\varkappa)} = \varphi \underset{v}{£}\, g_{\lambda\varkappa} = 0; \end{aligned}\right. \tag{2.13}$$

---

[1]) Cf. for more facts and more literature e.g. EISENHART 1926, 1 or 1949, 1, Ch. VI; 1933, 1, p. 208ff.; YANO 1949, 1, Ch. IV.

[2]) EISENHART 1933, 1, p. 212.

and this proves:

If two V_n's can be mapped on each other projectively, their groups of motions have the same number of parameters.[1])

Of course the groups need not be isomorphic because for two motions $u^\varkappa$ and $v^\varkappa$ in V_n we have in general

$$\underset{'v}{\pounds}\, 'u^\varkappa \neq \varphi\, 'g^{\varkappa\mu}\, g_{\mu\lambda} \underset{v}{\pounds}\, u^\lambda. \tag{2.14}$$

It is well known that the group of all motions in an S_n has exactly $\frac{1}{2}\, n\,(n+1)$ parameters. If we know that the group of all motions in a V_n has exactly $\frac{1}{2}\, n\,(n+1)$ parameters, $v^\varkappa$ and $v_{[\lambda\varkappa]}$ can be chosen arbitrarily at one point and the fields $v^\varkappa$, $v_\lambda^{\cdot\,\varkappa}$ can then be computed at all points. Hence the equations

$$\underset{v}{\pounds}\, \nabla_{\omega_k \ldots \omega_1} K_{\nu\mu\lambda\varkappa} = 0; \qquad k = 0, 1, 2, 3, \ldots \tag{2.15}$$

must be satisfied for all values of $v_{\lambda\varkappa}$ satisfying $v_{(\lambda\varkappa)} = 0$ and they should not contain $v^\varkappa$. This implies that in (2.2a) the first term in the right hand side must vanish

$$\nabla_\omega K_{\nu\mu\lambda\varkappa} = 0 \tag{2.16}$$

and that accordingly this equation can be written as

$$v^{\tau\sigma}\left(g_{\nu[\tau} K_{\sigma]\mu\lambda\varkappa} + g_{\mu[\tau} K_{|\nu|\sigma]\lambda\varkappa} + g_{\lambda[\tau} K_{|\nu\mu|\sigma]\varkappa} + g_{\varkappa[\tau} K_{|\nu\mu\lambda|\sigma]}\right) = 0 \tag{2.17}$$

valid for all values of $v^{[\tau\sigma]}$. But this is only possible if the expression between brackets is zero, hence, transvecting with $g^{\nu\tau}$

$$(n-1)\, K_{\nu\mu\lambda\varkappa} + 2 g_{[\nu|\lambda|} K_{\mu]\varkappa} = 0 \tag{2.18}$$

from which it follows for $n > 2$ immediately, after transvecting with $g^{\nu\varkappa}$, that the V_n is an S_n. For $n = 2$ the constancy of K does not follow from (2.18) but in this case it follows from (2.16). Hence[2])

The group of all motions of a V_n has $\frac{1}{2}\, n\,(n+1)$ parameters if and only if the V_n is an S_n.

Exercises.

VII 2,1[3]). An L_n with an integrable non-symmetric connexion is group space of a transitive n-parameter group if and only if $\nabla_\nu S_{\mu\lambda}^{\cdot\cdot\,\varkappa} = 0$.

VII 2,2[3]). An A_n has the (0)-connexion of an n-parameter group space if and only if there is a tensor field $S_{\mu\lambda}^{\cdot\cdot\,\varkappa}$ alternating in $\mu\lambda$ such that

$$R_{\nu\mu\lambda}^{\cdot\cdot\cdot\,\varkappa} = S_{\nu\mu}^{\cdot\cdot\,\sigma} S_{\sigma\lambda}^{\cdot\cdot\,\varkappa}; \qquad \nabla_\nu S_{\mu\lambda}^{\cdot\cdot\,\varkappa} = 0. \tag{VII 2,2 α}$$

[1]) KNEBELMAN 1930, 1; YANO 1949, 1.

[2]) FUBINI 1903, 1; 1904, 1; EISENHART 1933, 1, p. 215; YANO 1949, 1, p. 36.

[3]) EISENHART 1933, 1, p. 197; YANO 1949, 1, p. 21.

VII 2,3[1]). Let $\xi^\varkappa \rightarrow \xi^\varkappa + v^\varkappa dt$ be a motion in an L_n with an integrable connexion and let $\underset{h}{e^\varkappa}$ be n covariant constant vectorfields. Prove that the vectors $\underset{v}{\pounds}\, \underset{h}{e^\varkappa}$ are linear combinations of the $\underset{i}{e^\varkappa}$ with *constant* coefficients.

VII 2,4[2]). The fields $v^\varkappa$ and $\varrho\, v^\varkappa$ determine both a motion in A_n if and only if

VII 2,4 α) $$v^\varkappa \nabla_\mu \nabla_\lambda\, \varrho + 2 (\nabla_{(\lambda}\, \varrho) \nabla_{\mu)}\, v^\varkappa = 0 .$$

VII 2,5[3]). An L_n admits a one-parameter group of motions if and only if there exists a coordinate system such that either the $\Gamma_{\mu\lambda}^\varkappa$ are independent of ξ^1 or that the $\Gamma_{\mu\lambda}^\varkappa$ are homogeneous of degree -1 with respect to the coordinates.

VII 2,6 [cf. 1.8 or III 5.47].[4]) The field $v_\lambda^{;\varkappa}$ in L_n is covariant constant along every streamline of $v^\varkappa$. Streamlines, along which $v_\lambda^{;\varkappa} = 0$, are geodesics but this condition is only sufficient and not necessary. A necessary and sufficient condition is that $v^\varkappa$ lies at all points of the curve in an eigendirection of $v_\lambda^{;\varkappa}$.

VII 2,7. The motion in V_n defined by the field $v^\varkappa$ is a translation if and only if the angle between $v^\varkappa$ and any geodesic is constant along this geodesic.[5])

VII 2,8 (cf. IV § 8). The translations in the group space of a semi-simple group are the transformations of the first and second parameter-group.[6])

VII 2,9. Let $v^\varkappa$ be a displacement vector of a motion in an S_3. The equations of motion can be written in the form (cf. I 9.20)

VII 2,9 α) $$\nabla_\mu v_\lambda = 2 I_{\mu\lambda\varkappa} u^\varkappa; \quad \nabla_\mu u_\lambda = 2\varkappa\, I_{\mu\lambda\varkappa} v^\varkappa; \quad u^\varkappa \overset{\text{def}}{=} I^{\varkappa\mu\lambda} \nabla_\mu v_\lambda .$$

Prove that this system is totally integrable and that $u^\varkappa$ is the displacement vector of another motion.[7])

---

[1]) ROBERTSON 1932, 1; YANO 1949, 1, p. 15.

[2]) YANO 1949, 1, p. 17.

[3]) EISENHART 1933, 1, p. 232; YANO 1949, 1, p. 18. The booklet of YANO contains in Ch. III a great number of interesting propositions on affine motions in L_n and A_n. Note that A_n is used there in the sense of L_n and that an A_n is called an A_n with a symmetric linear connexion.

[4]) A. NIJENHUIS, personal communication.

[5]) EISENHART 1933, 1, p. 212; YANO 1949, 1, p. 32. As NIJENHUIS pointed out in 1952, 1, theses No. 4, the proof at these places of the lemma that the trajectories of two translations meet always under a constant angle is not sufficient and it seems probable that the lemma is not true.

[6]) CARTAN and SCHOUTEN 1926, 3, p. 811; EISENHART 1933, 1, p. 213; YANO 1949, 1, p. 33.

[7]) TONOLO 1950, 1.

§ 3. Deformation of subspaces. [1])

The deformation of a subspace can often be dealt with, using the method of transposition already explained in V § 9 and VII § 1. Here we give a detailed treatment of a V_m in V_n subjected to the infinitesimal transformation $\xi^\varkappa \rightarrow \xi^\varkappa + v^\varkappa dt$. Let a holonomic coordinate system (a); $a = 1, \ldots, \mathrm{m}$ in V_m be completed by basis vectors $\underset{y}{e}^\varkappa$; $y = \mathrm{m}+1, \ldots, \mathrm{n}$ perpendicular to V_m so that there results an anholonomic system (h); $h = 1, \ldots, \mathrm{n}$, defined over V_m. We suppose that the coordinate system (a) in V_m with its basis vectors $\underset{b}{e}^a$, and also the basis vectors $\overset{x}{e}_\lambda$, are dragged along over $v^\varkappa dt$. Further we suppose that the $\underset{y}{e}^\varkappa$ and $\overset{a}{e}_\lambda$ remain perpendicular to the $\underset{b}{e}^\varkappa$ and $\overset{x}{e}_\lambda$ respectively and that the equations $\overset{x}{e}_\lambda \underset{y}{e}^\lambda = \delta^x_y$; $\overset{a}{e}_\lambda \underset{b}{e}^\lambda = \delta^a_b$ remain valid. This implies that in general the $\underset{y}{e}^\varkappa$ and $\overset{a}{e}_\lambda$ are not dragged along, because the deformation may disturb orthogonality. In order to transpose the problem, the V_m and all basis vectors $\underset{i}{e}^\varkappa$, $\overset{h}{e}_\lambda$ are dragged back over $-v^\varkappa dt$ and at the same time the field $g_{\lambda\varkappa}$ is dragged along over $-v^\varkappa dt$. Then the $\underset{b}{e}^\varkappa$ and $\overset{x}{e}_\lambda$ return to their original positions but the fields $g_{\lambda\varkappa}$, $\underset{y}{e}^\varkappa$ and $\overset{a}{e}_\lambda$ over V_m will have been changed. As we have seen before, the absolute differentials $\overset{a}{D}\, dt$ of all quantities connected with V_m in the original problem equal the natural differentials $'\overset{n}{D}\, dt$ in the transposed problem. Hence

$$(3.1)\qquad \overset{a}{D}\, g_{\lambda\varkappa} = {}'\overset{n}{D}\, g_{\lambda\varkappa} = \underset{v}{\pounds}\, g_{\lambda\varkappa} = 2\nabla_{(\lambda} v_{\varkappa)} = 2v_{(\lambda\varkappa)}; \qquad v_\lambda^{\cdot\varkappa} = \nabla_\lambda v^\varkappa$$

$$(3.2)\qquad \overset{a}{D}\, B^\varkappa_b = {}'\overset{n}{D}\, B^\varkappa_b = 0$$

$$(3.3)\qquad \overset{a}{D}\, C^x_\lambda = {}'\overset{n}{D}\, C^x_\lambda = 0$$

$$(3.4)\qquad \left\{ \begin{array}{l} \overset{a}{D}\, 'g_{cb} = {}'\overset{n}{D}\, 'g_{cb} = B^{\mu\lambda}_{cb}\, {}'\overset{n}{D}\, g_{\mu\lambda} = 2B^{\mu\lambda}_{c\,b} v_{(\mu\lambda)} = 2B^\lambda_{(b} D_{c)} v_\lambda \\ \qquad\qquad\qquad\qquad (\text{cf. V § 7}). \end{array} \right.$$

---

[1]) General references: CARTAN 1916, 1 (in R_n); 1917, 1 (in C_n, $n \geq 5$); 1920, 1 (project.); 2; 1927, 5 (second var.); LEVI CIVITA 1926, 1 (second var.); SYNGE 1926, 1; 2 and 1934, 1 (second var.); SCHOUTEN 1927, 2 (V_m in V_n); MC CONNELL 1929, 1 (curves in V_n); HLAVATY 1929, 3 (curves in U_n); 1950, 1; 1952, 5 (families of V_m in V_n); BURSTIN 1931, 1; 2; HAYDEN 1931, 1 (curves in V_n); 1934, 2 (X_m in U_n); 1936, 1 (L_m in L_n); DIENES 1933, 4 and 1934, 2 (X_m in L_n); BOMPIANI 1935, 1; LAND 1936, 1; DAVIES 1936, 1 (V_m in V_n); 1938, 1 (curves in V_n); 1938, 2 (V_n^m); 1942, 1 (second var. V_m in V_n); YANO and MUTO 1936, 2 (second var.); DIENES and DAVIES 1937, 1 (X_m in V_n and A_n, literature); E II 1938, 2, p. 164 ff.; YANO and ADATI 1944, 7 (par. tang. and concirc.); YANO 1945, 9; 10; 1949, 2 (literature); YANO, TAKANO and TOMONAGA 1946, 10 and 1948, 2 (curves in L_n); CASTOLDI 1948, 2 (finite deform. of V_m in V_n); YANENKO 1949, 2 (V_m in V_n).

If $v^\varkappa$ is split up into $'v^\varkappa$ in V_m and $''v^\varkappa$ perpendicular to V_m, (3.4) leads to (cf. V 7.15)

$$\text{(3.5)}\qquad \boxed{\begin{aligned} &\text{a)}\ \overset{a}{D}\,'g_{cb} = 2D_{(c}\,'v_{b)} - 2\,''v_x\,H_{cb}^{\cdot\cdot\,x} \\ &\text{b)}\ \overset{a}{D}\,'g^{ab} = -2\,'g^{c(a}\,D_c\,'v^{b)} + 2\,''v_x\,H^{abx} \end{aligned}}\,.$$

From (3.1–4) and $B^a_\lambda = g_{\lambda\varkappa}\,B^\varkappa_b\,'g^{ba}$ we get

$$\text{(3.6)}\qquad \boxed{\overset{a}{D}\,B^a_\lambda = 2\,'g^{ab}\,B^\varkappa_b\,C^\mu_\lambda\,v_{(\mu\varkappa)}}$$

and accordingly

$$\text{(3.7)}\qquad \boxed{\overset{a}{D}\,B^\varkappa_\lambda = -\overset{a}{D}\,C^\varkappa_\lambda = 2\,'g^{\varkappa\varrho}\,C^\mu_\lambda\,v_{(\mu\varrho)}}\,.$$

Because of $\overset{a}{D} g_{\lambda\varkappa} = \underset{v}{\pounds}\, g_{\lambda\varkappa}$ we have (cf. 1.8)

$$\text{(3.8)}\qquad \overset{a}{D}\,\Gamma^\varkappa_{\mu\lambda} = \underset{v}{\pounds}\,\Gamma^\varkappa_{\mu\lambda} = \nabla_\mu\, v_\lambda^{\cdot\,\varkappa} + v^\sigma\, K_{\sigma\mu\lambda}^{\cdot\cdot\cdot\,\varkappa}$$

and from this we get for $\overset{a}{D}\,'\Gamma^a_{cb}$ (cf. 3.2, 6 and V 3.10)

$$\text{(3.9)}\qquad \left\{\begin{aligned} \overset{a}{D}\,'\Gamma^a_{cb} &= \overset{a}{D}\,(B^{\mu\lambda a}_{cb\varkappa}\,\Gamma^\varkappa_{\mu\lambda} + B^a_\varkappa\,\partial_c\,B^\varkappa_b) \\ &= B^{\mu\lambda}_{cb}\,\Gamma^\varkappa_{\mu\lambda}\,\overset{a}{D}\,B^a_\varkappa + B^{\mu\lambda a}_{cb\varkappa}\,\overset{a}{D}\,\Gamma^\varkappa_{\mu\lambda} + (\overset{a}{D}\,B^a_\varkappa)\,\partial_c\,B^\varkappa_b \\ &= B^{\mu\lambda a}_{cb\varkappa}\,\underset{v}{\pounds}\,\Gamma^\varkappa_{\mu\lambda} + (\overset{a}{D}\,B^a_\varkappa)\,(D_c\,B^\varkappa_b + '\Gamma^d_{cb}\,B^\varkappa_d) \end{aligned}\right.$$

hence

$$\text{(3.10)}\qquad \boxed{\overset{a}{D}\,'\Gamma^a_{cb} = B^{\mu\lambda a}_{cb\varkappa}\,\underset{v}{\pounds}\,\Gamma^\varkappa_{\mu\lambda} + H_{cb}^{\cdot\cdot\,\varkappa}\,\overset{a}{D}\,B^a_\varkappa}\,.^{1)}$$

It can be proved in the same way that

$$\text{(3.11)}\qquad \boxed{\overset{u}{D}\,H_{cb}^{\cdot\cdot\,x} = C^x_\varkappa\,B^{\mu\lambda}_{cb}\,\underset{v}{\pounds}\,\Gamma^\varkappa_{\mu\lambda}}\,.$$

By means of (3.8) and the CODAZZI equations (V 8.7) this equation can be transformed into

$$\text{(3.12)}\qquad \left\{\begin{aligned} &\overset{a}{D}\,H_{cb}^{\cdot\cdot\,x} = 'v^d\,D_d\,H_{cb}^{\cdot\cdot\,x} + 2H_{d(c}^{\cdot\cdot\,x}\,D_{b)}\,'v^d - H_{cb}^{\cdot\cdot\,y}\,''v_y^{\cdot\,x} + \\ &\qquad\qquad + ''v^y\,(C^{\nu x}_{y\varkappa}\,B^{\mu\lambda}_{cb}\,K_{\nu\mu\lambda}^{\cdot\cdot\cdot\,\varkappa} - H_{cd}^{\cdot\cdot\,x}\,H_{b\cdot y}^{\cdot d}) + D_c\,D_b\,''v^x\,; \\ &''v_y^{\cdot\,x} \overset{\text{def}}{=} C^{\lambda x}_{y\varkappa}\,v_\lambda^{\cdot\,\varkappa}. \end{aligned}\right.$$

In the exercises (VII 3,1; 2; 3) we give some other formulae concerning the V_m in V_n. [2])

[1]) DAVIES 1936, 1; cf. BORTOLOTTI 1928, 2; 3 (literature); HAYDEN 1934, 2.
[2]) Cf. for finite deformations of V_m in V_n CASTOLDI 1948, 2.

The V_n^m in V_n, the rigged L_m in L_n and the rigged L_n^m in L_n can be dealt with in the same way.[1])

The method of transposition cannot be used for a subspace imbedded in an X_n because in the X_n there is no field that could be dragged back. But a more general method that can also be used in this case, was developed by NIJENHUIS.[2]) Let a rigged X_n^m in X_n be subjected to the infinitesimal transformation $\xi^\varkappa \to \xi^\varkappa + v^\varkappa\, dt$ defined over X_n and let there be introduced an anholonomic coordinate system (h) such that before the transformation the $\underset{b}{e}^\varkappa$; $b = 1, \ldots, \mathrm{m}$ lie in X_n^m and the $\underset{y}{e}^\varkappa$; $y = \mathrm{m}+1, \ldots, \mathrm{n}$ in the rigging. Note that the tangent E_m and $E_{m'}$; $m' = n - m$, at $\xi^\varkappa$ after having been dragged along over $v^\varkappa\, dt$ do not coincide with the tangent E_m and $E_{m'}$ at $\xi^\varkappa + v^\varkappa\, dt$ before the transformation. This implies, that for a vector p^a of X_n^m, having components $p^\varkappa = B_b^\varkappa p^b$ with respect to $(\varkappa)$, a LIE derivative is as yet not defined because after the transformation the vector is no longer a vector in the tangent E_m. We now *define* the LIE derivatives for the fields p^a, q_b as the X_n^m-parts of the LIE derivatives of the fields $p^\varkappa$, q_λ considered as fields of X_n (cf. V 7.36):

$$(3.13)\quad \begin{cases} \text{a)} & \underset{v}{£}\, p^a \overset{\text{def}}{=} B_\varkappa^a \underset{v}{£}\, p^\varkappa \\ \text{b)} & \underset{v}{£}\, q_b \overset{\text{def}}{=} B_b^\lambda \underset{v}{£}\, q_\lambda . \end{cases}$$

The definition (3.13) implies that the LEIBNIZ rule remains valid at least for quantities. We *assume* that it holds always. If the X_n^m is holonomic (i.e. consisting of ∞^{n-m} X_m's) and if $v^\varkappa$ happens to be a vector of X_n^m, the vector $\underset{v}{£}\, p^a$ defined in the ordinary way with coordinates η^a in the X_m's, that is $v^b\, \partial_b\, p^a - p^b\, \partial_b\, v^a$, is the X_m-part of $\underset{v}{£}\, p^\varkappa$. This proves that $\underset{v}{£}\, p^a$ as defined in (3.13a) has the ordinary meaning in this special case. By the definition-formulae (3.13) the operator $\underset{v}{£}$ gets the properties of a v. D. WAERDEN-BORTOLOTTI operator (cf. V § 7). From (3.13) we get immediately (cf. II 9.10)

$$(3.14)\quad \begin{cases} \underset{v}{£}\, p^a = B_\varkappa^a (v^\mu\, \partial_\mu\, B_b^\varkappa\, p^b - p^b\, B_b^\mu\, \partial_\mu\, v^\varkappa) \\ \qquad \overset{*}{=} v^\mu\, \partial_\mu\, p^a - (\partial_b\, {}'v^a + 2 v^\mu\, \Omega_{\mu\lambda}^{a}\, B_b^\lambda)\, p^b ; \end{cases}$$

hence (cf. V 7.25, 26, 35)

$$(3.15)\quad \begin{cases} \text{a)} & \underset{'v}{£}\, p^a = {}'v^b\, \partial_b\, p^a - (\partial_b\, {}'v^a + 2\, {}'v^c\, {}'\overset{m}{\Omega}{}_{cb}^{a})\, p^b \\ \text{b)} & \underset{''v}{£}\, p^a = {}''v^y\, (\partial_y\, p^a + 2\Omega_{c\,y}^{a}\, p^c) = {}''v^y\, \mathfrak{D}_y\, p^a \\ \text{c)} & \underset{v}{£}\, p^a = \underset{'v}{£}\, p^a + \underset{''v}{£}\, p^a = \underset{'v}{£}\, p^a + {}''v^y\, \mathfrak{D}_y\, p^a \end{cases}$$

---

1) SCHOUTEN and v. KAMPEN 1933, 1; E II 1938, 2, p. 165ff.

2) Personal communication.

where $\mathfrak{D}_y$ is the invariant operator defined in V § 7 that is independent of any connexion in X_n. Note the resemblance of (3.15a) to the expression for the LIE derivative in anholonomic coordinates in exerc. II 10,6. In the same way we can derive

$$(3.16)\quad \begin{cases} \text{a)}\ \ \underset{'v}{\pounds}\, q_b = {}'v^c\, \partial_c\, q_b + (\partial_b\, {}'v^a + 2\, {}'v^c\, {}'\Omega_{cb}^{a})\, q_a \\ \text{b)}\ \ \underset{''v}{\pounds}\, q_b = {}''v^y\, (\partial_y\, q_b - 2\overset{m}{\Omega}{}_{b\,y}^{a}\, q_a) = {}''v^y\, \mathfrak{D}_y\, q_b \\ \text{c)}\ \ \underset{v}{\pounds}\, q_b \overset{*}{=} v^\mu\, \partial_\mu\, q_b + (\partial_b\, {}'v^a + 2 v^\mu\, \Omega_{\mu\lambda}^{a}\, B_b^\lambda)\, q_a = \underset{'v}{\pounds}\, q_b + \underset{''v}{\pounds}\, q_b \\ \qquad = \underset{'v}{\pounds}\, q_b + {}''v^y\, \mathfrak{D}_y\, q_b\,. \end{cases}$$

If we define

$$(3.17)\quad \begin{cases} \text{a)}\ \ \underset{v}{\pounds}\, r^x \overset{\text{def}}{=} C_\varkappa^x\, \underset{v}{\pounds}\, r^\varkappa \\ \text{b)}\ \ \underset{v}{\pounds}\, s_y \overset{\text{def}}{=} C_y^\lambda\, \underset{v}{\pounds}\, s_\lambda \end{cases}$$

for the fields r^x and s_y in the rigging (cf. 3.13) we get in the same way

$$(3.18)\quad \begin{cases} \text{a)}\ \ \underset{'v}{\pounds}\, r^x = {}'v^b\, \mathfrak{D}_b\, r^x; \\ \text{b)}\ \ \underset{''v}{\pounds}\, r^x = {}''v^y\, \partial_y\, r^x - (\partial_y\, {}''v^x + 2\, {}''v^z\, {}''\Omega_{zy}^{x})\, r^y \\ \text{c)}\ \ \mathfrak{D}_b\, r^x = \partial_b\, r^x + 2\overset{m'}{\Omega}{}_{yb}^{x}\, r^y \qquad \text{(cf. V 7.35c)} \\ \text{d)}\ \ \underset{v}{\pounds}\, r^x \overset{*}{=} v^\mu\, \partial_\mu\, r^x - (\partial_y\, {}''v^x + 2 v^\mu\, \Omega_{\mu\lambda}^{x}\, C_y^\lambda)\, r^y = \underset{'v}{\pounds}\, r^x + \underset{''v}{\pounds}\, r^x \end{cases}$$

$$(3.19)\quad \begin{cases} \text{a)}\ \ \underset{'v}{\pounds}\, s_y = {}'v^b\, \mathfrak{D}_b\, s_y \\ \text{b)}\ \ \underset{''v}{\pounds}\, s_y = {}''v^z\, \partial_z\, s_y + (\partial_y\, {}''v^x + 2\, {}''v^z\, {}''\Omega_{zy}^{x})\, s_x \\ \text{c)}\ \ \mathfrak{D}_b\, s_y = \partial_b\, s_y - 2\overset{m'}{\Omega}{}_{yb}^{x}\, s_x \qquad \text{(cf. V 7.35d)} \\ \text{d)}\ \ \underset{v}{\pounds}\, s_y \overset{*}{=} v^\mu\, \partial_\mu\, s_y + (\partial_y\, {}''v^x + 2 v^\mu\, \Omega_{\mu\lambda}^{x}\, C_y^\lambda)\, s_x = \underset{'v}{\pounds}\, s_y + \underset{''v}{\pounds}\, s_y\,. \end{cases}$$

All these formulae can also be derived easily from (Exerc. II 10,6). It is now easy to derive the general formula for a quantity lying with some indices in X_n, with other indices in X_n^m and with the remaining ones in the rigging, for instance

$$(3.20)\quad \begin{cases} \underset{v}{\pounds}\, P^{\varkappa a}_{\cdot\cdot y} \overset{*}{=} v^\mu\, \partial_\mu\, P^{\varkappa a}_{\cdot\cdot y} - P^{\mu a}_{\cdot\cdot y}\, \partial_\mu\, v^\varkappa - P^{\varkappa b}_{\cdot\cdot y}\, (\partial_b\, {}'v^a + 2 v^\mu\, \Omega_{\mu\lambda}^{a}\, B_b^\lambda) + \\ \qquad + P^{\varkappa a}_{\cdot\cdot x}\, (\partial_y\, {}''v^x + 2 v^\mu\, \Omega_{\mu\lambda}^{x}\, C_y^\lambda)\,. \end{cases}$$

For a density $\overset{(a)}{\mathfrak{p}}$ or $\overset{(x)}{\mathfrak{p}}$ of X_n^m or $X_n^{m'}$ with weight $'w$ or $''w$ respectively we find (cf. Exerc. II 10, 6δ)

$$(3.21)\quad \begin{cases} \text{a)}\ \ \underset{v}{\pounds}\, \overset{(a)}{\mathfrak{p}} \overset{*}{=} v^\mu\, \partial_\mu\, \overset{(a)}{\mathfrak{p}} + {}'w\, \overset{(a)}{\mathfrak{p}}\, (\partial_b\, {}'v^b + 2 v^\mu\, \Omega_{\mu\lambda}^{b}\, B_b^\lambda) \\ \text{b)}\ \ \underset{v}{\pounds}\, \overset{(x)}{\mathfrak{p}} \overset{*}{=} v^\mu\, \partial_\mu\, \overset{(x)}{\mathfrak{p}} + {}''w\, \overset{(x)}{\mathfrak{p}}\, (\partial_y\, {}''v^y + 2 v^\mu\, \Omega_{\mu\lambda}^{y}\, C_y^\lambda)\,. \end{cases}$$

For any *quantity* Φ (Ψ) (indices suppressed) of X_n^m ($X_n^{m'}$) we have (cf. V § 7)

$$\underset{''v}{£}\Phi = {}''v^y \mathfrak{D}_y \Phi; \qquad \underset{'v}{£}\Psi = {}'v^b \mathfrak{D}_b \Psi. \tag{3.22}$$

Of course $\mathfrak{D}_b$ ($\mathfrak{D}_y$) can not yet be applied to a quantity of X_n^m ($X_n^{m'}$) because so far no connexion has been given (cf. V § 7).

An X_n^m is invariant for a transformation $v^\varkappa dt$ if and only if every vector p^a of X_n^m remains a vector of X_n^m, that is if

$$\underset{v}{£} p^\varkappa = \underset{v}{£} B_a^\varkappa p^a = B_a^\varkappa \underset{v}{£} p^a + p^a \underset{v}{£} B_a^\varkappa \tag{3.23}$$

is a vector of X_n^m. Now the connecting quantity depending on $v^\varkappa$

$$\boxed{\underset{v}{H}{}_b^{\cdot\varkappa} \stackrel{\text{def}}{=} \underset{v}{£} B_b^\varkappa} \tag{3.24}$$

lies with the index $\varkappa$ in $X_n^{m'}$ because

$$B_\varkappa^a \underset{v}{H}{}_b^{\cdot\varkappa} = B_{\varkappa b}^{a\lambda} \underset{v}{£} B_\lambda^\varkappa = - B_{\varkappa b}^{a\lambda} \underset{v}{£} C_\lambda^\varkappa = - B_{\varkappa b}^{a\lambda} \underset{v}{£} C_y^\varkappa C_\lambda^y = 0. \tag{3.25}$$

Hence *the necessary and sufficient condition for the invariance of* X_n^m *is that the* LIE *derivative of* $B_b^\varkappa$ *is zero*: $\underset{v}{H}{}_b^{\cdot\varkappa} = 0$.

In the same way the connecting quantity

$$\boxed{\underset{v}{L}{}_y^{\cdot\varkappa} \stackrel{\text{def}}{=} \underset{v}{£} C_y^\varkappa} \tag{3.26}$$

lies with the index $\varkappa$ in X_n^m and *it is zero if and only if the rigging* $X_n^{m'}$ *is invariant.* From (3.24, 26) it follows easily that

$$\boxed{\text{a)}\quad \underset{v}{£} B_\lambda^a = - \underset{v}{L}{}_\lambda^{\cdot a}; \qquad \text{b)}\quad \underset{v}{£} C_\lambda^x = - \underset{v}{H}{}_\lambda^{\cdot x}} \tag{3.27}$$

and

$$\underset{v}{£} B_\lambda^\varkappa = - \underset{v}{£} C_\lambda^\varkappa = \underset{v}{H}{}_\lambda^{\cdot\varkappa} - \underset{v}{L}{}_\lambda^{\cdot\varkappa}. \tag{3.28}$$

For $\underset{v}{H}{}_b^{\cdot x}$ and $\underset{v}{L}{}_y^{\cdot a}$ it is easily proved that

$$\begin{cases} \text{a)}\quad \underset{v}{H}{}_b^{\cdot x} = C_\varkappa^x (v^\mu \partial_\mu B_b^\varkappa - B_b^\mu \partial_\mu v^\varkappa) \\ \text{b)}\quad \underset{v}{L}{}_y^{\cdot a} = B_\varkappa^a (v^\mu \partial_\mu C_y^\varkappa - C_y^\mu \partial_\mu v^\varkappa) \end{cases} \tag{3.29}$$

and these equations express that $\underset{v}{H}{}_b^{\cdot x}$ depends on the X_n^m only and not on the rigging, and in the same way $\underset{v}{L}{}_y^{\cdot a}$ depends on the $X_n^{m'}$ only.

If $v^\varkappa$ lies in X_n^m, $v^\varkappa = {}'v^\varkappa$, we get (cf. V 7.22)

$$\begin{cases} \underset{'v}{H}{}_b^{\cdot x} = C_\varkappa^x ({}'v^c \partial_c B_b^\varkappa - (\partial_b B_c^\varkappa)\, {}'v^c - B_c^\varkappa \partial_b {}'v^c) \\ \qquad = 2 C_\varkappa^x\, {}'v^c \partial_{[c} B_{b]}^\varkappa \stackrel{*}{=} - 2\, {}'v^c \Omega_{cb}^{x} \stackrel{*}{=} - 2\, {}'v^c \overset{m}{Z}{}_{cb}^{\cdot\cdot x} \end{cases} \tag{3.30}$$

and from this identity we see that *an* X_n^m *is invariant for all tangent deformations if and only if it is holonomic.*

If $v^\varkappa$ lies in $X_n^{m'}$, $v^\varkappa = ''v^\varkappa$, we get from (3.29) and (V 7.35 c)

$$(3.31)\qquad \begin{cases} \underset{''v}{H_b^{\cdot x}} = C_\varkappa^x\,(''v^z\,\partial_z\,B_b^\varkappa - (\partial_b\,C_z^\varkappa)\,''v^z - C_z^\varkappa\,\partial_b\,''v^z) \\ \qquad = -\,\partial_b\,''v^x - 2''v^z\,\overset{m'}{\Omega}{}_{zb}^{x} = -\,\mathfrak{D}_b\,''v^x \end{cases}$$

hence, from (3.30, 31) and the analogous reasoning for $X_n^{m'}$

$$(3.32)\qquad \begin{cases} \text{a)}\quad \underset{v}{H_b^{\cdot x}} = -\,\mathfrak{D}_b\,''v^x - 2'v^c\,\overset{m}{Z}{}_{cb}^{\cdot\cdot x} \\ \text{b)}\quad \underset{v}{L_y^{\cdot a}} = -\,\mathfrak{D}_y\,'v^a - 2''v^z\,\overset{m'}{Z}{}_{zy}^{\cdot\cdot a}. \end{cases}$$

If $u^\varkappa$ and $v^\varkappa$ are both vectors of X_n^m: $u^\varkappa = 'u^\varkappa$; $v^\varkappa = 'v^\varkappa$, we get from (3.30)

$$(3.33)\qquad \underset{'u}{\pounds}\,'v^\varkappa = \underset{'u}{\pounds}\,B_b^\varkappa\,'v^b = B_b^\varkappa\,\underset{'u}{\pounds}\,'v^b + 'v^b\,\underset{'u}{H_b^{\cdot\varkappa}} = B_b^\varkappa\,\underset{'u}{\pounds}\,'v^b - 2'u^c\,'v^b\,\overset{m}{Z}{}_{cb}^{\cdot\cdot\varkappa},$$

hence $\underset{'u}{\pounds}\,'v^\varkappa$ lies in X_n^m for every choice of $'u^\varkappa$ and $'v^\varkappa$ if and only if X_n^m is holonomic.

In Ch. II § 10 we found (Exerc. II 10,9)

$$(3.34)\qquad 2\underset{[u}{\pounds}\,\underset{v]}{\pounds} = \underset{(u,v)}{\pounds}\,; \qquad (u,v)^\varkappa \overset{\text{def}}{=} \underset{u}{\pounds}\,v^\varkappa = -\underset{v}{\pounds}\,u^\varkappa$$

for all quantities of X_n. But this simple rule does not hold for the operators $\underset{u}{\pounds}$ and $\underset{v}{\pounds}$ as defined in (3.13, 17), if applied to quantities of X_n^m or $X_n^{m'}$. Instead we get from (3.28)

$$(3.35)\qquad \boxed{\begin{array}{ll} \text{a)} & 2\underset{[u}{\pounds}\,\underset{v]}{\pounds}\,p^a = \underset{(u,v)}{\pounds}\,p^a + 2\underset{[u}{H_b^{\cdot x}}\,\underset{v]}{L_x^{\cdot a}}\,p^b \\ \text{b)} & 2\underset{[u}{\pounds}\,\underset{v]}{\pounds}\,q_b = \underset{(u,v)}{\pounds}\,q_b - 2\underset{[u}{H_b^{\cdot x}}\,\underset{v]}{L_x^{\cdot a}}\,q_a \\ \text{c)} & 2\underset{[u}{\pounds}\,\underset{v]}{\pounds}\,r^x = \underset{(u,v)}{\pounds}\,r^x + 2\underset{[u}{L_y^{\cdot a}}\,\underset{v]}{H_a^{\cdot x}}\,r^y \\ \text{d)} & 2\underset{[u}{\pounds}\,\underset{v]}{\pounds}\,s_y = \underset{(u,v)}{\pounds}\,s_y - 2\underset{[u}{L_y^{\cdot a}}\,\underset{v]}{H_a^{\cdot x}}\,s_x \end{array}}$$

hence, by applying (3.34, 35) to $B_b^\varkappa$ and $C_y^\varkappa$

$$(3.36)\qquad \begin{cases} \text{a)}\quad 2\underset{[u}{\pounds}\,\underset{v]}{\pounds}\,B_b^\varkappa = \underset{(u,v)}{H_b^{\cdot\varkappa}} - 2\underset{[u}{H_b^{\cdot x}}\,\underset{v]}{L_x^{\cdot\varkappa}} \\ \text{b)}\quad 2\underset{[u}{\pounds}\,\underset{v]}{\pounds}\,C_y^\varkappa = \underset{(u,v)}{L_y^{\cdot\varkappa}} - 2\underset{[u}{L_y^{\cdot a}}\,\underset{v]}{H_a^{\cdot\varkappa}} \end{cases}$$

and by transvection of these equations with $C^x_\varkappa$ and $B^a_\varkappa$ respectively

$$(3.37)\quad \begin{cases} \text{a)}\ \underset{(u,v)}{H_b^{\cdot x}} = 2\, C^x_\varkappa \underset{[u}{\pounds}\underset{v]}{\pounds}\, B^\varkappa_b = 2 \underset{[u}{\pounds}\underset{v]}{H_b^{\cdot x}} \\ \text{b)}\ \underset{(u,v)}{L_y^{\cdot a}} = 2\, B^a_\varkappa \underset{[u}{\pounds}\underset{v]}{\pounds}\, C^\varkappa_y = 2 \underset{[u}{\pounds}\underset{v]}{L_y^{\cdot a}}. \end{cases}$$

There is also a commutation formula for the operators $\underset{v}{\pounds}$ and $\mathfrak{D}_y$ (or $\mathfrak{D}_b$) that can be derived from (3.15, 16, 32, 35)

$$(3.38)\quad \begin{cases} \text{a)}\ (\mathfrak{D}_y \underset{v}{\pounds} - \underset{v}{\pounds}\, \mathfrak{D}_y)\, p^a = \underset{p}{\pounds}\underset{v}{L_y^{\cdot a}} + 2 \overset{m'}{Z}{}_{y x}^{\cdot\cdot a}\, \underset{v}{H_b^{\cdot x}}\, p^b \\ \text{b)}\ (\mathfrak{D}_b \underset{v}{\pounds} - \underset{v}{\pounds}\, \mathfrak{D}_b)\, r^x = \underset{r}{\pounds}\underset{v}{H_b^{\cdot x}} + 2 \overset{m}{Z}{}_{b a}^{\cdot\cdot x}\, \underset{v}{L_y^{\cdot a}}\, r^y. \end{cases}$$

So far no connexion or metric has been introduced either in X_n or in X^m_n. By the introduction of a connexion or a metric in the rigged X^m_n only, as was described in V § 7 and V § 9 respectively we get the rigged L^m_n and V^m_n in X_n. Introduction of a connexion or a metric in X_n leads to the rigged L^m_n in L_n and the V^m_n in V_n respectively. For all these special cases the formulae derived here for the rigged X^m_n in X_n remain valid, only new additional invariant operators and formulae arise.

Here we deal only with the rigged L^m_n in X_n arising by the introduction of a connexion $'\Gamma^a_{cb}$ in the X^m_n. Then, as we saw in V § 7 the operator $\mathfrak{D}_b$ can be extended to quantities of X^m_n (cf. V 7,35, 57)

$$(3.39)\quad \begin{cases} \text{a)}\ \mathfrak{D}_c\, p^a = D_c\, p^a = \partial_c\, p^a + {'\Gamma^a_{cb}}\, p^b \\ \text{b)}\ \mathfrak{D}_c\, q_b = D_c\, q_b = \partial_c\, q_b - {'\Gamma^a_{cb}}\, q_a \end{cases}$$

and the operator $\mathfrak{D}_\mu$ can be defined as

$$(3.40)\quad \mathfrak{D}_\mu \overset{\text{def}}{=} B^b_\mu\, \mathfrak{D}_b + C^y_\mu\, \mathfrak{D}_y.$$

Then we get from (3.15, 16) (cf. III 9.3 and III 5.48)

$$(3.41)\quad \begin{cases} \text{a)}\ \underset{v}{\pounds}\, p^a = v^\mu\, \mathfrak{D}_\mu\, p^a - {'v_b^{\cdot a}}\, p^b; & {'v_b^{\cdot a}} = D_b\, {'v^a} - 2\, {'S_{bc}^{\cdot\cdot a}}\, {'v^c} \\ \text{b)}\ \underset{v}{\pounds}\, q_b = v^\mu\, \mathfrak{D}_\mu\, q_b + {'v_b^{\cdot a}}\, q_a; & {'S_{cb}^{\cdot\cdot a}} = {'\Gamma^a_{[cb]}} + {'\Omega^a_{cb}}. \end{cases}$$

For $\underset{v}{\pounds}\, r^x$; $\underset{v}{\pounds}\, s_y$ there are no formulae of this kind as long as there is no connexion defined in $X^{m'}_n$.

We wish to find $\underset{v}{\pounds}\, {'\Gamma^a_{cb}}$. Although $\underset{v}{\pounds}$ and ∂_μ commute, $\underset{v}{\pounds}$ and ∂_c have this property if and only if the X^m_n is invariant for the transformation $v^\varkappa dt$, because, using the rule of LEIBNIZ (cf. p. 354) we have

$$(3.42)\quad \underset{v}{\pounds}\, \partial_c\, p^a = \underset{v}{\pounds}\, B^\mu_c\, \partial_\mu\, p^a = (\underset{v}{\pounds}\, B^\mu_c)\, \partial_\mu\, p^a + B^\mu_c\, \partial_\mu \underset{v}{\pounds}\, p^a = \underset{v}{H_c^{\cdot \mu}}\, \partial_\mu\, p^a + \partial_c \underset{v}{\pounds}\, p^a.$$

Hence, if the X_n^m is holonomic and if $v^\varkappa$ lies in X_n^m, it follows from (3.30) that $\underset{v}{£}$ and ∂_c are always commuting whether (c) is holonomic or anholonomic, as was to be expected (cf. II 10.17).

Applying (3.42) to $\underset{v}{£}\, \mathfrak{D}_c p^a$ we get

$$(3.43)\qquad \begin{cases} \underset{v}{£}\, \mathfrak{D}_c p^a = \underset{v}{£}\, (\partial_c p^a + {}'\Gamma^a_{cb} p^b) \\ \qquad = \underset{v}{H}{}_c^{\cdot\,\mu}\, \partial_\mu p^a + \partial_c \underset{v}{£}\, p^a + {}'\Gamma^a_{cb} \underset{v}{£}\, p^b + p^b \underset{v}{£}\, {}'\Gamma^a_{cb} \end{cases}$$

hence (cf. V 7.35a)

$$(3.44)\qquad p^b \underset{v}{£}\, {}'\Gamma^a_{cb} = (\underset{v}{£}\, \mathfrak{D}_c - \mathfrak{D}_c \underset{v}{£})\, p^a - \underset{v}{H}{}_c^{\cdot\,y}\, \mathfrak{D}_y p^a + 2\, \underset{v}{H}{}_c^{\cdot\,y}\, \overset{m}{\Omega}{}^a_{by}\, p^b .$$

Now it can be proved from (V 7.59–62) and (3.32a) that

$$(3.45)\qquad \begin{cases} (\underset{v}{£}\, \mathfrak{D}_c - \mathfrak{D}_c \underset{v}{£})\, p^a - \underset{v}{H}{}_c^{\cdot\,y}\, \mathfrak{D}_y p^a \\ \qquad = ({}'v^d\, {}^*\overset{m}{R}{}_{dcb}^{\cdot\cdot\cdot\,a} + {}''v^y\, {}^*\overset{mm'}{R}{}_{ycb}^{\cdot\cdot\cdot\,a} + \mathfrak{D}_c\, {}'v_b^{\cdot\,a})\, p^b , \end{cases}$$

hence

$$(3.46)\qquad \boxed{\underset{v}{£}\, {}'\Gamma^a_{cb} = {}'v^d\, {}^*\overset{m}{R}{}_{dcb}^{\cdot\cdot\cdot\,a} + {}''v^y\, {}^*\overset{mm'}{R}{}_{ycb}^{\cdot\cdot\cdot\,a} + \mathfrak{D}_c\, {}'v_b^{\cdot\,a} + 2\, \underset{v}{H}{}_c^{\cdot\,y}\, \overset{m}{\Omega}{}^a_{by}} .$$

In order to find $\underset{v}{£}\, \overset{m}{\Omega}{}^a_{cy}$ we start from $\underset{v}{£}\, \partial_y p^a$:

$$(3.47)\qquad \underset{v}{£}\, \partial_y p^a = \underset{v}{£}\, C^\mu_y\, \partial_\mu p^a = \underset{v}{L}{}_y^{\cdot\,\mu}\, \partial_\mu p^a + \partial_y \underset{v}{£}\, p^a$$

and apply this result to $\underset{v}{£}\, \mathfrak{D}_y p^a$ (cf. V 7.35a):

$$(3.48)\qquad \underset{v}{£}\, \mathfrak{D}_y p^a = \underset{v}{L}{}_y^{\cdot\,c}\, \partial_c p^a + \mathfrak{D}_y \underset{v}{£}\, p^a + 2 p^c \underset{v}{£}\, \overset{m}{\Omega}{}^a_{cy}$$

or

$$(3.49)\qquad (\underset{v}{£}\, \mathfrak{D}_y - \mathfrak{D}_y \underset{v}{£})\, p^a = \underset{v}{L}{}_y^{\cdot\,c}\, \mathfrak{D}_c p^a - \underset{v}{L}{}_y^{\cdot\,c}\, {}'\Gamma^a_{cb}\, p^b + 2 p^c \underset{v}{£}\, \overset{m}{\Omega}{}^a_{cy} .$$

Now from (3.15, 19) (cf. 3.45 and V 7.24, 58–61) it can be proved that

$$(3.50)\qquad \begin{cases} (\underset{v}{£}\, \mathfrak{D}_y - \mathfrak{D}_y \underset{v}{£})\, p^a - \underset{v}{L}{}_y^{\cdot\,c}\, \mathfrak{D}_c p^a \\ \qquad = (-\, \mathfrak{D}_b \underset{v}{L}{}_y^{\cdot\,a} + 2\, {}'S_{bc}^{\cdot\cdot\,a}\, \underset{v}{L}{}_y^{\cdot\,c} - 2\, \overset{m'}{Z}{}_{yx}^{\cdot\cdot\,a}\, \underset{v}{H}{}_b^{\cdot\,x})\, p^b ; \end{cases}$$

hence

$$(3.51)\qquad 2 \underset{v}{£}\, \overset{m}{\Omega}{}^a_{by} = -\, \mathfrak{D}_b \underset{v}{L}{}_y^{\cdot\,a} + \underset{v}{L}{}_y^{\cdot\,c}\, ({}'\Gamma^a_{cb} - 2\, {}'S_{cb}^{\cdot\cdot\,a}) - 2\, \overset{m'}{Z}{}_{yx}^{\cdot\cdot\,a}\, \underset{v}{H}{}_b^{\cdot\,x}$$

or

$$(3.52)\qquad \boxed{\underset{v}{£}\, \overset{m}{\Omega}{}^a_{by} = -\tfrac{1}{2}\, \partial_b \underset{v}{L}{}_y^{\cdot\,a} + \overset{m'}{\Omega}{}^z_{yb}\, \underset{v}{L}{}_z^{\cdot\,a} + {}'\Omega^a_{cb}\, \underset{v}{L}{}_y^{\cdot\,c} - \overset{m'}{Z}{}_{yx}^{\cdot\cdot\,a}\, \underset{v}{H}{}_b^{\cdot\,x}} .$$

Note that in the anholonomic case neither $\underset{v}{\pounds}\,'\Gamma^a_{cb}$ nor $\underset{v}{\pounds}\,\overset{m}{\Omega}{}^a_{by}$ is a quantity, as could have been foreseen, and that $\underset{v}{\pounds}\,\overset{m}{\Omega}{}^a_{by}$ does not depend on the connexion in L^m_n.

If an L^m_n is rigged in an L_n, there is induced a connexion in L^m_n and we have seen in (Exerc. V 7,1) that in this case $\overset{m}{{}^*R}{}_{dcb}^{\cdot\cdot\cdot a}$ as defined in (V 7.60) with respect to this induced connexion is identical with $\overset{m}{R}{}_{dcb}^{\cdot\cdot\cdot a}$ as defined in (V 7.39). $\underset{v}{\pounds}\,'\Gamma^a_{cb}$ can now also be computed by means of (cf. III 9.2; V 3.10)

$$'\Gamma^a_{cb} = B^{\mu\lambda a}_{cb\varkappa}\,\Gamma^\varkappa_{\mu\lambda} + B^a_\varkappa\,\partial_c\,B^\varkappa_b \tag{3.53}$$

using (3.24, 26, 27; III 5.47) with the result

$$\boxed{\begin{aligned}\underset{v}{\pounds}\,'\Gamma^a_{cb} \overset{*}{=} B^{\mu\lambda a}_{cb\varkappa}\,(\nabla_\mu\,v_{\cdot\lambda}^{\cdot\,\varkappa} + v^\sigma\,R_{\sigma\mu\lambda}^{\cdot\cdot\cdot\varkappa}) - \\ - \underset{v}{H}{}_b^{\cdot\,y}\,\overset{m}{L}{}_{c\cdot y}^{\cdot a} - \underset{v}{L}{}_x^{\cdot\,a}\,\overset{m}{H}{}_{cb}^{\cdot\cdot x} + \underset{v}{H}{}_c^{\cdot\,y}\,\Gamma^a_{yb}\end{aligned}}\,. \tag{3.54}$$

From (3.46) and (3.54) we get after some calculations the following relation between $\overset{mm'}{{}^*R}{}_{ycb}^{\cdot\cdot\cdot a}$ and $R_{\nu\mu\lambda}^{\cdot\cdot\cdot\varkappa}$,

$$\boxed{\begin{aligned}\overset{mm'}{{}^*R}{}_{ycb}^{\cdot\cdot\cdot a} \overset{*}{=} C^\nu_y\,B^{\mu\lambda a}_{cb\varkappa}\,R_{\nu\mu\lambda}^{\cdot\cdot\cdot\varkappa} - \\ - \mathfrak{D}_c\,(\overset{m}{L}{}_{b\cdot y}^{\cdot a} + 2\,S_{by}^{\cdot\cdot a}) - \overset{m}{H}{}_{cb}^{\cdot\cdot x}\,\overset{m'}{H}{}_{yx}^{\cdot\cdot a} + \overset{m}{L}{}_{c\cdot x}^{\cdot a}\,\overset{m'}{L}{}_{y\cdot b}^{\cdot x}\end{aligned}}\,. \tag{3.55}$$

Exercises.

VII 3,1. Prove that for a V_m in V_n

$$\text{VII 3,1}\,\alpha)\qquad \overset{a}{D}\,C^\varkappa_y = -'g^{\varkappa c}\,(D_c\,''v_y + H_{cdy}\,'v^d) - B^\varkappa_\lambda\,C^\mu_y\,\nabla_\mu\,v^\lambda.$$

VII 3,2. Prove that for a V_m in V_n

$$\text{VII 3,2}\,\alpha)\qquad \left\{\begin{aligned}\overset{a}{D}\,'\Gamma^a_{cb} = {}'v^d\,'K_{dcb}^{\cdot\cdot\cdot a} + D_c\,D_b\,'v^a + 2D_{(c}\,H_{b).x}^{\cdot\,a}\,''v^x - \\ - 'g^{ad}\,D_d\,H_{cbx}\,''v^x.\end{aligned}\right.$$

VII 3,3. Prove that for a V_m in V_n

$$\text{VII 3,3}\,\alpha)\qquad \left\{\begin{aligned}\overset{a}{D}\,'K_{dcba} = \underset{'v}{\pounds}\,'K_{dcba} + 4D_{[d}\,D_{[b}\,H_{c]a]x}\,''v^x - \\ - 2'K_{dc[b}^{\cdot\cdot\cdot e}\,H_{a]ex}\,''v^x.\end{aligned}\right.$$

VII 3,4. Prove that for an X_n^m in X_n

VII 3,4 α) $$\underset{v}{\pounds}\, \overset{m}{Z}{}_{cb}^{\cdot\cdot\varkappa} = \partial_{[c}\, \underset{v}{H}{}_{b]}^{\cdot\varkappa} + \underset{v}{H}{}_{a}^{\cdot\varkappa}\, '\Omega_{cb}^{a} + 2\overset{m'}{\Omega}{}_{y\,[c}^{\varkappa}\, \underset{v}{H}{}_{b]}^{\cdot y}.$$

VII 3,5. Prove that for an X_n^m in L_n

VII 3,5 α) $$\underset{v}{\pounds}\, H_{cb}^{\cdot\cdot\varkappa} = D_c\, \underset{v}{H}{}_{b}^{\cdot\varkappa} + B_{cb}^{\mu\lambda}\, C_{\varkappa}^{x}\, \underset{v}{\pounds}\, \Gamma_{\mu\lambda}^{\varkappa} - \underset{v}{H}{}_{c}^{\cdot y}\, \overset{m'}{L}{}_{y\cdot b}^{\cdot x}.$$

VII 3,6[1]). If the first and second fundamental tensor of a V_{n-1} in R_n admit the same infinitesimal transformation in V_{n-1} there exists at least one relation between the principal curvatures (cf. V § 6).

VII 3,7. (3.52) can also be derived by taking the LIE derivative of

VII 3,7 α) $$\overset{m}{\Omega}{}_{by}^{a} \overset{*}{=} B_b^{\mu}\, C_y^{\lambda}\, \partial_{[\mu} B_{\lambda]}^{a} \quad \text{(cf. V § 7)}.$$

VII 3,8. Prove that (3.54) is in accordance with (3.46) if for the $'\Gamma_{cb}^{a}$ in (3.46) we take the parameters of the induced connexion (use 3.38 and V 7.37).

§ 4. The holonomy group of an L_n.

Let C be a closed curve, in an $\mathfrak{N}(\underset{0}{\xi}{}^{\varkappa})$, through $\underset{0}{\xi}{}^{\varkappa}$ that can be contracted into a point and let $\underset{0}{u}{}^{\varkappa}$ be a vector at $\underset{0}{\xi}{}^{\varkappa}$. Then the components of $\underset{0}{u}{}^{\varkappa}$ after having been displaced parallel along the curve are (cf. VII § 1)

(4.1) $$\underset{*}{u}{}^{\varkappa} = \overset{C}{T}{}_{\cdot\lambda}^{\varkappa}\, \underset{0}{u}{}^{\lambda}$$

where $\overset{C}{T}{}_{\cdot\lambda}^{\varkappa}$ is a tensor depending on the choice of C but not on the choice of $\underset{0}{u}{}^{\varkappa}$ at $\underset{0}{\xi}{}^{\varkappa}$. The homogeneous linear transformations (4.1) for all choices of C form a group, the *holonomy group* of the L_n. If instead of parallel displacements we use CARTAN displacements (cf. III § 2) we get a non-homogeneous linear transformation of the points of the tangent E_n forming the *non-homogeneous holonomy group* of the L_n.[2])

[1]) TONOLO 1943, 1.

[2]) The homogeneous holonomy group in V_n was already used by SCHOUTEN 1918, 2 who remarked that the dimension of the R_N in which the V_n can be imbedded is n more than the number of parameters of this group. But the systematical treatment of the holonomy groups is due to CARTAN, 1924, 2; 1925, 1; 1926, 5. He introduced the name "groupe d'holonomie" though the group vanishes if the E_n moves "holonomically", that is if it always returns to its original position. Moreover the word holonomic or anholonomic gives the false impression that the group has something to do with an anholonomic coordinate system or with an X_n^m in L_n. The term "circuit group" would certainly be more suitable but the name given by CARTAN is so widely spread in literature that it seems better to leave it at that and to warn only against misinterpretation. Some american authors recently used the term "holonomic group" but this gives the false impression that the group itself has some property of holonomy. Of course holonomy groups can also be defined for the more general linear connexions mentioned in III § 2, footnote 1), p. 123 and especially for conformal and projective connexions, cf. CARTAN 1937, 1, p. 278 ff.

If C' is a closed curve in $\mathfrak{N}(\underset{0}{\xi^\varkappa})$ through another point $\underset{1}{\xi^\varkappa}$ and if $\overset{C'}{T}$ (matrix notation) is the transformation belonging to it, we may form a curve through $\underset{0}{\xi^\varkappa}$ by going first from $\underset{0}{\xi^\varkappa}$ to $\underset{1}{\xi^\varkappa}$ along some curve C'' in $\mathfrak{N}(\underset{0}{\xi^\varkappa})$, then along C' and finally back to $\underset{0}{\xi^\varkappa}$ along C''. If then U is the transformation along C'' from $\underset{0}{\xi^\varkappa}$ to $\underset{1}{\xi^\varkappa}$, we get for the whole transformation $U^{-1}\overset{C'}{T}U$ and this must be a transformation of the holonomy group defined at $\underset{0}{\xi^\varkappa}$. Hence the groups defined at $\underset{0}{\xi^\varkappa}$ and $\underset{1}{\xi^\varkappa}$ are isomorphic and this means that there is in fact only one holonomy group for all points of L_n. In fact, if $\overset{C'}{T}$ and $U^{-1}\overset{C'}{T}U$ are described with respect to two local coordinate systems that can be transformed into each other by parallel displacement along C'', the numbers in these descriptions are identical.

If for C we take the boundary of an infinitesimal facet $d f^{\nu\mu}$ at $\underset{0}{\xi^\varkappa}$, the vector $\underset{*}{u^\varkappa}$ is (cf. III § 4)

$$\underset{*}{u^\varkappa} = \underset{0}{u^\varkappa} - \tfrac{1}{2} R_{\nu\mu\lambda}^{\cdot\cdot\cdot\varkappa}\, \underset{0}{u^\lambda}\, d f^{\nu\mu}. \tag{4.2}$$

For every choice of the facet this is an infinitesimal transformation of the holonomy group. But this does not imply that these at most $\binom{n}{2}$ linearly independent infinitesimal transformations generate this group. For the non-homogeneous holonomy group we must use the transformations of the vectors together with the transformation of the point that originally coincided with the contactpoint and that after the displacement has the radiusvector (cf. III § 2)

$$- S_{\nu\mu}^{\cdot\cdot\varkappa}\, d f^{\nu\mu}. \tag{4.3}$$

In order to find all infinitesimal transformations of the homogeneous holonomy group we may use the results of VII § 1.[1]) With the notations introduced in that section we have for a curve arising from C by a displacement $\varepsilon v^\varkappa(t)$ with $v^\varkappa(t_0) = v^\varkappa(t_1) = 0$

$$\underset{*}{u^\varkappa} = \left(\left[{}^{t_1}_{t_0}\right]^\varkappa_\lambda + \varepsilon \overset{g}{D} \left[{}^{t_1}_{t_0}\right]^\varkappa_\lambda\right) \underset{0}{u^\lambda} \tag{4.4}$$

and accordingly the infinitesimal transformation belonging to these two curves is given by the tensor (cf. 1.36)

$$A^\varkappa_\lambda + \varepsilon \left[{}^{t_0}_{t_1}\right]^\varkappa_\sigma \overset{g}{D} \left[{}^{t_1}_{t_0}\right]^\sigma_\lambda = A^\varkappa_\lambda - \varepsilon \int_{t_0}^{t_1} \left[{}^{t}_{t_0}\right]^\tau_\lambda v^\nu \frac{d\xi^\mu}{dt} R_{\nu\mu\tau}^{\cdot\cdot\cdot\sigma} \left[{}^{t_0}_{t}\right]^\varkappa_\sigma dt. \tag{4.5}$$

[1]) There are many other methods. Cf. for instance KANITANI 1952, 1 where the transformations are developed in series.

Now we may take all curves through $\underset{0}{\xi}^\varkappa$ and all deformations $\varepsilon v^\varkappa$, hence every tensor of the form

$$(4.6)\qquad C_{\lambda}^{\cdot\,\varkappa} = \left[{}^{t}_{t_0}\right]_{\lambda}^{\tau} p^\nu \frac{d\xi^\mu}{dt} R_{\nu\mu\tau}^{\cdot\cdot\cdot\,\sigma} \left[{}^{t_0}_{t}\right]_{\sigma}^{\varkappa}$$

is a *generator* of the group (cf. IV § 3), i.e. $A_\lambda^\varkappa + C_\lambda^{\cdot\,\varkappa} dt$ is an infinitesimal transformation of the group, and the same holds for all derivatives of $C_\lambda^{\cdot\,\varkappa}$ with respect to t. In (4.6) $d\xi^\mu/dt$ and $R_{\nu\mu\tau}^{\cdot\cdot\cdot\,\sigma}$ are values at the point $\xi^\varkappa(t)$ and p^ν is an arbitrary vector at that point. Using (1.30, 31) we find for these derivatives

$$(4.7)\qquad \begin{cases} \dfrac{d}{dt} C_\lambda^{\cdot\,\varkappa} = \left[{}^{t}_{t_0}\right]_{\lambda}^{\tau} \left\{\dfrac{d\xi^\omega}{dt} \nabla_\omega p^\nu \dfrac{d\xi^\mu}{dt} R_{\nu\mu\tau}^{\cdot\cdot\cdot\,\sigma}\right\} \left[{}^{t_0}_{t}\right]_{\sigma}^{\varkappa} \\ \quad\vdots \\ \dfrac{d^k}{dt^k} C_\lambda^{\cdot\,\varkappa} = \left[{}^{t}_{t_0}\right]_{\lambda}^{\tau} \left\{\left(\dfrac{d\xi^\omega}{dt} \nabla_\omega\right)^k p^\nu \dfrac{d\xi^\mu}{dt} R_{\nu\mu\tau}^{\cdot\cdot\cdot\,\sigma}\right\} \left[{}^{t_0}_{t}\right]_{\sigma}^{\varkappa} \end{cases}$$

and in particular for $t = t_0$

$$(4.8)\qquad \begin{cases} C_\lambda^{\cdot\,\varkappa} = p^\nu \dfrac{d\xi^\mu}{dt} R_{\nu\mu\lambda}^{\cdot\cdot\cdot\,\varkappa} \\ \dfrac{d}{dt} C_\lambda^{\cdot\,\varkappa} = \dfrac{d\xi^\omega}{dt} \nabla_\omega p^\nu \dfrac{d\xi^\mu}{dt} R_{\nu\mu\lambda}^{\cdot\cdot\cdot\,\varkappa} \\ \quad\vdots \\ \dfrac{d^k}{dt^k} C_\lambda^{\cdot\,\varkappa} = \left(\dfrac{d\xi^\omega}{dt} \nabla_\omega\right)^k p^\nu \dfrac{d\xi^\mu}{dt} R_{\nu\mu\lambda}^{\cdot\cdot\cdot\,\varkappa}. \end{cases}$$

Conversely *all* generators depend linearly on the expressions (4.8) because the expressions (4.7) can be expanded in powers of $t - t_0$ and the coefficients in these series are the expressions (4.8) multiplied by constants. Hence the generators of the group are the quantities

$$(4.9)\qquad p^\nu q^\mu R_{\nu\mu\lambda}^{\cdot\cdot\cdot\,\varkappa}; \quad q^\omega \nabla_\omega p^\nu q^\mu R_{\nu\mu\lambda}^{\cdot\cdot\cdot\,\varkappa}; \ldots; (q^\omega \nabla_\omega)^k p^\nu q^\mu R_{\nu\mu\lambda}^{\cdot\cdot\cdot\,\varkappa}; \ldots$$

for all fields $q^\varkappa$ and all fields $p^\varkappa$ with $p^\varkappa = 0$ at $\underset{0}{\xi}^\varkappa$. That proves that the generators lie in the ${}^\varkappa_\lambda$-domain of the quantities

$$(4.11)\qquad R_{\nu\mu\lambda}^{\cdot\cdot\cdot\,\varkappa}; \quad \nabla_\omega R_{\nu\mu\lambda}^{\cdot\cdot\cdot\,\varkappa}; \ldots; \quad \nabla_{\omega_k \ldots \omega_1} R_{\nu\mu\lambda}^{\cdot\cdot\cdot\,\varkappa}; \ldots.$$

Moreover, as was proved by NIJENHUIS[1]) both domains coincide. Hence

The generators of the holonomy group of an L_n span the ${}^\varkappa_\lambda$-domain of the curvature tensor $R_{\nu\mu\lambda}^{\cdot\cdot\cdot\,\varkappa}$ and its covariant derivatives.

Let $r \leq n^2$ be the rank of this ${}^\varkappa_\lambda$-domain and let the domain be spanned by the r linearly independent tensors $C_{(b)\lambda}^{\quad\cdot\,\varkappa}$; $b = \dot{1}, \ldots, \dot{r}$. Here b is a dead index but it may be considered as a living one belonging to an auxiliary E_r in which every tensor of the domain is represented by a contravariant vector.[1]) Then $C_{b\lambda}^{\cdot\cdot\,\varkappa}$ is a connecting quantity

[1]) NIJENHUIS 1952, 1, Ch. II § 13.

between this E_r and the ${}^{\varkappa}_{\lambda}$-domain. The infinitesimal transformations of the holonomy group are

$$X_b = C_{b\lambda}^{\cdot\cdot\varkappa} x^\lambda \frac{\partial}{\partial x^\varkappa} \tag{4.12}$$

where the $x^\varkappa$ are rectilinear coordinates in the tangent E_n and from this we see that the E_r is identical with the tangent E_r at the point of group space corresponding to the identical transformation (cf. IV § 2). Now $R_{\nu\mu\lambda}^{\cdot\cdot\cdot\varkappa}$ and its covariant derivatives can be written in the form

$$(4.13)\quad \begin{cases} \text{a)} & R_{\nu\mu\lambda}^{\cdot\cdot\cdot\varkappa} = B_{\nu\mu}^{\cdot\cdot b} C_{b\lambda}^{\cdot\cdot\varkappa} \\ & \vdots \\ \text{b)} & \nabla_{\omega_k \ldots \omega_1} R_{\nu\mu\lambda}^{\cdot\cdot\cdot\varkappa} = \overset{k}{B}{}_{\omega_k \ldots \omega_1 \nu\mu}^{\cdot \ldots \cdot\cdot\cdot b} C_{b\lambda}^{\cdot\cdot\varkappa}; \qquad k = 1, 2, \ldots \end{cases}$$

with uniquely determined coefficients B and $\overset{k}{B}$.

To every point of L_n there now belongs also besides the ordinary tangent E_n a tangent E_r independent of this E_n, and covariant differentiation can be extended to the quantities of these E_r's by introducing the condition that the covariant derivative of the connecting quantity $C_{b\lambda}^{\cdot\cdot\varkappa}$ be zero for displacements in L_n.[1]) In fact, if we put

$$0 = \nabla_\mu C_{b\lambda}^{\cdot\cdot\varkappa} = \partial_\mu C_{b\lambda}^{\cdot\cdot\varkappa} - \Gamma_{\mu b}^a C_{a\lambda}^{\cdot\cdot\varkappa} - \Gamma_{\mu\lambda}^\varrho C_{b\varrho}^{\cdot\cdot\varkappa} + \Gamma_{\mu\sigma}^\varkappa C_{b\lambda}^{\cdot\cdot\sigma} \tag{4.14}$$

the $\Gamma_{\mu b}^a$ are determined uniquely because $C_{b\mu}^{\cdot\cdot\varkappa}$ has the maximum b-rank. If the $\Gamma_{\mu b}^a$ are found it follows from (4.13) that

$$(4.15)\quad \begin{cases} \nabla_\omega B_{\nu\mu}^{\cdot\cdot a} = \overset{1}{B}{}_{\omega\nu\mu}^{\cdot\cdot\cdot a} \\ \vdots \\ \nabla_\omega \overset{k}{B}{}_{\omega_k \ldots \omega_1 \nu\mu}^{\cdot \ldots \cdot\cdot\cdot a} = \overset{k+1}{B}{}_{\omega\omega_k \ldots \omega_1 \nu\mu}^{\cdot\cdot \ldots \cdot\cdot\cdot a}; \qquad k = 1, 2, \ldots \end{cases}$$

It is easily proved that the coefficients $\Gamma_{\mu b}^a$ transform in the following way if the coordinates in L_n and the (rectilinear) coordinates in E_r are transformed:

$$\Gamma_{\mu' b'}^{a'} = A_{\mu'}^{\mu} A_{b'a}^{b\,a'} \Gamma_{\mu b}^a + A_a^{a'} \partial_{\mu'} A_{b'}^a . \tag{4.16}$$

For the part $R_{\nu\mu b}^{\cdot\cdot\cdot a}$ of the extended curvature tensor (defined in the ordinary way) concerning the E_r we find

$$R_{\nu\mu b}^{\cdot\cdot\cdot a} = 2\partial_{[\nu} \Gamma_{\mu]b}^a + 2\Gamma_{[\nu|c|}^a \Gamma_{\mu]b}^c . \tag{4.17}$$

Because the covariant derivative of $C_{b\lambda}^{\cdot\cdot\varkappa}$ vanishes, the following identity holds

$$0 = 2\nabla_{[\nu}\nabla_{\mu]} C_{b\lambda}^{\cdot\cdot\varkappa} = - R_{\nu\mu b}^{\cdot\cdot\cdot a} C_{a\lambda}^{\cdot\cdot\varkappa} - R_{\nu\mu\lambda}^{\cdot\cdot\cdot\sigma} C_{b\sigma}^{\cdot\cdot\varkappa} + R_{\nu\mu\tau}^{\cdot\cdot\cdot\varkappa} C_{b\lambda}^{\cdot\cdot\tau}, \tag{4.18}$$

[1]) Cf. SCHOUTEN 1926, 1.

hence, according to (4.13)

(4.19) $$R_{\nu\mu b}^{\cdot\cdot\cdot a} C_{a\lambda}^{\cdot\cdot\varkappa} = B_{\nu\mu}^{\cdot\cdot c} (C_{c\lambda}^{\cdot\cdot\sigma} C_{b\sigma}^{\cdot\cdot\varkappa} - C_{b\lambda}^{\cdot\cdot\sigma} C_{c\sigma}^{\cdot\cdot\varkappa}).$$

Since $C_{b\lambda}^{\cdot\cdot\varkappa}$ has maximum b-rank, this implies that there exist uniquely determined coefficients $c_{cb}^{\cdot\cdot a}$ such that

(4.20) $$\begin{cases} \text{a)} \quad C_{c\lambda}^{\cdot\cdot\sigma} C_{b\sigma}^{\cdot\cdot\varkappa} - C_{b\lambda}^{\cdot\cdot\sigma} C_{c\sigma}^{\cdot\cdot\varkappa} = c_{cb}^{\cdot\cdot a} C_{a\lambda}^{\cdot\cdot\varkappa} \\ \text{b)} \quad R_{\nu\mu b}^{\cdot\cdot\cdot a} = B_{\nu\mu}^{\cdot\cdot c} c_{cb}^{\cdot\cdot a}. \end{cases}$$

It follows from (4.12, 20a) that

(4.21) $$(X_c X_b) = 2 C_{[c|\lambda|}^{\cdot\cdot\sigma} x^\lambda \frac{\partial}{\partial x^\sigma} C_{b]\mu}^{\cdot\cdot\tau} x^\mu \frac{\partial}{\partial x^\tau} = c_{cb}^{\cdot\cdot a} X_a$$

and this proves that the $c_{cb}^{\cdot\cdot a}$ are at each point of L_n the structural constants of the holonomy group at that point. In fact they satisfy the identities (cf. IV § 2)

(4.22) $$\text{a)} \quad c_{(cb)}^{\cdot\cdot\cdot a} = 0 \qquad\qquad \text{b)} \quad c_{[dc}^{\cdot\cdot e} c_{b]e}^{\cdot\cdot a} = 0$$

as can be checked easily by means of (4.20a). By covariant differentiation of (4.20a) we get because of (4.14)

(4.23) $$\nabla_\omega c_{cb}^{\cdot\cdot a} = 0.$$

The $c_{cb}^{\cdot\cdot a}$ are not in general constants throughout L_n as follows from the fact that the coordinates in the tangent E_r's are chosen quite arbitrarily at every point. But if, starting from $\underset{0}{\xi}^\varkappa$ we displace the local coordinate system in E_r parallel from $\underset{0}{\xi}^\varkappa$ along some curve to an arbitrary point $\xi^\varkappa$, the $c_{cb}^{\cdot\cdot a}$ remain constant. If the system returns to $\underset{0}{\xi}^\varkappa$ by some way or other, always moving parallel, the local coordinate system has suffered in the end a linear homogeneous transformation but the initial and the final values of $c_{cb}^{\cdot\cdot a}$ are the same. Hence the coordinate systems in the tangent E_r's can always be chosen in such a way that the $c_{cb}^{\cdot\cdot a}$ are constants over L_n, but it is not necessary to do so.

Now let C be a closed curve through $\underset{0}{\xi}^\varkappa$ and let the tangent E_n and the tangent E_r at $\underset{0}{\xi}^\varkappa$ be displaced parallel along C until they are back at $\underset{0}{\xi}^\varkappa$. Let a general vector $u^\varkappa$ at $\underset{0}{\xi}^\varkappa$ be changed by this process into

(4.24) $$'u^\varkappa = T_\lambda^{\cdot\varkappa} u^\lambda;$$

then T is a general transformation of the holonomy group and every transformation of this group can be obtained in this way. If $C_\lambda^{\cdot\varkappa}$ is any generator of the group at $\underset{0}{\xi}^{\varkappa}$, this quantity changes into

$$(4.25)\qquad 'C_{\lambda}^{\cdot\varkappa} = \overset{-1}{T}{}_{\lambda}^{\cdot\sigma}\, C_{\sigma}^{\cdot\varrho}\, T_{\varrho}^{\cdot\varkappa},$$

and this equation thus represents a transformation of the linear adjoint group, that is the adjoint group applied to the infinitesimal transformations of the holonomy group only (cf. IV § 2). Now for every choice of $C_{\lambda}^{\cdot\varkappa}$ we always have $C_{\lambda}^{\cdot\varkappa} = e^b C_{b\lambda}^{\cdot\cdot\varkappa}$ with uniquely determined constants e^b and the infinitesimal transformation represented by $C_{\lambda}^{\cdot\varkappa}$ is at the same time represented by the vector e^b in the tangent E_r. The quantity $C_{b\lambda}^{\cdot\cdot\varkappa}$ is covariant constant, i.e. it remains unchanged if it is displaced parallel along C. Hence we may write $'C_{\lambda}^{\cdot\varkappa} = {}'e^b C_{b\lambda}^{\cdot\cdot\varkappa}$ and then we have

$$(4.26)\qquad 'e^b\, C_{b\lambda}^{\cdot\cdot\varkappa} = \overset{-1}{T}{}_{\lambda}^{\cdot\sigma}\, e^b\, C_{b\sigma}^{\cdot\cdot\varrho}\, T_{\varrho}^{\cdot\varkappa}.$$

The b-rank of $C_{b\lambda}^{\cdot\cdot\varkappa}$ has the highest possible value r and therefore at least one quantity $\overset{-1}{C}{}^{a\varkappa}_{\cdot\cdot\lambda}$ must exist such that

$$(4.27)\qquad C_{b\varrho}^{\cdot\cdot\sigma}\, \overset{-1}{C}{}^{a\varrho}_{\cdot\cdot\sigma} = A_b^a.$$

By means of this quantity we get from (4.26)

$$(4.28)\qquad 'e^a = \overset{-1}{T}{}_{\lambda}^{\cdot\sigma}\, C_{b\sigma}^{\cdot\cdot\varrho}\, T_{\varrho}^{\cdot\varkappa}\, \overset{-1}{C}{}^{a\lambda}_{\cdot\cdot\varkappa}\, e^b$$

and this equation expresses how the vector e^a is transformed if the infinitesimal transformations of the holonomy group suffer a transformation T of the linear adjoint group.

In the same way as above it could now be proved that the generators of the linear adjoint group span the $\overset{a}{b}$-domain of the set

$$(4.29)\qquad R_{\nu\mu b}^{\cdot\cdot\cdot a};\qquad \nabla_{\omega} R_{\nu\mu b}^{\cdot\cdot\cdot a};\ \ldots;\ \nabla_{\omega_k \ldots \omega_1} R_{\nu\mu b}^{\cdot\cdot\cdot a};\ \ldots$$

and, according to (4.14, 15, 20b) also the $\overset{a}{b}$-domain of the set

$$(4.30)\qquad B_{\nu\mu}^{\cdot\cdot c}\, c_{cb}^{\cdot\cdot a};\qquad \overset{1}{B}{}_{\omega\nu\mu}^{\cdot\cdot\cdot c}\, c_{cb}^{\cdot\cdot a};\ \ldots;\ \overset{k}{B}{}_{\omega_k \ldots \omega_1 \nu\mu}^{\cdot\ \ldots\ \cdot\cdot\cdot c}\, c_{cb}^{\cdot\cdot a};\ \ldots.$$

But because the set of all B's has the maximum c-rank r, this proves that the generators of the linear adjoint group span the $\overset{a}{b}$-domain of $c_{cb}^{\cdot\cdot a}$ exactly, a result well known from IV § 2.

The holonomy group of an L_n is intimately connected with covariant constant fields of quantities. We prove the theorem[1]):

[1]) WAGNER 1943, 1; 1946, 2. This paper deals with general geometric objects. The proof is from NIJENHUIS 1952, 1, Ch. II § 13.

A covariant constant field of a quantity with a given kind of transformation in an L_n is possible if and only if there exists a quantity of this kind which is invariant for the holonomy group.

The proof is given here for a tensor $P^{\kappa}_{\cdot\lambda}$ of valence two. The generalization is obvious. The equation

$$(4.31)\qquad \nabla_\mu P^{\kappa}_{\cdot\lambda} = 0$$

has the first integrability condition

$$(4.32)\qquad R_{\nu\mu\sigma}^{\cdot\cdot\cdot\kappa} P^{\sigma}_{\cdot\lambda} - R_{\nu\mu\lambda}^{\cdot\cdot\cdot\varrho} P^{\kappa}_{\cdot\varrho} = 0$$

and the following integrability conditions are obtained by successive covariant differentiation of (4.32) (cf. III § 6):

$$(4.33)\qquad (\nabla_{\omega_k \dots \omega_1} R_{\nu\mu\sigma}^{\cdot\cdot\cdot\kappa}) P^{\sigma}_{\cdot\lambda} - (\nabla_{\omega_k \dots \omega_1} R_{\nu\mu\lambda}^{\cdot\cdot\cdot\varrho}) P^{\kappa}_{\cdot\varrho} = 0; \qquad k = 1, 2 \dots .$$

Because of (4.13) the conditions (4.32, 33) are equivalent to

$$(4.34)\qquad C_{b\sigma}^{\cdot\cdot\kappa} P^{\sigma}_{\cdot\lambda} - C_{b\lambda}^{\cdot\cdot\sigma} P^{\kappa}_{\cdot\sigma} = 0$$

and this equation expresses the invariance of $P^{\kappa}_{\cdot\lambda}$ for the holonomy group.[1])

If the L_n is a V_n special results can be obtained. Instead of (4.13 a) we get

$$(4.35)\qquad K_{\nu\mu\lambda\kappa} = B_{\nu\mu}^{\cdot\cdot b} C_{b\lambda\kappa};$$

hence, according to the fourth identity (cf. III § 5)

$$(4.36)\qquad B_{\nu\mu}^{\cdot\cdot b} C_{b\lambda\kappa} = B_{\lambda\kappa}^{\cdot\cdot b} C_{b\nu\mu}.$$

[1]) There are a great number of investigations on holonomy groups that leave certain geometric objects invariant. We give here some references. *Symmetric tensor of valence two in V_n:* LEVY 1925, 2; SAMUEL 1945, 1; LICHNEROWICZ 1950, 1; *Bivector in V_n:* IWAMOTO 1950, 1; LICHNEROWICZ 1950, 2. *Multivector in V_n:* LICHNEROWICZ 1951, 1; 2; IWAMOTO 1951, 1; *E_p-field in A_n and V_n:* ABE 1944, 1; 2; RUSE 1949, 1; 1950, 1; WALKER 1949, 1; 1950, 3; 4; YANO and SASAKI 1948, 3; WILLMORE 1951, 1; *Pointfield in V_n:* TACHIBANA 1949, 1; *Hyperquadric in a normal projective connexion:* SASAKI and YANO 1949, 1; ÔTSUKI 1950, 1; 1952, 1; *Point or hypersphere in a normal conformal connexion:* SASAKI 1943, 1; 2; 3; YANO and SASAKI 1947, 2. Connected with these investigations are a number of papers on classification of connexions by means of the holonomy group: SASAKI 1942, 1 (conform.); 1943, 1; 2; 3 (conform.); 1951, 1 (one-parameter groups); 1951, 2 (global); YANO 1943, 5 (conform.); YANO and SASAKI 1944, 13 (conform.); 1946, 4 (conform.); 1947, 2 (conform.); 1948, 3; WAGNER 1949, 2 (L_n); ÔTSUKI 1949, 1; 2 (V_4); 1950, 2 (conform.); 1950, 3 (W_4); 1951, 1 (conform.); LIBER 1949, 1 (one-parameter groups); DEBEVER 1950, 1 (conform.); KANITANI 1950, 1; 1952, 1 (projective); KUIPER 1951, 2; TAKIZAWA 1951, 1; HASHIMOTO 1951, 1; KURITA 1952, 1; WAKAKUWA 1952, 1; NIJENHUIS 1953, 1. Further literature is given in these papers.

But this implies that the $\nu\mu$-domain of $B_{\nu\mu}^{\cdot\cdot b}$ and of $C_{b\nu\mu}$ are identical. Hence there must exist equations of the form

$$B_{\nu\mu}^{\cdot\cdot a} = a^{ab} C_{b\nu\mu} \tag{4.37}$$

where a^{ab} is a uniquely determined *symmetric* tensor that can be computed from (4.35). Since the b-rank of $C_{b\nu\mu}$ is r, the rank of a^{ab} is equal to the a-rank of $B_{\nu\mu}^{\cdot\cdot a}$, and to the $\lambda\varkappa$-rank of $K_{\nu\mu\lambda\varkappa}$.

The rank of the symmetric tensor a^{ab} has the maximum value r if and only if the $\lambda\varkappa$-rank of $K_{\nu\mu\lambda\varkappa}$ equals r, or in other words, if the $\lambda\varkappa$-domain of $\nabla_\omega K_{\nu\mu\lambda\varkappa}$ lies in the $\lambda\varkappa$-domain of $K_{\nu\mu\lambda\varkappa}$.[1])

From (4.37) and (4.13) it follows that

$$K_{\nu\mu\lambda\varkappa} = a^{ab} C_{a\nu\mu} C_{b\lambda\varkappa} \tag{4.38}$$

and

$$\overset{k}{B}{}_{\omega_k\ldots\omega_1\nu\mu}^{\cdot\ \cdot\ \cdot\ \cdot\ \cdot a} = C_{b\nu\mu} \nabla_{\omega_k}\ldots{}_{\omega_1} a^{ab} \tag{4.39}$$

and from this latter equation we see that the a-rank of the set consisting of a^{ab} and all its covariant derivatives equals r.

As an example we consider a V_n with recurrent curvature tensor (cf. VIII § 9). This is a V_n with a vector field k_ω such that

$$\nabla_\omega K_{\nu\mu\lambda\varkappa} = k_\omega K_{\nu\mu\lambda\varkappa}; \quad k_\omega \neq 0. \tag{4.40}$$

From (4.38, 40) it follows that

$$\nabla_\omega a^{ab} = k_\omega a^{ab} \tag{4.41}$$

and from BIANCHI's identity, because a^{ab} has rank r

$$C_{b[\nu\mu} k_{\omega]} = 0. \tag{4.42}$$

But this equation expresses that for every choice of b the bivector C_b is simple and has a 2-direction containing the direction of k_ω. This implies that $C_{b\nu\mu}$ can be written in the form

$$C_{b\nu\mu} = 2u_{b[\nu} k_{\mu]} \tag{4.43}$$

in which $u_{b\nu}$ is uniquely determined to within an additive term of the form $u_b k_\nu$ with an arbitrary vector u_b. Now we may derive from (4.40)

$$2\nabla_{[\varrho\sigma]} K_{\nu\mu\lambda\varkappa} = 2(\nabla_{[\sigma} k_{\varrho]}) K_{\nu\mu\lambda\varkappa} \tag{4.44}$$

but on the other hand we have (cf. III 4.9 a)

$$2\nabla_{[\sigma\varrho]} K_{\nu\mu\lambda\varkappa} = -2K_{\sigma\varrho[\nu}^{\cdot\ \cdot\ \cdot\ \tau} K_{|\tau|\mu]\lambda\varkappa} - 2K_{\sigma\varrho[\lambda}^{\cdot\ \cdot\ \cdot\ \tau} K_{|\nu\mu\tau|\varkappa]}. \tag{4.45}$$

---

[1]) NIJENHUIS, personal communication.

Hence $\nabla_{[\sigma} k_{\varrho]}$ lies in the $\sigma\varrho$-domain of $K_{\sigma\varrho\lambda\varkappa}$, and thus in the $\sigma\varrho$-domain of $C_{b\sigma\varrho}$ also, and this implies that there is an equation of the form

(4.46) $$\nabla_{[\sigma} k_{\varrho]} = {'k}^b C_{b\sigma\varrho} = 2\,{'k}^b u_{b[\sigma} k_{\varrho]}.$$

This proves that k_λ is a gradient to within a scalar factor (cf. II § 7). Substituting (4.46) and (4.38) in (4.44, 45) we get

(4.47) $$\left\{\begin{aligned} {'k}^a C_{a\sigma\varrho} a^{cd} C_{c\nu\mu} C_{d\lambda\varkappa} = &- a^{ab} C_{a\sigma\varrho} C_{b}{}^{\cdot}{}_{[\nu}{}^{\cdot\tau} a^{cd} C_{|c\tau|\mu]} C_{d\lambda\varkappa} - \\ &- a^{ab} C_{a\sigma\varrho} C_{b}{}^{\cdot}{}_{[\lambda}{}^{\cdot\tau} a^{cd} C_{|c\nu\mu} C_{d\tau|\varkappa]} \end{aligned}\right.$$

from which it follows immediately by writing out

(4.48) $${'k}^{(a} a^{cd)} C_{a\sigma\varrho} C_{c\nu\mu} C_{d\lambda\varkappa} = 0$$

which is only possible if

(4.49) $${'k}^{(a} a^{cd)} = 0$$

that is, if ${'k}^a$ is zero. This proves *that k_λ is a gradientvector.*[1]) If we write $k_\lambda = -\nabla_\lambda \log \sigma$ we get from (4.41)

(4.50) $$\nabla_\mu \sigma a^{ab} = 0.$$

The $C_{b}{}^{\cdot}{}_{\lambda}{}^{\cdot\varkappa}$; $b = \dot{1}, \ldots, \dot{r}$ are the generators of the holonomy group, hence, from (4.20a) and (4.43)

(4.51) $$\left\{\begin{aligned} &(u_{c\lambda} k^\sigma - u_{c}{}^{\cdot\sigma} k_\lambda)(u_{b\sigma} k^\varkappa - u_{b}{}^{\cdot\varkappa} k_\sigma) - \\ &\quad - (u_{b\lambda} k^\sigma - u_{b}{}^{\cdot\sigma} k_\lambda)(u_{c\sigma} k^\varkappa - u_{c}{}^{\cdot\varkappa} k_\sigma) = c_{cb}^{\cdot\cdot a}(u_{a\lambda} k^\varkappa - u_{a}{}^{\cdot\varkappa} k_\lambda). \end{aligned}\right.$$

Now the r vectors $u_{b\lambda}$; $b = \dot{1}, \ldots, \dot{r}$ and k_λ are linearly independent because $C_{b\nu\mu}$ has b-rank r (cf. 4.43). Hence (4.51) is equivalent to

(4.52) $$\left\{\begin{aligned} &\text{a)}\quad k_\sigma k^\sigma u_{[c|\lambda|} u_{b]}{}^{\cdot\varkappa} = 0 \\ &\text{b)}\quad 2u_{[c}{}^{\cdot\sigma} u_{b]}{}^{\cdot\varkappa} k_\lambda k_\sigma = c_{cb}^{\cdot\cdot a} u_{a}{}^{\cdot\varkappa} k_\lambda. \end{aligned}\right.$$

If $k_\sigma k^\sigma \neq 0$ this implies that

(4.53) $$u_{[c|\lambda|} u_{b]}{}^{\cdot\varkappa} = 0; \qquad k_\sigma k^\sigma \neq 0$$

and if this substituted in (4.52b) we get $c_{cb}^{\cdot\cdot a} = 0$ because $u_{b}{}^{\cdot\varkappa}$ has b-rank r. Now for $r > 1$, (4.53) would imply that $u_{\dot{1}}{}^{\cdot\lambda}$ and $u_{\dot{2}}{}^{\cdot\lambda}$ were linearly dependent, hence $r = 1$. This proves[2])

[1]) WALKER 1950, 1, the proof given here is from NIJENHUIS (personal communication). Cf. RUSE 1946, 3; 1948, 2; 1949, 2. If the scalar curvature is not zero the proposition can also be proved by transvecting (4.40) with $g^{\nu\varkappa} g^{\mu\lambda}$.

[2]) NIJENHUIS, personal communication.

The holonomy group of a V_n with a recurrent curvature tensor is a one-parameter group if $k^\varkappa$ is not a nullvector.

Now let us suppose that $k^\varkappa$ is not a nullvector and that the 2-direction of the simple bivector $C_b^{\cdot\varkappa\lambda}$; $b=\dot{1}$ is not tangent to the nullcone. Then the holonomy group consists of all rotations in the tangent R_n around this 2-direction. This means that the 2-direction and the $(n-2)$-direction perpendicular to it are covariant constant because both are invariant for the holonomy group. But this implies that the V_n is a product $V_2 \times V_{n-2}$ (cf. V § 11). The V_2's are geodesic and parallel and their tangent R_2 is spanned by $u_b^{\cdot\nu}$ and k^ν (cf. 4.43). The V_{n-2}'s are also geodesic and parallel and their curvature tensor is the V_{n-2}-part of the curvature tensor in V_n (cf. V § 10, GAUSS). But according to (4.38) this part is zero. Hence

If in a V_n with a recurrent curvature tensor the vector k_λ is not a nullvector and if the simple bivector of the rotations of the one-parameter holonomy group is not tangent to the nullcone, the V_n is a product $V_2 \times R_{n-2}$.[1])

Note that accordingly every *ordinary* V_n with a recurrent curvature tensor is a product $V_2 \times R_{n-2}$.[2])

Exercise.

VII 4,1[3]). The non-homogeneous holonomy group of a V_n leaves a point invariant if and only if there exists a concurrent contravariant vector field (cf. VI § 9).

It leaves a direction invariant if and only if there exists a covariant constant contravariant vector field.

§ 5. Affine motions and the holonomy group in a symmetric A_n.

In a symmetric A_n (cf. III § 7) the covariant derivative of $R_{\nu\mu\lambda}^{\cdot\cdot\cdot\varkappa}$ vanishes and consequently the holonomy group is generated by the $\overset{\varkappa}{\lambda}$-domain of $R_{\nu\mu\lambda}^{\cdot\cdot\cdot\varkappa}$ only. This means that in the factorization (4.13a)

$$(5.1) \qquad R_{\nu\mu\lambda}^{\cdot\cdot\cdot\varkappa} = B_{\nu\mu}^{\cdot\cdot a} C_{a\lambda}^{\cdot\cdot\varkappa}$$

the a-rank of $B_{\nu\mu}^{\cdot\cdot a}$ and the a-rank of $C_{a\lambda}^{\cdot\cdot\varkappa}$ both are maximum and equal to the number r of parameters of the holonomy group. Hence the quan-

[1]) WALKER 1950, 1, p. 46. In this paper the author gave a detailed discussion of all other cases. He proved that the V_n can also be a product $V_3 \times R_{n-3}$ or $V_4 \times R_{n-4}$ where V_3 and V_4 have also recurrent curvature tensors. The proof given here is from NIJENHUIS (personal communication).

[2]) We mention some further recent publications on holonomy groups: LAPTEV 1950, 1; SASAKI 1951, 2 (symmetric spaces); KANITANI 1952, 1; BOREL and LICHNEROWICZ 1952, 2 (V_n).

[3]) YANO 1943, 3.

tities $\overset{k}{B}$ in (4.13b) vanish. The equations (4.13a), (4.14) and (4.16–23) remain valid but instead of (4.15) we have

$$\nabla_\omega B_{\nu\mu}^{\cdot\cdot a} = 0 \tag{5.2}$$

and there is another identity

$$C_{c\nu}^{\cdot\cdot\lambda} B_{\mu\lambda}^{\cdot\cdot a} - C_{c\mu}^{\cdot\cdot\lambda} B_{\nu\lambda}^{\cdot\cdot a} = c_{cb}^{\cdot\cdot a} B_{\nu\mu}^{\cdot\cdot b} \tag{5.3}$$

that can be derived from $\nabla_{[\omega}\nabla_{\nu]} B_{\mu\lambda}^{\cdot\cdot a} = 0$ because the a-rank of $B_{\nu\mu}^{\cdot\cdot a}$ is r.

The equations for the group of affine motions are now [cf. (2.1)]

$$\text{a)}\quad \nabla_\mu v_\lambda^{\cdot\varkappa} = -v^\nu R_{\nu\mu\lambda}^{\cdot\cdot\cdot\varkappa} \qquad \text{b)}\quad \nabla_\lambda v^\varkappa = v_\lambda^{\cdot\varkappa} \tag{5.4}$$

and the integrability conditions [cf. (2.2, 3)] reduce to

$$v_\nu^{\cdot\sigma} R_{\sigma\mu\lambda}^{\cdot\cdot\cdot\varkappa} + v_\mu^{\cdot\sigma} R_{\nu\sigma\lambda}^{\cdot\cdot\cdot\varkappa} + v_\lambda^{\cdot\sigma} R_{\nu\mu\sigma}^{\cdot\cdot\cdot\varkappa} - v_\varrho^{\cdot\varkappa} R_{\nu\mu\lambda}^{\cdot\cdot\cdot\varrho} = 0. \tag{5.5}$$

This equation expresses the fact that in the tangent E_n the tensor $R_{\nu\mu\lambda}^{\cdot\cdot\cdot\varkappa}$ is invariant for the infinitesimal transformation $x^\varkappa \to x^\varkappa + v_\lambda^{\cdot\varkappa} x^\lambda dt$. If (5.1) is substituted in (5.5) we get

$$(v_\nu^{\cdot\sigma} B_{\sigma\mu}^{\cdot\cdot a} + v_\mu^{\cdot\sigma} B_{\nu\sigma}^{\cdot\cdot a}) C_{a\lambda}^{\cdot\cdot\varkappa} + B_{\nu\mu}^{\cdot\cdot a} (v_\lambda^{\cdot\sigma} C_{a\sigma}^{\cdot\cdot\varkappa} - v_\varrho^{\cdot\varkappa} C_{a\lambda}^{\cdot\cdot\varrho}) = 0. \tag{5.6}$$

But because $C_{a\lambda}^{\cdot\cdot\varkappa}$ and $B_{\nu\mu}^{\cdot\cdot a}$ both have maximum a-rank, (5.6) can only be true if there exist uniquely determined coefficients $v_b^{\cdot a}$ such that

$$\text{a)}\quad v_\mu^{\cdot\varrho} B_{\varrho\lambda}^{\cdot\cdot a} + v_\lambda^{\cdot\varrho} B_{\mu\varrho}^{\cdot\cdot a} - v_b^{\cdot a} B_{\mu\lambda}^{\cdot\cdot b} = 0 \qquad \text{b)}\quad v_c^{\cdot a} C_{a\lambda}^{\cdot\cdot\varkappa} + v_\lambda^{\cdot\varrho} C_{c\varrho}^{\cdot\cdot\varkappa} - v_\sigma^{\cdot\varkappa} C_{c\lambda}^{\cdot\cdot\sigma} = 0. \tag{5.7}$$

The $v_b^{\cdot a}$ are the components of a tensor in the auxiliary tangent E_r's introduced in VII § 4 and (5.7) expresses the fact that in the tangent E_{n+r} spanned by the tangent E_n and the tangent E_r the quantities $B_{\mu\lambda}^{\cdot\cdot a}$ and $C_{c\lambda}^{\cdot\cdot\varkappa}$ are invariant for the infinitesimal transformation

$$x^\varkappa \to x^\varkappa + v_\lambda^{\cdot\varkappa} x^\lambda dt; \qquad x^a \to x^a + v_b^{\cdot a} x^b dt \tag{5.8}$$

in the tangent E_{n+r}. Moreover (5.7) is another form of the integrability conditions (5.5). As to $C_{a\lambda}^{\cdot\cdot\varkappa}$ this means that the holonomy group is invariant for all affine motions. Of course this is also true in a general A_n because a motion does not change the geometry in the space and consequently can not change the holonomy group. If p^a is arbitrary,

$p^a C_{a\lambda}^{\cdot\cdot\varkappa}$ is an arbitrary generator of the group. Taking the LIE derivative of this quantity we get according to (5.7b)

$$\underset{v}{£}\, p^a C_{a\lambda}^{\cdot\cdot\varkappa} = C_{a\lambda}^{\cdot\cdot\varkappa} (v^\mu \nabla_\mu p^a - p^c v_c^{\cdot a}) \tag{5.9}$$

and this proves in fact that the totality of all infinitesimal transformations of the holonomy group is invariant for affine motions.

If ∇_μ is applied to (5.7b) it follows, using (5.4a) and (4.18), that

$$\boxed{\nabla_\mu v_b^{\cdot a} = - v^\nu R_{\nu\mu b}^{\cdot\cdot\cdot a}} \,. \tag{5.10}$$

From (5.7) and (4.20a) it can be deduced that

$$\boxed{v_c^{\cdot d} c_{db}^{\cdot\cdot a} + v_b^{\cdot d} c_{cd}^{\cdot\cdot a} - v_d^{\cdot a} c_{cb}^{\cdot\cdot d} = 0} \tag{5.11}$$

and this equation expresses the fact that in the tangent E_r the quantity $c_{cb}^{\cdot\cdot a}$ is invariant for the infinitesimal transformation $x^a \rightarrow x^a + v_b^{\cdot a} x^b dt$.

The equations (5.4, 10) and (5.7) are equivalent to (5.4, 5) and thus (cf. VII § 2) form a totally integrable system with the unknowns $v^\varkappa$, $v_\lambda^{\cdot\varkappa}$ and $v_b^{\cdot a}$. In order to prove that the solutions of this system generate a group we can use the same way as was followed in VII § 2. Thus we extend the LIE operator $\underset{v}{£}$ to the X_{n+r} formed by the ∞^n tangent E_r's. Of course this can be done in several ways, but if we put the condition that $\underset{v}{£}\, C_{c\lambda}^{\cdot\cdot\varkappa} = 0$ we get uniquely for a vector of E_r [for instance for the contravariant vector p^a from (5.9)]

$$(5.12)\quad \begin{cases} \text{a)} & \underset{v}{£}\, p^a = v^\mu \nabla_\mu p^a - v_b^{\cdot a} p^b = v^\mu \partial_\mu p^a + v^\mu \Gamma_{\mu b}^a p^b - v_b^{\cdot a} p^b \\ \text{b)} & \underset{v}{£}\, q_b = v^\mu \nabla_\mu q_b + v_b^{\cdot a} q_a = v^\mu \partial_\mu q_b - v^\mu \Gamma_{\mu b}^a q_a + v_b^{\cdot a} q_a . \end{cases}$$

It is interesting to look at these equations from the point of view of the X_{n+r}. The general formula for the LIE-derivative of a vector u^A; $A = 1, \ldots, n, \dot{1}, \ldots, \dot{r}$ in this space with respect to the field v^A, resulting from $v^\varkappa$ and v^a, is

$$(5.13)\quad \begin{cases} \text{a)} & \underset{v}{£}\, u^\varkappa = v^\mu \partial_\mu u^\varkappa + v^c \partial_c u^\varkappa - u^\mu \partial_\mu v^\varkappa - u^c \partial_c v^\varkappa \\ \text{b)} & \underset{v}{£}\, u^a = v^\mu \partial_\mu u^a + v^c \partial_c u^a - u^\mu \partial_\mu v^a - u^c \partial_c v^a . \end{cases}$$

For a field u^A with $u^a = 0$, only dependent on the $\xi^\varkappa$, this gives the ordinary expression for $\underset{v}{£}\, u^\varkappa$. But for a field with $u^\varkappa = 0$ and u^a depending on the $\xi^\varkappa$ only, we get

$$(5.14)\qquad \begin{cases} \text{a)} & \underset{v}{\pounds}\, u^\varkappa = 0 \\ \text{b)} & \underset{v}{\pounds}\, u^a = v^\mu\, \partial_\mu\, u^a - u^c\, \partial_c\, v^a \end{cases}$$

because $v^\varkappa$ depends on the $\xi^\varkappa$ only. Now (5.14b) is in accordance with (5.12a) if

$$(5.15)\qquad v_b^{\cdot a} = \partial_b\, v^a + v^\mu\, \Gamma^a_{\mu b}.$$

This means that the LIE derivative $\underset{v}{\pounds}\, p^a$ in (5.12a) is the same as would have been found if the vector field $v^\varkappa$ were extended with a field v^a, dependent on the $\xi^\varkappa$ and the x^a, to $v^\varkappa, v^a$, in such a way that $\partial_b v^a$ were satisfying (5.15). This implies that

$$(5.16)\qquad \underset{v}{\pounds}\, \nabla_\mu - \nabla_\mu\, \underset{v}{\pounds} = 0$$

for all quantities of X_{n+r} but this equation can also be derived directly from (4.14, 22) and (5.1, 7, 11). It is very remarkable that v^a itself does not occur in (5.14b) but only its derivative $\partial_b v^a$. Hence, if $\underset{*}{v}^a$ is an arbitrary field not depending on the x^a, the field $v^a = \underset{*}{v}^a + (v_b^{\cdot a} - v^\mu\, \Gamma^a_{\mu b})\, x^b$ satisfies the condition (5.15).

The group property can now be proved by showing that $w^\varkappa \overset{\text{def}}{=} \underset{u}{\pounds}\, v^\varkappa$ is a solution of (5.4, 7, 10, 11) if $u^\varkappa$ and $v^\varkappa$ are solutions. This is done by substituting in these equations

$$(5.17)\qquad \begin{cases} \text{a)} & w^\varkappa \overset{\text{def}}{=} \underset{u}{\pounds}\, v^\varkappa = u^\mu\, \nabla_\mu\, v^\varkappa - v^\mu\, \nabla_\mu\, u^\varkappa = u^\mu\, v_\mu^{\cdot \varkappa} - v^\mu\, u_\mu^{\cdot \varkappa} \\ \text{b)} & w_\lambda^{\cdot \varkappa} \overset{\text{def}}{=} \nabla_\lambda\, w^\varkappa = \nabla_\lambda\, \underset{u}{\pounds}\, v^\varkappa = \underset{u}{\pounds}\, v_\lambda^{\cdot \varkappa} = u^\nu\, v^\mu\, R_{\nu\mu\lambda}^{\cdot\cdot\cdot\varkappa} + v_\sigma^{\cdot \varkappa}\, u_\lambda^{\cdot \sigma} - u_\sigma^{\cdot \varkappa}\, v_\lambda^{\cdot \sigma} \\ \text{c)} & w_b^{\cdot a} \overset{\text{def}}{=} \underset{u}{\pounds}\, v_b^{\cdot a} = u^\nu\, v^\mu\, R_{\nu\mu b}^{\cdot\cdot\cdot a} + v_c^{\cdot a}\, u_b^{\cdot c} - u_c^{\cdot a}\, v_b^{\cdot c}. \end{cases}$$

The equations (5.17) give the composition of two infinitesimal affine motions. If we drop the curvature tensor in (5.17b, c) these equations give the well known composition of affine motions in E_n where $u^\varkappa\, dt$, $v^\varkappa\, dt$ and $w^\varkappa\, dt$ are the displacements and $u_\lambda^{\cdot \varkappa}\, dt$, $v_\lambda^{\cdot \varkappa}\, dt$ and $w_\lambda^{\cdot \varkappa}\, dt$ the affine transformations of the tangent E_n's.[1]) From (5.17a) it follows that the affine motions leaving one arbitrary point $\underset{0}{\xi}^\varkappa$ invariant, form a group. This group is called following CARTAN the *group of isotropy with respect to* $\underset{0}{\xi}^\varkappa$.

We prove now that *the group of affine motions in a symmetric A_n is transitive and that it has at least $n+r$ parameters, where r is the number of parameters of the holonomy group.*

---

[1]) Some authors call them "affine rotations".

The equations (5.4, 7, 10, 11) are totally integrable, hence there exists a solution for every set of initial values $\underset{0}{v}^\varkappa$, $\underset{0}{v}_\lambda^{\cdot\varkappa}$ and $\underset{0}{v}_b^{\cdot a}$ at $\underset{0}{\xi}^\varkappa$, that satisfies (5.7, 11). But because $v^\varkappa$ does not occur in these conditions, $\underset{0}{v}^\varkappa$ can be chosen arbitrarily and this proves that the group is transitive. Now according to (4.20a, 22b; 5.3) the set defined by

$$(5.18)\qquad \underset{0}{v}_\lambda^{\cdot\varkappa} \overset{\text{def}}{=} \underset{0}{v}^c\, C_{c\lambda}^{\cdot\cdot\varkappa}; \qquad \underset{0}{v}_b^{\cdot a} \overset{\text{def}}{=} \underset{0}{v}^c\, c_{cb}^{\cdot\cdot a};$$

satisfies (5.7, 11) for every choice of the constants $\underset{0}{v}^a$. This proves that the group has at least $n+r$ parameters. The group with the parameters $\underset{0}{v}^\varkappa$, $\underset{0}{v}^a$ defined by (5.18) is an $(n+r)$-parameter subgroup of the group of affine motions. First we remark that (5.18) implies that equations of the same form

$$(5.19)\qquad \text{a)}\ v_\lambda^{\cdot\varkappa} = v^c\, C_{c\lambda}^{\cdot\cdot\varkappa}; \qquad \text{b)}\ v_b^{\cdot a} = v^c\, c_{cb}^{\cdot\cdot a}$$

with suitable values of v^c, hold at every point of A_n. In fact, according to (5.1, 4, 10) and (4.14, 20b) we have

$$(5.20)\quad \text{a)}\ C_{c\lambda}^{\cdot\cdot\varkappa}\, \nabla_\mu v^c = -v^\nu B_{\nu\mu}^{\cdot\cdot c}\, C_{c\lambda}^{\cdot\cdot\varkappa}; \qquad \text{b)}\ c_{cb}^{\cdot\cdot a}\, \nabla_\mu v^c = -v^\nu B_{\nu\mu}^{\cdot\cdot c}\, c_{cb}^{\cdot\cdot a}$$

or

$$(5.21)\qquad \nabla_\mu v^a = -v^\nu B_{\nu\mu}^{\cdot\cdot a}$$

and the integrability conditions of this equation are identically satisfied because according to (4.20b) and (5.3)

$$(5.22)\quad \begin{cases} 2\nabla_{[\nu}\nabla_{\mu]}v^a = -2(\nabla_{[\nu}v^\omega)\, B_{|\omega|\mu]}^{\cdot\cdot\ a} = -2v^c\, C_{c[\nu}^{\cdot\cdot\omega}\, B_{|\omega|\mu]}^{\cdot\cdot\ a} \\ \qquad = v^c\, c_{cb}^{\cdot\cdot a}\, B_{\nu\mu}^{\cdot\cdot b} = v^b\, c_{bc}^{\cdot\cdot a}\, B_{\nu\mu}^{\cdot\cdot c} = R_{\nu\mu b}^{\cdot\cdot\cdot a}\, v^b. \end{cases}$$

That proves that there is one and only one field v^a that satisfies (5.19, 21) and takes at $\underset{0}{\xi}^\varkappa$ the values $\underset{0}{v}^a$ satisfying (5.18).

Now we prove that the subgroup defined by (5.18) is an *invariant* subgroup of the group of affine motions (cf. VII § 2). Therefore we substitute (5.19) in (5.4, 10) and (5.7, 11). Then (5.7, 11) are satisfied identically as a consequence of (4.20a, 22b; 5.3), and (5.4, 10) reduces to the set

$$(5.23)\qquad \begin{cases} \nabla_\lambda v^\varkappa = v^c\, C_{c\lambda}^{\cdot\cdot\varkappa} \\ \nabla_\mu v^a = -v^\nu B_{\nu\mu}^{\cdot\cdot a} \end{cases}$$

which is of course totally integrable. If now $u^\varkappa$ is the vector of an affine motion not necessarily belonging to the subgroup it follows from (5.17, 23) that $\underset{u}{£}\, v^\varkappa$, $\underset{u}{£}\, v^a$ also satisfy (5.23) and this means that the subgroup is invariant.

Exercise.

VII 5,1. Prove (5.22) by writing out $2\nabla_{[\nu}\nabla_{\mu]}v_\lambda^{;\varkappa}$ in two different ways and using (4.18) (use the rule of LEIBNIZ for $\nabla_{[\nu}\nabla_{\mu]}$).

§ 6. CARTAN's method applied to the holonomy group and the symmetric A_n.

In order to give further examples of CARTAN's symbolical method we apply this method here independently of VII § 4, 5 to the homogeneous holonomy group. The equations of VII § 4, 5 will not be used in this section but results will be compared.

Starting from any point $\underset{0}{\xi}^\varkappa$ of an L_n we fix an anholonomic coordinate system (h) at *all* points of an $\mathfrak{R}(\underset{0}{\xi}^\varkappa)$ by parallel displacement of some given local coordinate system in the tangent E_n of $\underset{0}{\xi}^\varkappa$ along *any definitely chosen* curves to all points of $\mathfrak{R}(\underset{0}{\xi}^\varkappa)$. If this local system at $\underset{0}{\xi}^\varkappa$ suffers a linear homogeneous transformation, the local coordinate systems of (h) at all points are transformed in the same way and all these transformations are even numerically the same if they are expressed at all points with respect to (h). When this has been done we consider now at every point of $\mathfrak{R}(\underset{0}{\xi}^\varkappa)$ not only the local coordinate systems of (h) but *all* coordinate systems derivable from them by means of *all* transformations of the holonomy group. The whole set of all local coordinate systems obtained in this way will be called the *allowable local coordinate systems.*

If from this set we choose in an arbitrary way *one* local coordinate system at every point of $\mathfrak{R}(\underset{0}{\xi}^\varkappa)$ we get an *allowable anholonomic coordinate system* in that region.

It is clear that the set is independent of the choice of the curves used above. If the holonomy group is an r-parameter group, the local coordinate systems depend at each point on r parameters. If $\underset{1}{\xi}^\varkappa$ and $\underset{2}{\xi}^\varkappa$ are two points of $\mathfrak{R}(\underset{0}{\xi}^\varkappa)$ and if the tangent E_n at $\underset{1}{\xi}^\varkappa$ is displaced parallel from $\underset{1}{\xi}^\varkappa$ to $\underset{2}{\xi}^\varkappa$ along some arbitrary curve, the local coordinates of the points of E_n at $\underset{1}{\xi}^\varkappa$ with respect to some allowable local coordinate system and the local coordinates of the displaced points in the E_n at $\underset{2}{\xi}^\varkappa$ with respect to some other allowable local coordinate system *always* transform into each other by a transformation of the holonomy group.

Now let x^h be the radiusvector of a point of the tangent E_n of a point $\xi^\varkappa$ in $\underset{0}{\mathfrak{N}}(\xi^\varkappa)$ with respect to some allowable local coordinate system. The general infinitesimal transformation of the r-parameter holonomy group has the form (cf. IV § 3)

$$dx^h = \Xi_b^h(x^i)\,(d\eta)^b; \quad h = 1, \ldots, n; \quad b = \dot{1}, \ldots, \dot{r}. \tag{6.1}$$

But because this group contains linear homogeneous transformations only, dx^h must be linear homogeneous in x^h, hence there exist equations of the form

$$dx^h = C_{b\,i}^{\cdot\,\cdot\,h}\, x^i\,(d\eta)^b \tag{6.2}$$

in which the $C_{b\,i}^{\cdot\,\cdot\,h}$ are *constants*.

If a vector $v^\varkappa$ is displaced parallel from $\xi^\varkappa$ to $\xi^\varkappa + d\xi^\varkappa$ we have for the anholonomic components v^h (cf. III 2.10; 9.1; 10.1)

$$dv^h = -\Gamma_{\mu i}^h\, v^i\, d\xi^\mu \tag{6.3}$$

and because of the choice of the allowable local coordinate systems this transformation is a transformation of the holonomy group in the variables v^h. Hence there must be a relation of the form

$$\Gamma_{\mu i}^h = \tfrac{1}{2} C_{a\,i}^{\cdot\,\cdot\,h}\, \theta_\mu^a(\xi^\varkappa) \tag{6.4}$$

in which the θ_μ^a are *not* in general *constants*.

The infinitesimal transformations of the holonomy group are

$$X_b = C_{b\,i}^{\cdot\,\cdot\,h}\, x^i \frac{\partial}{\partial x^h} \tag{6.5}$$

and as we know that $(X_c X_b) = c_{c\,b}^{\cdot\,\cdot\,a} X_a$, it follows by substitution of (6.5) in this equation that (cf. 4.20a)

$$C_{c\,i}^{\cdot\,\cdot\,j} C_{b\,j}^{\cdot\,\cdot\,h} - C_{b\,i}^{\cdot\,\cdot\,j} C_{c\,j}^{\cdot\,\cdot\,h} = c_{c\,b}^{\cdot\,\cdot\,a} C_{a\,i}^{\cdot\,\cdot\,h}. \tag{6.6}$$

Now we substitute (6.4, 6) in the structural equations (III 10.4, 5) of an A_n. Then $S_{\mu\lambda}^{\cdot\,\cdot\,\varkappa} = 0$ and we get

$$\left\{\begin{array}{ll} & \text{a)}\quad \partial_{[\mu} A_{\lambda]}^h = \tfrac{1}{2} C_{a\,i}^{\cdot\,\cdot\,h} A_{[\mu}^i \theta_{\lambda]}^a \\ \complement & \text{b)}\quad [dA^h] = \tfrac{1}{2} C_{a\,i}^{\cdot\,\cdot\,h} [A^i \theta^a] \end{array}\right. \tag{6.7}$$

and

$$\left\{\begin{array}{ll} & \text{a)}\quad C_{a\,i}^{\cdot\,\cdot\,h} \partial_{[\nu} \theta_{\mu]}^a - \tfrac{1}{2} c_{c\,b}^{\cdot\,\cdot\,a} C_{a\,i}^{\cdot\,\cdot\,h} \theta_{[\nu}^c \theta_{\mu]}^b = R_{\nu\mu i}^{\cdot\,\cdot\,\cdot\,h} \\ \complement & \text{b)}\quad C_{a\,i}^{\cdot\,\cdot\,h} [d\theta^a] - \tfrac{1}{2} c_{c\,b}^{\cdot\,\cdot\,a} C_{a\,i}^{\cdot\,\cdot\,h} [\theta^c \theta^b] = R_i^{\cdot\,h} \end{array}\right. \tag{6.8}$$

or [cf. (4.13)]

$$\text{(6.9)} \qquad \begin{cases} \text{a)} & R_{\nu\mu i}^{\cdot\cdot\cdot h} = C_{ai}^{\cdot\cdot h} B_{\nu\mu}^{\cdot\cdot a} \\ \mathsf{C}\ \text{b)} & R_i^{\cdot h} = C_{ai}^{\cdot\cdot h} B^a \end{cases}$$

where

$$\text{(6.10)} \qquad \begin{cases} \text{a)} & B_{\nu\mu}^{\cdot\cdot a} \overset{\text{def}}{=} \partial_{[\nu} \theta_{\mu]}^a - \tfrac{1}{2} c_{cb}^{\cdot\cdot a} \theta_{[\nu}^c \theta_{\mu]}^b \\ \mathsf{C}\ \text{b)} & B^a \overset{\text{def}}{=} [d\theta^a] - \tfrac{1}{2} c_{cb}^{\cdot\cdot a} [\theta^c \theta^b]. \end{cases}$$

As an example we consider first the case of a *symmetric* A_n (cf. III § 7, VII § 5). Then we have $\nabla_\omega R_{\nu\mu\lambda}^{\cdot\cdot\cdot\varkappa} = 0$ and the components $R_{kji}^{\cdot\cdot\cdot h}$ have the same values at *all* points with respect to *all* allowable coordinate systems (h).

The infinitesimal transformation of the radiusvector x^h in the tangent E_n (cf. III 4.1)

$$\text{(6.11)} \qquad x^h \to x^h - \tfrac{1}{2} R_{kji}^{\cdot\cdot\cdot h} x^i \, d f^{kj}$$

is the transformation of the holonomy group belonging to the facet $d f^{kj}$. Hence there must exist relations of the form (cf. 4.13 a)

$$\text{(6.12)} \qquad R_{kji}^{\cdot\cdot\cdot h} = C_{ai}^{\cdot\cdot h} B_{kj}^{\cdot\cdot a}$$

in which now *not only the* $C_{ai}^{\cdot\cdot h}$ *but also the coefficients* $B_{kj}^{\cdot\cdot a}$ *are constants*. Substituting (6.12) in (6.8a) we get

$$\text{(6.13)} \qquad C_{ai}^{\cdot\cdot h} (\partial_{[\nu} \theta_{\mu]}^a - \tfrac{1}{2} c_{cb}^{\cdot\cdot a} \theta_{[\nu}^c \theta_{\mu]}^b - B_{kj}^{\cdot\cdot a} A_{[\nu}^k A_{\mu]}^j) = 0,$$

and this is equivalent to

$$\text{(6.14)} \qquad \begin{cases} \text{a)} & \partial_{[\nu} \theta_{\mu]}^a = \tfrac{1}{2} c_{cb}^{\cdot\cdot a} \theta_{[\nu}^c \theta_{\mu]}^b + B_{kj}^{\cdot\cdot a} A_{[\nu}^k A_{\mu]}^j \\ \mathsf{C}\ \text{b)} & [d\theta^a] = \tfrac{1}{2} c_{cb}^{\cdot\cdot a} [\theta^c \theta^b] + B_{kj}^{\cdot\cdot a} [A^k A^j] \end{cases}$$

because the a-rank of $C_{ai}^{\cdot\cdot h}$ is r.

We return to the case of a *general* A_n. The allowable local coordinate systems in A_n may be considered as depending on the general coordinates η^α; $\alpha = \dot{1}, \ldots, \dot{r}$ in the group space X_r of the holonomy group. To this end we have only to fix one allowable local coordinate system at every point of $\mathfrak{N}(\underset{0}{\xi^\varkappa})$ and to give these systems the coordinates $\underset{0}{\eta^\alpha}$ belonging to the identical transformation. Of course an allowable anholonomic system (h) can then be fixed by giving the η^α as functions of $\xi^\varkappa$. In the same way as in III § 10 the A_λ^h may be considered as the section of the A_B^h; $B = 1, \ldots, n, \dot{1}, \ldots, \dot{r}$ in the X_{n+r} of $\xi^\varkappa, \eta^\alpha$ with $A_\beta^h = 0$. But in contradistinction to III § 10 we have now $r \leq n^2$ and in the X_r we have not only the holonomic coordinates η^α but among others also the anholonomic system (a). Hence in X_{n+r} we have besides the holonomic system $(\varkappa, \alpha)$; $1, \ldots, n, \dot{1}, \ldots, \dot{r}$ the three anholonomic systems $(\varkappa, a)$; $1, \ldots, n, \dot{1}, \ldots, \dot{r}$ and (h, α); $1, \ldots, n, \dot{1}, \ldots, \dot{r}$ and (h, a); $1, \ldots, n, \dot{1}, \ldots, \dot{r}$.

According to (6.2, 5) the infinitesimal transformation of A_λ^h by the holonomy group is given by

$$d\eta^\beta\, \partial_\beta A_\lambda^h = C_{b i}^{\cdot\cdot h} A_\lambda^i (d\eta)^b \tag{6.15}$$

and from this equation we get

$$A_i^\lambda \partial_\gamma A_\lambda^h = A_\gamma^a C_{a i}^{\cdot\cdot h}. \tag{6.16}$$

Hence, if we write

$$\theta_\gamma^a \overset{\text{def}}{=} - A_\gamma^a \tag{6.17}$$

we get besides (6.7a) the equations

$$\begin{cases} \partial_{[\gamma} A_{\lambda]}^h = \frac{1}{2} C_{a i}^{\cdot\cdot h} A_{[\gamma}^i \theta_{\lambda]}^a \\ \partial_{[\gamma} A_{\beta]}^h = \frac{1}{2} C_{a i}^{\cdot\cdot h} A_{[\gamma}^i \theta_{\beta]}^a; \quad (\text{Note that } A_\beta^h = 0) \end{cases} \tag{6.18}$$

that can be gathered together with (6.7a) into the equations

$$\begin{cases} \text{a)} \ \partial_{[C} A_{B]}^h = \frac{1}{2} C_{a i}^{\cdot\cdot h} A_{[C}^i \theta_{B]}^a; \qquad B, C = 1, \ldots, n, \dot{1}, \ldots, \dot{r}; \\ \mathfrak{C} \ \text{b)} \ [d A^h] = \frac{1}{2} C_{a i}^{\cdot\cdot h} [A^i \theta^a]; \quad h, i = 1, \ldots, n; \quad a = \dot{1}, \ldots, \dot{r}. \end{cases} \tag{6.19}$$

Note *that* (6.19b) *has exactly the same form as* (6.7b), though (6.19b) contains much more.

Now we will prove that the equations (6.14) for the *symmetric* A_n can be generalized in the same way to

$$\begin{cases} \text{a)} \quad \partial_{[C} \theta_{B]}^a = \frac{1}{2} c_{c b}^{\cdot\cdot a} \theta_{[C}^c \theta_{B]}^b + B_{k j}^{\cdot\cdot a} A_{[C}^k A_{B]}^j \\ \mathfrak{C} \ \text{b)} \quad [d\theta^a] = \frac{1}{2} c_{c b}^{\cdot\cdot a} [\theta^c \theta^b] + B_{k j}^{\cdot\cdot a} [A^k A^j]. \end{cases} \tag{6.20}$$

We note that (6.20b) has the same form as (6.14b) but contains much more. We look at the $\Gamma_{\mu i}^h$ with respect to a definitely chosen allowable anholonomic coordinate system (h). If a transformation of the holonomy group is applied simultaneously to all local coordinate systems of (h) we get a new allowable anholonomic coordinate system (h') and the $A_h^{h'}$, $A_{h'}^h$ are constants. That means that in this case the transformation $\Gamma_{\mu i}^h \to \Gamma_{\mu i'}^{h'}$ is linear homogeneous (cf. 4.16). For an infinitesimal transformation of this kind it follows from (6.15) that

$$d\Gamma_{\mu i}^h = (C_{b j}^{\cdot\cdot h} \Gamma_{\mu i}^j - C_{b i}^{\cdot\cdot j} \Gamma_{\mu j}^h)\,(d\eta)^b \tag{6.21}$$

hence, according to (6.4, 6, 17)

$$(6.22)\quad \begin{cases} \partial_\delta \Gamma^h_{\mu i} = C_{\delta j}^{\cdot\cdot h} \Gamma^j_{\mu i} - C_{\delta i}^{\cdot\cdot j} \Gamma^h_{\mu j} \\ \qquad = \frac{1}{2}(C_{\delta j}^{\cdot\cdot h} C_{a i}^{\cdot\cdot j} - C_{\delta i}^{\cdot\cdot j} C_{a j}^{\cdot\cdot h})\, \theta^a_\mu = -\frac{1}{2} c_{c b}^{\cdot\cdot a} C_{a i}^{\cdot\cdot h} \theta^c_\mu \theta^b_\delta . \end{cases}$$

But on the other hand we have from (6.4)

$$(6.23)\qquad \partial_\delta \Gamma^h_{\mu i} = \tfrac{1}{2} C_{a i}^{\cdot\cdot h} \partial_\delta \theta^a_\mu$$

and because the a-rank of $C_{a i}^{\cdot\cdot h}$ is r, (6.22, 23) have as a consequence

$$(6.24)\qquad \partial_\gamma \theta^a_\lambda = c_{c b}^{\cdot\cdot a} \theta^c_\gamma \theta^b_\lambda .$$

As θ^a_γ depends here on the η^α only, we have $\partial_\lambda \theta^a_\gamma = 0$ and

$$(6.25)\qquad \partial_{[\gamma} \theta^a_{\lambda]} = \tfrac{1}{2} c_{c b}^{\cdot\cdot a} \theta^c_{[\gamma} \theta^b_{\lambda]} + B_{k j}^{\cdot\cdot a} A^k_{[\gamma} A^j_{\lambda]}$$

where the last term could be added because $A^k_\gamma = 0$. Now we have already two parts of (6.20) viz. (6.14) and (6.25). The last part is [cf. (6.17)]

$$(6.26)\qquad \partial_{[\gamma} A^a_{\beta]} = -\tfrac{1}{2} c_{c b}^{\cdot\cdot a} A^c_{[\gamma} A^b_{\beta]}$$

but this equation is equivalent to CARTAN's structural equations of the parameter groups of the holonomy group (IV 1,19), already proved in IV § 1.

In the X_{n+r} of $\xi^\varkappa, \eta^\alpha$ the A^h_B; $h = 1, \ldots, n$ and θ^a_B; $a = \dot{1}, \ldots, \dot{r}$ are $n+r$ covariant vector fields. They are linearly independent because an equation of the form

$$(6.27)\qquad \alpha_h A^h_B + \alpha_a \theta^a_B = 0$$

would be equivalent to

$$(6.28)\qquad \begin{cases} \text{a)}\quad \alpha_h A^h_\lambda + \alpha_a \theta^a_\lambda = 0 \\ \text{b)}\quad \alpha_a \theta^a_\beta = 0 \end{cases}$$

and these equations can only be true for $\alpha_a = 0$; $\alpha_h = 0$ because of (6.17). Hence the $n+r$ covariant vectors determine an anholonomic coordinate system in the X_{n+r} and the intermediate components A^X_B are

$$(6.29)\qquad \begin{cases} A^h_B; \quad A^a_B \overset{\text{def}}{=} -\theta^a_B; \quad X = 1, \ldots, n;\ \dot{1}, \ldots, \dot{r}; \\ \qquad\qquad h = 1, \ldots, n; \quad a = \dot{1}, \ldots, \dot{r}; \\ \qquad\qquad B = 1, \ldots, n;\ \dot{1}, \ldots, \dot{r}. \end{cases}$$

With this notation the equations (6.19, 20) of the *symmetric* A_n can be written together

$$(6.30)\qquad \begin{cases} \text{a)}\quad \partial_{[C} A^X_{B]} = \tfrac{1}{2} C^X_{Z Y} A^Z_{[C} A^Y_{B]} \\ \mathfrak{C}\ \text{b)}\quad [d A^X] = \tfrac{1}{2} C^X_{Z Y} [A^Z A^Y]; \quad X, Y, Z = 1, \ldots, n, \dot{1}, \ldots, \dot{r} \end{cases}$$

with *constant* coefficients C_{ZY}^{X}

$$(6.31)\qquad \begin{cases} C_{cb}^{a} \overset{\text{def}}{=} -c_{cb}^{a}; & C_{cb}^{h} \overset{\text{def}}{=} 0; & a, b, c = \dot{1}, \ldots, \dot{r} \\ C_{ic}^{a} = C_{ci}^{a} \overset{\text{def}}{=} 0; & C_{ic}^{h} \overset{\text{def}}{=} -C_{ci}^{h}; & h, i, j = 1, \ldots, n \\ C_{ji}^{a} \overset{\text{def}}{=} -B_{ji}^{\cdot\cdot a}; & C_{ji}^{h} \overset{\text{def}}{=} 0. & \end{cases}$$

But as was proved in IV § 2 this is only possible if

$$(6.32)\qquad C_{[ZY}^{U} C_{X]U}^{V} = 0.$$

Now if these latter equations are written out making use of (6.31), we see that they are in fact satisfied as a consequence of (III 5.4), (IV 1.28b), (6.6, 12) and $\nabla_\omega R_{\nu\mu\lambda}^{\cdot\cdot\cdot\varkappa} = 0$.

As we have seen in IV § 2, (6.32) has as a consequence that (6.30) are the structural equations after Cartan of an $(n+r)$-parameter group of point transformations in X_{n+r}. The structural equations of the same group after Lie are

$$(6.33)\qquad (A_Z A_Y) = -C_{ZY}^{X} A_X$$

or, written out,

$$(6.34)\qquad \begin{cases} (A_c A_b) = c_{cb}^{a} A_a; & A_b = A_b^{\mu} \partial_\mu + A_b^{\gamma} \partial_\gamma; \\ (A_c A_i) = -C_{ci}^{h} A_h; & A_j = A_j^{\mu} \partial_\mu + A_j^{\gamma} \partial_\gamma; \\ (A_j A_i) = B_{ji}^{\cdot\cdot a} A_a = -C_{ji}^{a} A_a. & \end{cases}$$

This proves that if a system of constants c_{cb}^{a}, C_{ci}^{h}, $B_{ji}^{\cdot\cdot a}$ is given, satisfying (6.31, 32) there always exists a *symmetric* A_n whose holonomy group depends on r parameters and has the c_{cb}^{a} for structural constants (cf. IV § 2). In this A_n there exists an anholonomic coordinate system (h) such that $\Gamma_{\mu i}^{h} = -\frac{1}{2} A_\mu^a C_{ai}^{h}$ and $R_{kji}^{\cdot\cdot\cdot h} = C_{ai}^{h} B_{kj}^{\cdot\cdot a}$.[1])

As was proved in IV § 2 there exists in X_{n+r} an $(n+r)$-parameter group leaving the $n+r$ covariant vector fields A_B^h (with $A_\beta^h = 0$) and A_B^a invariant. Hence, if $\xi^\varkappa \to \xi^\varkappa + v^\varkappa\, dt$ combined with $\eta^\alpha \to \eta^\alpha + v^\alpha\, dt$ is an infinitesimal transformation of this group, the Lie derivative of these $n+r$ fields must vanish:

$$(6.35)\qquad \begin{cases} \text{a)} & v^\mu \partial_\mu A_\lambda^h + v^\gamma \partial_\gamma A_\lambda^h + A_\mu^h \partial_\lambda v^\mu + (A_\gamma^h \partial_\lambda v^\gamma) = 0 \\ \text{b)} & (v^\mu \partial_\mu A_\beta^h) + (v^\gamma \partial_\gamma A_\beta^h) + A_\mu^h \partial_\beta v^\mu + (A_\gamma^h \partial_\beta v^\gamma) = 0 \\ \text{c)} & v^\mu \partial_\mu A_\lambda^a + (v^\gamma \partial_\gamma A_\lambda^a) + A_\mu^a \partial_\lambda v^\mu + A_\gamma^a \partial_\lambda v^\gamma = 0 \\ \text{d)} & (v^\mu \partial_\mu A_\beta^a) + v^\gamma \partial_\gamma A_\beta^a + A_\mu^a \partial_\beta v^\mu + A_\gamma^a \partial_\beta v^\gamma = 0 \end{cases}$$

---

[1]) This theorem was proved by E. Cartan in 1926, 1, p. 264 ff. for the case of a V_n.

(the terms in brackets vanish already because 1°. $A^h_\beta = 0$; 2°. the A^a_β are independent of the $\xi^\varkappa$ and 3°. the A^a_λ are independent of the η^α). From (6.35b) it follows that the $v^\varkappa$ are functions of the $\xi^\varkappa$ only. Hence $v^\varkappa\, dt$ represents an ordinary infinitesimal transformation of the X_n. With this transformation the fields $\overset{h}{e}_\lambda$ are dragged along and accordingly the changes of these fields are at every point of X_n equal to their negative Lie differentials. Apart from this transformation, $v^\alpha\, dt$ represents an extra infinitesimal transformation of these local vectors $\overset{h}{e}_\lambda$ and because in total the local fields $\overset{h}{e}_\lambda$ must be invariant, these two infinitesimal transformations must annihilate each other. This is just what the equation (6.35a) expresses. Moreover the A^a_λ are invariant, hence the $\Gamma^h_{\mu i} = -\frac{1}{2} A^a_\mu C^h_{ai}$ are invariant, and this means that the connexion in A_n is invariant. This proves that there exists in a *symmetric* A_n an $(n+r)$-parameter group of point transformations, which leaves the connexion invariant, and that this group has the $-C^X_{ZY}$ for structural constants.[1]) If from this group we take the subgroup that leaves invariant a point $\overset{0}{\xi}{}^\varkappa$, this subgroup transforms the vectors in the tangent E_n of $\overset{0}{\xi}{}^\varkappa$ in the same way as the holonomy group does.

VIII. Miscellaneous examples.

§ 1. The harmonic V_n.

Let $x^\varkappa$ be rectilinear coordinates in an ordinary R_n. Then the distance s of an arbitrary point $x^\varkappa$ from the origin is

$$s = (g_{\mu\lambda}\, x^\mu x^\lambda)^{\frac{1}{2}} \quad {}^{2)} \tag{1.1}$$

and by differentiation we get

$$\nabla_\nu s^p = p\, s^{p-2} g_{\mu\nu}\, x^\mu \tag{1.2}$$

$$\nabla^\nu \nabla_\nu s^p = p\,(n+p-2)\, s^{p-2}; \qquad \nabla^\nu \overset{\text{def}}{=} g^{\nu\mu} \nabla_\mu . \tag{1.3}$$

Hence $\nabla^\nu \nabla_\nu s$ is a function of s only, and s^{2-n} is for $n \neq 2$ a solution of the equation of Laplace

$$\nabla^\nu \nabla_\nu V = 0. \tag{1.4}$$

In a V_n (1.4) is an invariant equation but we can not be certain that solutions always exist. The V_n is called *harmonic*[3]) *at the point* $\overset{0}{\xi}{}^\varkappa$

[1]) This was proved by E. Cartan 1926, 1 for the case of a V_n.

[2]) From now we write in this section $d\sigma^2$ instead of ds^2 for the linear element.

[3]) Copson and Ruse 1940, 1, p. 117. Cf. for the affine generalization Ruse 1952, 1; Patterson 1952, 1.

if (1.4) has at least one solution $\varphi(s)$ which is a function of s *only*. s is here the distance of the point $\xi^\varkappa$ with respect to the point $\underset{0}{\xi}^\varkappa$. If this solution exists we have

$$(1.5)\qquad \begin{cases} 0 = \nabla^\nu \nabla_\nu \varphi(s) = \varphi' \nabla^\nu \nabla_\nu s + \varphi''(\nabla^\nu s)\nabla_\nu s; \\ \varphi' \overset{\text{def}}{=} \dfrac{d\varphi}{ds}; \quad \varphi'' \overset{\text{def}}{=} \dfrac{d^2\varphi}{ds^2} \end{cases}$$

hence

$$(1.6)\qquad \nabla^\nu \nabla_\nu s = (-\varphi''(\nabla^\nu s)\nabla_\nu s)/\varphi' = -\varphi''/\varphi'$$

and this is a function of s only. Conversely, if it is known that $\nabla^\nu \nabla_\nu s$ is a function of s *only:*

$$(1.7)\qquad \nabla^\nu \nabla_\nu s = \chi(s)$$

and if we write $V(s)$ for a still unknown solution of (1.4), we have

$$(1.8)\qquad \nabla^\nu \nabla_\nu V = V''(\nabla^\nu s)\nabla_\nu s + V'\chi(s) = 0$$

from which V can be found by quadratures as a function of s. Hence

A V_n is harmonic at a point $\underset{0}{\xi}^\varkappa$ if and only if the distance s with respect to $\underset{0}{\xi}^\varkappa$ satisfies an equation of the form (1.7). [1])

It is convenient to introduce the function $\Omega \overset{\text{def}}{=} \frac{1}{2}s^2$ [2]) and to write

$$(1.9)\qquad f(\Omega) \overset{\text{def}}{=} \nabla^\nu \nabla_\nu \Omega = s\,\chi(s) + 1 = (2\Omega)^{\frac{1}{2}}\chi\{(2\Omega)^{\frac{1}{2}}\} + 1.$$

According to Copson and Ruse a V_n that is harmonic at all points of the region considered is called *completely harmonic*. If it is harmonic at one point only, the term *centrally harmonic* will be used. In a completely harmonic V_n the functions $\chi(s)$ and $f(\Omega)$ are each the same for all points, as was proved by Lichnerowicz. [3])

Ruse [4]) has proved that an elementary solution of (1.4) is given by the integral

$$(1.10)\qquad \varphi(s) = A\int_a^s \varrho^{-1} s^{1-n}\,ds + B$$

where A, B and $a > 0$ are constants, s is the distance from $\xi^\varkappa$ to $\underset{0}{\xi}^\varkappa$ and where the integration has to be effected along the geodesic through

[1]) Copson and Ruse 1940, 1, p. 121.

[2]) Ω is a function of $\underset{0}{\xi}^\varkappa$ and $\xi^\varkappa$, and is often called the *characteristic function* of the V_n. Its covariant derivatives for fixed $\underset{0}{\xi}^\varkappa$ were derived by Synge 1930, 1. Cf. Copson and Ruse 1940, 1; Lichnerowicz 1944, 1, p. 158.

[3]) Lichnerowicz 1944, 1, p. 150; cf. Walker 1945, 1.

[4]) Ruse 1930, 2; 3; 1939, 1 for S_n; cf. Copson and Ruse 1940, 1.

$\underset{0}{\xi}^\varkappa$ and $\xi^\varkappa$. ϱ is RUSE's invariant

$$(1.11)\qquad \varrho \overset{\text{def}}{=} \mathfrak{g}^{\frac{1}{2}}(\xi^\varkappa)\, \mathfrak{g}^{\frac{1}{2}}(\underset{0}{\xi}^\varkappa) \Big/ \operatorname{Det}\left(\frac{\partial^2 \Omega}{\partial \xi^\mu \, \partial \underset{0}{\xi}^\lambda}\right);$$

it is a function of s only, if and only if the V_n is centrally harmonic at $\underset{0}{\xi}^\varkappa$. Its development in a power series of s and its geometric interpretation were given by WALKER.[1]) The solution (1.10) behaves like s^{2-n} at $\underset{0}{\xi}^\varkappa$ and on all null geodesics through this point, and it passes into s^{2-n} if the V_n is an R_n.[2])

A completely harmonic V_n with $\varrho = 1$ is called *simply harmonic.*[3]) WALKER proved that this condition is equivalent to $\chi(s) = (n-1)\, s^{-1}$.[4])

If the V_n is centrally harmonic at $\underset{0}{\xi}^\varkappa$ and if the geodesic joining $\xi^\varkappa$ and $\underset{0}{\xi}^\varkappa$ has a unit tangent vector at $\underset{0}{\xi}^\varkappa$ with components $t^{h'}$; $h' = 1', \ldots, \mathrm{n}'$ with respect to some fixed local orthogonal coordinate system (h') in the tangent R_n of $\underset{0}{\xi}^\varkappa$, we may introduce the coordinates

$$(1.12)\qquad \xi^{a'} \overset{\text{def}}{=} t^{a'}; \qquad \xi^{\mathrm{n}'} \overset{\text{def}}{=} s; \qquad a' = 1', \ldots, (\mathrm{n}-1)'.$$

Then the linear element takes the form

$$(1.13)\qquad d\sigma^2 \overset{*}{=} d\xi^{\mathrm{n}'}\, d\xi^{\mathrm{n}'} + g_{a'b'}\, d\xi^{a'}\, d\xi^{b'}$$

and for $\chi(s)$ we get

$$(1.14)\qquad \begin{cases} \chi(s) = \nabla^\mu \nabla_\mu s \overset{*}{=} (\partial_{h'} \mathfrak{g}^{i'h'})\, \partial_{i'} s + \overset{(h')}{\mathfrak{g}}{}^{-\frac{1}{2}}\, \partial^{j'} \overset{(h')}{\mathfrak{g}}{}^{\frac{1}{2}}\, \partial_{j'} s \\ \qquad = \overset{(h')}{\mathfrak{g}}{}^{-\frac{1}{2}}\, \partial^{\mathrm{n}'} \overset{(h')}{\mathfrak{g}}{}^{\frac{1}{2}}\, \partial_{\mathrm{n}'} s = \dfrac{1}{2} \dfrac{\partial}{\partial s} \log \overset{(h')}{\mathfrak{g}}, \end{cases}$$

from which by integration

$$(1.15)\qquad \tfrac{1}{2} \log \overset{(h')}{\mathfrak{g}} \overset{*}{=} \int \chi(s)\, ds + \Phi(\xi^{a'}); \qquad a' = 1', \ldots, (\mathrm{n}-1)'.$$

Instead of (h') we now introduce the coordinate system (h); $h = 1, \ldots, \mathrm{n}$, defined by

$$(1.16)\qquad \xi^a \overset{\text{def}}{=} \delta^a_{a'}\, s\, t^{a'}; \qquad \xi^{\mathrm{n}} \overset{\text{def}}{=} s\, t^{\mathrm{n}'}; \qquad a = 1, \ldots, \mathrm{n}-1$$

---

[1]) WALKER 1942, 1, p. 16.

[2]) LICHNEROWICZ 1944, 1, p. 148f. Mean value theorems for harmonic functions in a completely harmonic V_n were dealt with by WILLMORE 1950, 1. The more general equation $\nabla^\nu \nabla_\nu V = F(V)$ was discussed by NORDON 1944, 1.

[3]) WALKER 1942, 1; cf. PATTERSON 1951, 2; 1952, 1.

[4]) WALKER 1945, 1.

where $t^{\mathfrak{n}'}$ is a known function of the $t^{a'}$. Then

$$\log \operatorname{Det}(A_{h'}^{h}) \overset{*}{=} (n-1)\log s + \log t^{\mathfrak{n}'}(\xi^{a'}) \tag{1.17}$$

and

$$\tfrac{1}{2}\log \overset{(h)}{\mathfrak{g}} \overset{*}{=} \int \chi(s)\,ds - (n-1)\log s + \Phi(\xi^{a'}) - \log t^{\mathfrak{n}'}(\xi^{a'})\,. \tag{1.18}$$

But, for $s=0$, $\overset{(h)}{\mathfrak{g}}$ is independent of the $\xi^{a'}$ and has the value $+1$. Moreover in III § 7 we have seen that the transformation of two normal coordinate systems with respect to the same point is linear homogeneous with constant coefficients. Hence

In a centrally harmonic V_n with centrum $\underset{0}{\xi}{}^{\varkappa}$ the components $\overset{(h)}{\mathfrak{g}}$ of $\mathfrak{g}$ with respect to any arbitrary normal coordinate system (h) belonging to $\underset{0}{\xi}{}^{\varkappa}$ is a function of s only and satisfies the equation

$$\begin{cases} \dfrac{1}{2}\dfrac{d}{ds}\log \overset{(h)}{\mathfrak{g}} \overset{*}{=} \chi(s) - \dfrac{n-1}{s} \\[2ex] \qquad = \nabla^{\mu}\nabla_{\mu}s - \dfrac{n-1}{s}\ ^{1)}. \end{cases} \tag{1.19}$$

From (1.9) and (1.19) we get

$$\nabla^{\nu}\nabla_{\nu}\Omega \overset{*}{=} (n-1) + \frac{1}{2}s\frac{d\log \overset{(h)}{\mathfrak{g}}}{ds} + (\nabla^{\nu}s)\nabla_{\nu}s\ ^{2)}. \tag{1.20}$$

According to (III 3.19) it follows from (1.19) that for a positive definite fundamental tensor

$$\begin{cases} \Gamma_{ji}^{i} = \dfrac{1}{2}\partial_j \log \overset{(h)}{\mathfrak{g}} \overset{*}{=} \left(\chi(s) - \dfrac{n-1}{s}\right)\partial_j s \\[2ex] \quad = \dfrac{s\chi(s) - n + 1}{s^2}\partial_j\Omega = \dfrac{f(\Omega) - n}{2\Omega}\partial_j\Omega\ ^{3)}. \end{cases} \tag{1.21}$$

From the series (III 7.28) for Γ_{ji}^{h} we get for Γ_{ji}^{i}

$$\Gamma_{ji}^{i} \overset{*}{=} \xi^{k}\overset{4}{N}{}_{kji}^{\cdot\cdot\cdot i}(\underset{0}{\xi}) + \frac{1}{2!}\xi^{k_2}\xi^{k_1}\overset{5}{N}{}_{k_2k_1ji}^{\cdot\cdot\cdot\cdot i}(\underset{0}{\xi}) + \cdots. \tag{1.22}$$

Now in (1.21), $f(\Omega)$ can be developed in the series

$$f(\Omega) = f(0) + \Omega f'(0) + \tfrac{1}{2}\Omega^2 f''(0) + \cdots; \qquad f(0) = n. \tag{1.23}$$

[1]) LICHNEROWICZ 1944, 1, p. 155; the converse is also true. The proof given there is not quite satisfactory because the result seems to depend on the choice of the normal system, which in fact it does not.

[2]) LICHNEROWICZ 1944, 1, p. 150.

[3]) LICHNEROWICZ 1944, 1, p. 156.

Making use of the following identities in normal coordinates

$$(1.24)\qquad \partial_i s \overset{*}{=} g_{ij}\frac{\xi^j}{s}; \qquad \partial_i \Omega \overset{*}{=} g_{ij}\xi^j;$$

$$(1.25)\qquad (\partial_j \partial_i \Omega)_0 \overset{*}{=} (g_{ji})_0$$

we get by equating corresponding terms in the development of both sides of (1.21)

$$(1.26)\quad \begin{cases} \text{a)}\ \overset{4}{N}{}_{kji}^{\cdot\cdot\cdot\, i}(\underset{0}{\xi}) = \frac{1}{2} f'(0)\, g_{kj}(\underset{0}{\xi}) \\ \text{b)}\ \overset{5}{N}{}_{k_2 k_1 j i}^{\cdot\cdot\cdot\cdot\, i}(\underset{0}{\xi}) = 0 \\ \text{c)}\ \overset{6}{N}{}_{k_3 k_2 k_1 j i}^{\cdot\cdot\cdot\cdot\cdot\, i}(\underset{0}{\xi}) = \frac{3}{4} f''(0)\, g_{j(k_3}(\underset{0}{\xi})\, g_{k_2 k_1)}(\underset{0}{\xi}) \\ \text{d)}\ \overset{7}{N}{}_{k_4 k_3 k_2 k_1 j i}^{\cdot\cdot\cdot\cdot\cdot\cdot\, i}(\underset{0}{\xi}) = 0 \\ \vdots \end{cases}$$

known as the *equations of* COPSON *and* RUSE.[1])

Using the relations (III 7.36, 40; Exerc. III 7,5) between the normal tensors and the curvature tensor and its covariant derivatives, and the identities III (5.27, 30), the equations (1.26a, b) appear to be equivalent to

$$(1.27)\quad \begin{cases} \text{a)}\ \frac{1}{3} K_{\nu\mu\lambda}^{\cdot\cdot\cdot\,\lambda}(\underset{0}{\xi}) + \frac{1}{3} K_{\nu\lambda\mu}^{\cdot\cdot\cdot\,\lambda}(\underset{0}{\xi}) = -\frac{1}{3} K_{\nu\mu}(\underset{0}{\xi}) = \frac{1}{2} f'(0)\, g_{\nu\mu}(\underset{0}{\xi}) \\ \text{b)}\ \nabla_{(\omega} K_{\nu\mu)}(\underset{0}{\xi}) = 0; \end{cases}$$

hence (cf. III § 5)

$$(1.28)\qquad f'(0) = -\frac{2}{3n} K(\underset{0}{\xi}) = -\frac{2}{3}(n-1)\,\varkappa(\underset{0}{\xi}).$$

If the V_n is completely harmonic, (1.27a, b) hold for every choice of $\underset{0}{\xi}{}^\varkappa$. Then (1.27a) expresses that the V_n is an EINSTEIN space.[2]) (1.27b) is a consequence of (1.27a) for $n>2$ and expresses for $n=2$ that the V_2 is an S_2. As every EINSTEIN V_3 is an S_3 this proves that a completely harmonic V_n is always an S_n if $n=2$ or 3.[2]) Of course every S_n is completely harmonic because for every choice of $\underset{0}{\xi}{}^\varkappa$, $\nabla^\mu \nabla_\mu s$ is a function of s only. In the beginning of the investigations it seemed

[1]) COPSON and RUSE 1940, 1, p. 130; cf. LICHNEROWICZ 1944, 1, p. 156; RUSE 1945, 1. COPSON and RUSE went via the covariant derivatives of Ω in order to make use of SYNGE's relations in Synge 1930, 1.

[2]) COPSON and RUSE 1940, 1, p. 132. In the same paper it was proved that every SCHUR space [cf. V, p. 244, footnote 2)] is centrally harmonic with respect to the origin but that for $n>2$ not every centrally harmonic space is a SCHUR space.

quite probable that every completely harmonic V_n should be an S_n. Every conformally euclidean EINSTEIN space is for $n>2$ an S_n (cf. VI § 6) hence every completely harmonic space that is conformally euclidean is an S_n. But as WALKER proved by a counterexample[1]) a space can for $n>3$ be completely harmonic without being an S_n.

(1.26b) contains derivatives of the components of the curvature tensor. But for $n=2$ it follows by differentiation of (1.26a) that (1.26b) is identically satisfied. In an analogous way the second derivatives of the curvature tensor that occur in (1.26c), after expressing $\overset{6}{N}$ in terms of $\overset{4}{K}$ and $\overset{6}{K}$ (cf. III § 7), can be eliminated after differentiating the foregoing formulae of (1.26). Then we get after some calculation[2])

$$K_{\nu(\mu\lambda|\varkappa|}\, K^{\nu\cdot\cdot\varkappa}_{\cdot\sigma\varrho)} = -\,45 f''(0)\, g_{(\mu\lambda}\, g_{\sigma\varrho)}. \tag{1.29}$$

LICHNEROWICZ and WALKER[3]) used this equation to prove that a completely harmonic V_n is always an S_n if its index (cf. I § 8) is *1* or $n-1$. We give here a survey of results:

A completely harmonic V_n is an S_n if one of the following conditions is satisfied

a) $n=2$ or *3*;

b) *the V_n is conformally euclidean;*

c) *the index is 1 or $n-1$.*

If the V_n is simply harmonic it is an R_n if either one of these three conditions is satisfied or the condition[4])

d) *the index is 0 or n.*[5])

For any $n\geq 4$ there is a completely (simply) harmonic V_n which is not an $S_n(R_n)$, with any index not satisfying c (c or d).[6])

---

[1]) WALKER 1945, 1; 2.

[2]) LICHNEROWICZ 1944, 1, p. 159. RUSE 1945, 1, p. 156, gave the geometric interpretation of (1.29) for $n=4$ and he determined all algebraically possible V_4's satisfying (1.27) and (1.29). Cf. LICHNEROWICZ 1953, 2 also for literature on the harmonic V_n.

[3]) WALKER 1942, 1 for $n=4$; LICHNEROWICZ and WALKER 1945, 1 for the general case.

[4]) WALKER 1945, 1.

[5]) This is in accordance with T. Y. THOMAS and TITT 1939, 4, who proved that for the *definite* case s^{2-n} for $n>2$ and $\log s$ for $n=2$ can only be solutions of LAPLACE's equation if the V_n is an R_n. Cf. WALKER 1942, 1, p. 26; LICHNEROWICZ 1944, 1, p. 160.

[6]) WALKER 1945, 1; 2.

WALKER gave the following examples.[1]) The V_4 with the linear element with index 2

$$(1.30)\qquad \begin{cases} d\sigma^2 = C(\xi^2 d\xi^1 - \xi^1 d\xi^2)^2 + 2d\xi^1 d\xi^3 + 2d\xi^2 d\xi^4; \\ \qquad\qquad C = \text{const.} \neq 0 \end{cases}$$

is simply harmonic. It has a curvature tensor with the only non-zero component K_{1212}. Hence $K_{\mu\lambda}=0$, that means that the V_4 is a special EINSTEIN space. But it can not be an S_4 because an S_4 with $K_{\mu\lambda}=0$ would be an R_4. The V_4 is symmetric because $\nabla_\omega K_{\nu\mu\lambda\varkappa}=0$.

The V_4 with the linear element with index 2

$$(1.31)\qquad \begin{cases} d\sigma^2 = 2f^{-2}(d\xi^1 d\xi^2 + d\xi^3 d\xi^4) + \\ \qquad + 2Cf^{-2}(\xi^4 d\xi^1 - \xi^1 d\xi^4)(\xi^3 d\xi^2 - \xi^2 d\xi^3) \\ f = 1 + C(\xi^1\xi^2 + \xi^3\xi^4); \quad f>0; \quad C = \text{const.} \end{cases}$$

is completely but not simply harmonic. The curvature tensor satisfies the equation

$$(1.32)\qquad \begin{cases} K_{\nu\mu\lambda\varkappa} = 4C\, g_{[\nu[\lambda} g_{\mu]\varkappa]} + S_{\nu\mu\lambda\varkappa}; \\ S_{\nu\mu\lambda\varkappa} = -S_{\mu\nu\lambda\varkappa} = -S_{\nu\mu\varkappa\lambda} = S_{\lambda\varkappa\nu\mu}; \quad S_{\nu[\mu\lambda\varkappa]} = 0; \end{cases}$$

where $S_{\nu\mu\lambda\varkappa}$ is a tensor with the only non-zero components

$$(1.33)\qquad S_{1234} = Cf^{-3}; \quad S_{1342} = Cf^{-3}; \quad S_{1423} = -2Cf^{-3}.$$

The V_4 is symmetric but no S_4. A space with the same properties but a definite fundamental tensor can be derived from (1.31) by a non-real coordinate transformation.

In a later paper[2]) WALKER proved that a completely harmonic V_4 is always symmetric if its fundamental tensor is definite.

In the same paper he proved that the equations of a completely harmonic V_4 with a fundamental tensor of index two, only containing $g_{\lambda\varkappa}$, $K_{\nu\mu\lambda\varkappa}$ and $\nabla_\omega K_{\nu\mu\lambda\varkappa}$ [viz. the equations of COPSON and RUSE (1.27, 29)] admit solutions that do neither satisfy $\nabla_\omega K_{\nu\mu\lambda\varkappa}=0$ nor $\nabla_\omega K_{\nu\mu\lambda\varkappa}=k_\omega K_{\nu\mu\lambda\varkappa}$; $k_\omega \neq 0$. But, as WALKER remarks on p. 22 this does not yet prove that there exists a harmonic V_n that is neither symmetric nor of recurrent curvature.

LICHNEROWICZ[3]) proved from (1.27) and (1.29) that in a completely harmonic V_n the following inequality is valid

$$(1.34)\qquad f'^2(0) \leq -20(n-1)f''(0)$$

[1]) WALKER 1945, 1; 2; 1946, 1; cf. RUSE 1946, 3.

[2]) WALKER 1949, 2.

[3]) LICHNEROWICZ 1944, 1, p. 161ff; 2.

and that the sign = holds if and only if the V_n is an S_n. Moreover he proved in the same paper that a completely harmonic V_n is an R_n if there exists a number $k>1$ such that

$$f'^2(0) \geq -20k(n-1) f''(0). \tag{1.35}$$

This is a generalization of the result of THOMAS and TITT mentioned before. The proof of (1.34, 35) is not difficult but rather elaborate.

In 1938[1]) FIALKOW proved that an EINSTEIN V_n in an S_{n+1} with scalar curvature $\varkappa$ is either an S_n or a product (cf. V § 11) of an S_p and an S_{n-p} with scalar curvatures $\frac{n-2}{p-1}\varkappa$ and $\frac{n-2}{n-p-1}\varkappa$ respectively. LICHNEROWICZ proved that a non-flat completely harmonic V_n can never be a product space and that according to the theorem of FIALKOW every completely harmonic V_n in S_{n+1} must be an S_n.[2])

Global properties of harmonic spaces were discussed by WILLMORE.[3])

RENAUDIE[4]) dealt with properties of the group of isotropy (cf. VII § 5) in harmonic spaces.

Exercises.

VIII 1,1[5]). Every completely harmonic V_{n-1} in S_n is an S_{n-1}.

VIII 1,2[6]). The V_4 with the linear element

$$d\sigma^2 = C\,(\xi^2\, d\xi^1 - \xi^1\, d\xi^2)^2 + 2d\xi^1\, d\xi^3 + 2d\xi^2\, d\xi^4; \qquad C = \text{const.} \neq 0$$

is simply harmonic.

VIII 1,3[6]). In a centrally harmonic V_n, $\log\varrho$ is written as a power series in s:

$$\log\varrho = \frac{1}{2!}w_1 s + \frac{1}{3!}w_2 s^2 + \cdots. \tag{VIII 1,3 α}$$

Prove that

$$w_1 = 0; \qquad w_2 = \frac{1}{s^2} K_{\mu\lambda}\, x^\mu x^\lambda \tag{VIII 1,3 β}$$

§ 2. Connexions for hybrid quantities.

In I § 10 we defined hybrid quantities in E_n. In order to consider fields of such quantities in X_n it is necessary to make an agreement about functional dependency. Ordinary quantities were called analytic

[1]) FIALKOW 1938, 2, p. 782 also for literature.
[2]) LICHNEROWICZ 1944, 1, p. 163ff; 1944, 3.
[3]) WILLMORE 1953, 1.
[4]) RENAUDIE 1954, 1.
[5]) LICHNEROWICZ 1944, 3.
[6]) WALKER 1945, 1.

if their components with respect to some allowable coordinate system $(\varkappa)$ were analytic functions of the $\xi^{\varkappa}$ (cf. II § 2). It seems convenient to call a quantity of the second kind (cf. I § 10) analytic if its components with respect to $(\varkappa)$ are analytic functions of the $\xi^{\bar{\varkappa}}$. Then a quantity of the first or second kind is analytic if its complex conjugate is analytic. Accordingly a scalar is called analytic if it is an analytic function either of the $\xi^{\varkappa}$ or of the $\xi^{\bar{\varkappa}}$. Now it would be impossible to define analyticity of hybrid quantities in the same way, because such a definition would not be invariant for allowable coordinate transformations. Therefore we introduce for *all* fields the term *semi-analytic*. A field is called semi-analytic if its components with respect to some allowable coordinate system $(\varkappa)$ are analytic in the $2n$ variables $\xi^{\varkappa}$ and $\xi^{\bar{\varkappa}}$.[1]) Only in this case they are also analytic in the $2n$ variables $\eta^{\varkappa}, \zeta^{\varkappa}$ defined by $\xi^{\varkappa} = \eta^{\varkappa} + i\zeta^{\varkappa}$; $\xi^{\bar{\varkappa}} = \eta^{\varkappa} - i\zeta^{\varkappa}$. Obviously the definition of semi-analyticity has the desired invariance. In the following all fields are supposed to be semi-analytic (and perhaps but not necessarily analytic in $\xi^{\varkappa}$ or in $\xi^{\bar{\varkappa}}$) and $\partial_{\bar{\mu}}$ stands for $\partial/\partial\xi^{\bar{\mu}}$. Note that the $\xi^{\varkappa'}$ always remain *analytic* functions of the $\xi^{\varkappa}$.[2])

In order to establish a linear connexion[3]) for all quantities, hybrid quantities included, we start with the four conditions C I, II, III, V of Ch. III § 2. But instead of C IV we take the condition

C IV′ *The covariant differential of a quantity* Ψ (indices suppressed) *is linear homogeneous in* $d\xi^{\varkappa}$ *and* $d\xi^{\bar{\varkappa}}$:

$$\delta\Psi = \nabla_{\mu}\Psi\, d\xi^{\mu} + \nabla_{\bar{\mu}}\Psi\, d\xi^{\bar{\mu}}; \tag{2.1}$$

and as a sixth condition we introduce

C VI *The covariant differentials of complex conjugate quantities are complex conjugate.*

From these conditions we get in the same way as in III § 2 the most general form of the covariant differentials of contra- and covariant vectors of both kinds

$$\text{(2.2)} \quad \begin{cases} \text{a)} \quad \delta v^{\varkappa} = d v^{\varkappa} + \Gamma_{\mu\lambda}^{\varkappa} v^{\lambda}\, d\xi^{\mu} + \Gamma_{\bar{\mu}\lambda}^{\varkappa} v^{\lambda}\, d\xi^{\bar{\mu}} \\ \text{b)} \quad \delta w_{\lambda} = d w_{\lambda} - \Gamma_{\mu\lambda}^{\varkappa} w_{\varkappa}\, d\xi^{\mu} - \Gamma_{\bar{\mu}\lambda}^{\varkappa} w_{\varkappa}\, d\xi^{\bar{\mu}} \end{cases}$$

$$\text{(2.3)} \quad \begin{cases} \text{a)} \quad \delta v^{\bar{\varkappa}} = d v^{\bar{\varkappa}} + \Gamma_{\bar{\mu}\bar{\lambda}}^{\bar{\varkappa}} v^{\bar{\lambda}}\, d\xi^{\bar{\mu}} + \Gamma_{\mu\bar{\lambda}}^{\bar{\varkappa}} v^{\bar{\lambda}}\, d\xi^{\mu} \\ \text{b)} \quad \delta w_{\bar{\lambda}} = d w_{\bar{\lambda}} - \Gamma_{\bar{\mu}\bar{\lambda}}^{\bar{\varkappa}} w_{\bar{\varkappa}}\, d\xi^{\bar{\mu}} - \Gamma_{\mu\bar{\lambda}}^{\bar{\varkappa}} w_{\bar{\varkappa}}\, d\xi^{\mu} \end{cases}$$

---

[1]) SCHOUTEN 1929, 1.

[2]) GHOSH 1950, 1 considers the more general case where the $\xi^{\varkappa'}$ are semi-analytic.

[3]) SCHOUTEN and v. DANTZIG 1929, 5; 1930, 1; SCHOUTEN 1929, 1.

from which the covariant differentials of all other quantities can be derived. In these formulae $\Gamma_{\mu\lambda}^{\varkappa}$ and $\Gamma_{\bar{\mu}\lambda}^{\varkappa}$ are arbitrary but semi-analytic and $\Gamma_{\bar{\mu}\bar{\lambda}}^{\bar{\varkappa}}$ and $\Gamma_{\mu\bar{\lambda}}^{\bar{\varkappa}}$ are their complex conjugates. In the following we often suppress formulae that are the complex conjugates of other formulae already written out and write "conj." instead. The parameters of the connexion transform in the following way

$$
(2.4) \qquad \begin{cases} \text{a)} \quad \Gamma_{\mu'\lambda'}^{\varkappa'} = A_{\varkappa\mu'\lambda'}^{\varkappa'\mu\lambda} \Gamma_{\mu\lambda}^{\varkappa} + A_{\varkappa}^{\varkappa'} \partial_{\mu'} A_{\lambda'}^{\varkappa}; & \text{conj.} \\ \text{b)} \quad \Gamma_{\bar{\mu}'\lambda'}^{\varkappa'} = A_{\varkappa\lambda'\bar{\mu}'}^{\varkappa'\lambda\bar{\mu}} \Gamma_{\bar{\mu}\lambda}^{\varkappa}; & \text{conj.} \end{cases}
$$

from which we see that the $\Gamma_{\mu\lambda}^{\varkappa}$ transform in the ordinary way (cf. III 2.5) but that *the* $\Gamma_{\bar{\mu}\lambda}^{\varkappa}$ *are components of a hybrid tensor.*

It is sometimes convenient to introduce an auxiliary X_{2n} whose coordinates are the $\xi^{\varkappa}$ and $\xi^{\bar{\varkappa}}$ considered as $2n$ independent variables that can each take all complex values. Of course the $\eta^{\varkappa}, \zeta^{\varkappa}$ could also be used as (non allowable) coordinates. In this X_{2n} not all analytic transformations of the $\xi^{\varkappa}, \xi^{\bar{\varkappa}}$ are allowable but only those that can be written in the form

$$
(2.5) \qquad \xi^{\varkappa'} = f^{\varkappa'}(\xi^{\lambda}); \quad \text{conj.}
$$

But this implies that the X_{2n} contains two invariant sets of ∞^n X_n's with the equations $\xi^{\bar{\varkappa}} = \text{const.}$ and $\xi^{\varkappa} = \text{const.}$ respectively.[1]) In the X_n's of the first (second) invariant set the $\xi^{\varkappa}$, ($\xi^{\bar{\varkappa}}$) can be used as coordinates. Through every point of the X_{2n} there goes one X_n of the first and one X_n of the second set and two X_n's of different sets intersect exactly in one point. From this it follows that the points of each X_n of one set are in one to one correspondence with the points of every other X_n of the same set and that corresponding points lie in the same X_n of the other set. A figure in one X_n and the corresponding figure in another X_n of the same set are said to be *equipollent.*

The tangent space of the X_{2n} is an E_{2n} and in this E_{2n} the tangent n-directions of the two invariant X_n's through the point considered form with their parallels the two sets of invariant E_n's defined in I § 10. A general contravariant vector in X_{2n} has $2n$ components $v^{\varkappa}$ and $v^{\bar{\varkappa}}$ and here the $v^{\bar{\varkappa}}$ are in general not the complex conjugates of the $v^{\varkappa}$. The n-dimensional subspace of the X_{2n} with the equations

$$
(2.6) \qquad \xi^{\bar{\varkappa}} = \overline{\xi^{\varkappa}}
$$

is invariant and is called the *principal* X_n of the X_{2n}. It may be considered as the image of the original X_n. If the vector $v^{\varkappa}, v^{\bar{\varkappa}}$ belongs to

[1]) In SCHOUTEN and v. DANTZIG 1930, 1, p. 327 we called these X_n's "isotropic". Here the word isotropic has another meaning, cf. I § 9 and VIII § 3.

a point of the principal X_n it lies in this X_n if and only if $v^{\bar{\varkappa}} = \overline{v^{\varkappa}}$. In this way the original X_n appears as a subspace of the auxiliary X_{2n} and the quantities of the original X_n, hybrid quantities included, are those quantities of the X_{2n} at points of the principal X_n that satisfy the condition, that dropping a bar where there is one and inserting a bar where there is none changes every component into its complex conjugate.

Apart from the connexion in the original X_n defined in (2.2, 3) we now consider the most general linear connexion in the X_{2n}. This connexion has the form (cf. III § 2)

$$\text{(2.7)} \quad \begin{cases} \text{a)} \quad \delta v^{\varkappa} = d v^{\varkappa} + (\Gamma^{\varkappa}_{\mu\lambda} v^{\lambda} + \Gamma^{\varkappa}_{\mu\bar{\lambda}} v^{\bar{\lambda}})\, d\xi^{\mu} + (\Gamma^{\varkappa}_{\bar{\mu}\lambda} v^{\lambda} + \Gamma^{\varkappa}_{\bar{\mu}\bar{\lambda}} v^{\bar{\lambda}})\, d\xi^{\bar{\mu}} \\ \text{b)} \quad \delta v^{\bar{\varkappa}} = d v^{\bar{\varkappa}} + (\Gamma^{\bar{\varkappa}}_{\mu\lambda} v^{\lambda} + \Gamma^{\bar{\varkappa}}_{\mu\bar{\lambda}} v^{\bar{\lambda}})\, d\xi^{\mu} + (\Gamma^{\bar{\varkappa}}_{\bar{\mu}\lambda} v^{\lambda} + \Gamma^{\bar{\varkappa}}_{\bar{\mu}\bar{\lambda}} v^{\bar{\lambda}})\, d\xi^{\bar{\mu}} \end{cases}$$

where the $v^{\varkappa}$, $v^{\bar{\varkappa}}$ are $2n$ arbitrary components of a general contravariant vector of X_{2n} and where the Γ are $8n^3$ independent parameters not yet in any way connected with each other or with the Γ in (2.2, 3). Because the two sets of X_n's are invariant we may now *first* introduce the invariant condition that their tangent n-directions are parallel for *every* parallel displacement in X_{2n}. This gives immediately

$$\text{(2.8)} \qquad \Gamma^{\varkappa}_{\bar{\mu}\bar{\lambda}} = 0; \quad \Gamma^{\varkappa}_{\mu\bar{\lambda}} = 0; \quad \Gamma^{\bar{\varkappa}}_{\mu\lambda} = 0; \quad \Gamma^{\bar{\varkappa}}_{\bar{\mu}\lambda} = 0.$$

The geometrical meaning is that the X_n's of either set are geodesic and parallel. *Secondly* we introduce the invariant condition that the principal X_n is geodesic. Because for every displacement in this X_n we have $d\xi^{\bar{\varkappa}} = \overline{d\xi^{\varkappa}}$, the n.a.s. conditions are

$$\text{(2.9)} \qquad \left\{ \begin{array}{l} \text{a)} \quad \Gamma^{\bar{\varkappa}}_{\bar{\mu}\bar{\lambda}} = \overline{\Gamma^{\varkappa}_{\mu\lambda}} \\ \text{b)} \quad \Gamma^{\bar{\varkappa}}_{\mu\bar{\lambda}} = \overline{\Gamma^{\varkappa}_{\bar{\mu}\lambda}} \end{array} \right\} \quad \text{for} \quad \xi^{\bar{\varkappa}} = \overline{\xi^{\varkappa}}.$$

Hence *the linear connexion in X_{2n} for which the X_n's of either invariant set are geodesic and parallel and for which the principal X_n is geodesic, is at the same time a connexion in the original X_n of the kind* (2.2, 3), *that is a connexion satisfying the conditions* C I, C II, C III, C IV′, C V *and* C VI.[1])

The covariant derivatives belonging to this connexion are

$$\text{(2.10)} \qquad \begin{cases} \text{a)} \quad \nabla_{\mu} v^{\varkappa} = \partial_{\mu} v^{\varkappa} + \Gamma^{\varkappa}_{\mu\lambda} v^{\lambda}; \qquad \text{conj.} \\ \text{b)} \quad \nabla_{\mu} w_{\lambda} = \partial_{\mu} w_{\lambda} - \Gamma^{\varkappa}_{\mu\lambda} w_{\varkappa}; \qquad \text{conj.} \end{cases}$$

$$\text{(2.11)} \qquad \begin{cases} \text{a)} \quad \nabla_{\bar{\mu}} v^{\varkappa} = \partial_{\bar{\mu}} v^{\varkappa} + \Gamma^{\varkappa}_{\bar{\mu}\lambda} v^{\lambda}; \qquad \text{conj.} \\ \text{b)} \quad \nabla_{\bar{\mu}} w_{\lambda} = \partial_{\bar{\mu}} w_{\lambda} - \Gamma^{\varkappa}_{\bar{\mu}\lambda} w_{\varkappa}; \qquad \text{conj.} \end{cases}$$

[1]) Cf. footnote 3) on p. 389.

They can be looked upon as parts of the covariant derivatives of the vectors $v^\varkappa$, $v^{\bar\varkappa}$ and w_λ, $w_{\bar\lambda}$ in the auxiliary X_{2n}.

The curvature tensors belonging to the connexion (2.2, 3) may be considered as the only non-vanishing parts of the curvature tensor in X_{2n}. There are eight of them:

$$(2.12)\quad\begin{cases} \text{a)}\ R_{\nu\mu\lambda}^{\cdot\cdot\cdot\ \varkappa} = 2\partial_{[\nu}\Gamma^{\varkappa}_{\mu]\lambda} + 2\Gamma^{\varkappa}_{[\nu|\varrho|}\Gamma^{\varrho}_{\mu]\lambda};\quad \text{conj.} \\ \text{b)}\ R_{\bar\nu\mu\lambda}^{\cdot\cdot\cdot\ \varkappa} = \partial_{\bar\nu}\Gamma^{\varkappa}_{\mu\lambda} - \partial_\mu\Gamma^{\varkappa}_{\bar\nu\lambda} + \Gamma^{\varkappa}_{\bar\nu\varrho}\Gamma^{\varrho}_{\mu\lambda} - \Gamma^{\varkappa}_{\mu\varrho}\Gamma^{\varrho}_{\bar\nu\lambda};\quad \text{conj.} \\ \text{c)}\ R_{\nu\bar\mu\lambda}^{\cdot\cdot\cdot\ \varkappa} = -R_{\bar\mu\nu\lambda}^{\cdot\cdot\cdot\ \varkappa};\quad \text{conj.} \qquad\qquad \text{(isomer of (b))} \\ \text{d)}\ R_{\bar\nu\bar\mu\lambda}^{\cdot\cdot\cdot\ \varkappa} = 2\partial_{[\bar\nu}\Gamma^{\varkappa}_{\bar\mu]\lambda} + 2\Gamma^{\varkappa}_{[\bar\nu|\varrho|}\Gamma^{\varrho}_{\bar\mu]\lambda};\quad \text{conj.} \end{cases}$$

the other eight, $R_{\nu\mu\lambda}^{\cdot\cdot\cdot\ \bar\varkappa}$, $R_{\bar\nu\mu\lambda}^{\cdot\cdot\cdot\ \bar\varkappa}$, $R_{\nu\bar\mu\lambda}^{\cdot\cdot\cdot\ \bar\varkappa}$, $R_{\bar\nu\bar\mu\lambda}^{\cdot\cdot\cdot\ \bar\varkappa}$, conj. being zero. The tensors of asymmetry can in the same way be looked upon as the only non-vanishing parts of the tensor of asymmetry in X_{2n}. There are six of them:

$$(2.13)\quad\begin{cases} \text{a)}\ S_{\mu\lambda}^{\cdot\cdot\ \varkappa} = \Gamma^{\varkappa}_{[\mu\lambda]};\quad \text{conj.} \\ \text{b)}\ S_{\bar\mu\lambda}^{\cdot\cdot\ \varkappa} = \tfrac{1}{2}\Gamma^{\varkappa}_{\bar\mu\lambda};\quad \text{conj.} \\ \text{c)}\ S_{\mu\bar\lambda}^{\cdot\cdot\ \varkappa} = -S_{\bar\lambda\mu}^{\cdot\cdot\ \varkappa};\quad \text{conj.} \qquad \text{(isomer of (b))} \end{cases}$$

the other two, $S_{\mu\lambda}^{\cdot\cdot\ \bar\varkappa}$, conj. being zero.

A contravariant vector of an X_n of the first invariant set is characterized by $v^\varkappa \neq 0$; $v^{\bar\varkappa} = 0$. If this vector is displaced parallel in a direction lying in an X_n of the second invariant set, we have $d\xi^\varkappa = 0$, $d\xi^{\bar\varkappa} \neq 0$, hence

$$(2.14)\qquad d v^\varkappa = d\xi^{\bar\mu}\,\Gamma^{\varkappa}_{\bar\mu\lambda}\,v^\lambda.$$

But this proves that the transformation is an equipollence for every choice of $v^\varkappa$, if and only if $\Gamma^{\varkappa}_{\bar\mu\lambda} = 0$. Hence, if we introduce as a *third* invariant condition that the parallel displacement of a vector of an X_n of one invariant set in a direction lying in an X_n of the other set is always an equipollence, this leads to the equations

$$(2.15)\qquad \Gamma^{\varkappa}_{\bar\mu\lambda} = 0;\qquad \Gamma^{\bar\varkappa}_{\mu\bar\lambda} = 0.$$

Because $\Gamma^{\varkappa}_{\bar\mu\lambda}$ and $\Gamma^{\bar\varkappa}_{\mu\bar\lambda}$ are hybrid tensors as we have seen, the condition (2.15) is also invariant in the original X_n. Another geometrical meaning of (2.15) is that it is the n.a.s. condition for the existence of infinitesimal parallelograms in X_{2n} with two parallel sides in an arbitrary direction of an X_n of one invariant set and the other two in an arbitrary direction of an X_n of the other set. This is quite clear because $\Gamma^{\varkappa}_{\lambda\bar\mu}$ and $\Gamma^{\bar\varkappa}_{\bar\lambda\mu}$ are already zero according to (2.8), and (2.15) is therefore equivalent to the vanishing of the components $\Gamma^{\varkappa}_{[\bar\mu\lambda]}$, $\Gamma^{\bar\varkappa}_{[\mu\bar\lambda]}$ of the tensor of asymmetry in X_{2n}. So (2.15) is in fact a kind of condition of symmetry of the connexion in X_{2n} but only halfway.

If the third condition is also satisfied we have[1])

$$(2.16)\quad \begin{cases} \text{a)}\ \nabla_\mu v^\varkappa = \partial_\mu v^\varkappa + \Gamma^{\varkappa}_{\mu\lambda} v^\lambda; & \nabla_{\bar\mu} v^\varkappa = \partial_{\bar\mu} v^\varkappa; \quad \text{conj.} \\ \text{b)}\ \nabla_\mu w_\lambda = \partial_\mu w_\lambda - \Gamma^{\varkappa}_{\mu\lambda} w_\varkappa; & \nabla_{\bar\mu} w_\lambda = \partial_{\bar\mu} w_\lambda; \quad \text{conj.} \end{cases}$$

and this connexion differs from an ordinary connexion in L_n as dealt with in III § 2 only by its special equations for fields that are only semi-analytic.

From (2.16) it can be derived in the ordinary way via the n-vectors that for a density with the weights w and w' there is for ∇_μ a term $-w\,\Gamma^{\tau}_{\mu\tau}$ and for $\nabla_{\bar\mu}$ a term with $-w'\,\Gamma^{\bar\tau}_{\bar\mu\bar\tau}$. An X_n with the connexion (2.16) is called an $\tilde L_n$. It is the most general linear connexion for hybrid quantities that satisfies all invariant conditions and in this way the $\tilde L_n$ is analogous to the L_n in III § 2. The auxiliary X_{2n} of an $\tilde L_n$ is an L_{2n}.

In an $\tilde L_n$ only the curvature tensors $R_{\nu\mu\lambda}^{\cdot\cdot\cdot\varkappa}$; conj., $R_{\bar\nu\mu\lambda}^{\cdot\cdot\cdot\varkappa}$; conj. and $R_{\nu\bar\mu\lambda}^{\cdot\cdot\cdot\varkappa}$; conj. remain and (2.12b) takes the simpler form

$$(2.17)\qquad R_{\bar\nu\mu\lambda}^{\cdot\cdot\cdot\varkappa} = \partial_{\bar\nu} \Gamma^{\varkappa}_{\mu\lambda}; \quad \text{conj.}$$

From the tensors of asymmetry only $S_{\mu\lambda}^{\cdot\cdot\varkappa}$; conj. remain.

In an $\tilde L_n$ the quantities $R_{\nu\mu\lambda}^{\cdot\cdot\cdot\varkappa}$ and $S_{\mu\lambda}^{\cdot\cdot\varkappa}$ satisfy the identities (III 5.1), (III 5.2) and (III 5.19) and similar identities hold for their conjugates. The quantities $R_{\nu\mu\lambda}^{\cdot\cdot\cdot\varkappa}$, $R_{\bar\nu\mu\lambda}^{\cdot\cdot\cdot\varkappa}$, $R_{\nu\bar\mu\lambda}^{\cdot\cdot\cdot\varkappa}$, $S_{\mu\lambda}^{\cdot\cdot\varkappa}$ and their conjugates satisfy the following identities that can be easily derived from (III 5.1, 2, 19) if these equations are applied to the X_{2n}:

$$(2.18)\qquad R_{\bar\nu\mu\lambda}^{\cdot\cdot\cdot\varkappa} = -R_{\mu\bar\nu\lambda}^{\cdot\cdot\cdot\varkappa}; \quad \text{conj.} \quad \text{(first identity, cf. III 5.1)}$$

$$(2.19)\qquad R_{\bar\nu[\mu\lambda]}^{\cdot\cdot\cdot\varkappa} = \nabla_{\bar\nu} S_{\mu\lambda}^{\cdot\cdot\varkappa} = \partial_{\bar\nu} S_{\mu\lambda}^{\cdot\cdot\varkappa}; \quad \text{conj.} \quad \text{(second identity, cf. III 5.2)}$$

$$(2.20)\quad \begin{cases} \text{a)}\ 2\nabla_{[\omega} R_{\nu]\bar\mu\lambda}^{\cdot\cdot\cdot\varkappa} + \partial_{\bar\mu} R_{\omega\nu\lambda}^{\cdot\cdot\cdot\varkappa} = 2 S_{\omega\nu}^{\cdot\cdot\varrho} R_{\bar\mu\varrho\lambda}^{\cdot\cdot\cdot\varkappa}; \quad \text{conj.} \\ \text{b)}\ \nabla_{[\bar\omega} R_{\bar\nu]\mu\lambda}^{\cdot\cdot\cdot\varkappa} = S_{\bar\omega\bar\nu}^{\cdot\cdot\bar\varrho} R_{\mu\bar\varrho\lambda}^{\cdot\cdot\cdot\varkappa}; \quad \text{conj.} \\ \qquad \text{(identity of Bianchi, cf. III 5.19)} \end{cases}$$

---

1) Schouten 1929, 1, p. 460.

The Ricci tensor R and the bivector V in L_{2n} have the non vanishing components

$$
(2.21)\quad \begin{cases}
\text{a)}\quad R_{\mu\lambda} = R^{\cdot\cdot\cdot\nu}_{\nu\mu\lambda} + R^{\cdot\cdot\cdot\bar{\nu}}_{\bar{\nu}\mu\lambda} = R^{\cdot\cdot\cdot\nu}_{\nu\mu\lambda}; \quad \text{conj.}\\
\text{b)}\quad R_{\bar{\nu}\lambda} = R^{\cdot\cdot\cdot\mu}_{\mu\bar{\nu}\lambda} + R^{\cdot\cdot\cdot\bar{\mu}}_{\bar{\mu}\bar{\nu}\lambda} = R^{\cdot\cdot\cdot\mu}_{\mu\bar{\nu}\lambda}; \quad \text{conj.}\\
\text{c)}\quad V_{\nu\mu} = R^{\cdot\cdot\cdot\lambda}_{\nu\mu\lambda} + R^{\cdot\cdot\cdot\bar{\lambda}}_{\nu\mu\bar{\lambda}} = R^{\cdot\cdot\cdot\lambda}_{\nu\mu\lambda}; \quad \text{conj.}\\
\text{d)}\quad V_{\bar{\nu}\mu} = R^{\cdot\cdot\cdot\lambda}_{\bar{\nu}\mu\lambda} + R^{\cdot\cdot\cdot\bar{\lambda}}_{\bar{\nu}\mu\bar{\lambda}} = V'_{\bar{\nu}\mu} - V'_{\mu\bar{\nu}}; \quad \text{conj.};\\
\qquad V'_{\bar{\nu}\mu} \overset{\text{def}}{=} R^{\cdot\cdot\cdot\lambda}_{\bar{\nu}\mu\lambda} = \partial_{\bar{\nu}} \Gamma^{\lambda}_{\mu\lambda}
\end{cases}
$$

but from (2.18) *it does not follow that* $V'_{\bar{\nu}\mu} = -V'_{\mu\bar{\nu}}$, because $V'_{\mu\bar{\nu}}$ is according to our definitions the complex conjugate of $V'_{\bar{\mu}\nu}$ and thus equal to $R^{\cdot\cdot\cdot\bar{\lambda}}_{\mu\bar{\nu}\bar{\lambda}}$ and not equal to $R^{\cdot\cdot\cdot\lambda}_{\mu\bar{\nu}\lambda}$. From (2.19) we get the equation

$$
(2.22)\qquad R_{\bar{\nu}\lambda} + V'_{\bar{\nu}\lambda} = 2\,\partial_{\bar{\nu}} S^{\cdot\cdot\mu}_{\lambda\mu} = (n-1)\,\partial_{\bar{\nu}} S_{\lambda},
$$

which will be used hereafter. From (2.21) we see that $V_{\bar{\nu}\mu}$ (and *not* $V'_{\bar{\nu}\mu}$) is always a hybrid bivector and that $V'_{\bar{\nu}\mu}$ is hermitian if and only if $V_{\bar{\nu}\mu} = 0$. From (2.20) we derive by contraction over $\varkappa\lambda$

$$
(2.23)\quad \boxed{\begin{array}{ll}
\text{a)}\quad 2\nabla_{[\omega} V'_{|\bar{\mu}|\nu]} - \partial_{\bar{\mu}} V_{\omega\nu} = -2 S^{\cdot\cdot\varrho}_{\omega\nu} V'_{\bar{\mu}\varrho}; & \text{conj.}\\
\text{b)}\quad \nabla_{[\bar{\omega}} V'_{\bar{\nu}]\mu} = -S^{\cdot\cdot\bar{\varrho}}_{\bar{\omega}\bar{\nu}} V'_{\bar{\varrho}\mu}; \quad \text{conj.} &
\end{array}}
$$

from which, by means of (2.21 d)

$$
(2.24)\qquad 2\nabla_{[\bar{\omega}} V_{\bar{\nu}]\mu} + \partial_{\mu} V_{\bar{\omega}\bar{\nu}} = -2 S^{\cdot\cdot\bar{\varrho}}_{\bar{\omega}\bar{\nu}} V_{\bar{\varrho}\mu}; \quad \text{conj.}
$$

in accordance with (III 5.22). By contraction over $\varkappa\nu$ and $\varkappa\mu$ we get from (2.20a) and (2.20b)

$$
\boxed{\begin{array}{ll}
(2.25)\quad \nabla_{\omega} R_{\bar{\mu}\lambda} + \nabla_{\nu} R^{\cdot\cdot\cdot\nu}_{\bar{\mu}\omega\lambda} - \partial_{\bar{\mu}} R_{\omega\lambda} = 2 S^{\cdot\cdot\varrho}_{\omega\nu} R^{\cdot\cdot\cdot\nu}_{\bar{\mu}\varrho\lambda}; & \text{conj.}\\
(2.26)\quad \nabla_{[\bar{\nu}} R_{\bar{\omega}]\lambda} = -S^{\cdot\cdot\bar{\varrho}}_{\bar{\nu}\bar{\omega}} R_{\bar{\varrho}\lambda}; \quad \text{conj.} &
\end{array}}
$$

that could also be derived from (III 5.25). (2.23 b) and (2.26) are in accordance with (2.22).

An $\widetilde{L}_n$ with a symmetric connexion is called an $\widetilde{A}_n$. It is analogous to the A_n of III § 2. The auxiliary X_{2n} of an $\widetilde{A}_n$ is an A_{2n}. In an $\widetilde{A}_n$ we have instead of (2.19, 20, 22, 23, 24, 25, 26)

$$
(2.27)\qquad R^{\cdot\cdot\cdot\varkappa}_{\bar{\nu}[\mu\lambda]} = 0; \quad \text{conj.}
$$

$$
(2.28)\quad \begin{cases}
\text{a)}\quad 2\nabla_{[\omega} R^{\cdot\cdot\cdot\varkappa}_{\nu]\bar{\mu}\lambda} + \partial_{\bar{\mu}} R^{\cdot\cdot\cdot\varkappa}_{\omega\nu\lambda} = 0; \quad \text{conj.}\\
\text{b)}\quad \nabla_{[\bar{\omega}} R^{\cdot\cdot\cdot\varkappa}_{\bar{\nu}]\mu\lambda} = 0; \quad \text{conj.}
\end{cases}
$$

(2.29) $$R_{\bar{\nu}\lambda} = -V'_{\bar{\nu}\lambda} = -\partial_{\bar{\nu}} \Gamma^{\mu}_{\lambda\mu}; \quad \text{conj.}$$

(2.30) $$\begin{cases} \text{a)} & 2\nabla_{[\omega} V'_{|\bar{\mu}|\nu]} - \partial_{\bar{\mu}} V_{\omega\nu} = -2\nabla_{[\omega} R_{|\bar{\mu}|\nu]} + 2\partial_{\bar{\mu}} R_{[\omega\nu]} = 0; \quad \text{conj.} \\ \text{b)} & \nabla_{[\bar{\omega}} V'_{\bar{\nu}]\mu} = -\nabla_{[\bar{\omega}} R_{\bar{\nu}]\mu} = 0; \quad \text{conj.} \end{cases}$$

(2.31) $$2\nabla_{[\bar{\omega}} V_{\bar{\nu}]\mu} + \partial_{\mu} V_{\bar{\omega}\bar{\nu}} = 0; \quad \text{conj.}$$

(2.32) $$\nabla_{\omega} R_{\bar{\mu}\lambda} + \nabla_{\nu} R_{\bar{\mu}\omega\lambda}{}^{\nu} - \partial_{\bar{\mu}} R_{\omega\lambda} = 0; \quad \text{conj.}$$

(2.33) $$\nabla_{[\bar{\nu}} R_{\bar{\mu}]\lambda} = 0; \quad \text{conj.}$$

Of course the equations (III 5.2–4) remain valid, hence we have in an $\tilde{L}_n$:

(2.34) $$R_{[\nu\mu\lambda]}{}^{\varkappa} = 2\nabla_{[\nu} S_{\mu\lambda]}{}^{\varkappa} - 4S_{[\nu\mu}{}^{\varrho} S_{\lambda]\varrho}{}^{\varkappa}; \quad \text{conj.},$$

in a semi-symmetric $\tilde{L}_n$:

(2.35) $$R_{[\nu\mu\lambda]}{}^{\varkappa} = 2A^{\varkappa}_{[\nu} \nabla_{\mu} S_{\lambda]}; \quad \text{conj.},$$

and in an $\tilde{A}_n$:

(2.36) $$R_{[\nu\mu\lambda]}{}^{\varkappa} = 0; \quad \text{conj.}$$

Exercises.

VIII 2,1[1]). Let $v^{\varkappa}$ be a semi-analytic field and $v^{\bar{\varkappa}}$ its complex conjugate. This field is analytic if and only if the field $v^{\varkappa}, 0$ $(0, v^{\bar{\varkappa}})$ in X_{2n} is constant over each X_n of the second (first) invariant set.

A semi-analytic scalar field is analytic in $\xi^{\varkappa}$ $(\xi^{\bar{\varkappa}})$ if and only if it is constant in every X_n of the second (first) invariant set. It is the product of a scalar analytic in $\xi^{\varkappa}$ with a scalar analytic in $\xi^{\bar{\varkappa}}$ if and only if

VIII 2,1 α) $$\varphi\, \partial_{\bar{\nu}} \partial_{\mu} \varphi = (\partial_{\bar{\nu}} \varphi)\, \partial_{\mu} \varphi.$$

VIII 2,2[2]). An analytic transformation of X_{2n} is an analytic transformation of X_n if and only if it leaves invariant both invariant sets of X_n's and the principal X_n.

§ 3. Unitary connexions.

If a hermitian tensor field $a_{\lambda\bar{\varkappa}}$ of rank n is given in X_n we may ask for a connexion of the kind (2.2, 3) for which

(3.1) $$\nabla_{\mu} a_{\lambda\bar{\varkappa}} = 0; \quad \text{conj.}$$

or

(3.2) $$\partial_{\mu} a_{\lambda\bar{\varkappa}} - \Gamma^{\varrho}_{\mu\lambda} a_{\varrho\bar{\varkappa}} - \Gamma^{\bar{\sigma}}_{\mu\bar{\varkappa}} a_{\lambda\bar{\sigma}} = 0; \quad \text{conj.}$$

[1]) SCHOUTEN and v. DANTZIG 1930, 1, p. 328 f.; E II 1938, 2, p. 236.

[2]) SCHOUTEN and v. DANTZIG 1930, 1, p. 329.

From (3.2) we get immediately

$$\Gamma_{\mu\lambda}^{\varkappa} = a^{\bar{\varrho}\varkappa}\, \partial_\mu\, a_{\lambda\bar{\varrho}} - \Gamma_{\mu\bar{\varrho}}^{\bar{\sigma}}\, a_{\lambda\bar{\sigma}}\, a^{\bar{\varrho}\varkappa}; \quad \text{conj.} \tag{3.3}$$

which equation expresses $\Gamma_{\mu\lambda}^{\varkappa}$ in terms of $a_{\lambda\bar{\varkappa}}$ and its first derivatives and a hybrid tensor $\Gamma_{\mu\bar{\varrho}}^{\bar{\sigma}}$ that could be chosen arbitrarily and that is zero if we impose the condition that the X_n be an $\tilde{L}_n$. Then we get[1])

$$\left\{\begin{array}{lll} \text{a)} & \Gamma_{\mu\lambda}^{\varkappa} = a^{\bar{\varrho}\varkappa}\, \partial_\mu\, a_{\lambda\bar{\varrho}}; & \text{conj.} \\ \text{b)} & \Gamma_{\bar{\mu}\lambda}^{\varkappa} = 0; & \text{conj.} \end{array}\right. \tag{3.4}$$

Hence:

A hermitian tensor field $a_{\lambda\bar{\varkappa}}$ of rank n in an X_n determines one and only one $\tilde{L}_n$-connexion for which $a_{\lambda\bar{\varkappa}}$ is covariant constant.

Such a connexion is called *unitary* and an X_n with a unitary connexion is called a $\tilde{U}_n$.[2]) $a_{\lambda\bar{\varkappa}}$ is called the *fundamental tensor* of the $\tilde{U}_n$. It is analogous to the U_n of III § 3.

In the auxiliary X_{2n} of $\xi^\varkappa$ and $\xi^{\bar{\varkappa}}$ we now use indices $A, B, \ldots$ taking the values $1, \ldots, n, \bar{1}, \ldots, \bar{n}$. Then the $a_{\lambda\bar{\varkappa}}$ are the components of a symmetric tensor a_{BA} with $a_{\lambda\varkappa} = 0$, $a_{\bar{\lambda}\bar{\varkappa}} = 0$. Hence the invariant X_n's are isotropic (cf. I § 9) with respect to this fundamental tensor. The most general linear connexion in X_{2n} leaving a_{BA} invariant is (cf. III 3.5)

$$\Gamma_{CB}^{A} = \tfrac{1}{2}\, a^{AD}\, (\partial_C\, a_{DB} + \partial_B\, a_{DC} - \partial_D\, a_{CB}) + S_{\dot{C}\dot{B}}{}^{A} - S_{\dot{B}}{}^{A}{}_{\cdot C} - S_{\dot{C}}{}^{A}{}_{\cdot B} \tag{3.5}$$

where $S_{\dot{C}\dot{B}}{}^{A}$ is an arbitrary tensor alternating in CB, or

$$\left\{\begin{array}{ll} \text{a)} & \Gamma_{\mu\lambda}^{\varkappa} = a^{\bar{\varrho}\varkappa}\, \partial_{(\mu}\, a_{\lambda)\bar{\varrho}} + S_{\dot{\mu}\dot{\lambda}}{}^{\varkappa} - 2\, S_{(\mu}{}^{\varkappa}{}_{\cdot\lambda)} \\ \text{b)} & \Gamma_{\bar{\mu}\lambda}^{\varkappa} = a^{\bar{\varrho}\varkappa}\, \partial_{[\bar{\mu}}\, a_{\bar{\varrho}]\lambda} + S_{\dot{\bar{\mu}}\dot{\lambda}}{}^{\varkappa} - S_{\lambda}{}^{\varkappa}{}_{\cdot\bar{\mu}} - S_{\bar{\mu}}{}^{\varkappa}{}_{\cdot\lambda} \\ \text{c)} & \Gamma_{\mu\bar{\lambda}}^{\varkappa} = a^{\bar{\varrho}\varkappa}\, \partial_{[\bar{\lambda}}\, a_{\bar{\varrho}]\mu} + S_{\dot{\mu}\dot{\bar{\lambda}}}{}^{\varkappa} - S_{\bar{\lambda}}{}^{\varkappa}{}_{\cdot\mu} - S_{\mu}{}^{\varkappa}{}_{\cdot\bar{\lambda}} \\ \text{d)} & \Gamma_{\bar{\mu}\bar{\lambda}}^{\varkappa} = S_{\dot{\bar{\mu}}\dot{\bar{\lambda}}}{}^{\varkappa} - S_{\bar{\lambda}}{}^{\varkappa}{}_{\cdot\bar{\mu}} - S_{\bar{\mu}}{}^{\varkappa}{}_{\cdot\bar{\lambda}} \end{array}\right. \tag{3.6}$$

and four expressions for $\Gamma_{\bar{\mu}\bar{\lambda}}^{\bar{\varkappa}}$, $\Gamma_{\mu\bar{\lambda}}^{\bar{\varkappa}}$, $\Gamma_{\bar{\mu}\lambda}^{\bar{\varkappa}}$ and $\Gamma_{\mu\lambda}^{\bar{\varkappa}}$. Now these latter four expressions must be the complex conjugates of the expressions (3.6) in consequence of the condition that the principal X_n in the X_{2n} be geodesic. But this is possible if and only if $S_{\dot{\bar{\mu}}\dot{\bar{\lambda}}}{}^{\bar{\varkappa}}$, $S_{\dot{\mu}\dot{\bar{\lambda}}}{}^{\bar{\varkappa}}$ and $S_{\dot{\mu}\dot{\lambda}}{}^{\bar{\varkappa}}$ are the complex conjugates of $S_{\dot{\mu}\dot{\lambda}}{}^{\varkappa}$, $S_{\dot{\bar{\mu}}\dot{\lambda}}{}^{\varkappa}$ and $S_{\dot{\bar{\mu}}\dot{\bar{\lambda}}}{}^{\varkappa}$ respectively. If now we introduce the condition (2.8) expressing that the invariant X_n's of either set are geodesic and parallel and the condition (2.15) for the equipollence of

---

1) Cf. footnote 3) on p. 389.

2) In 1930, 3 we wrote U_n and in E II 1938, 2 we used the term unitary K_n instead of $\tilde{U}_n$. The notations $\tilde{L}_n$, $\tilde{A}_n$, $\tilde{U}_n$, $\tilde{V}_n$, $\tilde{S}_n$ and $\tilde{C}_n$ used now are in strict analogy to the L_n, A_n, U_n, V_n, S_n and C_n of Ch. III and VI.

parallel invariant X_n's we get

$$S_{\mu\lambda}^{\cdot\cdot\bar{\varkappa}} = 0; \quad S_{\bar{\mu}\lambda}^{\cdot\cdot\varkappa} = 0; \quad \text{conj.} \tag{3.7}$$

and

$$S_{\bar{\mu}\bar{\varkappa}\lambda} = \partial_{[\bar{\mu}} a_{\bar{\varkappa}]\lambda}; \quad \text{conj.} \tag{3.8}$$

or

$$S_{\mu\lambda}^{\cdot\cdot\varkappa} = a^{\bar{\varrho}\varkappa}\, \partial_{[\mu} a_{\lambda]\bar{\varrho}}; \quad \text{conj.} \tag{3.9}$$

hence

$$\begin{cases} \text{a)} \quad \Gamma_{\mu\lambda}^{\varkappa} = a^{\bar{\varrho}\varkappa}\, \partial_{\mu} a_{\lambda\bar{\varrho}}; & \text{conj.} \\ \text{b)} \quad \Gamma_{\bar{\mu}\lambda}^{\varkappa} = 0; & \text{conj.} \end{cases} \tag{3.10}$$

in accordance with (3.4). This proves that the connexion in a $\tilde{U}_n$ can be derived as a connexion in X_{2n} with the symmetric fundamental tensor a_{BA} with $a_{\lambda\varkappa}=0$, $a_{\bar{\lambda}\bar{\varkappa}}=0$, satisfying instead of the condition of symmetry other conditions concerning the invariant X_n's. Hence the auxiliary X_{2n} of a $\tilde{U}_n$ is a U_{2n}.[1]) In general the connexion of $\tilde{U}_n$ is not symmetric. If it is symmetric we call the $\tilde{U}_n$ a $\tilde{V}_n$. The $\tilde{V}_n$ is analogous to the V_n of III § 3 and its auxiliary X_{2n} is a V_{2n}. We prove that $S_{\mu\lambda}^{\cdot\cdot\varkappa}=0$; conj. if and only if there exists a semi-analytic scalar field φ such that

$$a_{\bar{\lambda}\varkappa} = \partial_{\bar{\lambda}}\, \partial_{\varkappa} \varphi \text{ [2])} \tag{3.11}$$

We have only to prove that (3.11) holds if $S_{\mu\lambda}^{\cdot\cdot\varkappa}=0$; conj. because the inverse is trivial.[3]) In order to prove this we remark that the equation

$$\partial_{\bar{\lambda}} \varphi_{\varkappa} = a_{\bar{\lambda}\varkappa} \tag{3.12}$$

with the unknowns $\varphi_{\varkappa}$ is totally integrable because of (3.2) and $S_{\bar{\mu}\bar{\lambda}}^{\cdot\cdot\bar{\varkappa}}=0$; conj. if the $\xi^{\varkappa}$ are considered as parameters in (3.12). Hence, if $\overset{0}{\varphi}_{\varkappa}$ is a solution of (3.12) the general solution has the form

$$\varphi_{\varkappa} - \overset{0}{\varphi}_{\varkappa}(\xi^{\varrho}, \xi^{\bar{\varrho}}) + f_{\varkappa}(\xi^{\varrho}) \tag{3.13}$$

[1]) Cf. ROZENFEL'D 1949, 1.

[2]) SCHOUTEN and v. DANTZIG 1930, 1 for a special case and KÄHLER 1933, 1 for the general case; cf. LEE 1942, 1; SASAKI 1949, 2; IWAMOTO 1950, 1, p. 125.

[3]) The symmetric unitary connexion or $\tilde{V}_n$ appeared first in SCHOUTEN and v. DANTZIG 1930, 1, p. 333. In 1933, 1 BERGMAN gave an application of this connexion to the theory of functions of two complex variables by means of his "kernel function". This application has been dealt with since in many publications of BERGMAN and others. In 1933, 1 KÄHLER independently found the same connexion and formulated the condition of symmetry by means of CARTAN's alternating differential forms. A $\tilde{V}_n$ is sometimes called "KÄHLER space" in the literature. Cf. for instance ECKMANN and GUGGENHEIMER 1949, 1; 2. HODGE 1951, 1; BOCHNER 1953, 1; GUGGENHEIMER 1953, 1. There are a great number of publications of these and other authors dealing with global properties of the $\tilde{V}_n$.

where the $f_\varkappa$ are arbitrary functions. This proves that φ is a solution of

$$\partial_\lambda \varphi = \overset{0}{\varphi}_\lambda + f_\lambda(\xi^\varkappa). \tag{3.14}$$

The integrability conditions of (3.14) are

$$\partial_{[\mu} \overset{0}{\varphi}_{\lambda]} + \partial_{[\mu} f_{\lambda]} = 0; \quad \text{conj.} \tag{3.15}$$

From (3.9, 12) and the vanishing of $S_{\mu\bar{\lambda}}^{\cdot\cdot\,\varkappa}$ it follows that

$$\partial_{\bar{\nu}} \partial_{[\mu} \overset{0}{\varphi}_{\lambda]} = \partial_{[\mu} \partial_{|\bar{\nu}|} \overset{0}{\varphi}_{\lambda]} = \partial_{[\mu} a_{\lambda]\bar{\nu}} = 0 \tag{3.16}$$

and this proves that $\partial_{[\mu} \overset{0}{\varphi}_{\lambda]}$ depends on the $\xi^\varkappa$ only and that accordingly f_λ can always be chosen such that (3.15) is satisfied identically.

If φ is a solution of (3.11), $\overline{\varphi}$ is also a solution and this proves that if (3.11) has solutions, there always exists a *real* solution.

In a $\widetilde{U}_n$ we have

$$\partial_{[\nu} \Gamma_{\mu]\lambda}^{\varkappa} + \Gamma_{[\nu|\sigma|}^{\varkappa} \Gamma_{\mu]\lambda}^{\sigma} = \partial_{[\nu} a^{\bar{\varrho}\varkappa} \partial_{\mu]} a_{\lambda\bar{\varrho}} + (a^{\bar{\varrho}\varkappa} \partial_{[\nu} a_{|\sigma\bar{\varrho}|}) a^{\bar{\tau}\sigma} \partial_{\mu]} a_{\lambda\bar{\tau}} = 0 \tag{3.17}$$

hence

$$R_{\nu\mu\lambda}^{\cdot\cdot\cdot\,\varkappa} = 0; \quad \text{conj.} \tag{3.18}$$

and the only remaining curvature tensors are $R_{\bar{\nu}\mu\lambda}^{\cdot\cdot\cdot\,\varkappa}$; conj. and $R_{\nu\bar{\mu}\lambda}^{\cdot\cdot\cdot\,\varkappa}$; conj. As in Riemannian geometry it is convenient to lower the last index of $R_{\bar{\nu}\mu\lambda}^{\cdot\cdot\cdot\,\varkappa}$. Then we get from (2.17) and (3.4a)

$$\text{(3.19)} \quad \begin{cases} \text{a)} \; R_{\bar{\nu}\mu\lambda\bar{\varkappa}} = \partial_{\bar{\nu}} \partial_\mu a_{\lambda\bar{\varkappa}} - a^{\bar{\varrho}\sigma} (\partial_{\bar{\nu}} a_{\sigma\bar{\varkappa}}) \partial_\mu a_{\lambda\bar{\varrho}}; \quad \text{conj.} \\ \text{b)} \quad R_{\bar{\nu}\lambda} = -\partial_{\bar{\nu}} a^{\bar{\varkappa}\mu} \partial_\mu a_{\lambda\bar{\varkappa}} \\ \qquad\quad = -\partial_{\bar{\nu}} \Gamma_{\mu\lambda}^{\mu} = -\partial_{\bar{\nu}} \partial_\lambda \log \mathfrak{a} + (n-1)\, \partial_{\bar{\nu}} S_\lambda; \quad \text{conj.} \\ \text{c)} \quad V'_{\bar{\nu}\mu} = \partial_{\bar{\nu}} \Gamma_{\mu\lambda}^{\lambda} = \partial_{\bar{\nu}} \partial_\mu \log \mathfrak{a} = V'_{\mu\bar{\nu}}; \quad \mathfrak{a} \overset{\text{def}}{=} |\operatorname{Det}(a_{\lambda\bar{\varkappa}})| \\ \text{d)} \quad V_{\bar{\nu}\mu} = 0; \quad \text{conj.} \end{cases}$$

The vanishing of $V_{\bar{\nu}\mu}$ is a consequence of the fact that the auxiliary X_{2n} is a U_{2n}. But as we have already seen this vanishing implies that $V'_{\bar{\nu}\mu}$ is hermitian and this is in accordance with (3.19c).

The following identities hold in a $\widetilde{U}_n$[1])

$$R_{\bar{\nu}\mu\lambda\bar{\varkappa}} = -R_{\mu\bar{\nu}\lambda\bar{\varkappa}}; \quad \text{conj.} \tag{3.20}$$

(first identity, cf. III 5.1; VIII 2.18)

$$R_{\bar{\nu}[\mu\lambda]\bar{\varkappa}} = \nabla_{\bar{\nu}} S_{\mu\lambda\bar{\varkappa}}; \quad \text{conj.} \tag{3.21}$$

(second identity, cf. III 5.2; VIII 2.19)

[1]) E II 1938, 2, p. 234–239.

(3.22) $$R_{\bar{\nu}\mu\lambda\bar{\varkappa}} = - R_{\bar{\nu}\mu\bar{\varkappa}\lambda}; \quad \text{conj.}$$
(third identity, cf. III 5.13)

(3.23) $$R_{\bar{\nu}\mu\lambda\bar{\varkappa}} - R_{\lambda\bar{\varkappa}\bar{\nu}\mu} = 2 \nabla_{\bar{\nu}} S_{\mu\lambda\bar{\varkappa}} - 2 \nabla_{\lambda} S_{\bar{\varkappa}\bar{\nu}\mu}; \quad \text{conj.}$$
(fourth identity, cf. III 5.15 IV)

(3.24) $$\nabla_{[\omega} R_{\nu]\bar{\mu}\lambda\bar{\varkappa}} = S_{\omega\nu}^{\cdot\cdot\cdot\varrho} R_{\bar{\mu}\varrho\lambda\bar{\varkappa}}; \quad \text{conj.}$$
(identity of Bianchi, cf. III 5.19, VIII 2.20)[1]

In consequence of (3.22) the equations (2.20a, b) lead only to one equation (3.24). The same holds for the equations (2.23a, b) which lead to one equation

(3.25) $$\nabla_{[\omega} V'_{|\bar{\mu}|\nu]} = - S_{\omega\nu}^{\cdot\cdot\cdot\varrho} V'_{\bar{\mu}\varrho}; \quad \text{conj.}$$

In (2.25) the term $-\partial_{\bar{\mu}} R_{\omega\lambda}$ drops out, (2.26) remains valid without simplifications and (2.34, 35) take the form

(3.26) $$2 \nabla_{[\nu} S_{\mu\lambda]}^{\cdot\cdot\cdot\varkappa} - 4 S_{[\nu\mu}^{\cdot\cdot\cdot\varrho} S_{\lambda]\varrho}^{\cdot\cdot\cdot\varkappa} = 0; \quad \text{conj.}$$

for the general $\widetilde{U}_n$ and

(3.27) $$\nabla_{[\mu} S_{\lambda]} = 0; \quad \text{conj.}$$

for the semi-symmetric $\widetilde{U}_n$, $n > 2$. Hence *in a semi-symmetric* $\widetilde{U}_n$, $n > 2$, S_λ *always is a gradient vector.*

In a $\widetilde{V}_n$ we have instead of (3.19b) [(cf. III 5.8) but consider the difference between $V_{\bar{\nu}\lambda}$ and $V'_{\bar{\nu}\lambda}$]

(3.28) $$R_{\bar{\nu}\lambda} = - \partial_{\bar{\nu}} \partial_\lambda \log \mathfrak{a} = R_{\lambda\bar{\nu}} = - V'_{\bar{\nu}\lambda}; \quad \text{conj.}$$

hence $R_{\bar{\nu}\lambda}$ is now also hermitian.[2] The identities (3.20–25) take the simpler form

(3.29)
$$\begin{cases} \text{a) } R_{\bar{\nu}\mu\lambda\bar{\varkappa}} = - R_{\mu\bar{\nu}\lambda\bar{\varkappa}}; \quad \text{conj.} & (1^\circ \text{ id.; cf. III 5.15 I}) \\ \text{b) } R_{\bar{\nu}\mu\lambda\bar{\varkappa}} = R_{\bar{\nu}\lambda\mu\bar{\varkappa}}; \quad \text{conj.} & (2^\circ \text{ id.; cf. III 5.15 II}) \\ \text{c) } R_{\bar{\nu}\mu\lambda\bar{\varkappa}} = - R_{\bar{\nu}\mu\bar{\varkappa}\lambda}; \quad \text{conj.} & (3^\circ \text{ id.; cf. III 5.15 III}) \\ \text{d) } R_{\bar{\nu}\mu\lambda\bar{\varkappa}} = R_{\lambda\bar{\varkappa}\bar{\nu}\mu}; \quad \text{conj.} & (4^\circ \text{ id.; cf. III 5.15 IV}) \end{cases}$$

(3.30) $$\nabla_{[\omega} R_{\nu]\bar{\mu}\lambda\bar{\varkappa}} = 0; \quad \text{conj.}$$

(3.31) $$\nabla_{[\omega} V'_{|\bar{\mu}|\nu]} = 0; \quad \text{conj.}$$

---

[1]) Hombu gave 1935, 1 a complete list of all identities containing first derivatives of R and S.

[2]) Schouten and v. Dantzig 1930, 1, p. 337; Kähler 1933, 1; Fuchs 1937, 1.

The complex conjugates of all equations are valid	$\tilde{L}_n$	$\tilde{A}_n$
Definition	$\Gamma_{\bar{\mu}\lambda}^{\kappa}=0;\quad \Gamma_{\mu\bar{\lambda}}^{\kappa}=0;\quad \Gamma_{\mu\lambda}^{\bar{\kappa}}=0$	$S_{\mu\lambda}{}^{\cdot\cdot\kappa}=0;$
Curvature quantities	—	—
	(2.12 a) $R_{\nu\mu\lambda}{}^{\cdot\cdot\cdot\kappa}=2\,\partial_{[\nu}\Gamma_{\mu]\lambda}^{\kappa}+2\,\Gamma_{[\nu\lvert\varrho\rvert}^{\kappa}\Gamma_{\mu]\lambda}^{\varrho}$	see $\tilde{L}_n$
	(2.17) $R_{\bar{\nu}\mu\lambda}{}^{\cdot\cdot\cdot\kappa}=\partial_{\bar{\nu}}\Gamma_{\mu\lambda}^{\kappa}$	see $\tilde{L}_n$
	(2.21 b) $R_{\bar{\nu}\lambda}=-R_{\bar{\nu}\mu\lambda}{}^{\cdot\cdot\cdot\mu}=-\partial_{\bar{\nu}}\Gamma_{\mu\lambda}^{\mu}$	(2.29) $R_{\bar{\nu}\lambda}=-\partial_{\bar{\nu}}\Gamma_{\lambda\mu}^{\mu}$
	(2.21 e) $V'_{\bar{\nu}\mu}=R_{\bar{\nu}\mu\lambda}{}^{\cdot\cdot\cdot\lambda}=\partial_{\bar{\nu}}\Gamma_{\mu\lambda}^{\lambda}$	see $\tilde{L}_n$
	(2.21 d) $V_{\bar{\nu}\mu}=V'_{\bar{\nu}\mu}-V'_{\mu\bar{\nu}}$	see $\tilde{L}_n$
Identities	(2.34) $R_{[\nu\mu\lambda]}{}^{\cdot\cdot\cdot\kappa}=2\,\nabla_{[\nu}S_{\mu\lambda]}{}^{\cdot\cdot\kappa}-4\,S_{[\nu\mu}{}^{\cdot\cdot\varrho}S_{\lambda]\varrho}{}^{\cdot\cdot\kappa}$	(2.36) $R_{[\nu\mu\lambda]}{}^{\cdot\cdot\cdot\kappa}=0$
	(2.35) $R_{[\nu\mu\lambda]}{}^{\cdot\cdot\cdot\kappa}=2\,A_{[\nu}^{\kappa}\nabla_{\mu}S_{\lambda]}$ for $S_{\mu\lambda}{}^{\cdot\cdot\kappa}=S_{[\mu}A_{\lambda]}^{\kappa}$	
	(2.18) $R_{\bar{\nu}\mu\lambda}{}^{\cdot\cdot\cdot\kappa}=-R_{\mu\bar{\nu}\lambda}{}^{\cdot\cdot\cdot\kappa}$	see $\tilde{L}_n$
	(2.19) $R_{\bar{\nu}[\mu\lambda]}{}^{\cdot\cdot\cdot\kappa}=\partial_{\bar{\nu}}S_{\mu\lambda}{}^{\cdot\cdot\kappa}$	(2.27) $R_{\bar{\nu}[\mu\lambda]}{}^{\cdot\cdot\cdot\kappa}=0$
	—	—
	—	—
	(2.22) $R_{\bar{\nu}\lambda}+V'_{\bar{\nu}\lambda}=(n-1)\,\partial_{\bar{\nu}}S_{\lambda}$	(2.29) $R_{\bar{\nu}\lambda}+V'_{\bar{\nu}\lambda}=0$
Identities with differentiated R	(cf. III 5.19) $\nabla_{[\omega}R_{\nu\mu]\lambda}{}^{\cdot\cdot\cdot\kappa}=2\,S_{[\omega\nu}{}^{\cdot\cdot\sigma}R_{\mu]\sigma\lambda}{}^{\cdot\cdot\cdot\kappa}$	(cf. III 5.21) $\nabla_{[\omega}R_{\nu\mu]\lambda}{}^{\cdot\cdot\cdot\kappa}=0$
	(2.20 a) $2\,\nabla_{[\omega}R_{\nu]\bar{\mu}\lambda}{}^{\cdot\cdot\cdot\kappa}+\nabla_{\bar{\mu}}R_{\omega\nu\lambda}{}^{\cdot\cdot\cdot\kappa}=2\,S_{\omega\nu}{}^{\cdot\cdot\varrho}R_{\bar{\mu}\varrho\lambda}{}^{\cdot\cdot\cdot\kappa}$	(2.28 a) $2\,\nabla_{[\omega}R_{\nu]\bar{\mu}\lambda}{}^{\cdot\cdot\cdot\kappa}+\nabla_{\bar{\mu}}R_{\omega\nu\lambda}{}^{\cdot\cdot\cdot\kappa}=0$
	(2.20 b) $\nabla_{[\bar{\omega}}R_{\bar{\nu}]\mu\lambda}{}^{\cdot\cdot\cdot\kappa}=S_{\bar{\omega}\bar{\nu}}{}^{\cdot\cdot\bar{\varrho}}R_{\mu\bar{\varrho}\lambda}{}^{\cdot\cdot\cdot\kappa}$	(2.28 b) $\nabla_{[\bar{\omega}}R_{\bar{\nu}]\mu\lambda}{}^{\cdot\cdot\cdot\kappa}=0$
	(2.23 a) $2\,\nabla_{[\omega}V'_{\lvert\bar{\mu}\rvert\nu]}-\partial_{\bar{\mu}}V_{\omega\nu}=-2\,S_{\omega\nu}{}^{\cdot\cdot\varrho}V'_{\bar{\mu}\varrho}$	(2.30 a) $\begin{cases}2\,\nabla_{[\omega}V'_{\lvert\bar{\mu}\rvert\nu]}-\nabla_{\bar{\mu}}V_{\omega\nu}=-2\,\nabla_{[\omega}R_{\lvert\bar{\mu}\rvert\nu]}+\\ \qquad +2\,\nabla_{\bar{\mu}}R_{[\omega\nu]}=0\end{cases}$
	(2.23 b) $\nabla_{[\bar{\omega}}V'_{\bar{\nu}]\mu}=-S_{\bar{\omega}\bar{\nu}}{}^{\cdot\cdot\bar{\varrho}}V'_{\bar{\varrho}\mu}$	(2.30 b) $\nabla_{[\bar{\omega}}V'_{\bar{\nu}]\mu}=-\nabla_{[\bar{\omega}}R_{\bar{\nu}]\mu}=0$
	(2.24) $2\,\nabla_{[\bar{\omega}}V_{\bar{\nu}]\mu}+\partial_{\mu}V_{\bar{\omega}\bar{\nu}}=-2\,S_{\bar{\omega}\bar{\nu}}{}^{\cdot\cdot\bar{\varrho}}V_{\bar{\varrho}\mu}$	(2.31) $2\,\nabla_{[\bar{\omega}}V_{\bar{\nu}]\mu}+\nabla_{\mu}V_{\bar{\omega}\bar{\nu}}=0$
	(2.25) $\nabla_{\omega}R_{\bar{\mu}\lambda}+\nabla_{\nu}R_{\bar{\mu}\omega\lambda}{}^{\cdot\cdot\cdot\nu}-\nabla_{\bar{\mu}}R_{\omega\lambda}=2\,S_{\omega\nu}{}^{\cdot\cdot\varrho}R_{\bar{\mu}\varrho\lambda}{}^{\cdot\cdot\cdot\nu}$	(2.32) $\nabla_{\omega}R_{\bar{\mu}\lambda}+\nabla_{\nu}R_{\bar{\mu}\omega\lambda}{}^{\cdot\cdot\cdot\nu}-\nabla_{\bar{\mu}}R_{\lambda\omega}=0$
	(2.26) $\nabla_{[\bar{\nu}}R_{\bar{\omega}]\lambda}=-S_{\bar{\nu}\bar{\omega}}{}^{\cdot\cdot\bar{\varrho}}R_{\bar{\varrho}\lambda}$	(2.33) $\nabla_{[\bar{\nu}}R_{\bar{\mu}]\lambda}=0$
	—	—
	—	—

$\tilde{U}_n$	$\tilde{V}_n$
$\nabla_\mu a_{\lambda\bar{\varkappa}} = 0; \quad S_{\mu\lambda}^{\cdot\cdot\varkappa} \neq 0$	$\nabla_\mu a_{\lambda\bar{\varkappa}} = 0; \quad S_{\mu\lambda}^{\cdot\cdot\varkappa} = 0$
(3.4a) $\Gamma_{\mu\lambda}^{\varkappa} = a^{\bar{\varrho}\varkappa}\,\partial_\mu a_{\lambda\bar{\varrho}}$	see $\tilde{U}_n$
(3.18) $R_{\nu\mu\lambda}^{\cdot\cdot\cdot\varkappa} = 0$	see $\tilde{U}_n$
(3.19a) $R_{\bar{\nu}\mu\lambda}^{\cdot\cdot\cdot\varkappa} = \partial_{\bar{\nu}}\, a^{\bar{\varrho}\varkappa}\,\partial_\mu a_{\lambda\bar{\varrho}}$	see $\tilde{U}_n$
(3.19b) $R_{\bar{\nu}\lambda} = -\partial_{\bar{\nu}}\,\partial_\lambda \log \mathfrak{a} + 2\,\partial_{\bar{\nu}} S_{\lambda\mu}^{\cdot\cdot\mu}$	(3.28) $R_{\bar{\nu}\lambda} = -\partial_{\bar{\nu}}\,\partial_\lambda \log \mathfrak{a} = R_{\lambda\bar{\nu}} = -V'_{\bar{\nu}\lambda}$
(3.19c) $V'_{\bar{\nu}\mu} = \partial_{\bar{\nu}}\,\partial_\mu \log \mathfrak{a} = V'_{\mu\bar{\nu}}$	see $\tilde{U}_n$
(3.19d) $V_{\bar{\nu}\mu} = 0$	see $\tilde{U}_n$
(3.26) $2\,\nabla_{[\nu} S_{\mu\lambda]}^{\cdot\cdot\cdot\varkappa} - 4\,S_{[\nu\mu}^{\cdot\cdot\cdot\varrho} S_{\lambda]\varrho}^{\cdot\cdot\varkappa} = 0$	—
(3.27) $\nabla_{[\mu} S_{\lambda]} = 0$ for $n > 2$ if $S_{\mu\lambda}^{\cdot\cdot\varkappa} = S_{[\mu} A_{\lambda]}^{\varkappa}$	—
(3.20) $R_{\bar{\nu}\mu\lambda\bar{\varkappa}} = -R_{\mu\bar{\nu}\lambda\bar{\varkappa}}$	(3.29a) $R_{\bar{\nu}\mu\lambda\bar{\varkappa}} = -R_{\mu\bar{\nu}\lambda\bar{\varkappa}}$
(3.21) $R_{\bar{\nu}[\mu\lambda]\bar{\varkappa}} = \nabla_{\bar{\nu}} S_{\mu\lambda\bar{\varkappa}}$	(3.29b) $R_{\bar{\nu}\mu\lambda\bar{\varkappa}} = R_{\bar{\nu}\lambda\mu\bar{\varkappa}}$
(3.22) $R_{\bar{\nu}\mu\lambda\bar{\varkappa}} = -R_{\bar{\nu}\mu\bar{\varkappa}\lambda}$	(3.29c) $R_{\bar{\nu}\mu\lambda\bar{\varkappa}} = -R_{\bar{\nu}\mu\bar{\varkappa}\lambda}$
(3.23) $R_{\bar{\nu}\mu\lambda\bar{\varkappa}} - R_{\lambda\bar{\varkappa}\bar{\nu}\mu} = 2\,\nabla_{\bar{\nu}} S_{\mu\lambda\bar{\varkappa}} - 2\,\nabla_\lambda S_{\bar{\varkappa}\bar{\nu}\mu}$	(3.29d) $R_{\bar{\nu}\mu\lambda\bar{\varkappa}} = R_{\lambda\bar{\varkappa}\bar{\nu}\mu}$
see $\tilde{L}_n$	see $\tilde{A}_n$
—	—
(3.24) $\nabla_{[\omega} R_{\nu]\bar{\mu}\lambda\bar{\varkappa}} = S_{\omega\nu}^{\cdot\cdot\cdot\varrho} R_{\bar{\mu}\varrho\lambda\bar{\varkappa}}$	(3.30) $\nabla_{[\omega} R_{\nu]\bar{\mu}\lambda\bar{\varkappa}} = 0$
(3.25) $\nabla_{[\omega} V'_{\vert\bar{\mu}\vert\nu]} = -S_{\omega\nu}^{\cdot\cdot\cdot\varrho} V'_{\bar{\mu}\varrho}$	(3.31) $\nabla_{[\omega} V'_{\vert\bar{\mu}\vert\nu]} = 0$
—	—
$\nabla_\omega R_{\bar{\mu}\lambda} + \nabla_\nu R_{\bar{\mu}\omega\lambda}^{\cdot\cdot\cdot\nu} = 2\,S_{\omega\nu}^{\cdot\cdot\cdot\varrho} R_{\bar{\mu}\varrho\lambda}^{\cdot\cdot\cdot\nu}$	$\nabla_\omega R_{\bar{\mu}\lambda} + \nabla_\nu R_{\bar{\mu}\omega\lambda}^{\cdot\cdot\cdot\nu} = 0$
see $\tilde{L}_n$	see $\tilde{A}_n$
—	(3.32) $\nabla_\mu G_\nu^{\cdot\mu} = 0; \quad G_\nu^{\cdot\mu} \overset{\text{def}}{=} R_\nu^{\cdot\mu} - R\,A_\nu^{\cdot\mu}$
—	$R \overset{\text{def}}{=} R_{\bar{\nu}\lambda}\, a^{\bar{\nu}\lambda}$

In (2.32) the term $-\partial_{\bar{\mu}} R_{\omega\lambda}$ drops out but (2.33) remains valid without simplifications. In a $\widetilde{U}_n$ a second contraction is possible. For a $\widetilde{V}_n$ this leads in (2.32) to $0=0$ but from (2.33) it follows that

$$(3.32) \qquad \boxed{\nabla_\mu G_\nu^{\cdot\,\mu} = 0;\ \text{conj.}}\ ; \quad G_\nu^{\cdot\,\mu} \overset{\text{def}}{=} R_\nu^{\cdot\,\mu} - R A_\nu^\mu;\ \text{conj.} \quad R \overset{\text{def}}{=} R_{\bar{\nu}\lambda} a^{\bar{\nu}\lambda}$$

which could be compared with (III 5.29).

On pp. 400, 401 there is a list of the most important formulae for $\widetilde{L}_n$, $\widetilde{A}_n$, $\widetilde{U}_n$ and $\widetilde{V}_n$.

If $\underset{i}{u}^\varkappa$ are n fields of unitvectors in a $\widetilde{U}_n$ that are mutually (unitary) perpendicular, if the reciprocal set (cf. I § 2) is $\overset{h}{u}_\lambda$ and if the fundamental tensor $a_{\lambda\bar{\varkappa}}$ is positive definite (cf. I § 10), this tensor can be written in the form

$$(3.33) \qquad a_{\lambda\bar{\varkappa}} = \sum_h \overset{h}{u}_\lambda \overset{h}{u}_{\bar{\varkappa}}$$

but the fields $\overset{h}{u}_\lambda$ need not to be analytic. In fact, it is not necessary that there exist analytic fields $\overset{h}{u}_\lambda$ satisfying (3.33). We prove that such fields exist if and only if

$$(3.34) \qquad \partial_{\bar{\nu}} \Gamma_{\mu\lambda}^\varkappa = 0$$ [1]

or in an equivalent form

$$(3.35) \qquad \partial_{\bar{\nu}} \partial_\mu a_{\lambda\bar{\varkappa}} = (\partial_{\bar{\nu}} a_{\lambda\bar{\varrho}})(\partial_\mu a_{\sigma\bar{\varkappa}}) a^{\bar{\varrho}\sigma}.$$

(3.34) expresses the fact that the $\Gamma_{\mu\lambda}^\varkappa$ are analytic in $\xi^\varkappa$. The condition is necessary. This follows immediately by differentiation of (3.33) if it is supposed that the $\overset{h}{u}_\lambda$ are analytic. In order to prove that it is also sufficient we consider the system of differential equations

$$(3.36) \qquad \begin{cases} \text{a)} & \nabla_\mu \overset{h}{u}_\lambda = \partial_\mu \overset{h}{u}_\lambda - \overset{h}{u}_\varrho a^{\bar{\sigma}\varrho} \partial_\mu a_{\lambda\bar{\sigma}} = 0 \\ \text{b)} & \nabla_{\bar{\mu}} \overset{h}{u}_\lambda = \partial_{\bar{\mu}} \overset{h}{u}_\lambda = 0. \end{cases}$$

The integrability conditions are

$$(3.37) \qquad \begin{cases} \text{a)} & \overset{h}{u}_\tau a^{\bar{\omega}\tau} (\partial_{[\nu} a_{|\varrho\bar{\omega}|}) a^{\bar{\sigma}\varrho} \partial_{\mu]} a_{\lambda\bar{\sigma}} + \overset{h}{u}_\varrho (\partial_{[\nu} a^{\bar{\sigma}\varrho}) \partial_{\mu]} a_{\lambda\bar{\sigma}} = 0 \\ \text{b)} & \overset{h}{u}_\varrho (\partial_{\bar{\nu}} a^{\bar{\sigma}\varrho}) \partial_\mu a_{\lambda\bar{\sigma}} + \overset{h}{u}_\varrho a^{\bar{\sigma}\varrho} \partial_{\bar{\nu}} \partial_\mu a_{\lambda\bar{\sigma}} = 0 \end{cases}$$

(3.37a) is satisfied identically and (3.36b) is a consequence of (3.35). Hence, if (3.35) is satisfied, (3.36) is totally integrable and this means that for every choice of a vector $\overset{0}{u}_\lambda$ at some point, there exists a covariant constant field having at this point exactly the value $\overset{0}{u}_\lambda$. Thus, if we

[1]) SCHOUTEN and v. DANTZIG 1930, 1; cf. E II 1938, 2, p. 236.

take n mutually perpendicular unitvectors at some point we get n fields of unitvectors that are mutually perpendicular at every point. The equation (3.34) implies that every analytic field remains analytic if it is displaced parallel. If we call such a connexion *analytic*, then (3.34) is the n.a.s. condition for the connexion to be analytic.

From (2.17) and (3.34) it follows that the last curvature tensors still remaining vanish and that the connexion is therefore *integrable* (cf. III § 4). Gathering results we have[1]):

In a $\tilde{U}_n$ the following three conditions are equivalent:

a) the connexion is integrable: $R_{\bar{\nu}\mu\lambda}^{\cdot\cdot\cdot\varkappa}=0$; conj.

b) the connexion is analytic: $\partial_{\bar{\nu}} \Gamma_{\mu\lambda}^{\varkappa}=0$; conj.

c) there exist sets of mutually perpendicular fields of unitvectors that are analytic.

If the connexion is integrable and moreover symmetric, we have from (3.9) and (3.36a)

$$(3.38) \qquad \partial_{[\mu} \overset{h}{u}_{\lambda]} = 0; \qquad \partial_{\bar{\mu}} \overset{h}{u}_{\lambda} = 0$$

and this means that the $\overset{h}{u}_{\lambda}$ are analytic gradientvectors, depending on the $\xi^\varkappa$ only. Hence in this case the scalars belonging to those fields are analytic in $\xi^\varkappa$ and if they are taken as coordinates, the linear element takes the form

$$(3.39) \qquad d s^2 \overset{*}{=} \sum_{\varkappa} d\xi^\varkappa \, d\xi^{\bar{\varkappa}}$$

because the $\overset{h}{u}_{\lambda}$ are mutually perpendicular unitvectors. From this form every linear coordinate transformation with constant coefficients leads to a form

$$(3.40) \qquad d s^2 \overset{*}{=} a_{\lambda\bar{\varkappa}} \, d\xi^\lambda \, d\xi^{\bar{\varkappa}}; \qquad a_{\lambda\bar{\varkappa}} \overset{*}{=} \text{const}.$$

Hence (cf. I § 10) [2])

If the connexion of a $\tilde{V}_n$ is analytic it is also integrable and the $\tilde{V}_n$ is an $\tilde{R}_n$.

Exercises.

VIII 3,1. Prove the third identity in $\tilde{U}_n$

$$\text{VIII 3,1 } \alpha) \qquad R_{\bar{\nu}\mu\lambda\bar{\varkappa}} = - R_{\bar{\nu}\mu\bar{\varkappa}\lambda}; \qquad \text{conj.}$$

VIII 3,2. Prove that $R \overset{\text{def}}{=} R_{\bar{\mu}\lambda} \, a^{\bar{\mu}\lambda}$ is always real in a $\tilde{V}_n$ but not always in an $\tilde{U}_n$.

[1]) SCHOUTEN and v. DANTZIG 1930, 1, p. 333, cf. E II 1938, 2, p. 237.

[2]) SCHOUTEN and v. DANTZIG 1931, 3; E II 1938, 2, p. 237f.

VIII 3,3[1]). In a $\widetilde{V}_n$ with $a_{\lambda\bar{\varkappa}} = \partial_\lambda \partial_{\bar{\varkappa}} \varphi$ is given that $e^\varphi f(\xi^\varkappa)\, g(\xi^{\bar{\varkappa}})$ is an ordinary density of weight $+2$. Prove that there exists a scalar field ψ such that

VIII 3,3 α) $$a_{\lambda\bar{\varkappa}} - V'_{\lambda\bar{\varkappa}} = \partial_\lambda\, \partial_{\bar{\varkappa}} \log \psi .$$

VIII 3,4. Let $L_{\mu\bar{\lambda}}$ be defined in a $\widetilde{V}_n$ by

VIII 3,4α) $$L_{\mu\bar{\lambda}} \overset{\text{def}}{=} - R_{\mu\bar{\lambda}} + \frac{2}{n+1} R\, a_{\mu\bar{\lambda}} .$$

Prove that

VIII 3,4) $$\begin{cases} \beta) & L_{\mu\bar{\lambda}} = L_{\bar{\lambda}\mu} \\ \gamma) & L \overset{\text{def}}{=} L_{\mu}^{\cdot\mu} = L_{\bar{\mu}}^{\cdot\bar{\mu}} = \text{real} \\ \delta) & \nabla_\nu L_{\mu}^{\cdot\nu} = \nabla_\mu L \qquad \text{(use 3.32)}. \end{cases}$$

VIII 3,5. Let $F_{\bar{\nu}\mu\lambda}^{\cdot\cdot\cdot\varkappa}$ be defined in a $\widetilde{V}_n$ by

VIII 3,5 α) $$F_{\bar{\nu}\mu\lambda}^{\cdot\cdot\cdot\varkappa} \overset{\text{def}}{=} R_{\bar{\nu}\mu\lambda}^{\cdot\cdot\cdot\varkappa} + \frac{2}{n-1} a_{\bar{\nu}\mu} L_{\lambda}^{\cdot\varkappa}$$

where $L_{\mu\bar{\lambda}}$ is defined in (Exerc. VIII 3,4α). Prove that

VIII 3,5 β) $$\nabla_\varkappa F_{\bar{\nu}\mu\lambda}^{\cdot\cdot\cdot\varkappa} = \nabla_\lambda L_{\bar{\nu}\mu} .$$

§ 4. The $\widetilde{V}_n$ of constant curvature.[2])

For a symmetric unitary connexion, $R_{\bar{\nu}\mu\lambda\bar{\varkappa}}$ is symmetric in $\mu\lambda$ (cf. 3.21, 29b). Now if

(4.1) $$R_{\bar{\nu}\mu\lambda\bar{\varkappa}} = \alpha\, a_{\bar{\nu}(\mu}\, a_{|\bar{\varkappa}|\,\lambda)} = \alpha\, a_{(\bar{\nu}(\mu}\, a_{\bar{\varkappa})\,\lambda)}$$

and therefore

(4.2) $$R_{\bar{\nu}\lambda} = -\tfrac{1}{2}(n+1)\, \alpha\, a_{\bar{\nu}\lambda}$$

the $\widetilde{V}_n$ is said to be of *constant curvature* and is called an $\widetilde{S}_n$. α is always *constant* and *real.*[3]) The scalar curvature (cf. III § 5) is also *constant* and *real* because

(4.3) $$\frac{1}{n(n-1)} R = -\frac{1}{2}\, \frac{n+1}{n-1}\, \alpha ; \qquad R \overset{\text{def}}{=} R_{\bar{\nu}\lambda}\, a^{\bar{\nu}\lambda} .$$

---

[1]) BERGMAN 1933, 1, p. 310.

[2]) SCHOUTEN and v. DANTZIG 1931, 3; FUCHS 1937, 1 also for a $\widetilde{V}_n$ with constant R; BOCHNER 1947, 1; E II 1938, 2, p. 239.

[3]) SATÒ 1950, 1.

If the connexion of a $\tilde{V}_n$ is transformed projectively (cf. VI § 1) we get a symmetric (but not necessarily unitary) connexion

$$(4.4)\qquad \begin{cases} \text{a)}\quad '\Gamma^{\varkappa}_{\mu\lambda} = \Gamma^{\varkappa}_{\mu\lambda} + 2p_{(\mu} A^{\varkappa}_{\lambda)}; & \text{conj.} \\ \text{b)}\quad '\Gamma^{\varkappa}_{\bar{\mu}\lambda} = \Gamma^{\varkappa}_{\bar{\mu}\lambda} = 0 \end{cases}$$

and the new curvature tensors are

$$(4.5)\qquad \begin{cases} \text{a)}\quad 'R_{\nu\mu\lambda}^{\cdot\cdot\cdot\varkappa} = -2p_{[\nu\mu]} A^{\varkappa}_{\lambda} + 2A^{\varkappa}_{[\nu} p_{\mu]\lambda}; & \text{conj.} \\ \text{b)}\quad 'R_{\bar{\nu}\mu\lambda}^{\cdot\cdot\cdot\varkappa} = \alpha\, a_{\bar{\nu}(\mu} A^{\varkappa}_{\lambda)} + 2(\nabla_{\bar{\nu}} p_{(\mu}) A^{\varkappa}_{\lambda)}; & \text{conj.} \\ \text{c)}\quad 'R_{\mu\bar{\nu}\lambda}^{\cdot\cdot\cdot\varkappa} = -'R_{\bar{\nu}\mu\lambda}^{\cdot\cdot\cdot\varkappa}; \quad \text{conj.} & p_{\mu\lambda} \stackrel{\text{def}}{=} -\nabla_\mu p_\lambda + p_\mu p_\lambda; \quad \text{conj.} \end{cases}$$

In order to make the new connexion integrable it is n.a.s. that for $n \geq 2$:

$$(4.6)\qquad \begin{cases} \text{a)}\quad \nabla_\mu p_\lambda = p_\mu p_\lambda; \quad \text{conj.} \\ \text{b)}\quad \nabla_{\bar{\mu}} p_\lambda = \partial_{\bar{\mu}} p_\lambda = -\tfrac{1}{2}\alpha\, a_{\bar{\mu}\lambda}; \quad \text{conj.} \end{cases}$$

but this implies that p_λ is a gradientvector:

$$(4.7)\qquad p_\lambda = \partial_\lambda \log \mu; \quad \text{conj.}$$

The integrability conditions of (4.6) are identically satisfied in consequence of (4.1) and the constancy of α. That proves

An $\tilde{S}_n$ can always be transformed into an $\tilde{R}_n$ by a projective transformation of the connexion.

In this $\tilde{R}_n$ there exist coordinate systems with respect to which $'\Gamma^{\varkappa}_{\mu\lambda} = 0$. If we use one of these coordinate systems in the $\tilde{S}_n$ we get

$$(4.8)\qquad \begin{cases} \text{a)}\quad \Gamma^{\varkappa}_{\mu\lambda} \stackrel{*}{=} -2A^{\varkappa}_{(\lambda}\partial_{\mu)} \log \mu; \quad \text{conj.} \\ \text{b)}\quad \Gamma^{\varkappa}_{\bar{\mu}\lambda} \stackrel{*}{=} 0; \quad \text{conj.} \end{cases}$$

where μ is a solution of (4.7). For this coordinate system we get

$$(4.9)\qquad \begin{cases} \text{a)}\quad \partial_\mu a_{\bar{\lambda}\varkappa} \stackrel{*}{=} -2a_{\bar{\lambda}(\varkappa} p_{\mu)}; & \text{conj.} \\ \text{b)}\quad \partial_\mu a^{\varkappa\bar{\lambda}} \stackrel{*}{=} a^{\varkappa\bar{\lambda}} p_\mu + A^{\varkappa}_{\mu} p^{\bar{\lambda}}; & \text{conj.} \\ \text{c)}\quad \partial_\mu p_\lambda \stackrel{*}{=} -p_\mu p_\lambda; & \text{conj.} \\ \text{d)}\quad \partial_\mu p^{\bar{\varkappa}} \stackrel{*}{=} p^{\bar{\varkappa}} p_\mu; & \text{conj.} \\ \text{e)}\quad \partial_\mu p_{\bar{\lambda}} \stackrel{*}{=} -\tfrac{1}{2}\alpha\, a_{\bar{\lambda}\mu}; & \text{conj.} \\ \text{f)}\quad \partial_\mu p^{\varkappa} \stackrel{*}{=} p^{\varkappa} p_\mu - \dfrac{1}{2}\alpha A^{\varkappa}_{\mu}\left(1 - \dfrac{2}{\alpha} p^{\bar{\lambda}} p_{\bar{\lambda}}\right); & \text{conj.} \end{cases}$$

Now it can easily be verified that

$$(4.10)\qquad \mu = 1 - \frac{2}{\alpha} p_\lambda p_{\bar{\varkappa}} a^{\lambda\bar{\varkappa}}$$

is a solution of (4.7) because α is real. From (4.10) we see that μ is also real. If we introduce the vector

$$z^\varkappa \stackrel{\text{def}}{=} -\frac{2}{\mu\alpha} p^\varkappa \tag{4.11}$$

we get from (4.9c, d) and (4.11)

$$\partial_\mu z^\varkappa \stackrel{*}{=} A_\mu^\varkappa; \qquad \partial_{\bar\mu} z^\varkappa \stackrel{*}{=} 0. \tag{4.12}$$

Hence the $z^\varkappa$ are analytic and by integration of (4.12) it follows that

$$\xi^\varkappa \stackrel{*}{=} z^\varkappa + c^\varkappa; \qquad c^\varkappa \stackrel{*}{=} \text{const}. \tag{4.13}$$

But this means that the coordinate system can be chosen such that

$$\xi^\varkappa \stackrel{*}{=} z^\varkappa = -\frac{2}{\mu\alpha} p^\varkappa. \tag{4.14}$$

From (4.6b) and (4.9a, c) it follows that

$$b_{\lambda\bar\varkappa} \stackrel{\text{def}}{=} \frac{\alpha\mu}{2} a_{\lambda\bar\varkappa} - \mu p_\lambda p_{\bar\varkappa} \tag{4.15}$$

is a hermitian tensor with *constant* components:

$$\partial_\mu b_{\lambda\bar\varkappa} \stackrel{*}{=} 0; \qquad \partial_{\bar\mu} b_{\lambda\bar\varkappa} \stackrel{*}{=} 0, \tag{4.16}$$

hence:

In an $\widetilde{S}_n$ the coordinate system can always be chosen such that in the region where the equation

$$\mu \stackrel{*}{=} 1 - \frac{\alpha\mu^2}{2} a_{\lambda\bar\varkappa} \xi^\lambda \xi^{\bar\varkappa} \tag{4.17}$$

has a real solution (that is where $1 + 2\alpha\, a_{\lambda\bar\varkappa} \xi^\lambda \xi^{\bar\varkappa} \geq 0$) *the fundamental tensor can be written in the form*

$$a_{\lambda\bar\varkappa} \stackrel{*}{=} \frac{2}{\alpha\mu} b_{\lambda\bar\varkappa} + \frac{\alpha\mu^2}{2} a_{\lambda\bar\varrho} a_{\sigma\bar\varkappa} \xi^\sigma \xi^{\bar\varrho}. \tag{4.18}$$

where $b_{\lambda\bar\varkappa}$ is a hermitian tensor with constant components.

But this form of $a_{\lambda\bar\varkappa}$ gives just the linear element of the hermitian non-euclidean geometry that was investigated by FUBINI and STUDY.[1]) Hence hermitian non-euclidean geometry in the sense of these authors is realized in an $\widetilde{S}_n$, in the same way as ordinary non-euclidean geometry is realized in an S_n.

---

[1]) FUBINI 1903, 2; STUDY 1905, 1; cf. COOLIDGE 1924, 1; CARTAN 1931, 1; E II 1938, 2, p. 243; BOCHNER 1947, 1, p. 185ff.

Exercise.

VIII 4,1[1]). Let the gradient vector p_λ be the vector of the transformation (4.4a) that transforms an $\widetilde{S}_n$ into an $\widetilde{R}_n$. μ is defined by (4.10). Prove that (use 4.9 and 4.15)

VIII 4,1 α) $$a_{\lambda\bar{\varkappa}} = -2\partial_\lambda \partial_{\bar{\varkappa}} \frac{\log \mu}{\alpha}$$

and

VIII 4,1 β) $$\mu \stackrel{*}{=} 1 - b_{\lambda\bar{\varkappa}} \xi^\lambda \xi^{\bar{\varkappa}}$$

with respect to the coordinate system (4.14).

§ 5. Imbedding in an $\widetilde{L}_n$.

Let an X_m in $\widetilde{L}_n$ be given by its parameter form

(5.1) $$\xi^\varkappa = f^\varkappa(\eta^a); \qquad a = 1, \ldots, \mathrm{m}; \qquad \text{conj.}$$

where the functions $f^\varkappa$ are analytic in the domain considered.[2]) If we define the connecting quantity $B_b^\varkappa$ in the ordinary way

(5.2) $$B_b^\varkappa \stackrel{\text{def}}{=} \partial_b \xi^\varkappa; \qquad \text{conj.}$$

the analyticity of $f^\varkappa$ ensures that the $B_b^\varkappa$ are analytic:

(5.3) $$\partial_{\bar{c}} B_b^\varkappa = 0; \qquad \text{conj.}$$

We suppose that the X_m is rigged and that the vectors $\underset{y}{e}^\varkappa$; $y = \mathrm{m}+1, \ldots, \mathrm{n}$ span the E_{n-m} of the rigging. *These vectors need not be analytic.* Besides the $\underset{y}{e}^\varkappa$ we have the vectors of $\widetilde{L}_n$ derived from the basis vectors $\underset{a}{e}^b$ in X_m

(5.4) $$\underset{a}{e}^\varkappa \stackrel{\text{def}}{=} B_b^\varkappa \underset{a}{e}^b; \qquad \text{conj.}$$

These vectors are analytic and they span the tangent E_m of X_m. All these vectors are given at points of X_m only. From the vectors $\underset{b}{e}^\varkappa$, $\underset{y}{e}^\varkappa$; conj. we form the reciprocal system $\overset{a}{e}_\lambda$; $\overset{x}{e}_\lambda$; conj. in the ordinary way. These covariant vectors need not be analytic. Now the formulae (V 7.1) and their conjugates hold over X_m. From them we need especially the definition of the second connecting quantity C_λ^x:

(5.5) $$C_\lambda^x \stackrel{\text{def}}{=} \underset{y}{e}^x \overset{y}{e}_\lambda; \qquad \text{conj.}$$

In contradistinction to $B_b^\varkappa$ this quantity need not be analytic.

---

[1]) Cf. Fuchs 1937, 1; Bochner 1947, 1.

[2]) Coburn also considered X_m's in $\widetilde{U}_n$ defined by semi-analytic functions, 1942, 1.

Now let p be a semi-analytic scalar field of X_m. Then we have the derivatives $\partial_b p$ and $\partial_{\bar b} p$ and if p is prolonged somehow in a neighbourhood of X_m we have also $\partial_\mu p$ and $\partial_{\bar\mu} p$ over X_m. The relations

$$(5.6) \qquad \partial_b p = B_b^\mu \partial_\mu p; \quad \text{conj.}$$

hold between them. If v^a is a vector field of X_m, the components of this field considered as a field of $\tilde{L}_n$ over X_m are $v^\varkappa = B_b^\varkappa v^b$. Hence a connexion in X_m can be defined by the equations

$$(5.7) \qquad \begin{cases} \text{a)} \quad {}'\nabla_c v^a \overset{\text{def}}{=} B_{c\varkappa}^{\mu a} \nabla_\mu v^\varkappa; & \text{conj.} \\ \text{b)} \quad {}'\nabla_{\bar c} v^a \overset{\text{def}}{=} B_{\bar c}^{\bar\mu} B_\varkappa^a \nabla_{\bar\mu} v^\varkappa; & \text{conj.} \end{cases}$$

where $B_\varkappa^a$ is defined as in (V 7.1). From (5.7) and the analyticity of $B_b^\varkappa$ we get the following equations for the parameters of this connexion:

$$(5.8) \qquad \begin{cases} \text{a)} \quad {}'\Gamma_{cb}^a = B_{cb\varkappa}^{\mu\lambda a} \Gamma_{\mu\lambda}^\varkappa - B_{c\lambda}^{\mu a} \partial_\mu B_b^\lambda; & \text{conj.} \\ \text{b)} \quad {}'\Gamma_{\bar c b}^a = B_{\bar c b\varkappa}^{\bar\mu\lambda a} \Gamma_{\bar\mu\lambda}^\varkappa - B_{\bar c\lambda}^{\bar\mu a} \partial_{\bar\mu} B_b^\lambda = 0; & \text{conj.} \end{cases}$$

and this proves that *the X_m with this induced connexion is an $\tilde{L}_m$*. From (5.8a) we see that if the connexion in $\tilde{L}_n$ is analytic, $\partial_{\bar\nu} \Gamma_{\mu\lambda}^\varkappa = 0$, it need not necessarily follow that the connexion in $\tilde{L}_m$ is also analytic, because $B_\varkappa^a$ need not be analytic. Alternation of (5.8a) gives

$$(5.8) \qquad \text{c)} \quad {}'S_{cb}^{\cdot\cdot a} = B_{cb\varkappa}^{\mu\lambda a} S_{\mu\lambda}^{\cdot\cdot\varkappa} - B_\varkappa^a \partial_{[c} B_{b]}^\varkappa = B_{cb\varkappa}^{\mu\lambda a} S_{\mu\lambda}^{\cdot\cdot\varkappa}; \quad \text{conj.}$$

and this means that *every X_m imbedded and rigged in an $\tilde{A}_n$ is itself an $\tilde{A}_m$ whatever the rigging may be.*

Instead of two curvature tensors with valence 3 as in (V 7.15, 16) here we get four with their complex conjugates

$$(5.9) \qquad \text{a)}\ H_{cb}^{\cdot\cdot\varkappa} \overset{\text{def}}{=} B_{cb}^{\mu\lambda} \nabla_\mu B_\lambda^\varkappa \qquad \text{b)}\ h_{\bar c b}^{\cdot\cdot\varkappa} \overset{\text{def}}{=} B_{\bar c b}^{\bar\mu\lambda} \nabla_{\bar\mu} B_\lambda^\varkappa; \quad \text{conj.}$$

$$(5.10) \qquad \text{a)}\ L_{\bar c\cdot\lambda}^{\cdot a} \overset{\text{def}}{=} B_{\bar c\varkappa}^{\bar\mu a} \nabla_{\bar\mu} B_\lambda^\varkappa \qquad \text{b)}\ l_{c\cdot\lambda}^{\cdot a} \overset{\text{def}}{=} B_{c\varkappa}^{\mu a} \nabla_\mu B_\lambda^\varkappa; \quad \text{conj.}$$

but from these $h_{\bar c b}^{\cdot\cdot\varkappa}$ and its complex conjugate are zero. In fact, according to the analyticity of $B_b^\varkappa$ we have

$$(5.11) \qquad h_{\bar c b}^{\cdot\cdot\varkappa} = B_{\bar c b}^{\bar\mu\lambda} \partial_{\bar\mu} B_\lambda^\varkappa = - B_{\bar c}^{\bar\mu} \partial_{\bar\mu} B_b^\varkappa = 0; \quad \text{conj.}$$

From (5.9a) it follows that

$$(5.12) \qquad \begin{cases} H_{[cb]}^{\cdot\cdot\varkappa} = - B_{cb}^{\mu\lambda} \partial_{[\mu} C_{\lambda]}^\varkappa + B_{cb}^{\mu\lambda} \Gamma_{[\mu\lambda]}^\varrho C_\varrho^\varkappa - B_{cb}^{\mu\lambda} \Gamma_{[\mu|\sigma|}^\varkappa C_{\lambda]}^\sigma \\ \qquad = C_\lambda^\varkappa \partial_{[c} B_{b]}^\lambda + B_{cb}^{\mu\lambda} S_{\mu\lambda}^{\cdot\cdot\varrho} C_\varrho^\varkappa = B_{cb}^{\mu\lambda} S_{\mu\lambda}^{\cdot\cdot\varrho} C_\varrho^\varkappa; \qquad \text{conj.} \end{cases}$$

and this proves that $H_{[cb]}^{\cdot\cdot\varkappa}$ is zero for every choice of the X_m and its rigging if the $\tilde{L}_n$ is an $\tilde{A}_n$ (cf. V 7.23).

An $\tilde{L}_m$ in $\tilde{L}_n$ will be called *geodesic* if every vector of $\tilde{L}_m$ remains in $\tilde{L}_m$ if it is displaced parallel in a direction of $\tilde{L}_m$. N.a.s. conditions are that for every field $v^\varkappa = B_b^\varkappa v^b$ the following equations should hold

$$(5.13)\quad \text{a) } v^\mu (\nabla_\mu v^\lambda)\, C_\lambda^\varkappa = 0;\quad \text{conj.};\qquad \text{b) } v^{\bar\mu} (\nabla_{\bar\mu} v^\lambda)\, C_\lambda^\varkappa = 0;\quad \text{conj.}$$

But these conditions are equivalent to

$$5.14)\quad \text{a) } H_{(cb)}^{\cdot\cdot\varkappa} = - B_{(cb)}^{\mu\lambda} \nabla_\mu C_\lambda^\varkappa = 0;\ \text{conj.}\quad \text{b) } v^{\bar c} v^b B_{\bar c b}^{\bar\mu\lambda} \nabla_{\bar\mu} C_\lambda^\varkappa = 0;\ \text{conj.}$$ [1]).

From these equations the first is equivalent to $H_{cb}^{\cdot\cdot\varkappa} = 0$ if the $\tilde{L}_n$ is an $\tilde{A}_n$ and the second is satisfied identically because $h_{\bar c b}^{\cdot\cdot\varkappa} = 0$ (cf. 5.11). It is easily proved that the transformation of the connexion in $\tilde{L}_n$

$$(5.15)\qquad {}'\Gamma_{\mu\lambda}^\varkappa = \Gamma_{\mu\lambda}^\varkappa + P_{\mu\lambda}^{\cdot\cdot\varkappa};\qquad P_{[\mu\lambda]}^{\cdot\cdot\varkappa} = 0;\qquad \text{conj.}$$

leaves invariant the set of all $\tilde{L}_m$'s which are geodesic according to this condition, if and only if $P_{\mu\lambda}^{\cdot\cdot\varkappa}$ has the form $2p_{(\mu} A_{\lambda)}^\varkappa$, that is, if the transformation is projective (cf. VI § 1) [2]).

For an $\tilde{L}_m$ in $\tilde{L}_n$ a D-symbolism may be introduced in the same way as for an L_m in L_n (cf. V § 7). Because the fields considered are all defined only over the $\tilde{L}_m$ here we need only D_μ and D_c and their conjugates. Using these symbols, the tensors $H_{cb}^{\cdot\cdot\varkappa}$, $L_{\bar c\cdot\lambda}^{\cdot a}$ and $l_{c\cdot\lambda}^{\cdot a}$ get the form (cf. V 7.15 b, 16 b)

$$(5.16)\qquad \text{a) } H_{cb}^{\cdot\cdot\varkappa} = D_c B_b^\varkappa;\quad \text{conj.}\qquad (\text{b)}\quad 0 = D_{\bar c} B_b^\varkappa;\ \text{conj.})$$

$$(5.17)\qquad \text{a) } L_{\bar c\cdot\lambda}^{\cdot a} = D_{\bar c} B_\lambda^a;\quad \text{conj.}\qquad \text{b) } l_{c\cdot\lambda}^{\cdot a} = D_c B_\lambda^a;\quad \text{conj.}$$

We see from this that the greek index in these three formulae always lies in the rigging and that (cf. V 7.15 a, 16 a)

$$(5.18)\qquad H_{cb}^{\cdot\cdot x} = C_\varkappa^x D_c B_b^\varkappa = - B_b^\varkappa D_c C_\varkappa^x;\qquad \text{conj.}$$

$$(5.19)\qquad \begin{cases} \text{a) } L_{\bar c\cdot y}^{\cdot a} = C_y^\lambda D_{\bar c} B_\lambda^a = - B_\lambda^a D_{\bar c} C_y^\lambda; & \text{conj.} \\ \text{b) } l_{c\cdot y}^{\cdot a} = C_y^\lambda D_c B_\lambda^a = - B_\lambda^a D_c C_y^\lambda; & \text{conj.} \end{cases}$$

---

[1]) If instead of (5.13) we take $(v^\mu \nabla_\mu v^\lambda + v^{\bar\mu} \nabla_{\bar\mu} v^\lambda)\, C_\lambda^\varkappa = 0$; conj. as the definition of the geodesic $\tilde{L}_m$, we get the same result (5.14) because $h_{\bar c b}^{\cdot\cdot\varkappa} = 0$. But this latter definition represents just the condition for the L_{2m} corresponding to the $\tilde{L}_m$ to be geodesic in the auxiliary L_{2n}.

[2]) Here is a difference with Bochner 1947, 1. He considers all geodesic subspaces in the auxiliary L_{2n} and requires that they all remain geodesic. Then it can be proved that this stronger condition leads to $P_{\mu\lambda}^{\cdot\cdot\varkappa} = 0$. But from our point of view the L_{2n} is only an auxiliary space in which we are only interested in what happens to subspaces in the principal X_n, and this leads to $P_{\mu\lambda}^{\cdot\cdot\varkappa} = 2p_{(\mu} A_{\lambda)}^\varkappa$.

If v^a is a field of $\widetilde{L}_m$ we have (cf. III 4.9b; V 8.4)

$$(5.20)\quad \begin{cases} {}'R_{\dot d c b}^{\cdot\cdot\cdot\,a} v^b - 2\,{}'S_{\dot d c}^{\cdot\cdot\,e}\,{}'\nabla_e v^a = 2\,{}'\nabla_{[d}\,{}'\nabla_{c]} v^a = 2 D_{[d} D_{c]} v^a = 2 D_{[d} B_{c]\varkappa}^{\mu a} \nabla_\mu v^\varkappa \\ \qquad = 2 H_{[\dot d c]}^{\cdot\cdot\,\mu} B_\varkappa^a \nabla_\mu v^\varkappa + 2 B_{[c}^{\mu} l_{\dot d]\cdot y}^{\cdot a} H_{\mu b}^{\cdot\cdot\,y} v^b + \\ \qquad + B_{d c \varkappa}^{\nu\mu a} R_{\nu\mu\lambda}^{\cdot\cdot\cdot\,\varkappa} v^\lambda - 2 B_{d c \varkappa}^{\nu\mu a} S_{\nu\mu}^{\cdot\cdot\,\varrho} \nabla_\varrho v^\varkappa , \end{cases}$$

hence

$$(5.21)\quad \begin{cases} \left({}'R_{\dot d c b}^{\cdot\cdot\cdot\,a} + 2 B_{[c}^{\mu} l_{\dot d]\cdot y}^{\cdot a} H_{\mu b}^{\cdot\cdot\,y} + B_{d c b \varkappa}^{\nu\mu\lambda a} R_{\nu\mu\lambda}^{\cdot\cdot\cdot\,\varkappa}\right) v^b \\ \qquad = 2 S_{\mu\lambda}^{\cdot\cdot\,\varrho} B_{d c \varrho \varkappa}^{\mu\lambda\nu a} \nabla_\nu v^\varkappa + 2 B_{d c}^{\mu\lambda} S_{\mu\lambda}^{\cdot\cdot\,\varrho} C_\varrho^\nu B_\varkappa^a \nabla_\nu v^\varkappa - \\ \qquad - 2 B_{d c \varkappa}^{\nu\mu a} S_{\nu\mu}^{\cdot\cdot\,\varrho} \nabla_\varrho v^\varkappa = 0 \end{cases}$$

or (cf. V 7.32; 8.4)

$$(5.22)\quad \boxed{{}'R_{\dot d c b}^{\cdot\cdot\cdot\,a} = B_{d c b \varkappa}^{\nu\mu\lambda a} R_{\nu\mu\lambda}^{\cdot\cdot\cdot\,\varkappa} + 2\, l_{[\dot d\cdot\,|y|}^{\cdot a} H_{c]b}^{\cdot\cdot\,y}; \quad \text{conj.}} \quad (\text{Gauss I})$$

In the same way we get (cf. V 7.32; 8.4)

$$(5.23)\quad \boxed{{}'R_{\dot{\bar d} c b}^{\cdot\cdot\cdot\,a} = B_{\bar d c b \varkappa}^{\bar\nu\mu\lambda a} R_{\bar\nu\mu\lambda}^{\cdot\cdot\cdot\,\varkappa} + L_{\dot{\bar d}\cdot x}^{\cdot a} H_{c b}^{\cdot\cdot\,x}; \quad \text{conj.}} \quad (\text{Gauss II})$$

So we have in $\widetilde{L}_n$ two Gauss equations and their conjugates.[1])

Proceeding in the same way we get after some calculations fiv Codazzi equations (cf. V 7.48, 51; 8.7, 8):

$$(5.24)\quad \boxed{2 D_{[d} H_{c]b}^{\cdot\cdot\,x} = - 2\,{}'S_{\dot d c}^{\cdot\cdot\,a} H_{a b}^{\cdot\cdot\,x} + B_{d c b}^{\nu\mu\lambda} C_\varkappa^x R_{\nu\mu\lambda}^{\cdot\cdot\cdot\,\varkappa}; \quad \text{conj.}} \quad (\text{Codazzi I})$$

$$(5.25)\quad \boxed{D_{\bar d} H_{c b}^{\cdot\cdot\,x} = C_\varkappa^x B_{c b \bar d}^{\mu\lambda\bar\nu} R_{\bar\nu\mu\lambda}^{\cdot\cdot\cdot\,\varkappa}; \quad \text{conj.}} \quad (\text{Codazzi II})$$

$$(5.26)\quad \boxed{2 D_{[d} l_{c]\cdot y}^{\cdot a} = - B_{d c \varkappa}^{\nu\mu a} C_y^\lambda R_{\nu\mu\lambda}^{\cdot\cdot\cdot\,\varkappa} - 2\,{}'S_{\dot d c}^{\cdot\cdot\,e} l_{e\cdot y}^{\cdot a}; \quad \text{conj.}} \quad (\text{Codazzi III})$$

$$(5.27)\quad \boxed{D_{[\bar d} L_{\bar c]\cdot y}^{\cdot a} = - {}'S_{\dot{\bar d}\bar c}^{\cdot\cdot\,\bar b} L_{\bar b\cdot y}^{\cdot a}; \quad \text{conj.}} \quad (\text{Codazzi IV})$$

$$(5.28)\quad \boxed{- D_{\bar d} l_{c\cdot y}^{\cdot a} + D_c L_{\bar d\cdot y}^{\cdot a} = C_y^\lambda B_{c \varkappa \bar d}^{\mu a \bar\nu} R_{\bar\nu\mu\lambda}^{\cdot\cdot\cdot\,\varkappa}; \quad \text{conj.}} \quad (\text{Codazzi V})$$

with their conjugates and two Ricci equations (cf. V 7.43; 8.6):

$$(5.29)\quad \boxed{\overset{m\,m'}{R}{}_{\dot d c y}^{\cdot\cdot\cdot\,x} = B_{d c}^{\nu\mu} C_{\varkappa y}^{x\lambda} R_{\nu\mu\lambda}^{\cdot\cdot\cdot\,\varkappa} + 2 H_{[\dot d|b|}^{\cdot\cdot\,x} l_{c]\cdot y}^{\cdot b}; \quad \text{conj.}} \quad (\text{Ricci I})$$

$$(5.30)\quad \boxed{\overset{m\,m'}{R}{}_{\dot{\bar d} c y}^{\cdot\cdot\cdot\,x} = B_{\bar d c}^{\bar\nu\mu} C_{y\varkappa}^{\lambda x} R_{\bar\nu\mu\lambda}^{\cdot\cdot\cdot\,\varkappa} - H_{c b}^{\cdot\cdot\,x} L_{\dot{\bar d}\cdot y}^{\cdot b}; \quad \text{conj.}} \quad (\text{Ricci II})$$

[1]) Cf. Hombu 1935, 1.

with their conjugates. Here $\overset{m\,m'}{R}{}_{\dot{a}cy}^{\cdot\cdot\cdot x}$ and $\overset{m\,m'}{R}{}_{\dot{a}cy}^{\cdot\cdot\cdot x}$ are the quantities defined by (cf. V 7.42; 8.5)

$$(5.31)\qquad 2D_{[d}D_{c]}r^x = \overset{m\,m'}{R}{}_{\dot{d}cy}^{\cdot\cdot\cdot x} r^y - 2\,'S_{\dot{d}c}^{\cdot\cdot e} D_e r^x$$

$$(5.32)\qquad 2D_{[\bar{d}}D_{c]}r^x = \overset{m\,m'}{R}{}_{\dot{\bar{d}}cy}^{\cdot\cdot\cdot x} r^y$$

with an arbitrary vector r^x in the $(n-m)$-direction of the rigging. Only the three CODAZZI equations (5.24, 26, 27) take a simpler form if the $\widetilde{L}_n$ is an $\widetilde{A}_n$.

In a $\widetilde{U}_n$ the induced fundamental tensor is

$$(5.33)\qquad 'a_{b\bar{a}} = B_{b\bar{a}}^{\lambda\bar{\varkappa}} a_{\lambda\bar{\varkappa}};\quad \text{conj.}$$

and the imbedding is fixed by $B_b^\varkappa$. Concomitants of $B_b^\varkappa$ and $a_{\lambda\bar{\varkappa}}$ are

$$(5.34)\qquad B_\lambda^a = {}'a^{\bar{b}a} B_{\bar{b}}^{\bar{\varkappa}} a_{\lambda\bar{\varkappa}};\quad \text{conj.}$$

and

$$(5.35)\qquad B_\lambda^\varkappa = B_{\lambda a}^{a\varkappa} = {}'a^{\bar{b}a} B_{\bar{b}}^{\bar{\varkappa}} a_{\lambda\bar{\varkappa}} B_a^\varkappa = a^{\bar{\lambda}\varkappa} B_{\bar{\lambda}}^{\bar{\varkappa}} a_{\lambda\bar{\varkappa}};\quad \text{conj.}$$

We easily derive from these equations that in a $\widetilde{U}_n$

$$(5.36)\qquad D_c\,'a_{b\bar{a}} = 0;\quad \text{conj.}$$

and

$$(5.37)\qquad \begin{cases} \text{a)}\quad L_{\bar{c}\,\cdot\lambda}^{\cdot\,\varkappa} = H_{\bar{c}\bar{\sigma}}^{\cdot\cdot\bar{\varrho}} a_{\lambda\bar{\varrho}} a^{\bar{\sigma}\varkappa}; & \text{conj.} \\ \text{b)}\quad l_{c\,\cdot\lambda}^{\cdot\,\varkappa} = h_{c\bar{\sigma}}^{\cdot\cdot\bar{\varkappa}} a^{\bar{\sigma}\varkappa} a_{\lambda\bar{\varkappa}} = 0; & \text{conj.} \end{cases}$$

From (5.36) we see that the imbedded X_m is a $\widetilde{U}_m$ with $'a_{b\bar{a}}$ as fundamental tensor. (5.37) expresses the facts that the quantities $l_{c\,\cdot\lambda}^{\cdot\,\varkappa}$ and $h_{\bar{c}\lambda}^{\cdot\cdot\varkappa}$ both vanish and that $H_{c\lambda}^{\cdot\cdot\varkappa}$ and $L_{c\,\cdot\bar{\lambda}}^{\cdot\,\bar{\varkappa}}$ are different components of one and the same quantity.

The first equation of GAUSS (5.22) drops out because both sides are zero and the second (5.23) takes the form

$$(5.38)\qquad \boxed{'R_{\dot{\bar{d}}cb}^{\cdot\cdot\cdot a} = B_{\bar{d}cb\varkappa}^{\bar{\nu}\mu\lambda a} R_{\bar{\nu}\mu\lambda}^{\cdot\cdot\cdot\varkappa} + H_{\bar{d}}^{\cdot a\bar{x}} H_{cb\bar{x}};\quad \text{conj.}}\qquad (\text{GAUSS}).$$

The third equation of CODAZZI (5.26) drops out, the first (5.24) is equivalent to the fourth (5.27):

$$(5.39)\qquad \boxed{D_{[d} H_{c]b}^{\cdot\,\cdot x} = -\,'S_{\dot{d}c}^{\cdot\cdot a} H_{ab}^{\cdot\cdot x};\quad \text{conj.}}\qquad (\text{CODAZZI I})$$

and the second (5.25) is equivalent to the fifth (5.28):

$$(5.40)\qquad \boxed{D_{\bar d} H_{c\bar b}^{\cdot\cdot\varkappa} = C_{\varkappa}^{x} B_{c\bar b\bar d}^{\mu\lambda\bar\nu} R_{\bar\nu\mu\lambda}^{\cdot\cdot\cdot\varkappa}; \quad \text{conj.}} \quad (\text{Codazzi II}).$$

The equations of Ricci take the form

$$(5.41)\qquad \left\{\begin{array}{l} \boxed{\overset{m m'}{R}{}_{\bar d c y}^{\cdot\cdot\cdot x} = 0; \quad \text{conj.}} \quad (\text{Ricci I}) \\ \boxed{\overset{m m'}{R}{}_{\bar d c y}^{\cdot\cdot\cdot x} = B_{\bar d c}^{\bar\nu\mu} C_{y\varkappa}^{\lambda x} R_{\bar\nu\mu\lambda}^{\cdot\cdot\cdot\varkappa} - H_{cb}^{\cdot\cdot x} H_{\bar d\cdot y}^{\cdot b}; \quad \text{conj.}} \quad (\text{Ricci II}). \end{array}\right.$$

Among all these equations only the Codazzi equation (5.39) contains $'S_{\bar d c}^{\cdot\cdot a}$ and it is only this equation that is simplified if the $\tilde U_n$ is a $\tilde V_n$.[1])

Exercise.

VIII 5,1[2]). Prove that for an $\tilde U_m$ in $\tilde U_n$

$$H_{c\lambda}^{\cdot\cdot\varkappa} = D_c B_\lambda^\varkappa$$

if $\xi^\varkappa$ is analytic in η^α. There is no analogous relation for a U_m in U_n.

§ 6. Curves in a $\tilde U_n$ with a positive definite fundamental tensor.

For $m = 1$ we write $\eta, \bar\eta, d_\eta, d_{\bar\eta}$ instead of $\eta^1, \eta^{\bar 1}, \partial_1, \partial_{\bar 1}$ and we suppose that nowhere $d_\eta \xi^\varkappa$ is a nullvector. Then we have

$$(6.1)\qquad B_1^\varkappa = d_\eta \xi^\varkappa; \quad \text{conj.}$$

$$(6.2)\qquad 'a_{1\bar 1} = a_{\lambda\bar\varkappa} (d_\eta \xi^\lambda)(d_{\bar\eta} \xi^{\bar\varkappa}) = \frac{ds^2}{d\eta\, d\bar\eta} = \left|\frac{ds}{d\eta}\right|^2; \quad 'a^{1\bar 1} = \left|\frac{ds}{d\eta}\right|^{-2}; \quad \text{conj.}$$

$$(6.3)\qquad '\Gamma_{11}^1 = 'a^{\bar 1 1}\, \partial_\eta\, 'a_{1\bar 1} = 2\,\partial_\eta \log\left|\frac{ds}{d\eta}\right|; \quad \text{conj.}$$

Here ds is defined by $ds^2 = a_{\lambda\bar\varkappa}\, d\xi^\lambda\, d\xi^{\bar\varkappa}$ and it is not a differential of a function of η or of η and $\bar\eta$. In fact $\tilde U_1$ is two-dimensional and $\int ds$ depends on the way of integration. From (5.16) and (6.3) it follows that

$$(6.4)\qquad H_{11}^{\cdot\cdot\varkappa} = \frac{d^2\xi^\varkappa}{d\eta^2} - 2\frac{d\xi^\varkappa}{d\eta}\,\partial_\eta \log\left|\frac{ds}{d\eta}\right| + \Gamma_{\mu\lambda}^\varkappa \frac{d\xi^\mu}{d\eta}\frac{d\xi^\lambda}{d\eta}; \quad \text{conj.}$$

---

[1]) Cf. for imbedding in $\tilde U_n$ Schouten 1929, 1; Schouten and v. Dantzig 1930, 1, p. 335ff.; Fuchs 1935, 1; 1937, 1; Rozenfel'd 1949, 2; Calabi 1953, 1. Imbedding in $\tilde R_n$ was dealt with by Bochner 1947, 1, p. 190ff.

[2]) Schouten and v. Dantzig 1930, 1; E II 1938, 2, p. 246.

If instead of ds we introduce a differential dz defined by $ds^2 = dz\, d\bar{z}$ [1]), dz is determined to within a factor $e^{i\varphi}$ with real φ. In general dz *is not a differential of a function and the expression* z *without a letter "d" has no meaning.* Using dz we have instead of (6.2–4) equations of exactly the same form but with dz instead of ds.

The tangent unitvector can now be defined by

$$(6.5)\qquad j^\varkappa \stackrel{\text{def}}{=} \frac{d\xi^\varkappa}{dz} = \sigma^{-1}\frac{d\xi^\varkappa}{d\eta}; \qquad \sigma \stackrel{\text{def}}{=} \frac{dz}{d\eta}; \quad \text{conj.}$$

It is defined to within a factor $e^{i\varphi}$ with φ real. This is not essentially new because the tangent unitvector of a real curve in V_n is also fixed to within a factor ± 1 because its sign depends on the choice of the sense on the curve in which s increases.

If (6.5) is substituted in (6.4) (for $d\eta \to dz$) we get

$$(6.6)\qquad \begin{cases} H_{\dot{1}\dot{1}}{}^{\varkappa} = \dfrac{d}{d\eta}\sigma j^\varkappa - 2\sigma j^\varkappa \partial_\eta \log|\sigma| + \sigma^2 \Gamma^\varkappa_{\mu\lambda} j^\mu j^\lambda \\ \qquad = \sigma^2 j^\mu \nabla_\mu j^\varkappa - \sigma^2 j^\mu j^\varkappa \partial_\mu \log\bar{\sigma}; \quad \text{conj.} \end{cases}$$

because $\partial/\partial_\eta = \sigma j^\mu \partial_\mu$. If the curvature vector of the curve is defined as in V § 9

$$(6.7)\qquad u^\varkappa \stackrel{\text{def}}{=} j^c j^b H_{cb}^{\cdot\cdot\varkappa}; \quad \text{conj.}$$

it follows that for a $\tilde{U}_1$ in a $\tilde{U}_n$ (cf. V § 9)

$$(6.8)\qquad u^\varkappa\; j^\mu \nabla_\mu j^\varkappa - j^\mu j^\varkappa \partial_\mu \log\bar{\sigma}; \quad \text{conj.}$$

in contradistinction to the case of a V_1 in V_n where $u^\varkappa$ equals $j^\mu \nabla_\mu j^\varkappa$. Note that $u^\varkappa$ gets a factor $e^{2i\varphi}$ if $j^\varkappa$ gets a factor $e^{i\varphi}$. The vector $j^\mu \nabla_\mu j^\varkappa$ is not perpendicular to $j^\varkappa$ because

$$(6.9)\qquad \begin{cases} a_{\lambda\bar{\varkappa}} j^\mu j^{\bar{\varkappa}} \nabla_\mu j^\lambda = -j^\mu j_{\bar{\varkappa}} \nabla_\mu j^{\bar{\varkappa}} = -\sigma^{-1} j_{\bar{\varkappa}} \partial_\eta j^{\bar{\varkappa}} \\ \qquad = -\sigma^{-1} j_{\bar{\varkappa}} \dfrac{d\xi^{\bar{\varkappa}}}{d\eta} \partial_\eta \bar{\sigma}^{-1} = j^\mu \partial_\mu \log\bar{\sigma}; \quad \text{conj.} \end{cases}$$

but $u^\varkappa$ is the part of $j^\mu \nabla_\mu j^\varkappa$ perpendicular to $j^\varkappa$ as follows from (6.8, 9)

$$(6.10)\qquad j_\lambda u^\lambda = j^\mu \partial_\mu \log\bar{\sigma} - j^\mu \partial_\mu \log\bar{\sigma} = 0; \quad \text{conj.}$$

and also from the fact that the $\varkappa$-region of $H_{\dot{1}\dot{1}}{}^{\varkappa}$ is perpendicular to $j^\varkappa$.

A geodesic $\tilde{U}_1$ in $\tilde{U}_n$ is characterized by

$$(6.11)\qquad \frac{d^2\xi^\varkappa}{d\eta^2} + \Gamma^\varkappa_{\mu\lambda}\frac{d\xi^\mu}{d\eta}\frac{d\xi^\lambda}{d\eta} = \alpha\frac{d\xi^\varkappa}{d\eta}; \quad \text{conj.}$$

[1]) Coburn 1941, 3.

Now the left hand side of this equation is equal to

$$(6.12)\quad \begin{cases} \sigma j^\mu \partial_\mu \sigma j^\varkappa + \sigma^2 \Gamma^\varkappa_{\mu\lambda} j^\mu j^\lambda = \sigma^2 j^\mu \nabla_\mu j^\varkappa + \sigma^2 j^\varkappa j^\mu \partial_\mu \log \sigma \\ \qquad = H_{\dot{1}\dot{1}}{}^\varkappa + 2\sigma^2 j^\varkappa j^\mu \partial_\mu \log|\sigma|; \quad \text{conj.} \end{cases}$$

hence the n.a.s. condition for a curve to be a geodesic is (cf. 5.14)

$$(6.13)\qquad H_{\dot{1}\dot{1}}{}^\varkappa = 0; \quad \text{conj.}$$

[equivalent to $u^\varkappa = 0$ according to (6.7)] and every geodesic satisfies the equation

$$(6.14)\qquad \frac{d^2 \xi^\varkappa}{d\eta^2} + \Gamma^\varkappa_{\mu\lambda} \frac{d\xi^\mu}{d\eta} \frac{d\xi^\lambda}{d\eta} = 2\sigma^2 j^\varkappa j^\mu \partial_\mu \log|\sigma|; \quad \text{conj.}$$

Note that the right hand side of this equation does not depend on the choice of the free factor in $j^\varkappa$ because $\sigma j^\varkappa$ and $|\sigma|$ are independent of this choice.

We now consider the special case where the free factor in dz can be chosen in such a way that dz is a differential of an analytic function $z(\eta)$. Because of (6.5) we then have

$$(6.15)\qquad \frac{d\xi^\varkappa}{d\eta} = j^\varkappa \frac{dz}{d\eta}; \quad \text{conj.}$$

and

$$(6.16)\qquad \frac{dz}{d\eta} = j_\varkappa \frac{d\xi^\varkappa}{d\eta} = j_\varkappa B^\varkappa_1 = j_1; \quad \text{conj.}$$

But (6.16) expresses the fact that the vector j_b in $\widetilde{U}_1$ is analytic. Conversely, if j_b is analytic, $dz/d\eta$ is analytic and z can be found by integration as an analytic function. It has already been proved in VIII § 3 that j_b is analytic if and only if the $\widetilde{U}_1$ is an $\widetilde{R}_1$. Because j_1 is determined to within a factor $e^{i\varphi}$ with φ real and analytic in η, it follows that φ must be a constant. Hence z is determined to within an affine transformation $'z = az + b$ with constant coefficients and with $|a| = 1$. We call z a *natural parameter* on the $\widetilde{R}_1$ in $\widetilde{U}_n$. Collecting results we have[1]:

If an X_1 in a $\widetilde{U}_n$ is given by the analytic functions $\xi^\varkappa(\eta)$ of a complex parameter η, this X_1 is always a $\widetilde{U}_1$. If and only if this $\widetilde{U}_1$ is an $\widetilde{R}_1$, that is, if and only if the induced connexion is analytic and thus integrable, it is possible to find a parameter $z = z(\eta)$ on this curve analytic in η such that $ds^2 = dz\, d\bar{z}$. This natural parameter is fixed to within an affine transformation $'z = az + b$ with constant coefficients and with $|a| = 1$.

Only these $\widetilde{R}_1$'s in $\widetilde{U}_n$ have an analytic tangent unitvector $d\xi^\varkappa/dz$. They are in many respects analogous to the V_1's in a V_n.[2]) For instance $u^\varkappa = j^\mu \nabla_\mu j^\varkappa$; (conj.) comes instead of (6.8).

---

[1]) SCHOUTEN and v. DANTZIG 1930, 1, p. 343; E II 1938, 2, p. 251. The theorem was proved there without the auxiliary differential dz of COBURN.

[2]) Cf. COBURN 1941, 3.

We see from (6.14) that the equation of a geodesic with respect to a natural parameter z (if existing) takes the simple form

$$\frac{d^2 \xi^\varkappa}{d z^2} + \Gamma^\varkappa_{\mu\lambda} \frac{d \xi^\mu}{d z} \frac{d \xi^\lambda}{d z} = 0; \quad \text{conj.} \tag{6.17}$$

There is still another kind of curves in $\tilde{U}_n$ that arises if we consider equations of the form $\xi^\varkappa = \xi^\varkappa(t)$ where t is a *real* parameter.[1] These curves are one-dimensional. It is possible to get them from the $\tilde{U}_1$'s in $\tilde{U}_n$ that are $\tilde{R}_1$'s, by considering only the points of such an $\tilde{R}_1$ where some definitely chosen natural parameter takes real values. Of course the natural parameter on such a curve is fixed to within transformations of the form $'z = z + b$ with real constant b.[2]

§ 7. Conformal transformation of a connexion in $\tilde{U}_n$.

If the fundamental tensor $a_{\lambda\bar{\varkappa}}$ of a $\tilde{U}_n$ is transformed into

$$'a_{\lambda\bar{\varkappa}} = \sigma\, a_{\lambda\bar{\varkappa}} \tag{7.1}$$

where σ is an arbitrary *real* semi-analytic scalar, the parameters $\Gamma^\varkappa_{\mu\lambda}$, $\Gamma^\varkappa_{\bar{\mu}\lambda}$ transform into

$$\left\{\begin{array}{ll} \text{a)} & '\Gamma^\varkappa_{\mu\lambda} = {}'\overset{-1}{a}{}^{\bar{\varrho}\varkappa}\, \partial_\mu\, 'a_{\lambda\bar{\varrho}} = \Gamma^\varkappa_{\mu\lambda} + \sigma_\mu A^\varkappa_\lambda; \quad \text{conj.} \\ \text{b)} & '\Gamma^\varkappa_{\bar{\mu}\lambda} = \Gamma^\varkappa_{\bar{\mu}\lambda} = 0; \quad \text{conj.} \qquad \sigma_\mu \overset{\text{def}}{=} \partial_\mu \log \sigma; \quad \text{conj.} \end{array}\right. \tag{7.2}$$

We call this transformation *conformal* (cf. VI § 5). From (7.2) we get

$$'\Gamma^\mu_{\mu\lambda} = \Gamma^\mu_{\mu\lambda} + \sigma_\lambda; \qquad '\Gamma^\lambda_{\mu\lambda} = \Gamma^\lambda_{\mu\lambda} + n\,\sigma_\mu; \quad \text{conj.} \tag{7.3}$$

and

$$\left\{\begin{array}{llll} \text{a)} & '\nabla_\mu v^\varkappa = \nabla_\mu v^\varkappa + \sigma_\mu v^\varkappa; & '\nabla_{\bar{\mu}} v^\varkappa = \nabla_{\bar{\mu}} v^\varkappa = \partial_{\bar{\mu}} v^\varkappa; & \text{conj.} \\ \text{b)} & '\nabla_\mu w_\lambda = \nabla_\mu w_\lambda - \sigma_\mu w_\lambda; & '\nabla_{\bar{\mu}} w_\lambda = \nabla_{\bar{\mu}} w_\lambda = \partial_{\bar{\mu}} w_\lambda; & \text{conj.} \end{array}\right. \tag{7.4}$$

The transformation (7.2a) is projective (cf. VI § 1) and leaves parallel fields parallel (cf. VI § 1). If and only if $\sigma_\lambda = 0$, $\Gamma^\varkappa_{\mu\lambda}$ and $'\Gamma^\varkappa_{\mu\lambda}$ can for $n > 1$ be symmetric at the same time. Hence

A conformal transformation of a $\tilde{U}_n$ is at the same time a restricted projective transformation[3]) *and σ_λ is the gradient of a* **real** *scalar. If a*

---

[1]) SCHOUTEN and v. DANTZIG 1930, 1, p. 343; 1931, 3; COBURN 1941, 1; 3. Cf. COBURN 1942, 2; SUGURI 1951, 1 for FRENET formulae for these curves. VARGA 1939, 1 dealt with curves in the elliptic $\tilde{R}_2$ and GOLIFMAN 1941, 1 considered one-dimensional curves in the hyperbolic $\tilde{R}_2$.

[2]) Congruences of curves of both kinds were studied by COBURN 1943, 1.

[3]) COBURN 1941, 1, p. 32f. Cf. VI § 1; cf. for projective transformations also DE MIRA FERNANDES 1950, 1.

conformal transformation leaves the connexion of a $\tilde{V}_n$, $n>1$, *symmetric,* $a_{\lambda\bar{\varkappa}}$ *gets a* **constant** *factor.*[1])

This latter statement implies that the only conformal transformation that transforms an $\tilde{S}_n$ into an $\tilde{S}_n$ is a transformation with $\sigma=$ const. We have seen in VI § 5 that an S_n can be transformed conformally into every other S_n. This is quite different in a unitary space. In VI § 5 we also saw that the connexion of a V_n is determined by its conformal and projective properties together. Now let

$$(7.5)\qquad '\Gamma^{\varkappa}_{\mu\lambda}=\Gamma^{\varkappa}_{\mu\lambda}+A^{\varkappa}_{\lambda}\,\partial_\mu \log\sigma$$

be a connexion derived from the $\Gamma^{\varkappa}_{\mu\lambda}$ of a $\tilde{U}_n$ by a conformal transformation. Then $\partial_\mu \log\sigma$ can be computed from $'\Gamma^{\varkappa}_{\mu\lambda}$ if $S_\mu=\frac{2}{n-1}\,S_{\mu\lambda}^{\cdot\cdot\lambda}$ is given. Hence

The connexion of a $\tilde{V}_n$ *is determined by its conformal properties only, but for the determination of the connexion of a* $\tilde{U}_n$ *we also need the field* S_μ.

$S_{\mu\lambda}^{\cdot\cdot\varkappa}$ and S_μ transform as follows (cf. 7.2)

$$(7.6)\qquad \begin{cases} \text{a)}\ \ 'S_{\mu\lambda}^{\cdot\cdot\varkappa}=S_{\mu\lambda}^{\cdot\cdot\varkappa}+\sigma_{[\mu}A^{\varkappa}_{\lambda]} \\ \text{b)}\ \ 'S_\mu=S_\mu+\sigma_\mu \end{cases}$$

hence

The connexion of a $\tilde{U}_n$ *can be transformed conformally into a symmetric connexion if and only if it is semi-symmetric and if* S_μ *is a gradient of a real scalar.*[2])

The transformation of the curvature quantities can be derived from (2.17, 21) and (7.2):

$$(7.7)\qquad \begin{cases} \text{a)}\ \ 'R_{\bar{\nu}\mu\lambda}^{\cdot\cdot\cdot\varkappa}=R_{\bar{\nu}\mu\lambda}^{\cdot\cdot\cdot\varkappa}+A^{\varkappa}_{\lambda}\,\partial_{\bar{\nu}}\sigma_\mu; & \text{conj.}\\ \text{b)}\ \ 'R_{\mu\bar{\nu}\lambda}^{\cdot\cdot\cdot\varkappa}=R_{\mu\bar{\nu}\lambda}^{\cdot\cdot\cdot\varkappa}-A^{\varkappa}_{\lambda}\,\partial_{\bar{\nu}}\sigma_\mu; & \text{conj.}\\ \text{c)}\ \ 'R_{\bar{\nu}\lambda}=R_{\bar{\nu}\lambda}-\partial_{\bar{\nu}}\sigma_\lambda; & \text{conj.}\\ \text{d)}\ \ 'V'_{\bar{\nu}\mu}=V'_{\bar{\nu}\mu}+n\,\partial_{\bar{\nu}}\sigma_\mu; & \text{conj.}\\ \text{e)}\ \ 'R=\sigma^{-1}R-\sigma^{-1}a^{\bar{\nu}\lambda}\,\partial_{\bar{\nu}}\sigma_\lambda; & \text{conj.} \end{cases}$$

A $\tilde{U}_n$ is called *conformally euclidean* or a $\tilde{C}_n$ if it can be transformed conformally into an $\tilde{R}_n$. From (7.6a, 7a) it follows:

A $\tilde{U}_n$ *is a* $\tilde{C}_n$ *if and only if there exists a real semi-analytic gradient field* σ_λ *such that*

$$(7.8)\qquad \text{a)}\ \ R_{\bar{\nu}\mu\lambda}^{\cdot\cdot\cdot\varkappa}=-A^{\varkappa}_{\lambda}\,\partial_{\bar{\nu}}\sigma_\mu;\ \ \text{conj.}\qquad \text{b)}\ \ S_{\mu\lambda}^{\cdot\cdot\varkappa}=A^{\varkappa}_{[\mu}\sigma_{\lambda]};\ \ \text{conj.}$$

---

1) Coburn 1941, 1, p. 31; 1942, 3, p. 140.

2) A gradient of a real scalar need not have real components.

For $n > 1$ it follows from $A^{\varkappa}_{[\mu}\, \sigma_{\lambda]} = 0$ that $\sigma_\lambda = 0$, hence

A $\tilde{V}_n$ is for $n > 1$ a $\tilde{C}_n$ if and only if it is an $\tilde{R}_n$.[1])

As a corollary we get that no $\tilde{S}_n$, $n > 1$, with non vanishing curvature is a $\tilde{C}_n$.[1])

From (7.8) we see that a $\tilde{C}_n$ is always semi-symmetric and that

$$R_{\bar{\nu}\mu\lambda}^{\cdot\cdot\cdot\varkappa} = A^{\varkappa}_{\lambda}\, \partial_{\bar{\nu}}\, S_\mu . \tag{7.9}$$

In VIII § 3 it has been proved that for a semi-symmetric $\tilde{U}_n$, $n > 2$, S_λ is always a gradient vector. Now if S_λ equals $\partial_\lambda \varphi$ it is also the gradient of the real scalar $\frac{1}{2}(\varphi + \bar{\varphi})$ if and only if $\partial_{\bar{\nu}} S_\lambda = \partial_\lambda S_{\bar{\nu}}$. Hence [2])

A $\tilde{U}_n$ can be transformed conformally into a $\tilde{V}_n$ if and only if for $n > 2$ the connexion is semi-symmetric and $\partial_{\bar{\mu}} S_\lambda = \partial_\lambda S_{\bar{\mu}}$, and for $n = 2$ S_λ is a gradient of a real scalar. A $\tilde{U}_n$ is a $\tilde{C}_n$ if and only if these conditions are satisfied and if moreover

$$0 = C_{\bar{\nu}\mu\lambda}^{\cdot\cdot\cdot\varkappa} \stackrel{\text{def}}{=} R_{\bar{\nu}\mu\lambda}^{\cdot\cdot\cdot\varkappa} - A^{\varkappa}_{\lambda}\, \partial_{\bar{\nu}}\, S_\mu \tag{7.10}$$

$C_{\bar{\nu}\mu\lambda}^{\cdot\cdot\cdot\varkappa}$ is called the *conformal curvature tensor* of the $\tilde{U}_n$.

Coburn has dealt with several problems concerning conformal transformations in $\tilde{U}_n$. He mostly considers the transformation not as a transformation of the connexion in one and the same space but as a transformation of a space $\tilde{U}_n$ into another space $'\tilde{U}_n$. But of course this interpretation is not essential. Since such a conformal transformation is always at the same time a restricted projective transformation he requires all those restricted projective transformations of this kind that transform a $\tilde{U}_n$ into a $\tilde{V}_n$.[3]) In the same paper[4]) he considers a *conformally symmetric* $\tilde{U}_n$, that is a $\tilde{U}_n$ with the fundamental tensor

$$a_{\lambda\bar{\varkappa}} = \beta\, \partial_\lambda\, \partial_{\bar{\varkappa}} \log \varphi ; \quad \text{conj.}; \qquad \beta = \text{real}; \quad \varphi - \text{real}, \tag{7.11}$$

subjected to conformal transformations of the form

$$'a_{\lambda\bar{\varkappa}} = \varphi^{\alpha}\, a_{\lambda\bar{\varkappa}} ; \quad \text{conj.}; \qquad \alpha = \text{real and const.} \tag{7.12}$$

and deduces invariants for these transformations.

---

[1]) Coburn 1942, 3, p. 140, only for $n > 2$. Remark that every S_n is a C_n (cf. VI § 5).

[2]) Coburn 1942, 3, p. 138.

[3]) Coburn 1941, 1, p. 32ff.

[4]) p. 37ff.

Exercises.

VIII 7,1[1]). In a $\widetilde{U}_n$ that can be transformed into a $\widetilde{U}_n$ with $R_{\bar{\nu}\mu\lambda}^{\cdot\cdot\cdot\varkappa}=0$ by the conformal transformation $a_{\lambda\bar{\varkappa}}\to\sigma\, a_{\lambda\bar{\varkappa}}$ the following identities hold

$$\text{VIII 7,1 } \alpha)\qquad R_{\bar{\nu}\mu\lambda}^{\cdot\cdot\cdot\varkappa}=\frac{1}{n}A_\lambda^\varkappa\,\partial_{\bar{\nu}}\,\partial_\mu\log\mathfrak{a};\qquad \mathfrak{a}=|\mathrm{Det}\,(a_{\lambda\bar{\varkappa}})|$$

$$\text{VIII 7,1 } \beta)\qquad R_\nu^{\cdot\nu}=\nabla_\nu\,\sigma^\nu;\qquad \sigma_\lambda\overset{\text{def}}{=}\partial_\lambda\log\sigma.$$

VIII 7,2. A $\widetilde{V}_n$, $n>1$, can be transformed conformally into a $\widetilde{U}_n$ with $R_{\bar{\nu}\mu\lambda}^{\cdot\cdot\cdot\varkappa}=0$, if it is an $\widetilde{R}_n$ (cf. the special case mentioned in the text).

VIII 7,3. If in a semi-symmetric $\widetilde{U}_n$ the vector S_λ is a gradient of a real scalar and if there exists a coordinate system $(\varkappa)$ such that

$$\text{VIII 7,3 } \alpha)\qquad \Gamma_{(\mu\lambda)}^\varkappa+A_{(\mu}^\varkappa\,S_{\lambda)}\overset{*}{=}0$$

it can be proved that the $\widetilde{U}_n$ is a $\widetilde{C}_n$ and that there exists an $\widetilde{R}_n$ such that the geodesics of $\widetilde{U}_n$ and $\widetilde{R}_n$ are in restricted correspondence.[2])

VIII 7,4. In a $\widetilde{U}_n$ the tensor $V'_{\bar{\mu}\lambda}+n\,R_{\bar{\mu}\lambda}$ is a conformal concomitant that vanishes if the $\widetilde{U}_n$ is a $\widetilde{C}_n$.

§ 8. Conformal unitary connexions[3]).

The quantity (cf. VI § 7)

$$(8.1)\qquad \mathfrak{A}_{\lambda\bar{\varkappa}}\overset{\text{def}}{=}\mathfrak{a}^{-\frac{1}{n}}\,a_{\lambda\bar{\varkappa}};\qquad \text{conj.}$$

is a hermitian tensor density of weight $-\frac{2}{n}$. It is conformally invariant and its determinant is $+1$. Analogous to VI § 7 and (VIII 3.4) we may define from it the conformal parameters

$$(8.2)\qquad \overset{c}{\Gamma}{}_{\mu\lambda}^\varkappa=\mathfrak{A}^{\bar{\sigma}\varkappa}\,\partial_\mu\,\mathfrak{A}_{\lambda\bar{\sigma}}=\Gamma_{\mu\lambda}^\varkappa-\frac{1}{n}A_\lambda^\varkappa\,\partial_\mu\log\mathfrak{a}=\Gamma_{\mu\lambda}^\varkappa-\frac{1}{n}\Gamma_\mu A_\lambda^\varkappa;\qquad \text{conj.}$$

They transform as follows

$$(8.3)\quad\begin{cases}\text{a)}\ \ \overset{c}{\Gamma}{}_{\mu'\lambda'}^{\varkappa'}=A_{\mu'\lambda'\varkappa}^{\mu\lambda\varkappa'}\,\overset{c}{\Gamma}{}_{\mu\lambda}^\varkappa+A_\varrho^{\varkappa'}\,\partial_{\mu'}A_{\lambda'}^\varrho-\frac{1}{n}A_{\lambda'}^{\varkappa'}\,\partial_{\mu'}\log\Delta;\qquad \text{conj.}\\ \text{b)}\ \ \overset{c}{\Gamma}{}_{\mu'\lambda'}^{\mu'}=A_{\lambda'}^\lambda\,\overset{c}{\Gamma}{}_{\mu\lambda}^\mu+\frac{n-1}{n}\,\partial_{\lambda'}\log\Delta;\qquad \text{conj.}\end{cases}$$

Now in contradistinction to the conformal parameters belonging to an ordinary metric dealt with in VI § 7 it is possible to derive a connexion

---

[1]) Cf. Exerc. VI 5,1.

[2]) COBURN 1941, 1 calls the correspondence between geodesics *restricted* if the vector of the projective transformation is a gradient of a real scalar.

[3]) COBURN 1942, 3, p. 129ff. This section is a free interpretation and at some points a simplification of the most important of his results.

from $\overset{c}{\Gamma}{}_{\mu\lambda}^{\varkappa}$ without using any auxiliary objects or invariant objects of higher order. In fact, if we define for $n > 1$

$$(8.4)\qquad \overset{c}{\Pi}{}_{\mu\lambda}^{\varkappa} \overset{\text{def}}{=} \overset{c}{\Gamma}{}_{\mu\lambda}^{\varkappa} + \frac{1}{n-1} A_{\lambda}^{\varkappa} \overset{c}{\Gamma}{}_{\varrho\mu}^{\varrho} = \Gamma_{\mu\lambda}^{\varkappa} - S_{\mu} A_{\lambda}^{\varkappa}; \qquad \text{conj.}$$

the $\overset{c}{\Pi}{}_{\mu\lambda}^{\varkappa}$ transform just like the $\Gamma_{\mu\lambda}^{\varkappa}$ because $A_{\lambda}^{\varkappa} S_{\mu}$ is a tensor, and these new parameters are conformally invariant because of (7.5, 6b). Denoting the covariant derivative belonging to them by $\overset{c}{\nabla}$ we have

$$(8.5)\qquad \begin{cases} \text{a)}\ \ \overset{c}{\nabla}_{\mu} a_{\lambda\bar{\varkappa}} = \nabla_{\mu} a_{\lambda\bar{\varkappa}} + S_{\mu} A_{\lambda}^{\sigma} a_{\sigma\bar{\varkappa}} = S_{\mu} a_{\lambda\bar{\varkappa}}; \quad \text{conj.} \\ \text{b)}\ \ \overset{c}{\nabla}_{\mu} \mathfrak{a} = \mathfrak{a} \overset{c}{\nabla}_{\mu} \log \mathfrak{a} = \mathfrak{a}\, a^{\bar{\varkappa}\lambda} \overset{c}{\nabla}_{\mu} a_{\lambda\bar{\varkappa}} = n\, \mathfrak{a}\, S_{\mu}; \quad \text{conj.} \end{cases}$$

hence

$$(8.6)\qquad \overset{c}{\nabla}_{\mu} \mathfrak{A}_{\lambda\bar{\varkappa}} = \overset{c}{\nabla}_{\mu} \mathfrak{a}^{-\frac{1}{n}} a_{\lambda\bar{\varkappa}} = 0; \quad \text{conj.}$$

We call $\overset{c}{\Pi}{}_{\mu\lambda}^{\varkappa}$ the parameters of the *conformal connexion* belonging to $\lfloor a_{\lambda\bar{\varkappa}} \rfloor$. From (8.4) it follows that the conformal connexion is identical with the connexion belonging to $a_{\lambda\bar{\varkappa}}$ if and only if $S_{\mu} = 0$, that is (cf. 3.9) if and only if $a_{\lambda\bar{\varkappa}}$ satisfies the differential equation

$$(8.7)\qquad \partial_{\mu} \log \mathfrak{a} = a^{\bar{\varrho}\lambda} \partial_{\lambda} a_{\mu\bar{\varrho}}; \quad \text{conj.}$$

This is always true in a $\widetilde{V}_n$, hence the conformal connexion of a $\widetilde{V}_n$ is the connexion of the $\widetilde{V}_n$ itself. In fact we have already seen that for a conformal transformation of a $\widetilde{V}_n$ into a $\widetilde{V}_n$ always $\sigma = \text{const}$. As a generalization we may mention the fact that for a conformal transformation of a $\widetilde{U}_n$ with $S_{\mu} = 0$ into a $\widetilde{U}_n$ with $S_{\mu} = 0$ also $\sigma = \text{const}$. as follows immediately from (7.2a; 6b).

The following conformally invariant tensors can be deduced from $\overset{c}{\Pi}{}_{\mu\lambda}^{\varkappa}$

$$(8.8)\qquad \begin{cases} \text{a)}\ \begin{cases} \overset{c}{R}{}_{\nu\mu\lambda}^{\cdot\cdot\cdot\varkappa} \overset{\text{def}}{=} 2\partial_{[\nu} \overset{c}{\Pi}{}_{\mu]\lambda}^{\varkappa} + 2\overset{c}{\Pi}{}_{[\nu|\varrho|}^{\varkappa} \overset{c}{\Pi}{}_{\mu]\lambda}^{\varrho} \\ \qquad = R_{\nu\mu\lambda}^{\cdot\cdot\cdot\varkappa} - 2A_{\lambda}^{\varkappa} \partial_{[\nu} S_{\mu]} = -2A_{\lambda}^{\varkappa} \partial_{[\nu} S_{\mu]}; \quad \text{conj.} \end{cases} \\ \text{b)}\ \ \overset{c}{R}{}_{\bar{\nu}\mu\lambda}^{\cdot\cdot\cdot\varkappa} \overset{\text{def}}{=} \partial_{\bar{\nu}} \overset{c}{\Pi}{}_{\mu\lambda}^{\varkappa} = R_{\bar{\nu}\mu\lambda}^{\cdot\cdot\cdot\varkappa} - A_{\lambda}^{\varkappa} \partial_{\bar{\nu}} S_{\mu} = C_{\bar{\nu}\mu\lambda}^{\cdot\cdot\cdot\varkappa}; \quad \text{conj.} \end{cases}$$

The conformal invariance of the first one follows from (7.6b) and the second one equals the conformal curvature tensor defined in (7.10). Both vanish in a $\widetilde{C}_n$.

The connexion $\overset{c}{\Pi}{}_{\mu\lambda}^{\varkappa}$ leaves invariant the hermitian tensor density $\mathfrak{A}_{\lambda\bar{\varkappa}}$ but not in general a hermitian tensor. If in an exceptional case a tensor $b_{\lambda\bar{\varkappa}}$ with rank n is invariant

$$\overset{c}{\nabla}_\mu b_{\lambda\bar{\varkappa}} = \partial_\mu b_{\lambda\bar{\varkappa}} - \overset{c}{\Pi}{}^{\varrho}_{\mu\lambda} b_{\varrho\bar{\varkappa}} = 0; \quad \text{conj.} \tag{8.9}$$

the first integrability conditions must be satisfied:

$$\text{(8.10)} \quad \begin{cases} \text{a)} \quad b_{\varrho\bar{\varkappa}} \overset{c}{R}{}_{\nu\mu\lambda}^{\cdot\cdot\cdot\varrho} = - b_{\lambda\bar{\varkappa}} \partial_{[\nu} S_{\mu]} = 0; \quad \text{conj.} \\ \text{b)} \quad R_{\bar{\mu}\mu\lambda}^{\cdot\cdot\cdot\varrho} b_{\varrho\bar{\varkappa}} - R_{\mu\bar{\mu}\bar{\varkappa}}^{\cdot\cdot\cdot\bar{\varrho}} b_{\lambda\bar{\varrho}} - b_{\lambda\bar{\varkappa}} \partial_{\bar{\mu}} S_\mu + b_{\lambda\bar{\varkappa}} \partial_\mu S_{\bar{\mu}} = 0; \quad \text{conj.} \end{cases}$$

From (8.10a) and (8.10b) transvected with $\overset{-1}{b}{}^{\bar{\varkappa}\lambda}$ (cf. 3.19d) it follows that S_μ must be the gradient of a *real* scalar (cf. VIII § 7). But then the last two terms in (8.10b) vanish and there remains only the algebraic equation

$$R_{\bar{\mu}\mu\lambda}^{\cdot\cdot\cdot\varrho} b_{\varrho\bar{\varkappa}} - R_{\mu\bar{\mu}\bar{\varkappa}}^{\cdot\cdot\cdot\bar{\varrho}} b_{\lambda\bar{\varrho}} = 0; \quad \text{conj.} \tag{8.11}$$

Because of (3.22) we have

$$R_{\bar{\mu}\mu\lambda}^{\cdot\cdot\cdot\varrho} a_{\varrho\bar{\varkappa}} = R_{\mu\bar{\mu}\bar{\varkappa}}^{\cdot\cdot\cdot\bar{\varrho}} a_{\lambda\bar{\varrho}}; \quad \text{conj.} \tag{8.12}$$

and this proves that (8.11) is satisfied if $b_{\lambda\bar{\varkappa}} = \sigma a_{\lambda\bar{\varkappa}}$, where σ is a *real* coefficient. In fact there exists such a solution if S_μ is a gradient of a *real* scalar because if in this case we take $\sigma_\mu = - S_\mu$ we get

$$\overset{c}{\nabla}_\mu \sigma a_{\lambda\bar{\varkappa}} = \sigma \sigma_\mu a_{\lambda\bar{\varkappa}} + \sigma S_\mu a_{\lambda\bar{\varkappa}} = 0; \quad \text{conj.} \tag{8.13}$$

and $\sigma a_{\lambda\bar{\varkappa}}$ is the fundamental tensor of a $\tilde{U}_n$ with parameters $\overset{c}{\Pi}{}^{\varkappa}_{\mu\lambda}$ in accordance with VIII § 3.

But there may exist other solutions. In order to find them we have to differentiate (8.11) and to eliminate the derivatives of $b_{\lambda\bar{\varkappa}}$ by means of (8.10). Proceeding in this way we get at last a set of algebraic equations for $b_{\lambda\bar{\varkappa}}$ that can be no more extended and that may have solutions with rank n (cf. the theory of sets of partial differential equations in II § 5). If there is no solution with rank n the $\overset{c}{\Pi}{}^{\varkappa}_{\mu\lambda}$ can not be the parameters of the connexion of a $\tilde{U}_n$.

Exercise.

VIII 8,1. If in a $\tilde{U}_n$ (cf. Exerc. VIII 7,1)

$$R_{\bar{\nu}\mu\lambda}^{\cdot\cdot\cdot\varkappa} = \frac{1}{n} A^{\varkappa}_{\lambda} \partial_{\bar{\nu}} \partial_\mu \log \mathfrak{a} \tag{VIII 8,1 α)}$$

and if $\tau = f(\xi^\varkappa)\, g(\xi^{\bar{\varkappa}})$ is an ordinary density of weight -2, the $\tilde{U}_n$ with the fundamental tensor

$$'a_{\lambda\bar{\varkappa}} = \tau^{-\frac{1}{n}} \mathfrak{A}_{\lambda\bar{\varkappa}}$$

has $'R_{\bar{\nu}\mu\lambda}^{\cdot\cdot\cdot\varkappa} = 0$.

§ 9. Spaces of recurrent curvature.[1]

In III § 2 a recurrent field was defined as a field that at every point and for every direction is proportional to its covariant differential and in VII § 4 we considered, from the point of view of their holonomy group, spaces with a recurrent curvature tensor.[2] A V_n is called a K_n if $K_{\nu\mu\lambda\varkappa}$ is recurrent but not covariant constant

$$(9.1) \qquad \nabla_\omega K_{\nu\mu\lambda\varkappa} = k_\omega K_{\nu\mu\lambda\varkappa}; \qquad k_\omega \neq 0$$

and it is said to be a K_n^* if it is either a K_n or symmetric

$$(9.2) \qquad \nabla_\omega K_{\nu\mu\lambda\varkappa} = 0$$

and if in this latter case there exists a vector k_ω such that

$$(9.3) \qquad k_{[\omega} K_{\nu\mu]\lambda\varkappa} = 0; \qquad k_\omega \neq 0.$$

If (9.2, 3) hold, k_ω need not be uniquely determined. For instance an R_n is a K_n^* with undetermined k_ω and an S_n with $K \neq 0$ is neither a K_n nor a K_n^*.

For $n=2$ we have (cf. III 5.31):

$$(9.4) \qquad K_{\nu\mu\lambda\varkappa} = -2\varkappa\, g_{[\nu[\lambda}\, g_{\mu]\varkappa]}$$

hence

$$(9.5) \qquad \nabla_\omega K_{\nu\mu\lambda\varkappa} = -2(\nabla_\omega \log\varkappa)\, K_{\nu\mu\lambda\varkappa}$$

and this proves that a V_2 is either an S_2 with $\varkappa \neq 0$ or a K_2.

In VII § 4 we proved that k_ω is a gradient in a K_n. In a symmetric K_n^* the integrability conditions of

$$(9.6) \qquad (\nabla_{[\omega} k) K_{\nu\mu]\lambda\varkappa} = 0$$

are identically satisfied in consequence of (9.2). Hence k_ω can always be chosen in such a way that it is a gradient.[3]

If for $n>2$ a K_n^* is a product $V_m \times V_{n-m}$ (cf. V § 11) the linear element can be written in the form

$$(9.7) \qquad \begin{cases} ds^2 \overset{*}{=} g_{\lambda_1\mu_1}(\xi^{\varkappa_1})\, d\xi^{\lambda_1} d\xi^{\mu_1} + g_{\lambda_2\mu_2}(\xi^{\varkappa_2})\, d\xi^{\lambda_2} d\xi^{\mu_2}; \\ \qquad\qquad\qquad \varkappa_1, \lambda_1, \mu_1 = 1, \ldots, m \\ \qquad\qquad\qquad \varkappa_2, \lambda_2, \mu_2 = m+1, \ldots, n \end{cases}$$

---

[1] This subject was introduced by RUSE 1948, 2; 1949, 2. His first investigations were connected with harmonic spaces and dealt especially with the cases $n=2, 3, 4$. WALKER 1950, 1 introduced the K_n^* and the simple K_n^* by means of which he and RUSE in 1951, 1; 2 succeeded in giving a classification.

[2] Spaces with recurrent RICCI tensor were considered by PATTERSON 1952, 2.

[3] WALKER 1950, 1, p. 46.

and from this we see that $\Gamma_{\mu\lambda}^{\varkappa}$ and $K_{\nu\mu\lambda\varkappa}$ have no components with indices from both ranges. But then it follows from (9.3) that

$$K_{\nu_1\mu_1\lambda_1\varkappa_1} k_{\omega_2} = 0; \quad K_{\nu_2\mu_2\lambda_2\varkappa_2} k_{\omega_1} = 0 \tag{9.8}$$

and this implies that either V_m or V_{n-m} or both are euclidean. The only non-trivial case is where only one of them, for instance V_{n-m} is euclidean. Because the rank of $g_{\lambda_2\mu_2}$ must be $n-m$, V_{n-m} must be an R_{n-m}. Because $k_{\omega_1} \neq 0$ and independent of the $\xi^{\varkappa_2}$ we have $k_{[\omega_1} K_{\nu_1\mu_1]\lambda_1\varkappa_1} = 0$ and this implies that the V_m is a K_m^*. Hence[1])

If a K_n^* is decomposable it is a product of a K_m^* and an R_{n-m}.

K_n^* is called the *flat extension* of K_m^*. Of course every flat extension of a K_m^* is some K_n^* and the same holds for K_m and K_n.[2])

If a K_n^* contains a parallel field of m-directions and a field of perpendicular $(n-m)$-directions, the first field forms a normal system of geodesic V_m's and the second a normal system of geodesic V_{n-m}'s perpendicular to the first. But then the linear element can be written in the form (9.7). Hence, the V_{n-m}'s are R_{n-m}'s and the V_m's are K_m^*'s or vice versa. If the V_{n-m}'s are R_{n-m}'s there exist in them $n-m$ mututally perpendicular non-null vector fields that are not only parallel but also covariant constant.[3]) Hence[4])

If a K_n^* admits exactly $n-m \geq 1$ mututally perpendicular non-null covariant constant vector fields it is a product of a K_m^* and an R_{n-m}.

This leads to the consideration of a K_n^* admitting exactly r mutually perpendicular covariant constant null vector fields. Now EISENHART[5]) proved that in a V_n of this kind $n-r$ is always $\geq r$ and that the linear element can be written in the form

$$\begin{cases} ds^2 \stackrel{*}{=} g_{\beta\alpha}\, d\xi^\beta\, d\xi^\alpha + 2\Sigma\, d\xi^\chi\, d\xi^{\chi+n-r}; \\ \qquad\qquad \alpha, \beta = 1, \ldots, n-r \\ \qquad\qquad \chi = 1, \ldots, r \end{cases} \tag{9.9}$$

where the $g_{\beta\alpha}$ depend on the ξ^α only. Such a V_n is called a *null-extension* of the V_{n-r} with the linear element $g_{\beta\alpha}\, d\xi^\beta\, d\xi^\alpha$. *This does not imply that the V_n is decomposable.* If this result is applied to a K_n^* with $n-m \geq 1$

[1]) WALKER 1950, 1, p. 38.

[2]) RUSE 1949, 2.

[3]) We differ here from RUSE and WALKER. We call a vector field covariant constant if its covariant derivative vanishes at all points, and parallel if at every point and for every direction it is proportional to its covariant differential (cf. III § 2).

[4]) WALKER 1950, 1, p. 39.

[5]) EISENHART 1938, 1; cf. WALKER 1949, 1; 1950, 3; 4.

mutually perpendicular covariant constant vector fields from which r are null and $n-m-r$ non-null, we get the linear element [1])

$$(9.10)\quad \begin{cases} ds^2 \stackrel{*}{=} g_{\beta\alpha}\, d\xi^\beta\, d\xi^\alpha + 2\Sigma\, d\xi^\chi\, d\xi^{\chi+m} + \sum_\omega \underset{\omega}{e}\, d\xi^\omega\, d\xi^\omega \\ \qquad \alpha, \beta = 1, \ldots, m \\ \qquad \chi = 1, \ldots, r \\ \qquad \omega = r+m+1, \ldots, n \\ \qquad \underset{\omega}{e} = \pm 1. \end{cases}$$

A V_n that admits $n-2$ independent covariant constant vector fields can be proved to be a K_n^*. Such a K_n^* is called *simple.* [2]) The classification of all simple K_n^*'s makes use of the following propositions due to RUSE

Every V_2 is a K_2^*. Every V_3 which is a null-extension of a V_2 is a K_3^*. Every V_4 which is a null-extension of a V_2 is a K_4^* and harmonic.

From this it can be proved that there are only three cases for the simple K_n^*. [2]) If there are exactly $n-2$ mutually perpendicular covariant constant vector fields of which exactly r are null vectors, we have

a) $r=0$: K_n^* is a flat extension of a K_2^*;

b) $r=1$: K_n^* is a flat extension of a K_3^* which is a null-extension of a K_2^*;

c) $r=2$: K_n^* is a flat extension of a K_4^* which is a null-extension of a K_2^*.

From the linear element (9.10) it can be proved by direct computation that $K_{\nu\mu\lambda\varkappa}$ in a simple K_n^* is the general product of two equal simple bivectors:

$$(9.11)\qquad K_{\nu\mu\lambda\varkappa} = u_{[\nu}\, v_{\mu]}\, u_{[\lambda}\, v_{\varkappa]}\,.$$

Conversely, if we know that the curvature tensor of a K_n^* has the form (9.11) the E_{n-2} spanned by u_λ and v_λ forms a parallel field. Hence a K_n^* is simple if and only if its curvature tensor has the form (9.11). [3])

RUSE [4]) has given the following classification of the simple K_n^*'s:

I. k_λ non-null; K_n^* simple

A. $K \neq 0$; $\quad K_n^* = K_2^* \times R_{n-2}$,

B. $K = 0$; $\quad K_{\mu\lambda} \neq 0$; $\quad K_n^* = k_3^* \times R_{n-3}$;

[1]) WALKER 1950, 1.

[2]) WALKER 1950, 1, p. 40ff.

[3]) WALKER 1950, 1, p. 44.

[4]) RUSE 1951, 1; 2.

II. k_λ null

A. $K \neq 0$; $K_n^* = K_2^* \times R_{n-2}$;

B. $K = 0$; $K_{\mu\lambda} \neq 0$; $K_n^* = k_3^* \times R_{n-3}$;

C. $K_{\mu\lambda} = 0$; $K_n^* = k_4^* \times R_{n-4}$;

where k_3^* and k_4^* are defined as K^*-spaces with the linear element

$$ds^2 \stackrel{*}{=} \psi(x, z)\, dx^2 + 2dx\,dy + dz^2 \tag{9.12}$$

and

$$ds^2 \stackrel{*}{=} \psi(x, z)\, dx^2 + 2dx\,dy + 2dz\,dt \tag{9.13}$$

respectively. WALKER[1]) proved that the linear element of a non-simple K_n^* for $n>3$ can always be written in the form

$$ds^2 \stackrel{*}{=} \psi\, d\xi^1\, d\xi^1 + 2d\xi^1\, d\xi^2 + k_{\alpha\beta}\, d\xi^\alpha\, d\xi^\beta; \qquad \alpha, \beta = 3, \ldots, n \tag{9.14}$$

where

$$\psi \stackrel{*}{=} \theta\, a_{\alpha\beta}\, \xi^\alpha \xi^\beta + \chi_\alpha \xi^\alpha; \quad a_{\alpha\beta} \stackrel{*}{=} \text{const.}; \quad k_{\alpha\beta} \stackrel{*}{=} \text{const.}; \quad |k_{\alpha\beta}| \neq 0 \tag{9.15}$$

and where $\theta \neq 0$ and the χ_α are functions of ξ^1. The K_n^* is symmetric if and only if $\theta = \text{const.}$

In the same paper WALKER also gave particulars about the symmetric K_n^*, the EINSTEIN K_n^* [2]) and the K_n^* which is a C_n. Moreover he proved some remarkable theorems on harmonic K_n^*'s, for instance that for $n>2$ every harmonic K_n^* is simply harmonic (cf. VIII § 1) and that for every $n>3$ there exists a simply harmonic K_n^* which is non-decomposable.

Exercises.

VIII 9,1. Prove that the V_n's with the linear elements

VIII 9,1 α) $ds^2 \stackrel{*}{=} \psi(x, z)\, dx^2 + 2dx\,dy + dz^2$

and

VIII 9,1 β) $ds^2 \stackrel{*}{=} \psi(x, z)\, dx^2 + 2dx\,dy + 2dz\,dt$

are both K_n^*'s.

VIII 9,2. Prove that the K_4^* with the linear element

$$ds^2 \stackrel{*}{=} \psi(x, y)\, dx^2 + 2\,\varphi(x, y)\, dx\,dy + \zeta(x, y)\, dy^2 + 2dx\,dz + 2dy\,dt$$

is simply harmonic (cf. Exerc. VIII 1,2).

VIII 9,3[3]). Every K_n^*, $n>2$, that is an EINSTEIN space, is a special EINSTEIN space.

[1]) WALKER 1950, 1, p. 51.
[2]) Cf. also MOGI 1950, 1.
[3]) WALKER 1950, 1.

Bibliography[1].

1944, 1, 2

ABE, M.: Sur la réductibilité du groupe d'holonomie I, Les espaces à connexion affine. Proc. Imp. Acad. Tokyo 20, 56–60; 177–182 . . . . *367*

1950, 1

ABRAMOV, B. M. (see also B. A. ROZENFEL'D): Topological invariants of Riemann spaces. Uspechi Mat. Nauk 5, 162–163 (russian) *269*

1951, 1

— On topological invariants of riemannian spaces obtained by the integration of tensor fields. Akad. Nauk SSSR. 81, 125–128 (russian) *269*

1951, 2

— On topological invariants of riemannian spaces obtained by the integration of pseudotensor fields. Akad. Nauk SSSR. 81, 325–328 (russian) *269*

1951, 1, 2

ADATI, T. (see also K. YANO): On riemannian spaces admitting a family of totally umbilical hypersurfaces I, II. Proc. Japan Acad. 27, 1–6, 49–54 *251, 272*

1951, 3–8

— On subprojective spaces I—VI. Tôhoku Math. J. 3, 159–173, 330–342, 343–358; Tensor I, 105–115, 116–129, 130–136 . . . . *321, 322, 324, 327, 329*

1933, 1

AGOSTINELLI, C.: Sui sistemi di velocità e le famiglie naturali di linee in uno spazio curvo. Atti Pont. Accad. Sci. Nuovi Lincei 86, 287–314 . . . *138, 253*

1933, 2

* — Parallelismo in un sistemi di S_k euclidei tangenti a una varietà V_n. Gili, Torino 31 pp.

1933, 3

— Sullo spostamento conforme in una varietà V_n. Atti Ist. Veneto. Sci. 92, 523–525 . . . . *315*

1952, 1

AKIVIS, M. A.: Invariant construction of the geometry of a hypersurface of a conformal space. Akad. Nauk SSSR. 82, 325–328 (russian) *315*

1925, 1

ALEXANDER, J. W.: On the decomposition of tensors. Ann. Math. 27, 421–423 *20*

1937, 1

ALLENDOERFER, C. B.: Einstein spaces of class one. Bull. Am. Math. Soc. 43, 265–270 . . . . *268*

1937, 2

— The imbedding of Riemann spaces in the large. Duke Math. J. 3, 317–333 *268*

1939, 1

— Rigidity for spaces of class greater than one. Am. J. Math. 61, 633–644 *268, 269, 275*

[1]) Titles marked with * are books or monographies.

1952, 1

* ANDELIC, T. P.: Tensorcalculus. Naučna Knjiga, Belgrade (yugoslavian)

1936, 1

ANDERSON, N. L., and L. INGOLD: Normals to a space V_n in hyperspace. Bull. Am. Math. Soc. 42, 429–435 . . . . *269, 275*

1935, 1

ANDREOLI, G.: Elementi di una geometria-differenziale di una V_1 in una V_2 isotrope cayleyane. Acc. Sc. Napoli Rend. (5) 4, 20–23 . . . . *227*

1864, 1

AOUST, L.: Théorie des coordonnées curvilignes quelconques. Ann. di Mat. 6, 64–87 . . . . *253*

1921, 1

BACH, R.: Zur Weylschen Relativitätstheorie und der Weylschen Erweiterung des Krümmungsbegriffs. Math. Zeitschr. 9, 110–135 . . . . *14, 147*

1943, 1

BECKENBACH, E. F., and R. H. BING: Conformal minimal varieties. Duke Math. J. 10, 637–640 . . . . *315*

1874, 1

BEEZ, R.: Über das Krümmungsmaß von Mannigfaltigkeiten höherer Ordnung. Math. Ann. 7, 387–395 . . . . *244*

1875, 1

— Zur Theorie des Krümmungsmaßes von Mannigfaltigkeiten höherer Ordnung. Zeitschr. f. Math. u. Phys. 20, 423–444; 21, 373–401 (1876) *244*

1934, 1

*BEHNKE, H., and P. THULLEN: Theorie der Funktionen mehrerer komplexer Veränderlichen. Springer, Berlin. 115 pp.; Chelsea, New York *61*

1868, 1

BELTRAMI, E.: Teoria fondamentale degli spazii di curvature costante. Ann. di Mat. (2), 232–255; Opere III, 406–409 . . . . *294*

1869, 1

— Zur Theorie des Krümmungsmaßes. Math. Ann. 1, 575–582; Opere II, 119 to 128 . . . . *245*

1933, 1

BERGMAN, S.: Über eine in der Theorie der Funktionen von zwei komplexen Veränderlichen auftretende unitäre Geometrie. Proc. Kon. Ned. Akad. Amsterdam 36, 307–313 . . . . *397, 404*

1922, 1

BERWALD, L.: Zur Geometrie einer n-dimensionalen Riemannschen Mannigfaltigkeit im $(n+1)$-dimensionalen euklidisch-affinen Raum. J. D. M. V. 31, 162–170 . . . . *237, 242*

1922, 2

— Die Grundgleichungen der Hyperflächen im euklidischen Raum gegenüber den inhaltstreuen Affinitäten. Monatshefte f. Math. u. Phys. 32, 89–106 *242*

1923, 1

*— Differentialinvarianten in der Geometrie. Enz. der Math. Wiss. III, Teil 3, D 11, 108 pp. *234*

1936, 1
BERWALD, L.: On the projective geometry of paths. Ann. of Math. 37, 8

1897, 1
BERZOLARI, L.: Sulla curvature delle varietà sopra una varietà qualunque. Acc. Accad. Torino 33, 693–700, 759–778 . . . . *271*

1888, 1
BIANCHI, L.: Sulle forme differenziali quadratiche indefinite. Mem. Acc. Linc. (4), 5, 539–603 . . . . *274*

1899, 1
*— Vorlesungen über Differentialgeometrie. Autorisierte deutsche Übersetzung von M. LUKAT, 1. Aufl. Teubner, Leipzig. 659 pp. *296*

1902, 1
— Sui simboli a quattro indice e sulla curvature di Riemann. Rend. d. Linc. (V), 11^{II}, 3–7 . . . . *146, 148*

1902, 2
*— Lezioni di geometria differenziale, 2° ed. Spoerri, Pisa I (1902), II (1903) *274, 312*

1918, 1
*— Lezioni sulla teoria dei gruppi continui finiti di trasformazioni. Spoerri Pisa. VI + 590 pp. *185, 348*

1922, 1
— Sul parallelismo vincolato di Levi-Civita nella metrica degli spazi curvi. Rend. Accad. Napoli (3), 28, 150–171 . . . . *230*

1930, 1
*BILIMOVIĆ, A.: Geometrical principles of the computation with dyads I.: Dyad and affinor (serbian), Belgrad.

BING, R. H. (see E. F. BECKENBACH).

1923, 1
*BIRKHOFF, G. D.: Relativity and modern physics. Harvard Univ. Press, Cambr. Mass. 283 pp. *155*

1920, 1
BLASCHKE, W.: Frenet's Formeln für den Raum von Riemann. Math. Zeitschr. 6, 94–99 . . . . *230*

1921, 1
*— Vorlesungen über Differentialgeometrie I. Springer, Berlin. 1. Aufl., 230 pp.. . . . . *229*

1923, 1
*— Vorlesungen über Differentialgeometrie und geometrische Grundlagen von Einsteins Relativitätstheorie. II. Affine Differentialgeometrie, bearbeitet von K. REIDEMEISTER. Springer, Berlin. IX + 259 pp. *234*

1928, 1
— Eine Umkehrung von Knesers Transversalensatz. Nieuw Archief v. Wisk. (2), 15, 202–204 . . . . *310*

1950, 1
*— Conferenze di geometria. Geometria di Riemann, geometria affine. La Edita Universitaria, Messina. 102 pp.

1950, 2

*BLASCHKE, W.: Geometria diferencial moderna (Modern differential geometry). Conferencias de Matematica, vol I. Instituto de Matemáticas "Jorge Juan", Madrid. I + 43 pp.

1946, 1

BLUM, R.: Anzahl der Identitäten von Bianchi in einer V_n. Acad. Roum. Bull. 28, 351–353 . . . . *147*

1946, 2

— Über die Bedingungsgleichungen einer riemannschen Mannigfaltigkeit, die in einer euklidischen Mannigfaltigkeit eingebettet ist. Bull. Math. Soc. Roumaine 47, 144–201 . . . . *268, 269*

1947, 1

— Sur les identités de Bianchi et Veblen. C. R. 224, 889–890 . . . *144, 147, 155*

1947, 2

— Sur les tenseurs dérivés de Gauss et Codazzi. C. R. 224, 708–710 . . *268, 269*

1947, 3

— Sur la classe des variétés riemanniennes. Bull. Math. Soc. Roumaine 48, 88–101 . . . . *268*

1910, 1

*BOCHER, M.: Einführung in die höhere Algebra. Teubner, Leipzig. 348 pp. *31*

1947, 1

BOCHNER, S.: Curvature in hermitian metric. Bull. Am. Math. Soc. 53, 179 to 195 . . . . *404, 406, 407, 409, 412*

1948, 1

*— and W. T. MARTIN: Several complex variables. Math. Ser. 10. Princeton, Univ. Press. IX + 216 pp. *61*

1950, 1

— Vector fields on complex and real manifolds. Ann. of Math. 52, 642–649 *66*

1951, 1

— Tensor fields with finite basis. Ann. of Math. 53, 400–411 . . . . *66*

1951, 2

— A new viewpoint in differential geometry. Canadian J. Math. 3, 460–470 *138*

1953, 1

*— Structure of complex spaces. Contributions to the theory of Riemann surfaces. Princeton Un. Press; Oxford Un. Press *397*

1950, 1

BOL, G.: Zur tensoriellen Behandlung der projektiven Flächentheorie. Math. Ann. 122, 279–295 . . . . *296, 301*

1914, 1

BOMPIANI, E.: Forma geometrica delle condizioni per la deformabilità delle ipersuperficie. Rend. d. Linc. (V), 23^{I}, 126–131 . . . . *244*

1921, 1

— Studi sugli spazi curvi. Atti Istituto Veneto (9), 5 (80), 355–386, 839–859, 1113–1145 . . . . *68, 256, 269, 274*

1924, 1

— Spazi riemanniani luoghi di varietà totalmente geodetiche. Rend. d. Palermo 48, 121–134 . . . . *269*

1935, 1

BOMPIANI, E.: Geometrie riemanniane di specie superiore. Mem. Akad. Ital. 6, 269–520 . . . . *352*

1938, 1

— Sulle varietà anolome. Atti d. Linc. (6), 27, 37–45; 45–52. . . . . . . *253*

1940, 1

— Le superficie emisotrope nello spazio euclideo a quattro dimensioni. Atti Accad. Italia 12, 1–23 . . . . *227*

1941, 1

— Intorno alla varietà isotrope. Ann. Mat. Pura Appl. 20, 21–58 . . . . . *227*

1943, 1

— Geometria proiettiva di elementi differenziali. Ann. Mat. Pura Appl. (4) 22, 1–32 . . . . *296, 301*

1945, 1

— Geometria degli spazi a connessione affine. Ann. Mat. Pura Appl. 24, 257 to 282 . . . . *123, 155*

1946, 1

— Le connessioni tensoriali. Atti d. Linc. 1, 478–482 . . . . *123, 138*

1950, 1

— Über das Theorem von F. Schur in der Riemannschen Geometrie. Math. Zeitschr. 52, 623–626 . . . . *148, 153, 269*

1950, 2

— Über die Gaußsche Krümmung der Überraumflächen. Math. Zeitschr. 53, 131–132 . . . . *269*

1950, 3

— Sopra una nozione di spazio osculatore ad una varietà introdotta da L. Berzolari. Boll. Un. Mat. Ital. 5, 213–218 . . . . *271*

1951, 1

— Significato del tensore di torsione di una connessione affine. Boll. Un. Mat. Ital. 6, 273–276 . . . . *127, 269*

1951, 2

— Géométries riemanniennes d'espèce superieure. Coll. Géom. Diff. Louvain. 1951, 123–156 . . . . *269, 277*

1951, 3

— Proprietà d'immersione di una varietà in uno spazio di Riemann. Rend. Sem. Mat. Milano 22, 3–26 . . . . *256, 269*

1952, 1

— Sulle connessioni affini non-posizionali. Arch. Math. 3, 183–186 . . . . *123*

1952, 2

— Sulle coordinate di Grassmann. Rend. d. Linc. (8) 13, 329–335 . . *23, 254*

1952, 1

BOREL, A., and A. LICHNEROWICZ: Espaces riemanniens et hermitiens symétriques. C. R. 234, 2332–2334 . . . . *163*

1952, 2

— — Groupes d'holonomie des variétes riemanniennes. C. R. 234, 1835–1837 *370*

1927, 1

Bortolotti, E.: Parallelismi assoluti nelle V_n riemanniane. Atti Veneto 86, 2, 455–465 . . . . *142*

1927, 2

— On metric connexions with absolute parallelism. Proc. Kon. Akad. Amsterdam 30, 216–218 . . . . *142*

1927, 3

— Spazi subordinati: equazioni di Gauss e Codazzi. Boll. Un. Mat. Ital. 6, 134–137 . . . . *254, 269*

1928, 1

— Parallelismo assoluto nella varietà a connessione affine e nuove veduti sulla relatività. Mem. Acad. Bologna (8) 6, 45–58 . . . . *125, 142*

1928, 2

— Varietà minime infinitesimale in une V_n riemanniana. Mem. Acad. Bologna 8, 5, 43–48 . . . . *353*

1928, 3

— Spostamento geodetico e sue generalizzatione. Giorn. M. Battaglini 3, 66, 153–191 . . . . *353*

1929, 1

— Stelle di congruenze e parallelisme assoluto: basi geometriche di una recente teoria di Einstein. Rend. d. Linc. (6) 9, 530–538 . . . . *125, 142*

1929, 2

— Sulle coordinate geodetiche lungo una linea I, II. Rend. d. Linc. (6) 10, 486–492; 544–548 . . . . *155, 166*

1929, 3

— Sulle varietà subordinate negli spazi a connessione affine e su di una espressione dei simboli di Riemann. Boll. Un. Mat. Ital. 7, 86–94 . . . . . . *232*

1930, 1

— Calcolo assoluto generalizzato di Pascal-Vitali etc. Atti. Soc. Ital. 19, 2, 15–16; Atti Ist. Ven. 90, 471–478. . . . . *58*

1930, 2

— Sulle forme differenziali quadratiche spezializzate. Rend. d. Linc. (6) 12, 541–547 . . . . *123, 133, 270*

1930, 3

— Sulla geometria della varietà a connessione affine. Teoria invariantiva delle trasformazione che conservano il parallelismo. Ann. di Mat. 8, 53–101 *123, 155, 287*

1930, 4

— Leggi di trasporto nei campi di vettori applicati ai punti di una curva o di una V_m in V_n riemanniana. Mem. Bologna (8) 7, 11–20 . . . . . . *166, 230*

1931, 1

— Trasporti rigidi e geometria della varietà anolonome. Boll. Un. Mat. Ital. 10, 5–12 . . . .*78, 253*

1931, 2

— Sulle varietà subordinate. Rend. Ist. Lombardo (2) 64, 441–463 *78, 232, 253, 300*

1931, 3

Bortolotti, E.: Calcolo assoluto rispetto a una forma differenziale quadratica specializzata. Rend. d. Linc. (6) 13, 19–25; 104–108 . . . . *123*

1931, 4

— Differential invariants of direction and point transformations. Ann. of Math. 32, 361–377 . . . . *123, 287, 288, 300*

1931, 5

— Directions concourantes et connexions dans les espaces courbes. Bull. Soc. Math. 59, 70–74 . . . . *322*

1931, 6

— Connessioni proiettive. Boll. Un. Mat. Italia 9 (1930), 288–294; 10, 28–34; 83–90 . . . . *301*

1932, 1

— Sulle connessioni proiettive. Rend. d. Palermo 56, 1–57 . . *78, 301, 302, 303*

1933, 1

— Connessione nella varietà luoga di spazi; applicazione alla geometrìa metrica differenziale della congruenze di rette. Rend. Univ. Cagliari 3, 11 pp. *78*

1933, 2

— Spazi proiettivamente piani. Ann. d. Bol. 4, 11, 111–134 . . . . *301*

1934, 1

— Riferimenti geodetici lungo più linee nelle varietà a connessione affine. Rend. d. Linc. 19, 556–560 . . . . *166*

1935, 1

— Trasporti di specie superiore. Rend. d. Linc. 21, 225–230 . . . . *123*

1936, 1

— Trasporti non lineare: geometria di un sistema di equazioni alle derivate parziale del 2e ordine. Rend. d. Linc. 22, 16–21; 104–110; 176–180. . . *123*

1936, 2

— and V. Hlavatý: Contributi alla teoria delle connessioni I. Ann. di Mat. (4) 15 (1936/37), 1–71 . . . . *123, 157, 301*

1936, 3

— Superficie anolonome complimentari. Ser. mat. off. a. L. Berzolari 553–576 *253*

1937, 1

— La méthode de G. Vitali; dans la récherche géométrique; répresentation fonctionelle et calcul absolu généralisé. Abh. Sem. Vector u. Tensor. Moskau 4, 269–288 . . . . *58*

1937, 2

— Geometria proiettiva differenziale delle superficie anolonome. Atti 1e congr. Un. Mat. Italia 305–311 . . . . *.78, 253*

1937, 3

— Varietà subordinata: connessione di specie superiore, il problema dell immersione. Atti 1e congr. Un. Mat. Italia 312–317 . . . . *.78, 123*

1939, 1

— Contributi alla teoria della connessioni. II. Connessioni di specie superiore, fondamenti analitici calcolo del Vitali generallizzato. Mem. Ist. Lombardo Mat. Nat. 24, 1–39 . . . . *123*

1940, 1

BORTOLOTTI, E.: Sulla geometria proiettiva differenziale di una superficie anolonoma nello spazio ordinario. Ann. Mat. Pura Appl. (4) 19, 315–325 *78*

1940, 2

— Coordinate normali ed "estensioni" nella geometria degli spazi a connessione lineare. Atti Accad. Italia (7) 2, 106–116 . . . . *155*

1941, 1

— Duale Verwandtschaften, anholonome Flächen im projektiven und im affinen Raume. J. D. M. V. 51, 151–169 . . . . *253*

1941, 2

— Varietà a connessione proiettiva e loro immagini tangenziali. Atti Accad. Italia (7) 2, 251–267 . . . . *296, 301*

1942, 1

— Vedute e problemi della teoria delle connessione. Univ. Roma e Ist. Naz. Alta Mat. 3, 241–281 . . . . *123*

1924, 1

*BOULIGAND, G.: Leçons de géométrie vectorielle, préliminaires à l'étude de la théorie d'Einstein. Vuibert, Paris. XIII + 365 pp.

1949, 1

*— Les principes de l'analyse géometrique I. Leçons de géométrie vectorielle, préliminaires à l'etude de la theorie d'Einstein. Vuibert, Paris. X + 436pp.

1947, 1

*BRAND, L.: Vector and tensor analysis. Wiley a. Sons, New York; Chapman a. Hall, London. XVI + 439 pp.

1935, 1

BRAUER, A., and H. WEYL: Spinors in n dimensions. Am. Journ. of Math. 57, 425–449 . . . . *50*

1951, 1

BRAUNER, H.: Orthogonalsysteme von riemannschen Hyperflächen der Klasse I. Anz. Öster. Akad. 1951, 29–36 . . . . *249, 269*

1946, 1

*BRILLOUIN, L.: Les tenseurs en mécanique et en élasticité. Dover Publications, New York. XI + 364 pp.

1923, 1

BRINKMAN, H. W.: On Riemann spaces conformal to Einstein spaces. Proc. Nat. Acad. 9, 172–174 . . . . *312, 314*

1923, 2

— On Riemann spaces conformal to euclidean spaces. Proc. Nat. Acad. 9, 1–3 *315*

1924, 1

— Riemann spaces conformal to Einstein spaces. Math. Ann. 91, 269–278 *312, 314*

1925, 1

— Einstein spaces which are mapped conformally on each other. Math. Ann. 94, 119–145 . . . . *312, 314, 330*

1906, 1

BROUWER, L. E. J.: Polydimensional vectordistributions. Proc. Kon. Ned. Akad. Amsterdam 9, 66–78 . . . . *85, 97*

1906, 2

BROUWER, L. E. J.: The force field of the non-euclidean spaces with negative curvature. Proc. Kon. Ned. Akad. Amsterdam 9, 116–133 . . . . . . *97, 122*

1919, 1

— Opmerkingen over meervoudige integralen. Versl. Kon. Ned. Akad. Amsterdam 28, 116–120 . . . . *97*

1943, 1

BRUCK, R. H., and T. L. WADE: The number of independent components of the tensors of given symmetry type. Bull. Am. Math. Soc. 49, 470–472 *15*

1944, 1

— The number of absolute invariants of a tensor. Am. J. Math. 66, 411–424 *15*

1948, 1

BUCHDAHL, H. A.: The hamiltonian derivatives of a class of fundamental invariants. Quart Journ. of Math. Oxford 19, 150–159 . . . . *114*

1951, 1

— Über die Variationsableitung von Fundamentalinvarianten beliebig hoher Ordnung. Acta Math. 85, 63–72 . . . . *114*

1951, 2

— On the hamiltonian derivatives arising from a class of gauge-invariant action principles in a W_n. Journ. London Math. Soc. 26, 139–149 . . . *114*

1951, 3

— An identity between the hamiltonian derivatives of certain fundamental invariants in a W_4. Journ. London Math. Soc. 26, 150–152 . . . . *114, 116*

1899, 1

*BUCHHOLZ, A.: Ein Beitrag zur Mannigfaltigkeitslehre. Mannigfaltigkeiten, deren Linienelemente auf die Form:

$$ds = f(\sum X_k^2)\sqrt{\sum dX_k^2}$$

gebracht werden können. Cohen, Bonn. 264 pp. *330*

1926, 1

BURSTIN, C., and W. MAYER: Das Formenproblem der l-dimensionalen Hyperflächen in n-dimensionalen Räumen konstanter Krümmung. Mon. f. Math. u. Phys. 24, 89–136 . . . . *277*

1929, 1

— Beiträge zur mehrdimensionalen Differentialgeometrie. Mon. f. Math. u. Phys. 36, 97–130 . . . . *277*

1931, 1

— Ein Beitrag zum Problem der Einbettung der Riemannschen Räume in euklidischen Räumen. Receuil. Math. Soc. math. (Math. Sbornik) Moscou 38, 74–85 . . . . *268, 352*

1931, 2

— Beiträge zum Problem der Verbiegung von Hyperflächen in euklidischen Räumen. Rec. Math. Soc. Math. (Math. Sbornik) Moscou 38, 86–93 . . *352*

1932, 1

— Zum Einbettungsproblem. Commun. Soc. Math. Kharkov et Inst. Sci. math. Ukraine (4) 5, 87–95 . . . . *268*

1937, 1

Burstin, C.: Ein Beitrag zur Theorie dés Klassenbegriffes der quadratischen Differentialformen. Abh. Sem. Vektor u. Tensor. Moskau 4, 121–136 . . *268*

1953, 1

Calabri, L.: Isometric imbedding of complex manifolds. Ann. of Math. 58, 3–23 . . . . *412*

1903, 1

*Campbell, J. E.: Introdüction treatise on Lie's theory of finite continuous groups. Oxford *203*

1894, 1

*Cartan, E.: Sur la structure des groupes de transformations finis et continus. Thèse, Nony, Paris; second edition Vuibert, Paris 1933 . . . . *117*

1899, 1

— Sur certaines expressions différentielles et le problème de Pfaff. Ann. École norm. sup. (III) 16, 239–332 . . . . *95*

1901, 1

— Sur l'intégration des systèmes d'équations aux différentielles totales. Ann. École norm. 18, 241–311 . . . . *117*

1916, 1

— La déformation des hypersurfaces dans l'espace euclidien a n dimensions. Bull. Soc. Math. 44, 65–99 . . . . *244, 352*

1917, 1

— La déformation des hypersurfaces dans l'espace conforme réel à $n \geq 5$ dimensions. Bull. Soc. Math 45, 57–121 . . . . *352*

1919, 1

— Sur les variétés de courbure constante d'un espace euclidien ou non euclidien. Bull. Soc. Math. d. France 47, 125–160 . . . .*39, 269*

1920, 1

— Sur la déformation projective des surfaces. C. R. 170, 1439; 171, 27; Ann. Éc. Norm. 37, 259–356 . . . . *352*

1920, 2

— Sur le problème général de la déformation. C. R. Congrès Strasbourg 397 to 406 . . . . *352*

1920, 3

— Sur les variétés de courbure constante d'un espace euclidien ou non euclidien. Bull. Soc. Math. d. France 48, 132–208 . . . . *269*

1922, 1

*— Leçons sur les invariants intégraux. Hermann, Paris. X + 210 pp. . . *89*

1922, 2

— Sur une généralisation de la notion de courbure de Riemann et les espaces à torsion. C. R. 174, 593–595 . . . . *127*

1922, 3

*— Sur les équations de la gravitation d'Einstein. Journ. d. math. p. et a. (9) 1, 141–203. Separate: Gauthier-Villars, Paris. 65 pp. *306*

1923, 1

— Sur un théorème fondamental de M. H. Weyl. Journ. d. Math. p. et a. (IX), 2, 167–192; cf. also C. R. 175, 82 85 . . . . *137*

1923, 2

CARTAN, E.: Sur les variétés à connexion affine et la théorie de la relativité généralisée. Ann. Éc. norm. 40, 325–412 . . . . *123*

1923, 3

— Les espaces à connexion conforme. Ann. Soc. polon. d. math. 2, 171–221 *315, 316*

1924, 1

— Sur les variétés à connexion projective. Bull. d. l. Soc. Mat. d. France 52, 205–241 . . . . *123, 302, 303, 316*

1924, 2

— La théorie des groupes et les récherches récentes de géometrie différentielle. Ens. Math. 1–18 . . . . *361*

1925, 1

— Sur les variétés à connexion affine et la théorie de la relativé généralisée. Ann. Éc. norm. 42, 17–88 . . . . *361*

1925, 2

*— La géométrie des espaces de Riemann. Mém. d. Sc. Math., fasc. 9.

1926, 1

— Sur une classe remarquable d'espaces de Riemann. Bull. Soc. Math. 54, 214–264; 55, 114–134 . . . . *163, 381*

1926, 2

— Sur les espaces de Riemann dans lesquels le transport par parallélisme conserve la courbure. Rend. d. Linc. (6) 3, 544–547 . . . . *163*

1926, 3

— and J. A. SCHOUTEN: On the geometry of the group manifold of simple and semi-simple groups. Proc. Kon. Ned. Akad. Amsterdam 29, 803–815 *142, 187, 351*

1926, 4

— — On riemannien geometries admitting an absolute parallelism. Proc. Kon. Ned. Akad. Amsterdam 29, 933–946 . . . . *142, 187*

1926, 5

— Les groupes d'holonomie des espaces généralisés. Acta Math. 48, 1–42 . . *361*

1927, 1

— La géométrie des groupes de transformations. Journ. de Math. 6 (9), 1–119 *163, 187, 201, 222*

1927, 2

— La géométrie des groupes simples. Ann. di Mat. 4, 209–256 . . . . . *142, 185*

1927, 3

— Sur les courbes de torsion nulle et les surfaces développables dans les espaces de Riemann. C. R. 184, 138–140 . . . . *229, 296*

1927, 4

— Sur la possibilité de plonger un espace riemannien donné dans un espace euclidien. Ann. d. l. Soc. Polon. de Math. 6, 1–7 . . . . *268*

1927, 5

— Sur l'écart géodésique et quelques notions connexes. Rend. d. Linc. (6a) 5 I, 609–613 . . . . *352*

1928, 1

*CARTAN, E.: Leçons sur la géométrie des espaces de Riemann. Gauthier-Villars, Paris. 270 pp. *149*

1930, 1

— Le troisième théorème fondamental de Lie. C. R. 190, 914–916; 1005–1007 *208*

1930, 2

*— La théorie des groupes finis et continus et l'analysis situs. Mém. Sc. Math. 42. Gauthier-Villars, Paris. 60 pp. *185*

1931, 1

*— Leçons sur la géométrie projective complexe. Gauthier-Villars, Paris. 322 pp. *406*

1932, 1

— Les espaces riemanniens symétriques. Verh. Int. Math. Kongr. Zürich I, 152–161 . . . . *163*

1934, 1

— Les espaces à connexion projective. Mem. du Semin. Géom. Diff. Tens. Moscou 4, 147–173. . . . . *301*

1935, 1

— Le calcul tensoriel projectif. Rec. Soc. Math. Moscou 42, 131–148 . . . *301*

1936, 1

*— La topologie des groupes de Lie. Act. Sc. et Ind. 358. Hermann, Paris. 28 pp. *185*

1937, 1

*— Leçons sur la théorie des espaces à connexion projective. Gauthier-Villars, Paris. 308 pp.. . . . . *128, 301, 361*

1937, 2

*— and J. LERAY: La théorie des groupes finis et continus. Gauthier-Villars, Paris. *208*

1937, 3

— L'extension du calcul tensorial aux géométries non affines. Ann. of Math. 38, 1–13 . . . . *301, 315*

1938, 1

*— Leçons sur la théorie des spineurs I 85 pp., II 96 pp. Hermann, Paris . . *21*

1939, 1

— Sur quelques familles remarquables de hypersurfaces. C. R. Congr. Sc. Math. Liège 30–41 . . . . *242, 269*

1939, 2

— Sur des familles remarquables d'hypersurfaces isoparamétriques dans les espaces sphériques. Math. Zeitschr. 45, 335–367 . . . . *242, 269*

1941, 1

— La notion d'orientation dans les différentes géométries. Bull. Soc. Math. France 69, 47–70 . . . . *5*

1941, 2

— Sur les surfaces admettant une seconde forme fondamentale donnée. C. R. 212, 825–828 . . . . *244*

1942, 1

CARTAN, E.: Sur les couples de surfaces applicables avec conservation des courbures principales. Bull. Sc. Math. (2) 66, 55–72, 74–85 . . . . *244*

1943, 1

— Les surfaces qui admettent une seconde forme fondamentale donnée. Bull. Sc. Math. (2) 67, 8–32 . . . . *244*

1946, 1

*— Leçons sur la géométrie des espaces de Riemann, 2nd éd. Gauthier-Villars, Paris. VIII + 378 pp. Reprinted 1951 . . . . *149, 175, 178*

1949, 1

*CARTAN, H.: Algebraic topology. Springer and Pollak, Cambridge, Mass. VI+169 pp. *85*

1946, 1

ĆASTOLDI, L.: Sopra alcune proprietà caratteristiche delle V_n totalmente geodetiche rispetto ad una V_n ambiente. Atti Acc. Naz. Linc. 1064–1069 . . *269*

1947, 1

— Sopra gli spazi a connessione semimetrica. Univ. Roma Ist. Naz. Alta. Mat. 6, 395–409 . . . . *145*

1948, 1

— Giaciture associate ad una serie di direzioni lungo una curva immersa in uno spazio a connessione affine. Atti Acc. Sc. Torino 81–82; 205–214 . . . . *232*

1948, 2

— Dislocazioni e deformazioni finite nella varietà sostanziali di Riemann. Ist. Naz. Alta Mat. 7, 373–392 . . . . *352, 353*

1949, 1

— V_m "canoniche" in V_n e loro caratterizzazione intrinseca. Ist. Lombardo Sc. Mat. 13, 267–279 . . . . *269*

1890, 1

CAYLEY, A.: Sur les surfaces minima. C. R. 111, 953–954; Coll. Papers XIII, 41–42 . . . . *274*

1925, 1

*CERDANZEV, J. A.: Grundlagen der Vektor- und Tensoranalysis. Moskau, Staatsverlag. 143 pp. (russian).

1937, 1

CHERN, S. S.: Sur la possibilité de plonger un espace à connexion projective donné dans un espace projectif. Bull. Sc. Math. (2) 61, 234–243 . . . . . *268*

1942, 1

— The geometry of isotropic surfaces. Ann. of Math. (2) 43, 545–559 . . . . *228*

1944, 1

— On a theorem of algebra and its geometrical application. J. Indian Math. Soc. (N. S.) 8, 29–36 . . . . *269*

1947, 1

— and H. WANG: Differential geometry in symplectic space I. Sci. Rep. Nat. Tsing Hua Univ. 4, 453–477 . . . . *179*

1946, 1

*CHEVALLEY, C.: Theory of Lie groups I. Princeton Math. Ser. vol. 8. Princeton Univ. Press. IX + 217 pp. *185*

1869, 1

CHRISTOFFEL, E. B.: Über die Transformation der homogenen Differentialausdrücke zweiten Grades. Crelle's Journal 70, 46–70; Ges. math. Abh. I, 352–377 . . . . *132, 164*

1932, 1

CHURCHIL, R. V.: On the geometry of the Riemann tensor. Trans. Am. Math. Soc. 34, 126–152 . . . . *146*

1932, 1

*CISOTTI, U.: Cenni sui fondamenti del calcolo tensoriale con applicazioni alla teoria dell'elasticità. Milano, Hoepli. 46 pp.

1940, 1

— Una notevole scomposizione dei tensori. Boll. Un. Mat. Ital. (2) 2, 210–215 *15*

1941, 1

— Invarianti quadratici dei tensori. Atti Accad. Italia 2, 511–516 . . . . *15*

1943, 1

— Invarianti ed eminvarianti cubici dei tensori. Atti Accad. Italia 4, 8–11 *15*

1944, 1

— Invarianti di vettori e di tensori doppi. Ist. Lombardo. Sc. Mat. Nat. Rend. 8, 253–258 . . . . *15*

1939, 1

COBURN, N.: V_m in S_n with planar points ($m \geqq 3$). Bull. Am. Math. Soc. 45, 774–783 . . . . *269, 272*

1940, 1

— Generalized Einstein hypersurfaces of spaces of constant curvature. J. Math. Phys. Mass. Inst. Tech. 19, 140–152 . . . . *242, 252, 269*

1940, 2

— A characterization of Schouten's and Haydens' deformation methods. J. London Math. Soc. 15, 123–136 . . . . *335*

1941, 1

— Conformal unitary spaces. Trans. Am. Math. Soc. 50, 26–39 *288, 415, 416, 417*

1941, 2

— A note on conformal geometry. Proc. Nat. Acad. Sci. U.S.A. 27, 57–60 *315*

1941, 3

— Unitary curves in unitary space. Univ. Nac. Tucumán Revista A 2, 159–167 *413, 414, 415*

1942, 1

— Semi-analytic unitary subspaces of unitary space. Am. J. Math. 64, 714–724 *407*

1942, 2

— Frenet formulae for curves in unitary space. J. Math. Phys. Mass. Inst. Tech. 21, 10–18 . . . . *415*

1942, 3

— Conformal geometry of unitary space. Univ. Nac. Tucumán Revista A 3, 125–140 . . . . *416, 417, 418*

1943, 1

— Congruences in unitary space. Trans. Am. Math. Soc. 53, 25–40 . . . *415, 418*

1931, 1
CONFORTO, F.: Parallelismo negli spazi funzionali continui. Rend. d. Linc. (6) 13, 173–178 . . . . *123*

1931, 2
— Formalismo matematico in uno spazio funzionale continuo retto da un elemento lineare di seconda spezie. Rend. d. Linc. (6) 13, 562–568 . . . *123*

1928, 1
CONNELL, A. J. MC.: Il trasporto parallelo di un vettore lungo un circuito finito. Rend. d. Linc. (6) 7, 208–213; 306–309 . . . . *138*

1928, 2
— The torsion of riemannian space. Atti Congr. Int. Mat. Bologna 6, 321–338 *126*

1928, 3
— Strain and torsion in riemannian space. Ann. di Mat. (4) 6, 207–231 . . *126*

1928, 4
— The derived directions of a vector defined along a curve in riemannian space. Proc. London Math. Soc. (2) 28, 501–509 . . . . *230*

1928, 5
— Schmidt's orthogonal ennuple and the Frenet formulas for a curve. Bull. Am. Math. Soc. 713–714 . . . . *230*

1929, 1
— The variation of curvatures in the deformation of a curve in riemannian space. Proc. Roy. Ir. Acad. 39 A, 1–9 . . . . *335, 352*

1931, 1
*— Applications of the absolute differential calculus. Blackie and Son, London. 330 pp.

1924, 1
*COOLIDGE, J. L.: The geometry of the complex domain. Clarendon Press, Oxford. 242 pp. *406*

1940, 1
COPSON, E. T., and M. S. RUSE: Harmonic riemannian spaces. Proc. Roy. Soc. Edinburgh 60, 117–133 . . . . *381, 382, 385*

1947, 1
COSSU, A.: Alcune osservazioni sulle varietà subordinate di una varietà a connessione affine asimmetrica. Atti Acc. Naz. Linc. (8) 3, 303–311 . . . . *232*

1949, 1
— Proprietà di curvatura di una particolare classe di varietà a connessione affine. Atti Acc. Naz. Linc. (6) 702–707 . . . . *144*

1899, 1
*COTTON, E.: Sur les variétés à trois dimensions. Ann. Fac. d. Sc. Toulouse (II) 1, 385–438; also Thèse Paris. 54 pp. *306, 316*

1940, 1
— Sur le calcul des invariants différentiels euclidiens d'une surface. J. Math. Pures et Appl. (9) 19, 211–220 . . . . *242*

1948, 1
— Sur un système de bivecteurs associé à un cycle. C. R. 226, 1564–1566 *232*

1943, 1
*CRAIG, H. V.: Vector and tensor analysis. McGraw-Hill, New York. XIV + 434 pp.. . . . . *57, 58*

1931, 1
*CUTLER, E. H.: On the curvature of a curve in Riemann space. Trans. Am. Math. Soc. 33, 832–838. Thesis Harvard 1930 . . . . *279*

1931, 2
— Frenet formulas for a general subspace of a Riemann space. Trans. Am. Math. Soc. 33, 839–850 . . . . *279*

1949, 1
CZECH, E.: Géométrie projective différentielle des correspondances entre deux espaces I. Časopis Pešt. Mat. Fys. 74, 32–48 (french, czech.) *301*

1950, 1, 2
— Géométrie projective différentielle des correspondances entre deux espaces II, III. Časopis Pešt. Mat. Fys. 75, 123–136; 137–158. (french, czech.) *301*

1932, 1, 2
DANTZIG, D. VAN (see also J. A. SCHOUTEN): Zur allgemeinen projektiven Differentialgeometrie I, II. Proc. Kon. Akad. Amsterdam 35, 524–534; 535–542 . . . . *105, 301, 303*

1932, 3
— Theorie des projektiven Zusammenhangs n-dimensionaler Räume. Math. Ann. 106, 400–454 . . . . *301, 303, 327*

1934, 1
— On the general projective differential geometry III. Proc. Kon. Acad. Amsterdam 37, 150–155 . . . . *301, 303*

1889, 1
DARBOUX, G.: Leçons sur la théorie générale des surfaces II. Gauthier-Villars, Paris 522 pp. *245*

1933, 1
DAVIES. E. T. (see also P. DIENES): On the infinitesimal deformations of a space. Ann. di Mat. (4) 12, 145–151 . . . . *335*

1935, 1
— On (r, r) subordination of a subspace in a Riemann space V_n. Journ. London Math. Soc 10, 216–232 . . . . *269*

1936, 1
— On the deformation of a subspace. Journ. London Math. Soc. 11, 295–301 *352, 353*

1937, 1
— On the second and third fundamental forms of a subspace. Journ. London Math. Soc. 12, 290–295. . . . . *335*

1938, 1
— Analogues of the French formulae determined by deformation operators Journ. London Math. Soc. 13, 210–216; 320 . . . . *110, 230, 335, 352*

1938, 2
— On the deformation of the tangent m-plane of a V_n^m. Proc. Edinburgh. Math. Soc. (2) 5, 202–206. . . . . *352*

1942, 1

DAVIES, E. T.: The first and second variations of the volume integral in riemannian space. Quart. J. Math. Oxford Ser. 13, 58–64 . . . . *352*

1953,1

— On the invariant theory of contact transformations. Math. Zeitschr. 57, 415–427 . . . . *253*

1950, 1

DEBEVER, R.: Le groupe d'holonomie des variétés de groupe de Lie à connexion conforme normale. Congr. Nat. des Sc. Bruxelles III, 56–58. . . . . *367*

1950, 1

*DELACHET, A.: Calcul vectoriel et calcul tensoriel. Presses Universit. de France, Paris. 128 pp.

1913, 1

DEMOULIN, A.: Sur une propriété caractéristique des familles de Lamé. C. R. 157, 1050–1053 . . . . *249*

1926, 1

*DICKSON, L. E.: Modern algebraic theories. Chicago (also german by E. Bodewig, Leipzig 1929) . . . . *.31, 34, 46, 53, 56*

1932, 1

DIENES, P.: On the fundamental formulae of the geometry of tensor submanifolds. Journ. d. Math. (9) 11, 255–282 . . . . *.78, 232, 253, 263, 278*

1933, 1

— Sur le déplacement d'un n-uple et sur une interprétation nouvelle des coefficients de rotation de Ricci. Rend. d. Linc. (6) 17, 119–122 . . *166, 171*

1933, 2

— Sur un théorème de M. Fermi. Rend. d. Linc. (6) 17, 369–372 . . . . *171*

1933, 3

— Sur la déformation des espaces à connexion linéaire générale. C. R. 197, 1084–1086 . . . . *335*

1933, 4

— Sur la déformation des sous-spaces dans un espace à connexion linéaire générale. C. R. 197, 1167–1169 . . . . *352*

1934, 1

— On curves in a space of general linear connection. Journ. London Math. Soc. 9, 259–266 . . . . *128, 131, 232*

1934, 2

— On the deformation of tensor manifolds. Proc. London Math. Soc. 37, 512–519 . . . . *352*

1937, 1

— and E. T. DAVIES: On the infinitesimal deformations of tensor submanifolds. Journ. de Math. 24, 111–150 . . . . *352*

1947, 1

*DIRAC, P. A. M.: The principles of quantum mechanics, 3rd ed. Oxford, Clarendon Press. 311 pp. *54*

1942, 1

DONDER, T. DE, and J. GÉHÉNIAU: Sur la dérivée covariante des tenseurs généralisés. Acad. Roy. Belgique Bull. 28, 630–633 . . . . *105*

1924, 1

DOUGLAS, J.: Normal congruences and quadruply infinite systems of curves in space. Trans. Am. Math. Soc. 26, 68–100 310

1925, 1

— A criterion for the conformal equivalence of a riemannian space to a euclidean space. Trans. Am. Math. Soc. 27, 299–306 306

1928, 1

— The general geometry of paths. Ann. of Math. (2) 29, 143–168 303

1931, 1

— Systems of K-dimensional manifolds in an N-dimensional space. Math. Ann. 105, 707–733 . . . . 155, 303

1941, 1

DOYLE, T. C.: Tensor decomposition with applications to the contact and complex groups. Ann. of Math. (2) 42, 698–722 15

1931, 1

*DROST, A. J.: Reeksontwikkelingen van algebraische vormen. Thesis. Groningen. 68 pp. 15

1941, 1

DUBNOV, Y. S.: Complete system of invariants of two affinors in centro-affine space of two or three dimensions. Abh. Sem. Vektor u. Tensor. 5, 250–270 (russian) 15

1927, 1

DUBOURDIEU, J.: Sopra le coordinate cartesiane lungo una curva. Rend. d. Linc., Rcma (5) 6, 654–655 155

1928, 1

— Quelques applications de la théorie des coordonnées géodésiques le long d'une courbe. Rend. d. Linc. (6) 7, 707–709 155

DUPARE, H. J. A. (see W. PEREMANS).

1930, 1

*DUSCHEK, A., and W. MAYER: Lehrbuch der Differentialgeometrie II: Riemannsche Geometrie. Teubner, Leipzig und Berlin. 245 pp. 68, 230, 242, 252, 272, 277, 285

1930, 2

— Parallelverschiebung in allgemeinen Räumen. J. D. M. V. 39, 42–44 . . 123

1935, 1

— Das Variationsproblem der F_m im Riemannschen R_n und eine Verallgemeinerung des Gauß-Bonnetschen Satzes. Math. Zeitschr. 40, 279–291 . . 269

1948, 1

*— and A. HOCHRAINER: Grundzüge der Tensorrechnung in analytischer Darstellung I. Springer, Wien. II 1950; III in printing.

1949, 1

ECKMANN, B., and H. GUGGENHEIMER: Formes différentielles et métrique hermitienne sans torsion. I. Structure complexe, formes pures. C. R. 229, 464–466 397

1949, 2

— and H. GUGGENHEIMER: Formes différentielles et métrique hermitienne sans torsion. II. Formes de classe K, formes analytique. C. R. 229, 489 to 491 397

1947, 1

EGOROV, I. P.: On the order of the group of motions of spaces with affine connection. Akad. Nauk. SSSR 57, 867–870 (russian) *346, 348*

1948, 1

— On collineations in spaces with projective connection. Akad. Nauk. SSSR 61, 605–608 . . . . *346, 348*

1949, 1

— On a strengthening of Fubini's theorem on the order of the group of motions of a riemannian space. Akad. Nauk. SSSR 66, 793–796 (russian) *346, 348*

1949, 2

— On the groups of motion of spaces with asymmetric affine connection. Akad. Nauk. SSSR. 64, 621–624 (russian) *346, 348*

1950, 1

— On groups of motion of spaces with general asymmetrical affine connection. Akad. Nauk. SSSR. 73, 265–267 (russian) *346, 348*

1951, 1

— Collineations of projectively connected spaces. Akad. Nauk SSSR. 80, 709 to 712 (russian) *348*

1952, 1

— A tensor characteristic of A_n of nonzero curvature with maximum mobility. Akad. Nauk SSSR. 84, 209–212 (russian) *348*

1952, 2

— Maximally mobile L_n with a semi-symmetric connection. Akad. Nauk SSSR. 84, 433–435 (russian) *348*

1950, 1

EHRESMANN, C.: Les connexions infinitésimales dans un espace fibré différentiable. Colloque de Topologie Bruxelles 29–55 . . . . *123*

1952, 1

— Structures locales et structures infinitésimales. C. R. 234, 587–589 . . . *67*

1952, 2

— Les prolongements d'une variété différentiable V. Covariants différentiels et prolongements d'une structure infinitésimale. C. R. 234, 1424–1425 *123*

1928, 1

EINSTEIN, A.: Riemann Geometrie mit Aufrechterhaltung des Begriffes des Fernparallelismus. Sitz. Preuß. Akad. 217–221 . . . . *142*

1945, 1

— A generalization of the relativistic theory of gravitation. Ann. of Math. (2) 46, 578–584 . . . . *179*

1946, 1

— and E. G. STRAUS: A generalization of the relativistic theory of gravitation II. Ann. of Math. 47, 731–741 . . . . *179*

1948, 1

— A generalized theory of gravitation. Rev. Modern Physics 20, 35–39 . . *179*

1950, 1

— The Bianchi identities in the generalized theory of gravitation. Canadian J. Math. 2, 120–128 . . . . *179*

1922, 1

EISENHART, L. P.: Ricci's principal directions for a Riemann space and the Einstein theory. Proc. Nat. Acad. 8, 24–26 . . . . *146*

1922, 2

— Spaces with corresponding paths. Proc. Nat. Acad. 8, 233–238 . . . . . *288*

1922, 3

— and O. VEBLEN: The Riemann geometry and its generalisation. Proc. Nat. Acad. 8, 19–23 . . . . *303*

1922, 4

— Fields of parallel vectors in the geometry of paths. Proc. Nat. Acad. 8, 207–212 . . . . *322*

1923, 1

— Symmetric tensors of the second order whose first covariant derivatives are zero. Trans. Am. Math. Soc. 25, 297–306 . . . . *285*

1923, 2

— Orthogonal systems of hypersurfaces in a general Riemann space. Trans. Am. Math. Soc. 25, 259–280 . . . . *246*

1923, 3

— Affine geometry of paths possessing an invariant integral. Proc. Nat. Acad. 9, 4–7 . . . . *144*

1924, 1

— Geometries of paths for which the equations of the paths admit a quadratic first integral. Trans. Am. Math. Soc. 26, 378–384 . . . . *293*

1925, 1

— Fields of parallel vectors in a riemannian geometry. Trans. Am. Math. Soc. 27, 563–573 . . . . *322*

1926, 1

*— Riemannian geometry. Princeton Univ. Press. 262 pp.
22, 156, 227, 229, 230, 242, 244, 245, 246, 263, 271, 278, 294, 307, 322, 348, 349

1927, 1

*— Non riemannian geometry. Am. Math. Soc. Coll. Publ. VIII. 184 pp.
123, 155, 166, 238, 289, 293, 296

1927, 2

— and M. S. KNEBELMAN: Displacements in a geometry of paths which carry paths into paths. Proc. Nat. Acad. 13, 38–42 . . . . *301*

1930, 1

— Projective normal coordinates. Proc. Nat. Acad. 16, 731–740 . . . . *155, 301*

1933, 1

*— Continuous groups of transformations. Princeton Univ. Press. 302 pp.
185, 215, 226, 349, 350, 351

1935, 1

— Groups of motions and Ricci directions. Ann. of Math. 36, 823–832 . . . *346*

1936, 1

— Simply transitive groups of motions. Mon. Math. Phys. 43, 448–462 . . *346*

1937, 1

— Riemannian spaces of class greater than unity. Ann. of Math. 38, 794–808
268, 281, 282, 283

1938, 1

EISENHART, L. P.: Parallel vectors in riemannian space. Ann. of Math. 39, 316–321 . . . . *322, 422*

1940, 1

*— An introduction to differential geometry. Princeton Math. Series, V, 3. Princeton Univ. Press. X + 304 pp.

1949, 1

*— Riemann geometry, 2nd printing. Princeton Univ. Press. VII + 306 pp. *153, 246, 349*

1951, 1

— Generalized Riemann spaces I. Proc. Nat. Acad. 37, 311–315 . . . . . *179*

1952, 1

— Generalized Riemann spaces II. Proc. Nat. Acad. 38, 505–508 . . . . . *179*

ENGEL, F. (see S. LIE).

1934, 1

*FABRICIUS BJERRE, FR.: Differentialgeometrische Untersuchungen, torsionsfreie Flächen in Räumen konstanter Krümmung. Thesis, Kopenhagen. 88 pp. (dansk.) *261, 269*

1936, 1

— Sur les variétés à torsion nulle. Acta Math. *66*, 49–77 . . . . . *253, 261, 269*

1922, 1

FERMI, E.: Sopra i fenomeni che avvengono in vicinanza di una linea oraria. Rend. d. Linc. I, 21–23, 51–52 . . . . *166*

1950, 1

H. E. FETTIS: A method for obtaining the characteristic equation of a matrix and computing the associated modal columns. Quart. Appl. Math. 8, 206–212 . . . . *30*

1938, 1

FIALKOW, A.: Einstein spaces in a space of constant curvature. Proc. Nat. Acad. 24, 30–34 . . . . *148*

1938, 2

— The riemannian curvature of a hypersurface. Bull. Ann. Math Soc. 44, 253–257 . . . . *242, 388*

1938, 3

— Hypersurfaces of a space of constant curvature. Ann. of Math. Princeton (2) 39, 762–785 . . . . *242*

1939, 1

— Totally geodesic Einstein spaces. Bull. Am. Math. Soc. 45, 423–428 *148, 312, 315*

1939, 2

— Conformal geodesics. Trans. Am. Math. Soc. 45, 443–473 . . *310, 315, 330, 332*

1940, 1

— The conformal theory of curves. Proc. Nat. Acad. 26, 437–439 . *310, 315, 330*

1942, 1

— Correction to "Totally geodesic Einstein spaces". Bull. Am. Math. Soc. 48, 167–168 . . . . *148, 312, 315*

1942, 2

FIALKOW, A.: The conformal theory of curves. Trans. Am. Math. Soc. 51, 435–501 . . . . *310, 315, 330*

1944, 1

— Conformal differential geometry of a subspace. Trans. Am. Math. Soc. 56, 309–433 . . . . *315*

1939, 1

FICKEN, F. A.: The riemannian and affine differential geometry of product spaces. Ann. of Math. 40, 892–913 . . . . *285*

1918, 1

*FINSLER, P.: Über Kurven und Flächen in allgemeinen Räumen. Thesis. Göttingen. 121 pp. *40, 41*

1940, 1

— Über die Krümmung der Kurven und Flächen. Reale Acc. d'Italia 9, Convegno Volta 18, 1–18 . . . . *269*

1902, 1

FINZI, A.: Le ipersuperficie a tre dimensioni che si possono rappresentare conformemente sullo spazio euclideo. Atti R. Istit. Veneto (VIII) 5 (= 62), 1049–1062 . . . . *306*

1921, 1

— Sulla representabilità conforme di due varietà ad n dimensioni l'una sull'altra. Atti R. Istit. Veneto 80^{II}, 777–789 . . . . *306*

1922, 1

— Sulle varietà in rappresentazione conforme con la varietà euclidea a più di tre dimensioni. Rend. d. Linc. (V) 31^{I}, 8–12 . . . . *306*

1923, 1

— Sulla curvatura conforme di una varietà. Rend. d. Linc. (V) 32^{I}, 215–218 *305*

1949, 1

*FINZI, B., and M. PASTORI: Calcolo tensoriale e applicazioni. Zanichelli, Bologna. VII + 427 pp.

1933, 1

FREUDENTHAL, H.: Ein Aufbau der Lieschen Gruppentheorie. J. D. M. V. 43, 26–39 . . . . *208*

1924, 1

FRIEDMANN, A., and J. A. SCHOUTEN: Über die Geometrie der halbsymmetrischen Übertragungen. Math. Zeitschr. 21, 211–223. . . . . *126, 127*

1925, 1

FRIESECKE, H.: Vektorübertragung, Richtungsübertragung, Metrik. Math. Ann. 94, 101–118 . . . . *123, 287*

1879, 1

FROBENIUS, G.: Über homogene totale Differentialgleichungen. Journ. f. d. r. u. a. Math. 86, 1–19 . . . . *92*

1903, 1

FUBINI, G.: Sugli spazi che amettono un gruppo continuo di movimenti. Ann. di Mat. (3) 8, 39–81 . . . . *348, 350*

1903, 2
FUBINI, G.: Sulle metriche definite da una forma Hermitiana. Atti Istit. Veneto (8) 6 (= 63), 501–513 . . . . *406*

1904, 1
— Sugli spazî a quattro dimensioni che ammettono un gruppo continuo di movimenti. Ann. di Mat. (3) 9, 33–90 . . . . *348, 350*

1905, 1
— Sulle coppie di varietà geodeticamenta applicabili. Rend. d. Linc. (V) 14^{I}, 678–683; 14^{II}, 315–322 . . . . *294, 296*

1909, 1
— Sulle rappresentazioni che conservano le ipersfere. Ann. di Matem. (III) 16, 141–160 . . . . *304*

1935, 1
FUCHS, B.: Über vollständig geodätische analytische Mannigfaltigkeiten einer vierdimensionalen. Mitt. Math. und Mech. Tomsk. 1, 168–174 . . . *269, 412*

1937, 1
— Über geodätische Mannigfaltigkeiten einer bei pseudokonformen Abbildungen invarianten Riemannschen Geometrie. Rec. Math. Moscow 2, 567 to 594 . . . . *399, 404, 407, 412*

1942, 1
GALVANI, O.: Sur la réalisation de certains espaces à parallelisme absolu par des congruences de droites. C. R. 214, 337–339 . . . . *125*

1942, 2
— Sur les connexions euclidiennes à courbure non nulle réalisable par des congruences de droites. C. R. 214, 733–735 . . . . *125*

1943, 1
— Sur la réalisation des connexions euclidiennes ponctuelles à deux dimensions les plus générales. C. R. 216, 23–25 . . . . *125*

1943, 2
— Sur la réalisation des connexions ponctuelles euclidienne et affines à n dimensions. C. R. 216, 519–521 . . . . *125*

1945, 1
— Sur la réalisation des espaces ponctuels à torsion en géométrie euclidienne. Ann. Ecole Norm. Sup. (3) 62, 1–92 . . . . *125*

1946, 1
— La réalisation des connexions ponctuelles affines et la géométrie des groupes de Lie. J. Math. Pures et Appl. 25, 209–239 . . . . *125*

1827, 1
*GAUSS, C. F.: Disquisitiones generales circa superficies curvas. Werke IV; german translation in Ostwald's Klassiker 5 (1889). 62 pp. *245*

GÉHÉNIAU, J. (see T. DE DONDER).

1949, 1
GEORGHIU, O. E.: Les lois de transformation des objets géométriques spéciaux linéaires de classe ν, avec une composante en X_1. C. R. 229, 611–613 *67*

1938, 1
GHOSH, N. N.: New methods in the geometry of a hypersurface. Bull. Calcutta Math. Soc. 30, 73–84 . . . . *242*

1940, 1

GHOSH, N. N.: The tortuosity of a variety. Bull. Calcutta Math. Soc. 32, 51–60 *269*

1941, 1

— The tortuosity of submanifolds of a variety. Bull. Calcutta Math. Soc. 33, 187–195 . . . . *269*

1950, 1

— On complex tensor calculus. Bull. Calcutta Math. Soc. 42, 65–72 . . . . *389*

1937, 1

GIVENS, J. W. (see also O. VEBLEN): Tensor coordinates of linear spaces. Ann. of Math. 38, 355–385 . . . . *23, 57*

1940, 1

— Factorization and signatures of Lorentz matrices. Bull. Am. Math. Soc. 46, 81–85 . . . . *50*

1930, 1

GOLAB, S. (see also J. A. SCHOUTEN): Über verallgemeinerte projektive Geometrie. Prace mat. fiz. Warszawa 37, 91–153. (Diss.) *3, 300, 303*

1931, 1

— Sopra le connessioni lineari generali. Estenzione d'un teorema di Bompiani nel caso più generale. Ann. di Mat. 4 (8) 283–291 . . . . *144*

1934, 1

— Sur l'ordre de planéité des espaces plongés. Ann. di Mat. (4) 13, 119–125 *232*

1934, 2

— Über die affinen Invarianten einer Kurve der X_p, die in einem L_n eingebettet ist. Abh. Sem. Vektor u. Tensor. Moskau 4, 360–362 . . . . *232*

1935, 1

— Ein Beitrag zur konformen Abbildung von zwei Riemannschen Räumen aufeinander. Ann. Soc. Pol. 13, 13–19 . . . . *315*

1938, 1

— Über die Klassifikation der geometrischen Objekte. Math. Zeitschr. 44, 104–114 . . . . *67*

1945, 1

— Sur la généralisation d'une formule de Lancret concernant l'uniformisation des équations de Frenet. Ann. Soc. Polon. Math. 18, 129–133 . . . . . *230*

1946, 1

— Sur la théorie des objets géométriques. Ann. Soc. Polon. Math. 19, 7–35 *67*

1947, 1

— Sur la théorie des objets géometriques. (Reduction des objets géométriques speciaux de première classe aux objets du type D.) Ann. Soc. Polon. Math. 26, 10–27 . . . . *67*

1948, 1

— Alcuni teoremi della teorema degli oggetti geometri. Atti Accad. Naz. Linc. 5, 120–122 . . . . *67*

1949, 1

— Généralisation des équations de Bonnet-Kowalewski dans l'espace à un nombre arbitraire de dimensions. Ann. Soc. Polon. Math. 22, 97–156 *242, 250*

1950, 1
GOLAB, S.: La notion de similitude parmi les objets géométriques. Bull. Int. Acad. Polon. 1–7 . . . . *67, 108*

1950, 2
— Sur les objets géométriques à une composante. Ann. Soc. Polon. Math. 23, 79–89 . . . . *67*

1951, 1
— and T. H. WRÖBEL: Courbure et torsion géodésique pour les courbes situées sur les hypersurfaces à n—1 dimensions plongées dans l'espace à n dimensions. Ann. Soc. Polon. Math. 24, 25–51 . . . . *242*

1941, 1
GOLIFMAN, R.: Sur les courbes à une dimension réelle dans l'espace hermitien hyperbolique. Bull. Soc. Roy. Sci. Liège 10, 57–67 . . . . *415*

1942, 1
*GONÇALVES-MIRANDA, M.: Tensor calculus. Anais Fac. Ci. Porto 27 (portuguese). 54 pp.

1922, 1
*GOURSAT, E.: Leçons sur le problème de Pfaff. Hermann, Paris. 386 pp. *83, 85*

1950, 1
GRAEUB, W.: Geometrische Deutung des Krümmungstensors. Arch. Math. 2, 148–151 . . . . *138*

1844, 1
*GRASSMANN, H.: Die lineale Ausdehnungslehre. Ein neuer Zweig der Mathematik. Ges. Schriften I, 139 pp. Teubner, Leipzig 1894 . . . . *36*

1862, 1
*— Die Ausdehnungslehre. Gesammelte Schriften II, 511 pp. Teubner, Leipzig 1896 . . . . *23, 366*

1934, 1
GRAUSTEIN, W. C.: The geometry of Riemann spaces. Trans. Am. Math. Soc. 36, 542–585 . . . . *253*

1925, 1
*GRISS, G. F. C.: Differentialinvarianten von Systemen von Vektoren, Diss. P. Noordhoff, Groningen. 61 pp. *142*

1951, 1
GUGGENHEIMER, H. (see also B. ECKMANN): Sur les variétés qui possèdent une forme extérieure quadratique fermée. C. R. 232, 470–472 . . . . *179*

1953, 1
— Geometria pseudo-kähleriana. Acc. Naz. Lincei (8) 14, 220–222 . . . . *397*

1935, 1
GUGINE, E.: Sul trasporto ciclico de un tensore di ordine qualunque. Rend. d. Linc. (6) 22, 295–299 . . . . *138*

1928, 1
*GUICHARD, C.: Les courbes de l'espace à n dimensions. Mém. d. Sc. Math. XXIX. Gauthier-Villars, Paris. 63 pp. *227*

1933, 1
GUREVIČ, G. B.: Über einige Integralaufgaben der Tensoranalysis. Abh. Sem. Vektor u. Tensor. Moskau 1, 143–187 . . . . *36, 87*

1933, 2

GUREVIČ, G. B.: Sur les formes canoniques d'un trivecteur dans l'espace à six dimensions. C. R. 197, 384–385 . . . . *36*

1934, 1

— Sur les trivecteurs dans l'espace à sept dimensions. C. R. Acad. Sc., URSS. 3, 564–569 . . . . *36*

1934, 2

— Sur quelques invariants d'un trivecteur et d'une ferme cubique. C. R. Acad. Sc. USSR. 3, 317–319 . . . . *36*

1934, 3

— Reduction d'un trivecteur à une forme canonique dans un cas spécial. C. R. Acad. Sc. USSR. 3, 320–321 . . . . *36*

1935, 1

— Classification des trivecteurs ayant le rang huit. C. R. Acad. Sc. USSR. 2, 353–356 . . . . *36*

1935, 2

— L'algèbre du trivecteur. Abh. Sem. Vektor u. Tensor. Moscow II, 51–113 . *36*

1950, 1

— Complete systems of symmetric and skew-symmetric tensors. Mat. Sbornik 27, 103–116 (russian). *36*

1950, 2

— On some linear transformations of symmetric tensors or polyvectors. Mat. Sbornik 26, 463–470 (russian) *35*

1950, 3

— Canonization of a pair of bivectors. Trudy Sem. Vektor, Tenzor. Analizu 8, 355–363 . . . . *35*

1950, 4

— On some affinors connected with trivectors of the eighth rank. Trudy Sem. Vektor, Tenzor. Analizu 8, 296–300 (russian) *35*

1940, 1, 2

*HAACK, W.: Differential-Geometrie I, II. Wolfenbütteler Verlagsanstalt, Wolfenbüttel and Hannover. 136 and 131 pp. *306, 342*

1935, 1

HAANTJES, J. (see also J. A. SCHOUTEN): Klassifikation der antilinearen Transformationen. Math. Ann. 112, 131–148 . . . . *53*

1937, 1

— On the projective geometry of paths. Proc. Edinburgh Math. Soc. 5, 103 to 115 . . . . *301, 302, 327*

1937, 2

— Conformal representations of an n-dimensional euclidean space with a non-definite fundamental form on itself. Proc. Kon. Ned. Akad. Amsterdam 40, 700–705 . . . . *312, 315*

1939, 1

— and W. WRONA: Über konformeuklidische und Einsteinsche Räume gerader Dimensionen. Proc. Kon. Ned. Akad. Amsterdam 42, 626–636 *148, 274, 305, 312*

1940, 1
HAANTJES, J.: Eine Charakterisierung der konformeuklidischen Räume. Proc. Kon. Ned. Akad. Amsterdam 43, 91–94 . . . . *305*

1941, 1
— Conformal differential geometry I. Curves in conformal euclidean spaces. Proc. Kon. Ned. Akad. Amsterdam 44, 814–824 . . . . *305, 315*

1942, 1
— Conformal differential geometry II. Curves in conformal two-dimensional spaces. Proc. Kon. Ned. Akad. Amsterdam 45, 249–255 . . . . . . *305, 315*

1942, 2
— Conformal differential geometry III. Surfaces in three-dimensional space. Proc. Kon. Ned. Akad. Amsterdam 45, 836–841 . . . . *305, 315*

1942, 3
— Conformal differential geometry IV. Surfaces in three-dimensional space. Proc. Kon. Ned. Akad. Amsterdam 45, 918–923 . . . . *305, 315*

1943, 1
— Conformal differential geometry V. Special surfaces. Proc. Kon. Ned. Akad. Amsterdam 52, 322–331 . . . . *305, 315*

1953, 1
— and G. LAMAN: On the definition of geometric objects. Indagationes Math. Proc. Kon. Ned. Akad. Amsterdam 15, 208–222 . . . . *67*

1938, 1
*HAIMOVICI, A.: Directions concourantes et directions parallèles sur une variété d'un espace quelconque. Thesis. Univ. Yassy. 70 pp. *315, 322*

1940, 1
— Sur la définition intrinsèque d'un élément plan normal à une variété non holonome. C. R. 210, 595–596 . . . . *253*

1943, 1
— Les espaces à connexion affine à parallélisme absolu et les systèmes homographiques de courbes. Ann. Sc. Univ. Jassy 29, 90–100 . . . . *142*

1945, 1
— Sur les familles de transformations ponctuelles simplement transitives. Acad. Roum. Bull. Sc. 28, 5 pp. *201*

1946, 1
— Sur une nouvelle espèce de connexions affines dans les espaces des familles simplement transitives de transformations de variables. Bull. Math. Soc. Roumaine 47, 35–48 . . . . *201*

1946, 2
— La géométrie des familles de transformations de variables dépendant de paramètres. C. R. 223, 969–971 . . . . *201*

1946, 3
— La géométrie intrinsèque des variétés Riemanniennes non holonomes. J. Math. Pures Appl. 25, 291–305 . . . . *253*

1947, 1
— Sur les espaces des familles simplement transitives de transformations de variables à courbure de III-e espèce nulle. Bull. Ecole Polytech. Jassy 2, 46–70 . . . . *201*

1948, 1
HAIMOVICI, A.: Sur la géométrie des familles de transformations ponctuelles simplement transitives. Ann. Univ. Jassy 1, 30 (1944–1947), 1–36 . . *201*

1948, 2
— La géométrie des familles de transformations de variables dépendant de paramètres. Disquisit. Math. Phys. 6, 81–128 . . . . *201*

1951, 1
HARTMANN, P.: On geodesic coordinates. Am. J. of Math. 73, 949–954 . . . *155*

1951, 1
HASHIMOTO, S.: A new proof of Liber's theorem. Kodai. Math. Sem. Rep. 118 to 119 . . . . *367*

1930, 1
HAYDEN, H. A.: On a generalized helix in a riemannian space. Proc. London Math. Soc. (2) 32, 337–345 . . . . *230*

1930, 2
— Asymptotic lines in a V_m in V_n. Proc. London Math. Soc. (2) 33, 22–31 *271, 275*

1931, 1
— Deformations of a curve in a riemannian n-space, which displace certain vectors parallely at each point. Proc. London Math. Soc. (2) 32, 321–336 *335, 352*

1932, 1
— Subspaces of a space with torsion. Proc. London Math. Soc. (2) 34, 27–50 *269*

1934, 1
— On some congruences associated with a sub-variety in a V_n. Quart. Journ. Math. Oxford (4) 33, 117–125 . . . . *275*

1934, 2
— Infinitesimal deformations of subspaces in a general metrical space. Proc. London Math. Soc. (2) 37, 416–440 . . . . *348, 352, 353*

1936, 1
— Infinitesimal deformations of an L_m in an L_n. Proc. Lond. Math. Soc. (2) 41, 332–336 . . . . *352*

1868, 1
HELMHOLTZ, H.: Über die Tatsachen, die der Geometrie zugrunde liegen. Gött. Nachr. 193–221; Wissensch. Abhandl. II, 618–639 . . . . *41*

1916, 1
HERGLOTZ, G.: Zur Einsteinschen Gravitationstheorie. Sitz. Leipzig 68, 199 to 203 . . . . *148, 153*

1949, 1
HESSELBACH, B.: Über die Erhaltungssätze der konformen Geometrie. Math. Nachr. 3, 107–126 . . . . *314*

1916, 1
HESSENBERG, G.: Vektorielle Begründung der Differentialgeometrie. Math. Ann. 78, 187–217 . . . . *123*

1924, 1
— Beispiele zur Richtungsübertragung. J. D. M. V. 33, 93–95 . . . . . . *143*

HIRAMATU, H. (see K. YANO).

1924, 1

HLAVATY, V. (see also E. BORTOLOTTI, J. A. SCHOUTEN): Sur le déplacement linéaire du point. Vestnik K. C. SP. N: 13, 8 . . . . *78*

1926, 1

— Les paramètres locaux dans une variété de Riemann. Rend. d. Linc. (6) 4, 5 pp. *155*

1926, 2

— Application des paramètres locaux. Ann. d. l. Soc. Polon. d. Math. 5, 19 *155*

1926, 3

— Deux notes sur la métrique. Rozpravy C. Ak. 16, 16 . . . . *155*

1926, 4

— Contribution au calcul différentiel absolu. Vestnik Kral. Nauk. Praha 2, 1 to 12 . . . . *232*

1927, 1

— Contact de deux courbes dans une V_n. Rend. d. Linc. (6) 5, 415–420 . . *155*

1927, 2

— Théorie des densités dans le déplacement général. Ann. di Mat. 5, 73–83 *129*

1927, 3

— Déplacements isohodoiques. Ens. Math. 26, 84–97 . . . . *322, 327*

1928, 1

— Sur les coefficients de Ricci. C. R. 186, 1691–1693 . . . . *78, 171*

1928, 2

— Bemerkung zur Arbeit von Herrn T. Y. THOMAS. Math. Zeitschr. 28, 142 to 146 . . . . *322*

1928, 3

— Ein Beitrag zur Theorie der Weylschen Übertragung. Proc. Kon. Ned. Akad. Amsterdam 31, 878–881 . . . . *231*

1929, 1

— Proprietà differenziali delle curve in uno spazio a connessione lineare generale. Rend. d. Palermo 53, 356–388. Ancora sulle proprietà differenziali delle curve in uno spazio a connessione lineare generale. Rend. d. Palermo 53, 389–410 . . . . *234*

1929, 2

— Le parallelisme de la connexion de M. Weyl. Ann. Ec. Norm. Sup. 46, 73 to 103 . . . . *133*

1929, 3

— Sur la déformation infinitésimale d'une courbe dans la variété métrique avec torsion. Bull. Soc. Mathém. d. France 56, 18–25 . . . . *352*

1930, 1, 2

— Sur la courbure des variétés non-holonomes. Rend. d. Linc. (6) 12, 566 to 574; 647–654 . . . . *78, 253*

1930, 3

— Sulle coordinate geodetiche. Rend. d. Linc. (6) 12, 502–509 . . . . . . *155*

1931, 1

HLAVATY, V.: Projektive Invarianten der Kurvenkongruenz und einer Kurve. Math. Zeitschr. 34, 58–73 . . . . *302*

1933, 1

— Connexion projective et déplacement projectif. Ann. di Mat. 12, 217–294 *301, 303, 322*

1934, 1

— Espaces de König. Abh. Sem. Vekt. u. Tensor. Moskau 4 (publ. 1937), 117–120 . . . . *123*

1934, 2

— Les courbes de la variété générale à n dimensions. Mém. d. Sc. Math., fasc. LXIII. Paris, Gauthier-Villars. 73 pp. *230, 232, 233, 234*

1934, 3

— Induzierte und eingeborene Konnexion in den (nicht) holonomen Räumen. Math. Zeitschr. 38, 283–300 . . . . *253*

1935, 1

— Espaces abstraits, courbes de König. Rend. d. Palermo 13, 1–39 . . . . *123*

1935, 2

— Zur Konformgeometrie I, II, III. Proc. Kon. Akad. Amsterdam 38, 281 to 286, 738–743, 1006–1011 . . . . *231, 315, 316*

1936, 1

— Systèmes des connexion de M. Weyl. Bull. Acad. des Sc. Boh. 37, 181–184 *315, 316*

1939, 1

— Contributo alla teoria delle connessioni. Reale Accad d'Italia, Fondazione Ales. Volta 479–506 . . . . *303*

1939, 2

*— Differentialgeometrie der Kurven und Flächen und Tensorrechnung. Noordhoff, Groningen-Djakarta. XI + 569 pp.

1949, 1

— Théorie d'immersion d'une W_m dans W_n. Ann. Soc. Polon. Math. 21, 196 to 206 . . . . *134, 269, 285*

1949, 2

— Affine embedding theory. I. Affine normal spaces. Proc. Kon. Ned. Akad. Amsterdam 52, 505–517; Indagationes Math. 11, 165–177 . . . . *285*

1949, 3

— Affine embedding theory. II. Frenet formulae. Proc. Kon. Ned. Akad. 52, 714–724; Indag. Math. 11, 244–254 . . . . *285*

1949, 4

— Affine embedding theory. III. Integrability conditions. Proc. Kon. Ned. Akad. Amsterdam. 52, 977–986; Indag. Math. 11, 365–374 . . . . *285*

1950, 1

— Deformation theory of subspaces in a Riemann space. Proc. Am. Math. Soc. 1, 600–617 . . . . *352*

1952, 1

— Embedding theory of a W_m in a W_n. Rend. d. Palermo 2, 403–438 *134, 269, 285*

1952, 2
HLAVATÝ, V.: The Einstein connection of the unified theory of relativity. Proc. Nat. Acad. 38, 415–419 . . . . *179*

1952, 3
— The elementary basic principles of the unified theory of relativity. Proc. Nat. Acad. 38, 243–247 . . . . *179*

1952, 4
— The elementary basic principles of the unified theory of relativity A. J. Rational Mech. Anal. 1, 539–562 . . . . *179*

1952, 5
— Intrinsic deformation theory of subspaces in a Riemann space. J. Rational Mech. Anal. 1, 49–72. . . . . *352*

1952, 6
— The Schrödinger final affine field laws. Proc. Nat. Acad. 38, 1052–1058 *179*

1953, 1
— The elementary basic principles of the unified theory of relativity B. J. Rational Mech. Anal. 2, 1–52 . . . . *179*

1953, 2
— Connections between Einstein's two unified theories of relativity. Proc. Nat. Acad. 39, 507–510 . . . . *179*

1953, 3
— and A. W. SAENZ: Uniquenes theorems in the unified theory of relativity. J. Nat. Mech. An. 2, 523–536 . . . . *179*

1953, 4
— The spinor connections in the unified Einstein theory of relativity. Proc. Nat. Acad. USA. 39, 501–536 . . . . *179*

1954, 1
— The elementary basic principles of the unified theory of relativity; C_1 Introduction; C_2 Applications I. J. Rat. Mech. Anal. 3, 103–146, 147–179 *179*

HOCHRAINER, A. (see A. DUSCHEK).

1951, 1
HODGE, W. V. D.: A special type of Kähler manifold. Proc. London Math. Soc. (3) 1, 104–117 . . . . *397*

1934, 1
HOKARI, S.: Über die Bivektorübertragung. Journ. Hokkaido Univ. 2, 103 to 117 . . . . *123, 138*

1935, 1
HOMBU, H.: Zur Theorie der unitären Geometrie. Journ. Fac. Sc. Hokkaido Univ. (1), 3, 27–42. *399, 410*

1936, 1
— Zur Theorie der Affinorübertragung. Proc. Imp. Acad. Japan 12, 109–111 *138*

1936, 1
— and M. MIKAMI: Parabolas and projective transformations in the generalized spaces of paths. Jap, J. Math. 17, 307–335 . . . . *300*

1941, 2

HOMBU, H. and Y. OKADA: On the projective theory of asymmetric connections. Proc. Phys. Math. Soc. Japan 23, 357–362 . . . . *322*

1942, 1

— and M. MIKAMI: Conics in the projectively connected manifolds. Mem. Fac. Sc. Kyūsyū Imp. Univ. 2, 217–239 . . . . *300*

1927, 1

HORAK, Z.: Die Formeln für allgemeine lineare Übertragung bei Benutzung von nichtholonomen Parametern. Nieuw Archief v. Wisk. (2) 15, 193–201 *78, 253*

1927, 2

— Eine Verallgemeinerung des Mannigfaltigkeitsbegriffes. Spisy Brno 86, 20 pp. *78, 253*

1928, 1

— Sur les courbures des variétés non holonomes. C. R. 187, 1273–1275 . . *253*

1929, 1

— Sur les dérivées par rapport aux affineurs. Ens. Math. 28, 212–214 . . . *72*

1948, 1

HUTCHINSON, L. C.: Notes on the member of leaves of an alternating tensor. Presented to the Am. Math. Soc. Dec. 28, 1948. The author was so kind to send me a manuscript *36*

1951, 1

ICHINOHE, A.: Sur la possibilité de plonger un espace à connexion conforme donné dans un espace conforme. Tôhoku Math. J. 2, 193–204 . . . . *268, 315*

IMAI, T. (see K. YANO).

1950, 1

INGRAHAM, R. L.: Contributions to the Schrödinger non-symmetric affine unified field theory. Ann. of Math. 52, 743–752 . . . . *179*

1951, 1

— Relativité conforme. C. R. 232, 1072–1074 . . . . *179*

1951, 2

— Sur une théorie de la "relativité conforme". C. R. 232, 938–940 . . . . *179*

1944, 1

IWAMOTO, H.: On Frenet's formulae. Tensor J. 44–49 (japanese) *277*

1950, 1

— On the structure of riemannian spaces whose holonomy groups fix a null-system. Tôhoku Math. J. 1, 109–135. . . . . *367, 397*

1951, 1

— On the relation between homological structure of riemannian spaces and exact differential forms which are invariant under holonomy groups I. Tôhuku Math. J. 3, 59–70 . . . . *367*

1934, 1

JACOBSTHAL, E.: Zur Theorie der linearen Ableitungen. Sitz. Berl. Math. Ges. 33, 15–34 . . . . *53*

1926, 1

JANET, M.: Sur la possibilité de plonger un espace riemannien donné dans un espace euclidien. Ann. d. l. Soc. Polon. d. Math. 5, 38–43 . . . . . *268*

1928, 1

*JÄRNEFELT, G.: Zur Affinoranalysis. Diss. Helsinki, 91 pp. Acta Sci. Fenniciae (A) 28, No. 9 . . . . *232, 263, 278*

1948, 1

JOHNSON, P. B.: The role of the directrix in Levi-Civita parallelism. Duke Math. J. 15, 389–405 . . . . *123, 131, 138, 340, 341*

1874, 1

JORDAN, C.: Sur la théorie des courbes dans l'espace à n dimensions. C. R. 79, 795–797 . . . . *230*

1948, 1

JORDAN, P.: Über den Riemannschen Krümmungstensor I. Einsteinsche Theorie. Z. Physik 124, 602–607 . . . . *179*

1948, 2

— Über den Riemannschen Krümmungstensor II. Eddingtonsche und Schrödingersche Theorie. Z. Physik 124, 608–613 . . . . *179*

1950, 1

— Einsteins neue Untersuchungen. Elektrotechn. Z. 71, 615–618 . . . . *179*

1921, 1

JUVET, G.: Les formules de Frenet dans un espace généralisé de Weyl. Bull. Soc. Neuchâteloise 46. Cf. also C. R. 172, 1647–1650 . . . . *230*

1922, 1

*— Introduction au calcul tensoriel et au calcul différentiel absolu. Blanchard, Paris. 100 pp.

1925, 1

— Sur le déplacement parallèle le plus général et sur l'étude des courbes tracées dans une multiciplité quelconque. Bull. Soc. Math. 53, 60–74 *230*

1930, 1

KAGAN, B.: Sur les espaces sous-projectifs. C. R. 191, 548–550 . . . . . . *317*

1933, 1

— Über eine Erweiterung des Begriffes vom projektiven Raume und dem zugehörigen Absolut. Abh. Sem. Vektor u. Tensor. Moskau 1, 12–101 *317, 318, 330*

1935, 1

— Der Ausnahmefall in der Theorie der subprojektiven Räume. Abh. Sem. Vektor u. Tensor. Moskau 2–3, 151–170 . . . . *317*

1933, 1

KÄHLER, E.: Über eine bemerkenswerte Hermitesche Metrik. Abh. Math. Sem. Hamburg 9, 173–186 . . . . *397, 399*

1934, 1

*— Einführung in die Theorie der Systeme von Differentialgleichungen. Hamb. Math. Einzelschr. Heft 16 . . . . *26, 74, 117*

1922, 1

*KÄMMERER, F.: Zur Flächentheorie im n-fach ausgedehnten Raume. Thesis. Gießen *269*

1927, 1

*KÄMMERER, W.: Die trilineare alternierende homogene Form in acht Veränderlichen. Mitt. Math. Sem. Gießen (Thesis) *36*

KAMPEN, E. R. VAN (see J. A. SCHOUTEN).

1941, 1

KANITANI, J.: Spaces with a linear connection reduced from a space with a projective connection. Tensor 4, 1–12 (japanese) *301*

1943, 1

— Sur les espaces répresentables géodésiquement sur l'espace projectif. Proc. Phys. Math. Soc. Japan 25, 87–100 . . . . *296, 301*

1943, 2

— Sur un espace à connexion projective renfermant des hyperquadriques. Proc. Phys. Math. Soc. Japan 25, 617–621 . . . . *296, 301*

1947, 1

— Sur l'espace à connexion projective majorante I. Jap. J. Math. 19, 343 to 361 . . . . *268, 296, 301*

1947, 2

— On a generalization of the projective deformation. Mem. Coll. Sc. Univ. Kyoto 25, 23–26 . . . . *301*

1948, 1

— Sur l'espace à connexion projective majorante II. Jap. J. Math. 19, 395–403 . . . . *268, 296, 301*

1949, 1

— On a method of plunging of R_2 with symmetric projective connexion into a projective space S_4. Mem. Coll. Sci. Univ. Kyoto 25, 87–91 *268, 296, 301*

1950, 1

— Sur le développement d'une courbe dans un espace à connexion projective. Mem. Univ. Kyoto 26, 31–43 . . . . *128, 367*

1950, 2

— Sur l'espace à connexion projective majorante III. Jap. J. Math. 20, 45–54 . . . . *268, 296, 301*

1952, 1

— Sur la transformation infinitésimale du groupe d'holonomie. Mem. Coll. Sc. Univ. Kyoto Math. 27, 75–93 . . . . *362, 367, 370*

1934, 1

KAPLAN, N.: V_3 in R_6. Proc. Nat. Acad. 20, 489–499 . . . . *269*

1934, 2

— V_m in R_{m+1} possessing an m-tuply orthogonal net along the lines of curvature. Journ. Math. Phys. Mass. 13, 426–432 . . . . *269*

1936, 1

— On certain three-dimensional manifolds in euclidean sixspace (V_3 in R_6). Journ. Math. Phys. Mass. 15, 219–248 . . . . *269*

1941, 1

— The problem of equivalence. Abh. Sem. Vektor. u. Tensor. 5, 284–289 (russian) *278*

1910, 1

KASNER, E.: The theorem of Thomson and Tait and natural families of trajectories. Trans. Am. Math. Soc. 11, 121–140 . . . . *310*

1913, 1
*KASNER, E.: Differential geometric aspects of dynamics. Princeton Colloquium lectures (delivered 1909). 117 pp. *310*

1921, 1
— Einsteins theory of gravitation: determination of the field by light signals. Amer. Journ. 43, 20–28 . . . . *314*

1922, 1
— The solar gravitational field completely determined by its light rays. Math. Ann. 85, 227–236 . . . . *314*

1932, 1
KAWAGUCHI, A.: Theory of connexions in the general Finsler manifold I. Proc. Imp. Acad. Tokyo 8, 340–343 . . . . *123*

1933, 1
— The foundation of the theory of displacement I. Proc. Imp. Acad. 9, 351–354 *123*

1933, 2
— Theory of connexions in the general Finsler manifold II. Proc. Imp. Acad. Tokyo 9, 351–354 . . . . *123*

1934, 1
— The foundation of the theory of displacements II, III. Proc. Imp. Acad. 10, 45–48; 133–136 . . . . *123*

1935, 1
— Differentialgeometrie in den verschiedenen Funktionalräumen. Journ. Hokkaido Univ. 3, 43–106 . . . . *123*

1939, 1
— Eine Verallgemeinerung von Extensoren. Monatsh. Math. Phys. 48, 329 to 339 . . . . *58*

1952, 1
— On the theory of nonlinear connections I. Introduction to the theory of general nonlinear connections. Tensor 2, 123–142 . . . . *123*

1952, 1
KAWAGUCHI, M.: A generalization of the extensor. Tensor 2, 59–66 . . . . . *58*

1885, 1
*KILLING, W.: Die nicht-euklidischen Raumformen in analytischer Behandlung. Teubner, Leipzig. XII + 264 pp. *244, 271*

1892, 1
— Über die Grundlagen der Geometrie. Journ. f. d. r. u. a. Math. 109, 121 to 186 . . . . *249, 348*

1893, 1
*— Einführung in die Grundlagen der Geometrie. Schönig, Paderborn. I, 357 pp.; II (1898), 361 pp. *244*

1943, 1
KIMPARA, M.: Theory of surfaces in a three-dimensional space with a projective connection. Tensor 6, 33–44 (japanese) *296, 301*

1872, 1
KLEIN, F.: Vergleichende Betrachtungen über neuere geometrische Forschungen. Erlangen. Reprinted Math. Ann. 43 (1893) 63–100 . . . . *64*

1926, 1

*KLEIN, F.: Vorlesungen über höhere Geometrie, 3. Aufl., bearbeitet von W. BLASCHKE. Springer, Berlin. 405 pp. *31, 316*

1951, 1

KLINGENBERG, W.: Zur affinen Differentialgeometrie. I. Über p-dimensionale Minimalflächen und Sphären im n-dimensionalen Raum. Math. Zeitschr. 54, 65–80 . . . . *232*

1951, 2

— Zur affinen Differentialgeometrie. II. Über zweidimensionale Flächen im vierdimensionalen Raum. Math. Zeitschr. 54, 184–216 . . . . *232*

1952, 1

— Über das Einspannungsproblem in der projektiven und affinen Differentialgeometrie. Math. Zeitschr. 55, 321–345 . . . . *232*

1930, 1

KNEBELMAN, M. S. (see also L. P. EISENHART): On groups of motions in related spaces. Am. J. Math. 52, 280–282 . . . . *350*

1945, 1

— On the equations of motion in a Riemann space. Bull. Am. Soc. 51, 682–685 *346*

1951, 1

— Spaces of relative parallelism. Ann. of Math. 53, 387–899 . . . . *123*

1939, 1

KOCZIAN, E.: Sopra i ds^2 di classe uno. Atti Ist. Veneto 98, 27–42 . . . . . *268*

KODAIRA, K. (see G. DE RHAM).

1920, 1

KÖNIG, R.: Beiträge zu einer allgemeinen Mannigfaltigkeitslehre. J. D. M. V. 28, 213–228 . . . . *123, 303, 316*

1932, 1

— Zur Grundlegung der Tensorrechnung. J. D. M. V. 41, 169–189 . . . . *123*

1934, 1

— and C. PESCHL: Axiomatischer Aufbau der Operationen im Tensorraum I, II, III. Ber. Sächs. Akad. 86, 129–154; 267–298; 383–410 . . . . . . *123*

1935, 1

— and K. H. WEISE: Axiomatischer Aufbau der Operationen im Tensorraum IV. Ber. Sächs. Akad. 87, 223–250 . . . . *123*

1946, 1

KOSAMBI, D. D.: Sur la différentiation covariante. C. R. 222, 211–213 . . . *108*

1947, 1

— Les invariants différentials d'un tenseur covariant à deux indices. C. R. 225, 790–792 . . . . *179*

1949, 1

— The differential invariants of a two-index tensor. Bull. Am. Math. Soc. 55, 90–94 . . . . *179*

1951, 1

— Series expansions of continuous groups. Quart. J. Math. Oxford 2, 244–257 *108, 346, 347*

1952, 1

KOSAMBI, D. D.: Path-spaces admitting collineations. Quart. J. Math. Oxford, Ser. 3, 1–11 . . . . *102, 137, 346*

1934, 1

KRAUS, A.: Hyperflächenstreifen im Riemannschen Raume. Mon. Math. Phys. 41, 424–438 . . . . *230*

1869, 1

KRONECKER, L.: Über Systeme von Funktionen mehrerer Variablen. Monatsber. Akad. Wiss. Berlin 159–193; 688–698; Werke I, 175–212; 213–226 *250*

1903, 1

KÜHNE, H.: Die Grundgleichungen einer beliebigen Mannigfaltigkeit. Arch. d. Math. u. Phys. (III) 4, 300–311 . . . . *230, 263, 278*

1904, 1

— Über die Krümmung einer beliebigen Mannigfaltigkeit. Arch. d. Math. u. Phys. (III) 6, 251–260 . . . . *256*

1950, 1, 2

KUIPER, N. H.: Einstein spaces an dconnections I, II. Kon. Ned. Akad. Amsterdam, Proc. 53, 1560–1567; 1568–1576 = Indagationes Math. 12, 505–512; 513–521 . . . . *148, 312, 314*

1951, 1

— Sur les propriétés conformes des espaces d'Einstein. Coll. Géométrie Différentielle, Louvain 1951, 165–166 . . . . *148, 312*

1951, 2

— On the holonomic groups of the vector displacement in riemannian spaces. Kon. Ned. Akad. Amsterdam. Proc. 54 = Indagationes Math. 13, 445–451 *367*

1939, 1

KULK, W. VAN DER (see also J. A. SCHOUTEN): Übertragungen mit alternierendem Krümmungsaffinor. Kon. Ned. Akad. Amsterdam. Proc. 42, 753 to 763 . . . . *144, 148*

1952, 1

KURITA, M.: Riemann space with two-parametric homogeneous holonomy group. Nagoya Math. J. 4, 35–42 . . . . *367*

1923, 1

*LAGRANGE, R.: Sur le calcul differential absolu. Thèse, Paris. 69 pp. . . . *306*

1926, 1

*— Calcul différentiel absolu. Mém. d. Sc. Math., fasc. 19 . . . . *126*

1941, 1

— Sur les invariants conformes d'une courbe. C. R. 242, 1123–1126 . . . . *315*

1941, 2

— Propriétés différentielles des courbes de l'espace conforme à n dimensions. C. R. 213, 551–553. . . . . *315*

LAMAN, G. (see J. HAANTJES).

1938, 1

LANCZOS, C.: A remarkable property of the Riemann-Christoffel tensor in four dimensions. Ann. of Math. (2) 39, 842–850 . . . . *146*

1936, 1

LAND, F. W.: On certain properties of non holonomic submanifolds of a non-riemannian space and the effect on them of a general deformation. Proc. London Math. Soc. (2) 41, 497–516 352

1938, 1

LAPTEV, B.: La dérivée de Lie des objets géométriques qui dépendent de point et de direction. Bull. Soc. Phys. Math. Kazan (3) 10, 3–39 (french summary) 105

1943, 1

— Sur une classe des géométries intrinsèques induites sur une surface dans un espace à connexion affine. C. R. Acad. Sci. URSS. 41, 315–317 268

1945, 1

— Sur l'immersion d'un espace à connexion affine dans un espace affine. C. R. Acad. Sci. URSS. 47, 531–534 268

1950, 1

— Differential connections of manifolds and their holonomy groups. Doklady Akad. Nauk SSSR. 71, 597–600 (russian) 370

1950, 1

*LASS, H.: Vector and tensor analysis. McGraw Hill, New York. XI + 347 pp.

1940, 1

LAURA, E.: Sopra un gruppo di condizioni necessarie affinchè un ds^2 sia di classe h. Ist. Veneto Sc. Mat. Nat. 99, 97–113 268

1942, 1

LEE, E. H.: On unitary geometry. Acad. Sinica Science Record 1, 49–54 . . 397

1943, 1

— A kind of even-dimensional differential geometry and its application to exterior calculus. Am. J. Math. 65, 433–438 179

1945, 1

— On even dimensional skew-metric spaces and their groups of transformations. Am. J. Math. 67, 321–328 179

1947, 1

— On skew-metric spaces and function groups. Am. J. Math. 69, 790–800 179

LEKKERKERKER, C. G. (see W. PEREMANS).

1926, 1

LENSE, J.: Über ametrische Mannigfaltigkeiten und quadratische Differentialformen mit verschwindender Diskriminante. J. D. M. V. 35, 281–294 . 227

1926, 2

— Über spezielle ametrische Mannigfaltigkeiten. Sitz. Akad. Wien (2a) 135, 29–32 227

1926, 3

— Über ametrische Mannigfaltigkeiten und quadratische Differentialformen mit verschwindender Diskriminante. J. D. M. V. 35, 280–294 227

1928, 1

— Über die Tangentialräume ametrischer Mannigfaltigkeiten. Math. Zeitschr. 29, 87–95 227

1932, 1
LENSE, J.: Über die Ableitungsgleichungen der ametrischen Mannigfaltigkeiten. Math. Zeitschr. 34, 721–736 . . . . *227*

1936, 1
— Über Kurven mit isotropen Normalen. Math. Ann. 112, 139–154 . . . *227*

1936, 2
— Über vollisotrope Flächen. Monatsh. f. Math. u. Phys. 43, 177–186 . . . *227*

1939, 1
— Über isotrope Mannigfaltigkeiten. Math. Ann. 116, 297–309 . . . . . *227*

1939, 2
— Beiträge zur Theorie der isotropen Mannigfaltigkeiten. Monatsh. Math. Phys. 48, 121–128 . . . . *227*

1940, 1
— Über die Ableitungsgleichungen einer Mannigfaltigkeit im mehrdimensionalen komplexen euklidischen Raum. Math. Zeitschr. 47, 78–84 *227, 269, 278*

1940, 2
— Längentreue Abbildung, isotrope Mannigfaltigkeiten vom Rang Null, Einbettungssatz. J. D. M. V. 50, 1–6 . . . . *227*

1944, 1
— Über einige Determinanten aus der Theorie der mehrdimensionalen Mannigfaltigkeiten. Math. Ann. 119, 216–220 . . . . *269*

LERAY, J. (see E. CARTAN).

1896, 1
LEVI-CIVITA, T. (see also G. RICCI): Sulle trasformazioni delle equazioni dinamiche. Ann. d. Mat. (II) 24, 255–300 . . . . *296*

1917, 1
— Nozione di parallelismo in una varietà qualunque e consequente specificazione geometrica della curvature Riemanniana. Rend. Palermo 42, 173 to 205 . . . . *122, 123*

1925, 1
*— Lezioni di calcolo differenziale assoluto. Stock, Roma *244*

1926, 1
— Sur l'écart géodésique. Math. Ann. 97, 291–320 . . . . *352*

1940, 1
— Formula di Green e di Stokes. Univ. Nac Tucumán Révista 1, 23–33 *97*

1935, 1
LEVINE, J. (see also T. Y. THOMAS): Conformal-affine connections. Proc. Nat. Acad. 21, 165–167 . . . . *316*

1936, 1
— Some new complete sets of identities for affine and metric spaces. Bull. Am. Math. Soc. 41, 497–501 . . . . *144, 147*

1937, 1
— Groups of motions in conformally flat spaces I. Bull. Am. Math. Soc. 42, 418–432 . . . . *346*

1939, 1
LEVINE, J.: Groups of motions in conformally flat spaces II. Bull. Am. Math. Soc. 45, 766–773 . . . 34

1942, 1
— A replacement theorem for conformal tensor invariants. Tohoku Math. J. 49, 69–86 . . . 16

1948, 1
— Invariant characterizations of two dimensional affine and metric spaces. Duke Math. J. 15, 69–77 . . . 133, 32

1949, 1
— Fields of parallel vectors in projectively flat spaces. Duke Math. J. 16, 23–32 . . . 291, 32

1950, 1
— Fields of parallel vectors in conformally flat spaces. Duke Math. J. 17, 15–20 . . . 291, 30

1950, 2
— Classification of collineations in projectively and affinely connected spaces of two dimensions. Ann. of Math. 52, 465–477 . . . 34

1951, 1
— Collineations in Weyl spaces of two dimensions. Proc. Am. Math. Soc. 2, 264–269 . . . 34

1951, 2
— Motions in linearly connected two-dimensional spaces. Proc. Am. Math. Soc. 2, 932–941 . . . 34

1925, 1
LEVY, H.: Ricci's coefficients of rotation. Bull. Am. Math. Soc. 31, 142–145 17

1925, 2
— Symmetric tensors of the second order whose covariant derivatives vanish. Ann. of Math. (2) 27, 91–98 . . . 285, 296, 36

1926, 1
— Forma canonica dei ds^2 per iquali si annulano i simboli di Riemann a cinque indici. Rend. d. Linc. (6) 3, 67–69 . . . 16

1926, 2
— Sopra alcune proprietà degli spazi per i quali si annullano i simboli di Riemann a cinque indici. Rend. d. Linc. (6) 3, 126–129 . . . 16

1930, 1
— Normal coordinates in the geometry of paths. Proc. Nat. Acad. 16, 492 to 496 . . . 15

1870, 1
LÉVY, M.: Mémoire sur les coordonnées curvilignes orthogonales. Journ. de l'Ec. Polyt. 26, 157–200 . . . 24

1941, 1
LIBER, A. E.: The first algebraic integrals of the equations of geodesy. C. R. Acad. Sci. URSS. (N.S.) 31, 840–841 . . . 29

1947, 1

Liber, A. E.: On the immersion of riemannian spaces of constant curvature in one another. C. R. Acad. Sci. URSS. 55, 291–293 . . . . *269*

1949, 1

— On spaces of linear affine connection with one-parameter holonomy groups. Akad. Nauk SSSR. 66, 1045–1046 (russian) *367*

1950, 1

— On the classification of an affine connection in a space of two dimensions. Mat. Sbornik 27, 249–266 (russian) *123*

1951, 1

— On comitants of geometric differential objects. Akad. Nauk SSSR. 80, 529–532 . . . . *67*

1941, 1

Lichnerowicz, A. (see also A. Borel): Les espaces à connexion semimétrique et la mécanique. C. R. 212, 328–331 . . . . *133*

1944, 1

— Sur les espaces riemanniens complètement harmoniques. Bull. Soc. Math. France 72, 146–168 . . . . . . *285, 286, 287, 382, 383, 384, 385, 386, 387, 388*

1944, 2

— Sur une inégalité relative aux espaces riemanniens complètement harmoniques. C. R. 218, 436–437 . . . . *387*

1944, 3

— Sur les espaces riemanniens complètement harmoniques. C. R. 218, 493 to 495 . . . . *388*

1945, 1

— and A. G. Walker: Sur les espaces riemanniens harmoniques de type hyperbolique normal. C. R. 221, 394–396 . . . . *386*

1950, 1

— Dérivation covariante et nombres de Betti. C. R. 230, 1248–1250 . . . *367*

1950, 2

— Sur les variétés riemanniennes admettant une forme quadratique extérieure à dérivée covariante nulle. C. R. 231, 1413–1415 . . . . *367*

1950, 3

*— Eléments de calcul tensoriel. Colin, Paris. 216 pp.

1951, 1

— Formes à dérivée covariante nulle sur une variété riemannienne. C. R. 232, 146–147 . . . . *367*

1951, 2

— Sur les variétés riemanniennes admettant une forme à dérivée covariante nulle. C. R. 232, 677–679 . . . . *367*

1953, 1

— Compatibilité des équations de la thèorie unitaire d'Einstein-Schrödinger. C. R. 237, 1383–1386 . . . . *179*

1953, 2

— Equations de Laplace et espaces harmoniques Premier colloque sur les èquations aux derivées partielles Louvain 17–19 Decembre 1953. 9–23 *386*

1872, 1

LIE, S.: Komplexe, insbesondere Linien und Kugelkomplexe mit Anwendung auf die Theorie partieller Differentialgleichungen. Math. Ann. 5, 145–246 *312*

1879, 1

— Über Minimalflächen. Math. Ann. 14, 331–416. Gesamm. Abh. II 1, 122 to 218 . . . . *274*

1888, 1

*— and F. ENGEL: Theorie der Transformationsgruppen, Vol. I. Teubner. *185, 205, 214, 215, 218, 220*

1890, 1

— Über die Grundlagen der Geometrie I, II. Ber. Sächs. Ges. d. Wiss. Leipzig 42, 284–321; 355–418 . . . . *41*

1922, 1

LIPKA, J.: Sui sistemi E nel calcolo differenziale assoluto. Rend. d. Linc. 31^{I}, 242–245 . . . . *51*

1869, 1

LIPSCHITZ, R.: Untersuchungen in Betreff der ganzen homogenen Funktionen von n Differentialen. Journ. f. d. r. u. a. Math. 70, 71–102; 72, 1–56 (1870) Excerpt in Monatsber. Akad. Berlin 1869, 44–53 . . . . *143*

1874, 1

— Extrait de six mémoires dans le journal de mathématiques de Borchardt. Bull. d. Sc. Math. 4, 97–110; 142–157; 212–224; 297–307; 308–320. . . . *143*

1874, 2

— Ausdehnung der Theorie der Minimalflächen. Journ. f. d. r. u. a. Math. 78, 1–45; cf. Monatsber. Akad. Berlin 1872, 361–367 . . . . *274*

1944, 1

LITTLEWOOD, D. E.: Invariant theory, tensors and group characters. Philos. Trans. Roy. Soc. London 239, 305–365 . . . . *15*

1950, 1

*— The theory of group characters and matrix representation of groups. Oxford. 310 pp. *15*

1948, 1

LONGO, C.: Sopra una classe di varietà che ammettono varietà subordinate quasi asintotiche. Atti Accad. Naz. Linc. 5, 19–21 . . . . *285*

1936, 1

LOPCHITZ, A. M.: Sugli spazi riemanniani contenenti un campo di giaciture parallele. Proc. Sem. Vector a Tensor. Moscow 2–3, 327–333. . . . . . *322*

1950, 1

— On the theory of a hypersurface in $(n+1)$-dimensional equiaffine space. Sem. Vektor Tensor An. 8, 273–285 (russian) *269*

1950, 2

— On the theory of a surface of n dimensions in an equicentroaffine space of $n+2$ dimensions. Sem. Vektor Tensor An. 8, 286–295 (russian) *269*

1934, 1

*LOVELL, C. A.: On certain associated metric spaces. Thesis Philadelphia. 19 pp. *148, 289*

1951, 1
LYRA, G.: Über eine Modifikation der Riemannschen Geometrie. Math. Zeitschr. 54, 52–64 . . . . *134*

1939, 1
MAEDA, J.: Einige Sätze in der Kurventheorie im n-dimensionalen affinen Raume. Tôhoku. Math. J. 45, 130–133 . . . . *232*

1940, 1
— Sur la podaire d'une hypersurface dans l'espace euclidien à n dimensions. Tôhoku. Math. J. 46, 360–384 . . . . *242*

MARTIN, W. F. (see S. BOCHNER).

1943, 1
MASINI VENTURELLI, L.: Sopra i ds^2 di Liouville di classe uno. Ist. Veneto, Cl. Sci. Mat. Nat. 102, 145–163 . . . . *268*

1943, 1
— Le ipersuperficie di rotazione del tipo di Liouville. Ist. Veneto, Cl. Sci. Mat. Nat. 102, 323–327 . . . . *268*

1950, 1, 2, 3
MATSUMOTO, M.: Riemann spaces of class two and their algebraic characterization I, II, III. J. Math. Soc. Japan 2, 67–76; 77–86; 87–92 . . . . . *269*

1951, 1
— On the special Riemann spaces of class two. Mem. Coll. Univ. Kyoto Math. 26, 149–157 . . . . *269*

1951, 2
— Affinely connected spaces of class one. Mem. Coll. Sci. Univ. Kyoto. Math. 26, 235–249 . . . . *269*

1951, 3
— Conformally flat Riemann spaces of class one. J. Math. Soc. Japan 3, 306 to 309 . . . . *305*

1952, 1
— The class number of embedding of the space with projective connection. J. Math. Soc. Japan 4, 37–58 . . . . *269*

1930, 1
MATTIOLI, G.: Su le ennuple di congruenze ortogonali a rotazioni costanti. Atti Istituto Veneto 89, 97–105 . . . . *171*

1930, 2
— Sulla determinazione della varietà riemanniane che ammettono gruppi semplicemente transitivi di movimenti. Rend. d. Linc. (6) 11, 369–371 *171*

1930, 3
— Sopra certi sistemi coordinati associati ad un ennupla di congruenze e applicazioni. Rend. Semenario Padova 1, 110–132 . . . . *155*

1888, 1
MAURER, L.: Über gewisse Invariantensysteme. Bayr. Akad. d. Wiss. 18, 103 to 150 . . . . *197*

1939, 1
MAXIA, A.: Varietà anolonome immerse in una varietà a connessione affine (X_n^{n-1} in E_n affine). Mem. Soc. Roy. Sci. Lett. Bohême. 18 pp. *253*

1939, 2

MAXIA, A.: Varietà anolonome associate ed una trasformazione dualistica. Ist. Lombardo Rend. 72, 437–454 . . . . *253*

1940, 1

— Geometria proiettiva differenziale dei complessi anolonomi dirette. Atti Sec. Congresso Un. Mat. Italia. Bologna 1940, 303–304 . . . . *253*

1928, 1

MAYER, W. (see also C. BURSTIN, A. DUSCHEK): Über das vollständige Formensystem der F_1 im R_n. Mon. f. Math. u. Phys. 35, 87–110 . . *232, 263, 277, 278*

1931, 1

— Beitrag zur Differentialgeometrie, l-dimensionaler Mannigfaltigkeiten die in euklidischen Räumen eingebettet sind. Sitz. Preuß. Akad. 27, 606–615 *277*

1933, 1

— Zum Tensorkalkül in Vektorräumen Riemannscher Mannigfaltigkeiten. Mon. Math. u. Phys. 40, 283–293 . . . . *269*

1935, 1

— and T. Y. THOMAS: Foundations of the theory of Lie groups. Ann. of Math. 36, 770–822 . . . . *185*

1935, 2

— Die Differentialgeometrie der Untermannigfaltigkeiten des R_n konstanter Krümmung. Trans. Am. Math. Soc. 38, 267–309 . . *232, 263, 269, 271, 277, 278*

1937, 1

— and T. Y. THOMAS: Fields of parallel vectors in non-analytic manifolds in the large. Comp. Math. 5, 198–207 . . . . *125*

1911, 1

*MEYER, W. F.: Über die Theorie benachbarter Geraden und einen verallgemeinerten Krümmungsbegriff. Teubner, Leipzig. 152 pp. *230*

1930, 1

MICHAL, A. D. (see also T. Y. THOMAS): Geodesic coordinates of order r. Bull. Amer. Math. Soc. 36, 541–546 . . . . *155*

1930, 2

— Scalar extensions of an orthogonal ennuple of vectors. Am. Math. Monthly 37, 529–533 . . . . *155*

1931, 1

— Notes on scalar extensions of tensors and properties of local coordinates. Proc. Nat. Acad. 17, 132–136 . . . . *155*

1931, 2

— An operation that generates absolute scalar differential invariants from tensors. Tôhoku Math. Journ. 34, 71–77 . . . . *155*

1945, 1

— First order differentials of functions with arguments and values in topological Abelian groups. Revista Ci. Lima 47, 389–422 . . . . *340*

1947, 1

*— Matrix and tensor calculus with applications to mechanics, elasticity and aeronautics. Wiley and Sons, New York. XIII + 132 pp. Chapman and Hall, London.

1941, 1

MIKAMI, M. (see also H. HOMBU): On the theory of non-linear direction displacements. Jap. J. Math. 17, 541–568 . . . . *123*

1941, 2

— On parabolas in the generalized spaces. Jap. J. Math. 17, 185–200 . . . *300*

1949, 1

— Projectively connected space and curves of the second order. Sūgaku Math. 1, 274–286 (japanese) *322*

1939, 1

MIKAN, M.: Die Räume von R. König, die den eingebetteten Mannigfaltigkeiten zugeordnet sind. Acad. Tchèque Sci. Bull. Math. Nat. 40, 218–221 . . . *253*

1950, 1

MIRA FERNANDES, A. DE: Le geodetiche degli spazi unitarii. Univ. Lisboa Revista Fac. Ci. A. Ci. Mat. 1, 173–186 . . . . *415*

1952, 1

MISHRA, R. S.: Sur certaines courbes appartenant à un sous espace d'un espace riemannien. Bull. Sci. Math. (2) 76, 77–84 . . . . *250*

1938, 1

MODESITT, V.: Conformal transformations between riemannian spaces. Amer. Journ. of Math. (2) 60, 325–336 . . . . *315*

1943, 1

MOGHE, D. N.: On isotropic spaces. J. Univ. Bombay (N.S.) 12, 1–3 . . . . *228*

1950, 1

MOGI, J.: A remark on recurrent curvature spaces. Kodai Math. Sem. Rep. 73–74 . . . . *424*

1930, 1

MOISIL, G. C.: Sur les propriétés affines et conformes des variétés nonholonomes. Bull. Cern. 4, 15–21 . . . . *253*

1940, 1

— Sur les géodésiques des espaces de Riemann singuliers. Bull. Math. Soc. Roumaine Sci. 42, 33–52 . . . . *133, 227, 270*

MOORE, C. L. E. (see E. B. WILSON).

1950, 1

MOREAU, J. J.: La symétrie de révolution en calcul tensoriel. C. R. 231, 1028 to 1029 . . . . *35*

1934, 1

MORINAGA, K.: An extension of the parallelism in X_n^{n-m} in X_n. Journ. Hiroshima Univ. (A) 4, 34, 177–183 . . . . *138, 140, 253*

1934, 2

— An extension of parallel displacements by matrices. Journ. of Sc. Hiroshima A 2, 127–140 . . . . *123, 131, 340*

1934, 3

— On parallel displacement in an n-dimensional space to which N-dimensional vectorspaces are attached. Journ. of Sc. Hiroshima A 5, 15–30 . . . . *123*

1899, 1
*MUTH, P.: Theorie und Anwendung der Elementarteiler. Teubner, Leipzig. 236 pp. 31

1939, 1
MUTÔ, Y. (see also K. YANO): On the equations of circles in a space with a conformal connection. Tensor 2, 50–52 (japanese) 315

1939, 2
— On some properties of hypersurfaces in a conformally connected manifold. Proc. Phys. Math. Soc. Japan 21, 615–625 315

1940, 1
— On some properties of umbilical points of hypersurfaces. Proc. Imp. Acad. Tokyo 16, 79–82 251

1940, 2
— On a family of curves on a hypersurface in the riemannian space. Tensor 3, 39–40 (japanese) 242

1940, 3
— On some properties of subspaces in a conformally connected manifold. Proc. Phys. Math. Soc. Japan 22, 621–636 315

1942, 1
— Theory of subspaces in a space with a conformal connection. Tensor 5, 31 to 46 (japanese) 315

1951, 1
— Some properties of a riemannian space admitting a simply transitive group of translations. Tôhoku Math. J. 2, 205–213 346

1924, 1
MYLLER, A.: Direzioni concorrenti sopra una superficie spiccanti dai punti di una curva. Rend. d. Lincei 5, 33, 339–341 322

1928, 1
— Directions concourantes dans une variété métrique à n dimensions. Bull. Soc. Math. 56, 1–6 322

1949, 1
NAGABHUSHANAM, K.: An extension of a theorem of Wintner. Proc. Nat. Inst. Sc. India 15, 35–36 89

1944, 1
NAKAE, T.: Das geometrische Objekt. Tensor 7, 1–5 (japanese) 67

1928, 1
NALLI, P.: Sopra le coordinate geodetiche. Rend. d. Linc. (6) 7, 195–198 155

1938, 1
NARLIKAR, V. V.: A note on riemannian tensors. Phil. Mag. VII, 26, 390–394 51

1948, 1
— and A. PRASAD: Canonical coordinates in general relativity. Bull. Calcutta Math. Soc. 40, 123–128 155

NEUMANN, J. v. (see O. VEBLEN).

1927, 1
NEWMAN, M. H. A.: A gauge-invariant tensorcalculus. Proc. Roy. Soc. 116, 603 to 623 135

1918, 1

NOETHER, E.: Invarianten beliebiger Differentialausdrücke. Göttinger Nachr. 37–44 . . . . *165*

1945, 1

NORDEN, A.: On pairs of conjugate parallel translations in n-dimensional spaces. C. R. Acad. Sc. URSS. 49, 625–628 . . . . *123, 132*

1945, 2

— On an interpretation of a space with degenerate metric. Akad. Nauk SSSR. 50, 57–60 (russian) *133*

1945, 3

— The affine connectivity on the surfaces of a projective and conformal space. C. R. Acad. Sc. URSS. 48, 539–541 . . . . *296, 301, 315*

1947, 1

— La connexion affine sur les surfaces de l'espace projectif. Rec. Math. 20, 263–281 (russian, french summary) *296, 301*

1948, 1

— The riemannian metric on surfaces of a projective space. Akad. Nauk SSSR. 60, 345–347 (russian) *296, 301*

1949, 1

— On the inner geometries of surfaces in a projective space II. Trudy Sem. Vektor, Tensor. An. 7, 31–64 (russian) *296, 301*

1950, 1

— On conjugate connections. Trudy Sem. Vektor, Tensor. An. 8, 93–105 (russian) *123*

1950, 2

*— Prostranstva affinnoï svyaznosti. (Affinely connected spaces.) Gosudarstv. Izdat. Tehn.-Teor. Lit., Moscow-Leningrad. 463 pp.

1944, 1

NORDON, J.: Sur la solution élémentaire d'une équation aux dérivées partielles associeé à un espace riemannien harmonique. C. R. 219, 436–438 . . . . *383*

1950, 1

NOZISKA, F.: Le vecteur affinonormal et la connexion de l'hypersurface dans l'espace affin. Časopis Pěst. Mat. Tys. 75, 179–209 (czech summary) . . *235*

1951, 1

NYENHUIS, A.: X_{n-1} forming sets of eigenvectors. Proc. Kon. Ned. Akad. Amsterdam 54 = Indagationes Math. 13, 200–212 . . . . *66, 248*

1951, 2

— An application of anholonomic coordinates. Math. Centrum Amsterdam. Rapp. Z. W. 1951, 017. 6.pp. *66*

1952, 1

*— Theory of the geometric object. Thesis. Univ. Amsterdam. 238 pp. (dutch summary) *64, 65, 67, 68, 102, 105, 107, 108, 109, 110, 114, 131, 340, 351, 363, 366*

1953, 1

— On the holonomy groups of linear connections IA, IB. Proc. Kon. Ned. Akad. Amsterdam = Indagationes Math. 15, 233–249 . . . . *367*

1936, 1

OBERTI, G.: Su una decomposizione carratteristica dei tensori doppi. Ist. Lombardo, Rend. II, 69, 397–402 . . . . *15*

OKADA, Y. (see H. HOMBU).

1949, 1, 2

OTSUKI, T.: On some 4-dimensional riemannian spaces I, II. Mem. Fac. Sc. Kyūsyū Univ. 4, 193–212; 213–217 . . . . *367*

1950, 1

— On projectively connected spaces whose groups of holonomy fix a hyperquadric. J. Math. Soc. Japan 1, 251–263 . . . . *128, 367*

1950, 2

— On the spaces with normal conformal connexions and some imbedding problem of riemannian spaces I. Tôhoku Math. J. 1, 194–224 . . . . *269, 315, 367*

1950, 3

— Classification of 4-dimensional analytic Weyl spaces by their holonomy groups. Mem. Fac. Sci. Kyūsyū Univ. A 5, 1–39 . . . . *367*

1951, 1

— On the spaces with normal conformal connexions and some imbedding problems of riemannian spaces II. Tôhoku Math. J. 2, 220–274. . . . . . *269, 367*

1952, 1

— On the spaces with normal projective connexions and some imbedding problems of riemannian spaces. Math. J. Okayama Univ. 1, 69–98. . . . . . *367*

1889, 1

PADOVA, E.: Sulle deformazioni infinitesimi. Rend. d. Linc. (IV) 5^{I}, 174–178 *146*

1884, 1

PAINLEVÉ, P.: Mémoire sur la transformations des équations de la dynamique. Journ. de Math. (4) 10, 5–92 . . . . *310*

1941, 1

PALATINI, A.: Sopra le varietà di classe uno. Ist. Lombardo, Mat. Nat. Rend. 5, 293–304 . . . . *268*

1943, 1

PANTAZI, A.: Sur la déformation projective des surfaces non holonomes de l'espace E_3. Bull. Math. Soc. Roumaine 45, 33–47 . . . . *253*

1947, 1

— Sur les variétés nonholonomes à asymptotiques rectilignes et leur déformation projective. Bull. Math. Soc. Roumaine 48, 3–26 . . . . *253*

1946, 1

PAPY, G.: Sur les formes cubiques alternées de rang inférieur à 8. Bull. Soc. Roy. Liège 15, 77–83 . . . . *36*

1946, 2

— and F. TOURNAY: Classification des formes cubiques alternées à six indéterminées par rapport au groupe canonique de degré trois. Bull. Soc. Roy. Sci. Liège 15, 513–523 . . . . *36*

PASTORI, M. (see B. FINZI).

1951, 1

PATTERSON, E. M.: On symmetric recurrent tensors of the second order. Quart. J. Math. Oxford 2, 151–158 322

1951, 2

— An existence theorem on simply harmonic spaces. J. London Math. Soc. 26, 238–240 383

1952, 1

— Simply harmonic Riemann extensions. J. London Math. Soc. 27, 102–107 381, 383

1952, 2

— Some theorems on Ricci-recurrent spaces. J. London Math. Soc. 27, 287 to 295 421

1952, 3

— and A. G. WALKER: Riemann extensions. Quart. J. Math. Oxford ser. (2) 3, 19–28 228

1937, 1

PAUC, C.: Etude géométrique d'un faisceau de transformations infinitésimales. C. R. 204, 1027–1029; 205, 101–102 253

1938, 1

— Extension aux variétés non holonomes V_n^{n-1}, de quelques propriétés des surfaces et des V_3^2. Atti Acad. Mat. Nat. (6) 27, 155–165 253

1938, 2

— Interprétation géométrique dans les variétés non holonomes des théories d'intégration des systèmes d'équations de Pfaff. C. R. 206, 885–888 253

1935, 1

PEARS, L. R.: Bertrand curves in riemannian space. Journ. London Math. Soc. 10, 180–183 343

1946, 1

PENSOV, Y. E.: Classification of differential geometric objects of the class v with one component. C. R. Acad. Sc. URSS. 54, 563–566 67

1948, 1

— The classification of geometric differential objects with two components. Akad. Nauk SSSR. 80, 537–540 67

1950, 1

— The classification of continuous pseudo-groups of Lie transformations in X_2 according to their characteristic objects. Trudy Sem. Vektor, Tenzor, An. 8, 382–413 (russian) 64

1952, 1

PEREMANS, W., H. J. A. DUPARC and C. G. LEKKERKERKER: A property of positive matrices. Kon. Ned. Akad. Amsterdam Proc. 55 = Indagationes Math. 14, 24–27 39

1934, 1

PEREPELKINE, D.: Sur les directions de courbure d'une V_m dans R_n. C. R. 199, 1089–1091 269

1935, 1

PEREPELKINE, D.: Sur les variétés paralleles dans un espace euclidien (ou riemannien). C. R. Ac. Sc. URSS. 593–598; french text 596–598 . . . . *269*

1935, 2

— Sur la transformation conforme et la courbure riemannienne normale intrinsèque d'une V_m dans V_n. C. R. 200, 513–515 . . . . *315*

1936, 1

— Sur la courbure et les espaces normaux d'une V_m dans R_n. Rec. Math. Moscow 42, 81–120 . . . . *269, 272, 275*

PESCHL, E. (see R. KÖNIG).

1948, 1

PETKANTSCHIN, B.: Allgemeine isotrope Regelscharen im euklidischen Raum. Ann. Univ. Sofia, Math. Phys. 44, 357–399 (bulgarian, german summary) *228*

1943, 1

PETRESCU, S. (see also K. YANO): Sur les invariants des hypersurfaces non holonomes V_n^{n-1}. Disquisit. Math. Phys. 3, 3–84 . . . . *253*

1943, 2

— Sur les hypersurfaces non holonomes V_{2p+1}^{2p}. Acad. Roum. Bull. Sect. Sc. 25, 249–253 . . . . *253*

1943, 3

— Sur les hypersurfaces non holonomes V_{2p+2}^{2p+1}. Acad. Roum. Bull. Sect. Sc. 25, 313–317 . . . . *253*

1944, 1

— Quelques propriétés des espaces non honolomes A_n^m, considérés dans un espace A_n, à connexion affine, admettant un parallelisme absolu. Bull. Math. Soc. Roumaine 46, 145–154 . . . . *253*

1945, 1

— Sur un problème d'existence des espaces non holonomes A_n^m totalement géodésiques dans un espace A_n à connexion affine symétrique. Disquisit. Math. Phys. 4, 117–130. . . . . *253*

1946, 1

— Quelques propriétés conformes des sous espaces V_l, dans un V_m de V_n. Bull. Sc. Ecole Polytech. Timisoara 12, 167–174 . . . . *315*

1948, 1

— Sur quelques propriétés conformes des espaces non holonomes V_n^m. Mathematica, Timisoara 23, 108–122 . . . . *253, 269, 315*

1948, 1

PETROV, A. Z.: On the curvature of Riemann space. Akad. Nauk SSSR. (N. S.) 61, 211–214 (russian). *138*

1886, 1

DELL-PEZZO, P.: Sugli spazii tangenti ad una superficie o ad una varietà immersa in uno spazio di più dimensioni. Rend. Acc. Napoli 25, 176–180 . . *275*

1932, 1

PINL, M.: Quasimetrik auf totalisotropen Flächen I, II. Proc. Kon. Akad. Amsterdam 35, 1181–1188; 36, 550–557 . . . . *227*

1932, 2

PINL, M.: Über Kurven mit isotropen Schiegräumen im euklidischen Raum von n Dimensionen. Mon. f. Math. u. Phys. 39, 157–172 227

1935, 1

— Quasimetrik auf totalisotropen Flächen III. Proc. Kon. Ned. Akad. Amsterdam 38, 171–180 227

1937, 1

— Zur dualistischen Theorie isotroper und verwandter Kurven im euklidischen Raum von n Dimensionen. Mon. f. Math. u. Phys. 44, 1–12 227

1937, 2

— Zur Existenztheorie und Klassifikation totalisotroper Flächen. Compositio Math. 5, 208–236 227

1940, 1

— W-Projektionen totalisotroper Flächen II. Časopis Pěst. Mat. Fys. 69, 23–35 (german, czech summary) 227

1940, 2

— Zur Theorie der halbisotropen Flächen in R_4. J. D. M. V. 50, 65–78 . . . 227

1940, 3

— Zur dualistischen Theorie isotroper Kurven. Mon. f. Math. u. Phys. 49, 261–278 227

1947, 1

— Über Flächen mit isotropem mittlerem Krümmungsvektor. Ber. Math. Tagung Tübingen 121–122 227

1948, 1

— Über Flächen mit isotropem mittlerem Krümmungsvektor. Monatsh. Math. 52, 301–310 227

1949, 1

— Binäre orthogonale Matrizen und integrallose Darstellungen isotroper Kurven. Math. Ann. 121, 1–20 227

1951, 1

— Geodesic coordinates and rest systems for general linear connections. Duke Math. J. 18, 557–562 . . . . 127, 155

1887, 1

POINCARÉ, H.: Sur les résidus des intégrales doubles. Acta mathematica 6, 321–380 97

1895, 1

— Analysis situs. Journ. de l'Éc. Polyt. II, 1, 1–123 97

PRASAD, A. (see V. V. NARLIKAR).

1953, 1

QUADE, W.: Beweis eines Satzes über positive Matrizen. Proc. Kon. Nederl. Akad. Amsterdam 56 (= Indag. Math. 15), 50–51 39

1949, 1

RACAH, G.: On the decomposition of tensors by contraction. Rev. Modern Physics 21, 494–496 15

1930, 1

RACHEVSKY, P.: Sur les espaces sous-projectifs. C. R. 191, 547–548 . . . *317*

1933, 1

— Caractères tensoriels de l'espace sous-projectif. Abh. des Seminars für Vektor und Tensoranalysis, Moskau 1, 126–140 . . . . *317, 329*

1949, 1

— Galois theory in fields of geometric objects. Trudy Sem. Vektor, Tenzor. An. 7, 167–186 (russian) *105*

1950, 1

— Symmetric spaces of affine connection with torsion I. Trudy Sem. Vektor, Tenzor An. 8, 82–92 (russian) *163*

1950, 2

*— Kurs differencial'noĭ geometrii (A course of differential geometry). 3rd ed. Gosudarstv. Izdat. Tehn.-Teor. Lit. Moscow-Leningrad. 428 pp.

1954, 1

RENAUDIE, J.: Un théorème sur les espaces harmoniques. C. R. 238, 199–201 *388*

1907, 1

*REICHEL, W.: Über trilineare alternierende Formen in 6 und 7 Veränderlichen. Diss. Greifswald. 59 pp. *36*

1950, 1

*RHAM, G. DE, and K. KODAIRA: Harmonic integrals. Inst. for Adv. Study. Princeton. N. J. 114 pp. *27*

1884, 1

RICCI, G.: Principii di una teoria delle forme differenziali quadratiche. Ann. di Mat. (II) 12, 135–167 . . . . *121, 143*

1886, 1

— Sui parametri e gli invarianti delle forme quadratiche differentiale. Ann. di Mat. (II) 14, 1–11 . . . . *121*

1886, 2

— Sui sistemi di integrali independenti di una equazione lineare ed omogenea a derivate parziali di 1° ordine. Rend. d. Linc. (IV) 2^{II}, 119–122, 190–194. Summary of 1887, 2 . . . . *121*

1887, 1

— Sulla derivazione covariante ad una forma quadratica differenziale. Rend. d. Linc. (IV) 3^{I}, 15–18 . . . . *121*

1887, 2

— Sui sistemi di integrali independenti di una equazione lineare ed omogenea a derivate parziali di 1° ordine. Ann. di Mat. (2) 15, 127–159 . . . . . . *121*

1888, 1

— Sulla classificazione delle forme differenziali quadratiche. Rend. d. Linc. (4) 41, 203–207 . . . . *121, 263, 278*

1888, 2

*— Delle derivazione covariante e contravariante e del loro uso nell'Analisi applicata. Studie editi della Università di Padova a commemorare l'ottavo Centenario delle origina della Università di Bologna, III, Padova. 30 pp. *121*

1889, 1

RICCI, G.: Sopra certi sistemi di funzioni. Rend. d. Linc. (IV) 5^1, 112–118 . . . *121*

1889, 2

— Di un punto della teoria delle forme differenziali quadratiche ternarie. Rend. d. Linc. (IV) 5^1, 643–651 . . . *121*

1892, 1

— Résumé de quelques travaux sur les systèmes variables des fonctions associés à une forme différentielle quadratique. Bull. Sc. Math. II 16, 167–189 . . *121*

1893, 1

— Di alcune applicazioni del calcolo differenziale assoluto alla teoria delle forme differenziale quadratiche binarie e dei sistemi a due variabili. Atti R. Istit. Veneto (VII) 4, 1336–1364 . . . *121*

1893, 2

— Dei sistemi di coordinate atti a ridurre la espressione del quadrato dell'elemento lineare di una superficie alla forma $ds^2 = (U + V)(du^2 + dv^2)$. Rend. d. Linc. (V) 2^1, 73–81 . . . *121*

1894, 1

— Sulla teoria delle linee geodetiche e dei sistemi isotermi di Liouville. Atti R. Istituto Veneto (VII) 5 (= 53), 643–681 . . . *121*

1894, 2

— Sulla teoria intrinseca delle superficie ed in ispecie di quelle di 2^0 grado. Atti R. Istituto Veneto (VII) 5 (= 53), 445–488 . . . *121*

1895, 1

— Dei sistemi di congruenze ortogonali in una varietà qualunque. Mem. Acc. Linc. (5) 2, 276–322 . . . *121, 171, 228, 244, 245*

1895, 2

— Sulla teoria degli iperspazi. Rend. d. Linc. (V) 4^{II}, 232–237 . . . *121*

1897, 1

*— Lezioni sulla delle superficie. Frat. Drucker, Verona-Padova. VIII + 416 pp. (Lithographic 1898) . . . *122*

1897, 2

— Sur les systèmes complètement orthogonaux dans un espace quelconque. C. R. 125, 810–811 . . . *122*

1897, 3

— Del teorema di Stokes in uno spazio qualunque a tre dimensioni ed in coordinate generale. Atti Istituto Veneto (7) 8 (= 56), 1536–1539, 1896–1897 *122*

1898, 1

— Sur les groupes continus de mouvements d'une variété quelconque à trois dimensions. C. R. 127, 344–346 . . . *122, 249*

1898, 2

— Sur les groupes continus de mouvements d'une variété quelconque. C. R. 127, 360–361 . . . *122, 249*

1901, 1

*— and T. LEVI-CIVITA: Méthodes de calcul différentiel absolu et leurs applications. Math. Ann. 54, 125–201, Berichtigungen 608. Reprinted without corrections. Collection de monographies scientifiques étrangères, publiée sous la direction M. G. Juvet, No 5. Blanchard, Paris *122, 165, 171, 244, 249*

1902, 1

RICCI, G.: Sui gruppi continui di movimenti in una varietà qualunque a tre dimensioni. Mem. d. Soc. Ital. (III) 12, 61–92 . . . . *122*

1902, 2

— Formole fondamentali nella teoria generale di varietà e della loro curvatura. Rend. d. Linc. (V) 11^{I}, 355–362 . . . . *122, 253*

1903, 1

— Sulle superficie geodetiche in una varietà qualunque e in particolare nella varietà a tre dimensioni. Rend. d. Linc. (V) 12^{I} 409–420 *122, 146, 148, 251, 256*

1904, 1

— Direzioni e invarianti principali di una varietà qualunque. Atti R. Istituto Veneto (VIII) 6 (= 63), 1233–1239 . . . . *122, 146*

1905, 1

— Sui gruppi continui di movimenti rigidi negli iperspazzi. Rend. d. Linc. (V) 14^{II}, 487–491 . . . . *122*

1910, 1

— Sulla determinazione di varietà dotate di proprietà intrinseche date a priori. Rend. d. Linc. (V) 19^{I}, 181–187; 19^{II}, 85–90 . . . . *122*

1910, 2

— Sulla determinazione di varietà che godone di proprietà intrinseche prestabilite. Atti Soc. Ital. 3, 477–480 . . . . *122*

1912, 1

— Di un metodo per la determinazione di un sistema completo di invarianti per un dato sistema di forme. Rend. Palermo 33, 194–200 . . . . *122*

1912, 2

— Della trasformazione delle forme differenziali quadratiche. Rend. d. Linc. (V) 21^{I}, 527–532. . . . . *122*

1918, 1

— Sulle varietà a tre dimensioni dotate di terne principali di congruenze geodetiche. Rend. d. Linc. (V) 27 I, 21–28, 75–87 . . . . *122*

1918, 2

— Della varietà a tre dimensioni con terne ortogonali di congruenze a rotazioni. Rend. d. Linc. (V) 27^{II}, 36–44 . . . . *122*

1928, 1

RICE, L. H.: Couche ranks in the general matrix. Journ. Math. a. Phys. Mass. 7, 93–96 . . . . *20*

1854, 1

*RIEMANN, B.: Über die Hypothesen, welche der Geometrie zugrunde liegen (Habilitationsschrift). Gött. Abh. 13 (1868) 1–20; Ges. Werke, 2. Aufl. (1892), 272–287. Neue Ausgabe mit Erläuterungen von H. WEYL. Springer 1921. 47 pp. *155, 158*

1861, 1

— Commentatio mathematica, qua respondere tentatur quaestioni ab III^ma^ Academia Parisiensi propositae etc. Gesamm. Werke, 2. Aufl. 1892, 391 to 423, mit Anmerkungen von DEDEKIND *143*

1904, 1

RIMINI, G.: Sugli spazi a tre dimensioni che ammettono un gruppo a quattro parametri di movimenti. Ann. R. Scuola Norm. Pisa, 9, 57 pp. *251*

1932, 1

ROBERTSON, H. P. (see also H. WEYL): Groups of motions in spaces admitting absolute parallelism. Ann. of Math. 33, 496–520 . . . . *351*

1949, 1

ROGOVOĬ, M. R.: On the projective differential geometry of nonholonomic surfaces in a three-dimensional space. Akad. Nauk SSSR. 66, 1055–1057 (russian) *253*

1950, 1

— On the projective differential geometry of nonholonomic surfaces in a three-dimensional space. Ukrain. Mat. Žurnal 2, 102–116 (russian) *253*

1940, 1; 1941, 1; 1943, 1

ROSENSON, N.: Sur les espaces riemanniens de classe 1, I, II, III. Bull. Acad. Sc. URSS. Sér. Math. 4, 181–192; 5, 325–351; 7, 253–284 (russian, french summary) *268*

1912, 1

ROTHE, H.: Über Komplexgrößen 2-ter und $(\nu-2)$-ter Stufe in einem Hauptgebiet ν-ter Stufe und die durch sie bestimmten linearen Komplexe. Sitz. Akad. Wien IIa, 121, 1015–1050 . . . . *36*

1921, 1

*— Einführung in die Tensorrechnung. Seidel & Sohn, Wien. 180 pp.

1945, 1

ROZENFEL'D, B. A.: Theory of surfaces in symmetrical spaces. Bull. Acad. Sci. URSS. Math. 9, 371–386 (russian, english summary) *268*

1947, 1

— The metric and affine connection in spaces of planes, spheres or quadrics. Akad. Nauk SSSR. 57, 543–546 . . . . *123, 125*

1947, 2

— Géométrie différentielle des familles de plans à plusieurs dimensions. Bull. Acad. Sci. URSS. Sér. Math. 11, 283–308 (russian, french summary) . . . *253*

1948, 1

— Differential geometry of figures of symmetry. Akad. Nauk SSSR. 59, 1057 to 1060 (russian) *123, 125*

1949, 1

— On unitary and stratified spaces. Trudy Sem. Vektor Tenzor. An. 7, 260 to 275 (russian) *397*

1949, 2

— Unitary-differential geometry of families of K_m in K_n. Mat. Sbornik 24, 53–74 (russian) *412*

1950, 1

— and A. A. ABRAMOV: Spaces with affine connection and symmetric spaces. Uspechi Matem. Nauk 5, 72–147 (russian) *163*

1950, 1

*RUBINOWICZ, W.: Wektory i tensory (vectors and tensors). Monogr. Matem. Tom XXII, Warszawa-Wroclaw. III + 170 pp.

1930, 1

RUSE, H. S. (see also E. T. COPSON): Normal covariant derivatives. Proc. London Math. Soc. (2) 33, 66–76 . . . . *155*

1930, 2

— On the "elementary" solution of Laplace's equation. Proc. Edinb. Math. Soc. (2) 2, 135–139 . . . . *382*

1930, 3

— Generalized solutions of Laplace's equation. Proc. Edinb. Math. Soc. (2) 2, 181–188 . . . . *382*

1931, 1

— On absolute partial differential calculus. Quart. Journ. Math. Oxford, Ser. 2, 190–202 . . . . *68*

1931, 2

— Normal covariant derivatives. Proc. London Math. Soc. (2), 33, 66–76 *155*

1931, 3

— Taylor's theorem in the tensor calculus. Proc. London Math. Soc. (2) 32, 87–92 . . . . *155*

1936, 1

— On the geometry of the electromagnetic field in general relativity. Proc. London Math. Soc. (2) 41, 302–322 . . . . *35, 146*

1939, 1

— Solutions of Laplace's equation in an n-dimensional space of constant curvature. Proc. Edinb. Math. Soc. (2) 6, 24–45 . . . . *382*

1944, 1

— On the line-geometry of the Riemann tensor. Proc. Roy. Soc. Edinburgh 62, 64–73 . . . . *146*

1944, 2

— Sets of vectors in a V_4 defined by the Riemann tensor. J. London Math. Soc. 19, 168–178 . . . . *146*

1945, 1

— The Riemann tensor in a completely harmonic V_4. Proc. Roy. Soc. Edinburgh 62, 156–163 . . . . *385, 386*

1946, 1

— A. G. D. Watson's principal directions for a riemannian V_4. Proc. Edinburgh Math. Soc. 7, 144–152 . . . . *146*

1946, 2

— The five-dimensional geometry of the curvature tensor in a riemannian V_4. Quart. J. Math. Oxford 17, 1–15 . . . . *146*

1946, 3

— On simply harmonic spaces. J. London Math. Soc. 21, 243–247 . . *369, 387*

1948, 1

— The self-polar Riemann complex for a V_4. Proc. London Math. Soc. 50, 75–106 . . . . *146*

1948, 2

RUSE, H. S.: On simply harmonic "kappa-spaces" of four dimensions. Proc. London Math. Soc. 50, 317–329 . . . . *369, 421*

1949, 1

— On parallel fields of planes in a riemannian space. Quart. J. Math. Oxford Ser. 20, 218–234 . . . . *322, 367*

1949, 2

— Three-dimensional spaces of recurrent curvature. Proc. London Math. Soc. 50, 438–446 . . . . *369, 421, 422*

1950, 1

— Parallel planes in a Riemann V_n. Proc. Roy. Soc. Edinburgh 63, 78–92 *322, 367*

1951, 1

— The Riemann complex in a four-dimensional space of recurrent curvature. Proc. London Math. Soc. 53, 13–31 . . . . *421, 423*

1951, 2

— A classification of K^*-spaces. Proc. London Math. Soc. 53, 212–229 *421, 423*

1952, 1

— Simply harmonic affine spaces of symmetric connection. Publ. Math. Debrecen 169–174 . . . . *381*

1948, 1

*RUTHERFORD, D. E.: Substitutional analysis. Edinburgh Univ. Press. 103 pp. *15*

1950, 1

RYŽKOV, V. V.: An imbedding theorem for riemannian geometries of higher order. Akad. Nauk SSSR. 75, 503–506 (russian) *268*

1950, 1

RŽEHNINA, N. F.: On the theory of fields of local curves in X_n. Akad. Nauk SSSR. 72, 461–464 (russian) *232*

SAENZ, A. W. (see V. HLAVATY).

1945, 1

SAMUEL, P.: Sur les tenseurs à dérivées covariantes nulles. C. R. 220, 160–162 *367*

1936, 1

SASAKI, S. (see also K. YANO): Some theorems of conformal transformations of riemannian spaces. Proc. phys. math. Soc. Japan (3) 18, 572–578 *311, 315*

1937, 1

— Contributions to the affine and projective differential geometries of space curves. Jap. J. Math. 13, 473–481 . . . . *302*

1939, 1

— A new proof of the theorem of J. A. Schouten on an umbilic surface in a riemannian space. Tensor 2, 25–29 (japanese) *251, 309*

1939, 2

— On the theory of curves in a curved conformal space. Sc. Rep. Tôhuku Imp. Univ. I, 27, 392–409 . . . . *315*

1939, 3

— On the theory of surfaces in a curved conformal space. Tensor 2, 52–53 (japanese) *315*

1940, 1

SASAKI, S.: On a remarkable property of umbilical hypersurfaces in the geometry of the normal conformal connexion. Sc. Rep. Tôhoku Imp. Univ. 29, 412–422 . . . . *309*

1940, 2

— Geometry of the conformal connexion. Sc. Rep. Tôhoku Imp. Univ. 29, 219–267 . . . . *315*

1940, 3

— On the theory of surfaces in a curved conformal space. Sc. Rep. Tôhoku Imp. Univ. 28, 261–285 . . . . *315*

1941, 1

— On conformal normal coordinates. Sc. Rep. Tôhoku Imp. Univ. 30, 71–80 *155*

1941, 2

— Geometry of conformal connections. Tensor 4, 13–24 (japanese) *315*

1942, 1

— On a relation between a riemannian space which is conformal with Einstein spaces and normal conformally connected spaces whose groups of holonomy fix a point or a hypersphere. Tensor 5, 66–72 (japanese) *312, 367*

1943, 1

— On the spaces with normal conformal connexions whose groups of holonomy fix a point or a hypersphere I. Jap. J. Math. 18, 615–622 . . . . *334, 367*

1943, 2

— On the spaces with normal conformal connexions whose groups of holonomy fix a point or a hypersphere II. Jap. J. Math. 18, 623–633 . . . . *334, 367*

1943, 3

— On the spaces with normal conformal connexion whose groups of holonomy fix a point or a hypersphere III. Jap. J. Math. 18, 791–795 . . . . *334, 367*

1948, 1

*— Geometry of conformal connection. Kawade-shobô, Tokyo. 3 + 3 + 265 pp.. . . . . *315*

1949, 1

— and K. YANO: On the structure of spaces with normal projective connexions whose groups of holonomy fix a hyperquadric or a quadric of $(n-2)$-dimensions. Tôhoku Math. J. 1, 31–39 . . . . *128, 367*

1949, 2

— On the real representation of spaces with hermitian connexion. Sc. Rep. Tôhoku Univ. 33, 53–61 . . . . *397*

1951, 1

— An alternative proof of Liber's theorem. Proc. Japan Acad. 27, 73–80 . . *367*

1951, 2

— On a theorem concerning the homological structure and the holonomy groups of closed orientable symmetric spaces. Proc. Japan Acad. 27, 81 to 85 . . . . *367, 370*

1950, 1

SATÒ, I.: Unitary connected spaces with constant curvature. Bull. Yamag. Univ. 2, 31–32 . . . . *404*

1871, 1

SCHLÄFLI, L.: Nota alla memoria del sig. Beltrami. Ann. di Mat. (2) 5, 170–193 *268*

1930, 1

SCHAPIRO, H.: Sur les espaces sous-projectifs. C. R. 191, 551–552 . . . . . *317*

1933, 1

— Über die Metrik der subprojektiven Räume. Abh. des Sem. Vektor u. Tensor., Moskau 1, 102–124 . . . . *317, 330*

1952, 1

SCHEIBE, E: Über einen verallgemeinerten affinen Zusammenhang. Math. Zeitschr. 57, 65–74 . . . . *287*

SCHILD, A. (see J. L. SYNGE).

1939, 1

SCHIROKOV, P.: Sur les directions concourantes dans les espaces de Riemann. Bull. Soc. Phys. Math. Kazan (3) 7, 77–87 . . . . *322*

1940, 1

— Symmetrisch-konformeuklidische Räume. Bull. Soc. Phys. Math. Kazan (3) 11, 19–28 . . . . *315*

1950, 1

— Projectively euclidean symmetric spaces. Trudy Sem. Vektor Tenzor, An. 8, 73–81 . . . . *163*

1928, 1

SCHLESINGER, L.: Parallelverschiebung und Krümmungstensor. Math. Ann. 99, 413–434 . . . . *123, 138, 340*

1929, 1

— Über Parallelverschiebung in der Weltgeometrie. Journ. f. d. r. u. a. Math. 161, 14–20 . . . . *78, 123*

1931, 1

— Ein Produktintegral mit Stetigkeitssprung. Journ. f. d. r. u. a. Math. 167, 73–78 . . . . *138, 340*

1931, 2

— Neue Grundlagen für einen Infinitesimalkalkül der Matrizen. Math. Zeitschr. 33, 33–61 . . . . *340*

1932, 1

— Weitere Beiträge zum Infinitesimalkalkül der Matrizen. Math. Zeitschr. 31, 485–501 . . . . *138, 340*

1949, 1

SCHMIDT, H.: Winkeltreue und Streckentreue bei konformer Abbildung Riemannscher Räume. Math. Zeitschr. 51, 700–701 . . . . *315*

1902, 1

*SCHOUTE, P. H.: Mehrdimensionale Geometrie I. Göschen, Leipzig. 295 pp. *51*

1918, 1

SCHOUTEN, J. A. (see also E. CARTAN): Die direkte Analysis zur neueren Relativitätstheorie. Verh. Kon. Akad. Amsterdam, 12, Nr. 6. 95 pp. *58, 59, 122, 123, 304*

1918, 2

— On the number of degrees of freedom of the geodetically moving systems and the enclosing euclidean space with the least possible number of dimensions. Proc. Kon. Akad. Amsterdam 21, 607–613 . . . . *81, 268, 361*

1919, 1

SCHOUTEN, J. A. and D. J. STRUIK: On n-tuple orthogonal systems of $(n-1)$-dimensional manifolds in a general manifold of n dimensions. Proc. Kon. Akad. Amsterdam 22, 596–605; 684–695 . . . . . . *58, 81, 171, 244, 245, 246*

1921, 1

— Über die konforme Abbildung n-dimensionaler Mannigfaltigkeiten mit quadratischer Maßbestimmung auf eine Mannigfaltigkeit mit euklidischer Maßbestimmung. Math. Zeitschr. 11, 58–88. . . . . *58, 251, 306, 307, 309, 311*

1921, 2

— and D. J. STRUIK: Über das Theorem von Malus-Dupin und einige verwandte Theoreme in einer n-dimensionalen Mannigfaltigkeit mit beliebiger quadratischer Maßbestimmung. Rend. Palermo 45, 313–331 . *58, 244, 249, 310*

1921, 3

— and D. J. STRUIK: On curvature and invariants of deformation of a V_m in V_n. Proc. Kon. Akad. Amsterdam 24, 146–161 . . . . . *58, 256, 269*

1921, 4

— and D. J. STRUIK: On some properties of general manifolds relating to Einstein's theory of gravitation. Am. Journ. of Math. 43, 213–216 . . . *305, 314*

1922, 1

— and D. J. STRUIK: Einführung in die neueren Methoden der Differentialgeometrie. Christiaan Huygens I, 333–353; II, 1–24; 155–171; 291–306 (1923) *24, 58, 59, 256, 269, 271*

1922, 2

— and D. J. STRUIK: Über Krümmungseigenschaften einer m-dimensionalen Mannigfaltigkeit, die in einer n-dimensionalen Mannigfaltigkeit mit beliebiger quadratischer Maßbestimmung eingebettet ist. Rend. Palermo 46, 165–184 . . . . *58, 256, 269, 271*

1922, 3

— Über die verschiedenen Arten der Übertragung, die einer Differentialgeometrie zugrunde gelegt werden können. Math. Zeitschr. 13, 56–81. Addendum 15, 168 . . . . *60, 123*

1922, 4

— and D. J. STRUIK: Berichtigung zur Mitteilung über das Theorem von Malus-Dupin. Rend. Palermo 46, 346 . . . . *310*

1923, 1

— Über die Einordnung der Affingeometrie in die Theorie der höheren Übertragungen. Math. Zeitschr. 17, 161–182, 183–188 *8, 123, 144, 235, 237, 238, 239, 253, 256*

1923, 2

— Über die Bianchische Identität für symmetrische Übertragungen. Math. Zeitschr. 17, 111–115. . . . . *147*

1923, 3

— and D. J. STRUIK: Note on Mr. Harward's paper on the identical relations in Einstein's theory. Philos. Mag. 47, 584–585 . . . . *147*

1923, 4

— and D. J. STRUIK: Un théorème sur la transformation conforme dans la géometrie différentielle à n dimensions. C. R. 176, 1597–1600 . . . *310, 312, 315*

1924, 1

*SCHOUTEN, J. A.: Der Ricci-Kalkül. Springer, Berlin. 312 pp. Here referred to as RK 1924, 1 *22, 123, 133, 153, 232, 235, 246, 269, 291, 292, 295, 305, 309, 310, 312*

1924, 2

— On the place of conformal and projective geometry in the theory of linear displacements. Proc. Kon. Akad. Amsterdam 27, 407–424 . *123, 302, 303, 316*

1924, 3

— Sur les connexions conformes et projectives de M. Cartan et la connexion linéaire générale de M. König. C. R. 178, 2044–2046 . . . . . *123, 303, 316*

1925, 1

— On the conditions of integrability of covariant differential equations. Trans. Am. Math. Soc. 27, 441–473. . . . . *57, 87*

1925, 2

— Projective and conformal invariants of half-symmetrical connexions. Proc. Kon. Akad. Amsterdam 29, 334–336 . . . . *289, 306*

1926, 1

— Erlanger Programm und Übertragungslehre. Neue Gesichtspunkte zur Grundlegung der Geometrie. Rend. Palermo 50, 142–169 . . . . . *123, 364*

1927, 1

*— Vorlesungen über die Theorie der halbeinfachen kontinuierlichen Gruppen. Univ. of Leiden. 285 pp. (opolographic) *3, 187*

1927, 2

— On infinitesimal deformations of V_m in V_n. Proc. Kon. Ned. Akad. Amsterdam 31, 208–218 . . . . *68, 352*

1927, 3

— Über n-fache Orthogonalsysteme in V_n. Math. Zeitschr. 26, 706–730 *246, 247, 307*

1928, 1

— On non holonomic connections. Proc. Kon. Akad. Amsterdam 31, 291–299 *78, 123, 253*

1928, 2

— Über die Umkehrung eines Satzes von Lipschitz. Nieuw Archief v. Wisk. (2) 15, 97–102 . . . . *310*

1929, 1

— Über unitäre Geometrie. Proc. Kon. Akad. Amsterdam 32, 457–465 *52, 389, 391, 393, 396, 412*

1929, 2

— and V. HLAVATÝ: Zur Theorie der allgemeinen linearen Übertragung. Math. Zeitschr. 30, 414–432 . . . . *11, 129, 135*

1929, 3

— Über nicht-holonome Übertragungen in einer L_n. Math. Zeitschr. 30, 149 to 172 . . . . *78, 253, 261, 263, 270*

1929, 4

— Zur Geometrie der kontinuierlichen Transformationsgruppen. Math. Ann. 102, 244–272 . . . . *142, 187*

1929, 5

SCHOUTEN, J. A. and D. v. DANTZIG: Über die Differentialgeometrie einer hermiteschen Differentialform und ihre Beziehungen zu den Feldgleichungen der Physik. Proc. Kon. Akad. Amsterdam 32, 60–64 . . . . . . *389, 391, 396*

1930, 1

— and D. VAN DANTZIG: Über unitäre Geometrie. Math. Ann. 103, 319–346 *52, 389, 390, 395, 397, 399, 402, 403, 412, 414, 415*

1930, 2

— and E. R. VAN KAMPEN: Zur Einbettungs- und Krümmungstheorie nichtholonomer Gebilde. Math. Ann. 103, 752–783 *3, 78, 253, 254, 261, 263, 278*

1930, 3

— and S. GOLAB: Über projektive Übertragungen und Ableitungen I. Math. Zeitschr. 32, 192–214; II. Ann. Math. Pure Appl. (4) 8, 141–157 *301, 303, 322*

1931, 1

— Klassifizierung der alternierenden Größen dritten Grades in 7 Dimensionen. Rend. d. Palermo 55, 137–157 . . . . *36*

1931, 2

— and E. R. VAN KAMPEN: Über die Krümmung einer V_m in V_n; eine Revision der Krümmungstheorie. Math. Ann. 105, 144–159 . . . . . . *277, 278, 280*

1931, 3

— and D. VAN DANTZIG: Über unitäre Geometrie konstanter Krümmung. Proc. Kon. Akad. Amsterdam 34, 1293–1314 . . . . *403, 404, 415*

1932, 1

— and D. VAN DANTZIG: Generelle Feldtheorie I: Zum Unifizierungsproblem der Physik. Skizze einer generellen Feldtheorie. Proc. Kon. Akad. Amsterdam 35, 642–656 . . . . *303*

1932, 2

— and D. VAN DANTZIG: Generelle Feldtheorie III. Z. f. Phys. 78, 639–667 *303*

1933, 1

— and E. R. VAN KAMPEN: Beiträge zur Theorie der Deformation. Prac. Matematyczno-Fizycznych, Warszawa 41, 1–19 *102, 103, 104, 107, 274, 335, 336, 337, 345, 346, 354*

1933, 2

— and D. VAN DANTZIG: General field theory VI: On projective connexions and their applications to the general field theory. Ann. of Math. 34, 271–312 *301*

1934, 1

— and D. VAN DANTZIG: Was ist Geometrie? Abh. Sem. Vektor u. Tensor., Moskau 2–3, 15–50 . . . . *65*

1935, 1

*— and D. J. STRUIK: Einführung in die neueren Methoden der Differentialgeometrie, Vol. I by J. A. SCHOUTEN, NOORDHOFF, Groningen, here referred to as E I 1935, 1. Russian translation Moscow 1939 *31, 36, 52, 54, 56, 57, 59, 60, 68, 99, 102, 103, 107, 126, 135, 137, 142, 143, 148, 153, 155, 165, 166, 169, 171, 227, 246, 265, 327, 335, 336, 340*

1935, 2

— and J. HAANTJES: Konforme Feldtheorie und Spinraum. Ann. di Pisa 2, 4. 175–189 . . . . *53*

1935, 3
SCHOUTEN, J. A. and J. HAANTJES: Über allgemeine konforme Geometrie in projektiver Behandlung I, II. Proc. Kon. Akad. Amsterdam 38, 706–708; 39, 27 . . . . *316*

1936, 1
— and J. HAANTJES: Zur allgemeinen projektiven Differentialgeometrie. Compositio Math. 3, 1–51. . . . . *296, 301, 303, 327*

1936, 2
— and J. HAANTJES: Beiträge zur allgemeinen (gekrümmten) konformen Differentialgeometrie I, II. Math. Ann. 112, 594–629; 113, 568–583 (cf. 1935, 3) *314, 315, 316*

1937, 1
*— and J. HAANTJES: On the theory of the geometric object. Proc. London Math. Soc. (2) 42, 356–376 . . . . *67*

1938, 1
— Über die geometrische Deutung von gewöhnlichen p-Vektoren und W-p-Vektoren und den korrespondierenden Dichten. Proc. Kon. Akad. Amsterdam 41, 709–716 . . . . *27*

1938, 2
*— and D. J. STRUIK: Einführung in die neueren Methoden der Differentialgeometrie, Vol. II by D. J. STRUIK, NOORDHOFF, Groningen, here referred to as E II 1938, 1. Russian translation Moscow 1948
102, 171, 227, 230, 232, 234, 242, 244, 245, 246, 249, 250, 253, 261, 265, 268, 271 272, 274, 275, 277, 278, 290, 291, 292, 296, 302, 307, 309, 310, 311, 312, 313, 314 317, 318, 329, 330, 335, 348, 352, 354, 395, 396, 398, 402, 403, 404, 406, 412, 414

1940, 1
— and D. VAN DANTZIG: On ordinary quantities and W-quantities. Classification and geometrical application. Comp. Math. 7, 447–473 . . . . . *27, 97*

1940, 2
— Über Differentialkomitanten zweier kontravarianter Größen. Proc. Kon. Ned. Akad. Amsterdam 43, 449–452 . . . . *67*

1947, 1
— Regular systems of equations and supernumerary coordinates. Scriptum VI. Math. Centre of Amsterdam. Here referred to as R. S. 1947, 1 *64, 72, 74, 75*

1949, 1
*— and W. V. D. KULK: Pfaffs problem and its generalizations. Clarendon Press, Oxford. Here referred to as P. P. 1949, 1 . . . . *4, 22, 23, 26, 34, 35, 38, 64, 67, 72, 74, 75, 76, 79, 80, 81, 82, 85, 88, 89, 102, 107, 110, 214, 291*

1949, 2; 3; 4; 1950, 1
— On the geometry of spin spaces I, II, III, IV. Proc. Kon. Ned. Akad. Amsterdam 52, 597–609; 687–695; 938–948; 53, 261–272. Indagationes Math. 11, 178–190; 217–225; 336–346; 12, 41–52 . . . . *53, 291*

1949, 5
*— Lie's differential operator. Math. Centrum. Amsterdam. Rapport Z.W. 1949–010. 7 pp. (dutch) *104*

1951, 1
*— Tensor analysis for physicists. Clarendon Press, Oxford. X + 275 pp. *27, 44, 53, 54, 60, 98, 101, 102, 107, 112*

1951, 2

SCHOUTEN, J. A.: Sur les tenseurs de V_n aux directions principales V_{n-1}-normales. Colloque de Géometrie Différentielle, Louvain, 67–70 . . . . . . *248*

1953, 1

— On subprojective affine connections. Proc. Kon. Nederl. Akad. Amsterdam 56 (= Indag. Math. 15) 72–79 . . . . *319, 320, 321, 324*

1929, 1

SCHRENZEL, E.: Über Kurven mit isotropen Normalen. Sitz. Akad. Wien 138, 439–446 . . . . *227*

1947, 1; 1948, 1; 2

SCHRÖDINGER, E.: The final affine field laws I, II, III. Proc. Roy. Irish Acad. 51, 163–171; 205–216; 52, 1–9. *114, 179*

1950, 1

*—Space-time structure. Cambr. Univ. Press. 119 pp. *114, 179*

1951, 1

— On the differential identities of an affinity. Proc. Roy. Irish Acad. 54, 79 to 85 . . . . *114, 117*

1951, 2

— Studies in the non-symmetric generalization of the theory of gravitation I. Com. Dublin Inst. Adv. Studies 6, 28 pp. *179*

1886, 1

SCHUR, F.: Über den Zusammenhang der Räume konstanten Krümmungsmaßes mit den projektiven Räumen. Math. Ann. 27, 537–567 . . . *153, 244*

1886, 2

— Über die Deformation der Räume konstanten Riemannschen Krümmungsmaßes. Math. Ann. 27, 163–176 . . . . *244*

1887, 1

— Über die Deformation eines dreidimensionalen Raumes in einem ebenen vierdimensionalen Raume. Math. Ann. 28, 343–353 . . . . *244*

1941, 1

SCHWARTZ, A.: The Gauß-Codazzi-Ricci equations in riemannian manifolds. J. Math. Phys. Mass. Inst. Tech. 20, 30–79 . . . . *269, 279*

1946, 1

— On higher normal spaces for V_m in S_n. Am. J. Math. 68, 660–666 . . *268, 275*

1944, 1; 1945, 1; 1946, 1

SEN, R. N.: On parallelism in riemannian space, I, II, III. Bull. Calcutta Math. Soc. 36, 102–107; 37, 153–159; 38, 161–167 . . . . *132*

1948, 1; 1949, 1; 2

— Parallel displacement and scalar product of vectors I, II, III. Proc. Nat. Inst. Sci. India 14, 45–52; Bull. Calcutta Math. Soc. 41, 41–46; 113–120 . *132*

1950, 1

— Corrections to my paper on "Parallel displacement and scalar product of vectors III". Bull. Calcutta Math. Soc. 42, 56 . . . . *132*

1950, 2

— On an algebraic system generated by a single element and its application in riemannian geometry. Bull. Calcutta Math. Soc. 42, 1–13 . . . . . . *132*

1942, 1

SHABBAR, M.: One parameter groups of deformations in riemannian spaces. J. Indian Math. Soc. 6, 186–191 . . . . *108, 335*

1950, 1

SHANKS, E. B.: Homothetic correspondences between riemannian spaces. Duke Math. J. 17, 299–311 . . . . *346*

1934, 1

SIBATA, T.: An extension of the definition of vector and parallel displacement. J. Sc. Hirosh. Univ. 5, 83–97 . . . . *123*

1905, 1

SINIGALLIA, L.: Sugli invarianti differenziali. Rend. Palermo 19, 161–184 . *22*

1927, 1

SLEBODZINSKI, W.: Sur une classe d'espaces riemannien a trois dimensions. Ann. Soc. Polon. 6, 54–82 . . . . *248*

1930, 1

— Sur une généralisation des coefficients de Ricci. C. R. Varsovie 23, 147–150 *171*

1931, 1

— Sur les équations de Hamilton. Bull. Acad. Roy. d. Belg. (5) 17, 864–870 *102*

1932, 1

— Sur les transformations isomorphiques d'une variété a connexion affine. Prac. Matem. Fiz. Warszawa 39, 55–62 . . . . *346*

1951, 1

*SOKOLNIKOFF, I. S.: Tensor analysis theory and applications. Wiley and Sons, New York; Chapman and Hall, London. IX + 335 pp.

1950, 1

SOURIAU, J. M.: Le calcul spinoriel et ses applications. Recherche Aéronautique 1950, 14, 3–8 . . . . *30*

1950, 1

SPRINGER, C. E.: Union curves of a hypersurface. Canadian J. Math. 2, 457 to 460 . . . . *250*

1951, 1

STELLMACHER, K. L.: Geometrische Deutung konforminvarianter Eigenschaften des Riemannschen Raumes. Math. Ann. 123, 34–52 . . . . . . *308, 309*

1949, 1

STRAUS, E. G. (see also A. EINSTEIN): Some results in Einstein's unified field theory. Rev. Modern Physics 21, 414–420 . . . . *179*

1941, 1

STRUBECKER, K.: Differentialgeometrie des isotropen Raumes I. Theorie der Raumkurven. Sitz. Akad. Wiss. Wien 150, 1–53 . . . . *228*

1942, 1

— Differentialgeometrie des isotropen Raumes. II. Die Flächen konstanter Relativkrümmung $K = rt - s^2$. Math. Zeitschr. 47, 743–777 . . . . *228*

1942, 2

— Differentialgeometrie des isotropen Raumes. III. Flächentheorie. Math. Zeitschr. 48, 369–427 . . . . *228*

1944, 1

STRUBECKER, K.: Differentialgeometrie des isotropen Raumes. IV. Theorie der flächentreuen Abbildungen der Ebene. Math. Zeitschr. 50, 1–92 . . . . . *228*

1922, 1

*STRUIK, D. J. (see also J. A. SCHOUTEN): Grundzüge der mehrdimensionalen Differentialgeometrie in direkter Darstellung. Springer, Berlin. 198 pp. *58, 59, 81, 230, 245, 249, 250, 251, 256, 263, 274, 305*

1928, 1

— On the geometry of linear displacement. Publ. Mass. Inst. of Techn. II 135, 522–564 . . . . . *123*

1934, 1

*— Theory of linear connections. Erg. der Math. III, 2. Springer, Berlin. 68 pp.

1950, 1

*— Lectures on classical differential geometry. Addison-Wesley Press. Inc Cambridge Mass. VIII + 221 pp.

1905, 1

STUDY, E.: Kürzeste Wege im komplexen Gebiet. Math. Ann. 60, 321–378 *406*

1943, 1

SU, B.: The projective differential geometry of a nonholonomic hypersurface. Duke Math. J. 10, 575–586 . . . . . *253*

1949, 1

— Lie derivation in the geometry of conformal connections. Acad. Sinica Sc. Record 2, 331–339 . . . . . *346*

1951, 1

SUGURI, T.: Theory of curves in the unitary K_n-connected spaces. Mem. Fac. Sc. Kyūsyū Univ. 6, 31–40 . . . . . *415*

1937, 1

SUN, T.: On the complete set of identities for affine metric space. Tôhoku Math. Journ. 43, 261–266; 44, 32–38 . . . . . *144*

1922, 1

SYNGE, J. L.: On principal directions in a riemannian space. Proc. Nat. Acad. 8, 198–207 . . . . . *146*

1924, 1

— Parallel propagation of a vector around an infinitesimal circuit in an affine-connected manifold. Ann. of Math. 25, 181–184 . . . . . *138*

1924, 2

— Normals and curvatures of a curve in a riemannian manifold. Proc. Int. Math. Congr. Toronto I (publ. 1928), 857–862 . . . . . *229, 230*

1926, 1

— The first and second variations of the length integral in riemannian space. Proc. London Math. Soc. (2) 25, 247–264 . . . . . *352*

1926, 2

— On the geometry of dynamics. Trans. Roy. Soc. London A, 226, 33–106 *352*

1928, 1

— Geodesics in non-holonomic geometry. Math. Ann. 99, 738–751 . . . *78, 253*

1928, 2
SYNGE, J. L., and A. J. MCCONNELL: Riemannian null-geometry. Phil. Mag. (7) 5, 241–263 227

1930, 1
— A characteristic function in riemannian space and its application to the solution of geodesic triangles. Proc. London Math. Soc. (2) 32, 241–258 382, 385

1934, 1
— On the deviation of geodesics and nullgeodesics, particularly in relation to the properties of spaces of constant curvature and indefinite line-element. Ann. of Math. 35, 705–713 352

1949, 1
*— and A. SCHILD: Tensor calculus. Math. Expos. No. 5, University of Toronto Press, Toronto. XII + 324 pp.

1932, 1
SYPTÁK, M.: Sur les hypercirconférences et hyperhélices dans les espaces euclidiens à p dimensions. C. R. 195, 298–299 230

1934, 1
— Sur les hypercirconférences et hyperhélices généralisées dans les espaces euclidiens à p dimensions. C. R. 198, 1665–1667 230

1949, 1
TACHIBANA, S.: On normal coordinates of a Riemann space, whose holonomy group fixes a point. Tôhoku Math. J. 1, 26–30 . . . . 128, 367

1951, 1
— On pseudo-parallelism in Einstein spaces. Tôhoku Math. J. 2, 214–219 148

1951, 2
— On concircular geometry and Riemann spaces with constant scalar curvatures. Tôhoku Math. J. 3, 149–158 334

TAIT, P. G. (see W. THOMSON).

TAKANO, K. (see K. YANO).

1953, 1
TAKASU, T.: Connexion spaces in the large I–VI. Yokohama Math. Journ. 1, 1–87 . . . . 123, 253

1942, 1
TAKENO, H.: On the geometry of m-vectors. J. Sc. Hiroshima Univ. 12, 109–123 138

1951, 1
TAKIZAWA, S.: On generalized spaces which admit given holonomy groups. Mem. Coll. Sc. Univ. Kyoto Math. 26, 199–209 367

1950, 1
TASHIRO, Y.: Sur la dérivée de Lie de l'être géométrique et son groupe d'invariance. Tôhoku Math. J. 2, 166–181 . . . . 105, 107

1952, 1
— Note sur la dérivée de Lie d'un étre géométrique. Math. J. Okayama Univ. 1, 125–128 110

1941, 1
*THALER, H.: Isotrope Projektion isotroper Kurven. These Innsbruck, 27 pp. 228

1925, 1
THOMAS, J. M. (see also O. VEBLEN): Conformal correspondence of Riemann spaces. Proc. Nat. Acad. 12, 352–359 . . . . *315*

1925, 2
— Note on the projective geometry of paths. Proc. Nat. Acad. 11, 207–209 *300*

1926, 1
— Asymetric displacement of a vector. Trans. Am. Math. Soc. 28, 658–670 *127, 287*

1926, 2
— On normal coordinates in the geometry of paths. Proc. Nat. Acad. 12, 58 to 63 . . . . *155*

1926, 3
— Conformal invariants. Proc. Nat. Acad. 12, 389–393 . . . . *315*

1927, 1
— On systems of total differential equations. Ann. of Math. 28, 379–385 . . *88*

1922, 1
THOMAS, T. Y. (see also W. MAYER, O. VEBLEN): On the equi-projective geometry of paths. Proc. Nat. Acad. 8, 592–594 . . . . *300*

1925, 1
— On Weyl's treatment of the parallel displacement of a vector around an infinitesimal closed circuit in an affinely connected manifold. Ann. of Math. 27, 25–28 . . . . *138*

1925, 2
— Note on the projective geometry of paths. Bull. Am. Math. Soc. 31, 318 to 322 . . . . *155*

1925, 3
— On the projective and equi-projective geometries of paths. Proc. Nat. Acad. 11, 198–203 . . . . *300, 301, 302, 303*

1926, 1
— The identities of affinely connected manifolds. Math. Zeitschr. 25, 714–722 *78*

1926, 2
— A projective theory of affinely connected manifolds. Math. Zeitschr. 25, 723–733 . . . . *155, 165, 300, 303*

1926, 3
— On conformal geometry. Proc. Nat. Acad. 11, 722–725 . . . . *315*

1927, 1
— Appell on tensor calculus. Bull. Am. Math. Soc. 33, 493–495 . . . . . . *172*

1927, 2
— and D. MICHAL: Differential invariants of affinely connected manifolds. Ann. of Math. (2) 28, 196–236 . . . . *155, 165*

1927, 3
— The replacement theorem and related questions in the projective geometry of paths. Ann. of Math. (2) 28, 549–561 . . . . *155, 165, 300*

1928, 1
— A theorem concerning the affine connections. Am. Journ. of Math. 50, 518 to 520 . . . . *155*

1929, 1

THOMAS, T. Y.: Determination of affine and metric spaces by their differential invariants. Math. Ann. 101, 713–728 . . . . *165*

1931, 1

*— The elementary theory of tensors with applications to geometry and mechanics. McGraw-Hill, London. 122 pp.

1932, 1

— Conformal tensors. Proc. Nat. Acad. 18, 103–112; 189–193 . . . . . *315, 316*

1934, 1

— and J. LEVINE: On a class of existence theorems in differential geometry. Bull. Am. Math. Soc. 40, 721–728 . . . . *87, 133*

1934, 2

*— The differential invariants of generalized spaces. Cambr. Univ. Press. 251 pp.. . . . . *87, 144, 156, 159, 160, 165, 300, 303, 315, 316*

1935, 1

— Algebraic characterization of complex differential geometry. Trans. Am. Math. Soc. 38, 501–514; Bull. Am. Math. Soc. 41, 349 . . . . *133*

1936, 1

— Fields of parallel vectors in the large. Compositio Math. 3, 453–468 . . . *125*

1936, 2

— On the metric representation of affinely connected spaces. Bull. Am. Math. Soc. 42, 77–78 . . . . *133*

1936, 3

— On normal coordinates. Proc. Nat. Acad. 22, 309–312 . . . . *155*

1936, 4

— Riemann spaces of class one and their characterization. Acta Math. 67, 169–211 . . . . *268, 269*

1938, 1

— New theorems in Riemann-Einstein spaces. Rec. Math. Moscow 3, 331 to 340 . . . . *148*

1938, 2

— Extract from a letter by E. Cartan concerning my note: On closed spaces of constant mean curvature. Am. Journ. Math. 59, 793–794 . . . . . . *251*

1938, 3

— Arcs in affinely connected spaces. Ann. o. Math. Princeton (2) 38, 120–130 *232*

1939, 1

— Reducible Riemann spaces and their characterization. Mon. f. Math. u. Phys. 48, 228–292 . . . . *285, 286*

1939, 2

— The decomposition of Riemann spaces in the large. Mon. f. Math. u. Phys. 47, 388–418 . . . . *285, 286*

1939, 3

— Imbedding theorems in differential geometry. Bull. Am. Math. Soc. 45, 841–850 . . . . *269*

1939, 4

THOMAS, T. Y., and E. W. TITT: On the elementary solution of the general linear differential equation of the second order with analytic coeffients. J. Math. Pures et Appl. 18, 217–248 . . . . *386*

1879, 1

*THOMSON, W., and P. G. TAIT: Treatise on natural philosophy. Cambr. Univ. Press *310*

THULLEN, P. (see H. BEHNKE).

1923, 1

TIETZE, H.: Über die Parallelverschiebung in Riemannschen Räumen. Math. Zeitschr. 16, 308–317 . . . . *138*

TITT, E. W. (see T. Y. THOMAS).

TOMONAGA, Y. (see K. YANO).

1939, 1

TOMPKINS, C.: Isometric imbedding of flat manifolds in euclidian space. Duke Math. Journ. 5, 58–61 . . . . *269*

1941, 1

— A flat Klein bottle isometrically embedded in euclidean 4-space. Bull. Am. Math. Soc. 47, 508 . . . . *268, 269*

1949, 1

TONNELAT, M. A.: Théorie unitaire du champ physique. I. Les tenseurs fondamentaux et la connection affine. C. R. 228, 368–370 . . . . *179*

1950, 1

— Résolution des équations fondamentales d'une théorie unitare purement affine. C. R. 230, 182–184 . . . . *179*

1950, 2

— Théorie unitaire affine. I. Choix des tenseurs de base et obtention de l'équation fondamentale. C. R. 231, 470–472 . . . . *179*

1950, 3

— Théorie unitaire affine. II. Résolution rigoureuse de l'équation fondamentale. C. R. 231, 487–489 . . . . *179*

1950, 4

— Théorie unitaire affine. III. Les équations du champ. C. R. 231, 512–514 *179*

1951, 1

— Théorie unitaire affine du champ physique. Journ. Phys. Radium (8) 12, 81–88 . . . . *179*

1951, 2

— Compatibilité des équations de la théorie unitaire des champs. C. R. 232, 2407–2409 . . . . *179*

1951, 3

— Les équations du champ unitaire et leurs approximations. C. R. 233, 513 to 516 . . . . *179*

1951, 4

— Les équations électromagnétique déduites d'une théorie unitaire des champs. C. R. 233, 555–557 . . . . *179*

1941, 1

TONOLO, A.: Determinazione di una classe di varietà riemanniane normali a tre dimensioni. Ist. Lombardo Sci. Mat. Nat. Rend. 5, 113–124 . . . . *248*

1941, 2

— Sulle equazione di Weingarten relative ai sistemi tripli ortogonali di superficie isostatiche. Univ. Roma e Ist. Naz. Alta Mat. 2, 170–192 . . *248*

1943, 1

— Sulle ipersuperficie V_n le cui due forme fondamentali amettono una stessa trasformazione infinitesima $X(f)$. Boll. Un. Mat. Ital. (2) 5, 137–140 . . *361*

1949, 1

— Sulla varietà riemanniane normali a tre dimensione. Pont. Acad. Sc. Acta 13, 29–53 . . . . *248*

1949, 2

— Sulla varietà riemanniane normali a tre dimensione. Atti d. Linc. (8) 6, 438–444 . . . . *248*

1949, 3

— Sopra una classa di deformazione finite. Ann. Mat. Pura Appl. (4) 29, 99 to 114 . . . . *248*

1950, 1

— Sopra un sistema di equazioné differenziali relativo ai moti rigidi delle varietà riemanniane a tre dimensioni a curvatura costante. Rend. Sem. Mat. Univ. Padova 19, 250–272 . . . . *335, 351*

1953, 1

— Sopra un problema di Darboux relativo alla meccanica dei mezzi continui. Atti IV Congr. Un. Mat. Ital. 2, 578–582 . . . . *248*

TOURNAY, F. (see G. PAPY).

1931, 1

TUCKER, A. W.: On generalized covariant differentation. Ann. of Math. (2) 32, 451–460 . . . . *68*

1935, 1

— Non-riemannian subspaces. Ann. of Math. 36, 965–968 . . . . *123*

1950, 1

UDESCHINI, P.: Le equazioni di prima approssimazione nella nuova teoria relativistica unitaria di Einstein. Atti d. Linc. Fis. Mat. Nat. 9, 256–261 *179*

1951, 1

— Le equazioni di seconda approssimazione nella nuova teoria relativistica unitaria di Einstein I. Atti d. Linc. 8, 21–24 . . . . *179*

1951, 2

— Le equazioni di seconda approssimazione nella nuova teoria relativistica unitaria di Einstein II. Atti d. Linc. 10, 121–123 . . . . *179*

1951, 3

— Sulle mutue azioni fra campo gravitazionale e campo elettromagnetico. Atti d. Linc. (8) 10, 390–394 . . . . *179*

1940, 1

URABE, M.: Geometry of parallel displacement making a volume invariant. J. Sc. Hiroshima Univ. 10, 95–108 . . . . *123, 155*

1947, 1

URBAN, A.: L'équation différentielle des courbes sur une V_{n-1} spéciale dans V_n. Acad. Tcheque. Mat. Nat. 48, 57–61 . . . . *242*

1948, 1

— Differential equations of curves on a special V_{n-1} in V_n. Rozpravy II Trědy Ceské Akad. 57, no. 9, 12 pp. (czech) *242*

1934, 1

VANDERSLICE, J. L.: Non holonomic geometries. Am. Journ. 56, 153–192 *123, 253*

1934, 2

— Conformal tensorinvariants. Proc. Nat. Acad. 20, 672–676 . . . . *315, 316*

1939, 1

VARGA, O.: Über die Integralinvarianten, die zu einer Kurve in der hermitischen Geometrie gehören. Acta Hungar. 9, 88–102 . . . . *415*

1943, 1

— Über eine Charakterisierung der Riemannschen Räume konstanter Krümmung. Mat. Fiz. Lapok 50, 34–39 (hungarian, german summary) *148*

1950, 1

— Normalkoordinaten in allgemeinen differentialgeometrischen Räumen und ihre Verwendung zur Bestimmung sämtlicher Differentialinvarianten (hungarian and german). C. R. first congress of hungarian mathem. 131–162 *155*

1951, 1

VASILIEV, A. M.: General invariant methods in differential geometry. Akad. Nauk SSSR. 79, 5–7 (russian) *67*

1922, 1

VEBLEN, O. (see also L. P. EISENHART): Normal coordinates for the geometry of paths. Proc. Nat. Akad. 8, 192–197 . . . . *147, 155, 300, 303*

1922, 2

— Projective and affine geometry of paths. Proc. Nat. Acad. 8, 233–238 . . *300*

1923, 1

— and T. Y. THOMAS: The geometry of paths. Trans. Am. Math. Soc. 25, 551 to 608 . . . . *155, 159, 300, 303*

1923, 2

— Equiaffine geometry of paths. Proc. Nat. Acad. 9, 3–4 . . . . *144*

1924, 1

— and T. Y. THOMAS: Extensions of relative tensors. Trans. Am. Math. Soc. 26, 373–377 . . . . *129, 155, 160*

1925, 1

— and J. M. THOMAS: Projective normal coordinates for the geometry of paths. Proc. Nat. Acad. 11, 204–207 . . . . *155*

1926, 1

— and J. M. THOMAS: Projective invariants of affine geometry of paths. Ann. of Math. 27, 278–296 . . . . *87, 300, 302*

1927, 1

*— Invariants of quadratic differential forms. Cambr. Tracts No 24, 102 pp. *155*

1928, 1

— Projective tensors and connections. Proc. Nat. Acad. 14, 154–166 *300, 303*

1928, 2

VEBLEN, O.: Differential invariants and geometry. Atti del Congress Int. Bologna 6, 181–189 . . . . *316*

1928, 3

— Conformal tensors and connexions. Proc. Nat. Acad. 14, 735–745 . . *315, 316*

1929, 1

— Generalized projective geometry. Journ. London, Math. Soc. 4, 140–160 *300*

1930, 1

— A generalization of the quadratic differential form. Quart. Journ. Oxford 1, 60–76 . . . . *315*

1932, 1

*— and J. H. C. WHITEHEAD: The foundations of differential geometry. Cambr. Tracts. No 29, 96 pp. Reprinted 1953 . . . .*61, 62, 64*

1935, 1

— Formalism for conformal geometry. Proc. Nat. Acad. 21, 168–173 *315, 316*

1936, 1

*— and J. v. NEUMANN and J. W. GIVENS: Geometry of complex domains. Princeton Seminar, stencilled. *53, 57*

1949, 1

VERBICKII, L. L.: On the metric differential geometry of hypersurfaces of the second order. Trudy Sem. Vektor Tenzor., An. 7, 319–340 (russian) *242*

1950, 1

— On the equations for imbedding riemannian spaces of class 2 in euclidean spaces. Trudy Sem. Vektor, Tenzor, An. 8, 425–429. . . . . *269*

1951, 1

— A tensor criterion of conformally euclidean spaces of class 1. Akad. Nauk SSSR. 81, 133–136 . . . . *315*

1951, 2

— Construction of a conformally euclidean space of class 1 within a containing euclidean space. Akad. Nauk SSSR. 81, 333–336 (russian) . . . *315*

1923, 1

VITALI, G.: I fondamenti del calcolo assoluto generalizzato. Giorn. d. Matem. 61, 159–202 . . . . *57*

1924, 1

— Una derivazione covariante formata coll'ausilio di *n* sistemi covarianti del 1° ordine. Atti Pavia 248–253 . . . . *142*

1946, 1

VOLTA V. DALLA: Sulla torsione di una varietà a tre dimensioni dotati di una connessione affine. Univ. Roma Ist. Naz. Mat. e Appl. (5) 5, 271–282 . . *123*

1946, 2

— Sulle connessioni affine asimmetriche. Univ. Roma Ist. Naz. Mat. e Appl. (5) 5, 315–326 . . . . *123*

1887, 1

VOLTERRA, V.: Sui fondamenti della teoria delle equazioni differenziali lineari. Mem. Soc. Ital. 40, III 6, No 8, 107 pp. *340*

1889, 1

— Sulle funzioni conjugate. Rend. d. Linc. (IV) 5^{I}, 599–611 . . . . *85*

1880, 1

VOSS, A.: Zur Theorie der Transformation quadratischer Differentialausdrücke und der Krümmung höherer Mannigfaltigkeiten. Math. Ann. 16, 129–178 *146, 250, 256, 263*

1926, 1

VRANCEANU, G.: Sur les espaces non holonomes. C. R. 183, 852–854 *78, 253, 261*

1926, 2

— Sur le calcul différentiel absolu pour les variétés non holonomes. C. R. 183, 1083–1085 . . . .*78, 253, 261*

1927, 1

— Studio geometrico meccanico dei sistemi anolonomi. Ann. di Mat. (4) 5, 275–278 . . . .*78, 253*

1927, 2

— Sur l'écart géodésique dans les espaces non holonomes. Ann. Sci. Univ. Jassy 15, 7–24 . . . .*78, 253*

1928, 1

— Sur quelques tenseurs dans les variétés non holonomes. C. R. 186, 995–997 *78, 253, 261*

1928, 2

— Seconda forma quadratica fondamentale di una varietà anolonoma ed applicazioni. Rend. d. Linc. (6) 8, 669–673 . . . . *78, 253*

1928, 3

— Sullo scostamento geodetico nella varietà anolonome. Rend. d. Linc. (6) 7, 134–137 . . . .*78, 261*

1928, 4

— Note sur l'écart géodésique dans les espaces non-holonomes. Ann. Univ. Jassy 15, 307–310 . . . . *78*

1928, 5

— Parallelisme et courbure dans une variété non holonome. Atti Congr. Int. Mat. Bologna 6, 61–67 . . . . *253*

1929, 1

— Les trois points de vue dans l'etude des espaces non-holonomes. C. R. 188, 973–976 . . . . *78, 253*

1929, 2

— Studio geometrico dei sistemi anolonomi. Ann. di Mat. (4) 6, 9–43 . . *78, 253*

1929, 3

— Sur les espaces de Riemann ayant leurs coefficients de rotation constants. C. R. 189, 386–389 . . . . *171*

1930, 1

— Sur les groupes d'applicabilité des variétés non-holonomes. C. R. 190, 1100 to 1102 . . . .*78, 253*

1930, 2

— Sulle condizioni di rigidità di una V_m in un S_n. Rend. d. Linc. (6a) 11, 385 to 389 . . . . *268*

1931, 1

— Sur quelques théorèmes relatifs aux variétés non-holonomes et aux systèmes de formes de Pfaff. C. R. 192, 721–724 . . . . *78, 253*

1932, 1

VRANCEANU, G.: Sur quelques points de la théorie des espaces non-holonomes. Bull. fac. sci. Cernauti 5, 177–205 . . . . *78, 170, 253*

1933, 1

— Connexions affines liées aux systèmes de Pfaff. Bull. fac. sci. Cernauti 6, 188–196 . . . . *78*

1934, 1

— Étude des espaces non holonomes. Journ. de Math. (9) 13, 113–174 *78, 253*

1936, 1

*— Les espaces non holonomes. Mém. d. Sc. Math., fasc. LXXVI, Gauthier-Villars. Paris. 70 pp. *78, 253*

1938, 1

— Invariants de deux surfaces non holonomes complementaires. C. R. Roum. 459–465 . . . . *253*

1940, 1

— Sur les espaces à connexion conforme. Disquisit. Math. Phys. 1, 63–81 *315*

1942, 1

— Sur les invariants des espaces de Riemann singuliers, Disquisit. Math. Phys. 2, 253–281 . . . . *133, 270*

1942, 2

— Sur les espaces à connexion affine et projective. Bull. Math. Soc. Roumaine Sci. 44, 85–139 . . . . *253*

1943, 1

— Sur la théorie des espaces à connexion conforme. Bull. Math. Soc. Roumaine Sci. 45, 3–31 . . . . *353*

1947, 1

*— Leçons de géometrie différentielle, Vol. I. Congruences, formes de Pfaff, groupes continus, invariants et équivalence, espaces à connexion affine, espaces de Riemann, espaces à connexion projective. Bucarest. 422 pp. *185, 348*

1947, 2

— Sur les espaces partiellement projectifs. Bull. Math. Soc. Roumaine Sci. 48, 43–64 . . . . *317, 326*

1949, 1

— La géométrisation des espaces symplétiques. C. R. 229, 336–338 . . . *179, 348*

1949, 2

— Sur les espaces à connexion à groupe maximum des transformations en eux-mêmes. C. R. 229, 543–545 . . . . *348*

1950, 1

— On Kagan spaces (romanian, russian and french summaries). Rom. Acad. 2, 299–304 . . . . *317*

1950, 2

— On metric Kagan spaces. (romanian, russian and french summaries). Rom. Acad. 2, 6 pp.. . . . . *317*

1951, 1

*— Lectii de geometrie diferentiala, II (romanian) *348*

1951, 2

VRANCEANU, G.: Asupra grupurilor etc. Com. Ac. R.P.R. 1, 137; cf. Stud. S. Cerc. Mat. 2, 57 (romanian) *348*

1951, 3

— Groupes de mouvement des espaces à connexion. Ac. R. P. R. 2, 387–433 (romanian with a french summary) *348*

1953, 1

— Sur les groupes de mouvement d'un espace de Riemann à quatre dimensions. Stud. S. Cerc. Mat. 4, 121–153 (romanian with a french summary) *348*

1941, 1

WADE, T. L. (see also R. H. BRUCK): Tensor algebra and Young's symmetry operators. Am. J. Math. 63, 645–657 . . . . *15*

1943, 1

— The factorization of rank tensors. Duke Math. J. 10, 493–498 . . . . .*15, 34*

1944, 1; 1945, 1

— Tensor algebra and invariants I, II. Nat. Math. Mag. 19, 3–10; 20, 5–10 *15*

1927, 1

WAERDEN, B. L. VAN DER: Differentialkovarianten von n-dimensionalen Mannigfaltigkeiten in Riemannschen m-dimensionalen Räumen. Abh. Math. Sem. Hamburg 5, 153–160 . . . .*68, 254, 269*

1931, 1

*— Moderne Algebra II. Springer, Berlin. 216 pp. *31*

1936, 1

WAGNER, V.: Sur la géométrie différentielle des multiplicités anholonomes. Proc. Sem. Vect. a. Tens. Moscow 2–3, 269–318 . . . . *253*

1938, 1

— Über V_3^2 von der Krümmung Null in R_3. Rec. Math. Moscow 4 (46) 2, 333 to 338 . . . . *253*

1938, 2

— On the geometric interpretation of the curvature vector of a non-holonomic V_3^2 in the three-dimensional euclidean space. Rec. Math. Moscow 4 (46) 2, 339–356 . . . . *253*

1940, 1

— Differential geometry of non-linear non holonomic manifolds in the three-dimensional euclidean space. Rec. Math. (Mat. Sbornik) N. S. 8 (50), 3–39 (english, russian summary) *253*

1941, 1

— The geometry of an $(n-1)$-dimensional nonholonomic manifold in an n-dimensional space. Proc. Sem. Vect. and Tensor 5, 173–225 (russian) *253*

1942, 1

— On the Cartan group of holonomicity for surfaces. C. R. Acad. Sc. URSS. 37, 6–8 . . . . *128*

1942, 2

— Differential geometry of the family of R_k's in R_n and of the family of totally geodesic S_{k-1}'s in S_{n-1} of positive curvature. Rec. Math. (Mat. Sbornik) 10 (52) 165–212 (english, russian summary) *269*

1943, 1

WAGNER, V.: The absolute derivative of field of local geometric object in a compound manifold. C. R. Acad. Sc. URSS. 40, 94–97 . . . . *123, 366*

1943, 2

— The inner geometry of non-linear non-holonomic manifolds. Rec. Math. (Mat. Sbornik) 13 (55) 135–167 (english, russian summary) *253*

1945, 1

— The theory of geometric objects and the theory of finite and infinite continuous groups of transformations. C. R. Acad. Sc. URSS. 46, 347–349 *67*

1945, 2

— The generalization of Ricci's and Bianchi's identities for a connexion in the compound manifold. C. R. Acad. Sc. URSS. 46, 303–305 . . . . *123, 147*

1946, 1

— On a sufficient condition in the problem of Lagrange for multiple integrals. C. R. Acad. des Sc. Moscow 54, 479–482 . . . . *232*

1946, 2

— Constant fields of local geometric objects in compound manifolds with a linear connexion. C. R. Acad. Sc. URSS. 53, 183–186 . . . . *366*

1949, 1

— Classification of simple differential-geometric objects. Akad. Nauk SSSR. 69, 293–296 (russian). *67*

1949, 2

— Classification of linear connections in a composite manifold $X_{n+(1)}$ according to their holonomy group. Trudy Sem. Vektor Tenzor, An. 7, 205–226 (russian) *367*

1950, 1

— On the theory of pseudogroups of transformations. Akad. Nauk SSSR. 72, 453–456 . . . . *64, 67, 105*

1952, 1

WAKAKUWA, H.: On Riemann spaces whose homogeneous holonomy groups are integrable. Tôhoku Math. J. 4, 96–98 . . . . *367*

1934, 1

*WALBERER, P.: Orthogonale Kurvengewebe in Riemannschen Räumen. Dissertation Hamburg. The three parts also appeared in Abh. Math. Sem. Hamburg 10, 136–151; 152–168; 169–179 . . . . *246*

1942, 1

WALKER, A. G. (see also E. M. PATTERSON): Note on a distance invariant and the calculation of Ruse's invariant. Proc. Edinburgh Math. Soc. (2) 7, 16 to 26 . . . . *383, 386*

1944, 1

— Completely symmetric spaces. J. London Math. Soc. 19, 219–226 . . . *163*

1944, 2

— Complete symmetry in flat space. J. London Math. Soc. 19, 227–229 *163*

1945, 1

— A particular harmonic riemannian space. J. London Math. Soc. 20, 93–99 *382, 383, 386, 387, 388*

1945, 2

WALKER, A. G.: On completely harmonic spaces. J. London Math. Soc. 20, 159–163 . . . . *386, 387*

1946, 1

— Symmetric harmonic spaces. J. London Math. Soc. 21, 47–57 . . . . . *387*

1949, 1

— On parallel fields of partially null vector spaces. Quart. J. of Math. Oxford Ser. 20, 135–145 . . . . *322, 367, 422*

1949, 2

— On Lichnerowicz's conjecture for harmonic 4-spaces. J. London Math. Soc. 24, 21–28 . . . . *387*

1950, 1

— On Ruse's spaces of recurrent curvature. Proc. London Math. Soc. 52, 36 to 64 . . . . *153, 322, 369, 370, 421, 422, 423, 424*

1950, 2

— On parallel fields of partially null vector spaces. Quart. J. of Math. Oxford 20, 135–145 . . . . *228*

1950, 3

— Canonical form for a riemannian space with a parallel field of null planes. Quart. J. of Math. Oxford (2) 1, 69–79 . . . . *228, 322, 367, 422*

1950, 4

— Canonical forms II. Parallel partially null planes. Quart. J. Math. Oxford (2) 1, 147–152 . . . . *228, 322, 367, 422*

1943, 1

WANG, H. CH.: On a projective invariant of a nonholonomic surface. Ann. of Math. (2) 44, 562–571 . . . . *253*

1943, 2

— A projective invariant of a nonholonomic surface. Duke Math. J. 10, 565–574. (This paper appeared also Ann. of Math. (2) 44, 562–571.) . . . *253*

1927, 1

*WEATHERBURN, C. E.: Differential geometry of three dimensions. Cambr. Univ. Press. 268 pp. *306, 342*

1933, 1

*— The development of multidimensional differential geometry. Report Australian and New Zealand Assoc. for Advancement of Science 21, 12–28.

1938, 1

*— An introduction to Riemann geometry and the tensor calculus. Cambr. Univ. Press. Cambridge. 191 pp.

1900, 1

*WEBER, E. v.: Vorlesungen über das Pfaffsche Problem und die Theorie der partiellen Differentialgleichungen erster Ordnung. Teubner, Leipzig. XI + 622 pp. *81*

1934, 1

WEISE, K. H. (see also R. KÖNIG): Beiträge zum Klassenproblem der quadratischen Differentialformen. Math. Ann. 110, 522–570 . . . . *268*

1935, 1

— Autoparallele Mannigfaltigkeiten im affin zusammenhängenden Raum. Math. Zeitschr. 40, 208–220 . . . . *232*

1908, 1

*WEITZENBÖCK, R.: Komplex-Symbolik. Leipzig *23, 58*

1922, 1

*— Neuere Arbeiten der algebraischen Invariantentheorie. Differentialinvarianten. Encycl. Mathem. Wissensch. III D 10 (E 1). 71 pp.

1923, 1

*— Invariantentheorie. Noordhoff, Groningen. 408 pp. *15, 23, 26, 97, 142, 147, 165*

1928, 1

— Differentialinvarianten in der Einsteinschen Theorie des Fernparallelismus. Sitz. Preuß. Akad. 466–474 . . . . *142*

1932, 1

— Über den Reduktionssatz bei affinem und projektivem Zusammenhang. Proc. Kon. Ned. Akad. Amsterdam 35, 1220–1229 . . . . *302*

1937, 1

— Bemerkungen über Trivektoren. Proc. Kon. Ned. Akad. Amsterdam 40, 312–315 . . . . *36*

1937, 2

— Über Trivektoren I, II, III. Proc. Kon. Ned. Akad. Amsterdam 40, 676 to 681; 704–710; 815–857 . . . . *36*

1938, 1

— Über Trivektoren IV–VII. Proc. Kon. Ned. Akad. Amsterdam 41, 108–116; 238–245; 362–369; 454–461 . . . . *36*

1939, 1

— Zur Theorie der Komplexgrößen a_{ijk}. Monatsh. Math. Phys. 48, 129–140 *15*

1918, 1

WEYL, H. (see also A. BRAUER): Reine Infinitesimalgeometrie. Math. Zeitschr. 2, 384–411 . . . . *133, 304, 306*

1921, 1

*— Raum, Zeit, Materie, 4. Aufl. Springer, Berlin. 300 pp. *40, 41, 136*

1921, 2

— Zur Infinitesimalgeometrie: Einordnung der projektiven und der konformen Auffassung. Göttinger Nachrichten 1921, 99–112 . . . . . *148, 289, 294, 304*

1922, 1

*— Mathematische Analyse des Raumproblems. Springer, Berlin. 117 pp. *31, 33*

1922, 2

— Die Einzigartigkeit der Pythagoreischen Maßbestimmung. Math. Zeitschr. 12, 114–146 . . . . *136, 137*

1922, 3

— Zur Infinitesimalgeometrie; p-dimensionale Fläche im n-dimensionalen Raum. Math. Zeitschr. 12, 154–160 . . . . *232, 242*

1929, 1

WEYL, H. and H. P. ROBERTSON: On a problem in the theory of groups arising in the foundations of infinitesimal geometry. Bull. Am. Math. Soc. 35, 686–690 . . . . *300, 303*

1929, 2

— On the foundations of general infinitesimal geometry. Bull. Am. Math. Soc. 35, 716–725 . . . . *300, 303*

1939, 1

*— The classical groups, their invariants and representations. Princeton Univ. Press. Princeton N. J. 302 pp. *15*

1937, 1

WEYSSENHOFF, J. VAN: Duale Größen, Großrotation, Großdivergenz und die Stokes-Gaußschen Sätze in allgemeinen Räumen. Ann. Soc. Pol. 16, 127 to 144 . . . . *97*

1929, 1

WHITEHEAD, J. H. C. (see also O. VEBLEN): On a class of projectively flat affine connexions. Proc. London Math. Soc. 2, 32, 93–114 . . . . . *300, 301, 315*

1930, 1

— and B. V. WILLIAMS: A theorem of linear connections. Ann. of Math. (2) 31, 150–157 . . . . *155*

1931, 1

— On linear connexions. Trans. Am. Math. Soc. 33, 191–209 . . . . *155*

1931, 2

— The representation of projective spaces. Ann. of Math. 32, 327–360 . . . *302*

1932, 1

— Affine spaces of paths which are symmetric about each point. Math. Zeitschr. 35, 644–659 . . . . *163*

1932, 2

— Note on Maurer's equations. Journ. London Math. Soc. 7, 223–227 . . . *197*

WILLIAMS, B. V. (see J. H. C. WHITEHEAD).

1950, 1

WILLMORE, T. J.: Mean value theorems in harmonic riemannian spaces. J. London Math. Soc. 25, 54–57 . . . . *383*

1951, 1

— Les plans parallèles dans les espaces riemanniens globaux. C. R. 232, 298 to 299 . . . . *367*

1953, 1

— Some properties on harmonic riemannian manifolds. Int. Congress Venice Sept. 1953 . . . . *388*

1951, 1

WINOGRADSKI, J.: Sur la connexion des espaces affines. C. R. 232, 936–938 *123*

1940, 1

WONG, Y. CH.: On the principal directions of a tensor. J. London Math. Soc. 15, 172–183 . . . .*35, 269*

1940, 2

WONG, Y. CH.: On the Frenet formulae for a V_m in a V_n. Quart. J. of Math. Oxford 11, 146–160 . . . 277

1940, 3

— On a certain matrix occuring in the theory of helices. J. London Math. Soc. 15, 168–172 . . . 230

1941, 1

— Generalized helices in an ordinary V_n. Proc. Cambridge Philos. Soc. 37, 14 to 28 . . . 230

1941, 2

— On the generalized helices of Hayden and Sypták in an N-space. Proc. Cambridge Philos. Soc. 37, 229–243 . . . 230

1943, 1

— A note on complementary subspaces in a riemannian space. Bull. Am. Math. Soc. 49, 120–125 . . . 272, 286

1943, 2

— Some Einstein spaces with conformally separable fundamental tensors. Trans. Amer. Math. Soc. 53, 157–194 . . . 287, 312, 315

1943, 3

— Family of totally umbilical hypersurfaces in an Einstein space. Ann. of Math. 44, 271–297 . . . 251, 272, 312

1945, 1

— A note on the first normal space of a V_m in an R_n. Bull. Am. Math. Soc. 51, 997–1000 . . . 269

1945, 2

— Quasi-orthogonal ennuple of congruences in a riemannian space. Ann. of Math. (2) 46, 158–173 . . . 228, 246

1946, 1

— Some theorems on Einstein 4-space. Duke Math. J. 13, 601–610 . . . 272

1895, 1

*WOODS, F. S.: Über Pseudominimalflächen. Diss. Göttingen. 73 pp. . . . 274

1922, 1

*— Higher geometry. Ginn & Co., Boston. 423 pp. . . . 316

1929, 1

WOUDE, W. VAN DER: Over de invoering van het begrip kromtestraal. N. Archief v. Wisk. (2) 13, 396–397 . . . 229

1908, 1

*WRIGHT, J. E.: Invariants of quadratic differential forms. Cambridge Tracts 9. 90 pp. . . . 296

WRÖBEL, F. H. (see S. GOLAB).

1941, 1

WRONA, W. (see also J. HAANTJES): Eine Verallgemeinerung des Schurschen Satzes. Proc. Kon. Ned. Akad. Amsterdam 44, 943–946 . . . 148, 153

1947, 1

— Conditions nécessaires et suffisantes qui déterminent les espaces einsteiniens conformément euclidiens et de courbure constante. Ann. Soc. Polon. Math. 20, 28–80 . . . 148, 153

1948, 1

WRONA, W.: On multivectors in a V_n I. Proc. Kon. Ned. Akad. Amsterdam 51, 1291–1301 = Indagationes Math. 10, 435–445 . . . . *146, 305*

1948, 2

— Conditions nécessaires et suffisantes qui déterminent les espaces einsteiniens, conformément euclidiens et de courbure constante. Ann. Soc. Polon. Math. 20, 28–80 . . . . *305, 312*

1949, 1

— On multivectors in a V_n II. Proc. Kon. Ned. Akad. Amsterdam 52, 29–36 = Indagationes Math. 11, 61–68 . . . . *305*

1950, 1

— Sur les multivecteurs dans V_n III. Casopis Pest. Mat. Fys. 74, 273–280 (polish, french summary) *305*

1929, 1

WUNDHEILER, A.: Sur un déplacement généralisé dans les espaces riemanniens. Rend. d. Linc. 9, 387–389 . . . . *123*

1937, 1

— Objekte, Invarianten und Klassifikation der Geometrien. Abh. Sem. Vektor u. Tensor., Moscow 4, 366–375 (Congress Diff. Geom. Moscow 1934) *67*

1949, 1

YANENKO, N. N.: The geometric structure of surfaces of small type. Akad. Nauk SSSR. 64, 641–644 (russian) *269*

1949, 2

— On some necessary conditions for the deformability of surfaces in a multidimensional euclidean space. Akad. Nauk SSSR. 65, 449–452 (russian) *352*

1953, 1

— Some questions of the theory of imbedding of riemannian metrics in euclidean spaces. Uspedi Matem. Nauk (N. S.) 8, 21–100 (russian) *269*

1935, 1; 1936, 1

YANO, K. (see also S. SASAKI) and Y. MUTÔ: On the connexion in X_m associated with the points of Y_m. Proc. Phys. Math. Soc. Japan III 17, 379 to 390; 18, 1–9 . . . . *123*

1936, 2

— and Y. MUTÔ: Notes on the derivation of geodesics and the fundamental scalar in a riemannian space. Proc. Phys. Math. Soc. Japan III 18, 142–152 *352*

1937, 1

— Sur les espaces non holonomes totalement géodésiques. C. R. 205, 9–12 *253*

1937, 2

— Sur le changement des coefficients d'une connexion projective. C. R. 205, 637–639 . . . . *302*

1937, 3

— Sur les équations des géodésiques dans une variété à connexion projective. C. R. 205, 829–831 . . . . *302, 303*

1938, 1

— La relativité non holonome et la théorie unitaire d'Einstein et Mayer. Math. Cluj. 14, 124–132 . . . . *253*

1938, 2

YANO, K.: Remarques relatives à la théorie des espaces à connection conforme. C. R. 206, 560–563 . . . . *315, 316*

1938, 3

— Sur les conférences généralisées dans les espaces à connection conforme. Proc. Imp. Acad. 14, 329–332 . . . . *315, 330*

1938, 4

*— Les espaces à connection projective et la gèométrie projective des "paths". Thése Univ. Yassy. *303*

1938, 5

— Sur l'espace projectif de M. D. van Dantzig. C. R. 206, 1610–1612 . . . *303*

1938, 6

— The non holonomic representation of projective spaces. Proc. Phys. Math. Soc. Japan (3) 20, 442–450 . . . . *303*

1939, 1

— Theory of non-holonomic spaces. Tensor 2, 13–24 (japanese) *253*

1939, 2

— Sur les équations de Gauss dans la géométrie conforme des espaces de Riemann. Proc. Imp. Acad. Tokyo 15, 247–252. . . . . *311, 315*

1939, 3

— Sur les équations de Codazzi dans la géométrie conforme des espaces de Riemann. Proc. Imp. Acad. Tokyo 15, 340–344 . . . . *311, 315*

1939, 4

A projective treatment of a conformally connected manifold. Proc. Phys. Math. Soc. Japan 21, 270–286 . . . . *315*

1939, 5

— Sur la théorie des espaces à connexion conforme. J. Imp. Univ. Tokyo 4, 1–59 . . . . *315*

1940, 1

— Conformally separable quadratic differential forms. Proc. Imp. Acad. Tokyo 16, 83–86 . . . . *287, 315*

1940, 2

— and S. PETRESCU: Sur les espaces métriques non holonomes complémentaires. Disquisit. Math. Phys. 1, 191–246 . . . . *253*

1940, 3

— Sur quelques propriétés conformes de V_1 dans V_m dans V_n. Proc. Imp. Acad. Tokyo 16, 173–177 . . . . *285, 311, 315*

1940, 4

— Sur la connexion de Weyl-Hlavaty et ses applications à la géométrie conforme. Proc. Phys. Math. Soc. Japan (3) 22, 595–621 . . . . *303, 316, 334*

1940, 5

— Concircular geometry I. Concircular transformations. Proc. Imp. Acad. Tokyo 16, 195–200 . . . . *322, 330, 331, 334*

1940, 6

— Concircular geometry II. Integrability conditions of $p_{\mu\nu}=\Phi g_{\mu\nu}$. Proc. Imp. Acad. Tokyo 16, 354–360 . . . . *322, 330, 331, 332*

1940, 7

YANO, K.: Concircular geometry III. Theory of curves. Proc. Imp. Akad. Tokyo 16, 442–448 . . . . *322, 330, 334*

1940, 8

— Concircular geometry IV. Theory of subspaces. Proc. Imp. Acad. Tokyo 16, 505–511 . . . . *322, 330, 332*

1940, 9

— and Y. MUTÔ: Sur la théorie des hypersurfaces dans un espace à connexion conforme. Proc. Imp. Acad. Tokyo 16, 266–273 . . . . *315*

1941, 1

— and Y. MUTÔ: Sur la théorie des espaces à connexion conforme normale et la géométrie conforme des espaces de Riemann. Proc. Imp. Acad. Tokyo 17, 87–94 . . . . *315*

1941, 2

— and Y. MUTÔ: Sur la théorie des espaces à connexion conforme normale et la géométrie conforme des espaces de Riemann. J. Imp. Univ. Tokyo 4, 117–169 . . . . *315*

1941, 3

— and Y. MUTÔ: On the conformal arc length. Proc. Imp. Acad. Tokyo 17, 318–322 . . . . *315*

1941, 4

— and Y. MUTÔ: On the generalized loxodromes in the conformally connected manifold. Proc. Imp. Acad. Tokyo 17, 455–460 . . . . *315*

1941, 5

— and Y. MUTÔ: Sur la théorie des hypersurfaces dans un espace à connexion conforme. Jap. Journ. of Math. 17, 229–288 . . . . *315*

1941, 6

— The variable u^0 of T. Y. Thomas and the homogeneous coordinates of D. van Dantzig. Tensor 4, 38–48 (japanese) *303*

1942, 1

— and Y. MUTÔ: Sur le théorème fondamental dans la géométrie conforme des sous-espaces riemanniens. Proc. Phys. Math. Soc. Japan (3) 24, 437–449 *278, 311, 315*

1942, 2

— On the fundamental theorem of conformal geometry. Tensor 5, 51–59 (japanese) *278, 315*

1942, 3

— Concircular geometry V. Einstein spaces. Proc. Imp. Acad. Tokyo 18, 446–451 . . . . *334*

1942, 4

— Les espaces d'éléments linéaires à connexion projective normale et la géométrie projective générale des paths. Proc. Phys. Math. Soc. Japan 24, 9–24 . . . . *301, 303*

1943, 1

— Conformal and concircular geometries in Einstein spaces. Proc. Imp. Ac. Tokyo 19, 444–453 . . . . *148, 312, 334*

1943, 2

YANO, K.: Sur les équations fondamentales dans la géométrie conforme des sous-espaces. Proc. Imp. Acad. Tokyo 19, 326–334 . . . . *311, 315*

1943, 3

— Sur le parallélisme et la concourance dans l'espace de Riemann. Proc. Imp. Acad. Tokyo 19, 189–197 . . . . *322, 333, 334, 370*

1943, 4

— Sur une application du tenseur conforme C_{jk} et de la scalaire conforme C. Proc. Imp. Acad. Tokyo 19, 335–340 . . . . *315*

1943, 5

— On Sasaki's study of holonomy groups of spaces with normal conformal connections. Tensor 6, 68–85 (japanese) *367*

1944, 1, 2

— and K. TAKANO: Sur les coniques dans les espaces à connexion affine ou projective I, II. Proc. Imp. Acad. Tokyo 20, 410–417; 418–424 . . . . *298, 300*

1944, 3

— Equations of paths in O. Veblen's projective space. Tensor 7, 65–72 (japanese) *302, 303, 327*

1944, 4

— Projective parameters in projective and conformal geometries. Proc. Imp. Acad. Tokyo 20, 45–53 . . . . *302, 315*

1944, 5

— On the torse-forming directions in riemannian spaces. Proc. Imp. Acad. Tokyo 20, 340–345 . . . . *322, 332*

1944, 6

— Subprojective transformations, subprojective spaces and subprojective collineations. Proc. Imp. Acad. Tokyo 20, 701–705 . . . . *327, 328, 330*

1944, 7

— and T. ADATI: Parallel tangent deformation, concircular transformation and concurrent vectorfield. Proc. Imp. Acad. Tokyo 20, 123–127 *330, 333, 352*

1944, 8, 9, 10

Sur les espaces à connexion affine qui peuvent représenter les espaces projectifs des paths I, II, III. Proc. Imp. Acad. Tokyo 20, 631–639; 21, 16–24; 97–103 . . . . *303*

1944, 11

— Projective parameters on paths in D. van Dantzig's projective space. Proc. Imp. Acad. Tokyo 20, 210–215 . . . . *303*

1944, 12

— Über eine geometrische Deutung der projektiven Transformationen nichtsymmetrischer affiner Übertragungen. Proc. Imp. Acad. Tokyo 20, 284 to 287 . . . . *322*

1944, 13; 1946, 4

— and S. SASAKI: Sur les espaces à connexion conforme normale dont les groupes d'holonomie fixent une sphère à un nombre quelconque de dimensions I, II. Proc. Imp. Acad. Tokyo 20, 525–535; 22, 225–232 . . . . *367*

1945, 1

YANO, K.: Lie derivatives in general space of paths. Proc. Japan Acad. 21, 363–371 102

1945, 2, 3

— Sur la théorie des espaces à hyperconnexion euclidienne I, II. Proc. Japan Acad. 21, 156–163; 164–170 123

1945, 4

— On the fundamental differential equations of flat projective geometry. Proc. Japan Acad. 21, 392–400 . . . . 303, 305

1945, 5, 6; 1946, 2, 3

— On the flat conformal differential geometry I, II, III, IV. Proc. Japan Acad. 21, 419–429; 454–465; 22, 9–19; 20–31 305

1945, 7

— Bemerkungen über infinitesimale Deformationen eines Raumes. Proc. Japan Acad. 21, 171–178 335

1945, 8

— Sur la déformation infinitésimale des sous-espaces dans un espace affine. Proc. Japan Acad. 21, 248–260 335

1945, 9

— Sur la déformation infinitésimale tangentielle d'un sous-espace. Proc. Japan Acad. 21, 261–268 . . . . 335, 352

1945, 10

— Quelques remarques sur un article de M. N. Coburn intitulé "A characterization of Schouten's and Hayden's deformation methods". Proc. Japan Acad. 21, 330–336 . . . . 335, 352

1946, 1

— and Y. MUTÔ: Note sur le théorème fondamental dans la géométric conforme des sous-espaces riemanniens. Proc. Jap. Acad. 22, 338–342 . . . . 311, 315

1946, 5, 6, 7, 8, 9; 1947, 3

— Quelques remarques sur les groupes de transformations dans les espaces à connexion linéaires I, II, III, IV, V, VI. Proc. Japan Acad. 22, 41–47; 69 to 74 (with K. TAKANO); 167–172; 173–183 (with Y. TOMONAGA); 275–283 (with Y. TOMONAGA); 23, 143–146 . . . . 104, 346

1946, 10

—, K. TAKANO and Y. TOMONAGA: On the infinitesimal deformations of curves in the spaces with linear connexion. Proc. Japan Acad. 22, 294–309 . . . 352

1947, 1

— On the flat projective differential geometry. Jap. J. Math. 19, 385–440 303

1947, 2

— and S. SASAKI: On the structure of spaces with normal conformal connection whose holonomy group leaves invariant a sphere of arbitrary dimension. Sugaku (Math.) 1, 18–28 (japanese) 367

1947, 4

*— Setsuzoku no kikagaku (Geometry of connections). Kawade shobô, Tokyo. 2 + 4 + 2 + 185 pp.

1948, 1

YANO, K.: Union curves and subpaths. Math. Japonicae 1, 51–59 . . . . . *250*

1948, 2

—, K. TAKANO and Y. TOMONAGA: On infinitesimal deformations of curves in spaces with linear connection. Jap. J. Math. 19, 433–477 . . . . *298, 343, 352*

1948, 3

— and S. SASAKI: Sur la structure des espaces de Riemann dont le groupe d'holonomie fixe un plan à un nombre quelconque de dimensions. Proc. Japan Acad. 24, 7–13 . . . . *367*

1949, 1

*— Groups of transformations in generalized spaces. Akad. Press. Comp. Ltd. Tokyo. 70 pp. *104, 346, 349, 350, 351*

1949, 2

— Sur la théorie des déformations infinitésimales. J. Univ. Tokyo 6, 1–75 *298, 335, 352*

1949, 3

— and K. TAKANO: Conics in D. van Dantzig's projective space. Proc. Japan Acad. 21, 179–187 . . . . *300*

1949, 4

— and S. SASAKI: On the structure of spaces with normal projective connexions whose groups of holonomy fix a hyperquadric or a quadric of $(N-2)$-dimension. Tôhoku Math. J. (2) 1, 31–39 . . . . *303*

1950, 1

— Note on the conformal theory of curves. Tensor 1, 6–13 . . . . . . *310, 315*

1950, 2

— and T. IMAI: On affine collineations in projectively related spaces. J. Math. Soc. Japan 1, 287–288 . . . . *346*

1951, 1

— and H. HIRAMATU: Affine and projective geometries of systems of hypersurfaces. J. Math. Soc. Japan 3, 116–136 . . . . *125*

1951, 2

— On groups of homothetic transformations in riemannian spaces. Journ. Ind. Math. Soc. 15 B, 105–117. *316*

1953, 1

— On n-dimensional riemannian spaces admitting a group of motions of order $n(n-1)/2+1$. Trans. Amer. Math. Soc. 74, 260–279 . . . . . . *348*

1948, 1

YEN, CHIH-TA: Sur une connexion projective normale associée a un système de variétés à k dimensions. C. R. 227, 461–462 . . . . *301*

1937, 1

ZAREMBA, S. K.: Démonstration élémentaire d'un théorème de M. Golab sur la représentation conforme. Ann. Soc. Polon. Math. 14, 142–144 . . . . *315*

Index.

Symbols.

Zeitfracht Medien GmbH
Ferdinand-Jühlke-Straße 7
99095 Erfurt, Deutschland
produktsicherheit@kolibri360.de